Emil Bretschneider

History of European Botanical Discoveries in China

SE**V**ERUS
Verlag

Bretschneider, Emil: History of European Botanical Discoveries in China
Hamburg, SEVERUS Verlag 2011.
Nachdruck der Originalausgabe von 1898.

ISBN: 978-3-86347-165-1
Druck: SEVERUS Verlag, Hamburg 2011

Der SEVERUS Verlag ist ein Imprint der Diplomica Verlag GmbH.

Bibliografische Information der Deutschen Nationalbibliothek:
Die Deutsche Nationalbibliothek verzeichnet diese Publikation in der
Deutschen Nationalbibliografie; detaillierte bibliografische Daten sind
im Internet über http://dnb.d-nb.de abrufbar.

SEVERUS Verlag

Preface.

The botanical exploration of Eastern Asia by European travellers and botanists has for a long time attracted the author's attention, and the greater part of the materials for the present work were brought together, many years ago, from various sources of information, frequently unprinted, some of which were only obtainable in China.

It has been attempted in the following pages to supply the want of a work of reference, in which botanists dealing with Chinese plants preserved in European herbariums might find some particulars regarding the history of these collections, of which the labels affixed to the herbarium specimens generally give only an imperfect account.

The botanists of the last century, who received exotic collections of plants did not pay much attention to the localities, collectors or dates of their specimens. In determining and describing new exotic plants, they were content to know that they had been sent from China, India etc., or the native countries were even still more loosely indicated: Asia, Africa, America. Even so late as the first part of the present century, a great carelessnes prevailed in the great botanical institutions of Europe with respect to the registring of botanical collections, especially those received from distant countries. The original letters of the collectors, which no doubt, accompanied the collections, were not preserved, and the original labels written by the collectors are generally not found in these old collections *).

Besides giving a survey of botanical travels performed in China and botanical collections made in that Empire, we wish also to record in this work the services rendered by professional botanists and other promoters of botanical science, who, although they had never been in China, have contributed to our

---

*) We beg to refer to a paper on the subject, we published in the Journ. of Bot., 1894, pp. 292—299.

knowledge of the vegetable productions of that country by determining and describing or by introducing and cultivating interesting Chinese plants.

The scope of our investigations is not confined to China Proper, as might be supposed from the title of my book, but it comprehends also besides Manchuria, the vast territories of Mongolia, Eastern Turkestan, Tibet etc., the so called «Vassal States», and, following the plan adopted in Mr. Hemsley's Index Florae Sinensis, we include even Corea and the Liu kiu and Bonin I-s.

The early period of the progress of botanical discoveries in China, concluding with Linnaeus, has received detailed notice in a paper published by the present writer under the title of «Early European Researches into the Flora of China», Shanghai, 1881 (Journ. N. China Br. Asiat. Soc.). Our first plan was to publish the present work as a continuation of the Researches, commencing with the time after Linnaeus, but, as our former paper required some revision and additions, we decided to reproduce it here in a new, partly abbreviated form, omitting amongst other things all Sinological details referring to Chinese names of plants, but adding, however, much new information.

It is obvious that the present work could not be accomplished solely from the printed sources accessible to the author. From the beginning of its preparation he has been conscious that it would be necessary to apply for assistance to such persons as might be able to elucidate special important questions. We are now glad to state that none of the many letters of inquiry addressed to friends and frequently to people personally unknown to the author, have remained unanswered. We, therefore consider it a duty as well as a pleasure to express here our great obligations to those who have placed at our disposal much valuable information regarding the subjects dealt with in the following pages.

In the first place we may mention Mr. W. B. HEMSLEY of the Royal Kew Gardens, the able author of the «Index Florae Sinensis», who in the most liberal manner has favoured us with much interesting information, derived from the Kew herbarium and other unprinted sources, with respect to old and recent collections of Chinese plants. We have likewise gratefully to acknowledge many kindnesses received from another distinguished botanist, Mr. A. FRANCHET of the Muséum d'Histoire Naturelle, Paris. He has furnished us with a complete list of the collections of Chinese plants of the last and present centuries preserved in the herbarium of the Museum, besides other important extracts from unpublished documents. To our valued late friend Mr. TH. SAMPSON, long resident in Canton, we owe a mass of valuable matter regarding early botanical collectors in South China and interesting biographical and other notices relating to the great explorer of the Chinese Flora, the late Dr. H. F. Hance, with whom he was most intimately acquainted. Nor should we omit to name here Mr. CHARLES

FORD, Director of the Botanical Gardens, Hong kong, who has kindly assisted us in elucidating dubious questions concerning European travellers and botanical collectors in South China. The valuable services of many others who have kindly lent us assistance will be acknowledged in the proper places.

It was at first our intention not to conclude this work before Mr. Hemsley's Index Florae Sinensis, which constitutes one of the chief sources for our investigations, has been completed at least to the end of Phanerogams (for all the Filices received at Kew from China, have been determined and published by Mr. Baker in the «Synopsis Filicum» and in various periodicals, and are found enumerated in his «New Ferns», v. s. 817, 818); but we cannot wait any longer for it is not to be foreseen when Mr. Hemsley's admirable work interrupted more than four years ago, will be brought to an end.

St. Petersburg, September 1898. Moïka, 64.

E. B.

NOTE ON MY MAPS OF CHINA.

The author has thought it useful to accompany his History of Botanical Discoveries in China with a Map of that country drawn up with special reference to the explorations which form the subject of that work, in order that the reader may be able to trace the localities mentioned in the text.

In consequence we published, in the spring of 1896, a MAP OF CHINA PROPER, in four sheets, scale 71,4 Engl. miles to an inch, which, however, owing to the rapid progress of geographical explorations in recent times, already requires corrections.

During the last and present year we prepared FIVE SUPPLEMENTARY MAPS representing parts of China on a larger scale, which are now ready, viz:

I. PART OF NORTHERN CHILI. Scale 12 miles to an inch.

II. THE MOUNTAINS WEST OF PEKING. $5^1/_2$ miles to an inch *).

III. PART OF MID CHINA AND THE YANG TZE RIVER, in two sheets, A and B. 35,7 miles to an inch.

IV. THE GREAT RIVERS OF THE KUANG TUNG PROVINCE. $23^1/_2$ miles to an inch **).

V. SKETCH MAP OF THE REGIONS IN YÜN NAN, WHERE FROM 1882 TO 94 FATHER J. M. DELAVAY HAS MADE VAST BOTANICAL COLLECTIONS.

These maps are sold at Mr. Edward Stanford's, 26 and 27, Cock spur Street, Charing Cross, London.

---

*) We consider it a duty to confess that sheets I and II are for the greater part based upon the excellent maps of Baron v. Richthofen and Dr. O. v. Moellendorff (v. s. pp. 947 and 949). We have occasionally added our own observations.

**) The names of places are generally given in the Canton dialect.

Table of Contents.

Part I. THE PRE-LINNEAN PERIOD.

From the Middle-Ages down to about the middle of the 18th Century.

I. Marco Polo.

The illustrious traveller MARCO POLO, who lived many years in China during the latter part of the 13th century and to whom we owe the first authentic account of that mighty empire, notices from his own observation many of the vegetable productions of China used for economical purposes. Let me extract from the late Colonel Henry Yule's, splendid edition of Marco Polo (The Book of Ser Marco Polo, the Venetian, 2nd edition, 1875) that traveller's notes on Chinese plants, and add some explanations.

"Province of Manzi (China south of the Yellow River). They have also *Wheat* and *Rice* and other kinds of corn". (II, 27).

"Most of the people of Cathay (North China) drink Rice-Wine. It is a liquor which they brew of rice with a quantity of spice, in such fashion that it makes better drink than any other kind of wine; it is not only good, but clear and pleasing to the eye". (I, 427, also II, 47, 53, 99, 184, 185).

Compare regarding Chinese Wine my Botany of the Chinese Classics, p. 154 seq.

"In the city called Unken, in the kingdom of Fuju (present province of Fu-kien) there is an immense quantity of *sugar* made. From this city the Great Kaan gets all the sugar for the use of his Court, a quantity worth a great amount of money. And before this city came under the Great Kaan these people knew not how to make fine sugar; they only used to boil and skim the juice, which when cold left a black paste. But after they came under the Great Kaan some men of Babylonia who happened to be at the Court proceeded to this city and taught the people to refine the sugar with the ashes of certain trees". (II, 208).

The sugar cane, although not indigenous in China, was known to the Chinese in the 2nd cent. B. C. It is largely cultivated in the southern provinces.

"The city of Taianfu (T'ai yüan fu in Shansi). There grow here many excellent *Vines*, supplying great plenty of wine; and in all Cathay (North China) this is the only place where wine is produced. It is carried hence all over the country". (II, 9, 47).

"City of Kinsay (Hang chou fu in Chekiang). Neither *Grapes* nor *Wine* are produced there, but very good raisins are brought from abroad and wine likewise. The natives, however, do not much care about wine, being used to that kind of their own made from rice and spices". (II, 184).

The vine, now largely cultivated in the northern part of China, was introduced to that country from Western Asia in the 2nd cent. B. C. See my Botanicon Sin. I, 25.

"There are in the city of Kinsay (Hang chou-fu in Chekiang) particular certain *Pears* of enormous size, weighing as much as ten pounds a piece, and the pulp of which is white and fragrant like a confection; besides *Peaches* in their season both yellow and white, of every delicate flavour". (II, 184).

Large pears are now-a-days produced in Shantung and Manchuria, but they are rather tasteless and coarse. I am inclined to suppose that Polo's large pears were Chinese quinces, *Cydonia chinensis*, Thouin, this fruit being of enormous size, sometimes one foot long, and very fragrant. The Chinese use it for sweet-meats.

"In the province of Acbalec Manzi there grows such a great quantity of *Ginger*, that is is carried all over the region of Cathay, and it affords a maintenance to all the people of the province, who get great gain thereby". (II, 27).

"The kingdom of Fuju (Fu kien province) produces *Ginger* and *Galingale* in immense quantities". (II, 207).

According to the Chinese authors the provinces of Sze ch'uan and Han chung (Southern Shensi) were in ancient times famed for their Ginger. Ginger is still exported in large quantities from Han k'ou. It is known also to be grown largely in the southern provinces. — Galingale is the lesser or Chinese Galanga of commerce, *Alpinia officinarum* Hance.

"Province of Caindu. They have *Ginger* and *Cinnamon* in great plenty. There grows also in this country a quantity of *Clove*. The tree that bears it is a small one, with leaves like laurel but longer and narrower, and with a small white flower like the clove". (II, 47).

Polo's Caindu seems to answer to the present Ning yüan fu in Southern Sz'ch'uan. Polo does not state explicitly that Cinnamon grows there. As is known, now-a-days, the Cassia bark or Chinese Cinnamon of commerce is produced in the provinces of Kuang si and Western Kuang tung. Col. Yule (l. c. 50) may be right in supposing that Polo's Cloves were Cassia buds.

"The ships in Manzi (China) are of *Fir* timber". (II, 231).

Pine is still the staple timber for ship-building in China. It is the *Pinus sinensis*, so common along the coast of China.

"City of Ghiuju (province of Chekiang). Here you find the largest and longest *Canes* that are in all Manzi; they are full four palms in girth and 15 paces in length". (II, 203).

"They (the Chinese near the Kiang river) take those great canes and split them from end to end into many slender strips, and then they twist these strips together as to make a rope of any length they please. And the ropes so made are stronger than if they were made of hemp". (II, 156).

"The Kaan's palace at Chandu is built of *Cane*. The roof, like the rest, is formed af canes, covered with varnish. These canes are a good 3 palms in girth, and from 10 to 15 paces in length. They are cut across at each knot, and then the pieces are split so as to from from each two hollow tiles and with these the house is roofed . . . . . . These canes serve also for a great variety of other useful purposes". (I, 290).

"Province of Tebet (Polo includes in Tibet the western part of the present Sze ch'uan) .In this country you find quantities of Canes, full 3 palms in girth and 15 paces in length, with some 3 palms' interval between the joints. Merchants and travellers through this country are wont at nightfall to gather these canes and make fires of them; for as they burn they make such loud reports that the lions (tigers) and bears and other wild beasts are greatly frightened and make off as fast as possible". (II, 33).

By Canes we have to understand Bamboos. Of this useful plant China produces many species and varieties, some of them of considerable dimensions. The provinces of Chekiang and Szech'uan are especially rich in bamboos. In corroboration of Polo's statement regarding the explosions produced when burning bamboos, I may adduce Sir Joseph Hooker's Himalayan Journals, edition of 1891, p. 100, where, in speaking of the fires in the jungles, he says: "Their triumph is in reaching a great bamboo clump, when the noise of the flames drowns that of the torrents, and as the great stem-joints burst, from the expansion of the confined air, the report is as that of a salvo from a park of artillery".

"The city and great haven of Zayton (Tsʿüan chou fu in Fu kien). You travel by mountains and valleys (between the latter place and Fu chou fu) and in some places by great forests in which are many of the trees which give *Camphor*". (II, 217).

The *Laurus (or Cinnamomum) Camphora*, a large timber tree, grows abundantly in Fu kien.

"The kingdom of Tajanfu (Northern Shansı). There is a great deal of silk here, for the people have great quantities of *Mulberry-trees* and silk-worms". (II, 9).

"When you leave the city of Cachanfu (Pʿu chou fu in S. W. Shansi) and travel westward you meet with beautiful gardens and fine plains planted with *Mulberries*, which are the trees on the leaves of which the silkworms do feed". (II, 18).

Morus alba is largely grown in North China for feeding silkworms.

Polo states (I, 409) "that the Great Kaan causeth the bark of Mulberry-trees, made into something like paper, to pass for money". He seems to be mistaken. Paper in China is not made from mulbery-trees but from the bark of *Broussonetia papyrifera*, which latter tree belongs to the same order of Moraceae. The same fibres are used also, in some parts of China for making cloth, and Marco Polo alludes probably to the same tree when stating (II, 108) "that in the province of Cuiju (Kuei chow) they manufacture stuffs of the bark of certain trees which form very fine summer clothing".

"Merchant Ships of Manzi (China). They take some lime and some chopped hemp, and these they knead together with a certain *Wood-oil*; and when the three are thoroughly amalgamated, they hold like any glue. And with this mixture they do pay their ships". (II, 232).

What goes under the name of "wood-oil" to-day in China is the poisonous oil obtained from the nuts of *Elaeococca verrucosa*. It is much used for painting and caulking ships.

"Kingdom of Fuju (province of Fukien). They have a kind of fruit resembling *Saffron*, and which serves the purpose of saffron just as well". (II, 207).

Col. Yule concludes that the fruit of a *Gardenia*, which yields a yellow colour, is meant. But Polo's vague description might just as well agree with the Bastard Saffron, *Carthamus tinctorius*, a plant introduced into China from Western Asia in the 2nd cent. B. C. and since then much cultivated in that country.

"All over the mountains of the province of Tangut *Rhubarb* is found in great abundance, and thither merchants come to buy it, and carry it thence all over the world». (I, 219).

"In the mountains belonging to the city of Su ju (Su chou fu in Kiang su) *Rhubarb* and *Ginger* grew in great abundance". (II, 166).

Polo is correct in giving Tangut as the native country of Rhubarb *(Rheum palmatum)*, but no species of Rheum has hitherto been gathered by our botanists as far south as Kiang su, indeed, not even in Shantung.

Polo further states in speaking (l. c.) of the Tangutan mountains: "Travellers however, dare not visit those mountains with any cattle but those of the country, for a certain plant grows there which is so poisonous that cattle which eat it loose their hoofs. The cattle of the country know it and eschew it".

This poisonous plant seems to be the *Stipa inebrians* described by the late Dr. Hance in the Journal of Bot. 1876, p. 211, from specimens sent to me by Belgian Missionaries from the Ala shan Mts, west of the Yellow River.

In speaking of the great Haven of Zayton (Ts῾üan chou fu) Polo notices the duties paid at this place on *Pepper, Lignaloes, Sandal wood* articles then largely imported from southern countries, just as at the present time. (II, 217, 186).

It is a strange fact that Polo never mentions the use of *Tea* in China, although he travelled through the Tea districts in Fu kien, and tea was then as generally drunk by the Chinese as it is now. It is mentioned more than four centuries earlier by the Mohammedan merchant Soleyman, who visited China about the middle of the 9th. cent. He states (Reinaud, Relation des Voyages faits par les Arabes et les Persans dans l'Inde et à la Chine, 1845, I, 40): "The people of China are accustomed to use as a beverage an infusion of a plant, which they call *sakh*, and the leaves of which are aromatic and of a bitter taste. It is considered very wholesome. This plant (the leaves) is sold in all the cities of the empire".

II. Early Sea-Trade of the Portugueuse with China.

After the memorable travels of Marco Polo and the wanderings of Odoric and Marignolli in Eastern Asia, in the 14th century, China was for a long time shut up to European access. It had been altogether forgotten in Europe, when in 1516 the Portuguese first arrived in China, which therefore was considered a new discovery. The Portuguese, having in 1511

conquered Malacca, Captain RAPHAEL PESTRELLO, in 1516, took passage
on board a junk to China and brought back some scanty information about
that country. Most probably he had visited Canton. In the next year, in
1517, before Pestrello had returned, a small Portuguese fleet under the
command of FERD. ANDRADE left Malacca for China, entered the Gulf of
Canton and anchored off the island of Sancian (St. John's I.), S. W. of
Macao. Andrade was well received by the Chinese authorities, and even
allowed to proceed with two ships to Canton, with which important port
he obtained permission to trade. Meanwhile G. MASCARENHAS, who had
come with Andrade in his own ship, visited the eastern coast of China in a
native junk. He no doubt touched at some ports in the provinces of Fu kien
and Chekiang and paved the way for his countrymen to a profitable mercan-
tile intercourse. Some years later the Portuguese succeeded in forming a
factory at Liampo (Ningpo), and subsequently, about 1544, we find them
also established near Chin chew (Chang chou fu in Fu kien) and, it seems,
in Amoy. It is known from ancient Portuguese records, that in 1537 there
were on the islands in the Gulf of the Canton River, three Portuguese
settlements, where they carried on trade with the Chinese. Of these Macao,
on account of its favourable situation, was the most important, and soon
became the base and the starting point for the commercial enterprises of
the Portuguese in Eastern Asia. Here they monopolized for 70 or 80 years
European trade with China. They brought from Europe to China woollens;
from India amber, corals, elephant's teeth, sandal wood, but above all a
great quantity of pepper. As the ancient reports state they exported from
China silkstuffs, gold, musk, pearls, precious stones, china ware, sugar and
a variety of trifles (Comp. "Ljungstedt, Hist. Sketch of the Portuguese
settlements in China", 1836, p. 1—9, 82).

The only vegetable product specified in the above ancient list of articles exported from China is sugar. There can however be no doubt that
many other valuable Chinese products of vegetable origin were carried to
Europe in Portuguese ships in the 16th century, especially medicines.
Among the latter I may mention *China root (Smilax glabra)*, which acquired great celebrity in Europe, since about the year 1535. The Emperor
Charles V is reported to have been cured of gout by this drug (Comp.
Flückiger & Hanbury, Pharmacographia, p. 648). Tea was not exported
from China during this period, although its use by the Chinese was known
to the Portuguese.

The Portuguese claim the honour of having introduced the *Sweet
Orange* from China to Portugal. The valiant Juano de Castro, Vice-roy of

India, from 1545 to his death in 1548, according to tradition, sent a living tree of the Chinese Orange to Lisbon. (Comp. "Galesio, Traité du Citrus", 1829, p. 297). The Jesuit missionary Le Comte in his Account of China, 1696, vol. I, 173 (see further on) states, that the first Orange tree, which had been introduced by the Portuguese from China, still existed at that time (end of the 17th cent.) in the garden of Count St. Laurent in Lisbon.

Garcia ab Orta, First Physician to the Portuguese Vice-roy of India at Goa, and a resident in India for thirty years, published in 1563 at Goa his COLOGIOS DOS SIMPLES E DROGAS etc., of which a Latin version is found in Clusius' Exotica, p. 145—252. In this interesting account of Indian spices and medicines, the author mentions also several Chinese drugs:

Camphor (p. 161). "There are two sorts of Caphura (Camphor), one is produced in Bornea, the other is brought from China. The first is of a superior quality and a hundred times more costly than the Chinese sort. It is never seen in Europe. The camphor of China, which is alone exported to Europe, appears in compressed orbicular cakes, 5 inches in diameter, in which it is probably mixed with other medicines".

Cassia bark (p. 169, 170). Garcia states that Canella, Cassia lignea and Cinnamomum are different names for the same spice. The best is produced in Ceylon. That brought in large quantities by Chinese ships to Ormuz and thence carried to Alexandria and Europe, is known to the Persians under the name of *darchini* or Chinese wood, for they suppose it to come from China. But this is an error, as the Chinese buy in Ceylon all the Cinnamon sold to the Persians. — Garcia is evidently mistaken as to the place of production of Cassia bark. Flückiger and Hanbury, Pharmacographia, p. 468, regard the ancient Cinnamon to have been Chinese Cassia lignea, which the Chinese from very early times used to bring to the ports of Western Asia.

Rhabarbarum (p. 207). "This drug, which is produced only in China, is shipped thence in Chinese vessels to India. But rhubarb is also conveyed fron China overland through Tatary to Ormuz and Aleppo from whence it reaches Alexandria and Venice. When coming this way the drug is less subject to become wormeaten".

Radix Chinae (p. 208). Garcia gives a long account of this Chinese drug and its medicinal virtues, stating that in its native country it is celebrated as a remedy in venereal and cutaneous diseases. China root became first known in India in A. D. 1535. The Chinese call the plant *lampatam*. — The latter name given by Garcia seems to be a corruption of *leng*

fan tŭ an, which according to the Chinese Herbal is another name for the
t'u fu ling or China root (*Smilax glabra* and other species).

Galanga (p. 211). There are two kinds of this root (both figured by
Garcia). The Galanga minor is produced in China from whence it is exported
to Lusitania. It is fragrant. The natives (Chinese) call it *luan don*. The
other sort, the Galanga major is a native of Java. — *Alpinia officinarum*,
Hance, is the source of Chinese Galangal. Its Chinese name is *kao-liang-
kiang* or *liang-kiang*. The rhizome of *Alpinia Galanga* Willd., a plant of
Java, constitutes the drug known as Rad. Galangae majoris.

Zedoaria (p. 213, 214). Garcia believes that this drug (the root of
Curcuma zedoaria) is produced in China. This is confirmed by Roxburgh,
who states (Fl. Ind. ed. 1874, p. 8) that this plant is a native, not only of
Bengal, but also of China.

Garcia, in describing (p. 156) the various sorts of Benzoin, mentions
Roça malha, which, he says, is the name given by the Chinese to liquid
Storax. Hanbury, Science Pap. p. 143, confirms Garcia's statement that the
drug Rose Malloes is identical with liquid Storax, but the latter is mistaken
as to the Chinese origin of the name, Rose Malloes being the Javanese
name for the tree *Liquidambar altingia*.

III. Early Botanical Information with respect to China supplied by Catholic Missionaries, and especially the Jesuits.

Thirty six years after the rediscovery of China by the Portuguese the
first Christian missionaries made their appearance in that country. In 1552
FRANCISCUS XAVIER, a Jesuit started from Goa in a royal vessel bound to
Malacca. Then he embarked in the same ship and arrived at the island of
Sancian (St. John), S. W. of Macao, where the Portuguese had a settlement.
But he was disappointed in his hope of beginning his missionary work among
the Chinese, and died on the same island a few months after his arrival.
Nearly thirty years elapsed before a new attempt was made by the Jesuits
to gain a footing in China. From 1581 to 1583 they sent successively four
missionaries to Macao. One of them was MATTHAEUS RICCI, who holds one
of the most conspicuous places in the history of Chinese missions. By perse-
vering efforts he obtained permission to reside at Peking, where he arrived
in 1600. At the time of Ricci's death, in 1610, the number of Jesuit
missionaries in China had already considerably increased, and we find them

working in many parts of the empire, namely (besides Peking) at Canton, Chao king fu, Shao chou fu (all in Kuang tung province), Nanking, Shanghai, Su chou fu and Sung kiang fu (all in the province of Kiang su), in which they then had altogether 90 churches; at Hang chou fu (in Chekiang), Nan ch'ang fu (in Kiang si). In the provinces of Hu kuang and Sze ch'uan they had also built many churches, and it appears that at that time there were missionaries also in Fu chou fu and in some parts of the province of Shan si. The Jesuit missionaries were not only assiduously labouring to learn the Chinese language and preach the gospel, but also occupied themselves in acquiring knowledge of the customs of the people and the literary works of the Chinese, and directed their attention to the features of the country, its natural productions etc. The Jesuit missionaries have always had the well-merited reputation of great learning and of a classical and scientific education; and it seems that those who were sent to convert the Chinese, had been especially trained with the object of convincing the latter, by means of striking experiments, of the superiority of western science, and of demonstrating to them the accuracy of Europeans in observing natural phenomena, and their ingenuity in making the laws of nature serviceable to the purposes of industry, economy and the arts. The early success enjoyed by the propaganda of the Jesuits in China was principally due to the great authority they had acquired at the Court of Peking on account of their skill in astronomy, physics, chemistry etc. Many of these distinguished scholars used to take great pleasure in investigating objects of natural history, and thus we find in the collections of the letters and memoirs of the Jesuits in China a great number of articles treating of mineralogy, zoology and botany, and supplying a mass of most valuable information. The circumstances in which they lived amony the natives, gave them many more facilities for gathering information than can be obtained by travellers or naturalists of the present time, who are looked upon with suspicion, constantly watched and often molested by the people. There are still in the interior of China some plants known to us only from the descriptions of the Jesuits. I need hardly say, that the accounts left by the early missionaries, concerning Chinese botany, have for the greater part no claim to be considered scientific papers in our modern sense. Their descriptions, however, of the plants applied by the natives to economic or other useful purposes, and also of wild-growing medicinal, and other remarkable plants are generally quite satisfactory and popularly correct. The Chinese names are generally added. In most of the cases the plants described are easily recognizable.

Before proceeding to a chronological survey of these publications of the ancient Jesuit fathers, one of the earliest accounts of China, written before their advent, deserves to be noticed here.

J. Gonzales de Mendoza's HISTORY OF THE GREAT AND MIGHTY KINGDOM OF CHINA.

This book was first printed in Spanish, in 1585, in Rome. An English version was published in 1853 by the Hakluyt Society. Mendoza, an Augustine monk, had himself never been in China. He depends mainly upon the accounts furnished by MARTIN DE HERRADA, a monk of the same order, who had been taken, in 1575, by a Spanish ship from Manilla to the Chinese port of Ts'üan chou fu (prov. of Fu kien), where he was allowed to spend three months. The vegetable productions observed there by him were *Chestnuts*, large *Melons*, a kind of plum called *leechias (Nephelium Litchi)* of an excellent gallant taste. Of cereals cultivated there he notices, besides Wheat, Barley, Millet, "the plant called *Maize*, which constitutes the principal food of the Indians in Mexico". This latter statement, made at so early a date, has a peculiar interest for us, for Maize is not indigenous to China and has been introduced to that country since the discovery of America.

Joannis Petri Maffei. Bergamatis, e Soc. Jesu, HISTORIARUM INDICARUM LIBRI XVI. 1589. Florentiae.

Maffeus, who never visited India himself, compiled his account of that country from the information supplied by the Jesuit missionaries residing there. On p. 103 we find a short notice of China, Sinarum regio, in which he speaks of the vegetable products of that empire in the following terms: "Saltus et colles *Pinu Vitibus*-que: campi et planities *Oryza, Hordeo, Frumento*, caeterisque vegetibus nitent, quamquam e vitibus more nostro non exprimunt merum. Uvas quodam condimenti genere in hyemem adservare mos est. Caeterum ex herba quadam expressus liquor admodum salutaris, nomine *Chia* calidus hauritur, ut apud Japonios: cujus maxime beneficio, pituitam, gravedinem, lippitudinem nesciunt; vitamque bene longam, sine ullo terme languore traducunt". — Here we find, it seems, the earliest mention made by European writers of Tea (*cha* in Chinese).

Trigault. DE CHRISTIANA EXPEDITIONE APUD SINAS suscepta ab Soc. Jesu. Ex P. Matthaei Ricci commentariis. 1615.

Nicolas Trigault, of Belgian origin, born in 1577, having entered the Order of the Jesuits, arrived in China in 1610, and died at Hang chou fu in 1628. Under the above title he published the matter compiled by Ricci concerning China and the history of the mission. In Chap. III we find the following particulars regarding the vegetable productions of China:

They have in China all our principal fruits with the exception of Olives and Almonds. Besides these they have a great variety of fruits unknown in Europe as f. i. in the Canton province the *Licy* and the *Longan (Nephelium Litchi, N. Longan)* not found elsewhere. *Indian nuts (Cocoa nuts)* are also met with there.

There is a peculiar very common fruit which the Portuguese use to call *Chinese Figs.* It is of a red colour of the size of a peach and without kernels. It is also eaten in a dried state. *(Diospyros Kaki, D. Schitse.* The dried fruit resembles very much in appearance and in taste our figs).

Their *Oranges*, of which they have many varieties, are superior in sweetness to the same fruits cultivated in other countries.

In the four southern provinces of China is found the famous leaf which the Indians call *betre*, and the *Arqueira* tree. *(Betel pepper* leaf and the *Areca* nut).

Olive oil is unknown to them. But they use largely *Sesam* oil, which is very fragrant, and is found everywhere.

The *Vine* is rare there (i. e. in Southern China) and they do not use grapes for making wine. Their wines, which are inferior to ours are made of rice and other ingredients.

Flax is unknown in China. For their clothes they use *Cotton.* The cotton plant is grown in great abundance there, but it is not indigenous in China. The seeds were brought to China by foreigners, 400 years ago.

As far as timber is concerned, the *Oak* is very rare there. But they have a precious wood which the Portuguese call *Pao de ferro* (Iron wood). It is very hard and has also the colour of iron. *(Baryxylum rufum* according to Loureiro).

You met there also with *Cedars*, which trees are considered fateful. The timber is used for coffins.

They have a kind of reed called *Bambu* by the Portuguese. It is almost as hard as iron. The largest kind is scarcely encompassed with two hands. It is hollow inside and presents many joints outside. The Chinese use it for pillars, shafts of lances and for 600 other domestic purposes.

Of drugs China produces: *Rhubarb* and *China wood* (root) as the Portuguese term it. Others call it Holy wood. It is not cultivated but is brought from desert places.

Sugar and Honey are produced in abundance.

The *White Wax* (of China) is superior in quality to ordinary wax, for it is less sticky and (when manufactured into candles) burns with a brighter flame. It is produced by certain insects which live upon trees planted for this purpose. (Here the well known Chinese Insect wax is meant, produced by the Coccus pela upon the branches of *Fraxinus* and *Ligustrum*).

They use yet another kind of wax which is obtained from the fruit of a certain tree. It is also used for candles. (This is the Vegetable Tallow yielded by the seeds of *Stillingia sebifera*).

There is in China a shrub the leaves of which are boiled into a famous beverage much used by the Chinese, Japanese and other nations conterminous to China. They call it *Cia*. Its use among the Chinese does not seem to be of very old date; for in their most ancient books no hieroglyphic character is found to designate the cia. The leaves are gathered in spring and dried in the shade. They drink this decoction almost continually. *(Tea.* Compare above Maffeus).

Cinnamon and *Ginger* are also produced in China.

Alvarus de Semedo, RELATIONE DELLA GRANDE MONARCHIA DELLA CHINA. 1643.

Semedo, a Portuguese, born in 1585, entered the Society of Jesus, arrived in China in 1613, and died at Macao in 1658. He wrote his book about 1633. In his description of the provinces of China he notices the excellent *Peaches* and *Grapes* of Shensi, the fine *Apricots* produced in Honan, and the large *Pears* of Shan tung. He gives a good description of the so-called *Chinese Figs (Diospyros)* already mentioned, of the fruit of *Myrica sapida* in Che kiang, of *Litchis* and *Longans* (v. supra 11) produced in the Fu kien and Canton provinces. He further tells us that the island of Hainan is famed for its precious woods such as the fragrant *Eagle wood (Aloexylon Agallochum)*, the wood called *Rose wood* by the Portuguese, *hua li mu* by the Chinese. Formosa produces *Pepper, Cinnamon, Camphor* trees, *China root.* The *Rhubarb* plant is found in the province of Shensi, the precious *Ginseng* root in the province of Liao tung (Earliest mention of this plant by an European author). Semedo speaks also of the *Tea* plant and its use, and of the *Varnish* tree which yields the famous Chinese varnish *(Rhus vernicifera);* he notices Chinese *lamp-wicks* made from the pith of a rush *(Juncus effusus)*. Among garden flowers he mentions the fragrant *la mei (Chimonanthus fragrans)* and the *Air plant (Aërides, Vanda)*. — For further particulars I beg to refer to my Early Europ. Res. into the Flora of China, p. 4—7.

Martini's Atlas Sinensis. 1655.

Martinus Martini was born in 1614 at Trent in S. Tyrol. He arrived in China in 1643, returned in 1653 to Europe, published his work in Vienna and returned to China. He died at Hang chou fu in Chekiang in 1661 or 1662. The Atlas Sinensis is a short geographical description of the provinces of China. It contains a mass of interesting information regarding the vegetable productions of the different provinces. As in my Earl. Europ. Res. etc. I have extracted all these passages (p. 7—20), I can restrict myself here to give a list of the plants spoken of by Martini, and which can be easily recognized from his descriptions.

Garden fruits: Apples, Pears, Plums, Pomegranates, Persimons *(Diospyros)*, Grapes, common Figs, various Oranges, the Pampelmoes *(Citrus decumana)*, the fingered Citron or Budha's hand, Chestnuts. In the southern provinces Litchis, Longans (v. supra 11), Pineapples, the Jackfruit, Cocoa nuts, and Areca nuts.

Useful waterplants: *Nelumbium speciosum*, *Trapa bispinosa*, *Elaeocharis tuberosa*.

The *Tea* plant is treated of in detail, and the *Sugar cane* is mentioned as being cultivated in the prov. of Szech'uan, Cinnamon produced in Kuang si, Pepper in Yünnan. A tree *kuang lang* in Kuang si from the pith of which Sago is obtained *(Caryota)*.

Trees: the *Mulberry* tree cultivated for feeding silkworms. Allusion to the *Oaks* of Shan tung upon which the wild silkworms live, the Willow *(Salix babylonica)*. — *Bamboos* are produced in the province of Chekiang. The island of Hainan is famed for its *Eagle wood, Rose wood, Ebony, Brasil wood (Caesalpinia sappan)*. The *Iron wood* grows in the province of Kuang-si. The *Rotang (Calamus rotang* and other species) is found in Kuang tung and Hainan.

Martini mentions the Chinese *Varnish* tree *(Rhus vernicifera)* the *Tallow* tree *(Stillingia sebifera)*, Chinese *White wax* (Insect wax).

Of textile plants he notices: *Cotton, Bochmeria nivea, Pueraria Thunbergiana*, a red *Musa* with textile fibres.

Ornamental plants: *Paeonia Moutan, Hibiscus mutabilis, Jasminum Sambac, Olea fragrans.*

Medicinal plants: *Ginseng, Rhubarb, China root (Smilax), Pachyma Cocos.*

Michael Boym's Flora Sinensis. 1656.

M. Boym, a Pole, born in 1612. He left Europe as a Jesuit missionary for China in 1643, returned to Europe in 1652, reembarked for China in 1656, and died in the province of Kuang si in 1659. His Flora Sinensis was originally printed in Vienna in Latin. A French translation of it is found in «Thevenot's Relation des Voyages», 1696, sec. partie, p. 15—30. Boym gives an account of 22 plants, of which the greater part are rather plants of the Indian Archipelago; 21 of them are represented by engravings. The plants noticed there are the following:

Cocos nucifera.	*Psidium Guyava.*
Areca Catechu.	*Artocarpus integrifolia.*
Betel pepper.	*Diospyros Kaki.*
Carica papaya.	*Anona squamosa.*
Musa sapientium.	*Durio zibethinus.*
Anacardium occidentale.	*Cynometra cauliflora.*
Nephelium Litchi.	*Piper nigrum.*
Nephelium Longan.	*Radix Sinae.*
Eugenia Jambos.	*Rhabarbarum.*
Ananassa sativa.	*Cinnamomum.*
Mangifera indica.	*Zingiber.*
Eriobotrya japonica.	

For further particulars sea my Earl. Res. p. 21—24.

Michael Boym wrote also a Treatise on Chinese Medicine, which was published in 1682 by ANDREAS CLEYER with the title: SPECIMEN MEDICINAE SINICAE. One section of it is entitled: MEDICAMENTA SIMPLICIA QUAE A CHINENSIBUS AD USUM MEDICUM ADHIBENTUR. It is nothing more than an enumeration of 289 Chinese drugs, for the greater part of vegetable origin with the Chinese names added. The medicinal properties of each drug are explained.

For further particulars see Earl. Europ. Res. p. 24.

Gabriel de Magalhaens. NOUVELLE RELATION DE LA CHINE. 1688.

Magalhaens, a Portuguese, born in 1609, arrived in China in 1640, and died at Peking in 1677. The MS was written in Portuguese, in 1668, and after the author's death translated into French and published in 1688 under the above title. On p. 173 we find an interesting account of the Chinese *white Insect wax* (v. supra) produced in the provinces of Shan tung and Hu kuang and of the trees on which the insects live.

Le Comte, NOUVEAUX MÉMOIRES SUR L'ÉTAT DE LA CHINE. Paris 1696, 2 vols.

Louis Le Comte, a Frenchman, born in 1655, joined the Jesuit mission in China in 1687, and died at Bordeaux, in 1729.

In his book he notices (I, 173) the introduction of the *Orange* tree of China in to Portugal by the Portuguese (compare supra   ). He saw in the province of Shensi small yellow *Melons* and large *Watermelons* with red and white pulp (I, 132). — A good description of the *Tea plant,* its culture etc., in the province of Fu kien is found I, 368. The author explains that the Chinese call the plant properly *Cha,* and that only in the dialect of Fu kien the name sounds *Te.* — Le Comte reports (I, 168) that *Tobacco* is cultivated in the neighborhood of Peking and in the provinces of Shan si, Shensi, Sz'ch'uan. He seems to be the first European who notices Tobacco in China. — The *black* and *yellow Peas,* to which Le Comte refers (I, 168) as used in North China for feeding horses, are varieties of the Soy bean, *Soja hispida.* — The *hoa tsiao,* or Chinese Pepper tree described I, 177, is *Zanthoxylum Bungei,* — the tree *ou-tom-chu* with large leaves and seeds borne on the edges of the leaves (I, 268) with a drawing, is *Sterculia platanifolia.* — Le Comte devotes also some pages to the celebrated *Ginseng* (I, 377). — Further particulars see Early Europ. Res. p. 26—28.

George Joseph Kamel, or latinized Camellus, a Jesuit father born in 1661 at Brünn in Moravia, entered the order in 1683 and proceeded to the Marianne Islands and thence to the Philippines. Having acquired some knowledge of botany and pharmacy, he opened at Manila a pharmaceutical shop for distributing medicines to the poor. Being interested in botany he sent herbarium specimens and botanical drawings to Europe, partly to Ray and Petiver. — This short notice, found in A. de Backer's Bibliothèque des écrivains de la Compagnie de Jésus, IV (1858) 89 (quoted by Prof. Flückiger), is all we know regarding Kamel's life. His scientific exertions appear in the SYLLABUS STIRPIUM IN INSULA LUZONE PHILIPPINARUM NASCENTIUM A REV. PATRE G. J. CAMELLO OBSERVATARUM ET DESCRIPTARUM etc. It forms the Appendix to the third volume of Ray's Historia. Plantarum, 1704. More than 90 drawings of the plants sent by Kamel were published by Petiver in his «Gazophylacium Naturae et Artis» 1702—1709.

Among the Luzon plants specified by Kamel in the above article there seem to occur many Chinese plants cultivated by the Chinese in that island, or found there also in a wild state. Of these Kamel frequently gives the Chinese names (some times corrupted), and the latter may in some cases prove serviceable to recognize the plants intended. Let me quote a few instances:

P. 29, n. 3. Cyperus esculentus sinicus seu *Maa tchi* vel *Pee tchi*. . . .
Folium habet surrectum, teres tubulosum: radicem fuscam, fibrosam, de
qua sessiles aliquot liris cincti, deforis subruffi, interne candidi ut Apulit,
castanea majores, aut pares, et nuce indica cocco, ac amygdalis suaviores,
esu, ac dulciores, bulbosive dependent glandes. Florem non observavi, quia
inundatione periere plantae. — This is, I have no doubt, the Chinese Water-
chestnut, *Scirpus tuberosus*, Roxburgh, Flora Ind. (ed. 1874) p. 70, in
Chinese *pi tsʻi* or *maa tai*. Comp. also Earl. Europ. Res. p. 11.

P. 33, n. 4. Visco aloës Luzonis prima, ... Sinis *Tiao lan*, visci
modo arboribus ut reliquae species innascitur. Loco aëre perflato suspensum
diutissime vivit, floretque. Folium habet succulentum, crassum. Flores gra-
tiosi, candidi, suavissimo olentes, et roseis variegatis maculis.

P. 34, n. 14. Visco Aloës nona, Sinarum *Lan hoa*. Planta est quasi
bulbosa, arboribus innascens. Sinae hanc ob floris suaveolentiam in fictilibus
magna cura adolescere cogunt. Ex innumeris radicum fibris quasi tuberosum habet caput, ex hoc folia crebra et plantae assurgunt multae, ut dense
stipatum ex Narcissi, aut Hyacinthi tuberosi foliis, cespitem effingant, cras-
siora autem et contumaciora sunt folia quam Narcissi. Figured in Gazophyl.
Tab. CIII, fig. 7.

The names *lan hua* and *tiao lan* in China are applied to several fragrant Orchids, *Aërides*, *Vanda*, *Dendrobium*. Compare my Botanicon Sin.
II, p. 229.

P. 52, n. 13. Osa, Usao, Hispanis Lechias, Sinis *Lichi* Arbor. Fructus parvo ovo gallinaceo saepe compar verrucoso-membranaceo cortice in-
tectus, primum viret, maturus ruber, desiccatus fusci est coloris, continens
buccellam carnis candidae, diaphanae et gratissimi acido-dulcis saporis, in
qua nucleus oblongo-teres, laevis, glandiformis, rubre fuscus latet. Abundat
in montibus Batan, Palicpican, in Zebu, Bohol et Basilan. Huc spectat fructus a Clusio pictus et descriptus, fol. 36, n. 5. — The above description
agrees well with *Nephelium Litchi*, the well known Chinese fruit. Kamel
refers to Clusius' Exotica p. 36, n. V. The fruit represented there seems
indeed to be a Litchi.

P. 52, n. 14. Boa, Boboa, Boabas, Sinis *Lung yen*, id est "Oculus
Draconis", Hispanis Longanes, Lanzones. Arbor ceraso similis, sed foliis
longius, est annue bis fructicans. Fructus ex melioribus et sanioribus, inno-
cuus, etiamsi liberalius esitetur, recens saporis suavis, acidulo-dulcis, passus
uvas passas sapit, racematim de caudice arboris proveniens, figurae et coloris
pruni Brunensis habitioris, putamine verrucoso-membranaceo coopertus,
quod in sicco fragile est et rubre fuscum. Caro niveo-pellucida est, ossiculo

obvoluta, quod sphaericum, avellanae compar, nigrum, politum et splendens, cujus nucleus amarus et lumbricos fugans. — This is the *Nephelium longan*, likewise a common fruit in China.

P. 54, n. 5. *Zapotl de China*, seu Sinense *xi cu*, Hispanis Chicoy, Lusitanis Ficocaque. Arbor procera. Poma pomi ordinario sunt magnitudinis, quasi conica, alia tamen nuces Juglandes vix superant, suntque exossa. Caricarum modo ad solem desiccantur, suntque ita curata itineribus transmarinis et mensis secundis aptissima, aqua scilicet prius macerata, mox vino, saccharo, et canela pulvere conspersa apponuntur. Recentia duplici insident caliculo, deforis tenui, cinnabarino, et membranaceo teguntur putamine, immatura acredine morsicant vehementer, matura carne farciuntur molli, lutescente, grati saporis, cui oblongo-compressa, cornea uncialia, tersa, coloris terrei interjecta sunt ossicula. — Kamel here describes the fruits of Chinese Persimons, *Diospyros Kaki*, *D. Shitse*. Comp. Earl. Europ. Res. p. 5.

On. p. 57. Kamel enumerates 17 sorts of *Oranges*, amongst which n. 25. Ochan, Aurantium sinense seu Olysiponense vocatum, cortice tenui dulci et sature croceo rubente.

n. 26. Aurantium japonicum pyriforme.

n. 36. Aurantium sinicum pumilum. Poma magnitudine nucis juglandiae.

P. 39, n. 1. De Faba aegyptia..... Sinae *Lien* vocant, ejus flores *Hu hua*, id est Florem Lacus: radicem edulem *ngeu*, fructum *Lien zu* et *Lien po*. In China communiter in omnibus arte factis stagnis, lacubus ac piscinis feliciter adolescit sed et hortorum incolam curiosi Sinae arte et industria esse cogunt eam in vastae amplitudinis fictilibus, luto tenaci ex palustribus allato, et aqua diluto, repletis serendo........ Radix decocta et cruda tabidos reficet et consumptos. — This is *Nelumbium speciosum*. Comp. Earl. Europ. Res. p. 12.

P. 79, n. 3. (see also p. 48, n. 7). *Anisum sinicum*, Chabraeo indicum, Nierembergio Philippinicum. Est fructus quem Chabraeus (1677) vocat Anisum indicum, Clusius: Damur, in Luzone Anisum sinicum, Sinae *Pui ka*, alii *Zingi*. Hispani Anis de Sanglay: arboris Pansi pansi, quae abundat in montibus Boholanis, folia ferentis odorata. —

Nieremberg, Historia naturae, 1635. Clusius, Rariorum Plantarum Historiae, 1601, p. 202, states that the fruit of a peculiar kind of Anise quite different from the common Anise, was brought from the Philippine Islands to England by an Englishman, Thomas Candi, who had circumnavigated the globe and returned a few years before (In Flückiger & Hanbury's

Pharmacographia, p. 21, he is called Candish and stated to have brought
the Star-anise about 1588). Clusius obtained specimens of it in London from
the apothecary Morgan and the druggist Garet, and described and figured
the fruit capsule with the seeds. According to Candi it was called Damor
in the Philippines. *Pui ka* is perhaps a corruption of *pa küe hiang* (eight-
horned fragrance) as the Chinese call Star-anise. In Bauhin's Hist. Plant.
(1650), I, 485 we read, *Zingi*, fructus stellatus s. Anisum indicum.

P. 83, n. 2. Zhampacae species et arbor Alanguilang zibuanorum, Sinarum *In yao hao*. — The tree of which the fragrant flowers yield the oil
known under the name of Alanguillan is the *Cananga (Uvaria odorata)* of
the Philippines. (Flückiger. Pharm. Journ. May 14 1881). According to
Lamarck it also occurs in Southern China. But the Chinese name *ying chao
hoa* (hawk's claw flower) is generally applied to another fragrant Anonacea,
the *Artabotrys odoratissimus*.

P. 87, n. 1—8. Description of various sorts of *Aloes-wood* which the
Chinese are accustomed to bring to Manila, with the Chinese names added.

P. 6, n. 18. Telephium Sempervivum seu Sinarum *Kalan chau huy*.
Figured in the Gazophyl. Tab. XCV, 384. — I can make nothing of the
above, which is evidently a corrupted Chinese name. Adanson, in his Fa-
milles des Plantes, 1763, II, 248 transformed it into *Kalanchoe*, a new
genus name. The plant intended by Kamel is Linnaeus' *Cotyledon laciniata*,
De Candolle's *Kalanchoe laciniata*, a native of Southern China.

Martyn, in the 9th edition of Miller's Gardener's Dictionary, 1797,
states that Linnaeus named the Genus *Camellia* in honour of the Jesuit
G. J. Kamel or Camellus. Sir J. E. Smith in Rees' Cyclopaedia says the
same. They may be right, although no corroboration of these statements is
found in Linnaeus' writings. The name Camellia first appears in his Systema
naturae, 1735, without any explanation. J. Dryander, Linnaeus' country-
man and contemporary, in the Trans. Linn. Soc. I, p. 172, note, (1789)
states: — "In a letter by Kamel to Petiver, preserved in the Brit. Mus. he
signs himself "Kamel". The plant named from him ought therefore to be
called "Kamelia" instead of Camellia. The Abbé Berlèse, however, in his
"Monographie du genre Camellia", 1837, reports, without giving the source
of his information, that Linnaeus named the Genus Camellia in testimony
of the exertions of the Jesuit Father *Camelli,* who in 1739 introduced the
Camellia japonica from Japan to Europe. Whether Berlèse's Camelli is identical with Kamel and whether the latter had ever visited Japan, I am not
prepared to say. Poiret in Enc. Bot. VIII, 749, asserts that Linnaeus named
the genus *Camellia* in honour of the botanist Camelli, who at the end of the

17th cent. collected plants in America. In Aiton's Hortus Kewensis, first ed. II, 460, we read that Camellia japonica was cultivated by Rob. James Lord Petre, before 1739.

LETTRES ÉDIFIANTES ET CURIEUSES ÉCRITES DES MISSIONS ÉTRANGÈRES. —

This collection of letters written by the ancient Jesuit missionaries to their Superiors or friends in Europe is a vast mine of information regarding China in the 17th and 18th centuries. The following names of Jesuit missionaries appear there in connection with articles on Chinese plants:

Joannes Laureati, an Italian, born in 1666, joined the Chinese mission in 1697, and died in 1727 at sea. He wrote a letter dated Fokien, July 26, 1714, to the Baron de Zea, in which he gives some accounts of the vegetable productions of China, especially of *Tea*. He notices *Tobacco*, which in the beginning of the 18th century was largely cultivated in Fo kien, and clothes made of the fibres of a nettle *(Boehmeria nivea)*.

Francis Xavier d'Entrecolles, a Frenchman, born in 1662, joined the Chinese mission in 1698, and died in Peking in 1741. Two of his letters, dated 1727 and 1736, referring to Chinese plants have been preserved. In the first he gives an interesting account of the manufacture of artificial flowers, which the Chinese make of the pith of the plant *tong tsao (Aralia papyrifera)*. He further speaks of the *Fingered Citron* or Buddha's hand. In the other letter he notices the Chinese fruit *shi tse (Diospyros Kaki, D. Shitse)*, the tree *hoaï shu,* a kind of Acacia the flowers of which are used for dyeing yellow, *(Sophora japonica)*, the Chinese Willow *(Salix babylonica)* the wool of which is used like cotton, the Belvedère *(Kochia scoparia)* recommended as a substitute for food in times of scarcity. Finally d'Entrecolles gives an account of the Chinese *Camphor tree,* and the method used by the Chinese to obtain Camphor from it.

Dominicus Parennin, a Frenchman, born 1665, came to China in 1698, and died at Peking in 1741.

In 1723 Parennin sent a few Chinese drugs to the Academy of sciences in Paris, furnishing some explanatory remarks on them in an accompanying letter.

The first of these drugs, in Chinese *hia tsao tong tchong,* meaning "a plant in summer, an insect in winter", is produced in Tibet and Szech'uan.

Réaumur described and figured it in the Mémoires de l'Académie, 1726, p. 302, tab. 16. The drug in question is a Fungus, *Cordyceps sinensis (Sphaeria sinensis)*.

The drug next described, is the *san tsi*. The plant from which it is derived is still unknown to botanists. Another drug of which P. gives a short account, is an aromatic root from Szech'uan, probably a species of *Ligusticum*.

During 18 years Parennin had accompanied the Emperor Kang hi on all his frequent travels and hunting expeditions generally directed to Southern Mongolia and Manchuria. He gives in the same letter a slight sketch of the botany of these regions. The *ulana* a great red cherry produced on a dwarf stem is the *Prunus humilis*. Bge. — In another letter Parennin mentions the climbing plant *teng lo* with beautiful violet flowers hanging down in large bunches. This is *Wistaria chinensis*.

Petrus Jartoux, a Frenchman, born in 1668, joined the Chinese mission in 1701, and died at Peking in 1720. The fathers Jartoux and Regis had been intrusted by the Emperor Kanghi with the survey of Manchuria, and on this occasion Jartoux had opportunity to visit the very country where the famous *Ginseng* plant is produced. In a letter, dated Peking, 12 April 1711 he gives the first authentic account of this plant accompanied with a drawing made by himself from nature.

Gaspar Chanseaume, a Frenchman, born in 1711, joined the Chinese mission 1746, and died in the province of Kiangsi in 1761. He has left a very interesting Memoir on *Chinese Insectwax* written about 1750, and published among the Lettres édifiantes etc.

For further particulars regarding D'Entrecolles, Parennin, Jartoux and Chanseaume and the Chinese plants mentioned by them I beg the reader to refer to my Earl. Europ. Res. p. 28—32.

Du Halde. DESCRIPTION DE L'EMPIRE DE LA CHINE. 1735.

Joh. Bapt. Du Halde, 1674—1743 a Jesuit, who had never visited China, compiled this valuable work, of which several editions were published in French, and which has been translated into many other languages, from the letters of the Jesuit missionaries in China, partly addressed to him, and not included in the collection of the Lettres édifiantes etc. Du Halde devotes several chapters to Chinese natural history, agriculture, and medicine, and gives translations from the well known *Pen ts'ao kang mu*, the Chinese standard work on Natural History and Materia Medica. A list of all the Chinese plants spoken of in Du Halde with their identifications has been given in my Earl. Europ. Res. p. 33—35.

Aug. Pyram. De Candolle has commemorated the name of the learned Jesuit father in a Chinese plant of the order Compositae. In the Prodromus Regni veget. V (1836) p. 366 sub *Duhaldea chinensis* we read: "Nomen in memoriam J. B. Duhalde qui medio seculo XVII Chinam adiit et descripsit". The great botanist was mistaken. Du Halde was born in the second half of the 17th cent. and never visited China in person.

IV. Sea-Trade of the Dutch with Eastern Asia in the 17th Century.

The first appearance of the Dutch in the Indian Archipelago dates from the end of the 16th century. In 1595 a squadron consisting of 4 ships under the command of C. Houtman sailed from Holland to the East Indies. Having visited in the next year several places in the island of Java and the adjacent islands, they returned to Holland with a rich cargo of spices. This first attempt was followed by other Dutch commercial expeditions to India and the Archipelago, which also proved successful. In 1602 the Dutch East India Company was established. The war which then issued between the Dutch, Spaniards and Portuguese for the possession of the Spice islands lasted till 1610, when the Dutch remained masters of these seas and monopolized the lucrative trade there. The seat of the Dutch government was first established in the island of Amboyna, but in 1619 it was transferred to the newly founded city of Batavia in Java, from which year may be dated the formation of the Dutch East Indian Empire.

Twelve years before the first commercial expedition of the Dutch to the Archipelago took place, **J. H. Van Linschoten**, a studious young Dutchman had visited India. Having obtained a place in the suite of the newly appointed Portuguese Archbishop of Goa, he reached that place in 1583, and spent 5 years there. On his return to Europe he began to compile a book with the title: NAVIGATIO ET ITINERARIUM IN INDIAM ORIENTALEM, which was first printed in 1599, and published also in Dutch. In 1885 the Hakluyt Society edited an English version of it. This book contains interesting accounts of India and other countries visited by the author, or of which he had gathered information from other persons. There are also a number of chapters devoted to the natural productions of India especially those of vegetable origin, fruits, drugs etc., accompanied with drawings. L. also speaks of some Chinese drugs, as Rhubarb, China root, China Camphor, and notices the Chinese Lac or Varnish. But his notes on plants are for the greater part borrowed from Garcia ab Orta.

It was in 1601, a year before the Dutch E. I. Company was founded, that Dutch ships made their first appearance in the Chinese waters. In this year the Admiral J. C. Van Neck, who had been sent out to the East with a small trading fleet by Dutch merchants, made sail for China. In Sept. 1601 he found himself, without knowing it, with his two ships off Macao. Some of the crew, who were sent on shore, were hung by the Portuguese, whereupon the admiral made haste to return to the Archipelago.

In 1604 the first attempt was made by the Company to obtain a footing for their trade in China. W. Van Warwijk was sent with a commission to open friendly commercial intercourse with the Chinese, but owing to the influence of the Portuguese of Macao, this was refused.

In 1622, J. P. Koen, Governor of the Dutch settlements, sent out from Batavia, which city he had founded in 1619, a fleet of 8 vessels under the command of Bontekoe Van Hoorn and C. Reijersz, to attempt the expulsion of the Portuguese from Macao. This place having been bombarded without any success and with considerable loss on the part of the assailants, the Dutch fleet sailed to the Pescadore I° situated between Formosa and the mainland and the Dutch established themselves on one of the islands, and built a fort there. In the same year they visited Taivan in Formosa, Amoy and Chin chew (Chang chou fu), but were not allowed by the Chinese authorities to carry on trade in the Chinese ports. In 1624 the Dutch removed from the Pescadore I° to Formosa and built the fort of Zeelandia on an island at the entrance of the Bay of Taivan.

In 1655 the Company at Batavia resolved on despatching an embassy to Peking to the Emperor Shun chi in order to obtain free trade in some Chinese ports. P. de Goijer and J. de Keyzer were appointed envoys. The narrative of this embassy was published by J. Nieuhof, who had accompanied them as steward of the mission, with the title: "Legatio Batavica ad Magnum Tartariae Chanum Sung Teium, Sinae Imperatorem", 1665. Comp. my Earl. Europ. Res. p. 25. The embassy started from Canton on March 17th 1656 in Chinese boats, followed up the Pe kiang or North River to Shao chou fu, then ascended an affluent of it which comes from the N. E., to Nan hiung chow. Then they had to cross the mountain range separating the province of Kuang tung from that of Kiang si. At Nan an fu in Kiang si they came again to a river, where they embarked. Then descending the

Kan River they reached Nan ch'ang fu, sailed across Lake Po yang, entered the Yangtze kiang, reached at Yang chou fu the Grand Canal, which led them to T'ien tsin. Peking was reached on July 17th. The envoys were admitted to an audience, prostrated themselves before the Emperor, but did not obtain permission to trade. On Oct. 16th 1656 they set out on their return, and travelling by the same river way they had come, the embassy reached Canton on Jan. 28th 1657.

After occupying for about 38 years a large part of Formosa, the Dutch were expelled from the island, in Jan. 1662, by Cheng Ch'eng kung (Koxinga of the Portuguese), a powerful Chinese pirate, and thus lost their footing in China.

In Japan the Dutch had been more successful in their commercial enterprises. According to Kaempfer, History of Japan, the first Dutch vessels visited Japan in 1600. Nine years later, in 1609, the Dutch E. I. Comp. sent several small vessels to Firando (N. W. of Nagasaki), where they were well received by the Japanese, and, in 1611, a formal edict in favour of their trade was obtained. A Dutch factory was established at Firando. The Dutch trade was opened in Japan by Jac. Spex sent by the Company in the quality of an envoy and subsequently chief of the factory. He left Japan in 1620, and in 1629 was appointed Governor-General of the Netherlands India. Subsequently Japan was closed against foreigners (Portuguese, Spaniards) with the exception of the Dutch and Chinese. From 1834, however, the Dutch trade with Japan has been limited to the island of Decima (Deshima) in Nagasaki. After the year 1653 the Chinese pirate Koxinga began to harass with his fleet the forces of the new Tartar (Manchu) dynasty in China, and ravaged and plundered the coast of the Fu kien province. He even established himself in the islands of Amoy and Quemoy and built fortifications there. After the expulsion of the Dutch from Formosa, the Council at Batavia decided to send a fleet to China, and to propose to the Tartars to operate conjointly against Koxinga. On June 20th 1662 B. Bort, in the capacity of envoy left Batavia with a squadron of twelve vessels, and reached Hoksieu (Fu chou fu) on Aug. 14th. The vice roy Sing la mong and the commander-in-chief of the Tartar forces Lipui, then living in the interior of the province, induced the Dutch envoy to send two of his officers to the vice roy's camp, to arrange concerning operations. These officers performed the journey to this camp, at a place called Sinksien (?), in 11 days, conferred with the commander and returned to Fu chou fu. After leaving this port the squadron visited several ports on the coast, and amongst them Swatow, and on April 11th 1663 reached Batavia. In a few months Bort was again despatched to China with a reinforcement of 16 other ships. Meanwhile Koxinga had died and his son prosecuted his father's depredations. The Dutch fleet, in conjunction with the Tartars, attacked the pirates and succeeded in expelling them from Amoy and Quemoy. The Dutch were then permitted to trade with Canton, Fu chou and Chin chew (Chang chou fu).

In 1666 the council at Batavia despatched for the second time an embassy to the court of Peking. Pieter Van Hoorn was appointed ambassador and received instructions to petition for free trade and permission to erect factories. He landed at Hoksieu (Fu chou fu) on Aug. 6th 1666, but did not start for Peking till the 20th of January 1667. The narrative of this embassy as well as that of Bort's mission, in 1662, are found in Dapper's "Gedenkw. Bedryf d. Nederl. O. I. Maetschappye in het keizerrijk van Taising of Sina, 1670". We find there a detailed account of the journey and the route followed, which we can easily trace as the geographical names are not much corrupted. They proceeded from Fu chou in boats up the Min River to Kien ning fu, and then ascended an affluent of this river on which the cities of Kien yang fu and P'u ch'eng are situated. From the latter place they went by land, and having to cross the mountain chain forming the boundary between the provinces of Fu kien and Che kiang they came to a navigable river where they embarked again. Descending this river they passed by Lan k'i hien, Yen chou fu and entered the great river (Ts'ien t'ang kiang) which flows past Hang chou fu. The rest of the journey was performed on the Grand Canal, and Peking was reached on June 21st 1667. At the end of August they left the capital, and returned to Fu chou by the same route. Hoorn was admitted to an audience and had to perform the act of prostration before the emperor. None of the privileges he solicited for the Company were granted.

After this summary outline of the early commercial intercourse of the Dutch with Eastern Asia, derived from various authentic sources, I shall now proceed to examine what advantages resulted to botanical science from it, and what the Dutch in the 17th century contributed to our knowledge of Chinese plants:

Clusius, in his "Exotica", 1605, p. 36, describes and figures a collection of fruits brought home by the Dutch expedition (Van Neck) sent to the East Indies, in 1597. In fig. V we may easily recognise the Chinese *Litchi* fruit. — In the same book, p. 82, Clusius states that the first specimen of *Gummi Gutti* or *Gambogia* was brought to Europe from China by the Dutch admiral Van Neck, in 1603. But these articles had hardly been procured in China. Comp. supra p. 22.

Let me further notice that we owe to the Dutch the first authentic account of the useful *Tea* plant. Jac. Bontius a Dutch physician, for many years a resident in Batavia, in his "Historia naturalis et med. Indiae orientalis", written in 1631, p. 87, gives a short notice of the Tea shrub: "de Herba seu Frutice quam Chinenses *The* dicunt, unde potum suum ejusdem nominis conficiunt". B. states that no European has ever seen the Chinese Tea plant, and that he is indebted for all information about it to the General Jac. Spex, who resided several years in Japan (v. supra p. 23) and saw it growing there. G. Piso, likewise a Dutch physician, who in 1658 published Bontius' writings, adds a more detailed and quite correct description of the Tea shrub, with a figure which, as he states, had been drawn from nature in Japan, and presented to him by D. Caron, "olim in Japonia praefectus" *).

It has also been asserted and even admitted as a fact, that Tea (as a beverage, dried leaves) was first introduced into Europe by the Dutch E. I. Comp. J. C. Lettsom, in his Monograph on Tea, 1772, states that the Dutch first imported Tea in the beginning of the 17th century, and suggests that it was then brought not from China, but from Japan, for the Dutch trade with China, at that time, was trifling compared with that carried on in Japan. I have not been able to make out the earliest or original source of this assertion, nor has Professor Schlegel found any allusion to such an early Tea trade of the Dutch in ancient Dutch records. I may observe that in Valmont de Bomare's "Dictionnaire d'Hist. nat.", 1791, article 'Sauge', a curious statement referring to the Dutch Tea trade is found. We read there (source of information not given) that our common Sage *(Salvia officinalis)* is higly valued by the Chinese and Japanese and that the Dutch use (or used) to gather this plant in South Europe and take it to the Chinese, accepting in return the Chinese Tea. For one chest of Sage they receive from two to three chests of Green Tea. I have in vain tried to find a corroboration for this statement elsewhere.

Although it cannot be proved that the Dutch in the first half of the 17th century first introduced Tea as an article of commerce into Europe, it is, nevertheless, a fact that they were the first to introduce the Tea

---

*) My respected friend, Professor Dr. G. Schlegel of Leyden, to whom I owe many interesting notices regarding Dutch intercourse with Japan and China, drawn from Dutch documents, kindly informed me that François Caron was sent to Yeddo by the Dutch E. I. Comp. in 1638, and in 1639 was appointed chief of the Dutch factory at Decima.

plant from Japan into the gardens of Holland, and to recommend Tea as a beverage.

N. Tulpius, a Dutch physician of Amsterdam, in his "Observationes medicae", 1641, lib. IV, cap. 60, details the medical effects of the "Herba Thée". But he confesses that Tea was unknown to him, and that all he writes about it was communicated to him by persons who had visited the native countries of the plant. — Piso (l. c.), however, states (in 1658) that the Magnates in Europe were accustomed to drink Tea: "unde fit ut magna copia et magno pretio ipsa folia exsiccata et in capsulis plumbeis recondita, undiquaque nunc distrahantur".

Jac. Breyn (1637—1716) a merchant of Danzig, a distinguished botanist, repeatedly visited, in the last quarter of the 17th century, the flower gardens in Holland and described the remarkable plants he met with there in his Prodromus Plantarum rariorum in hortis celeberrimis Hollandiae observatarum, which appeared in two parts, published respectively in 1680 and 1689. In Part II, 98, he describes a shrub "Styraci et Evonymo media affinis, *The* Sinensium sive *Tsia* Japonensibus, flore simplice, nobis", and says: "in horto Amstelodamensi saltem conspeximus". Thus Linnaeus was not the first, as is generally admitted, who cultivated the Tea shrub in Europe.

Two famous Chinese gardenflowers, the *Azalea indica* and the *Chrysanthemum sinense* were likewise first cultivated in Europe in the flower gardens of Holland, and we may therefore presume that they had been imported in Dutch ships. In Breyn, Prodr. I (1680) p. 24, we read: "Chamaerhododendron exoticum amplissimis floribus liliaceis. Frutex spectabilis elegans. In horto Beveringiano. Flores purpurei elegantissimi." This is, according to Linnaeus, *Azalea indica*. Breyn does not say from what country this ornamental shrub had been brought, but P. Hermann, who in his Catal. Horti Ludg. Batav. (1687), p. 152, describes the same plant under the name of "Cistus indicus Ledi alpini foliis, floribus amplis," and figures it on tab. 153, reports that it was introduced from Jaccatra (Batavia, Java), where it was probably found cultivated by the Chinese. Comp. Flore des Serres, 1853, p. 77. *Rhododendron Breynii,* Planchon.

The *Chrysanthemum sinense* Sab. is described by Breyn in Prodr. II (1689), p. 66 as Matricaria japonica maxima, flore roseo sive suave rubente, pleno elegantissimo, nobis. Kychonophane Japonensibus (kiko no fanna in Thbg. Fl. jap. 320). Breyn enumerates no less than six varieties then cultivated in the gardens of Holland, viz: floribus suave rubentibus, fl. candidissimis, fl. purpureis, fl. luteo-obsoletis, fl. carneis, fl. phoeniceis. These plants were subsequently lost in the Dutch gardens, for no further account of them, except that given by Breyn, can be traced, nor were the gardeners of Holland acquainted with them when again introduced into Europe, in 1789.

In the above mentioned garden of H. Beveringk, Breyn saw also, in 1678, a fine specimen of the *Camphor tree* which had been introduced from Japan. When ten years later he visited the same garden, it was a large tree showing rudiments of flowers. See Prodr. I, p. 4; II, p. 16. In his "Exoticarum Centuria prima", 1678, p. 11, 12, Breyn describes and figures a branch of the "Arbor camphorifera japonica" which had been sent to him from Japan by Dr. W. Rhyne in 1674.

G. E. Rumphius, 1627—1706, a physician and a merchant in the service of the Dutch E. I. Company, was Deputy Governor of Amboyna, on which island he passed the greater part of his life, from 1655 till his death, devoting himself ardently to the exploration of its natural productions. His chief work the Herbarium Amboinense was finished in 1690. A short time before he died he added an Auctuarium. In the whole he describes about 1750 different plants, of which 1060 are figured. The MS of this work, after the author's death in the possession of the E. I. Company, was only published half a century later, from 1741—1754 by J. Burmann of Amsterdam. In this elaborate illustration of the Flora of Amboyna, Rumphius notices and figures also a considerable number of Chinese plants, either cultivated in the island or regarding which he had derived information from the Chinese settlers there. The Chinese names of the plants are frequently misspelt, but, as Rumphius generally gives the meaning of the sounds, they can be recognized. The scope of the present work does not permit us to extract all the interesting notes on Chinese plants found scattered in the Herbarium amboinense. But let us here mention one Chinese plant which Rumphius himself says that he sent to Amsterdam, Herb. amb. I, p. 92, "Arbor calappoides sinensis, Sinensibus *titsju,* h. e. arbor ferrea, dum ejus folia instar pinnae ferreae pungant". This is the *Cycas revoluta,* Thbg., a small palm like tree of China and Japan, in Chinese *t'ie shu* (iron tree) or *t'ie tsiao* (iron banana). The Chinese authors say that when the tree is about to die, it will thrive again, if burned with hot iron nails, and hence its name. The same is reported by Bontius (1631), l. c. p. 85, and by Kaempfer, Amoen. exot. 897, sub *tessio:* characterem ferreum ingreditur, quia ab infixis clavis melius nutritur.

E. Kaempfer, 1651—1716, gives in his AMOENITATES EXOTICAE, published in 1712, an admirable account of Japanese plants, the fruit of his observations during a stay of about two years, 1690—92, in Japan, when in the service of the Dutch E. I. Company. Among these plants he notices the following as natives of China and introduced from that country into Japan.

P. 608, *Thea,* Japonibus *Tsjaa,* Sinensibus *Theh* dicta etc.

P. 817, *Rjugan* vel *Djugan niku,* i. e. oculus serpentis. Frutex sinensis etc. — This is *Nephelium Longan.*

P. 818, *Sju sin, Nindsin* etc. Ob polychrestam radicem, omnium Orientis celebratissima planta; quae ex Coraea Japoniae illata, in hortis colitur. In Katajae (provinciae Sjamsai), Coraeaeque montibus rigidiori Jove nata etc. Kaempfer speaks of the *Ginseng* plant, but what he figures under this name, p. 819, fig. 1, 2, 3 is *Sium Ninsi;* only fig. 4 is a correct drawing of the Ginseng root.

P. 831, *Kantats* or *Too Tsisa,* i. e. Lactuca sinensis. — According to Thunberg, Fl. jap. 300 this is *Lactuca sativa.*

P. 834, *Sjokkuso,* vulgo *Too Kibbi,* Milium sinense dictum, quod inde antiquitus delatum: caule foliisque arundinaceis magnis; spica ramosa erecta, granis flavescentibus. — Thunberg, Fl. jap. 37, identifies this with Zea Mays, but judging from the Chinese characters given by Kaempfer, the latter means *Sorghum.* Matsumura, in his Nomencl. Japan. Plants, gives *To kibi* as the Japanese name for *Sorghum saccharatum.* — *Zea Mays* is noticed by Kaempfer, p. 835, under the name of *Kjokuso,* vulgo *Nan ban Kiwi,* i. e. Milium populi septentrionalis (but *nan* means South), quia a Lusitanis ex India primum invectum. Frumentum indicum, juba fastigii divulsa. Thunb. l. c. 51 identifies Kaempfer's Nan ban Kiwi erroneously with Cynosurus coraeanus.

P. 884, *Nankin Ssugi,* Juniperus Barmudiana, Herm. H. Bat. quae ex regni Sinae provincia Nankin invecta, ob pulchritudinem colitur. According to Thunberg, l. c. 264, this is Juniperus barbadensis, but Thunberg's Jun. barb. is reduced by Siebold to *Juniperus chinensis.*

P. 878, *Obai* seu *Robai,* cum icone. Frutex Nanquino primum adductus, in hortulis ob pulchritudinem colitur. — This is *Chimonanthus fragrans.*

V. Early English Intercourse with China. First Botanical Collections made in that Country.

The English first visited the islands of the Indian Archipelago some years after the Dutch, i. e. in 1602, but it was not till 35 years later that they made their appearance in Chinese waters. In 1637 an English fleet under the command of Captain John Weddell, sent to China by Charles I, King of England, to open up trade with that country, anchored in the roads of Macao and Weddell proceeded even towards Canton. But owing to the intrigues of the Portuguese his negotiations with the Chinese mandarins at Canton were frustrated. Finally, after Weddell had bombarded and forced one of the Chinese forts at the entrance of the Canton River, the Chinese agreed to supply the English ships with cargoes. No other attempt to open up trade was made till 1664, when the English East-India Company, who then had factories at Bantam in Java and at Madras, sent a ship to Macao, which, however, had to return without having sold any goods. This did not discourage the Company, who made inquiries respecting other favourable Chinese ports and at length succeeded, in about 1677, in starting a trade with the son of the famous pirate Koxinga (v. supra), who was then in possession of Formosa (Zelandia) and Amoy. The Company was allowed to establish factories at these places and the trade there was continued for several years with considerable profit. In 1681, however, the Company ordered their factories at Amoy and in Formosa to be withdrawn, and in 1684 succeeded in obtaining, notwithstanding the intrigues of the Portuguese, a footing at Canton. In 1685 trade was renewed at Amoy. The next attempt to enlarge the commercial relations with China was made by the E. I. Company in 1701, when they obtained permission from the Chinese government to open up trade in Chusan, the largest of the islands of the Archipelago opposite to Ningpo. The Company's factory there was not, however continued later than 1703, the trial being found unsatisfactory.

This last mentioned expedition to Chusan, in 1701, led to the first botanical exploration of that island by J. Cunningham, of whose exertions we are now about to give a brief account. But let us in the first place notice, that he was not, as is generally believed, the first European collector of Chinese plants, for there is evidence from the statements of Petiver and Ray, that previous to the year 1688 some dried plants collected at Amoy had reached England.

In my Earl. Europ. Res. p. 46—52 full notices have been given of

the Chinese plants described in J. Petiver's Museum. The latter comprises
ten centuriae, written successively from 1692 to 1703, but it is not stated
to what year each centuria is to be referred. The Chinese specimens there
described were gathered chiefly by Cunningham. His name, however, appears only in a few cases. On p. 9 we read: "Since my 4th and 5th centuriae, Mr. Keir and Mr. Barclay, surgeons, presented me with some
plants they collected in China". In his account of Cunningham's Chusan
plants, published in the Phil. Trans. XXIII, 1703, Petiver states sub. n. 90,
Ricinus chinensis sebifera: "Mr. Sam. Brown*) first sent me this some years
ago from China, since which I have received it from Emoy and Chusan".

In the first centuria of the Museum the following three Chinese Ferns
are mentioned, and two of them as having been gathered at Hamuy (Amoy).

Adiantum nigrum chinense, tenuiter divisum, pinnulis minimis obtusis
bifidis. Chinese black Maidenhair with blunt forked leaves Hamoy in China.
Petiver quotes: Raii Historia plant. (II, 1688), p. 1854 and Plukenet's
Almagestum (1696) p. 10. Both Ray and Plukenet say that they received
it from Petiver. Figured in Pluk. Phytographia (1691) tab. 4, fig. 1. —
This is *Adiantum tenuifolium*, Lamarck, Enc. Bot. I, 44, or *Davallia chinensis*, J. E. Smith (1798) and Swartz Syn. Fil. (1806).

Adiantum nigrum lanuginosum chinense, Hoary Black Chinese Maiden-
hair. Hamoy in China. Petiver quotes: Raii Hist. pl. (II, 1688) p. 1854,
Dryopteris Chinensis lanuginosa, Pluk. Almagestum p. 11, Phytogr. tab. 4,
fig. 2. Ray and Plukenet received it from Petiver.

Filix pyramidalis chinensis. Filix e China mollis, auricula ad pinnulae
basim superne producta, summo folio longius mucronata. Our China Steeple
Fern. Petiver quotes: Raii Hist. plant. II, (1688) p. 1853, and Plukn. Almagestum (1696). 151, Phytogr. (1691) tab. 30, fig. 2.

Ray states that he saw it in Petiver's collection.

A fourth Chinese Fern is noticed by Ray l. c. p. 1853, as Adiantum
folio Coriandri ramosum etc. Ramulum siccum e China communicavit nobis
D. Petiver. The same in Plukn. Almagestum, p. 11, Adiantum Chinese,
perelegans ramosum. Phytogr. tab. 4, fig. 3. Petiver does not mention
this species.

In the Philos. Trans. XXVII, 1710—12, p. 417, Rare plants grown
in the Physic Garden of Chelsea etc., Petiver notices the China *Shunda* or
Thorny Nightshade, Shunda chinensis folio pannoso minore, Nobis. Received
from China. He quotes: Plukenet Phytographia (1691), tab. 62 fig. 1, So-

---

*) Sam. Brown, mentioned by Dillenius, in Hort. Elth. 1732, 345 as having introduced
Senecio Pseudochina from India.

lanum incanum chinense minus, spinosum, floribus parvis fere umbellatis. Plukenet says: a D. Petiver accepi. — Shunda is a Malabar name for Solanum. See Rheede Malab. II, p. 69. — Dunal in DC. Prodr. XIII, I, p. 373 named this plant *Solanum Chinense*. Comp. Forbes & Hemsley, Ind. Florae Sin. II, 171, sub *Solanum indicum*.

Plukenet mentions yet another plant received from China in the last quarter of the 17th cent., the *Artemisia chinensis,* cujus mollugo *moxa* dicitur, which is found figured in the Phytographia (1692) tab. 15, fig. 1, and described in the Almagestum (1696) p. 50. This is *A. vulgaris,* var. *indica,* according to Linn. Hort. Cliff. 404, n. 6.

James Cunningham. 1701.

Very little is known regarding the life-history of this English naturalist, who in the beginning of the 18th century sent from China to England a rich herbarium of Chinese plants.

According to the Dict. of Nation. Biogr., Cunningham was a Scotchman. He entered the service of the E. I. Company and first went out in 1698, as surgeon to the Comp.'s factory at Emuy (Amoy) in China. He seems to have returned to England in the next year. In 1699 he was elected a Fellow of the Royal Society. In 1700 he made a second voyage to China. From two of his letters, written in 1701 and published in the Philos. Trans. XXIII, 1702, p. 1201 sqq., we learn that the ship in which he went out, touched at Borneo, on the 17th July 1701, and reached the coast of China on August 13th. On the last of August they anchored at the Crocodile Islands [probably Dogs Islands, S. E. of the mouth of the Min River (Fu chou). Alligator I of modern sea charts is a rock in the sea farther N. E.]. On October first they reached the Island of Chusan. His second letter is dated Chusan, 22nd November 1701. In February 1703 he is reported to have been sent to the Company's station at Pulo Condore, to try and open up trade with Cochinchina, but this attempt proved a failure. In 1707 he proceeded to Batavia, and soon after embarked for England, but there is some reason to believe that he died on the voyage. His last letter to Sloane and Petiver is dated Jan. 4th, 1709 and was received in August 1709.

In my Early European Researches into the Flora of China, p. 38—43, I have reproduced Cunningham's letters and his account of the Island of Chusan. His botanical collections were made at Amoy, in the Dogs Islands, but chiefly in Chusan. He distributed his plants among his friends Sloane, Petiver and Plukenet.

Sloane, Sir Hans. This enlightened and munificent patron of botanical science, collector and successful physician, was born April 16th, 1660 at Killileagh, Country Down, Ireland. He went to London to study medicine, and directed his attention assiduously to botany. After having travelled through France, spending some time at Paris and Montpellier, he took his M. D. degree at the University of Orange (S. France) and returned to London. In 1685 S. was elected into the Roy. Society and two years later became Fellow of the College of Physicians. In 1687 he took the opportunity of proceeding to Jamaica as physician in the suite of the Duke of Albemarle, who died soon after landing. Sloane's visit to that island lasted only 15 months, but during that time he got together about 800 new species of plants, the island then being virgin ground to the botanist. Of these he published an elaborate Catalogue, and 1707—25 two folio volumes, "Natural History of Jamaica," on the experiences of his visit.

Sloane was a distinguished physician and had a large and lucrative practice among the upper classes of London. In 1714 King George I made him Surgeon-General of the army, and in 1727 appointed him his Physician-in-Ordinary. In 1716 he had already been created Baronet.

In 1693 S. became Secretary to the Roy. Society and edited the Transactions for 20 years. In 1727 he succeeded Newton in the presidential chair of the Roy. Soc., which post he held till 1740, when, at the age of eighty, he resigned and retired to his manor-house of Chelsea, where he died 11 Janv. 1753.

On purchasing the great estate of Chelsea, in about 1720, S. presented the Apothecaries' Comp. with the fee simple of the garden which they had already established there, in 1673, on the condition that it should for ever continue a botanic garden, on certain conditions that have ever since been observed.

But Sloane's memory survives principally through the bequest of his immense collections of objects of natural history and artificial curiosities, library and rare manuscripts etc. to the English nation. His extensive herbarium comprised all Petiver's and Plukenet's (including Cunningham's) collections which Sloane had purchased after their death. He had also got possession of E. Kaempfer's important collections comprising dried specimens and drawings of Japanese plants. All these valuable treasures S. bequeathed to the public on condition of a payment of 20,000 pounds to his heirs. They formed the basis of the "British Museum", that great national repository for the establishment of which the nation is entirely indebted to Sloane. For the reception of these collections was purchased, after Sloane's death, by Act of Parliament, the noble mansion built in 1680 by Ralph, first Duke of Montague. At first it was divided into three departments, Books, Manuscripts, Natural History. In 1827 a special department — that of Botany — was created in consequence of the bequest by Sir Jos. Banks of his botanical collections.

Petiver, James, born c. 1658 at Hill Morton, Warwickshire, Apothecary to the Charterhouse, London, Fellow Roy. Soc. 1695, was an active collector of objects of natural history, and had correspondents in most parts of the world, who sent him plants, animals etc. He was a friend of the botanist Ray and of Sir Hans Sloane. Petiver belonged to the Apothecaries' Company (supra p. 32) and took a great interest in their garden at Chelsea. He described many new exotic plants raised there. P. died at London, April 20th 1718. His botanical collections, acquired by Sloane, are now in the British Museum.

In a paper published in the Philos. Trans. vol. XXIII, in 1703, Petiver describes about 70 Chinese plants, supplied principally by Cunningham. Previously he had published in his Museum, Rariora Naturae etc, 1692—1703, short characteristics of 1000 exotic plants, scattered amongst which we find about a hundred from China, the greater part of which are not mentioned in the Philos. Trans. — Besides these he issued the Gazophylacii Naturae Decades decem, 1702—1709, descriptions and figures of a hundred exotic plants, about one fifth of which were selected from Cunningham's Chinese collection.

It appears from Cunningham's letter (Earl. Res. 43), and from Petiver's quotations, that the latter had received from Cunningham a collection of Chinese drawings representing plants. Petiver repeatedly speaks of «Herbarium nostrum sinense pictum». Mr. F. B. Forbes, who in 1883 inquired after these drawings in the British Museum, found only, in Sloane's MSS, n. 2376 ff. 82—110, a «Catalogus plantarum quarum icones in China delineatae sunt», which turned out to be a brief enumeration of tabulae 1 to 43, each with 18 numbers, making a total of 774 numbers. As Petiver's handwriting appears in this catalogue, it can hardly be doubted that it refers to the Herb. sin. pict. (See Journ. Bot. 1883, 12).

Plukenet, Leonard, a distinguished English botanist, born in 1642, was a M. D., but it is unknown where he took this degree. He was Queen's Botanist to Mary II and had charge of the Royal Garden at Hampton Court. His own botanic garden was at Old Palace Yard, Westminster. He died

July 6, 1706. See Mr. B. D. Jackson's "Notice of Plukenet", in Journ. Bot. 1882, 338.

Plukenet has described and figured a great number of new plants from America, the East Indies and China. Almost all the Chinese plants noticed by him are from Cunningham's herbarium.

One of Plukenet's principal works, the Amaltheum Botanicum seu Stirpium Indicarum alterum Copiae Cornu, issued in 1705, contains descriptions of about 400 Chinese plants, intermixed with E. Indian and American species and plants from the Cape of G. H. Nearly one half of the Chinese plants are figured on the plates appended to the Amaltheum. In my Earl. Res. 44, I erroneously referred these plates, tab. 351 to 454 to Plukenet's Phytographia. But he gives this name only to the first series of his delineations of plants, tab. 1 to 328, issued in 4 parts, from 1691—1696, the text to which was furnished by the Almagestum, 1696.

Plukenet's MSS and Herbarium of 8000 plants are in the Herb. Sloane, in the British Museum. We learn from Mr. J. Britten (Journ. Bot. 1882, 249) that Cunningham's collections of Chinese plants still exist in the Brit. Mus. scattered through various volumes of the Sloane Herb., and that most of the specimens could be readily identified by any one familiar with Chinese botany; they are mostly localised under the general title of "China;" but some are from "Emuy" (Amoy), and others from "Chusan".— Some of Cunningham's Chusan plants have also found their way to the Herbarium of the Imp. Academy, St. Petersburg. They were received before 1768.

In my Earl. Res., 46—88, I have extracted at full length all the notices given by Petiver and Plukenet of Cunningham's Chinese plants, nearly 600 species in all. I may therefore restrict myself, in the present work, to mention briefly only those which can be easily recognized from the ancient descriptions or figures, or which have been determined by Linnaeus and since his time by other botanists.

Clematis minor maritima Sinensium. Pluk. Amalth. 60. — Earl. Res. 71. Identified in DC. Syst. I (1818), 136, with *Clematis minor*, Lour.

Aquilegiae corniculis, Moschatellinae foliis planta pusilla. Chusan. Pluk. Amalth. 19, t. 360, f. 3. — Earl. Res. 64. — Lamarck, Enc. Bot. III, 99, species of Helleborus or Isopyrum. In DC. Syst. I (1818) 324, it is named *Isopyrum adoxoides*. — Ind. Fl. Sin. I, 18.

Aconitum minus autumnale cheusanense. Sin: *Tsou u.* Flor. Sept. Oct. Pluk. Amalth. 5. — Early Res. 63. — Perhaps *Aconitum autumnale*, Lindl., which was brought by Fortune from Chusan in 1846.

La boe Chinensibus. Arbor flore luteo, foliis acutis binis decussatis. Petiv. Mus. 937. — Earl. Res. 49. — *Chimonanthus fragrans.*

Cocculi orientalis frutex convolvulaceus, orbiculatis foliis, prona parte villosis. Ins. Crocodil. Pluk. Amalth. 61, t. 384, fig. 6. — Earl. Res. 71. — Linnaeus, Sp. Pl. 1468 named it *Menispermum orbiculatum.*

Thea chinensis Pimentae jamaicensis folio, *Swa Tea* seu *Cha hoa,* flore rosaceo simplici. Petiv. Gaz. t. 33, f. 4; Petiv. Phil. Trans. 93; detailed description. — Earl. Res. 51, 62. — Arbor indica cheusan.... Swa Tea Sinens., flore pleno albo, rubro et variegato. Species Theae Indorum. Pluk. Amalth. 21. — Earl. Res. 64. — Linn. Sp. 982, *Camellia japonica.*

Thea Chusan, floribus majoribus, folio Alaterni serrato. Petiv. Mus. 983; Gazoph. t. 21, f. 10; Petiv. in Phil. Trans. 92, Thea chinensis vera potulenta. — Earl. Res. 40, 41, 49, 51, 61. — Pluk. Phytogr. (1691) t. 88, f. 6, Styraci et Euonymo media affinis, *The* Sinensium seu *Tja* Japonensium, flore simplici; Pluk. Almagest. 139, Amalth. 79. — Linnaeus Sp. Pl., first ed. 515, *Thea sinensis.*

Alcea indica Sinarum. Frutex arboreus flore amplo luteo malvaceo, fundo purpureo etc. Folia fert ut Populus alba. Cunningh. Pluk. Amalth. 6, t. 355, f. 5. — Earl. Res. 63. — Alcea forte fruticosa Chusan. Leaves like our Aspentree. Petiv. Phil. Trans. 42. — Early Res. 54. — Linn. Sp. 976, *Hibiscus tiliaceus.*

Alcea sinica Manihot, fructu pyramidato hirsuto 5 capsulari, seminibus reniformibus, foliis digitis longioribus. Sin: *Tchu whei.* Cunningh. Pluk. Amalth. 7, t. 355, f. 2. — Earl. Res. 63. — Chusan Musk Mallow, Abelmoschus chusanensis folio palmato, fructu piloso. Cunningham brought it in seed from Chusan, from which it was raised in Chelsea. Petiv. Phil. Trans. XXVII, 1710, 417. — Linnaeus, Sp. Pl. 980: *Hibiscus Manihot.*

Ou tum chu, P. Le Comte, folio trifido, petalis bacciferis. Detailed description of the tree. Petiver in Phil. Trans. 82, Gazoph. tab. 27. — Earl. Res. 60, 51. — "Tetraglottis baccifera Sinarum seu Ligularia arbor sinensis, Gossypii 5 fidi amplioribus foliis, quatuor ligulas foliaceas, longo pedunculo ex alis insidentes, pro floribus ferens, e ligularum margine baccifera. Ou tom chu Sinarum". Pluk. Amalth. 199, tab. 444, fig. 4. — Earl. Res. 85. — The above Chinese name refers to *Sterculia platanifolia* Linn. fil.

Alcea olitoria s. Corchorus longiori folio Sinarum. Pluk. Amalth. 7.— Earl. Res. 63. — Probably *Corchorus capsularis.* Lin.

Hemsley, Ind. Florae Sin., I, 94, *Corchoropsis crenata.* S. & Z. quotes Cunningham in herb. Sloane XCIV 146.

Hemsl. 1. c. I, 100, *Impatiens Balsamina*. L., quotes Cunningham, Amoy, in herb. Sloane CCLXVII p. 20.

Fagara Chusan, Rhois virginianae folio, caule alato. Petiv. Gazoph. tab. 36, fig. 8. Idem Petiv. Phil. Trans., 74. Sinice *What chaw*, cultivated by Sloane. — Earl. Res. 51, 59. — Arbuscula baccifera spinosa etc. Wha tchaw Sinensibus dicta. Pluken. Amalth. 33, tab. 362, fig. 1. — Earl. Res. 67. — *Hua tsiao* is the Chinese name for *Zanthoxylum Bungei*. Hemsley, Ind. Fl. Sin. I, 106, found this in Cunningham's collection.

"Euonymo affinis aromatica s. Zanthoxylum spinosum Fraxinellae foliis cheusanicum ......" detailed description. Sinice *Hoa tchaw*. Pluken. Amalth. 78, tab. 393, fig. 2. — Earl. Res. 73. — According to Hooker & Arnott. Voy. Beech. 175, this is *Zanthoxylum nitidum*. DC.

Euonymo affinis aromatica, s. Zanthoxylum spinosissimum Fraxini angustiore folio punctatum. Cheusan. Pluken. Amalth. 76, tab. 392, fig. 1.— Lamarck, Enc. Bot. II, 39, identifies this with *Zanthoxylum rhoifolium* of India. — Mr. Hemsley discoverd Cunninghams original plant in Hb. Sloane, British Museum and found that it was *Z. emarginellum,* Miq. Prol. Jap. 210, Japan. The plant has been gathered also by Ford in Formosa. See I. F. S., I, 108, and Annals of Botany 1895, 149.

There are several other species of *Zanthoxylum* in Cunningham's Chinese collection. Comp. Petiv. Mus 640, Fagara Emuyca Cardamomi sapore etc., Earl. Res. 47, — and Petiv. Phil. Trans. 71, Euonymo affinis, Chusan, Fraxini folio, semine nigro, 72, Fagara Emuyaca Fraxini folio. Detailed descriptions. — Earl. Res. 88. — Hemsl. Ind. Fl. Sin., I, 107, notices *Z. schinifolium,* Sieb. & Zucc. among Cunningham's Chinese plants.

Hemsl. Ind. Fl. sin., I, 118, mentions *Euonymus Bungeanus*. Maxim. in Cunningham's collection. This is perhaps the: Euonymo nostrati similis sed multo latioribus foliis etc in Pluk. Amalth. 75. — Earl. Res. 72.

Paliurus Emuyaca major folio rotundiore. Petiv. Philos. Trans. 83. — Earl. Res. 60. This is probably *Paliurus ramosissimus*. Poir. common in Mid and South China.

Frutex sinensis Majoranae minoribus foliis ..... lineolis transcurrentibus elegantissime delineatis. Chusan. Cnghm. Pluk. Amalth. 100, tab. 407, fig. 4. — Early. Res. 76. — This is identified in Lamarck, Enc. Bot. IV, 473 with the *Rhamnus lineatus* of Linnaeus, *Berchemia lineata,* DC.

Berchemia racemosa, Sieb. & Zucc. is mentioned in the Ind. Fl. Sin., I, 127, among the specimens of Cunningham's collection.

Ibidem I, 131, *Sageretia theezans,* Brogn. Cunningham's collection.

Pluk. Amalth. 211 mentions the "Vitis vinifera sylvestris", the "Vitis vulpina dicta Virginiana alba", and the "Vitis agrestis sinica minor", or "the least Fox grape", all sent by Cnghm from Chusan. — Earl. Res. 86. — Wild vines with edible fruit seem to be meant, probably *Vitis labrusca.* L. (V. ficifolia, Bge.) and *V. bryoniaefolia* Bge., both common in Northern and Central China.

Arbor Sinensium Lauri folio *Lei chi,* i. e. Oculum Draconis, fructum ferens; et aliquando *Lung yen* indigenis audit. Pluk. Amalth. 25, tab. 365, fig. 6. — Earl. Res. 65. — *Nephelium Litchi* Camb. and *N. Longan.* Camb. The second is called Lung yen (oculum draconis) by the Chinese.

Acer forte Chusan, folio A. Monspessulani minore trifido. Petiv. Philos. Trans. 57. — Earl. Res. 56. — Arbuscula cheusanensis Aceri Monspessulani folio, subtus rore coeruleo tincto, longo pediculo insidente. Pluk. Amalth. 32, tab. 366, fig. 3. — Earl. Res. 67. — *Acer trifidum.* Hook. & Arn. mentioned by Hemsley, Ind. Fl. Sin., ı, 142, in Cunningham's collection.

Rhus quinquefolia Sinarum lactescens, rachi medio alata, foliis molli hirsutie pubescentibus. Pluk. Amalth. 183. — Earl. Res. 83. — Rhus chinensis latifolia pinnis alatis, nobis, Petiv. in Philos. Trans. XXVII (1710) p. 417. Petiver quotes Amalth. 183 and adds: "The leaves resemble our Ash but are broader and deeper serrated, and very soft underneath. Each stalk is composed of 3 or 4 wings, oftner than 2. James Cunningham brought large specimens with ripe berries from China. From these berries living plants were raised in the Chelsea Garden. — Perhaps the Rhus Emuyca folio serrato subtus molli, rachi alato etc., previously described by Petiver in the Philos. Trans. 89 (Earl. Res. 61) is the same. Miller, Gard. Dict. 1768 named Plukenet's Rhus Sinarum lactescens, *Rhus chinensis* (n. 7), which in the Ind. Fl. Sin., ı, 147 is united with *Rhus semialata,* Murr.

Planta sativa *Co* dicta unde efficitur *Co-pou* pro vestibus aestivus. Cnghm. Pluk. Amalth. 212. — Earl. Res. 87. — This is the *Pachyrhizus Thunbergianus (Pueraria).* — Ind. Fl. Sin., ı, 191.

Hai hoa Chinensibus, flore albo, siliquis gummosis articulatis. Petiv. Mus. 930. Earl. Res. 48. — Most probably *Sophora japonica,* Lin.

Frutex sinensis Senae sylvestris folio angustiore, nodosa siliqua rostro longiore donata. Pluk. Amalth. n. 18, tab. 451, fig. 10. — Earl. Res. 87. — Hemsley, Ind. Fl. Sin., ı, 201, thinks that the figure represents *Sophora flavescens.* Ait.

Clematis arborea, summo folio bicorni. Chusan. Hort. Malab. VIII, tab. 30, 31 nagamuvalli. Pluk. Amalth. 60. — Earl. Res. 71. — The tab. quoted in the Hort. Malab. represents *Bauhinia scandens.* Lin., which is not met with in China. But Hemsley, Ind. Fl. Sin., I, 212, enumerates seven other Chinese species of Bauhinia. — Comp. also Petiver in Phil. Trans. 80, Mandaru forte Chusan, folio acuminato, alte bifido. — Earl. Res. 59. — Mandaru is a Malabar name for Bauhinia.

Rubus, Chusan, folio Corchori. Description. Petiv. Philos. Trans. 43. — Earl. Res. 54. — Perhaps *Rubus corchorifolius*, Linn. fil., which is common in Mid and South China.

Rosa Chusan glabra Juniperi fructu. Received from Chusan and China. Petiver Gazoph. tab. 35, fig. 11. — Earl. Res. 51. — Linnaeus, Sp. 705, named this *Rosa indica*. But Lindley, Monogr. Ros. 130 refers Petiver's figure to *R. microcarpa*.

Mespilus oxycantha, Chusan, foliis mucronatis et serratis, fructu longiore, summis ramulis innascente. Pluk. Amalth. 216, tab. 453, fig. 3. — Earl. Res. 88. — Loureiro, Fl. cochin. 392 refers this to Linnaeus' *Mespilus pyracantha* or *Crataegus pyracantha*, Pers. which is common in Central China.

Arbor sinica *Pipa* dicta. Cunningham. Detailed description. Pluk. Amalth. 26, tab. 371, fig. 2. — Earl. Res. 65. — This is *Eriobotrya japonica*, Lindl.

Jasminum japonicum. Description. Sinice: *Seu kiu hoa*. Cnghm. Pluk. Amalth. 122. — Earl. Res. 78. — *Hydrangea*.

Arbuscula sinensis Cisti minoris folio etc. Detailed description. Cheusan. Cnghm. Pluk. Amalth. 32, tab. 368, fig. 2. — Earl. Res. 67. — R. Brown in Abel's Journey, China (1818) p. 375 identifies this with his *Hamamelis chinensis*.

Arbor sinensis Canellae folio minore, trinervi, prona parte villoso, fructu Caryophylli aromatici majoris, villis obducto. Ins. Crocodil. Pluk. Amalth. 21, tab. 372, fig. 1. — Earl. Res. 65. — Aiton, Hort. Kew (1789) II, 159, identifies this with his *Myrtus tomentosa*.

Petiver, Philos. Trans. 77. Hedera arborea. C. B. 305, 1. Petiver says it is our common Ivy *(Hedera Helix.* L.*)*. — Earl Res. 59. — Hedera arborea cheusanensis, vulgari similis sed foliis perangustis. Pluk. Amalth. 114, tab. 415, fig. 5. — Earl. Res. 77. — Hedera Helix is not uncommon in Mid China. Cantor gathered it in Chusan.

Periclymenum scandens hirsutum sinicum subrotundis foliis binis, oppositis. Caprifolium flore minore ex albo flavescente etc. Sinice *kin in hoa,* i. e. flos auro argenteus. Cnghm. Pluk. Amalth. 167, tab. 435, fig. 1. — Earl. Res. 82. — The above Chinese name is applied to *Lonicera japonica*. Thbg.

Um ki Chinensibus. Frutex Cynobasti fructu alato tinctorio, barbulis longioribus coronatis, etc. Petiv. Mus. 498. — Earl. Res. 46. — Arbuscula sinensis Myrti majoris folio, vasculo seminali hexagono, ad singulos angulos alis foliaceis munito quae porrectae vasculi coronam efformant. Umki Sinensibus dicta. Flos rosaceus albus hexapetalus. Semina tinctoribus inserviunt etc. Cnghm. Pluk. Amalth. 29, tab. 448, fig. 4. — Earl. Res. 66. — Loureiro Fl. cochin. 183 identifies this with *Gardenia florida,* L.

The plant called *Sam tan guy* or Flammula in Chinese. It grows about a yard and a half high, into many branches, bearing at the top scarlet Jasmin-like flowers. Kam. Ray. App. p. 7. pl. 23. Petiv. Gazoph. tab. 59, fig. 6. — Earl. Res. 52. — *Shan tan* (the red of the mountains) is the Chinese name for *Ixora stricta,* Roxb. Comp. also Rumph. Amb. IV p. 106, tab. 46. Flamma sylvarum. Sinenses dicunt in ipsorum patria quoque crescere, Ipsisque dicitur *San tan hoa.* — The plant described and figured by Rumph is *Ixora coccinea* L. — Hemsley unites I. stricta and coccinea with *I. chinensis.* Lam.

Lycium Myrsinitis Sinarum, spinis longioribus bijugis, foliis lanceolatis laevibus, ex adverso binis absque pediculis, bacca rubente coronata, duo vel tria semina continente. Ins. Pu to (Chusan Archip.). Colitur. Sinice: *hu tsu.* Cnghm. Pluk. Amalth. 136, tab. 426, fig. 3. — Earl. Res. 79. — In Japan as well as in China the name *hu tz̆e* (tiger's spines) is applied to *Damnacanthus indicus.* Gaert. — Ind. Fl. Sin., I, 387.

Rubiae facie planta Crocodyl. cordato folia. Petiv. Mus. 970. — Earl. Res. 49. — Rubia quadrafolia aspera, baccis numerosis singularibus, nigris, succulentis semine unico, rotundo, umbilicato foetis. Cnghm. Pluk. Amalth. 185. — Earl. Res. 85. — Probably *Rubia cordifolia.* L.

Aster cheusanensis Tripoli nostratis aemulus et forte idem Pluk. Amalth. — 40. Earl. Res. 68. — *Aster tripolium,* L, is common in Northern and Mid China.

Cichorio affinis Lampsana sinica, Mentastri foliis, calyce fimbriato hispido, flore luteo. Sinice *Hi him tsaw.* Detailed description. Cnghm. Pluk. Amalth. 58, tab. 380, fig. 2. — Earl. Res. 70. — Linnaeus, Sp. 1269 identified this with his *Siegesbeckia orientalis,* the Chinese name of which is *hi hien ts̆ao.*

Matricaria Chusan, flore luteo minore simplici, *keuk hoa* Chinensibus, etc. Petiv. Philos. Trans. 36. — Earl. Res. 54. — Matricaria sinensis, minore flore, petalis et umbone ochroleucis. Pluk. Amalth. 142, tab. 430, fig. 3. — Earl. Res. 80. — Linnaeus identifies this with his *Chrysanthemum indicum.* — Hemsley, Gard. Chron. 1889, II, 522, reports that among Cunningham's specimens preserved in the British Mus. there are two small double Chrysanthemums. One of them is that figured by Plukenet.

Absinthium maritimum Sinarum, Lavandulae folio, pulchrioribus corymbis inodorum, sapore aromatico etc. Cnghm. Chusan. Pluk. Amalth. 3, tab. 353, fig. 5. — Earl. Res. 63. — Linnaeus, Sp. 1190, identifies this with his *Artemisia chinensis,* same as *Crossostephium artemisiodes* Less., or *Tanacetum chinense,* A. Gray.

Tussilago chusanensis ramosa, folio rotundo glabro etc. Petiv. Philos. Trans. 31. — Earl. Res. 53. — Doronicum Tussilaginis folio, flore magno singulari. Chusan. Pluk. Amalth. 71, tab. 390, fig. 6.— Earl. Res. 71.— Linnaeus, Mant. 113, identifies this with his *Tussilago japonica,* which is the *Ligularia Kaempferi,* Sieb. & Zucc.

Xeranthemum sesamoides flore albo, Ericae foliis cauli tomentoso adstrictis etc. Ins. Crocodyl. Pluk. Amalth. 213, tab. 449, fig. 5. — Earl. Res. 87. — This is identified in Lam. Enc. Bot. III, 239 with *Xeranthemum heterophyllum,* a native of the Cape of G. H. No species of Xeranthemum has been found in China in modern times. Loureiro's X. chinense (Fl. cochin. 498) most probably does not belong to this genus.

Vaccinia forte Chusan, Laurocerasi folio flore tubuloso. Petiver Gazoph. tab. 35, fig. 7. — Earl Res. 51. — Probably *Vaccinium bracteatum* Thunb. which has been gathered in Chusan by modern collectors.

Androsace chusanensis Cortusae Matthioli folia. Petiv. Mus. 858, Gazoph. tab. 33, fig. 8., Phil. Trans. 52. — Earl. Res. 47, 51, 55.— Auriculae Ursi affinis, Androsace dicta, sinensis. Saxifragae aureae foliis, pediculis longis insidentibus. Pluk. Amalth. 43, tab. 450, fig. 6. — Earl. Res. 69. — Probably *Androsace saxifragaefolia,* Bunge.

Zapotl Chinens. fructu cinnabarino, *xi cu* Sinensibus, Figocaque Lusitan. Fruits dried in the sun, as they do figs. Petiv. Gazoph. tab. 45, fig. 9—11.— Earl. Res. 52. — *Diospyros Kaki* L., *D. Schitze,* Bge.

Arbor Lauri folia, floribus minimis ex albo flavescentibus odoriferis, monopetalis, 4 fidis etc. A Sinis *Quei hoa* nuncupatur. Cnghm. Pluk. Amalth. 27. — Earl. Res. 66. — This is the *Olea fragrans.* Thbg., in Chinese *kui hua.*

Lycium Chusan, Pruni minoris folio. The twigs end in a thorn. Berries (when dried) black, wrinkled like pepper, and with two kernels. Petiver, Philos. Trans. 79. — Earl. Res. 59. — Most probably *Lycium chinense* Mill., gathered in Chusan by Dr. Cantor about 58 years ago.

Clematis cheusanensis, Bucananthi majoris fere foliis solitariis, ad margines spinosis serris, spica florum ampliori, ex foliorum alis. Planta repens flore purpureo violaceo. Cnghm. Pluk. 60, tab. 384, fig. 1. — Earl. Res. 71. — Hemsl. in Ind. Fl. Sin., II, 195 identifies this with *Calorhabdos axillaris* Benth. & Hook.

Euphrasia sinica Parietariae foliis, Rubiae modo spicata. Flos albens personatus etc. Cnghm. Pluk. 82, tab. 392, fig. 4. — Earl. Res. 74. — Linnaeus, Sp. 22, identifies this with his *Justicia procumbens*.

Ocymum Melissophyllum sinicum, foliis mucronatis, serratis suavem Menthae odorem spirantibus. Flos albus etc. Sinice *ssu tse*. Cnghm. Pluk. 159, tab. 429, fig. 7. — Earl. Res. 80. — Judging from the Chinese name this seems to be *Perilla ocymoides*. L.

Mentha cheusan, verticillata odoris Menthae, flore ex albo coerulescente. Sinice: *po ho*. Cnghm. Pluk. 144. — Earl. Res. 80. — *Mentha arvensis*, L., the only species of this genus met with in China.

Scutellaria sinica Betonicae folio, floribus albis. Pluk. Amalth. 190, tab. 441, fig. 1. — Earl. Res. 84. — Linnaeus, Sp. 836 identifies this with his *Scutellaria indica*.

Clinopodium parvum sinicum hirsutis Majoranae foliis coronatum, flore coerulescente labiato etc. Cnghm. Pluk. Amalth. 61, tab. 380, fig. 4. — Early Res. 71. — Linnaeus, Syst. 163, identifies this with his *Cometes alterniflora,* which, however, is an Indian plant, hitherto not gathered in China by modern collectors.

Cadelari siciliana folio acuto, Amaranthus siculus spicatus. Petiv. Philos. Trans. 27. — Earl. Res. 53. — This is, according to Lamarck, *Achyranthes argentea* Lam. Enc. Bot. I, 545, a Sicilian plant not recorded from China by other botanists, and therefore, as it has been frequently confounded with *A. aspera* L., Cunningham's plant may be the latter species, which is common in China.

Fagotritici similis spinosa minor sinica cauliculis erectis. Persicaria calycis floris rubente, foliis lanceolatis auritis caulibus et ramibus spinosis, semine nigro splendente triquetro. Cnghm. Pluk. Amalth. 87, tab. 398, fig. 1. — Earl. Res. 74. — Lamarck, Enc. Bot. VI 152, identifies this with *Polygonum perfoliatum*. Lin.

Camphora officinarum, the Camphor tree. It is very well figured and amply described in Breynius centuria (1678) and first Prodromus. — Petiv. Phil. Trans. 65. — Earl. Res. 57.

Arbuscula sinica Anonae dulcis folio minore non splendente. Flos est stamineus, minimus, calyce multifido. Fructus compressus, instar seminis Malvae, multicapsularis, plurima semina rubra etc. Cnghm. Pluk. Amalth. 35, tab. 368, fig. 1. — Earl. Res. 68. — Gaertner, Fruct. II (1792) iden-

tified this with his *Bradleia sinica* which is, according to Hemsl., Ind.
Fl. Sin., II, 425, *Glochidium obscurum,* Blume.

Ricinus chinensis sebifera, Populi nigrae folio. Chinese Tallow tree.
Petiver Mus. 965, Gazoph. tab. 34, fig. 3. — Earl. Res. 49, 51. — Petiv. in
Philos. Trans. 90. Detailed description of the tree. Petiver cultivated it in
the Chelsea Garden in 1705. Philos. Trans. XXIII, p. 1427. — Earl. Res.
61. — Cunningham's letter, Tallow tree. — Earl. Res. 42. — Arbor sinensis sebifera *kieu-yeu.* Detailed description. Pluk. Amalth. 25, tab. 390,
fig. 2. — Earl. Res. 65. — Linnaeus, Sp. 1425, named it *Croton sebiferum,* = *Stillingia sebifera,* Michx.

Lupulus Chusan minor Rubi folio. Petiv. Mus. 944. Earl. Res. 49. —
Lupulus foemina cheusanensis florens solummodo et non fructum ferens.
Pluk. Amalth. 131. — Earl. Res. 79. — *Humulus japonicus,* Sieb. & Zucc.
gathered also by Dr. Cantor in Chusan.

Urtica racemifera maxima Sinarum, foliis subtus argentea lanugine
villosis. Textile plant. Cunningham. Plukenet Amalth. 212, tab. 429, fig.
2. — Earl. Res. 87. — Linnaeus, Sp. 1398, identified this with his *Urtica nivea,* = *Boehmeria nivea.* Hook. & Arn.

Prunifera Elaeagni foliis, sinice *Yang muy,* fructu hispido fraga redo-
lènte. In vino destillato toto anno conservant. Pluk. Amalth. 178. — Earl.
Res. 82. — *Yang mei* is the Chinese name for *Myrica sapida.* Wall.

Quercus Chusan Castaneae folio pubescente. Underside of the leaves
hoary. Its catkin or julus is round and echinated; acorn small and smooth.
Petiv. Philos. Trans. 87. — Earl. Res. 60. — Quercus cheusan Castaneae
minoris folio prona parte candicante. Pluk. Amalth. 180. — Earl. Res.
83. — *Quercus Bungeana,* Forbes, has leaves resembling those of the chestnut. Mr. F. B. Forbes found, among Cunningham's Chusan plants in the
British Museum, the *Quercus thalassica,* Hance. (Journ. Bot. 1884, 86).

Castanea Chusan folio fere serrato, subtus glauco. Petiv. Philos. Trans.
66. — Earl. Res. 57.

Yang diu (liu) Chinensibus. Arbor Salicis folio ramulis pendulis. Frequently painted on Japan work. Of the wood they make arrows. Petiv. Mus.
997. — Earl. Res. 50. — *Salix babylonica.* L.

Juniperi folia arbuscula cheusanensis conifera foliis variis Cupressi
squamosis et Juniperinis. Cnghm. Pluk. Amalth. 125. — Earl. Res. 78. —
Carrière, Conif. 151, identifies this with *Glyptostrobus heterophyllus* Endl.
But is seems to me that this short description agrees equally well with
Juniperus chinensis Lin., which is common all over China, whilst Glypto-
strobus is only known from South China.

Cunningham in Pluk. Amalth. 102 (Earl. Res. 76) notices also a Frutex cheusanensis conifer, foliis Juniperi planis et teretibus Cupressi.

Cupressus chusanensis Abietis folio. Leaves triangular, carinated, stiff, and stand off from the stalk; seed brown and small, resembling buckwheat. Petiv. Gazoph. tab. 6, fig. 3. The same Petiv. Philos. Trans. 70. — Earl. Res. 50, 58. — Cupressus chusanensis Juniperinis arcuatis foliis, clavis galbulorum eleganter cristatis. Pluk. Amalth. 69, tab. 386, fig. 3. — Earl. Res. 71. — In DC. Prodr. XVI, 2. p. 438 this is identified with *Crypto-meria japonica*. Don.

Abies major sinensis pectinatis Taxi foliis, subtus caesiis, conis gran-dioribus sursum rigentibus, foliorum et squamarum apiculis spinosis. Pluk. Amalth. 1, tab. 351, fig. 2. — Earl. Res. 62. — This is the tree, first discovered by J. Cunningham in China, upon which R. Brown founded the Genus *Cunninghamia*, and of which the type specimen is in the British Mu-seum. The tree was first described, in 1803, by Lambert as *Pinus lanceo-lata*. Salisbury in 1805, named it *Belis jaculifolia*. Lambert described it from specimens gathered by G. Staunton in China. In a note, in P. P. King's Survey of the Coasts of Australia, V (1827), appendix, Botany, p. 564, R. Brown states that he requested Richard, who then was writing a monograph on Conifers to change the genus name of *Belis* into *Cunninghamia* in com-memoration of the merits of James Cunningham, an excellent observer in his day, by whom this plant was discovered, and in honour of Mr. Allan Cunningham, a very deserving botanist, who accompanied Mr. Oxley in his first expedition into New South Wales, and Capt. King in all his Voyages during his Survey of the Coast of New Holland. The name *Cunninghamia sinensis* thus appears first in Richard's Conifers (1826) p. 80.

Frutex convolvulaceus spinosus sinicus rotundiore nervoso folio, flori-bus parvis, umbellatis, claviculis ligneis binatim donatus. Pluk. Amalth. 101, tab. 408, fig. 1. — Earl. Res. 76. — Linnaeus, Sp. 1459, identifies this with *Smilax China*.

Sagittaria chinensis foliis ternis longissimis. Herb. nost. Chin. tab. 12, fig. 3. *Sa heo chaw* indigenis. Petiv. Gazophyl. tab. 19, fig. 5. — Earl. Res. 50. — Linnaeus, Sp. 1410 identifies this with his *Sagittaria trifolia*.

Gramen cyperoideum capillaceis foliis, pusillum. Ins. Cheusan. Pluk. Amalth. 110, tab. 417, fig. 8. — Early Res. 77. — Willdenow, Sp. pl. I, 272, identifies this with his *Cyperus nanus*. But Roem. & Schultes, Syst. Veget. II, 167, mean that Plukenet's figure represents a variety of *Cy-perus tenellus*. Lin.

Muscus clavatus erectus crispatis foliolis, Spongiolae imitamentum.

China. Pluk. Amalth. 149, tab. 431, fig. 3. — Earl. Res. 80. — Linnaeus identifies this, Sp. 1506, with his *Lycopodium cernuum*.

Muscus denticulatus Emuyacus minor erectus. Cunningham. Petiver Mus. 536. — Earl. Res. 47. — Muscus denticulatus minor. Cheusan. Pluk. Amalth. Append. tab. 453, fig. 9. — Earl. Res. 88. — Dillenius, Hist. Muscorum (1741), p. 470, tab. 65, fig. 7. Lycopodiodes denticulatum, erectum, filicinum et argutius. — Not identified by Linnaeus. According to Spring, Monogr. Lycopod. (1842) 109, it is *Selaginella uncinata*.

Filix Cochine. Branched Comb Fern. Petiv. Gazoph. t. 70 fig. 12. Cunningham sent it from Cochin China. Linnaeus, Sp. 1524, identifies it with his *Acrostichum dichotomum*.

Filix Adianto nigro officinarum similis, pediculo viridi, pinnulis magis eleganter incisis. Cheusan. Pluk. Amalth. 91, tab. 403, fig. 2. — Earl. Res. 75. — Swartz, Synopsis Filic. (1806) p. 125, identifies this with his *Adiantum pallens*, from Plukenets figure alone. Comp. Hook. Spec. Filicum II, 55.

Linnaeus describes an *Adiantum chusanum*, Sp. 1558, evidently from Cunningham's collection, but he does not quote either Petiver or Plukenet. He says only: Hab. in China.

Adiantum nigrum chinense, tenuiter divisum, pinnulis minimis obtusis bifidis, Pluken. Almag. 29, tab. 4, fig. 1. Raji hist. pl. 1854. Chinese black Maidenhair with blunt forked leaves. Hamoy in China. Mus. Petiv. cent. I. — Lamarck, Enc. Bot. I, 44, identifies this with his *Adiantum tenuifolium*.

Adiantum chinense perelegans, ramosum, folio flabelliformi cum rubedine perfuso. Pluken. Almag. 11, tab. 4, fig. 3. — Linnaeus named this *Adiantum flabellulatum*.

Filicula cheusanica s. Hemionitis multifido folio tenuissime serrato, ad margines seminifera. Pluk. Amalth. 94, tab. 407, fig. 2. — Earl. Res. 75. — Swartz, Syn. Filic. (1806) 97, identifies this with *Pteris serrulata* Linn.

Filix chinensis Lonchitidis facie cujus lanago radicis *Poco sempe* vocatur. *Kim kow ja* (i. e canis aurei crines). Herb. nostr. Chin. pict. tab. 10, fig. 8. Petiv. Mus. 554. — Earl. Res. 47. — Judging from the Chinese name this seems to be Linnaeus' *Polypodium Barometz*.

VI. Botanical Collections made by the Jesuit Missionaries in China in the first half of the 18th century.

The learned Jesuit fathers, who in the 17th and the beginning of the 18th century furnished so much valuable information regarding the vegetable productions of China, as has been shown in a previous chapter, commenced from the beginning of the 18th century to send botanical collections, seeds and herbarium specimens, to Europe, which for the greater part came from Peking and were transmitted overland, via Siberia; for in the 18th century Peking was not unfrequently visited by Russian caravans and couriers, and in the first half of that period even by Russian embassies and diplomatic agents.

In 1705, Tournefort described and figured, in the "Memoirs of the Academy", Paris p. 264, n. 5, a plant Caryophyllus sinensis supinus, leucoji folio, flore unico, of which he states that it had been sent, about three years before from China to the Abbé Bignon. In Miller's Gard. Dict. it is said that the French missionaries had sent the seeds to Paris. This is the *Dianthus chinensis*. L., since that time much cultivated in our gardens.

Tilli, in the "Catalogus Plantarum horti Pisani", 1723, p. 49, t. 22, describes and figures a Cucurbita sinensis fructu longo, anguino, flore candido capillamentis tenuissimis, of which the seeds had been brought to Florence from China by the abbot Josephus, Ignatius **Cordero**. The plant was cultivated by Tilli. This is Linnaeus' *Trichosanthes anguina*, known previously from India. In the "Catalogue of the Jesuit Missionaries in China", published at Shanghai, 1873, the name *Cordero* appears sub n. 443. The time of his sojourn in China is stated to be unknown.

Professor J. Amman describes and figures in the "Comment. Acad. Petropol." X. 1739 p. 305, tab. 7, a plant Bermudiana floribus maculatis etc. raised from seeds sent from Peking by the Jesuit missionaries together with other botanical collections. — Linnaeus identifies this with *Ixia chinensis*, same as *Pardanthus chinensis.*

Father Antoine **Gaubil,** a Frenchman, from 1723—1759 in Peking, Honorary member of the Academy of St. Petersburg, 1739, had sent to Stephan Krasheninnikov, Professor of Botany at St. Petersburg, the seed of a plant, which was cultivated at Peking, and used for preparing Indigo. The plant raised from these seeds, Krasheninnikow described and figured in the Nov. Comment. Petropol. I, 1747 p. 375, tab. 13: Persicaria foliis ovatis utrinque incanis etc. — Linnaeus identified it with his *Polygonum* *barbatum*. But this plant is not met with in North China, and the plant which is now cultivated at **Peking** for manufacturing Indigo is *Polygonum tinctorium*.

Thouin in Enc. méthod., Dict. d'Agriculture, I, 710, states that seeds of the *Aster chinensis* or "Reine Marguerite" were first sent from China about 1728, by d'Incarville to Antoine de Jussieu (1686—1758), Professor of Botany. These seeds sown in the Royal Garden Paris produced several varieties of the Reine Marguerite. I quote this from Grosier, la Chine, III, 129. I have not seen Thouin's original article. There seems to be some misapprehension in the above statement, for d'Incarville did not arrive in China before 1740. The seeds may have been first sent by some other missionaries. Dillenius, in the Ht. Eltham. 1732, t. 34, fig. 38, first figured the plant, stating on the plate, that it was introduced from North China... He calls it Aster chenopodiifolius, annuus, flore ingenti specioso*).

Ph. Miller, in the 8th edition of the Gardener's Dictionary, 1768, describes the *Rhus chinense*, n. 7, and states that plants of it were raised from seeds received from China in the Royal Garden, Paris, before 1737, for in this year Miller received one of these plants, which grew well in the open air at Chelsea. — As we have seen, Mr. Hemsley identifies the Rhus chinese with *Rhus semialata*, Murr.

D'Incarville, Pierre, 1740—1756.

The name of this Jesuit Father has been repeatedly inscribed in the annals of botanical science, for he sent to Paris towards the middle of the 18th cent. an interesting collection of dried plants and seeds of the Peking Flora, from which a number of beautiful new Chinese plants were raised in Europe. D'Incarville was a Frenchman, born in 1706. In 1740 he joined the Chinese mission of the Jesuits and died in 1757 at Peking, where he seems to have laboured during the greater part of his sojourn in China. He was a pupil of the great French botanist Bernard de Jussieu (1699—1776), Professor of Botany at the Royal Garden, Paris, to whom he used to send his collections. In a memoir by Incarville on Chinese natural productions, of which we shall speak further on (article Bambou) he calls Jussieu "mon maître en botanique," and speaks of a botanical collection made for him at Macao. In another place (article Loup-marin) he states that he was in Ca-

---

*) *Callistephus (Aster) chinensis* is indeed a native of North China. In the herbarium of the Botan. Gard. St. Petersb. there are wild growing specimens of the plant brought from the Po hua shan by Kirolov and Tatarinov. The present writer has gathered it on the same mountain. The wild plant has much smaller flowers than the cultivated one.

nada before he went to China. He was a Corresponding Member of the Paris Academy.

D'Incarville's name has been commemorated in the new genus *Incarvillea*, established by Antoine Laurent de Jussieu (nephew of Bernard) in his Genera Plantarum, 1789, p. 138. There we read, that the character of this genus was founded upon dried specimens found in a herbarium, which Father d'Incarville, in 1743, had transmitted to Bern. de Jussieu, and in which many other new species were contained, as for example *Aster chinensis*. — *Incarvillea sinensis* is a beautiful Bignoniaceous plant with large scarlet flowers, frequently met with in the Peking plain and in the mountains; it flowers towards the end of summer.

When in 1881, in my Early Europ. Researches into the Flora of China, I ventured to give an account of Incarville's life and his merits as a botanist (p. 120—124), very little was known regarding the collections forwarded by him, after about two years' residence at Peking, to his instructor Bern. de Jussieu. I had only been able to ascertain the existence of Incarville's herbarium, consisting of Peking plants and specimens from South China, in the Museum d'Hist. natur., in the Jardin des Plantes in Paris. Only a few of these plants then had been described by botanists of the last and the first half of the present century. The following can be traced out from various botanical works as having been noticed previous to 1882, when the whole collection was determined by Mr. A. Franchet (v. infra):

Incarvillea sinensis. Jussieu defined only briefly the distinctive character of the genus. The plant, then the only known species of it, was first fully described by Lamarck in the Enc. Botan. III, 1789, p. 243.

Fumaria caudata. Lamarck, l. c. II, 1786, 569. Peking, d'Incarv. — DC. Prodr. I, 127 considers this to be a variety of *Corydalis longiflora*, Pers. Franchet determined the two species of Corydalis found in Incarville's collection, both from Peking as *C. Bungeana*, Turcz. and *C. solida*, Smith.

Fumaria spectabilis, L. Lam. l. c. II, 571. Chine, d'Incarv. — This is *Dicentra spectabilis*, DC.

Geranium sibiricum, L. Lam. l. c. II, 653. Chine, d'Incarv.

Cycas revoluta, Thbg. Lamarck, l. c. II, 232, Chine, d'Inc.

Dalechampia parvifolia, Lam. l. c. II, 258. Chine, d'Inc.

Fagara Avicennae, Lam. l. c. II, 445. Chine, d'Inc. — *Zanthoxylum Avicennae* DC.

Hypecoum erectum, Lin., Lam. l. c. III, 1789, 161. Peking, d'Inc.

Artocarpus Jaca, Lin., Lam. l. c. III, 209. Chine, d'Inc.

Ailantus glandulosa, Desf., Lam. l. c. III, 416. Chine, d'Inc.

Ionidium heterophyllum, Ventenat, Jard. Malm. 1803, p. 27.

Grewia nitida, A. L. de Jussieu, Ann. Mus. d'Hist. nat., IV, 1804, p. 90. E China. — Jussieu does not mention Incarville, but from Franchet we learn that this species is in Incarville's collection. — Cult. in the Jardin des Plantes. Desfontaines' catal., 1804, p. 151.

Syringa villosa, Vahl, Enum. Plant. I, 1805, 38. In montibus circa Pekin, d'Incarv.

Ficus hirta, Vahl, l. c. II, 1805, 201. Chine, d'Incarv.

Taenitis chinensis, Desvaux, Journ. de Botan. III, 1813, 270. Chine, d'Incarv.

Cocculus ovalifolius, DC, Syst. Regni vegetab., I, 1818, 526. China, d'Incarv.

Cedrela sinensis, first described from a specimen collected by d'Incarville near Peking, by Adrien de Jussieu (1797—1853) son of Ant. Laurent, in Mém. Mus. Paris, XIX, 1830, 294.

Plants introduced into the gardens of Europe by Incarville, raised from seeds sent by him from China:

Ailantus glandulosa, first described and figured by Desfontaines in the Hist. de l'Acad. d. Sc. Paris, 1786, (resp. 1783), p. 265, tab. 8. He says that this tree had been cultivated for a long time in France, and that it was generally believed, from the resemblance of the leaves to be the "Grand Vernis du Japon" or Rhus succedanea. But when it first flowered and produced fruit it turned out to be a new and quite different tree. L'Héritier, Stirp. nov. II, (1784), 179 and Lamarck, Enc. Bot. III, 1789, 416, state that Ailantus glandulosa had been introduced by d'Incarville from China. Lamarck saw this tree in Jussieu's garden. In "Miller's Gard. Dict." 9th edition, 1797 we read that it is the Rhus sinense, foliis alatis etc. described by J. Ellis in Philos. Trans. XLIX, 1757, 870 (not to be confounded with Miller's Rhus chinense) and that it was first raised in England by Ph. Miller and Ph. C. Webb, in about 1751 from seeds sent over by Father d'Incarville. Ellis, l. c. asserts that, according to Incarville, the tree grows at Nanking. But I have little doubt that Incarville sent the seeds of A. glandulosa from Peking, where it is one of the most common trees.

Sophora japonica, L. — Desfontaines, Hist. d. Arbres etc. II, 1809, p. 258, states that this tree was raised in Paris from seeds sent over from China by Incarville in 1747, and from thence its cultivation soon spread over the whole of Europe. It was for a long time unknown to what Genus the tree belonged, till it first flowered near Paris in 1779. According to Hort. Kew, first ed. 1789, II, 45, S. japonica was first introduced to England by J. Gordon, 1753 (I presume from France).

Polygonum tinctorium. Lour. J. Castera in his French translation of Staunton's Account of Lord Macartney's Embassy, 1798, vol. III, 206, states that Bern. de Jussieu cultivated this plant successfully as a garden plant from seeds sent over by d'Incarville, under the name of *siao lan.* (Still the Chinese name for Polygonum tinctorium in Peking). See also Grosier, "la Chine", III, 276.

Thuja orientalis, L., was most probably also introduced into Europe by Incarville. Miller, Gard. Diction. 6th edition, 1752, reports that this tree was raised from seeds sent, not long ago, by the missionaries, and in the 8th edition, 1768, we read (n. 2) that the seeds were sent to Paris from North China by missionaries.

Regarding some other Northern Chinese plants, the introduction of which is I suspect, due to Incarville, namely *Koelreuteria paniculata,* Laxm. (first cultivated in St. Petersburg), *Zizyphus chinensis,* Lam., *Caragana Chamlagu,* Lam. *Gleditschia sinensis,* Lam., *Vitex incisa,* Lam., *Lycium chinese,* Mill., particulars will be found in my Earl. Res. 120—122.

I. Chr. Hebenstreit, Academician, Professor of Botany, St. Petersburg, in 1761 published in the Nov. Comment. Acad. Sc. Imp. Petrop., VIII, p. 331 seqq., an article on some rare plants. In this he notices a peculiar kind of Cabbage and another of Mustard, both raised from seeds received, in 1756 from the Jesuit missionaries in Peking, who, he adds, for a long time used to send seeds of rare Chinese plants to the Academy by the merchants of the Russian caravans returning from Peking. One of these plants proved to be the Brassica foliis ovalibus subintegerrimis etc. in Linn. Cent. I, pl. Upsal. 1755 n. 54, or *Brassica chinensis* of Linnaeus. The other was a *Sinapis,* the roots of which are said to be eaten in China together with radish roots. There can be no doubt that these seeds had been sent by Incarville, for in his catalogue of Chinese natural productions (see further on) he notices the same plants and gives their Chinese names, from which we can infer that he means the Mustard with large tuberous roots, much cultivated at Peking for its roots, and which I introduced into the Jard. d'Acclim., Paris, eighteen years ago. It is a variety of *Sinapis juncea.* L.

There is in the Library of the Asiatic Museum (Imp. Academy of Sc.) at St. Petersburg an interesting MS. Memoir written in Incarville's own handwritting entitled: Catalogue alphabétique des Plantes et Drogues simples, que j'ai vues en Chine, avec quelques observations que

J'AI FAITES DEPUIS 15 ANS QUE JE SUIS DANS LE PAYS. As Incarville arrived in China in 1740 we may conclude that the MS, which bears no
date, was written in 1755. It is an alphabetical list of Chinese natural
productions and drugs, for the greater part Peking plants, beasts and minerals. The Chinese name of each object is given in nicely written Chinese
characters. About 260 Chinese plants are noticed. Incarville identifies or
compares them with European plants quoted under their popular French
names. Occasionally he adds some particulars regarding the medicinal or
economical uses of the drugs or plants, which generally can easily be identified from the Chinese names appended. Incarville concludes his Catalogue
with the following words: "J'ai écrit ce catalogue un peu à la hâte de peur
de ne l'avoir fini à temps. Tel qu'il est, il suffit pour donner quelque idée
de l'histoire naturelle de Chine. Je serais faché que rien de ce que j'envoye
(à l'Académie de St.-Pétersbourg) parût en public; je risquerais à me faire
des affaires. On est jaloux en France. On m'a marqué que les Ministres en
seraient choqués. J'aurais pu donner une autre forme à cet écrit; mais je
l'ai mis exprès dans un style simple. Je l'ai fait en français, Monsieur le
Chirurgien de la caravane m'ayant dit que tout le monde à la Cour de St.
Pétersbourg parlait français".

From other allusions in d'Incarville's paper it can be inferred that he
used to keep up a correspondence with some of the learned academicians in
St. Petersburg. He refers to a letter Krasheninnikov (Academician, Prof. of
Botany St. Petersburg) had written to him about some seeds received from
Peking. Incarville's collections destined for Jussieu were likewise intrusted
to the care of the Russian caravans from Kiakhta (resp. Moscow), which
every three years visited Peking. Jussieu seems to have forwarded European plants to Incarville by the same way, for in one instance the latter
speaks of some bulbs and seeds sent by Jussieu in 1748, and of the delay
in their transmission from Kiakhta by the Russian caravan.

The alphabetical Catalogue is followed by an Appendix in which
d'Incarville gives an interesting account of the mode of preparation of the
Chinese Wine, or rather *White Beer*, and *Vinegar* from the grains of
cereals.

D'Incarville's alphabetical Catalogue of Chinese natural productions
was published in 1812 and 1813, from the original MS, in the "Mémoires
des Naturalistes de Moscou", tome III and IV. In an introductory note it
is suggested that this catalogue ·had probably been drawn up by Incarville
at the request of Bern. de Jussieu. But as we have seen, in the MS, Incarville states positively that he wrote it for the Academy of St. Petersburg.

In the Mém. des Natur. the Chinese characters in the Catalogue are omitted. The Appendix was not published. Dr. Fischer, Inspector of the Botan.
Garden at Gorenki (subsequently Director of the Botan. Garden St. Petersburg) tried to identifiy some of the plants mentioned in Incarville's catalogue, but his botanical commentaries are very poor. At that time the Flora
of Peking was almost completely unknown in Europe.

Koch, "Dendrology", II, 307, relates that Adrien de Jussieu, after his
death (1853) left a MS by Incarville, describing his voyage to China and
a collection of 4010 Chinese drawings, representing plants and animals,
which he had bequeathed to the Muséum d'Hist. nat. Whether they still
exist there I know not. Cavanilles, Diss. bot. 5 (1788), p. 287, tab. 143,
fig. 1, describes and figures a new plant, *Sterculia lanceolata*, from this
collection of drawings "Pulcherrimae icones vivis coloribus quas olim misit
ex China Pater d'Incarville ad D. de Jussieu".

There are two other articles written by d'Incarville on botanical matters, which have been published, viz:

Mémoire sur le Vernis de la Chine, published after d'Incarville's
death in the Mém. Mathém. et Phys. Acad. Sc. Paris, vol. III (1760), p.
117—142, with several plates.

Mémoire sur les Vers à soie sauvages, in which he describes two
remarkable trees of Northern China which he terms "Frêne puant" (Stinking
Ash) and "Frêne odorant" (odoriferous Ash). The first is *Ailantus glandulosa*, the other *Cedrela sinensis*. In the same article he speaks of an Oak
of North China with Chestnut leaves, evidently *Quercus Bungeana* Forbes
(*Q. chinensis* Bge). This paper was published a long time after Incarville's
death by Father Cibot in the Mém. conc. Chinois, vol. II, 1777, p. 583.
It is reproduced in Stan. Julien's Culture des Mûriers, 1837, p. 187.

After my first notices of Father d'Incarville and his collections had
appeared in my Earl. Res., Mr. A. Franchet succeeded in finding, in the
Muséum d'Hist. nat. Paris the whole collection of Chinese plants, which
Incarville about 140 years earlier had transmitted to Bern. de Jussieu, and
he then published a most interesting article on the subject in the Bulletin
de la Soc. Botan. de France, Jan. 1883. Franchet met with no difficulty
in the determination of the specimens. I extract the following particulars
from Franchet's paper and reproduce the list of the plants.

D'Incarville's collection, after Bern. de Jussieu's death, in 1776, came
into the possession of his nephew Ant. Laurent de Jussieu, and when the
latter died, in 1836, his son Adrien inherited his collections. Adrien died
in 1853, and in 1857 his heirs generously presented his botanical collec-

tions to the Muséum d'Hist. nat. Incarville's herbarium consists of 149 plants gathered at Peking and in the neighbouring mountains, and 144 specimens from Macao, where Incarville seems to have spent some time *) before he proceeded to the north of China. Besides herbarium specimens Incarville also sent seeds. Some of them are still in the Museum, while others were no doubt sown, and became the origin of the Chinese plants cultivated for more than a century at the gardens, this being the case with *Polygonum tinctorium*, some varieties of *Callistephus (Aster) chinensis*, *Gleditschia sinensis* and probably also with *Sophora japonica*.

Dr. Bunge, who in 1830 visited Peking, was hitherto considered to have first explored the Flora of the Peking district. But the types of most of the genera recognised and described by Bunge, had existed in a French collection since 1743 and to the learned Jesuit must be referred the discovery of *Orychophragmus*, *Xanthoceras*, *Myripnois*, *Bothriospermum* and others.

List of Incarville's Peking Plants.

Incarville on the labels attached to the specimens distinguishes between plants gathered in Peking (including probably those collected in the plain which surrounds the capital) and plants from the Peking mountains. (Abbreviated P. and P. m. in our list.).

Clematis angustifolia, Jacq. — P. m.

Atragene macropetala, Led. — P. m.

Thalictrum petaloideum, L. — P. m.

Anemone chinensis, Bge. — P.

Ranunculus hydrophilus, Bge. — P. m.

Ranunculus Cymbalariae, Pursh. — P.

Aquilegiae sp. (prob. variety of *A. vulgaris*, L.) — P. m.

Menispermum dauricum, DC. — P. m.

Berberis sinensis, Def. — P. m.

Chelidonium majus, L. — P. m.

Hypecoum erectum, L. — P. m.

Dicentra spectabilis, DC. — P. m. (I have myself found this plant in a wild state in the Peking mountains).

Corydalis solida, Smith.

Cooydalis Bungeana, Turcz.

Erysimum cheiranthoides, L. — Inc.'s label has Macao. But. this is evidently a mistake.

Dontostemon dentatus, Bge. — P. m.

Orychophragmus sonchifolius, Bg. — P.

Thlapsi Bursa pastoris, L. — P.

Lepidium latifolium, L. — The label has Macao. This also I consider to be a mistake.

Viola Patrinii, DC. β. *chinensis*, Ging. — P.

Viola pinnata, L. — P. m.

Polygala sibirica, L. var. *tenuifolia*, Reg. — P. m.

Stellaria nemorum, L.

---

*) In the Hist. Acad. Sc. année 1741, Obs. p. 35, mention is made of a letter written by d'Incarv. to M. Geoffroy and dated 15 Jan. 1741, Canton.

Malva sylvestris, L. — P.
Sida Abutilon, L. — P.
Hibiscus ternatus, Cav. — P. m.
Tribulus terrestris, L. — P.
Geranium sibiricum, L. — P.
Erodium Stephanianum, Willd. — P.
Ailantus glandulosa, Def. — P.
Cedrela sinensis, Adr. Juss. — P.
Xanthoceras sorbifolia, Bge. — P.
Rhus Cotinus, L. — P. and P. m.
Melilotus parviflora, L. — P.
Indigofera Bungeana, Steud.—P. m.
Caragana frutescens, DC. — P.
Güldenstaedtia pauciflora, Fisher.—P.
Güldenstaedtia multiflora, Bge. — P.
    and P. m.
Oxytropis hirta, Bge. — P. m.
Glycyrrhiza echinata, L. I suspect
    from Peking.
Lespedeza trichocarpa, Pers. P.?
Sophora flavescens, Ait. — P. m.
Albizzia Julibrissin, Boiv. — P.?
Spiraea trilobata, L. — P. m.
Potentilla chinensis, Ser. — P.
Potentilla fragarioides, L. — P. m.
Potentilla supina, L. — P.
Saxifraga sarmentosa, L. — P.?
Deutzia parviflora, Bge. — P. m.
    » grandiflora, Bge. — P. m.
Lagerstroemia indica, L. — P.
Siler divaricatum, Benth. & Hook. —
    P. m.
Sambucus racemosa, L. — P. m.
Viburnum fragrans, Bge. — P.
Rubia cordifolia, L. — P.
Aster altaicus, Willd.— P.
Aster (Calimeris) integrifolius, Turcz.
    — P.
Inula Britannica, L. — P.

Xanthium strumarium, L. — P.
Artemisia annua, L. — P.
    » Scoparia, W. & K.— P. m.
    » Sieversiana, Willd. — P.
    » indica, L. — P.
Senecio glabellus, DC. — P.
    » Kirilowii (Cineraria Kiril.,
    Turcz.). — P. m.
Cnicus (Cirsium) segetum, Bge. — P.
Rhaponticum uniflorum, DC. — P. m.
Myripnois dioica, Bge. — P. m.
Anandria Bellidiastrum, DC.—P. m.
Lactuca denticulata, Maxim. — P.
    » versicolor, Maxim. — P.
Rhododendron micranthum, Turcz. —
    P. m.
Androsace saxifragaefolia, Bge. —
    P. m.
Lysimachia barystachys, Bge.—P. m.
Jasminum Sambac, L. — P.
Syringa villosa, Vahl. — P. m.
    » amurensis, Rupr. -- P. m.
Fraxinus rhynchophylla, Hance. —
    P. m.
Apocynum venetum, L. — P.?
Periploca sepium, Bge. — P. m.
Vincetoxicum sibiricum, Dcne. — P.
Cynanchum pubescens, Bge. (C. chi-
    nense, R. Br.) — I suppose from
    Peking.
Gentiana squarrosa, Led. — P. m.
Tournefortia Arguzia, L. — P. m.
Eritrichium pedunculare, A. DC.—P.
Bothriospermum chinense, Bge. — P.
Convolvulus arvensis, L., var. sagit-
    taefolius, Led. — P.
Solanum nigrum, L. — P.
    » septemlobum, Bge. — P.
Physalis Alkekengi, L. — P

Lycium chinense, Mill. — P.

Rehmannia glutinosa, Lib. — P.

Orobanche ammophila, Mey. — P. m.

Incarvillea sinensis, Lam. — P.

Vitex incisa, Lam. — P. m.

Perilla ocimoides, L. — P.

Salvia miltiorrhiza, Bge. — P. m.

Marrubium incisum, Benth. — P.

Leonurus sibiricus, L. — P.

Amarantus ascendens, Lois. — P.

Chenapodium viride, L. — P.

Kochia scoparia, Schrad. — P.

Salsola Kali, L. — P.

Polygonum orientale, L. — P.

Diarthron linifolius, Turcz. — P. m.

Elaeagnus angustifolia, L. — P.

Euphorbia humifusa, Willd. — P.

    »   lunulata, Bge. — P. m.

Andrachne chinensis, Bge. — P. m.

Acalypha pauciflora, Horn. — P.

Canabis sativa, L. — P.

Broussonetia papyrifera, Vent. — P.

Cypripedium macranthum, Sw. — Pm.

Iris oxypetala, Bge. — Pm.

Pardanthus dichotomus, Led. — P.

Asparagus trichophillus, Bge. — P. m.

Polygonatum chinense, Kth. — P. m.

    »   officinale All.? — P. m.

Funkia subcordata, Spr. — P.

Lilium tenuifolium, Fischer. — P. m.

Commelina communis, L. — P.

Carex stenophylla, Wahl. — P.

Paspalum villosum, Thbg. P.

Panicum (Setaria) viride, L. — P.

    »   (») glaucum, L. — P.

    »   Crus Galli, L. var. muticum.
— P.

    »   miliaceum, L. — P.

    »   italicum, L. — P.

Sorghum vulgare, — P.

Hierochloe dahurica. Trin. — P.

Melica scabrosa, Trin. — P.

Poa pilosa, L. — P.

    » megastachys, Link. — P.

Selaginella mongolica, Rupr. — P. m.

    »   Stauntoniana, Spring. —
    P. m.

Plants gathered by Incarville near Macao.

Anona muricata, L.

Cocculus ovalifolius, DC.

Ionidium heterophyllum, Vent.

Actinidia chinensis, Planch.

Sida acuta, Burm.

    » humilis, Willd.

Urena lobata, L.

Hibiscus mutabilis, L.

    »   Rosa sinensis, L.

Helicteres angustifolia, L.

Grewia nitida, Juss.

Triumfetta Lappula, L.

Corchorus acutangulus, L.

Oxalis corniculata, L.

Averrhoa Carambola, L.

Ruta angustifolia, L.

Zanthoxylon Avicennae, DC.

Murraya exotica, L.

Atalantia monophylla, DC.

Brucea sumatrana, Roxb.

Melia Azedarach, L.

Sageretia, theezans, Brongn.

Cardiospermum Halicacabum, L.

Nephelium Litchi, Camb.

    »   Longanum, Hook.

Rhus succedanea, L.

Mangifera indica, L.
Crotalaria albida, Heyne.
Indigofera hirsuta, L.
Tephrosia purpurea, Pers.
Desmodium gangeticum, L.
     »   polycarpum, DC.
     »   pulchellum, Benth.
     »   latifolium, DC.
     »   triquetrum, DC.
Uraria crinita, Desv.
Alysicarpus vaginalis, DC.
Cajanus bicolor, DC.
Atylosia scarabacoides, Benth.
Guilandina Bondurella, L.
Cassia Tora, L.
     »   mimosoides, L.
Tamarindus indica, L.
Acacia Farnesiana, Willd.
Eriobotrya japonica, Lindl.
Kandelia Rheedii, Arnolt.
Psidium Guijava, L.
Eugenia Jambos, L.
Syzygium sp., near S. oblatum, Wall.
Jussieua villosa, Lam.
Coccinia grandis, Cogn.
Paratropia cantoniensis, Hook. & Arn.
Oldenlandia paniculata, L.
Mussaenda pubescens, Ait.
Ixora stricta, Roxb.
Psychotria Reevesii, Wall.
Paederia foetida, L.
Borreria stricta, L.
Elephantopus scaber, L.
Vernonia cinerea, Less.
     »   chinensis, Less.
Eupatorium Reevesii, Wall.
Conyza aegyptiaca, L.
Carpesium abrotanoides, L.
Siegesbeckia orientalis, L.

Eclipta alba, L.
Wedelia calendulacea, Less.
Wollastonia biflora, DC.
Bidens pilosa, L.
Pyrethrum indicum, Cass.
Artemisia annua, L.
Gynura pseudochina, DC.
Ipomoea septans, Poir.
     »   sp. near I. chryseides, Ker.
Solanum Melongena, L.
Capsicum conoides, Mill.
Herpestis Monnieria, H. & Kth.
Vandellia sp.
Siphonostegia chinensis, Benth.
Dicliptera cardiocarpa, Nees.
     »   cuneata, Nees.
Ocimum Bacilicum, L.
     »   sanctum. L.
Perilla ocimoides, L.
Mentha arvensis, L.
Scutellaria indica, L.
Anisomeles ovata Br. var. molissima,
    Benth.
Celosia cristata, Moq.
Amaranthus ascendens, Lois.
     »   melancolicus, L.
Cyathula prostrata, Blume.
Aerva lanata, Juss.
Achyranthes aspera, L.
Alternanthera sessilis, Rob. Br.
Polygonum chinense, L.
Cassytha filiformis, L.
Euphorbia Tirucali, L.
     »   pilulifera, L.
Acalypha pauciflora, Horn.
Dalechampia parvifolia, Lam.
Fatoua pilosa, Gaud. var. subcordata,
    Bur.
Ficus erecta, Thbg.

Ficus pertusa, L.
» sp. F. pumilae Thbg. affinis.
» sp. F. rufescenti Vahl. aff.
» sp. F. indicae, L. aff.
Artocarpus Jaca, Lam.
Pouzolzia indica, Gaud. forma *micro-phylla*.
Biota orientalis, Endl.
Cycas revoluta, Thbg. (See Lamarck, Enc. Bot. II, 232).

Amomum Zingiber, L
Alpinia Galanga, Sw.
Asparagus lucidus, Lindl.
Commelyna benghalensis, L.
Pothos sandens, L.
Cyperus Iria, L.
» *distans*, L.
Isachne sp. near I australis, Rob. Br.
Panicum compositum, L.
» *sanguinale*, L.

Oryza sativa, L.
Andropogon Schoenanthus, L.
Apluda mutica, L.
Chloris caudata, Trin.
Eleusine indica, L.
Dactyloctenium aegyptiacum, Willd.
Poa tenella, L.
» *unioloides*, Retz.
Bambusae spec.

Angiopteris evecta, Hoffm.
Lygodium japonicum, Sw.
Blechnum orientale, L.
Adiantum caudatum, Hook.
» *flabellulatum*, L.
Pteris semipinnata, L.
Asplenium sp.
Aspidium molle, Sw.
Polypodium adnascens, Sw.
Taenitis blechnoides, Sw.
Davallia tenuifolia, Sw.

Part II. THE LINNEAN PERIOD.

From about the middle of the 18th century to 1793.

This period is marked by the general adoption of the Linnean botanical system. The first outlines of the new method of classifying animals, plants and minerals are contained in Linnaeus' "Systema Naturae", published in 1735, but the plants known to him were first described according to the binary nomenclature, genera and species, in his "Species Plantarum", 1753, of which the second enlarged edition appeared ten years later, 1762—63. Although the great Linnaeus died in 1778, and his son in 1783, I make this period extend to the year 1793, when, with the first botanical exploration of the interior of China, a new era for European botanical investigations was inaugurated.

It must be remarked that from about the year 1755 the Chinese government had restricted all the foreign trade to the port of Canton. In 1755 Messrs Flint and Harrison, a British firm, attempted to trade at Ning po. But in 1759 their factory there was demolished by the Chinese. It is therefore obvious that all the botanical collections from China, herbarium specimens, seeds, living plants which during this period reached Europe came from Canton or Macao (belonging to the Portuguese), all the other Chinese ports being shut up from European access. We may, however, except a few seeds and perhaps also dried plants occasionally brought overland from Peking to St. Petersburg by the Russian caravans, which seem to have been sent by the Jesuit missionaries.

I. The First Swedish Botanical Collectors in China.

In 1731 a Swedish East India Trading Company was founded in Göteborg. The trade with China is recorded to have been opened in 1732, and
during the first 15 years 22 Swedish ships were sent to Canton, of which
4 were lost. (Chin. Repos. II, 295). The great Linnaeus, always anxious
to enrich his collections of natural objects with exotic specimens, never neglected to interest the captains or the chaplains of the ships despatched to
distant countries in his scientific pursuits. Almost all the Chinese plants
which Linnaeus describes in the "Species Plantarum" seem to have been
brought to him in Swedish ships. Occasionally he mentions the donor by
name. Thus in an article on the Phaseolus zeylanicus (radiatus) published
in the Acta Stockholm, 1742, p. 202, Linnaeus records that the plant was
brought to him from Canton by Admiral Ankarcrone. But the most important collection of Chinese plants he received in 1752 from Osbeck.

Peter Osbeck. 1751.

P. Osbeck, a Swede and a pupil of Linnaeus, was born in 1723. In
1750 he set out on a voyage to China, as chaplain to a Swedish E. India
man, the "Prince Charles", which left Göteborg 18 Nov. 1750, and arrived
at Cadiz 4 Jan. 1751. After a stay of 10 weeks they left this place, 20
March, sailed round the Cape, without landing there, in the latter half of
May. They passed St. Paul June 12th, anchored in the harbour of Angeri
(Java, Sunda Str.) July 15th, and left on the 17th. Here Osb. was able to
collect some plants. On Aug., 25th 1751, the "Prince Charles" anchored
at *Whampoa,* near Canton, and remained there more than four months,
weighing anchor on the 6th of Jan. 1752. On his way home Osb. again
collected Javanese plants and beasts in New Bay, where the Prince Charles
stopped a few days. In April Osb. made some collections on the island of
Ascension, and on June 26th got back to Göteborg.

Osbeck was a zealous naturalist and brought home a rich collection of
natural objects, chiefly Chinese specimens, for during his long stay at Whampoa and Canton he had ample opportunities of visiting the neighbourhood
of these places. Osbeck's original account of his voyage appeared in 1757
in his native language with the title: "Dagbok öfver en Ostindisk resa."
He published also some botanical articles in the Acta acad. Holm. 1762,
1765 and 1769. He was a Member of the Academy of Stockholm and of
the Soc. of Sciences, Upsala, and died as Praepositus of Hasloff and Wox-
torp, in 1805.

A German translation of Osbeck's narrative was made by J. G. Georgi, revised and completed by Osbeck himself, and published in 1765. From this German version J. R. Forster translated Osbeck's book into English and published this translation in 1771, adding a Faunula and Florula sinensis. For further particulars see my Earl. Eur. Res. p. 88 seqq.

Osbeck's notes on his plants collected in the neighbourhood of Canton and Whampoa are found scattered in his diary. He enumerates in all 244 Chinese plants, often describing them under their Linnean names, or giving them new ones first proposed by himself. Eleven new Chinese plants are figured there. (see Earl. Res. 91). It is generally believed that Osbeck placed all his Chinese collections in the hands of Linnaeus, but Linnaeus determined and described only a small portion of Osbeck's Chinese plants. In the first edition of the Species Plantarum, which appeared in 1753, a year after Osbeck's return and 4 years before he published his diary, Linnaeus describes only 37 plants brought by Osbeck from China. We may suppose that 18 more Chinese plants noticed there were likewise collected by Osbeck, for these plants appear in Osbeck's list, but Linnaeus omitted to mention his name. Almost all the Chinese plants from Osbeck's collection, described by Linnaeus, are found in the first edition of the Spec. Plant. Only *Clerodendron fortunatum* is first described in Amoen. Acad. IV, 320 (1756) and *Rhamnus theezans* in the Mantissa II, 207 (1771).

Besides the Osbekian plants described by Linnaeus there are in Osbeck's list 155 more plants mentioned as growing near Canton, for the greater part previously named and described by Linnaeus as Indian plants. But there appear also 21 new Chinese plants for which Osbeck proposes the following names:

Clematis chinensis (Earl. Res. 91).   *Convallaria chinensis* (l. c. 110).
Urena chinensis (l. c. 93).   *Commelina chinensis* (l. c. 110).
Buxoides aculeata (l. c. 94). Perhaps   *Cyperus dichotomus* (l. c. 111).
    Atalantia buxifolia, Oliv. see Ind.   *Scirpus chinensis* (l c. 111).
    Fl. sin. i, 110.   *Saccharum chinense* (l. c. 111).
Mimosa chinensis (l. c. 97).   »   *pluviatile* (l. c. 111).
Solidago chinensis (l. c. 101).   *Briza elegans* (l. c. 113).
Nyctanthes orientalis (l. c. 102).   *Lycopodium varium* (l. c. 113).
Columnea chinensis (l. c. 105).   *Jungermannia chinensis* (l. c. 114).
Monarda chinensis (l. c. 106).   *Lichen chinensis* (l. c. 114).
Achyranthes chinensis (l. c. 107).   *Agaricus chinensis* (l. c. 114).
Cryptanthus chinensis (l. c. 115).

Linnaeus in the second edition of his Species Plantarum, published

1762—63, does not make any mention of these names, although he knew of Osbeck's diary, for the quotes it repeatedly as Osb. iter. A full account of Osbeck's collection of Chinese plants is given in my Earl. Res. p. 91 seqq. where his plants are enumerated together with the Chinese plants known to Linnaeus.

According to De Candolle, "Phytographie", 1880, p. 438 Osbeck's Herbarium is now kept in the Government Museum at Stockholm. Joh. Müller, Arg. in the "Linnaea", XXXIV (1865) p. 107, states that the type specimen of his *Croton tomentosus,* which is described there, was gathered by Osbeck in China, and is now in the Stockholm Herbarium.

Linnaeus named the genus *Osbeckia* after Osbeck; see sub *Osbeckia chinensis,* Sp. Plant. first ed. 345: "In honorem Petr. Osbeck, qui 1751 plantas Chinae et Javae periculoso itinere adiit, legit, examinavit, descripsit, communicavit."

Olaf Toreen, 1751—52.

O. Toreen was Chaplain to the "Gothic Lion", a Swedish E. India man which visited Canton at the same time that Osbeck was there. This ship anchored at Whampoa, 7 July, 1751, and left 4 Jan. 1752. Toreen presented to Linnaeus, his instructor, a collection of Indian plants chiefly from Suratte, where he had made a stay of more than 5 months, and some also from China. The letters addressed to Linnaeus by Toreen during his voyage were published as an appendix to Osbeck's narrative along with some short biographical notes. The 5th and 7th letters contain some particulars regarding Canton, and the cultivation of Rice, and Batates, but are of no interest concerning Chinese botany. Toreen died in 1753, a year after his return to Sweden. Linnaeus dedicated to his memory the genus *Torenia*. Although he says that T. brought the *Torenia asiatica* from India, it is more probable that he gathered it near Canton. Only one plant, *Conyza chinensis* is stated by Linnaeus to have been gathered by Toreen in China.

Carolus Gustavus Eckeberg. 1743—71.

Eckeberg was Captain to several ships of the Swedish E. I. Company. Between 1743 and 1771 he visited India and China (Canton) eight times. On his first voyage to China, in 1743, his ship, the "King of Sweden" met with a heavy storm and was forced to seek shelter in the large island of *Aynam* (Hainan), 400 miles from Canton. Eckeberg has left an account of this island. "Its mountains are covered with forests producing precious woods as Rose-wood, Aloe-wood. In the valleys the Chinese settled there cultivate rice, sugarcane, beans, tobacco, batates, Pisang, Betel, cocoa nuts, bamboos, ananasses etc." Eckeberg spent some time in this island and made excursions into the interior. In March 1743 they sailed for Macao.

In 1754 Eckeberg presented to the Swedish Academy of Sciences a memoir on CHINESE HUSBANDRY, which was published in 1757 as an appendix to Osbeck's narrative (v. supra). A detailed account of this paper is found in my Earl. Res. p. 116—118. D. de Blackford translated it into French, in 1771.

As Linnaeus states in the Syst. nat., 12th edition, 1767, Eckeberg was the first to bring from China a living *Tea plant*, which safely reached Linnaeus, 3 Oct. 1763.

Another of Eckeberg's voyages to China is mentioned by his cousin A. Sparrmann (see further on), who accompanied the former on his voyage to China in 1766. Eckeberg was then Captain to the "Navarcha", which arrived at Macao Aug. 24, 1766.

Eckeberg seems to have made his last voyage to China in 1770. The narrative of it is contained in three letters addressed by E. to the Secretary of the Swedish Academy. These letters were published together with 5 plates of which pl. 3—5 represent the Bocca Tigris, Whampoa and Canton. In the same memoir there is also the account of Eckeberg's first voyage, in 1743, above referred to. I have seen a German translation of it by Bernouilli, Berlin, 1784. From this we learn that Eckeberg was a Member of the Swedish Academy. In 1770 he was commander of the Royal Swedish Ship "Finland" which together with the "Prince Gustav" was sent on an expedition to China. In the beginning of September the "Finland" anchored at Whampoa near Canton, and remained there till Dec. 20, 1770.

Sparrmann has dedicated to the memory of his friend the genus *Ekebergia* (Meliaceae, Cape of G. H.), Act. Holm. 1779 p. 282.

It seems that Eckeberg has also made botanical collections in China. Lessing, Linnaea, 1831 p. 220 sub *Crossostephium artemisioides* says: "e Cantone Chinensium ab Eckeberg. In herbario Bergiano". (P. J. Berg, 1730—1790, a pupil of Linnaeus, professor at Stockholm). — Lindley, Rosar. Monogr. 108, sub *Rosa semperflorens:* Hab. in China, Eckeberg, v. v. c. et sp. 5. herb. Banks.

Andreas Sparrmann, 1766.

A. Sparrmann, a Swedish naturalist, born in 1747, died as Professor of Natural History at Upsala, in 1787. In 1766 he visited China, accompanied Captain James Cook on his first circumnavigation of the world, went to the Cape of G. H., where he lived, 1772—75, and travelled in the interior of Africa.

A brief account of his voyage to China, and of the collections of natural objects which he made there, is found in the Amoen. acad. VII, 1768, p. 497—506, Sparrmann, Iter in China. He states that Captain Eckeberg had invited him to accompany him on his ship the "Navarcha". The latter passed by Macao, Aug. 24, 1766, and anchored not far from the city of Canton, Aug. 26. Sparrmann enumerates the following plants gathered there:

Oryza sativa.	*Barleria cristata.*
Arundo saccharifera.	*Hedyotis fruticosa.*
Thea Bohea.	Planta dubia, *Spermacoce?*
Ixora coccinea.	*Sinapis brassicata.*
Rhamnus lineatus.	*Polygonum chinense.*
Baeckia frutescens.	*Torenia asiatica.*
Triumfetta Barthramia.	*Aralia chinensis,* sed tota glabra.
Urena procumbens.	*Verbesina calendulacea.*
Hedysarum biarticulatum.	

The following fruits were collected

Bromelia Ananas.	*Citrus aurantium.*
Cucurbita Citrullus.	*Citrus decumana.*
Tamarindus indica.	

Sparrmann has written many botanical articles in the Acta Acad. Holm. and in Nova Acta Soc. Upsal. — Thunberg dedicated to his memory the genus *Sparmannia* (Tiliaceae, Cape of G. H.).

Magnus von Lagerstroem, 1696—1759.

I may finally mention an ardent Swedish naturalist, M. v. Lagerstroem, a friend of Linnaeus, who, although he had never visited Asiatic countries, has contributed much to the knowledge of the natural history of India and China. His position as Director of the Swedish E. I. Company at Göteborg enabled him to procure many rare objets of natural history from India and China, which he used to present to Linnaeus.

There is in the Amoen. acad. IV, p. 230—266 a paper by J. L. Odhello, written in 1754, devoted to the *Chinensia Lagerstroemiana*. The author reports that Lagerstroem had obtained dried specimens of plants from India and China, and had also succeeded in introducing living plants from those distant countries into the botanic garden at Upsala, amongst

others *Arum chinense*, a plant not mentioned by Linnaeus. — Toreen in
his letters repeatedly speaks of Lagerstroem. Linnaeus received first from
him the beautiful Chinese plant, which he named, in honour of his friend,
Lagerstroemia indica. (Linnaeus makes no difference between India and
China, see further on). Besides this Linnaeus notices two Chinese Artemisias,
A. chinensis and *A. minima* as having been sent to him by Lagerstroem.

Odhello mentions a *Botanicon Chinense*, written in Chinese characters,
in 36 volumes, which was in the possession of Lagerstroem. Strindberg,
"Relations de la Suède avec la Chine, Revue de l'Extr. Orient," I, 515, has
proved that this is the well known Chinese Materia Medica "Pen ts'ao kang
mu". — Lagerstroem had moreover received from China a collection of
about 1000 Chinese drugs.

II. Chinese Plants known to Linnaeus, -+ 1778, and his Son, -+ 1783.

We propose in the present chapter to give a complete account of all
the Chinese plants described by Linnaeus father and son with the quotations
from earlier authors regarding some of these plants. Linnaeus in the earlier
days of his scientific career, cultivated in Clifford's Garden *) 10 Chinese
plants raised from seeds received from China, and described them in 1737
in the "Hortus Cliffordianus". Two of them *(Aster chinensis, Urena lobata)*
had been previously cultivated in James Sherard's garden at Eltham (in
Kent) by Linnaeus' older friend Dillenius, who described and figured them,
in 1732, in the "Hortus Elthamiensis". The same plants were subsequently
cultivated by Linnaeus in the botanic garden at Upsala. In the "Hortus
Upsaliensis", published in 1748, two new Chinese plants are noticed as
cultivated in that garden: *Brassica chinensis* and *Brassica violacea*. When
the younger Linnaeus died, in 1783, about 160 Chinese plants had been
determined by his father and himself. Of these nearly 100 had been named
and described in the first edition of the Species Plantarum published in
1753. In his Systema Naturae, 10th edition, 1759 he gives the diagnoses
of 9 more Chinese plants, and the descriptions of 19 new Chinese plants

---

*) J. E. Smith in Rees' Cyclop., art. Linnaeus, says: «When Linnaeus was in Holland
(1735), Boerhave introduced him to Clifford an opulent banker, whose garden at Hartecamp,
was one of the richest in the world. For two years L. was superintendent of this garden, and
in 1737 published the «Hortus Cliffordianus». He styles Clifford: Dr. utriusque juris and states
that Hartecamp lies in Holland at the distance of one hour's journey from Harlem, and three
hours from Leyden.

appear in the 2d edition of the Species Plantarum, 1762—63, to which 20 more are added in the Mantissa, 1767—71, and 10 in the Supplementum Plant. Linn. filii et patris, 1781.

Linnaeus the father is generally very careless in his statements regarding the native countries of exotic plants. He seems to have had a very confused idea with respect to the geographical position of China, for he identifies it not unfrequently with India. *Daphne indica,* thus named by Linnaeus, was known to him from China only. It is the same with his *Rosa indica,* as he states himself, a Chinese plant. On the other hand he first named *Ixia chinensis, Poa chinensis, Osbeckia chinensis, Polygala chinensis, Dolichos chinensis, Citrus chinensis, Sphaeranthus chinensis, Bidens chinensis,* but gives as the native country of all thes plants only India. Many of Osbeck's Chinese plants appear in the Species Plantarum as plants collected by Osbeck in India. But as we have seen Osbeck never visited India. Linnaeus even does not always distinguish between India orientalis and India occidentalis. Thus, in the Species Plantarum, first ed. 63, second' ed. 94, we read: "*Agrostis indica,* habitat in India". It is only from the quotation, Sloane, Jam. we can conclude that Linnaeus means India occidentalis. Linnaeus, in describing new plants he had received from foreign countries deems it generally superfluous to notice the names of the collectors. With respect to Chinese plants he mentions occasionally Osbeck, in a few cases Toreen, Eckeberg and Gmelin. The latter had sent him a few Chinese plants or seeds received from China (Peking). But for about one half of the Chinese plants in his herbarium he does not give the names of the donors.

Abbreviations of the titles of some of Linnaeus botanical works frequently quoted in the subsequent account:

Sp. I = Species Plantarum, first edition, 1753.

Sp. II = Species Plantarum, second edition, 1762—63.

S. X = Systema Naturae, vol. II, Vegetab., 10th edition, 1759.

S. XII, Systema Naturae, vol. II, Vegetab. 12th ed., 1767.

S. XIII, Systema Vegetabilium, 12th ed., cur. Murray, 1774.

Illicium anisatum, S, X, 1050, Sp. II, 664. Illicium floribus flavescentibus. — Somo vulgo Skimmi, Kaempf. Amoen. exot. 880, t. 881 — Habitat in Japonia, China.

Planta a me non visa, fide Kaempferi recepta, forte Anisum stellatum officinarum, quod adjectum Tetraodonti occelari (Globe-fish) ejus auget venenum.

Linn. Mat. Med. (1749) app. n. 510 (sub dubiis): Anisum stellatum. Loc. nat. Tartaria, Chinae, Philippinae.

Badianifera, Evonymo affinis Philippinarum insularum Anisum spirans, nuculas in capsulis stellaformiter coaggestis proferens, Pluken. Almagest. 140. —

Anisum peregrinum, C. A. Bauhin, Pinax Theat. Bot. (1623) 159. —

Clusius, Plant. rar. hist. (1601) II, 202: Anisum Philippinarum insularum. Ante paucos annos allatum a Nauclero Anglo Thomas Candi, cum e sua navigatione circum orbem rediisset. Accipiebam semina et umbellarum particulum. Odor Anisi vulgari. Damor appelatur. Folia, flores non vidi. (Comp. also supra p. 17, Kamel, Luzon.).

Linnaeus, who had seen only the fruit of the Chinese Star Anise, identifies it erroneously with Kaempfer's Skimmi of the Japanese or *Illicium religiosum*. The plant which yields the Chinese Star Anise was first described as *Illicium verum* by Sir Jos. Hooker, Bot. Mag. t. 7005 (1888) from living plants sent by Mr. Ch. Ford.

Anona hexapetala. Supplem. 270. A. foliis elliptico-oblongis, acutis, glabris; petalis spathulatis, aequalibus acutis. — Hab. in China, colitur in India orient.

According to Aiton, Hort. Kew. 1789, p. 253, this plant was introduced into England in 1758. The name does not appear in the Ind. Florae Sin. — Dunal Monogr. Anonaceae, 1817, p. 106 identifies Linnaeus' A. hexapetala with *Anona uncinata*, Lam. Enc. Bot. II, 127, which is the *Artabotrys odoratissimus* R. Br. See Ind. Fl. Sin., I, 26.

Menispermum orbiculatum, Sp. I, 340; Sp. II, 1468. M. foliis orbiculatis, subtus villosis. Hab. in insula Crocodilorum Asiae. Cocculi orientalis frutex convolvulaceus orbiculatis foliis prona parte villosis, Pluken. Amalth. 61, t. 384, fig. 6 (comp. supra p. 35.). — Thbg. Fl. Jap. 194. Japan.

D. C. Syst. Veget. I, 523, refers it to *Cocculus*. Not mentioned in the Ind. Fl. Sin.

Fumaria spectabilis, Sp. I, 699, — Sp. II, 983. F. floribus postice bilobis. Corollae magnit. extimi articuli pollicis, postice in duos lobos aequales rotundatos divisae. Planta eximia floribus speciosissimis maximis. Habitus F. bulbosae. — Hab. in Sibiria. D. Demidoff.

Amoen. acad. II, 336: Gregor Demidoff, nobilis, discipulus Linnaei,

1761. Plantas a Stellero et Lercheo maxime lectas ruthenicas jam antea Linnaeo dedit.

Linnaeus first mentions this plant in his "Plantae camtchatkenses rar.", 1750, n. 18 (Amoen. acad. II, 327): Inter Flores Sibiriae speciosos et maxime singulares est etiam quaedam Fumaria, bulbosis affinis floribus condecorata in suo genere maximis.

Linnaeus is mistaken with respect to the native country of the plant, which is not Siberia but China. S. G. Gmelin, in his Flora Sibirica, IV p. 68, states that this extremely beautiful plant was brought to him (when he was travelling in Siberia, 1733—43) from China (probably Peking), by the surgeon Heuke. At about the same time d'Incarville (supra p. 47.) sent herbarium specimens of the plant from Peking to Paris.

Fumaria spectabilis was first figured by Alex. de Karamysheff, who, when student at the Moscow University, in 1766, sent to Linnaeus a memoir, "Necessitas Historiae naturalis Rossiae", which was published in the Amoen. acad. VII. We read there, p. 457: Memini de caetero me vidisse domi varias plantas omnium pulcherrimas utpoti Hypecoum erectum, Fumariam spectabilem aliasque quae dignissimae essent quae inserentur hortis Magnatum summorum, ob illarum pulchritudinem. — The plant is represented on tab. 7.

D. C. Syst. Veg. II (1821) 110, named it *Diclytra spectabilis,* Miquel, in 1866, *Dicentra spectabilis.*

Brassica chinensis, Sp. II, 932, B. foliis ovalibus subintegerrimis, floribus amplexicaulibus lanceolatis, calycibus ungue petalorum longioribus. Flores ut in B. oleracea lutei, siliquae parum compressae. — Hab. in China.

First described in Amoen acad. IV p. 281, Cent. I, n. 52 (1755). Osbeck attulit e China.

Comp. Earl. Res. p. 92 and supra p. 49. I may notice that Boerhave, Index alter Lugd. Bat., 1727, II, 12, describes a *Brassica sinensis* folio lactucae, flore luteo, which then was cultivated at Leyden.

Brassica violacea. Sp. I, 667; — Sp. II, 932.

Linn. Hort. Upsal (1748), 191: B. foliis lanceolato-ovatis glabris, indivisis, dentatis. Flores violacei magni. — Habitat in China. Misit Cel. Gmelinus.

Hemsley, Ind. Fl. Sin., I, 47, states that this plant is not represented in the Linnaean herbarium (now in the possession of the Linnean Soc., London) and is quite indeterminable from the description. — Gmelin received most probably the seeds from Peking. (through Heuke?) I also, am not prepared to say which Peking (cultivated) plant is to be understood by Linnaeus' Br. violacea.

Sinapis juncea, Sp. I, 668; Sp. II, 934. Habitat in Asia, China.

Linn. Hort. Upsal (1748) 191, n. 2. Sinapis ramis fasciculatis, foliis summis lanceolatis integerrimis. Sinapi sinensi acanthifolio, Gmelin. — Habitat in China.

Hermann, Paradisus Batavus (1698), 230, t. 230, Sinapi indicum maximum, lactucae folio.

I. F. S., I, 47.

Sinapis brassicata, S. XII, vol. III, App. 231. S. foliis obovatis denticulatis laevibus. Statura Brassicae aut Lactucae, sed calyx sinapeos. Flores flavi. Siliqua sim. Br. oleraceae. — Hab. in China. —

This seems to have been brought to Linnaeus by Sparrmann from Canton. V. supra p. 62. — Hemsley, Ind. Fl. Sin., I, 47, unites Linnaeus' S. brassicata with S. juncea.

Sinapis chinensis, Mant. I, 95. S. siliquis laevibus subarticulatis, patulis, foliis lyrato-runcinatis subhirtis. Flores albi, parvi. — Habitat in China? Arduini.

P. Arduini, Animadv. botanicarum Specimen (1759), I, p. 23, tab. 10. Sinapi chinensis, folio acanthi, siliquis glabris. Radix annua crassiuscula, candida, parum ramosa. Semina hujus plantae ad nos transmissa sunt ab amicis nostris (Gmelin?) per epistolam signata nomine S. chinense folio acanthi.

Boerhave, Index alter Plant. hti Lugd. Batav. II (1727) p. 13. Sinapi chinense folio acanthi, a Vaillant, (1669—1721).

DC. Syst. II, 613 says e herbario Vaillant.

Hemsley, l. c. reduces this also to *Sinapis juncea.*

Raphanus sativus, var. γ *Raphanus chinensis* annuus, *oleiferus,* Sp. II, 935. — Hab. in China.

Seeds of this had been brought from China by Eckeberg. See Earl. Res. p. 116. Linnaeus cultivated it at Upsala.

Polygala chinensis. Sp. I, 704; II, 989. P. floribus imberbibus spica-

tis, axillaribus, caulibus suffruticosis, foliis ovalibus. Caules pedales lig-
nosi. — Habitat in India.

P. arborea, fol. lanceol.-ovat. etc. Browne, Jam. 287.

This may serve, once more, as an example for the carelessness and
the contradictions so frequently met with in Linnaeus' writings. He names
the plant P. chinensis, gives India as its native country, and from the
quotation of Browne's natur. history of Jamaica we may conclude that he
means West India. — Hemsley, Ind. Fl. Sin., ɪ, 59, states that P. chinensis, according to the Flora of British India, is found in Tropical Asia and
Australia. Whether it grows in China is uncertain.

Dianthus chinensis. Sp. I, 411; II, 588. D. floribus solitariis: squamis
calycinis subulatis patulis tubum aequantibus. — China.

Hort. Clifford. (1737) p. 164, n. 4 and Hort. Upsal. (1748), 104,
n. 3. — China.

Philos. Trans. XXVIII (1713) p. 62: Petiver, Rare plants raised in
Chelsea. China Pink, Coryophyllus sinensis flore pulcherrimo, Boerh. 88, 23.

Tournefort, Act. (Mémoires de l'Acad. Paris) 1705, p. 264, fig. 5:
Caryophyllus sinensis supinus, leucoji folio, flore unica.

Tournefort states that about three years ago, Abbé Bignon received
this (probably seeds) from China.

Hypericum monogynum. Sp. II, 1107. H. floribus monogynis, staminibus corolla longioribus, calycibus coloratis, caule fruticoso. —

China. (Brought by Osbeck, it seems. Comp. Earl. Res. 93).

S. X, 1184, *Hypericum chinense*. (Diagn. eadem).

Miller, icon. t. 151, f. 2.

Amoen. acad. VIII (1776) p. 328. H. chinense, monogynum. Dissert.
Linn. — Helenius.

I. F. S., ɪ, 72.

Camellia japonica. Sp. I, 698; II, 982. — Hab. in Japonia, China.

Thea chinensis pimentae jamaicensis folio, flore roseo. Petiv. Gazoph.
t. 33, f. 4. (SVIII). V. supra p. 35.

Tsubakki montanus s. sylvestris, flore roseo simplici, Kaempfer, Amoen.
exot. 850, t. 851.

    var. β. (Sp. I). Tsubakki hortensis, flore pleno maximo Rosae hortensis.
Kaempf. l. c. 852.

Rosa chinensis, G. Edward, Natur. Hist. (1758—64) II, 67, t. 67.

Thea sinensis. Sp. I, 515; — *Thea Bohea,* Sp. II, 735. Vidi flores in aliis hexapetalos, in aliis enneapetalos; an ejusdem speciei judicent, qui possunt vivam inspicere. — Loc. nat. Japonia et China.

S XII, XIII: Nobis attulit vivam primus Carl. Gust. Eckeberg, navis ad Chinenses gubernator, d. 3. Oct. 1763. Obtinetur, si semina ad navis recessum e China in nave terrae mandentur, ut in itinere serius germinent. (Comp. also Earl. Res. 93).

Hort. Clifford (1737) 204: Thea. The following authors quoted.

Thee, Kaempf. Jap. (Amoen. exot.) p. 605, t. 606.

Breynius Centuria Plant. rar. (1678), p. 111, t. 112, (Description of the Tea plant from Japan). In the Appendix 17, t. 3, an article on Japanese Tea by Dr. W. Rhyne.

Breynius, Prodromus II, Plant. rar. in hort. Hollandiae (1689) p. 98: Styraci et Evonymo medio affinis *The* Sinensium sive *Tsia* Japonensibus, flore simplice, nobis. In horto Amstelodamensi saltem conspeximus. *The* fruticis notae: In omnibus cum Styracis conveniunt nisi quod 1, nuculae in fructu intersepimentis Evonymi de more distinctae, — 2, floris petala majora rotunda, obtusa, — 3, stamina apicibus minoribus ornata sunt. — Ibidem: Styraci et Evonymo media affinis, *The* Sinensium sive *Tsja* Japon., flore niveo pleno, nobis. A praecedente differt flore decem petalis duplice serie composito, Rosae Damascenae flore pleno et facie et magnitudine. Hujus ramos simul cum variis plantis aliis Japonensibus, Cl. D. Cleyer nobis ex India transmisit. (Comp. p. 26.)

Raji Hist. Plant. (1688). 1619: *The* Sinensium sive *Tsja* Japonensibus. In Japonia et China crescit.

Bontius, Hist. nat. Indiae orient (1631), 87, t. 88, de herba seu frutice quam Chinenses *The* dicant etc. (v. supra p. 25).

C. Bauhini Pinax Theatri Botanici (1623) p. 147: *Chaa* herba in Japonia ex cujus pulvere decoctum preciosum parant et hospitibus dignioribus propinant: et olla in qua hujus herbae decoctio facta, in eo apud ipsos precio, in qua apud nos Adamantes sunt. Linscot. par. 2, Ind. Orient. cap. 28. (Bauhin quotes from Linschoten's Voyage to the East-Indies, end of 16th cent. See Engl. transl. Hakluyt Soc. 1885, I, p. 157, the island of Japan.).

Pluken. Amalth. 79. Evonymo affinis *tsia* Japonensibus dicta. (v. supra p. 35.).

Pluken. Almagest. p. 139: Evonymo affinis arbor orientalis nucifera, flore roseo. Phytogr. tab. 88, fig. 6. Quoted Breynius, Bontius (v. supra), J. B. III, lib. 27, 5. (Joh. Bauhin. Historia plant. univers. 1650, notice of Tea, copied from C. Bauh. Pin. Th. Bot.).

J. Hill, Exotic Botany (1759), t. 22.

Linn. Mantissa, II, 402. Thea Bohea. Semina terrae mandanda debent prius macerari paululum in aqua calida quam digitus immissus potest fere tolerare, ut alia indica. Perfidi Chinenses saepe pro Theae frutice offerunt Camelliam, qua leviter versatus absque floribus facile seducitur, non observatis attente petiolis. Multi ostenderunt alium fruticem pro Thea, sed, ni fallor nullus vivam habuit in Europa ante me.

Thea viridis. Sp. II, 735. Thea floribus enneapetalis. Folia hujus longiora, illius (i. e Theae Bohea) breviora. Plantam non vidi.

Hab. in China. — Hill, Exot. Bot. (1759), t. 22.

S. XIII, 412. Florum Theae viridis libram integram aliquando accepi; esse Theam viridem ex Hillio dedici: stylos 3 conglutinatos esse, nec simplicem. — Jo Coakly Letsom. (J. C. Lettsom, Natural History of the Tea tree etc. 1772).

Linn. Amoen. acad. VII, p. 237, t. 4 (1765). Thea Bohea et viridis. Thea crescit, quantum adhuc notum, in China et Japonia tantum, ad latera collium et ad ripas inprimis fluviorum. A Canton usque ad Pechinum in China reperitur, quod mirandum. Urbs Pechini aequali a polo longitudine ac Roma distat. Frigidus Pechini est multo acerbius ac Stockholmiae.

I need not mention that tea is not grown at Peking. The extreme line of its cultivation in China does not go farther north than the 31° of lat. — Botanists are now generally agreed that the two supposed Chinese species are nothing more than varieties of one and the same species for which Sims in 1807 proposed the name *Thea chinensis;* Link, 1822, *Camellia Thea.* See Ind. Fl. Sin., 1, 82.

Vatica chinensis. Mantissa II, 242. — Hab. in China. Statura accedit ad Toberam, Kaempf. Amoen. exot. 797 (Pittosporum Tobira Ait.) aut Citrum. Rami striati seu angulati, subtomentosi. Folia alterna petiolata, cordato-ovata, integerrima, utrinque glabra, venosa, alterne nervosa, spithamaea, latitudine palmaria. Flores paniculati, praecipue terminales, magnitudine Citri.

Represented in J. E. Smith's Icones ined. herb. Linn. (1789—91) t. 36. In Hooker's Fl. Indica I, 302 it is identified with *V. Roxburghiana,* a species apparently restricted to the western peninsula of India and Ceylon. (Ind. Fl. Sin., 1, 83.)

Malva verticillata. Sp. I, 689; — Sp. II, 970. M. caule erecto, foliis angulatis, floribus axillaribus glomeratis. — China.

Viridarium Cliffordianum (1737), 68.

Hortus Clifford. (1737) append. p. 502, n. 7. Diagnosis as above.

Hort. Upsal. (1748) 200. In this and the preceding habitat not given.

Royen ht. Lugd. Batav. (1740) 356: "Malva caule erecto, floribus verticillatis".

Boerhave, Ind. alter Plant. hti. Lugd. Batav. (1727) II, 268. Malva sinensis erecta, flosculis albis minimis.

Morison, Hist. Plant. Oxon. I (1680) 520. Malva annua rotundifolia, floribus omnium minimis albis pentaphyllis verticillatim genicula ambientibus. — Habitat not given.

Raji Hist. Plant. (1686) 598. Habitat not given.

I. F. S., I, 84.

Urena lobata. Sp. I, 692; Sp. II, 474. Urena foliis angulatis. — China.

Hort. Cliff. (1737) p. 348, n. 1. — India orientalis. Quoted:

Dillenius, Hort. Elth. (1732) p. 430, t. 319, fig. 412. Urena sinica xanthii facie. — Dillenius reports that the seeds came from China by an Ostend ship, and were communicated by Egmont van Nyenborg, a Fleming nobleman.

Hort. Upsal. (1748) 200.

Breyn, Centuria exot. etc. (1678) p. 82, t. 35. Trifolio adfinis Indiae orientalis, xanthii facie.

Osbeck gathered U. lobata in China. Earl. Res. 93.

I. F. S., I, 86.

Urena procumbens. Sp. I, 692; II, 975. U. foliis hastato subcordatis indivisis, serratis, caule procumbente. — In China monticulis, Osbeck.

Collected also by Sparrmann in China. Earl. Res. 93, 119.

I. F. S., I, 86.

Hibiscus mutabilis. Sp. I, 694, — Sp. II, 977. Hibiscus foliis cordato quinqungularibus obsolete serratis, caule arboreo. — Habitat in India.

Hort. Clifford (1737) p. 349, n. 3. Locus nat. China, Japonia, Malabar.

Hort. Upsal. (1748) 205; the same.

In the Hort. Cliff. the following quotations are found:

Joh. Bapt. Ferrari, Senensis, Soc. Jes. Flora seu de Florum cultura, 1632, p. 479—502, tab. 497: Decem ante annos a nostris hominibus occi-

dentali ex India missum semen est floris fabulosi quem barbari Indi nomine
Fu yo, alii a genitali solo Malvam indicam, Japonicam alii vocant, nos cum
iis qui Romam ex India miserunt semina, *Sinensem Rosam* nominabimus,
sinensis enim floris satio et culture Senensi cultori debebatur. — *Fu yo* is
the Japanese name for Hibiscus mutabilis.

Morison, Hist. Plant. Oxon. II (1680) 530, Sec. 5, tab. 18, fig. 2.
Althaea arborea Rosa sinensis nobis.

Breyn. Prodr. Rar. Plant. Holland. II (1688) 10, Alcea arborescens
japonica, pampineis foliis subasperis.

Hina-parite, Rheede, Malab. VI, p. 69, t. 38—41.

Flos Horarius, Rumph. Amb. IV, p. 27, t. 9. Rumphius says, a native of China.

Tournefort, Instit. Rei Herb. (1694), Ketmia sinensis, fructu subro-
tundo, flore simplici.

I. F. S., I, 87.

Hibiscus Rosa sinensis. Sp. I, 694; — Sp. II, 977. H. foliis ovatis acuminatis serratis, glabris, caule arboreo. Variat flore pleno. — Habitat in India.

Linn. Flora zeylan. (1747) n. 260.

Breyn. Centur. exot. 121, t. 56. Alcea javanica arborescens, flore
pleno rubicundo.

Hermann, Catal. hti Lugd. Batav. (1687), 22. Althaea arborea Rosa
sinensis.

Raji Hist. Plant. (1688), 1068 n. 11, Alcea.

Sheru paritu, Rheede Malab. II, p. 25, t. 16.

Flos festivalis, Rumph. Amb. IV, p. 24, t. 8. India and China.

Tournefort, Institut. Rei Herb. (1694): Ketmia sinensis fructu sub-
rotundo, flore pleno.

I. F. S., I, 87.

Hibiscus Manihot. Sp. I, 696; — Sp. II, 980. H. foliis palmato-digi-
tatis septem-partitis. — Habitat in Indiis.

Hort. Clifford. (1737), 450, n. 7. Diagnosis as above. — Locus nat.
China, America.

Hort. Upsal. (1748) p. 206, n. 5.

The following quotations are found in the Hort. Clifford:

Ketmia folio Manihot serrato, flore amplo sulphureo, Dillen. Hort.
Eltham. (1732), 189, t. 156, fig. 189. Dillenius, has the following quotations:

Alcea sinica Manihot stellato folio etc. Pluken. Amalth. 7, t. 355, fig. 2, and Abelmoschus chusanensis folio palmato, fructu piloso, Herb. Sloane, Petiver n. 1. (v. supra p. 35).

Hibiscus Manihot does not appear in the Ind. Fl. Sin.

Hibiscus tiliaceus. Sp. I, 694; Sp. II, 976. H. foliis cordatis subrotundis indivisis acuminatis crenatis, caule arboreo, calycum exteriore decemcrenato. — Hab. in India.

Lin. Flora zeyl. 118.

Alcea malabarica abutili folio, flore minore ex albo flavescente exterius aspero, Raji hist. pl. (1688) 1070.

Pariti s. Talipariti, Rheede, Malab. I, p. 53, t. 30.

Novella, Rumph. Amb. II, 218, t. 73. In Sina quoque arbor vulgaris est.

Alcea indica Sinarum flore luteo malvaceo etc. Pluken. Amalth. 6, tab. 355, fig. 5. (v. supra p. 35).

I. F. S., ı, 88.

Hibiscus simplex. Sp. II, 977. H. foliis cordatis trilobis repandis integerrimis, caule arboreo simplicissimo. Planta mihi triennis, caule indiviso, recto, crassiusculo. Folia integerrima, cordata, triloba, obtusiuscula, glabra, margine inaequali, nervis subtus poro mellifero. Flores non vidi. — Habitat in Asia.

Sterculia plantanifolia. Suppl. 423, St. hermaphrodita, foliis cordatis, lobatis, floribus paniculatis. — Habitat in India. Ex horto Patavino communicavit A. Murray, Prof. Anat. Upsal. Colitur in horto Upsal.

Cavanilles, Dissert. Monadelph. 5 (1788), p. 288, tab. 145, *Sterculia plantanifolia,* Lin., identifies it with the Tetraglottis baccifera Sinarum Pluken. Amalth. 199, tab. 444, fig. 4 (v. supra p. 35), and states that it is a native of China.

Vahl, Symbol. Bot. I (1790) p. 80, sub. *Sterculia platanifolia,* Lin., observes: Folia hujus sub nomine *Hibisci simplicis* in herbario Linnaei asservati.

Vahl quotes as a synonym *Culhamia,* Forskal (+ 1763), Flora aegypt. arab. p. 96. Floret quotannis sub dio Pataviae. (DC. Prodr. I, 483 says: Syn. dub. ex Salisb.).

G. Marsili, in Act. Acad. de Padova I, 1786 p. 106, tab. 1, 2 named it *Firmania.*

Sterculia platanifolia is a native of China and Japan, and it not found in India.

I. F. S., ı, 90.

Helicteres angustifolia. Sp. I, 963, II, 1366. H. foliis lanceolatis integerrimis, fructu ovato recto. — Hab. in China; Osbeck Iter 232, t. 5. (See Earl. Res. 94).

I. F. S., ɪ, 90.

Impatiens chinensis. Sp. I, 937; Sp. II, 1328. I. pedunculis unifloris solitariis, foliis oppositis ovatis, nectariis arcuatis crassis. Flos purpureus.— Habitat in China.

Collected by Osbeck. See Earl. Res. 94.

I. F. S., ɪ, 100.

Citrus aurantium. Sp. I, 783; Sp. II, 1101. Citrus petiolis alatis. — "Malus aurantia major", C. Bauhini Pin. Theatri Bot. (1623) 436. (This is the bitter or Seville Orange). — Hab. in India.

Linnaeus notices two varieties of the Orange:

β. var. *sinensis.* He quotes C. Bauh. l. c. "Malus aurantia cortice dulci edule". "Aurantia in Indiae insula Zeilan cortice dulci et suavi", Scaliger Ct. Lugd. (It seems Commentarii in Theophrastum, Lugduni 1566).

Bauhin further quotes: "Aurea malus eduli cortice, Clusius, Rar. Plant. Hist. (1601) p. 6, and: in China poma aurea quae dulcedine saccharum vincant, Par 2. Ind. or. c. 25" *).

γ. var. *grandis.* In the second edition of the Spec. Pl. Linnaeus calls this = *Citrus decumana* and records it from Jamaica only. — But the tree is indigenous in China and mentioned in the Chinese Classics. See Botanicon Sinicon II, p. 311.

Salacia chinensis. Mantissa II, 293. Frutex ramis angulatis, laevibus divaricatissimis. Folia peltata, alterna, remota, ovalia (Pruni facie), integerrima, acutiuscula, laevia. Flores axillares, plures e singulis gemmis; pedunculis unifloris petiolo brevioribus. — Habitat in China.

This is a dubious plant. The specimen in the herbarium of Linnaeus is too bad to allow any determination. See Ind. Fl. Sin., ɪ, 125.

Rhamnus lineatus. Sp. II, 281. Habitat in China, elevatis. Osbeck iter, 219, t. 7. (Earl. Res. 95, 119).

---

*) Bauhin evidently means Linschotten's Voyage to the East Indies. See Engl. Transl., Hakluyt Soc., I, 130.

First described in Linn. Amoen. acad. IV (1756), p. 308. Rh. inermis floribus hermaphrod., foliis ovatis lineatis repandis subtus reticulatis.

This is *Berchemia lineata* DC. Prodr. II, 23. — I. F. S., ɪ, 127.

Rhamnus theezans Mantissa, II, 207. Rh. spinis terminalibus, foliis ovatis, serrulatis, ramis divaricatis. Arbuscula sarmentosa. Spicae terminales in panicula nuda. Flores sessiles minuti. — Hab. in China pauperibus Theae succedaneum. Rhamnus Thea Osbeck iter, 232, 161. (See Earl. Res. 95).

Sageretia theezans, Brongn. in Ann. Sc. Nat. Ière sér. X (1827) p. 360. — I. F. S., ɪ, 131.

Sapindus chinensis. Supplem. 228. Sap. foliis pinnatis, foliolis laciniatis. Linn. Syst. veget. ed. XIII, (1774) p. 315.

Koelreuteria paniculata, Laxmann in Nov. Comment. Petrop. XVI (1772) p. 561, tab. 18. — Habitat in China.

The later name is now generally adopted. (V. supra p. 49). — I. F. S., ɪ, 138.

Rhus succedanea. Mantissa II, 221. Rh. foliis pinnatis integerrimis perennantibus lucidis, petiolo integro aequali. Distinguenda omnino a Rhoe vernice cum qua affinis, sed magnitudine foliorum imprimis diversa. Folia hujus rigidula, utrinque nitida, rarius utroque margine aequalia, Rh. vernicis vero opaca viridia. Fructus hujus magnitudine cerasi nec albus; illius vero magnitudine pisi, omnino albus. — Habitat in Japonia, China. —

Arbor vernicifera spuria sylvestris angustifolia, Kaempf. Amoen. exot., 794, t. 795.

I. F. S., ɪ, 47.

Rhus javanica. Sp. I, 265; Sp. II, 380. Osbeck, Java. It is now reduced to *Rhus semialata,* first described in 1784 by Murray from a specimen received from Macao. — Osbeck states, that he gathered Rh. javanica near Canton. (Earl. Res. 96). — I. F. S., ɪ, 146.

Crotalaria chinensis. Sp. II, 1003. — China.

S. X, 1158: C. foliis simplicibus ovatis subpetiolatis, stipulis minutissimis. Corolla lutea.

I. F. S., ɪ, 151.

Crotalaria sessiliflora. Sp. II, 1004. C. foliis simplicibus lanceolatis subsessilibus, floribus sessilibus lateralibus coeruleis. — Habitat in China.

     I. F. S., ɪ, 152.

Astragalus chinensis. Sp. II, 1066.

     First described and figured in Linnaei fil. Plant. rar. hti Upsal. Decas I (1762) 1, tab. 3. A. caulescens strictus glaber, floribus racemosis pendulis, leguminibus ovatis inflatis utrinque mucronatis. — Habitat in China unde anno 1760 inter varia alia semina advenit.

     I. F. S., ɪ, 165.

Astragalus sinicus. Mantissa, I, 103. A. caulescens prostratus, capitulis pedunculatis, leguminibus prismaticis erectis triquetris apice subulatis. Folio pinnata, foliolis 7—9. — Habitat in China. — Cult. in horto Upsal.

     Mill. Act. Angl. (1765) p. 138. (Philos. Trans. LIV, 1764 p. 138 n. 2059. A. sinicus cultivated by Ph. Miller in the Chelsea Garden under the erroneous name of *A. chinensis).*

     Pallas. Spec. Astragalus, 1800, n. 106 named it *Astr. lotoides,* to avoid its being confused with A. chinensis. DC. Prodr. II, 282, erroneously attributes the specific name *A. lotoides* to Lamarck, Enc. Bot. 1783, I, 316. But l. c. we find only the Linnaean name A. sinicus. The same error is repeated in Bunge's Genus Astragalus (1868) n. 10, Hemsl. Ind. Fl. Sin., ɪ, 166.

Phaca trifoliata. Mantissa, II, 270. Ph. foliis ternatis ovalibus obtusis. Racemus terminalis pedicellis geminatis unifloris. Legumina semiorbiculata ventricoso inflata. — Habitat in China.

     J. E. Smith, in Rees' Cycl. says: "Of this a specimen from China is in the Linnaean herbarium. It resembles Hedysarum, Glycine, Flemingia".

     DC. II, 275. — Ind. Florae Sin., ɪ, 167. An obscure plant.

Hedysarum lagopodioides. Sp. I, app. p. 1198, n. 34; Sp. II, 1057. H. foliis ternatis ovatis obtusis, spicis hirsutis. Corollae parvae. Legumina 1 sperma. — Habitat in China, Osbeck. (Early Res. 96).

     Uraria lagopoides, DC. Prodr. II, 324. — Ind. Fl. Sin., ɪ, 178.

Phaseolus radiatus. Sp. I, 725; Sp. II, 1018. Phas. caule erecto tereti, floribus capitatis, leguminibus cylindricis horizontalibus. —

     Habitat China, Zeylonia.

Linn. Flora Zeylan. (1747), 281. Decoctum seminum in colica calculosa a Chinensibus commendatur.

Hort. Upsal. p. 213, n. 4

Act. Stockholm, 1742, p. 202, t. 7, fig. 2. Linnaeus on Phas. zeylanicus. It was brought to him from Canton by Admiral Ankarcrone.

Dillenius, Hort. Eltham. (1732) 315, t. 235, fig. 304. Phas. zeylanicus siliquis radiatim digestis.

Phas. minimus. Rumph. Amb. V, p. 386, t. 139, fig. 2. (from Mant. II).

Dolichos sinensis. Sp. II, 1018. — Habitat in India. — (Earl. Res. 97. Osbeck, Canton).

Linn. Amoen. acad. IV (1756) p. 326, D, volubilis, pedunculis multiflóris erectis, leguminibus pendulis cylindricis torulosis, communiter in singulo pedunculo duo vel tria. Semina esculenta a nautis emuntur in China pro alimento et viatico.

Dolichos sinensis, Rumph. Amb. V, p. 375, t. 134.

Vigna sinensis, Hassk. Ind. Fl. Sin., ı, 193.

Gleditschia sinensis, first described by Lamarck, in 1786, was known to Linnaeus. According to J. E. Smith (Rees Cyclop.) he had it at Upsala but never described it. —

I. F. S., ı, 209.

Rubus parvifolius. Sp. I, app. p. 1197; Sp. II, 707. R. foliis ternatis subtus tomentosis, caule hirto petiolisque aculeis recurvis. — Hab. in India, Osbeck. (Osbeck gathered it at Canton. Earl. Res. 98).

R. moluccanus parvifolius Rumph. Amb. V, p. 88, t. 47, fig. 1.

I. F. S., ı, 235.

Rosa indica. Sp. I, 492; II, 705. R. germ. ovatis pedunculisque glabris, caule subinermi, petiolis aculeatis, foliolis quinis subtus tomentosis, impari majore, stipulis obsoletis. Fructus magnitudine Sorbi aucupariae. — Habitat in China.

Rosa cheusan glabra, juniperi fructu, Petiver Gazoph. 56, t. 35, fig. 11. (V. supra p. 38).

Lindley, Monogr. Rosarum, 1820, p. 106 says: "It is now perhaps too late to inquire what was really intended by Linnaeus for R. indica, since his specific character and description will agree with no species from China at present known, and the figure of Petiver which he quotes to this belongs to a widely different plant nearly allied to R. Banksiae", which Lindley named *R. microcarpa*. — Lindley has, however, examined the R. indica in Linnaeus herbarium, which no doubt belongs to this species. From the Ind. Flor. Sin., I, 250, it would appear that Linnaeus' R. indica has subsequently been gathered in China.

Rosa sinica. S. XIII, 394. R. germinibus subglobosis glabris, pedunculis aculeatis hispidis, caule petiolisque aculeatis, calycinis foliolis lanceolatis subpetiolatis.

Lindley, l. c., who saw the specimen in Linnaeus' herbarium, pronounces it to be a monstrous state of *R. indica*. The *R. sinica*. Ait. Hort. Kew ed. 2 III 261 is different, = *R. laevigata*, Michx. — I. F. S., I, 250.

Crassula pinnata. Suppl. 191. C. foliis pinnatis, caule arboreo. — Habitat in China.

Calanchoe pinnata. Persoon, Syn. Pl. I, (1805) 446.

Bryophyllum calycinum Salisb. Parad. Lond. (1806) tab. 3.

See also Ind. Fl. Sin., I, 280.

Baeckia frutescens. Sp. I, 358; Sp. II, 515. Frutex habitu Abrotani ramis virgatis. Folia linearia acuta glabra integerrima. Flores axillares solitarii. Perianthum infundibiliforme 5 dentatum. Petala 5.

D. D. Abraham Baeck, S. R. Mitis Sueciae archiater et amicus noster vere sincerus, e cujus horto sicco plurimas rariores stirpes obtinuimus, etiam hanc primus nobiscum communicavit. — Habitat in China (Sp. I). — In Sp. II, Linnaeus quotes Osbeck iter 231, tab. 1. Chinensibus tjongina. (Earl. Res. 98, 119).

I. F. S., I, 295.

Osbeckia chinensis. Sp. I, 345; Sp. II, 490. In honorem Petri Osbeck etc. (v. supra p. 60).

Habitus Melastomae (sic Linnaeus!). Folia angusto-lanceolata trinervia opposita scabra. Flores terminales aliquot sessiles, cincti, foliis 4, flore longioribus patentibus. — Habitat in India, Osbeck iter 213, t. 2. (Osbeck gathered it at Canton, Earl. Res. 98).

Echinophora maderaspatana sideritis non serratis nervosis foliis, fructu capsulari etc. Pluken. Almag. 132, t. 173, fig. 4 (a false reference according to Benth. Fl. hgk. 115).

I. F. S., I, 298.

Melastoma octandra. Sp. I, 391; Sp. II, 560. M. foliis integerrimis trinerviis ovatis glabris margine hispidis. — Hab. in India (Sp. I). — In SX, 1022, and Sp. II, Linnaeus quotes Osbeck iter 213. (See Earl. Res. 98).

Linn. Fl. Zeylan. n. 173.

Burmann, Fl. Zeylan. 154, t. 72. Melastoma scabra trinervia etc.

This is *Osbeckia octandra,* DC. Prodr. III, 142, a native of Ceylon and Decan. According to Richter, Codex Linn., the above reference to Osbeck is erroneous. Not mentioned in Ind. Fl. Sin. But Osbeck mentions Mel. octandra for Canton. — J. E. Smith in Rees' Cyclop. has Melastoma octandra, Ceylon and China.

Ammania baccifera, Sp. I, 120; Sp. II, 175. A. foliis subpetiolatis, capsulis calyce majoribus coloratis. Planta tenera. — Hab. China, Osbeck. (See Earl. Res. 98).

Anonymos linariae folio orientalis, galii lutei flore, herba capsularis verticillata, Pluken. Alm. 33, t. 136, f. 2?

Cornelia verticillata, Arduino Animadv. bot. spec. (1759) 2, p. 9, t. 1. Ard. states: E. China facta Italica. Sponte nascitur in Patavinis agris. (Observ. Richter Cod. Linn.: sed specie differt italica, A verticillata, Lamarck).

I. F. S., I, 302.

Lythrum fruticosum. S. X, 1045; — Sp. II, 641. L. foliis oppositis subtus tomentosis, floribus decandris solitariis, corollis calyce calyceque genitalibus brevioribus. — Habitat in China.

Roxb. Pl. Coromand. (1795) t. 31 identified this with his *Grislea tomentosa.* Same in Willd. Sp. pl. II, 321.

Salisb. Parad. Lond. 1806, tab. 42, named the plant *Woodfordia floribunda.* — Ind. Fl. Sin., I, 304.

Lagerstroemia indica. S. X, 1076; — Sp. II, 634. Arbor. Folia alterna subsessilia oblonga integerrima glabra, floralia subrotunda. Thyrsus terminalis laxus, pedicellis trifidis seu trifloris. — China.

Sibi, Kaempf. Amoen. exot. 855. (from Mantissa II).

Tsjin kin, Rumph. Amb. VII, p. 61, t. 28.

I. F. S., I, 306.

Munchhausia speciosa. Mantissa II, 243. Arbuscula ramis teretibus laevibus. Folia alterna petiolata ovata acuminata glabra. Racemus terminalis. Petala 6 obovata patentia unguiculata.

First described in the "Hausvater", a periodical Journal published by baron Otto von Münchhausen (who used to sent plants to Linnaeus), V (1765) p. 357, t. 356.

See also A. Murray, Prodrom. design. stirp. Goetting. 1770, prefamen: Munchhausia speciosa described and figured. Murray says it flowered in Upsala Garden, 1749.

Habitat in China. (J. E. Smith in Rees' Cycl. observes that Linnaeus' herbarium specimen is marked as coming from India).

Lagerstroemia Flos-Reginae, Retz. Obs. V. 25. — *L. speciosa* Pers. See Ind. Fl. Sin., ɪ, 305.

Trapa bicornis. Supplem. 128. Trapa nucibus bicornibus. — In Chinae aquis.

Plumier, Icon. posth. plant. American (ed. Burmann 1755—60) t. 67, fig. inf. Siliqua fusca chinensis. Plumier figures only the fruit and states: per nautam nostratem olim ex Sina delata fuit.

The Chinese Water Caltrop is *Trapa bispinosa,* Roxb. which in Ind. Florae Sin., ɪ, 311, is reduced to the common *Tr. natans.*

Trichosanthes anguina. Sp. I, 1008; — Sp. II, 1432. T. pomis teretibus oblongis incurvis. Folia cordata. Corolla alba 5 partita ciliis albis. — Habitat in China.

Tilli, Catal. Plant. hti Pisani (1723), p. 49, t. 22. Cucurbita sinensis fructu longo anguino vario, flore candido capillamentis tenuissimis, ex Indiae seminibus Florentiam allatis a D. Abbate Josepho Ignatio *Cordero* (a Jesuit missionary in China, v. s. p. 45), Eminentissimi Cardinalis de Turnon comite itineris; dein Illustr. ac Clar. Senator D. Philippus Bonarrote bonarum artium hujus Cucurbitae semen ad me misit. — After this he quotes Bauhinius and Tournefort.

P. A. Michelius, Nov. Gen. Plant. (1729), 12, t. 9. Anguina sinensis, flore albo elegantissimo, fructu oblongo intorto. Cucurbita sinensis Hort. Pisani t. 22. Ex seminibus e Sina a D. Abbate Ign. Cordero transmissis. Cum alia Cucurbitae specie in hortis nostris germinavit. Utriusque junioribus fructibus Sinenses vescuntur. Prima ab ipsis *patula,* altera *pipingai* appellantur. (These are not Chinese but rather Indian names. The first seems to refer to Luffa).

Royen, plant. hti Lugd. Batav. (1740), 262.

Miller, Dict. t. 32. (Linn. means Miller's Icones, 1760, t. 32).

Hort. Clifford. (1737), 450. The above authors quoted.

A figure and description of the same plant was published by Breynius, in Ephemer. cur. nat. ann. IV (1688), obs. 137. He received the plant from Ceylon.

Trichos. anguinea is not a native of India. See Ind. Flor. Sin., i, 313.

Cucumis acutangulus. Sp. I, 1011; — Sp. II, 1436. C. foliis rotundato-angulatis, pomis angulis decem acutis. Flores lutei in pedunculis multifloris. Poma oblonga, angulis 10 compressis saepe dentatis. Semina nigra quasi masticata ut in Anguria. — Habitat in Tataria, China.

Cucumis longus indicus, Grew, mus. Regalis Societatis (1681) 229, t. 17, fig. 2.

Cucumis indicus striatus operculo donatus: corticoso putamine tectus, Pluken. Alm. 123, t. 172, fig. 1.?

Petola, Rumph. Amb. V, p. 408, t. 149.

Luffa acutangula. Roxb. Hort. Beng. — I. F. S., i, 314.

Momordica cylindrica. Sp. I, 1009; — Sp. II, 1434. M. pomis cylindricis longissimis. Folia cucumerina angulata basi serrato-dentata, angulis acutis. Flores lutei, semina nigra. Cirrhi ex axillis. — Habitat Zeylonia, China.

Hermann, Catal. hti Lugd. Batav. (1687), 482. Pepo indicus reticulatus seminibus nigris. Zeylonia.

Hermann, Parad. Batav. (1698), app. 11. Pepo reticulatus sulcatus semine nigro.

Raji hist. plant. Supp. (1704) 332.

Pluken. Almag. 286, Pepo indicus reticulatus seu Luffa Arabum Veslingii.

Luffa cylindrica, Roemer Synops. Peponif. (1846) II, 63. — I. F. S., i, 315.

Hydrocotyle chinensis. Sp. I, 234; Sp. II, 339. H. foliis linearibus, umbellis multifloris. — Habitat in China.

See Earl. Res. 99. Osbeck, Canton.

According to the I. F. S., i, 326, an obscure plant, but supposed to be the N. American *Crantzia lineata.* (Sprengel in Schultes Syst. Veg. VI (1820) 355).

Sium Sisarum. Sp. I, 251; — Sp. II, 361. S. foliis pinnatis, floralibus ternatis. — Hab. in China?

Hort. Clifford (1737) 98, n. 2. Sium foliis pinnatis, floralibus ternatis.
Crescendi locus naturalis nobis latet, videtur e facie sinensis vel proxima.

Hort. Upsal. (1748), 62. Diagn. as above. Habitat forte in China?
Hospitatur in hortis oleraceis; perennis.

C. Bauhini Pinax Theatri Bot. (1623) 155. Sisarum Germanorum.

Sium Sisarum has not been observed in China by recent botanists.
Comp. Maximovicz, Mél. biol. IX, p. 17. It is a native of N. Persia, the
Altai mountains in Siberia and Japan (perhaps only cultivated).

Sium Ninsi. Sp. I, 251; — Sp. II, 361. S. foliis serratis pinnatis:
rameis ternatis. — Habitat in China.

Linn. Materia Med. (1749), 117, Radix Ninsi.

Kaempf. Amoen. exot. 817, t. 818. Sisarum montanum coraeensc,
radice non tuberosa.

This plant is known from Japan, Maximowicz l. c. 18; Linnaeus is
the only botanist who notices it from China.

Athamanta chinensis. Sp. I, 245; — Sp. II, 353. A. seminibus mem-
branaceo-striatis, foliis supra decompositis laevibus multifidis. Statura Se-
leni Monnieri. — Habitat in China. Chinensem dixit Barthram qui semina
misit e Virginia.

The late Asa Gray observes (Journ. Bot. 1881, 325) that Linnaeus
was mistaken. Barthram sent the seeds from *Genesee* a river in New York,
which L. took for China. But Athamanta chinensis, the same as *Selinum
Monnieri,* Linn. or *Cnidium Monnieri* Cuss., is a common plant also in
China. See I. F. S., I, 332.

Aralia chinensis. Sp. I, 273; — Sp. II, 293. A. caule petiolisque
aculeatis, foliolis inermibus villosis. — Habit. in China, Osbeck. (See Earl.
Eur. Res. 100, and supra p. 62, Sparrmann).

Frutex aquosus mas, Rumph. Amb. IV, 103, tab. 44.

Same as *Aralia spinosa* Linn., Sp. I, 273; II, 392, Virginia. See
I. F. S., I, 338.

Zanthoxylum trifoliatum. Sp. I, 270; — Sp. II, 1455. Z. foliis ternatis. Frutex aculeo recurvo sub basi et apice petiolorum. Folia solitaria
ternata, ad flores terna: foliolis ovatis obtusis. Umbellae pedunculatae hemi-

sphaericae simplices. Flores stylis tribus, nec quinque ut nequeat Araliis sociari. — Habitat in China, Osbeck. (See Earl. Res. 100).

Panax aculeatum, Ait. in DC. Prodr. IV, 252, = *Acanthopanax aculeatum*, Seem. Journ. Bot. 1867, p. 238. — I. F. S., ɪ, 339.

Lycium foetidum. Suppl. 150. Lycium foliis oppositis ovato-lanceolatis, stipulis interfoliaceis setaceo-spinescentibus, floribus axillaribus sessilibus.— *Austella* Chinens. (Linnaeus seems to mean that this is its Chinese name).— Habitat in Japonia, Thunberg, — in China Thouin, — in Madeira, Koenig. — Cl. Thouin misit illam nomine *Spermacoces fruticosae,* Jussieu.

In A. L. de Jussieu's Genera plantarum 1789, p. 232 we read sub *Serissa* Commers: Frutex idem cum Spermacoce fruticosa, Hort. Reg. Paris, et Lycio foetido, Linn.

Serissa foetida, Commers., DC. Prodr. IV, 575. — I. F. S., ɪ, 391.

Rubia cordifolia. Mantissa, II, 197. R. foliis perennantibus quaternis cordatis. Flores paniculati ex apice caulis et ramorum, pauci, albi subcampanulati 4partiti patuli apice anguiculati. Fructificationem non vidi. — Hab. in Sibiria, China.

Cruciata daurica scandens, smilacis folio aspero, flore luteolo, fructu majori rubro, Amman Ruth. (1739) 19.

I. F. S., ɪ, 393.

Eupatorium chinense. Sp. I, 837; — Sp. II, 1172. E. foliis ovatis petiolatis serratis. Corymbi terminales flosculis quinis. China, Osbeck. (Not mentioned by Osbeck).

A dubious plant, perhaps *E. Lindleyanum.* See I. F. S., ɪ, 404.

Aster chinensis. Sp. I, 877; — Sp. II, 1232. A. foliis ovatis angulatis dentatis petiolatis, cal. terminalibus patentibus foliosis. Plena est eximia. — Habitat in China?

Dillenius, Hort. Eltham. (1732), p. 38, t. 34, fig. 38. Aster chenopodiifolius, annuus flore ingenti specioso. Speciosae hujus plantae semina benevolente communicavit Adr. v. Royen, Acad. Lugd. Professor. — On the plate we read: E sept. Chinae delatus.

Hort. Cliff. (1737) 407. Aster foliis ovatis angulatis dentatis chenopodii folio, annuus. Crescit?

Royen, Plant. hti Lugd. Batav. (1740), 169. He quotes only Hort. Eltham. and Hort. Cliff.

Hort. Upsal. (1748). Diagnosis as above. Habitat forte in China. Variat radio albo et violaceo.

Callistephus hortensis, Cassin. (1825). — *C. chinensis,* Nees (1832).— I. F. S., ɪ, 407. — Comp. s. p. 46.

Aster indicus. Sp. I, 876; — Sp. II, 1230. A. foliis ovato-oblongis serratis, floralibus ovali-lanceolatis integerrimis, ramulis unifloris. — Habitat in China. (See Earl. Res. 101. Osbeck).

Aster conyzoides Indiae orientalis ramosior, caulibus sparsis. Pluken. Almag. 57, t. 149, fig. 4.

Linnaeus' specimens came, it seems, from China. Aster indicus is found not only in China but also in Japan, the Malay Archipelago and Burma. — *Boltonia indica* in Benth. Flora hongk. 174. — I. F. S., ɪ, 413.

Conyza chinensis. Sp. I, 862; — Sp. II, 1208. C. foliis lanceolato-ovatis reflexo serratis subtus tomentosis, floribus terminalibus congestis. Raro plures quam tres flores simul congesti. — Habitat in China, Toreen. (Osbeck gathered it near Canton. Earl. Res. 101).

Senecio amboinicus, Rumph. Amb. VI, 36, t. 14, fig. 2.

Blumea chinensis, DC. Prodr. V, 444. — I. F. S., ɪ, 420.

Conyza hirsuta. Sp. I, 863; — Sp. II, 1209. C. foliis ovalibus integerrimis scabris subtus hirsutis. Flores in racemis. — Habitat in China. (Osbeck. See Earl. Res. 101).

Lessing, Linnaea 1831, p. 150, named this plant *Pluchea hirsuta.*

Hemsley, I. F. S., ɪ, 422, thinks that Conyza (Pluchea) hirsuta, recorded from China also by Loureiro, Fl. cochin., 606, does not occur in that country.

Sphaeranthus chinensis, Mantissa I, 119. Sph. foliis sessilibus pinnatifidis. — Habitat in India.

DC, Prodr. V, 371, justly asks: cur chinensis, cum ipse cel. auctor ex India ortam dicat?

Carpesium abrotanoides. Sp. I, 860; — Sp. II, 1204. C. floribus lateralibus, sparsis saepe solitariis. Semina oblonga nuda glabra balsamo quasi illita. — Habitat in China. Osbeck iter tab. 10. (See Earl. Res. 101). — I. F. S., ɪ, 430.

Verbesina chinensis. S. I, 901; — Sp. II, 1270. V. foliis alternis petiolatis ovato-lanceolatis serratis. Corolla lutea. — Habitat in China, Osbeck. (Earl. Res. 101).

Anisopappus chinensis, Hook & Arnott, Bot. Beechey (1841), 196. — I. F. S., I, 431.

Xanthium orientale. Sp. II, 1400. X. caule inermi, foliis cuneiformi-ovatis subtrilobis. Facies X. strumarii, sed planta magis scabra. — Habitat in China, Japonia Zeylana.

Linn. filius, Plant. rar. hti Upsal., Decas II, (1763) tab. 17, Xanthium orientale. Description and habitat as above. Horto Upsaliensi inter alia semina chinensia allata fuere aliquot fructus hujus anno 1761.

Linnaeus in Sp. II, X. orientale, refers to Sp. I, 987, X. strumarium, var. β. elatius, americana. Europa, Canada, Japonia.

Linnaeus' X. orientale is now generally reduced to our common *Xanthium Strumarium, Xanthium indicum,* Roxb.

Bidens pilosa, var. β. *chinensis.* Mantissa II, 281, 463. Bidens pilosa (Sp. I ed., Sp. II) is an American plant according to Linnaeus, but is now also recorded from China.

Var. β. *chinensis.* "Simillima sed foliola distincta et semina semper 4 aristata (in B. pilosa vulgari 2 seu 3). Flores radiati, radius etiam floris albus. — Habitat in India orientali".

Agrimonia molucca, Rumph. Amb. VI, p. 38, t. 15.

Culta in Hort. Upsal.

I. F. S., I, 435.

Chrysanthemum indicum. Sp. I, 889. Chr. foliis simplicibus ovatis sinuatis angulatis serratis acutis. — Habitat in India. (Earl. Res. 101. Osbeck, Canton).

Matricaria sinensis minore flore, petalis et umbone ochroleucis. Pluk. Amalth. 142, t. 430, fig. 2. (V. supra p. 39).

Matricaria sinensis, Rumph. Amb. V, p. 259, t. 91, fig. 1.

Tsietti pu, Rheede, Malab. X, t. 44.

Var. β. India.

Chrysanthemum maderaspatanum oxycanthae foliis caesiis ad marginem spinosis, calyce argenteo. Pluken. Almag. 101, t. 160, fig. 6.

Matricaria indica latiore folio, flore pleno, Morison, Hist. Plant. Oxon. III, (1680) 33.

Matricaria sinensis, flore monstruoso, S. Vaillant in Act. Paris. 1720, p. 368.

Linn. Flor. Zeylan. (1747), n. 421.

Matricaria zeylanica hortensis, flore pleno, Rajus. Suppl. (1704), 224.

Chrysanthemum indicum and the variety β, as described in Linnaeus' Sp. Plant., are not represented in his herbarium. Sabine, in Trans. Hort. Soc. IV, (1821), 330, thinks that Linnaeus' genuine from is single flowered and he identified it with a poor specimen of a plant from China found in Banks' herbarium, of which he gives a figure. But Lindley claims the name given by Linnaeus for a different single flowered small yellow Chrysanthemum, introduced about 1821, it seems from China, by Brookes of Ball's Pond, which Lindley figured and described as Ch. indicum, Linn. in Bot. Reg. t. 1287 (1829). For the plant given by Sabine under this name, and of which Lindley possessed perfect specimens, from Macao, he proposes the name Chr. Sabini. The late Maximowicz considered both Sabine's and Lindley's plants as Ch. indicum. See Hemsley, Chrysanthemum in Gard. Chron. 1889, II, 522.

Linnaeus' var. β. is, according to Sabine, l. c., the double-flowered form of Chr. indicum. The so called Chinese Chrysanthemums with large double flowers, Ch. sinense, Sabine seem to have been unknown to Linnaeus.

Hooker in Flora indica, III, 314, states that Chrysanthemum indicum in India is known only in a garden state. It is a native of China.

Ind. Fl. Sin., I, 437.

Artemisia minima. Sp. I, 849; — Sp. II, 1190. A. foliis cuneiformibus repandis, caule procumbente, floribus axillaribus sessilibus solitariis. Planta omnium minima. — Habitat in China. Lagerstroem.

Sphaeromorphaea centipeda DC. Prodr. VI, 140. *Myriogyne minuta*, Lessing in Linnaea VI (1831) 219. The plant was cultivated, previous to 1783 in the Royal Garden, Paris. Lam. Enc. Bot. I, 265. — Ind. Fl. Sin., I, 440.

Artemisia chinensis. Sp. I, 849; — Sp. II, 1190. A. foliis simplicibus tomentosis obtusis lanceolatis, inferioribus cuneiformibus trilobis. — Habitat in China, Lagerstroem. Sibiria.

Absinthium maritimum Sinarum lavandulaefolio pulchrioribus corymbis, inodorum, sapore aromatico. Pluken. Amalth. 3, t. 153, fig. 5.

Artemisia foliis radicalibus a caulinis diversis, Gmelin, Flora Sibir. II, 127, tab. 61, fig. 1, 2.

The Chinese plant, reference to Plukenet, is *Crossostephium artemisioides*, Lessing in Linnaea VI, (1831) p. 220 or *Tanacetum chinense*, A. Gray, Maxim. Mél. VIII (1872) 520, — the Siberian plant, reference to Gmelin, is *Artemisia lagocephala* Fischer, Maxim. l. c. 532.

The "*Artemisia chinensis* cujus mollugo Moxa dicitur", Pluken. Almag. 50, t. 15, fig. 1, is identified in the Hort. Cliff. (1737) 404, n. 6, with the Artemisia foliis pinnatif. planis laciniatis *(A. vulgaris)* var. γ. It does not appear in the Spec. Plant. = A. indica Willd. Sp. III, 1846. — Moxa in China is indeed made of the leaves of A. vulgaris.

Ethulia tomentosa. Mantissa I, 110, E. suffruticosa, foliis linearibus integerrimis tomentosis. Fruticulus facie Lavandulae. — Habitat in India.

Ethulia tomentosa is according to J. E. Smith (Rees' Cyclop.) the same as *Artemisia chinensis*.

But in DC. Prodr. VI, E. toment. is identified with *Artemisia lavandulaefolia,* considered by Maxim. l. c. 534, the same as *A. vulgaris*.

Senecio divaricatus. Sp. I, 866; — Sp. II, 1215. Senecio corollis nudis, foliis lanceolatis dentatis scabris, ramulis floriferis divaricatis. — Habitat in China, Osbeck. (See Earl. Res. 102.)
Gynura divaricata, DC. VI, 301. — I. F. S., I, 447.

Cacalia sonchifolia. Sp. I, 835; — Sp. II, 1169. C. caule herbaceo, foliis lyratis dentatis. — Habitat Zeylona, China.
Kleinia caule herb., folio lyrato. Linn. Flor. Zeyl. n. 305.
Sonchus amboin., Rumph. Amb. V, 297, t. 103, fig. 1.
Muel-Shavi, Rheed. Malab. X, p. 135, t. 68.
Senecio maderaspat. sinapios folio, floribus parvis luteis, Pluk. Amalth. 192, t. 444, fig. 1.
Tagolina Luzonum flore purpureo, Petiv. Gazoph. t. 80, fig. 13, Camell. Luzon. III, p. 3, n. 8.
Emilia sonchifolia, DC. Prodr. VI, 302. — I. F. S., 449.

Lobelia zeylanica. Sp. I, 932; — Sp. II, 1322. L. caulibus repentibus procumbentibus, foliis ovatis serratis, inferioribus obtusis, superioribus acutis, pedunculis unifloris, capsulis subvillosis. —
Habitat in China, Osbeck (Sp. I), — China Aethiopia (Sp. II). (See Earl. Res. 102, Osbeck.)

Campanula zeylanica senecionis folio, flore purpureo, A. Seba, Thesaurus Rer. nat. (1734), I, p. 37, t. 22, fig. 12?

Lobelia affinis Wall. See DC. Prodr. VII, 360, I. F. S., ii, 2.

Azalea indica. Sp. I, 150; — Sp. II, 214. A. floribus subsolitariis, calycibus pilosis. — Habitat in India.

S. XII, n. 1, adds: Chamaerhododendron exoticum amplissimis floribus liliaceis, Breyn, Prodr. plant. rar. (1680) I, p. 24 (V. supra p. 26).

P. Hermann, Catal. hti Ludg. Batav. (1687) 152, tab. 153. Cistus indicus Ledi alpini foliis, floribus amplis. Speciosissimus hic Cistus ex Jaccatra (Batavia, Java) advectus. Flores eximiae pulchritudinis, modo liliacei, monopetali 5 partiti colore fulgentis dilutioris cocci. Odor deest.

Raji Hist. Plant. (1688) 1895. (He quotes only Breyn and Hermann).

Tsutsusi, Kaempf. Amoen. exot. 845, t. 846.

Azalea indica is a Chinese plant, not found in a wild state in India. — I. F. S., ii, 25, *Rhododendron indicum*, Sweet.

Pergularia tomentosa. Mantissa I, 53; II, 522. P. foliis cordatis tomentosis. — Hab. Arabia, Forskähl.

Asclepias cordata, Burm. Fl. ind. 72, t. 27, fig. 2.

Edwards, Botan. Reg. 1819, tab. 412 sub *Pergularia odoratissima*, Smith states: Linnaeus had in his herbarium a Chinese specimen marked *tomentosa* with a note on the back, signifying that the Catholic clergy at Macao prepare from its milky juice a medicine for dysentery. He cultivated it in his stove and described it. (from J. E. Smith in Rees' Cyclop.) — I. F. S., ii, 114.

Asclepias carnosa. Supplem. 170. A. foliis ovatis carnosis glaberrimis. — Habitat in China. — Hujus vidi tantum folia duo et umbellam floriferam quam astuti Chinenses pro Gummi Gutta dederunt.

According to J. E. Smith, Exotic Botany (1804) t. 70 this plant was received from China by Linnaeus through his friend the merchant *Bladh* (see further on).

Rob. Brown in Mém. Wern. Soc. I (1808) p. 27, named this plant *Hoya carnosa.* — I. F. Sin., ii, 115.

Convolvulus biflorus. Sp. II, 1668. C. foliis cordatis pubescentibus, pedunculis geminis, corollis lobis trifidis. — China.

This is *Ipomoea biflora,* Persoon, Syn. Pl. I (1805). 183. — I. F. S., II. 157.

Convolvulus obscurus. Sp. II, 220 *(Convolvulus hederaceus var. δ, Sp. I, 154).* C. foliis cordatis indivisis, caule subpubescente, pedunculis incrassatis unifloris, calycibus glabris. — China, Batavia, Zeylona, Surinam.

Convolvulus flore minore lacteo, fundo atrorubente, Dillen. Hort. Elth. (1732) 98, t. 93, fig. 95. Grown from seeds received from Batavia.

Ipomoea obscura. Ker in Bot. Reg. t. 239 (1817). — I. F. S., II, 161.

Convolvulus hirtus. Sp. I, 159; — Sp. II, 226. C. foliis cordatis, hastatis glabris acuminatis angustis rotundatis, caule petiolisque villosis. — Habitat India, Osbeck. (See Earl. Res. 103, Osbeck, Canton.)

Hort. Cliff. (1737) p. 496, n. 16? C. foliis cordato-hastatis glabris etc. Crescit e seminibus americanis adhuc dum tenela.

Royen, Flor. Leyd. et pl. hti Lugd. Bat. (1740), 429. (He quotes only the Hort. Cliff.).

Choisy, in DC. Prodr., IX, 412, identifies C. hirtus doubtfully with *Conv. farinosus,* Lin., Graecia, Madeira, Cape of G. H. etc.

Solanum aethiopicum. Sp. II, 265. — Habitat in China. This is first described in Amoen. Acad. IV (1756), p. 307. S. caule inermi herbaceo, foliis ovatis dentato angulatis, pedunculis fertilibus unifloris. Aethiopia, Burserus (+ 1649, his herbarium in Upsala).

Loureiro, Fl. cochin. 160, mentions S. aethiopicum as cultivated in China. — I. F. S., II, 169.

Lycium barbarum. Sp. I, 192; — Sp. II, 278. L. foliis lanceolatis crassiusulis, calycibus subbifidis. — Habitat in Asia, Africa, Europa.

Mill. Dict. n. 6, L. foliis lanceolatis acutis.

"Jasminoides sinensis, halimifolio longiore et angustiore", Duhamel, Arbres et Arbustes (1755) I, 306, tab. 121, fig. 4.

Osbeck mentions L. barbarum for Canton. Earl. Res. 104.

This is *Lycium chinense* Mill. Dict. 8th edit. (1768) n. 5. — I. F. S., II, 175.

Datura ferox. Sp. II, 255. D. pericarpiis spinosis erectis ovatis, spinis inferioribus minoribus, superioribus maximis convergentibus. — Habitat in China.

First described in Linn. Amoen. Acad. III (1753), Datura fastuosa. Diagnosis as above. — China.

Zanoni, Hist. Bot. (1675), ed. latin., I, 76, tab. 162, Stramonium ferox ob fructus aculeos dictum, Daturae spinosissimae Cochinensis nomine. Ex Orientali India delatam ferunt: quandoquidem illius semen ad Zanonium misit is qui Mediolani iter Carmelitas Excalceatos pharmaceuticam faciebat, Michael a S. Elisaeo, qui a religioso hominem ejusdem ordinis prope Cochinum Malabariae urbem lectum narrabat.

Boccone, Icones et Descript. Rar. Plant. Siciliae, Melitae etc. (1674), 50. Stramonium ferox seu Datura. Lutetiae vidi. Cum icone.

D. ferox is common in North China. Not mentioned in Hooker's Flora indica. — I. F. S., ii, 176.

Nicotiana fruticosa. Sp. II, 258. Ad. Cap. B. Sp., China. — Simillima N. Tabacco.

First described, S. X, 932, N. foliis lanceolatis subpetiolatis amplexicaulibus; floribus acutis, caule fruticoso.

Mill. Dict., N. foliis lineari lanceolatis acuminatis semiamplexicaulibus, caule fruticoso. Coast of Guinea, cultivated in Brazil. Icon, 185, fig. 1.

In Ait. Hort. Kew, first ed. 1789, I, 241, Nicotiana fruticosa is recorded as a Chinese plant. But Sir Jos. Hooker, Bot. Mag. 6207 (1876), does not doubt that this plant which he considers to be a variety of N. Tabacum, is a native of South America and not of Africa or China.

Scrophularia chinensis. Mantissa, II, 250. S. foliis ovato-oblongis serratis pubescentibus, figura Tanaceti Balsamitae. Racemi terminales longissimi simplices. Flores solitarii, capsulae compressae. — Habitat in China.

J. E. Smith, in Rees' Cyclop: Scroph. chinensis in Linnaeus' herb. consists of an imperfect specimen of what seems to be an *Ocymum*, accompanied by a still more imperfect branch of what may be a *Celsea* or *Verbascum*, but neither of them has anything to do with Scrophularia nor was Linnaeus at all satisfied about them.

Bentham in DC. Prodr. X, 317, states that Linnaeus' Scrophularia chinensis is *Celsia coromandeliana* (first described in Vahl, Symb. Bot. III (1794) 79) and *Salvia plebeja*. — I. F. S., ii, 177.

Gerardia glutinosa. Sp. I, 611; — Sp. II, 849. G. foliis ovatis serratis, bracteis linearibus hispidis. — China, Osbeck iter tab. 9. (See Earl. Res. 104).

Bentham, Scrophular. Ind. (1835) p. 21, named this plant *Ptero-stigma grandiflorum*, and subsequently *Adenosma grandiflora*. See I. F. S., II, 185.

Torenia asiatica. Sp. I, 619, India, Toreen; — Sp. II, 862, Osbeck iter, 210. (See Earl. Res. 104, 119).

Amoen. Acad. III (1751) p. 25. Only the name.

Habitat in India unde allata a D. Toreen nostrate qui hanc plantam cum variis aliis collegit et communicavit.

Euphrasiae affinis pusilla planta, Pluken. Amalth. 85, t. 373, fig. 2.

Asarinae foliis et facie capsula bivalvi lignosa, Pluken. Amalth. 19, t. 360, fig. 3?

Kaka-pu, Rheede Malab. IX, 103, tab. 53.

Osbeck brought this plant from Canton, and Toreen gathered it most probably at the same place. According to Hooker, Fl. Ind. IV, 277, it grows in the Nilgherry, Tenasserim, Ceylon, Java, China.

Capraria crustacea. Mantissa I, 87. C. repens, foliis oppositis ovatis subpetiolatis crenatis. — Habitat in Amboina, China.

Caranasi minus, Rumph. Amb. V, p. 461, t. 170, fig. 3.

Bentham, Scrophular. Ind. (1835), p. 35 named this plant *Vandellia crustacea*. — I. F. S., II, 189.

Buchnera asiatica. Sp. I, 630; — Sp. II, 879. B. foliis integerrimis linearibus alternis, calycibus scabris. — Zeylon, China.

Bentham in DC. Prodr. X, 499: Buchnera asiatica Linn. = Strigae species complures, ibid. 502 = *Striga densiflora*, 503 = *Striga hirsuta*. — I. F. S., II, 201.

Utricularia bifida. Sp. I, 18; — Sp. II, 26. U. scapo nudo bifido, flores alterni lutei. — Habitat in China, Osbeck iter, 243, t. 3, fig. 2 (from S. X, 851.). (See Earl. Res. 105). — I. F. S., II, 222.

Ruellia crispa. Sp. I, 635; — Sp. II, 886. R. foliis subcrenatis lanceolato-ovatis, capitulis ovatis foliosis hispidis, caule repente. Corolla flava. — Habitat in India. Osbeck iter, 240. (Earl. Res. 105, Osbeck, Canton).

Adhatoda luzanensis, spica plana, Petiv. Gazoph. t. 73, fig. 6?

Not recorded, after Osbeck, from China.

Barleria cristata. Sp. I, 636; — Sp. II, 887. B. foliis oblongis integerrimis, calycis foliolis duobus latioribus ciliatis duobusque linearibus
acutis. Corolla coerulea. — Habitat in India (Sp. I), Osbeck iter, 225,
t. 8 (Sp. II). (Earl. Res. 105, 119).

Melampyro cognata maderaspatensis quam ipse habuit Morison, Hist.
Pl. Oxon. III (1680) 425, S. 11. t. 53, fig. 7.

I. F. S., ii, 242.

Justicia procumbens. Sp. I, 15; — Sp. II, 22. J. foliis lanceolatis
integerrimis, spicis terminalibus lateralibusque alternis, bracteis setaceis,
caule procumbente. Lin. Fl. Zeylan. 19. — Zeylania.

Euphrasia alsines angustiore folio, Rubiae modo spicata, Pluken.
Almag. 142, tab. 56, fig. 3.

Euphrasia sinica Parietariae foliis, Rubiae modo spicata, Pluken.
Amalth. 83, tab. 392, fig. 4. (supra p. 41). (This reference only in
Sp. II.).

Just. procumbens noticed by Osbeck, Canton. (Earl. Res. 105). The
plant is common in China. I. F. S., ii, 246.

Justicia chinensis. Sp. I, 16; — Sp. II, 22. J. herbacea, foliis ovatis,
floribus lateralibus, pedunculis trifloris. — Habitat in China.

Burmann Fl. indica (1768) 8, t. 4, fig. 1. J. chinensis, in urbe Batavia circa mures.

Dicliptera chinensis, Nees in DC. Prodr. XI, 477. — I. F. S., ii, 247.

Justicia purpurea. Sp. I, 16; — Sp. II, 23, — S. X, 850. — J. foliis
ovatis utrinque mucronatis integerrimis glabris, caule geniculato, spicis
secundis. — Habitat in China, Osbeck iter 230 (See Earl. Res. 105).

Folium tinctorium, Rumph. Amb. VI, p. 51, t. 22, fig. 1. (from S. X).

Nees ab Esenb., DC. Prodr. XI, 438, named the plant *Rostellularia
diffusa.* See also I. F. S., ii, 245: *Justicia diffusa,* Willd.

Clerodendron fortunatum. Sp. II, 889, Cl. foliis simplicibus lanceolatis integerrimis. — Habitat in India, Osbeck iter, 228, tab. 11. (Earl.
Res. 105, Osbeck, Canton).

Amoen. Acad. IV (1756), 320. Clerodendron fortunatum first noticed. — S. X, 1158.

Cl. fortunatum is a Chinese plant not met with in India (Hooker, Fl.
Br. India IV, 596) — I. F. S., ii, 260.

Ocimum capitellatum. Supplem. 276. O. herbaceum, foliis ovatis, floribus aggregatis, petiolis lateralibus. — Habitat in China. Colitur in hto Upsal.

Bentham, in Wallace pl. as. rar. (1831) named it *Acrocephalus capitatus.* DC. Prodr. XII, 47. — I. F. S., II, 269.

Hyssopus Lophanthus. Sp. I, 569, habitut in Sibiria; — Sp. II, 769, hab. in China septentrionali.

Hort. Upsal. (1748), 162, n. 2. Hyssopus corollis transversalibus, staminibus inferioribus corolla brevioribus. Lophanthus vulgo. Habitat ubique in Sibiria, inventore Cl. Gmelino.

Haller, Hort. Goetting. (1753), 338, Cataria floribus inversis. Ex Sinarum regione.

Comment. Goetting. II (1753), 344, tab. 14, Haller: Catariae species quam ex Sinensi semine in horto alimus. Odor in toto planta Catariae satis vehemens.

Although this plant is found in the Altai montains and in Dahuria (Ledebour, Flora ross. III, 372). Gmelin does not mention it in his "Flora sibirica". He received the seeds evidently from China (Peking, Heuke, v. s. 66) and communicated them to Linnaeus and Haller to whom he used to send dried specimens and seeds (Gmel. Fl. sib. III, 144).

Bentham in Bot. Reg. vol. XV (1829, sub n. 1282, described a plant, which he named *Lophanthus chinensis.* Fischer, Director of the Bot. Gard. St. Petersb., who had received it from Dahuria, had sent it to Bentham. The latter identifies it with Linnaeus' Hyssopus Lophanthus. — L. chinensis has not been recorded from China but in 1835, in Index Semin. Hti Petrop. 30, Fischer and Meyer, described a new species, *Lophanthus rugosus,* which they had raised from seeds received from China (Peking). This plant is commonly cultivated at Peking, and was I should think, the Hyssopus Lophanthus of Linnaeus. — I. F. S., II, 288.

Scutellaria indica. Sp. I, 600; — Sp. II, 836. S. foliis subovatis obtusis crenatis petiolatis, racemis nudiusculis. — Habitat in China (Sp. I), — Osbeck iter 244 (Sp. II). (See Earl. Res. 106).

Serratula amara, Rumph. Amb. V, 459, t. 170, fig. 1, 2.

Scutellaria sinica betonicaefolia, Pluken. Amalth. 190, t. 441, fig. 1. (v. supra p. 41).

I. F. S., II, 295.

Leonurus sibiricus. Sp. I, 584; — Sp. II, 818. L. foliis tripartitis multifidis linearibus obtusiusculis. — Habitat in Sibiria, China.

Hort. Upsal. (1748), 170. Diagnosis as above. Quoted: Amman. Ruth. 48, t. 8: Ballota inodora, foliis Coronopi. Sibiria ad Udam fluv.

I. F. S., ii, 302.

Plantago asiatica. Sp. I, 113; — Sp. II, 163. P. foliis ovatis glabris, scapo angulato, spica flosculis distinctis. — Habitat in China, Sibiria.

Planta ita refert Plantaginem majorem, differt spica longiore, floribus remotis et distantibus, foliis saepe basi subdentatis, scapo angulato.

Pl. asiatica is now generally considered only a variety of *Plantago major.* — I. F. S., ii, 316.

Valeriana chinensis. Sp. I, 33; — Sp. II, 47. V. floribus triandris, foliis omnibus cordatis repando-lobatis. — Habitat in China, Osbeck. (See Earl. Res. 106).

S. XII: Burmann Fl. ind. (1768), t. 6, fig. 3. China, Coromandel.

Willdenow, Sp. pl. I (1797) p. 22, identified this plant with his *Boerhavia repanda.* — I. F. S., ii, 317.

Celosia argentea. Sp. I, 205; — Sp. II, 296. C. foliis lanceolatis, stipulis subfalcatis, pedunculis angulatis, spicis scariosis. — Habitat in America (Sp. I), in China (Sp. II). (Earl. Res. 106. Osbeck, Canton).

Hort. Clifford. (1737) Celosia foliis lineari-lanceolatis. America.

Boerhave, Index alter hti Lugd. Batav. (1727) II, 98, Amarantus spicatus argenteus, americanus.

Rheede, Malab. X, p. 77, t. 39. Tsiera adeca manjen. (from Mantissa II).

I. F. S., ii, 318.

Amaranthus (sic Linn.!) *tristis.* Sp. I, 989; — Sp. II, 1404. A. glomerulis triandris rotundatis subscpicatis, foliis ovato-cordatis. — Habitat in China. — (See Earl. Res. 106. Osbeck, Canton).

Blitum indicum secundum, Rumph. Amb. V, p. 231, t. 82, fig. 1. Same as *A. gangeticus* Linn. See Ind. Fl. Sin., ii, 319.

Amaranthus cruentus. Sp. II, 1406. — Habitat in China.

First described in S. X, 1269. A. racemis pentandris decompositis remotis patulo-nutantibus, foliis lanceolato-ovatis. Spicae sanguineae patentissimae parum curvatae.

Amaranthus sinensis foliis variis, Martyn, Hist. Pl. rar. cent. (1728) p. 6, tab. 6.

Same as *A. paniculatus* and *A. sanguineus*. Lin. See I. F. S., ii, 320.

Chenopodium Scoparia. Sp. I, 221; — Sp. II, 321. Ch. foliis lineari-lanceolatis planis integerrimis. — Habitat in Graecia, Japonia, China. (China only in Sp. II).

Hort. Clifford. (1737), p. 86, n. 10. Chenopodium lini folio villoso, Tournef: Inst. 506. Other quotations:

Scoparia seu Belvedere Italorum, Raji hist. pl. I, (1686) p. 210.

Linaria Belvedere dicta, J. Bauhin. hist. pl. III, (1650) p. 462.

Tsisu, Kaempf. Amoen. exot. 885. Japan.

Hort. Upsal. (1748), 55.

Schrader, in 1809, named this plant *Kochia scoparia*. — I. F. S., ii, 328.

Basella alba. Sp. I, 272; — Sp. II, 390. B. foliis ovatis undatis, pedunculis simplicibus folio longioribus. — Habitat in Syria (Sp. I), in China, Amboina (Sp. II).

Rumph. Amb. V, 417, Gandola alba.

Pluken. Almag. 252, t. 63, fig. 1. Mirabili peruv. affinis tinctoria betae folio, scandens.

Kaempf. Amoen. exot. 784 murasakki, Japan.

Thran, Ind. plant. hti Carlsruhe (1733) 11. loc. nat. Syria?

Same as *Basella rubra*, Linn. See I. F. S., ii, 331.

Polygonum barbatum. Sp. I, 362; — Sp. II, 518. P. floribus hexandris trigynis, spicis virgatis, stipulis truncatis setaceo-ciliatis, foliis lanceolatis. Caulis herbaceus rufus. — Habitat in China. (Earl. Res. 107, Osbeck, Canton).

Nov. Comment. Petropol. I, (1747) p. 375, tab. 13. Stephan Krasheninnikow, Descript. rar. plant: Persicaria foliis ovatis utrinque incanis non in Sibiria sed in Sinarum regno provenit. Comp. supra p. 49.

Velutta-modela-mucu, Rheede Malab. XII, 145, t. 77, non 76.

Pol. barbatum (Ind. Fl. Sin., ii, 334) is not met with in North China, but the plant of which the Chinese there extract a beautiful Indigo is *Pol. tinctorium*. — I. F. S., ii, 334, 351.

Polygonum chinense. Sp. I, 363; — Sp. II, 520. P. floribus octandris trigynis, pedunculis scabris, foliis ovatis. Flores in capitulis rotundis. — Habitat in India, China. (Earl. Res. 107, 119. Osbeck, Sparrmann, Canton). I. F. S., II, 335.

Rheum palmatum. Sp. II, 531.

First described in S. X, 1010: Rheum foliis palmatis acuminatis Sp. II: Gemma vernans non rubescens sed flavescens. Folia scabriuscula, foliorum laciniae oblongae. — Habitat in China ad murum. Radicem hujus misit D. D. David de Gorter, Imperatricis Ruthenorum Medicus ordinarius, 1762.

Linn. fil. Plant. rar. Upsal descrip. et figurae, fasc. 7 (1767) tab. 4. Rh. palmatum.

Act. Angl. (Philos. Trans. LV) 1765, p. 292, t. 12. (article on Rh. palmatum by Prof. Dr. Hope, who cultivated the plant in 1763).

Seeds of the true officinal Rhubarb plant were first received at Kiakhta in 1750, and the plants raised from them in Russia and in many parts of Western Europe were described by Linnaeus as *Rh. palmatum.* — I. F. S., II, 354.

Rheum undulatum. Sp. II, 531. Rh. foliis subvillosis undulatis, petiolis aequalibus. — Habitat in China, Sibiria.

In Sp. I, 372, Linnaeus named this plant incorrectly *Rheum Rhabarbarum*, Rh. foliis subvillosis petiolibus aequalibus. China ad murum, Sibiria. He was under the impression that it was the plant yielding the officinal root, but subsequently, when aware of his error, changed the name into *Rh. undulatum.*

Hort Upsal. (1748) 98. Rhabarbarum folio longiore hirsuto, crispo, florum thyrso longiori et tenuiori, J. Amman, Stirp. rar. Ruth. (1739), p. 7, n. 9. (Amman says: Semina hujus Moscua primum pro vero et genuino Rhabarbaro missa sunt. Postea quoque e China accepimus. Verum plantae exinde enatae nondum floruerunt.)

J. Amman. l. c. p. 160, n. 226. Acetosa montana, folio cubitale oblongiore crispo etc. Messerschmidt, Rhabarbarum sibiricum Ruthenorum, Dauria.

Linn. Materia Medica (1749), 147. Rhabarbarum sinense folio crispo, flagellis rarioribus et minoribus; Amman herb. 206.

Linn. Amoen. acad. III (1752) 215—218, tab. 4. *Rheum undulatum.* Hort. Ups. 98. — Habitat in Udae fluvii montanis (Transbaicalia) et circa lacum in Dauria, Messerschmidt. In China ad murum vel in circumjacentibus terris, Tataria chinensi loco lutoso frequens.

Rh. undulatum has not been gathered in China. I. F. S., II, 355.

Rheum compactum. Sp. II, 513. Rh. foliis sublobatis obtusissimis glaberrimis argute denticulatis lucidis. Folia magis quam reliqua, coriacea seu compacta, lobis rotundatis, obsoletioribus crenatis, margine cartilaginea, acutis denticulis, utrinque glaberrimis. Paniculae rami nutantes. — Habitat in Tataria, China.

Rheum foliis cordatis glabris marginibus sinuatis, spicis divisis nutantibus. Miller's Dict. tab. 218.

Rheum compactum is known from Siberia and North China.

Cassytha filiformis. Sp. I, 35; — Sp. II, 531. Habitat in India, Osbeck iter 243. (Earl. Res. 108. Osbeck, Canton). Habitus et omnia Cuscutae sed fructificatio diversissima.

Cuscuta altera s. major, Camel Luz. 1, p. 1, n. 1.

Petiver Gazoph. 77, t. 49, fig. 12.

Cuscuta baccifera barbadensis a maritimis. Pluken. Almag. 126, t. 172, fig. 2.

Acatsia valla, Rheede Malab. VII, p. 83, t. 44.

Cussuta, Rumph. Amb. V, p. 491, tab. 184, fig. 4.

I. F. S., II, 393.

Daphne indica. Sp. I, 357. D. capitulo terminali pedunculato, foliis oppositis oblongo ovatis glabris. — Habitat in China. Osbeck iter 246. (Earl. Res. 108. Osbeck, Canton).

C. A. Meyer, Bull. Phys. mathem. Acad. Petrop. I, (1843) p. 358, named this plant *Wikstroemia indica* and means that *Capura purpurata*, Linn. Mant. II, 225 is the same.

According to Hooker, Flora Indica, V, 195, *Wickstroemia indica* is found in Chittagong, Tenasserim, Singapore, China. — I. F. S., II, 398.

Loranthus scurulla. Sp. II, 472. Loranthus pedunculis unifloris congestis, foliis obovatis — Habitat in China.

In Sp. I, 110, Scurrula parasitia, China.

Viscum vitici innascens Camell. Luzon in Raji Suppl. III, p. 3, n. 36.— Petiv. Gazoph. t. 63, fig. 8.

I. F. S., II, 407.

Andrachne fruticosa. Sp. I, 1014; — Sp. II, 1440. A. erecta arborea, ramis alternis ad ramulos compressis. Folia ovata integerrima, petiolata, glabra. — Flores axillares conferti. — Habitat in China. Osbeck iter, 228. (Earl. Res. 108, Osbeck, Canton). — I. F. S., ii, 427, *Breynia fruticosa,* Hook. fil.

Agyneia impubes. Mant. II, 296. A. foliis utrinque glabris alternis bifariis subpetiolatis ellipticis integerrimis pollicariis. Flores axillares plures congesti. Pedunculi uniflori tenuissimi. Masculi flores inferiores minores glabri; feminei flores superiores majores pedunculis tomentosis. Frutex erectus. — Habitat in China.

Phyllanthus puberus Müller Arg. var. γ. *impubes* in DC. Prodr. XV, 2, p. 307. — I. F. S., ii, 426: *Glochidion obscurum,* Blume.

Agyneia pubera, Mant. II, 296. A. foliis subtus tomentosis, ovalioblongis alternis bifar. integerrimis sesqui pollicaribus subtus tomentosis. Flores ut in praecedenti. Frutex erectiusculus. — Habitat in China.

Phyllanthus puberus var. β. *genuinus* Müller Arg. in DC. Prodr. l. c.— According to Ind. Fl. Sin., l. c. same as preceding.

There is in the Linnaean herbarium a specimen of *Croton tomentosus,* Müller Arg. brought by Osbeck from Canton. DC. Prodr. XV, 2, 588.

Croton sebiferum. Sp. I, 1004; — Sp. II, 1425. S. foliis rhombeo-rotundatis utrinque mucronatis integerrimis glabris, figura Populi nigrae.— Habitat in Chinae humidis, Osbeck iter, 245. (Earl. Res. 108, Osbeck, Canton).

Ricinus chinensis sebifera, Populi nigrae folio, Petiv. Gazoph. 53, t. 34, fig. 3.

Evonymo affinis Sinarum, Populi nigrae folio, tricapsularis, granis nigris candidissima substantia obductis sebifera, Pluken. Amalth. 76, t. 390, fig. 2.

Arbor sebifera chinensis, kiu yeu, Martini, Le Compte. (Earl. Res. 18, 26.)

P. Amman, hort. Bosianus, 1686, 375.

Stillingia sebifera Mich. *Excaecaria sebifera* Müller Arg. in DC. Prodr. XV, 2, 1210. — I. F. S., ii, 445.

Morus alba. Sp. I, 986; — Sp. II, 1398. M. foliis oblique cordatis laevibus. — Habitat in China.

Hort. Cliff. (1737) 441. Morus fructu albo Bauhin. Crescit in Italia.

Hort. Upsal. (1748), 283. Morus fructu albo, Bauhin, Pin. 459. China, Italia. — Ibidem: Morus fructu nigro, Bauhin, 1. c. China, Italia. Morus alba expetitur a Bombyce prae sequenti (M. nigra).

C. Bauhini, Pinax Theatri Botan. (1623) 459, says nothing regarding the native country of M. alba and nigra.

Morus alba is said to have been brought into Tuscany from the Levant in 1434. Its native country is no doubt China. M. nigra is unknown in China and Linn. in Spec. Plant gives only Italy as its native country.

I. F. S., II, 455.

Ficus pumila. Sp. I, 1060; — Sp. II, 1515. F. foliis oblongo-ovatis acutis integerrimis subtus reticulatis, caule repente. — Habitat in China, Japonia.

Linn. Amoen. acad. I (1744). Diagnosis as above. Quoted Kaempf. Amoen. exot. 803, tab. 804. Ficus sylvestris procumbens folio simplici.

Varinga repens, Rumph. Amb. III, p. 134, t. 85.

Urtica nivea. Sp. I, 985; — Sp. II, 1398. U. foliis alternis suborbiculatis utrinque acutis subtus tomentosis. — Habitat in Chinae muris. (See Earl. Res. 109, Osbeck, Canton).

Hort. Clifford. (1737). Diagnosis as above. Crescit in China. Facies tamen americana est.

Urtica racemifera maxima Sinarum, foliis subtus argentea lanugine villosis. Pluken. Amalth. 212. (v. supra p. 42).

Ramium majus, Rumph. Amb. V, 214, t. 79, fig. 1.

Boehmeria nivea, Hook. & Arn. Voy. Beech. (1840) p. 214.

Thuya orientalis. Sp. I, 1002; — Sp. II, 1422. Th. strobulis squarrosis, squamis acuminatis reflexis. — Habitat in China. (Earl. Res. 109, Osbeck, Canton).

Hort. Clifford. (1737) 449, n. 2. Thuya strobilis uncinatis, squamis reflexo-acuminatis. Crescit forte in China. Communicata per Cl. Royenum qui hujus se daturum esse historiam pollicitus est.

Royen, Plant. ht. Lugd. Batav. (1740) p. 87. Description of the tree. Not stated from whence it came.

Biota orientalis, Endlicher Conif. (1847) p. 47.

Juniperus chinensis. Mant. I, 127, II, 519. J. foliis decurrentibus imbricato patentibus, confertis: caulinis ternis, rameis quaternis. Folia patula utrinque viridia, magis quam in reliquis conferta, basi adnata, vix pungentia distinctissima densitate foliorum. Culta in hto Upsalensi. — Habitat in China.

Epidendrum ensifolium. Sp. I, 954; — Sp. II, 1352. E. caule tereti laevi, foliis ensiformibus striatis, petalis lanceolatis glabris, labio recurvato latiore, nectario integro. Planta terrestris nec parasitica. Flos fragrantissimus. — Habitat in China, Osbeck. (See Earl. Res. 110).

Cymbidium ensifolium, Swartz, Nova Acta Upsal. VI, 77.

Ixia chinensis. Sp. I, 36; — Sp. II, 52. Ixia foliis ensiformibus, floribus pedunculatis, panicula dichotoma. — Habitat in India.

Hort. Upsal. (1748), 16. Ixia foliis ensiformibus, floribus remotis. Facies Ireos. Flores lutei maculis fulvis. Hanc debemus amicissimo D. Gmelino. Habitat in China.

Bermudiana radice carnosa, floribus maculatis, seminibus pulpa obductis, Amman, Act. Petrop. XI, p. 308, t. 7. — Linnaeus refers to Comment. Acad. Sc. Imp. Petrop. XI (1739) p. 305, Amman, Descriptio et Icon (tab. VII) novae Bermudianae speciei: Inter res naturales plures quas R. R. P. P. Soc. Jesu in Pekino sinensis Imperii capite ad Academiam Petropolitanam ante aliquot annos miserunt, semina etiam erant globosa et aterrima *yen tchi* titulo insignita. Haec terrae in horto Academiae commisi. Brevi temporis spatio exinde enascebantur plantae Iridum junioribus admodem similes. . . . . Description and good drawing.

Linnaeus further quotes:

Bermudiana Iridis folio majori, flore croceo eleganter punctato, Kraus hort. 25, t. 25 (i. e Rolof, Index plant. hti Krausiani, Berolini, 1747). — Nothing said regarding the origin of the plant.

Balemcanda shularmandi, Rheede, Marab. XI, p. 73, t. 37.

Trew, Plantae selectae pinxit G. D. Ehret, London (1750) seqq.) 23, t. 52. (Quoted in S. XII).

Moraea chinensis in S. XIV, p. 93, Murray.

Pardanthus chinensis, Ker in Konig and Sims, Ann. Bot. I (1805) 246.

Belamcanda chinensis in Benth. & Hook. Gen. Plant. III, 697.

Smilax China. Sp. I, 1029; — Sp. II, 1459. S. caule aculeato teretiusculo, foliis inermibus ovato-cordatis obtusis cum acumine. Petioli bidentati. — Habitat in China, Japonia. (See Earl. Res. 110, Osbeck, Canton.)

Linn. Mat. med. (1749), locus nat. China.

S. minus spinosa, fructu rubicundo, rad. virtuosa China dicta seu San-kiva. Kaempf. Amoen. exot. 781, t. 782.

Fruticulus convolvulaceus spinosus sinicus floribus parvis umbellatis, Pluken. Amalth. 101, t. 408, fig. 1. (V. supra p.   ).

Rumph. Amb. V t. 161 (quoted in S. X, 1292), V t. 162, VII t. 30 (quoted in Amoen. acad. IV, 133, 137).

C. Bauhin Pinax Theatri Bot. (1625) p. 296. China suis sarmentis Smilaci asperae absimilis non est. Dicitur China et Chinaea radix a China Indiae regione in quo copiose nascitur. Est autem duplex, alba enim ex India occidentali, Hispania nova et Peru quae colore magis rufo est: alia ex India orientali ex regione Sinarum (ubi ab indigenis Lampatam vocatur) adfertur. Verum Gracias (he means Garcias ab Orta, who calls this drug Lampatam, v. supra p. 7) et Acosta (qui integram plantam depingit et ex eo Lugd. hist.) Chinam orientalem aliter describunt quam Monardes suam occidentalem. At radix haec Indiae primum A. Christi, 1535, innotuit. — Cina, alias China, Matthiolus (+ 1577).

Chinae radix fruticis more excrescit, foliis paucis Aurantiorum forma. In China arbor Lampatou dicitur: et radix quo nigrior eo melior. Linscot. par. 4. Ind. Orient. c. 33. (Bauhin quotes Linschotten Voyage to the East Indies 1598. See Engl. transl. Hakluyt Soc. 1885, II, 107).

Hemerocallis fulva. Sp. II, 462. H. corollis. fulvis. — Habitat in China.

In Sp. I, 324 = *H. lilioasphodelus,* var. β. *(fulvus).*

Hort. Clifford. (1737) 128. H. lilioasphodelus, β. Crescit in Ungaria, de alterius loco nil scimus.

Hort. Upsal. (1748). H. lilioasph. var. α. Habitat in uliginosis pratis Hungariae.

Clusius, Rar. Plant. Hist. (1601), 137. Lilio-asphodelus puniceus. Habitat in Austria.

C. Bauhin, Pin. Theatri Bot. (1623) 80. "Lilium rubrum Asphodeli radice".

J. E. Smith, in Rees' Cyclop., says: "H. fulca, supposed by Linnaeus to be a native of China. It is an old and very hardy inhabitant of our gardens". — Linnaeus is right in considering the plant to be indigenous in China. It is also found in a wild state in Mid Russia.

Sagittaria trifolia. Sp. I, 993; — Sp. II, 1410. S. foliis ternatis. — Habitat in China. (See Earl. Res. 111, Osbeck, Canton).

Sagittaria chinensis foliis ternis longissimis, Petiv. Gazoph. 29, t. 19, fig. 3. (V. supra p. 43).

According to Kunth, Enum. III, 157, this is perhaps *S. chinensis*, Sims in Bot. Mag. t. 1631, (1814).

Cyperus Iria. Sp. I, 45; — Sp. II, 67. C. culmo triquetro seminudo, umbella foliosa decomposita, spiculis alternis, granis distinctis. — Habitat in India, Osbeck. (See Earl. Res. 111, Osbeck, Canton).

Gramen Cyperoides Indiae orientalis elatius, panicula sparsa pallescente, Pluken. Almag. 179, t. 191, fig. 7.

Iria seu Balari. Rheede Malab. XII, p. 105, t. 56.

C. Iria is common in India and China.

Panicum alopecuroides. Sp. I, 55. S. spica tereti involucris setaceis fasciculatis unifloris flosculo quadruplo longioribus. — Habitat in China. (Osbeck gathered this species near Canton. See Earl. Res. 112.)

Gramen geniculatum brevifolium crispum, spica purpureo-sericea, Pluken. Almag. 177, t. 119, fig. 1.

In Sp. II, 82, P. alopecoroides, the above reference to Pluken. Almag. is given with a? and Gramen indicum alopecuroides holosericeum majus, Pluken. Almag. 177, t. 92, fig. 5 is quoted. Linnaeus notices it as a plant from Jamaica only.

In Linn. Mant. altera, 322, P. alopecuroides figures with a new description. Habitat in India orientalis.

In S. XIII, 92 the same plant is called *Alopecurus indicus.*

In Roem. & Schultes Syst. Veg. II (1817), 498, it is *Penicillaria cylindrica,* in Kunth, Enum. Pl. I, 163, *Pennisetum Linnaei.*

Ischaemum aristatum. Sp. I, 1049; — Sp. II, 1487. Isch. seminibus aristatis, spica bipartita. Semina arista intorta flosculis longiore. — Habitat in China, Osbeck. (See Earl. Res. 112).

Holcus latifolius. S. X, 1305; — Sp. II, 1486. H. glumis trifloris: flosculo primo inermi; duobus margine aculeatis, foliis subovatis valde latis. — Habitat in India, Osbeck iter 247. (See Earl. Res. 113). J. E. Smith (in Rees' Cyclop.) says that this is identical with *Cenchrus lappaceus,* Lin. of the East-Indies.

According to Kunth, Enum. I, 366 (resp. Trinius) it is the same as Linnaeus' *Cenchrus lappaceus* or *Centhoteca lappacea,* Desv. But in Kunth, I, 510 the H. latifolius of Lin. is named *Andropogon latifolius.*

Poa chinensis, Sp. I, 69; — Sp. II, 100. P. paniculae ramis simplicissimis, floribus sessilibus, seminibus imbricatis, culmu erecto. — Habitat in India, Osbeck. (See Earl. Res. 113. Osb. Canton).

Burmann, Flora ind. (1768) 27, t. 11, fig. 3, *Poa chinensis.* India. — Poiret, Enc. Bot. V, 80, 89 proves, that the P. chinensis of Linn. is not identical with Burmann's plant, which latter he names *Poa sessilis,* same as *Leptochloa tenerrima* Roem. & Schlt. (See Kth, enum. I, 270) and *Leptochloa chinensis,* Nees (See Benth. Fl. hgkg. 430.)

Lycopodium cernuum. Sp. I, 1103; — Sp. II, 1506. L. foliis sparsis curvatis, caule ramosissimo, spicis nutantibus. — Habitat in India.

Lycopodium ceylanicum erectum ramosissimum, Burm. Zeyl. (1737) 144.

Bellan-Patsja, Rheede Malab. XII, p. 73, t. 39. Muscus zeylanicus erectus, perpetua virens in arboribus proceritatem excrescens, Pluken. Almag. 247, tab. 47, fig. 9.

Muscus clavatus erectus crispatis foliolis, Spongiolae imitamentum, China, Pluken. Amalth. 149, tab. 431, fig. 3. (V. supra p. 43).

Osbeck gathered L. cernuum near Canton. See Earl. Res. 113.

Acrostichum punctatum. Sp. II, 1524. A. frondibus cordato-lingula ts acuminatis integerrimis, supra punctatis. — Habitat in China. J. Fothergill. Chinensibus officinalis.

According to Hooker & Baker, Synopsis Filicum, 1874, p. 360, this is the same as the *Polypodium irioides,* Lamarck, Enc. Bot. V, 513. N. India, Chusan, Mascarenes, Guinea etc. — Swartz, Syn. Fil. 30. *Polypodium lingulatum.*

Acrostichum dichotomum. Sp. I, 1068; — Sp. II, 1524. A. nudum, dichotomum spicis secundis adscendentibus reflexis compressis. — Habitat in China.

Filix Cochine, Petiv. Gazoph. t. 70, fig. 12. Cochine branched Comb. Fern, Cat. 305. One of the most elegant amongst the numerous tribe of Capillars. Mr. James Cunningham first discovered it, and sent it from Cochinchina. (V. supra p. 44).

This is *Schizaea dichotoma,* Swartz, Syn. Filicum, 1806, p. 150, Hooker & Baker Syn. Fil. 1874, p. 430.

Blechnum orientale. Sp. I, 1077; — Sp. II, 1535. B. frondibus pinnatis: pinnis linearibus alternis. — Habitat in China, Osbeck. (Earl. Res. 114).

In Sp. I, Bl. orientale is erroneously given as an American species and Bl. occidentale as Chinese.

Bentham, Fl. hgk. 444. Bl. orientale, India, Ceylon, Archipelago, China.

Adiantum flabellulatum. Sp. I, 1095; — Sp. II, 1557. A. frondibus decompositis: pinnis alternis rhombeis rotundatis multifloris, stipitibus supra pubescentibus. — Habitat in China, Osbeck. (See Earl. Res. 114).

Adiantum chinense perelegans ramosum, folio flabelliformicum rubedine perfuso, Pluken. Almag. 11, t. 4, fig. 3.

A. foliis Coriandri ramosum. Raji Hist. Pl. (1688) p. 1853. Bentham, Fl. hgk. 447. India, Ceylon, China.

Adiantum chusanum. Sp. I, 1095; — Sp. II, 1558. A. frondibus decompositis: pinnis alternatis pinnatifidis: lobis inaequalibus. — Habitat in China.

Probably from Cunningham's collection. J. E. Smith in Rees' Cyclop. refers it to his *Davallia chinensis,* first described in 1798, as does also Swartz, Syn. Fil. 133. See also Hook. Spec. Fil. I, 187.

Pteris vittata. Sp. I, 1074; — Sp. II, 1532. P. frondibus pinnatis: pinnis linearibus rectis basi rotundatis. — Habitat in China, Osbeck (Sp. I)— in China, Osb. iter t. 4, Jamaica (Sp. II). (See Earl. Res. 114).

Filix odorata luzonica, Petiv. Gazophyl. t. 93, fig. 10.

Fragrant Luzon Fern with plain willow leaves and seed edged lists used in Luzon instead of Melilot. Raji Hist. III, (1704), Suppl. Kamel Luzon p. 2, n. 10.

Lonchitis major pinnis longis angustissimis, Sloane Jamaica (1707) I, p. 79, t. 34.

P. t. *vittata* occurs in India. Roxbg. Fl. Ind. ed. 1874, p. 757.

Pteris semipinnata. Sp. I, 1076; — Sp. II, 1534. P. frondibus subpinnatis, foliolis lateralibus loboque infimo semipinnatifidis. — Habitat in China. Osbeck iter, t. 3, fig. 1. (See Earl. Res. 114).

Bentham, Fl. hgk. 448, Pteris semipinnata, India, Ceylon, Archipelago, China, Japan.

Polypodium varium. Sp. I, 1090; — Sp. II, 1551. P. frondibus lateralibus bipinnatis, foliolo infimo pinnatifido. — Habitat in China, Osbeck. (Earl. Res. 114).

Swartz, Syn. Filicum, 1806, 51, calls it *Aspidium varium.* See also Hooker, Species Fil. IV (1862) p. 30.

Polypodium Barometz. Sp. I, 1092; — Sp. II, 1553. Polypodium (?) frondibus bipinnatis: pinnis pinnatifidis lanceolatis serratis. Radix decumbens crassa vellere molissimo, densissimo intense flavo obvestita. — Habitat in China, unde habeo cum radice. (Osbeck gathered this plant near Canton. Earl. Res. 114).

Linnaeus seems to have believed that this was the plant which gave rise to the tale of the "Agnus scythicus", a plant with the appearance of a lamb, the shaggy caudex of the fern having indeed some resemblance to a small wolly animal. The first account of the "Tartarian lamb" was given by Herberstein, the German ambassador to the Grand Prince Vassily III of Moscow, 1517 and 1526, and who reports in his Rerum Moscovit. Comment., 1549 (Engl. edit. Hakl. Soc. II, 74), from hearsay, that beyond the Volga there grows a curious plant resembling a lamb, and called "baranetz", a lambkin, by the Russians. Now-a-days the name baranetz in Russia is applied to Lycopodium. Linnaeus' Polyp. Barometz, however, is a tropical plant and a native of China, and Cochinchina. See Lour. Fl. cochin. 667. It is now more generally known under the names of *Cibotium Barometz* J. Sm. and *Dicksonia Barometz.*

Trichomanes chinense. Sp. I, 1099; — Sp. II, 1562. T. frondibus supradecompositis, foliis pinnisque alternis lanceolatis: pinnis laciniis cuneiformibus. — Habitat in China. Osbeck iter, 222, t. 6. (Earl. Res. 114).

Swartz, Synops. Filicum (1806) 133 refers it to *Davallia chinensis.* (v. supra). See also Hooker Species Filicum I (1846), 187.

Fucus Tendo. Sp. I, 1162; — Sp. II, 1631. F. filiformis simplex tenuissimus subdiaphanus. Fucus hic crassitie setae suillae, pallidus, longitudine 6 vel 7 pedum, teres, tenacissimus instar tendinis, basi angustissimus et fere capillaris, apice vero crassiore terminatus granulo subovato mucilaginoso. Singulare est productum maris, et quasi ab ipsa natura hominibus

loco fili destinatum, quo et Chinensibus summa cum utilitate utuntur, quod
in fila contorquent triplicata, tanti roboris, ut manibus vix ac ne vix quidem
a vallidissimo queant rumpi, quorum specimina simul missa in Museo
asservantur. Planto hoc tempore botanicis minus est cognita. — Habitat
in China.

The above details are from Linn. Amoen. acad. IV (1751—53) cum
icone t. 3, fig. 2, with the following synonyms:

Fucus indicus teres, setam piscatoriam referens, longissimus. Pluken.
Almag. 160, t. 184, fig. 3.

Gramen spartium setas equinas referens etc. J. Bauhin, Prodrom.
Hist. Plant. (1619) 11; — C. Bauhin, Pinax Th. Botan (1623) 5. E Java
insula allatum.

Esper, Icones Fucorum, I (1797) p. 45 means that Linnaeus' F. tendo
is the same as *Fucus filum*. But C. A. Agardh, Species Algarum, 1821,
p. 163 says: "Fucus tendo Lin. non ad regnum vegetabile pertinet".

In the copy of Esper's Icon. Fung. in the library of the Bot. Garden,
St. Peterb., l. c., there is a marginal MS note by a former possessor of
the book, stating, that Turner had informed him, that the Fucus tendo in
the Linnaean herbarium is the intestine of an animal.

Phallus mokusin. Suppl. 452.

Acta Petrop. XIX, (Nov. Comment. Acad. Petropol. 1775) p. 373,
t. 5. (Article by Father Cibot, Peking: Fungus Sinensium mo ku sin.).

Habitat in China prope Pekin in Mori radicibus putridis. Chinensibus
raro edulis, cum saepe venenatus et verminosus, sed laudant ut praestans
remedium in ulceribus cancrosis.

E. Fries, Systema mycologicum, II (1821) p. 286 calls this Fungus, *Lysurus mokusin*. He knows it only from Cibot's description and
figure.

Boletus favus. Sp. II, 1645. — First described in S. X, 1349.
B. acaulis subpulvinatus scaber: setis erectis ramosis, poris angulatis patu-
lis. — Fungus planus fuscus, vix pulvinatus, pagina superiore undique
hispida quasi Lichene rangiferino parvo, e setis compressis ramosis fuscis.
Pagina inferior instar favi apum poris magnis angulatis. — Habitat in
China.

E. Fries, l. c. I (1821) p. 345 describes this fungus of which he had
seen a specimen, as *Polyporus favus*, var. *sinensis*.

III. Philip Miller and his Gardener's Dictionary, 1731—1807.

This celebrated Gardener and botanist was born in 1691, near London. He was Superintendent of the Physic Garden at Chelsea belonging to the Apothecaries' Company, to which appointment he had succeeded his father in 1722. Regarding the foundation of the Chelsea Garden v. supra p. 32. In 1731 Miller published the first edition of his GARDENER'S DICTIONARY, which during his life time went through 8 editions:

First edition, 1731.			5th edition, 1747.		
2nd	»	1733.	6th	»	1752.
3d	»	1737—39.	7th	»	1759.
4th	»	1741.	8th	»	1768.

Seven editions were written on Tournefortian principles, in the 8th edition Miller adpted the nomenclature of Linnaeus. A much more ample edition of the work, in 2 enormous volumes, was published after Miller's death 1797—1807, under the care of the Rev. Thomas Martyn (1736—1825); Professor of Botany, Cambridge University, with the title: "THE GARDENER'S AND BOTANIST'S DICTIONARY BY THE LATE PH. MILLER, to which are now first added a complete enumeration and description of all plants hitherto known". This is still a very useful book.

In 1755 Miller began to publish: "FIGURES OF THE MOST BEAUTIFUL, USEFUL AND UNCOMMON PLANTS, described in the Gardener's Dictionary, to which are added the descriptions". It was completed in 1760; 300 coloured plates, in folio.

Miller was a Fellow of the Royal Society and enriched the Transactions with several papers. Most of these were catalogues of the annual collection of the 50 plants, which were required to be sent to that learned body from the Chelsea Garden, by the rules of its founder. These herbarium specimens are still preserved in the British Museum. Miller was by his infirmities finally obliged to resign the charge of the Garden and died soon after, in 1771.

Chinese plants cultivated by Miller in the Chelsea Garden:

Brassica violacea, Lin., Diction. 1768, n. 5.

Sinapis juncea, Lin. Cult. 1731. Dict. ed. 1, n. 3. In ed. 8 (1768) n. 5. "China, whence the seeds are frequently brought to England".

Raphanus chinensis oleifera, Lin., Dict. ed. 8 (1768). See Ind. Fl. Sin., I, 50.

Dianthus chinensis, Lin. — Miller, Dict. 1768, n. 10; Icon. tab. 81, fig. 2. Caryophyllus sinensis, Tournef. China. The French missionaries sent seeds to Paris, in 1705. The flowers first produced in European gardens were single till about 1719, when many double flowers appeared in Paris.

Hypericum monogynum. Miller, Icon. t. 151, fig. 2. Hyp. floribus monogynis, staminibus corolla longioribus, calycibus coloratis, caule fruticoso. The plant grows naturally in China, from whence the seeds were brought in 1753 to His Grace, the Duke of Northumberland, and were sown in His Grace's curious garden at Stanwick. The Chelsea Garden was also furnished with it.

Alcea rosea. Dict. ed. 8 (1768): "I many years ago saw some plants with variegated flowers in the garden of the late Lord Burlington, London, raised from seeds which came from China. Linnaeus, Sp. pl. 966 refers it to Siberia" (Linn. says only: in Oriente). "It grows naturally in China from whence I have often received seed".

Malva chinensis. Dict. 5th ed. 1747. It was formerly sent from China as a potherb and cultivated in England. — Dict. ed. 8 (1768) n. 6. In the last edition it is called *Malva verticillata,* Lin. (n. 29).

Urena lobata Lin. Cult. Chelsea 1731. Chelsean plants presented to the Roy. Soc. n. 474. See Hort. Kew, ed. 2, 222. — Diction. ed. 8 (1768) n. 9. China, America.

Hibiscus Manihot Lin. Cult. 1712 in Chelsea Garden. Philos. Trans. n. 333, p. 417, n. 64. — See Hort. Kew. first ed. II, 457.

Rhus chinense. Dict. ed. 8 (1768) n. 7, Rhus foliis pinnatis, foliolis ovatis obtuse serratis, petiolo membranaceo villoso. Rhus Sinarum lactescens, costa foliorum alata, Pluken. Amalth. 183. (v. supra p. 37). "It grows naturally in the East. Seeds of this were sent to the Royal Garden, Paris where they succeeded, and from thence I received the plant, which grew very well in the open air at Chelsea, three years, but the severe winter in 1740 destroyed it". — Hemsley, I. F. S., ı, 147, reduces Miller's Rhus chinense to *Rhus semialata.* Murr. It is not to be confounded with *Rhus sinense,* Ellis, in Philos. Transactions (v. supra p. 48), which plant is *Ailantus glandulosa.*

Another species of Rhus was cultivated by Miller under the name of *Toxicodendron* altissimum before 1768. Dict. ed. 8, n. 10, where it is stated that it is the Fasinoki, arbor vernicifera spuria, Kaempf. Amoen. exot. 794. Raised in Chelsea from seed received by Miller from China. This is *Rhus succedanea,* Lin.

Astragalus chinensis. Mill. Dict. ed. 8 (1768) n. 21. Native of China.

Cult. 1763 in Chelsea, Philos. Trans. Plants presented to the Royal Soc. n. 2059. According to Hort. Kew, ed. 2, IV. 368, Miller's Astr. chinensis is Linnaeus' *Astrag. sinicus*. But they are different.

Rosa sinica. Hort Kew, ed. 1. II, 203. China, Cultivated by Miller 1759.

Ammania baccifera. Lin., Miller's Dict. 1768. China. Cult. in Chelsea Garden.

Trichosanthes anguinea Lin. Cult. in Chelsea 1755. Figured in Miller's Icon. t. 32, under the name of snake-gourd, Trichosanthes. The seeds had been received from China.

Conyca patula. Ait. Hort. Kew, ed. 1. III, 184. Figured in Miller's Icon. t. 247. Serratula foliis oblongo-ovatis, obtuso dentatis, caule ramoso patulo, calycibus subrotundis, mollibus. The seeds were given to Miller by J. Browning of Lincoln's Inn, who received them from the northern part of China, where the plant grows naturally. — *Vernonia chinensis*, Less. — I. F. S., I, 401.

Conyza hirsuta, Lin. Cult., 1767, Chelsea. Dict. ed. 8, n. 18. Philos. Trans. Plants from Chelsea presented to the Royal Soc. n. 2258.

Aster chinensis, Lin., Dict. ed. 8, n. 30. "Native of China from whence the seeds were sent to France by the missionaries, where the plant was first raised in Europe. In 1731 I received seeds of them, from which I raised plants with red, and some with white flowers. In 1736 I received seeds of the blue flowered, but all were single. *Reine marguerite* was the title by which they came. In 1752 I received seeds of the double flowers, both red and blue, and in 1753 seeds of the double white sort." — This is *Callistephus hortensis*, Cass. — I. F. S., I, 407.

Carpesium abrotanoides, Lin. Cult. at Chelsea before 1768. Dict. ed. 8, n. 2.

Xanthium chinense. Mill. Dict. ed. 8 (1768), n. 4. We have often received seeds from China. — *Xanthium strumarium*, Lin. — I. F. S., I, 433.

Siegesbeckia orientalis, Lin. Cult. Chelsea before 1768. Dict. ed. 8.

Matricaria indica. Miller Dict. ed. 8 (1768). Miller received it from Nimpu (probably Ning po in China). Chinese *Chrysanthemum*, Sabine in Trans. Hort. Soc. IV, 326—354. Sabine saw Miller's herbarium specimen, which still exists in the British Mus. Small double flowerheads. Hemsley in Gard. Chron. 1889, II, 522.

Artemisia minima. Lin. Cult. before 1768, Chelsea. Dict. ed. 8, n. 5. Comp. supra p. 86.

Lycium chinense. Mill. Dict. ed. 8 (1768) n. 5. L. foliis ovatis lanceolatis, ramis diffusis, floribus solitariis patentibus alaribus, stylo longiori. Grows naturally in China, from whence the seeds were brought to England a few years past, and the plants were raised in several gardens and by some were thought to be Tea.

Lycium halimifolium. Mill. Dict. ed. 8, n. 6. L. foliis lanceolatis acutis. Jasminoides sinense halimifolio, Du Hamel, Arbres et Arbustes (1755), 306, t. 121, fig. 4. Native of China, from whence seeds were brought to the Royal Garden, Paris. Seeds were sent me by Dr. Bernh. Jussieu. — In the last edition of Mill. Dict., both L. chinense and L. halimifolium are referred to *Lycium barbarum,* China. Seeds brought to England in 1759.— I. F. S., II, 175, *Lycium chinense.*

Datura ferox, Lin. Mill. Dict. ed. 8 (1768). n. 4. China. Cult. Chelsea by Miller in 1731, Dict. ed. 1. Strammonium 3. See Hort. Kew first ed. I, 238.

Vitex chinensis. Mill. Dict. ed. 8 (1768) n. 5. Introduced lately into English gardens from Paris, where it has been raised from seeds sent by the missionaries from China.

Mill. Icon. tab. 275 figures, of Vitex: Fig. 1. Vitex foliis ternatis quinatisque pinnato-incisis, spicis verticillatis, floribus coeruleis, fig. 2, idem floribus albis. — In the last edition of Mill. Dict. these and the V. chinensis are all referred to *Vitex incisa,* Lamarck, Enc. Bot. II, 612.

Vitex Negundo, Lin. Mill. Dict. ed. 8 (1768), n. 4. From North China. Cult. at Chelsea. Destroyed in England during the severe winter of 1740.

Celosia coccinea, Lin., Mill. Dict. ed. 8 (1768). Miller received the seeds from China and cultivated the plant at Chelsea.

Amarantus cruentus, Lin., Mill. Dict. ed. 8 (1768) n. 17. The seeds were brought from China and the two first years produced beautiful heads of flowers, but afterwards they degenerated. The plant was cultivated in the Chelsea garden before 1747; it is mentioned in the 5th edition of Mill. Dict. as *Amarantus sinensis,* a name first given to the plant by Martyn in his Hist. Plant. rar. (1728), tab. 6.

Amarantus tristis, Lin. — Hort. Kew, ed. 1, III 348. China. Cult. by Miller in 1759. Dict. ed. 7, n. 3.

Rheum, Lin., Rhabarbarum, Tournef., Miller Icon. t. 218: Rheum foliis cordatis, glabris, marginibus sinuatis, spicis divisis nutantibus. The seeds of this sort were sent me from St. Petersburg for the true Rhubarb. But Linnaeus has supposed another species to be the true Rhubarb: Rheum foliis subvillosis, petiolis aequalibus. Lin. Sp. pl. 372. The seeds of this were sent to Miller in 1734 by the late Dr. Boerhave from Leyden by the title of *Rhabarbarum chinense verum*. — This is the *Rh. rhabarbarum* of the first edition of Linnaeus Sp. pl. 372, *Rh. undulatum* of. ed. 2, p. 531. But tab. 218 in Mill. icon. represents *Rheum compactum*, Lin. — Comp. above p. 96.

Ficus pumila, Lin., Hort. Kew ed. 2, V. 487. China, Japan. Cult. 1759 by Miller. Dict. ed. 7, n. 8.

Urtica nivea, Lin. Miller, Dict. ed. 8 (1768) n. 9. China, were is is called Pea ma. Cult. Chelsea. — In the last edition of the Dict. Urtica nivea, n. 52, it is stated that Miller cultivated it in 1759.

Thuya orientalis, Lin., Mill. Dict. ed. 8 (1768) n. 2. Native of the northern part of China. The seeds were first sent to Paris by the missionaries. Dict. ed. 6 (1752) says: "not long ago seeds were sent by the missionaries".

Ixia chinensis, L. Cult. 1759 by Miller. Dict. ed. 7. Ixia n. 1.

Smilax China. L. Cult. 1759 by Miller. Dict. ed. 7. n. 5.

IV. Swedish Botanical Collectors in Southern China in the last third of the 18th Century.

Bladh, Peter John, Supercargo to the Swedish E. I. Company and resident at Canton. He presented the plants collected by him in the neighbourhood of Canton and Macao to Thunberg, Retz, Sir Joseph Banks and it seems also to Linnaeus filius. Bladh seems to have returned to Europe about 1779.

C. P. Thunberg (1743—1822), the explorer of the Flora of Japan, 1775—76, and subsequently Professor of Botany at Upsala, in his "Nova Genera plantarum", 1781, p. 7, established the genus *Bladhia* and writes there: "Nomen in honorem et memoriam amici Domini P. J. Bladh in Canton Chinae mercatoris celeb. et botanici indefessi, inclyti". Compare also Miller's Gard. Dictionary, last edition, sub Bladhia.

In Thunberg's Flora Japonica, published in 1784, three plants from China are mentioned, two of which had been sent by Bladh:

p. 257, *Vitex ovata*, Japonia. "E Macao Chinae mihi missa fuit a Dn. P. J. Bladh."

p. 301. *Prenanthes chinensis*. Japonia. E China misit D. Bladh.

p. 138. *Anthericum japonicum*. Japonia, China, Java.

This and several species of *Gardenia* from Macao, described by Thunberg in his Dissertatio de Gardenia, 1780, may likewise have been transmitted to him by Bladh.

The late C. Maximowicz found in Thunberg's herbarium two species of *Hedyotis*, *H. lancea* and *H. linearis,* and a plant named by Thunberg *Ilex asiatica,* which plants, as was written on the labels, had been collected at Macao by Bladh. The Ilea asiatica turned out to be a male specimen of *Zanthoxylum vitidum.* DC. (Mél. biol. Acad. Petrop. XI; 781, 784; Mém. Acad. Petr. XXIX, n. 3, 67).

G. v. Martens, Tange, Preuss. Exped. Ost Asien, 1866, describes several Chinese Algae found in Thunberg's herbarium.

Fourteen plants gathered by Bladh near Canton and almost all new, were described by Professor Retz. (See further on).

J. E. Smith, Exotic Botany (1804) t. 70, *Asclepias carnosa* Lin. fil., tells us that the original specimen of this plant in Linnaeus' herbarium, was received from China through his friend Bladh.

Canton plants collected by Bladh are also met with in the Banksian herbarium, now in the British Museum.

Bot. Reg. 469 (1820) *Strophanthus dichotomus,* β. *chinensis.* Bladh. Macao, Banksian Herb.

Lindley, Collect. Bot. (1821) tab. 3, *Raphiolepis rubra.* Bladh, China. Banks. Herb.

Lindley, Monogr. Rosarum (1820), p. 126. *Rosa sinica.* Bladh, China. Banks. Herb.

DC. Prodr. XIII. 1, p. 179. *Solanum Osbeckii.* Bladh. Herb. Banks.

I. F. S., ɪ, 109, *Murraya exotica.* Bladh, China. — Ibidem I, 134, *Vitis japonica,* Bladh, China. — Ibid. I, 447. *Gynura divaricata.* Bladh, near Canton.

Retz, A. J. (Retzius), Professor of Natural History at Lund in Sweden, in his "Observationes Botanicae", which appeared in 6 fascicles, 1779—91, described 36 Chinese plants, most of them new, transmitted to him by Bladh, Osbeck, **Tranchell, Wennerberg** and other collectors, not named. Wennerberg seems to have been chaplain to a ship. He brought also plants from Java and Sumatra.

Retz's herbarium (he received also plants from India, sent by Koenig) is preserved at Lund. (De Candolle, Phytogr. 1880, p. 443).

The following is a list of the Chinese plants described in the 6 fascicles of Retz' Observ. Botan:

Clematis chinensis, n. sp. II, 18, tab. 2. E China misit Bladh. —

Macartney, a few years later, from Dane's Island. — DC. Syst. I (1818) 137. — Ind. Fl. Sin., ɪ, 3.

Hypericum chinense Retz, (non Lin.), V, 27. E China adduxit D. Bladh. — *Hyp. biflorum*, Lam. Enc. bot. IV, 170. *Cratoxylon polyanthum*, Kort. — Ind. Fl. Sin., ɪ, 74.

Sida chinensis. Retz. IV, 29. E China, Wennerberg. — *Sida rhombifolia*, Lin. — I. F. S., ɪ, 85.

Hedysarum triflorum. Lin., Retz, IV, 29. E China, Wennerberg. — *Desmodium triflorum*, DC. — I. F. S., ɪ, 176.

Dolichos sinensis, Lin., Retz, IV, 30. Not stated where collected.

Hedyotis fruticosa Lin.? Retz, II, 8. E China specimen accepimus. — Probably *Hedyotis consanguinea*, Hance. — I. F. S., ɪ, 373.

Hedyotis hispida, Retz, IV, 23. E Canton adduxit Wennerberg. — I. F. S., ɪ, 374.

Gardenia jasminoides, Retz, II, 13. E China ramulum dedit cel. D. P. J. Bladh. — *Gard. scandens*, Thbg, DC. Prodr. IV, 383. A doubtful plant. — I. F. S., ɪ, 383.

Hieracium heterophyllum (Prenanthes repens Lin.*)*. Retz. II, 24. E China misit Bladh. — *Lactuca repens*, Benth., Maxim. Mél. biol. IX, 364.

Scrophularia chinensis, Lin.? Retz, V, 5. Not stated where gathered and by whom. — Comp. supra p. 90.

Phlomis chinensis, Retz, II, 19. E China accepi. — *Leucas chinensis*, DC. Prodr. XII, 525. — I. F. S., ɪɪ, 303.

Polygonum barbatum, Linn., Retz, IV, 25. In China lectum.

Curcuma longa. Lin. (communicated by Koenig). Retz, III, 72. Colitur in hortis Sinensium copiossime Malaccae, rarissime semina profert.

Languas chinensis, Retz, III, 65. Misit Koenig (ex India). Maleys: Sina languas. Retz, VI, 18, *Heritiera chinensis* or *Languas chinensis*, Koenig Descr. Monandr. Plants. Colitur in hortis Sinensium pro usu medico. — This is *Alpinia chinensis*, Roscoe, Linn. Trans. VIII. 346.

Canna juncea, Retz, I, 9. E China allatum exemplar dedit D. Bladh.

Cyperus compactus. Retz, V. 10. E China allatum communicavit Rev. D. D. Osbeck. — Kth, enum. plant. II, 92, *Cyp. dilutus*, Vahl.

Cyperus albidus, Retz, VI, 21. E China redux dedit D. Bladh. — Knth, enum. pl. II, 101, considers this a dubious plant.

Scirpus junciformis, Retz, VI, 19. E China dedit D. Bladh. — Kth enum. II, 175.

Carex chinensis, Retz, III, 42. E China misit Bladh. — Kth enum. II, 517. Benth. Fl. hongk. 402.

Paspalum hirsutum, Retz, II, 7. E China misit Bladh. — Kth enum. I, 57.

Panicum setigerum, Retz, IV, 15. E China misit Bladh. — Kth enum. I, 90.

Panicum ciliare, Retz, IV, 16. E Java et China adtulit Wennerberg. — Kth enum. I, 82.

Panicum radicans, Retz, IV, 18. Ad Canton lectum, Wennerberg, — Kth enum. I, 126.

Panicum aristatum, Retz, IV, 17. E China adduxit plur. rever. D. Wennerberg. — Kth enum. I, 146, *Oplismenus lanceolatus,* Kth.

Rottboellia sanguinea, Retz, III, 25. Ramulum misit e China D. Bladh. — Kth enum. I, 468.

Rottboellia compressa, Linn., Retz, III, 12. In China lectum attulit D. Tranchell. — Kth enum. I, 465. *Hemarthria compressa,* Brown.

Ischaemum ciliare, Retz, VI, 36. E China accepi. — Kth enum. I, 513.

Andropogon Bladhii, Retz, II, 27. In China lectum, cum aliis misit D. Bladh. — Kth enum. I, 498.

Aira chinensis, Retz, III, 10. E China misit D. Bladh. — Kth enum. I, 291.

Poa chinensis, Retz, (non Linn.), V, 19. Not stated from whence received. — A dubious plant.

Poa panicea, Retz, III, 11. (similis Poae chinensi, Burm. Ind. tab. 11). E China misit Bladh. — Accord. to Poiret, Enc. Bot. V, 80, = *Poa chinensis,* Lin.

Adiantum fuscum, Retz, II, 28, figured tab. V. China. This is according to Swartz, Syn. Filic. 121, *Adiantum flabellulatum* Lin.

Pteris nigra, Retz, VI, 38. Circa Canton lecta, dedit Wannerberg. According to Swartz, l. c. 108, this is the *Adiantum tenuifolium,* Lam.

Trichomanes chinense, Lin. Retz, VI, 40. Osbeck dedit. — V. supra p. 105.

Björkegren. Maximowicz found this name on a label, affixed to *Hedyotis ovata,* Thbg, in Thunberg's herbarium, where it was stated that Bj. brought this plant from the island of Hainan.

Fagraeus. In Acta Holm. IX (1788) there is an article on the genus Turraea by C. N. Hellenius, Professor of Botany, Abo. On p. 309 it is stated that *Turraea pubescens,* a new plant, had been collected in Hainan by the assistant Fagraeus together with other plants found in the collection of natural objects belonging to H. Chr. Pentz, councillor in Alingsas. The plant is described and figured.

Waenman. G. Wahlenberg (1780—1851), Professor of Botany, Upsala, in his article on Carices, Act. Holm. XXIV (1803) p. 149, describes a new species, *Carex cruciata*, gathered near Canton by Waenmann. Ex herbario Bergiano. (P. J. Berg, Professor at Stockholm, + 1790, friend of Linnaeus).

Dahl. Martin Vahl (1749—1804) a Swede, pupil of Linnaeus, and Professor of Botany at Copenhagen, in his "Symbolae Botanicae", 1790—94, describes two new Chinese plants, *Croton laevigatus* (II, 97) and *Ardisia humilis* (III, 40). Both are stated to have been brought from the island of Hainan by Dr. Dahl.

V. Botanical Researches of the Jesuit Missionaries in China, in the last quarter of the 18th Century.

Cibot, Pierre, Martial, a Frenchman, born, 1727. He entered the Order of the Jesuits, was sent to China in 1759, and died at Peking in 1784. A complete list of his numerous papers on Chinese, especially Peking plants, published in the Mémoires concernant les Chinois, 1777—86, has been given in my Early Res. Flora Chin. p. 124—126, to which I beg to refer. Like Gaubil and d'Incarville, Cibot kept up a correspondence with the botanists of the Academy of Sc. in St. Petersburg. In 1767 he was elected Corresponding Member of that learned body. In the Novi Commentarii Acad. Petrop. XIX, 1875, p. 375 seqq., there is an article in Latin by Cibot, "*Fungus sinensis mokusin*", cum icone, tab. 5. (Comp. supra p. 106). The same article in French appeared 4 years later, in the Mém. conc. Chin. IV, 500.

Prof. J. Gaertner, St. Petersburg, described and figured in the Novi Comment. Acad, Petrop. XIV, 1, (1770) p. 544, tab. 20, a new plant, *Digitalis glutinosa*, of which he says: habitat versus Chinam. It seems that Gaertner had before him a living plant. This is the *Rehmannia glutinosa,* Libosch., a common plant in North China, especially at Peking, but not met with beyond the Great Wall. I feel little doubt that the seeds of this plant had been sent so Gaertner from Peking, and probably by Cibot.

Loureiro, Ioannis de, Portuguese, born, 1715. He proceeded in 1735 as a missionary to Cambodja, and, in 1743 went to Cochinchina, where he spent 36 years. He studied the Flora of the country, and collected nearly 1000 species there. In 1779 he established himself at Canton, and entered the Order of the Jesuits. At this place he continued his botanical researches during three years, and then embarked with his botanical treasures for Europe, in 1782. On his way home he visited the island of Mozambique, where he made a stay of three months, enriching his herbarium with many rare specimens. After having reached his native country, Loureiro was taken up during several years, in Lisbon, with the description of his plants. In 1790 his FLORA COCHINCHINENSIS was brought out. The preface is dated 1788, Oct. Three years later Willdenow edited it anew. Loureiro died in 1794 or 1796.

In 1774 Loureiro sent 60 species of plants from Cochinchina, via Canton, to England and Sweden, and in 1779 transmitted 230 species more from Canton to London. This collection is still in the British Museum. The whole of Loureiro's herbarium was kept by the Academy of Lisbon, but in 1808, when Napoleon I had taken possession of Portugal, a part of it was transferred to Paris, where it still exists in the Muséum d'hist. nat., 120 species, as M. Franchet informs me; nearly all have been identified. M. De Candolle in his "Phytographie", 1880, p. 429, referring to a letter from M. Gomes fil. 1877, states that Loureiro's herbarium at Lisbon has been destroyed.

In his Flora Cochinchinensis Loureiro notices 539 Chinese plants. As a full account of them has been given in my Earl. Res. p. 134—184, I need not revert to his subject here.

Grosier's DESCRIPTION GÉNÉRALE DE LA CHINE.

In 1785 Abbé Grosier published a book with the above title, forming the 13th volume of De Mailla's "Histoire générale de la Chine". The learned Father Grosier, who was "Bibliothécaire de Son Altesse Royale, Monsieur", had himself never visited China. He compiled his book from the reports of the missionaries in that country. A considerable part of it is devoted to natural history; for instance 108 pages to botany. In a new and much enlarged edition of the work which appeared from 1818—20, in 7 volumes, nearly three of them treat of natural history; vol. II and III, 658 pages, deal with botany. It supplies a mass of most valuable information with respect to Chinese plants. Grosier endeavours to give a list of the Chinese plants described by European botanists. But besides Loureiro's

Chinese plants, of which he generally gives a full account, his enumeration is far from being complete.

VI. Poivre, Pierre. 1740—73.

This indefatigable traveller and naturalist was born at Lyons in 1719. He was sent to the Congregation of the Missions Étrangères at Paris, where he was prepared for missionary work, but at the same time his studies were directed to natural history. In 1740 the Missions Étrangères dispatched him to Cochinchina and China, where he learned the languages of the two countries. Soon after his arrival at Canton Poivre was, in consequence of a mistake, thrown into prison, but he defended himself so well before the mandarins, that he was honorably discharged, and the Vice-roy was even so much pleased with him that he allowed him to travel in the interior of the province. On his return to Europe, P. had the misfortune to lose an arm and be made prisoner in an engagement between the French ship which he was on board of and an English vessel, in the Straits of Banca. He was, however, soon released and sent to Pondicherry to his countrymen. From his place he got to Isle de France (Mauritius) and continued his voyage home. But when in sight of the coast of France he was for a second time made prisoner by the English, but released in consequence of the peace between France and England, in 1745. The loss of his arm obliged Poivre to renounce the ecclesiastical profession. He presented, after his return to France, the results of his observations during his voyage, to the French E. I. Company and proposed amongst other things to open up a direct trade between France and Cochinchina, and to transplant to Isle de France the precious spice trees of the Moluccas. Accordingly he was entrusted with the execution of these projects, and in 1749 departed for Cochinchina as an envoy of the King of France. The King of Cochinchina received him well, and made no objection to the establishment of a French factory at *Tai fo* on the Bay of Turan. P. discovered in this country the "upland rice" grown there in the mountains without irrigation, and introduced it subsequently into Isle de France. He noticed in Cochinchina an Indigo plant, called *tsai*. (Rondot, Vert de China p. 21). After leaving Cochinchina he went to China for a second time, visited the Philippines and the Moluccas, and from the latter islands took with him living spice trees. He paid finally a visit to the island of Madagascar. On his way home he was for a third time made prisoner by the English. It was only in 1757 that he again saw his native country. As a tribute to the value of his scientific explorations, the Academy of Lyons elected him their Member. After this he spent nearly ten years near Lyons in complete devotion to agriculture, when in 1767, the Duke of Choiseul, then Prime Minister, invited him to accept the post of Intendant of the French colonies Isle de France and Bourbon. P. embarked the same year and successfully administrated these colonies for 6 years. He formed there in the celebrated garden of "Monplaisir", a nursery of every kind of useful tropical plants, naturalised the clove, the nutmeg, the breadfruit and also many Chinese plants which it was not difficult to obtain, for, as Poivre states (Voyages d'un philosophe, v. infra), all the French ships carrying on trade with India and China used to call at Isle de France, which has two excellent harbors. He also sent many exotic plants and herbarium specimens to Paris. Poivre returned to France in 1773, and died at Lyons in 1786. He left numerous manuscripts, which have never been published. In 1768 there appeared a small book with the title: "Voyages d'un Philosophe". The author was not named, but it is well known that it was written by Poivre. It went through several editions, that of the year 1797 begins with Poivre's biography by Dupont de Nemours. This little memoir is not, as might be presumed, a narrative of his travels, but it consists of short accounts, without any date, of the countries visited by him, viz: Cape of G. H., Madagascar, Bourbon, Isle de France, Coromandel, Siam, Cambodia, Cochinchina, China. As to China he speaks only of Canton and its neighborhood. The author pays special attention to the state of agriculture in the countries visited by him. — Ph. Commerson, the well known French botanist and traveller, contemporary of Poivre, named the genus *Poivrea* (Combretaceae) in honour of his friend.

Poivre presented his herbarium of exotic plants to Bern. de Jussieu. Mr. A. Franchet informs me that about 100 Chinese plants gathered by Poivre are found in Jussieu's herbarium, now in the Muséum d'hist. nat. It seems that only two plants of this collection have been described:

Hedysarum lutescens, Poiret, Enc. Bot. VI, 417. Croît à la Chine. Cette plante a été envoyée par M. Poivre. V. s. in herb. Jussieu. — This plant has not been gathered after Poivre. See Ind. Fl. Sin., ɪ, 174, sub *Desmodium lutescens*.

Selaginella uncinata, Spring, Monogr. Lycopod. p. 109. Quoted Poivre, China, Herb. Jussieu.

The following is a list of Chinese plants which are recorded to have been naturalised in the second half of the last century in Isle de France.

Their introduction seems to be due in most of the cases to the exertions of Poivre and the Jesuit missionaries in China.

Poiret, Enc. Bot. VIII, 753, in his biographical notice of Poivre, states that he introduced from China to Isle de France the *Tea* plant, the Chinese Tallow tree *(Stillingia sebifera)*, the "Savonnier de Chine" *(Sapindus chinensis.* Linn., or *Koelreuteria paniculata?* Perhaps Poiret means another tree, *Gymnocladus, Gleditschia?).*

Grosier, la Chine II, 531 notices the introduction of *Chinese Cinnamon* by Poivre, into Isle de France, before 1776, whilst Poiret, l. c. says that he introduced the "Cannellier de Cochinchine".

Grosier, II, 401, speaks of a kind of Tea sent from China by Poivre under the name of *tcha hoa,* and gives a description of the plant (probably a *Camellia).*

Euphoria punicea, or *Litchi chinensis.* Lamarck Enc. Bot. III, 573. Cultivé à l'Isle de France, introduit par Poivre. — See also Grosier, II, 478.

Euphoria longana, Lam. l. c. Chine méridionale, maintenant cultivé à l'Isle de France.

Mespilus japonica, Thbg., Lamarck Enc. Bot. IV, 444. Chine et Japon. Les Chinois le nomment *lou-koet.* Cultivé à l'Isle de France.

Cookia punctata, Sonnerat., Poiret, Enc. Bot. VIII, 327. "Chine. Cultivé à l'Isle de France.

Litsea chinensis. Faux cérisier de la Chine. Lamarck, Enc. Bot. III, 574. Cet arbre croît naturellement à la Chine et est cultivé à l'Isle de France, où, par la faculté qu'il a de resister aux vents, on l'emploie comme en charmille pour former des abris contre les ouragans. Ses baies ont un goût de camphre et une odeur de Lierre. Les oiseaux s'en nourrissent. — This is the *Tetranthera laurifolia,* Jacq. or *Litsea sebifera,* Pers. See I. F. S., II, 385.

Dryandra oleifera, Lam. Enc. Bot. II, 330. Cultivé au Jardin du Roi à l'Isle de France. Les fruits de cet arbre sont de la grosseur d'une noix munie de son brou; on retire de leurs amandes une huile pour les lampes et qu'on nomme "huile de bois, *mou yeou* des Chinois". — This tree, the *Elaeococca verrucosa* of Adr. Juss., is common in Mid and South China and in Japan. It was probably introduced from China.

Rhus vernicifera L., the Chinese Varnish tree was, according to Cossigny, introduced to Isle de France, from China, by a Jesuit missionary, in 1766.

Latania borbonica, a palm observed by Commerson in the island of

Bourbon. Lam. Enc. Bot. III, 427. The same tree was, in 1788, introduced by Jacquin from Mauritius. Jacq. Fragm. botan. p. 16, tab. 11, fig. 1. Jacquin named it *Latania chinensis,* for in Mauritius this palm was known under the name of "Latanier de Chine". R. Brown named it subsequently *Livistona chinensis.*

    Aletris chinensis, Lam. Enc. Bot. I, 79, (ibid. II, 324 *Dracaena terminalis* Linn.). Chine. Cultivé an Jardin du Roi à Paris et à l'Isle de France.

    Hortensia opuloides, Lam. Enc. Bot. III, 136. Japon, Chine. Cultivé à l'Isle de France. — *Hydrangea Hortensia,* DC. Prodr. IV, 15. — I. F. S., ı, 273.

VII. Sonnerat, Pierre. 1776.

    This distinguished French traveller and naturalist, who visited China in about 1776, was born at Lyons, in 1745. He entered the navy, and early commenced to take an interest in natural history. A great part of his life was spent in travelling. In 1768 he went in the quality of commissaire de la marine to Isle de France (Mauritius) where his relative Poivre was then Intendant of the French Colonies, and remained there for six years in company with Commerson, another celebrated French naturalist, who died there in 1773. During Sonnerat's sojourn at Isle de France, Poivre dispatched a second expedition to the Moluccas to bring living spice trees to the island. Sonnerat embarked with them. They executed their commission, and also visited the Philippines. Sonnerat returned to Isle de France in 1772, along with Commerson, visited Madagascar, and in 1774 embarked for France. After a short stay in his native country he was again despatched by the French government to continue his observations on Asiatic countries. He visited Ceylon, Malabar, the Gulf of Cambay, Coromandel, Malacca, China. Then he returned to Coromandel, travelled two years in India, went to Pondicherry, which then belonged to the French, but was captured in 1778 by the English. Sonnerat finally returned to Isle de France and on his way home to Europe sojourned some time at the Cape of G. H. He reached France in 1781 and in 1782 published a work in two volumes with the title: VOYAGE AUX INDES ORIENTALES ET À LA CHINE, 1774—1781, with 140 plates. Having finished this work and arranged his collections, Sonnerat sailed again for India. He settled in Pondicherry, which in 1783 had been rendered to the French, and lived there till 1803, when he re-

turned to France. He died at Paris, in 1814. — As Poiret states in Enc.
Bot. VIII, 756, Sonnerat brought home very rich zoological and botanical
collections from the countries he had visited. The plants he presented principally to Jussieu and Lamarck. Linnaeus filius named in his honour the
genus *Sonneratia* (Myrtaceae). Lamarck and Poiret described many of his
Indian and Chinese specimens in the Encycl. Botan. Mr. Franchet writes
me that the herbariums of Lamarck and Jussieu, now in the Muséum d'hist.
nat., contain some of Sonnerats specimens, but not a single one from China,
and that in general his plants are very rare in the French collections.

The following Chinese plants, gathered by Sonnerat near Canton or
Macao, were described by Lamarck and Poiret:

Uvaria odorata, Lam. Enc. Bot. I, 595. Alanguilan de la Chine, Cananga, Rumph. Amb. II, 195, t. 65. Molluques, Java, Chine. Sonnerat. —
I. F. S., I, 27. *Cananga oderata,* Hook. fil. & Thoms. No Chinese specimens in the London herbaria.

Hypericum chinense, Lin. — Lam. Enc. Bot. IV, 144. Cultivé au
Jardin des Plantes. Sonnerat l'a cueilli en Chine.

Hibiscus populneus, Lin. — Lam. Enc. Bot. III, 353. Indes orientales, partie australe de la Chine, Isle de France. Communiqué par M.
Sonnerat.

Cookia punctata, Sonnerat, Voyage II, p. 181, tab. 130. — Poiret,
Enc. Bot. VIII, 327. Cette plante croît naturellement à la Chine, elle est
cultivée à l'Isle de France. (V. s. in herb. Lamarck). — I. F. S., I, 110,
Clausena Wampi, Oliv.

Citrus buxifolia, Poiret in Lam. Enc. Bot. IV, 581, 382; Suppl. III,
594. Cette plante est originaire de la Chine et y a été observée par Sonnerat, qui en a communiqué des exemplaires au citoyen Lamarck (v. s.).
Marsana buxifolia, Sonn. Voyage II, p. 245, tab. 139. *Murraya exotica,*
Linn. — I. F. S., I, 109.

Rhamnus lineatus, Linn. — Poiret in Lam. Enc. IV, 473. Cette plante
croît à la Chine et aux Indes. Sonnerat en a communiqué plusieurs individus
à Lamarck.

Euphoria punicea, Lam. Enc. Bot. III, 573. (Comp. supra p. 11).
Litchi chinensis, Sonnerat, Voyage II, p. 230, tab. 129. Cet arbre croît
en abondance à la Chine et à la Cochinchine. Cultivé à l'Isle de France
(Poivre).

Euphoria longana, Lam. l. c. (Comp. supra p. 11). Cet arbre croît à
la Chine, dans les provinces méridionales. Cultivé à l'Isle de France. Nous
en possédons des rameaux communiqués par Sonnerat et Jos. Martin.

Melastoma repens, Desrousseaux in Lam. Enc. Bot. IV, 54. Chine. Communiqué par Sonnerat. — I. F. S., ı, 300.

Cephalanthus chinensis, Lam. Enc. Bot. I, 678. Cet arbre croît à la Chine, aux Isles Philippines et dans les Moluques. Sonnerat nous en a communiqué des rameaux chargés de fleurs. — Poiret, l. c. IV, 435, suspects that this may be the same as *Nauclea orientalis* Linn., but DC, Prodr. IV, 346, identifies it with *Nauclea purpurea,* Roxb., which, according to Hemsl., I. F. S., ı, 370, is not a native of China.

Ixora chinensis, Lam. Enc. Bot. III, 344. Cette espèce nous a été communiquée par Sonnerat comme provenant de la Chine. Elle croît aussi dans l'isle de Java. — I. F. S., ı, 385.

Cordia sinensis, Lamarck. Poiret, E. B. VII, 49. Chine, Herb. Lamarck. — Lam. Illustr. Genres, n. 1914. Chinese, Sonnerat. — Hemsley, Ind. Fl. Sin., ıı, 143, states, that this species has not been identified by subsequent writers.

Capsicum sinense. Jacq. Hort. Vindob. III, tab. 67. (Jacquin had it cultivated from Martinique). Poiret, Enc. Bot. V, 326: Cette plante croît naturellement à la Chine. Des exemplaires de cette espèce communiqués par Sonnerat à Lamarck.

Bignonia chinensis, Lam. Enc. Bot. I, 423. Cet arbrisseau croît à la Chine. Il nous a été communiqué par Sonnerat. — DC, Prodr. IX, 223: *Tecoma (Bignonia) grandiflora,* Delaun. — I. F. S., ıı, 235.

Litsea chinensis, Lam. Enc. Bot. III, 574 (v. supra 119). Nous avons reçu des rameaux de cet arbre par Sonnerat, Stadman et Jos. Martin.

Phyllanthus lucens, Poiret, Enc. Bot. V, 296. Sonnerat, Chine. I. F. S., ıı, 427, *Breynia fruticosa,* Hook. fil. (v. s. 98).

Phyllanthus villosus, Poiret, l. c. 297. Sonnerat, Chine. — I. F. S., ıı, 423: incompletely known.

Croton sebiferum, Lin. — Lam. Enc. Bot. II, 209. Chine. Sonnerat et Commerson en ont rapporté des branches chargées de fleurs et de fruits. On en cultive des jeunes pieds au Jardin du Roi. — *Stillingia sebifera,* the Chinese Tallow tree, v. supra p. 98.

Ficus heterophylla, Lam. Enc. Bot. II, 499. Indes Orientales, Chine. Cette espèce nous a été communiquée par Sonnerat. — Hooker, Fl. Ind. V, 518.

Cupressus japonica, Thbg. — Lam. Enc. Bot. II, 244. Cet arbre croît au Japon. Il se trouve aussi à la Chine, d'ou Sonnerat nous en a rapporté des branches dépourvues de frutification. — DC, Prodr. XVI, 2, p. 438, *Cryptomeria japonica,* Don.

Aletris chinensis, Lam. Enc. Bot. I, 79 v. supra p. 120). Chine. Sonnerat en a apporté des spécimens d'herbier.

Poa subsecunda. Lam. Enc. Bot. V, 84. Croît à la Chine, d'où elle a été rapporté par Sonnerat. Herb. Lamarck.

Selaginella uncinata, Spring, Monogr. Lycopod. p. 109. (v. supra p. 44). Sonnerat, China. Herb. Delessert, Herb. Palissot-Beauvois.

Adiantum flabellatum, Lin. — Lam. Enc. Bot. I, 42. Chine. Communiqué par Sonnerat.

Pteris crenata, Swartz. Poiret, Enc. Bot. V, 715. Chine. Herb. Lamarck. Communiqué par Sonnerat.

VIII. The Royal Garden at Paris and the Great French Botanists who laboured there in the 18th Century.

The starting point of this celebrated Garden, now generally known under the name of «Jardin des Plantes», dates from the year 1626, when Hérouard, First Physician to Louis XIII, obtained from the King letters patent authorizing the establishment of a Botanical Garden in the suburb St. Victor, of which Hérouard became Superintendent. When he died, in 1633, Bruvard, the new physician to the king was intrusted with the superintendency of the garden, and La Brosse was appointed Director of it. In 1640 three professors of the Faculty of Medicine were appointed to lecture at the Garden on Botany, Pharmacology, and Chemistry. They were called «demonstrateurs», and the garden received the name «Jardin Royal des Plantes médicinales». A good many exotic plants, especially from Canada, which then belonged to France, were cultivated there.

Tournefort, Joseph Pitton de, born at Aix, in 1656, early devoted himself ardently to the study of botany, and soon occupied an eminent position as a botanist. In 1683 he accepted the chair of botany at the Royal Garden and with him begin the days of glory in the history of the garden. Before he entered it, he had much travalled in France, Savoy and explored the Pyrenees, and after his appointment at the garden travelled again in Spain, Portugal, Holland, England. In 1691 he was elected Member of the French Academy of Sc.' In 1700 he was commissioned by the King to explore Greece, Asia minor, Armenia. He returned after two years and brought home many new plants. Tournefort died 28 Sept. 1708.

He established a classification of plants and explained his system first in his Éléments de Botanique, 1694, 3 vols, and afterwards more in detail in his Institutiones Rei Herbariae, 1700.

Jussieu, Antoine de, born at Lyons in 1686, Tournefort's pupil. He succeeded the latter as Professor of Botany at the Royal Garden and filled this place till his death in 1758.

Jussieu, Bernard de, younger brother of Antoine, born at Lyons in 1699. He studied medicine at Montpellier and took his degree of M. D. in 1720. In 1722 he was appointed Demonstrator at the Royal Garden, which post he held till his death. With characteristic modesty he declined the office of professor in 1770, when Le Monnier, the successor of Antoine Jussieu, had given up his duties; he continued his labours as manager of the cultural department and director of the herborizing excursions. From 1758 Bernh. de Jussieu was Superintendent of the Royal Garden at Trianon, where he arranged the plants according to a new system. He died in 1777. — Genus *Jussiaea*, Lin. — According to Poiret, Enc. Bot. VI (1804) 511, the magnificent old Cedar of Lebanon in the Jardin des Plantes, was planted (in 1735) by Bernh. de Jussieu. It is still an object of admiration there.

Jussieu, Antoine Laurent de, nephew of Bernard (son of Bernard's brother Joseph) born 1748 at Lyons. In 1770 he succeeded Le Monnier as Professor of Botany at the Royal Garden. He resigned this position in 1785, died in 1836. In 1773 he had been elected Member of the Paris Academy of Sc. — Bernh. de Jussieu's System was first made known by A. L. de Jussieu: Genera Plantarum secundum ordines naturales disposita, juxta methodum in horto regio Parisiensi exaratam anno 1774. Published 1789.

Jussieu, Adrien de, son of Ant. Laur., the last of that illustrious family, born at Paris in 1797. In 1826 he was appointed Professor of Rural Botany at the Muséum d'hist. nat. (Royal Garden). Member of the Academy. He died 29 June 1853.

Thouin, André, born in 1747, a pupil of Bern. Jussieu. He was 17 years old when his father, Chief Gardener at the Royal Garden, died. Jussieu then did not hesistate to entrust him, notwithstanding his youthful age, with the direction of the cultures in the garden, and this place of

Chief Gardner he held from 1764 till his death in 1823. He was at the same time Professor of Horticulture there, from 1793. He was Member of the French Institute and Foreign Member of the Linnean Soc., London. Whilst Lamarck and Poiret disclosed the treasuries of the French botanical collections, especially those made in distant countries, their contemporary Thuin in the Royal Garden successfully cultivated many exotic plants, and amongst them not a few natives of China. — Genus *Thouinia,* Linn. fil. (1781).

Desfontaines. René Louiche, 1750—1833. From 1785 Professor of Botany at the Royal Garden, Member of the French Institute. In 1804 he published a small book with the title TABLEAU DE L'ÉCOLE DE BOTANIQUE DU MUSÉUM D'HISTOIRE NATURELLE, in which a list of the plants then cultivated in the Jardin des Plantes is given. More than a hundred names of Chinese plants are met with in this catalogue. A second edition was issued in 1815, a third in 1829. — Genus *Desfontainea,* Ruiz & Pavon (1794).

Lamarck, Jean Bapt. Antoine, Pierre Monet, chevalier de. This eminent French naturalist was born in 1744. He was originally destined for the Church and sent to the Jesuits at Amiens, where he remained till his fathers death. In 1760 he entered the army. Subsequently he went to Paris, and began to study medicine. He early became interested in natural sciences, and threw his main strength into botany. In 1778 he published the «Flore française». In 1779 the great naturalist Buffon obtained for him an appointment as Botanist at the Academy of Sc. In 1781 and 1782 Lamarck travelled with Buffon's son in Holland, Germany, Hungary. On his return he commenced that elaborate work on which his reputation in botany principally rests. He took upon himself to write the botanical part of the great ENCYCLOPÉDIE MÉTHODIQUE par ordre des matières, which Panckouck & Agasse began to publish in 1783, and of with the section BOTANIQUE forms 8 large volumes and 5 vols Supplements, 1783—1817. Lamarck issued only the first four volumes of the work, 1783—1796. The Abbé **Poiret,** Jean Louis Marie, born in 1755, ᵻ 1834, Professor of Natural History at Paris, published the rest. From 1791—1823 appeared the ILLUSTRATION DES GENRES (des plantes), 3 volumes with 1000 plates and 2 vols of text. Of these I and II (1791—93) tab. 1—445, were edited by Lamarck; vol. III (1823) tab. 446—1000 by Poiret.

From 1788 Lamarck had a botanical appointment in the Royal Garden, and was keeper of the Herbarium there. In 1793, when the «Mu-

séum d'histoire naturelle» was created in the Royal Garden, Lamarck was
presented to a zoological chair and called on to lecture on invertebrate
animals. About this time he was elected Member of the French Institute.
For the last seventeen years of his life he was afflicted with total blindness.
He died in 1829. — Genus *Lamarkea,* Persoon (1805).

The excellent descriptions found in the Enclopédie, Botanique, of all
the plants then known to botanists, were based upon the rich botanical
collections in the Royal Garden and the living plants cultivated there.
Lamarck, who himself possessed large collections (now in the Muséum
d'hist. nat.) had also access to various private herbariums of French travellers and botanists. When noticing Chinese plants, Lamarck and Poiret
frequently do not give the names of the collectors and even do not state
from what part of China these plants had been sent, but for the greater
part they can be traced to the collections of d'Incarville, Poivre and Sonnerat, already spoken of.

L'Héritier de Brutelle, Charles Louis, a distinguished French botanist,
born at Paris in 1746, Member of the French Institute. He was assassinat-
ed in 1800. His friend W. Aiton, Director of the Kew Gardens, named in
his honour the genus *Heritiera* (1789). Of l'Héritier's botanical publications I may mention here two, in which are found descriptions and illustrations of Chinese plants:

STIRPES NOVAE AUT MINUS COGNITAE DESCRIPTIONIBUS ET ICONIBUS
ILLUSTRATAE (coloured). Paris. In 6 fascicles: I and II, 1784; III, 1785;
IV, 1788; V, 1789; VI, after 1789, — altogether 184 pages, 84 plates.
Pritzel, Thes. Lit. Bot. observes that in De Candolle's library there are 28
unpublished plates of the Stirpes novae, forming the fascicles VII and VIII.

Robinia Chamlagu, l. c. p. 161, t. 77. In horto Paris. accepta, gigni-
tur in China, eamque dicunt "Chamlagu", nomine vulgari Sinarum. Multis
abhinc annis cicur apud nos, sed nondum fructus perfecit. Ineunti vere floruit
floribus albo-luteis, aliis sanguineis sat cospicuis.

Ailantus glandulosa, Desf. — L'Hérit. l. c. 179, t. 84. E Peking
introducta. D'Incarville.

W. Aiton in the Hort. Kew., refers to three plates of the Stirpes
novae, representing Chinese plants, not found among the published
ones, viz:

t. 93, *Sterculia platanifolia.*

t. 99, *Panax aculeatum.*

t. 100, *Rhapis flabelliformis.*

In 1786 l'Héritier went to London and was staying there for 18 months. He made a study of the Kew collections, which he subsequently utilised in another iconographical work which he published in 1788 with the title: SERTUM ANGLICUM seu Plantae rariores quae in Hortis juxta Londinum, imprimis in Horto Regio Kewensi excoluntur, ab ann. 1786—87. The volume contains 36 pages text and 34 uncoloured plates. The latter were the work of Redouté, the celebrated French botanical artist.

The following Chinese plants are figured and described in the Sertum Anglicum:

Chloranthus inconspicuus (Swartz in Philos. Trans. LXXVIII, 2, (1787) p. 359, t. 15). L'Hérit. Sert. Angl., p. 1, t. 2, and p. 35. Sinensibus *tchin tchu lan* seu *tchu lan*. Habitat in China. Planta viva allata in Angliam a Jac. Lind, M. D., anno 1781, floruit dum navis Atlas adveheret. Sinensibus miscere foliis Theae, Theamque suavius olere Dr. Lind narrabat. Sed Chloranthus omnino inodorus est.

Amaryllis aurea, sp. nov. L'Hér. Sert. Angl. 14, t. 15 bis, China.

Koelreuteria paullinoides, L'Hér. l. c. 18, t. 19. China. (Same as *K. paniculata*, Laxm.)

Limodorum Tankervillae, L'Hér. l. c. 28. China. No figure.

In 1739 the great naturalist **Buffon**, George Louis Leclerc, comte de, born 1707, was appointed Director of the Royal Garden, and superintended it till his death, 1789.

He was succeeded in the superintendency by **Bernardin de St. Pierre,** Jacques Henri, born 1837. After France had been proclaimed a Republic, Sept. 25, 1792, the name of the «Jardin Royal» was changed into that of «Muséum d'histoire naturelle», which name this important Institution bears to this day. In the next year, at the motion of Bernardin, the animals of the royal menagerie of Versailles were transferred to the Jardin des plantes and a Library was founded in the Museum. But soon after wards the Directorship there was abolished, and Bernardin retired. The Natural History Department was reorganized, 12 chairs were established, and the body of Professors henceforth represented the Museum.

CHINESE PLANTS KNOWN AT THE END OF THE 18TH CENTURY IN THE JARDIN DES PLANTES.

The following is a list of Chinese plants cultivated in the Jardin des Plantes, according to Lamarck, Poiret and Defontaines, and of Chinese herbarium specimens described as new, or new for China in Lamarck and Poiret's Encyclop., Botanique, but without mentioning the names of the collectors. I omit the plants described by them from the collections made by d'Incarville, Poivre, Sonnerat, of which notice has already been given in the preceding chapters.

Calycanthus praecox, Lin. — Lam. Enc. Bot., II, (1783) 565. Japon, Chine. — Desf. Cat. (1804) 180. Cult. Jardin d. Pl. — *Chimonanthus fragrans*, Lindl.

Magnolia obovata, Thbg.
» *discolor*, Vent. (variety of the preceding)
» *purpurea*, Curtis. (same as the preceding)
» *pumila*, Andr.

The above four are mentioned by Desf. l. c., 150, as Chinese species cultivated in the Jard. d. Pl. They had been received from England. Cultivated at Kew: M. obovata since 1790, M. pumila since 1786. — Ind. Fl. Sin., I, 23, 24.

Brassica chinensis, Lin. This is mentioned in Aiton's Hort. Kew. (1789) II, 402, as having been introduced (into France), in 1770, by M. Richard (Gardener to the King at Versailles). It is not noticed among the plants cultivated at the Jard. d. Pl., either by Lamarck or Desfontaines. — I. F. S., I, 46.

Sinapis chinensis. Linn. According to Aiton, Hort. Kew, 1789, II, 403, introduced (into France) by M. Thouin in 1782. It is however not noticed among the plants cultivated in the Jard. d. Pl. either by Lamarck or Desfontaines.

Sinapis juncea, Lin. — Lam. Enc. Bot. IV (1796), 342. Cult. Jard. d. Pl. Desf. (1804) 128. — DC. Syst. veg. II, 612, states, that he saw an herbarium specimen of this plant, from Cheusan (China) in the herbarium of Jussieu.

Sinapis brassicata, Lin. — Cult. Jard. Pl. Lam. l. c., Desf. 128.

Raphanus sativus, var. *oleifera*, Lin. — Cult. Jard. Pl. Desf. (1804) 128.

Viola heterophylla, Poiret, Enc. Bot. VIII (1808), 646. China, herb. Jussieu. (Incarville Macao, v. supra p. 54). *Ionidium heterophyllum*, Ventenat, Jard. Malmaison (1803) p. 27. Pluken. tab. 120, fig. 8.

Dianthus chinensis. Linn. First cult. in Jard. Pl. 1705. V. supra p. 68. Lam. Enc. Bot. IV, 520.

Lychnis grandiflora, Jacq. — Lam. Enc. Bot. III (1789). Chine, Japon. Cult. au Jard. du Roi.

Hypericum chinense Lin. — Lam. Enc. Bot. IV (1796), 144. Chine, Cult. Jard. Pl.

Camellia japonica. Lin. — Lam. Enc. Bot. I (1783), 572. Chine, Japon. Cult. Jard. du Roi.

Thea Bohea and *Thea viridis* Lin., Desf. (1804) 138, cult. Jard. d. Pl.

Alcea chinensis. Hort. Reg., Lam. Enc. Bot. I, (1783), 77. Chine. Cult. Jard. du Roi. — Cavanilles, Monad. Diss. 2, (1790) p. 92, tab. 29, fig. 3, *Althaea sinensis.* — Ind. Fl. Sin., ɪ, 83: not identified.

Malva verticillata. Lin. — Lam. Enc. Bot. III (1789) 751. Chine. Cult. Jard. du Roi.

Malva sinensis, Cavanilles, l. c. Diss. 10, p. 77. Colitur in horto regio, Paris. — According to DC, Prodr. I, 432, this is a variety of *M. Mauritiana*, Lin.

Urena lobata Lin. Chine. — Poiret, Enc. Bot. VIII, 252. Cult. Jard. Pl. — Cult. in England as early as 1723. (V. supra p. 71).

Sterculia platanifolia, Lin. fil. — Cavanilles, Monad. Diss. 5 (1788), p. 288, tab. 145. Cult. in the Royal Garden, Paris.

Limonia trifoliata Lin. — Desf. 1804: cult. Jard. d. Pl. — This plant was first introduced from China into the Kew Gardens in 1787. Ait. Hort. Kew. ed. alt. III, 43.

Citrus sinensis Lin. — Desf. 1804, 138: Cult. Jard. d. Pl.

Ailantus glandulosa. Desf. — Lam. Enc. Bot. III (1789) 416. Cultivé dans nos jardins. Comp. supra p. 48.

Trichosanthes anguinea. Lin. Chine. — Desf. 1804, 208, cult. Jard. d. Pl.

Zizyphus sinensis, Hort. Reg., Lam. Enc. Bot. III (1789), 317. On cultive cette espèce depuis plusieurs années au Jardin du Roi. On la dit originaire de la Chine. — Ind. Fl. Sin., ɪ, 126, same as *Z. vulgaris*, Lam.

Rhamnus theezans, Lin. — Desf. cat. (1804) 202, cult. Jard. d. Pl.

Koelreuteria paniculata. Laxm. — Poiret, Enc. Bot. VI (1804). Chine, Cult. Jard. d. Pl. — V. supra p. 49.

Caragana Chamlagu, Lamarck, Enc. Bot. I (1783) 616. Chine. Cult. Jard. du Roi. — Robinia Chamlagu, L'Héritier, Stirp. nov. II (1783), 161, tab. 77. — According to Hort. Kew. first ed. III, 54, it was introduced (to Kew) in 1776, by Richard, gardener to the King at Versailles. — Comp. my Earl. Res. 122.

Phaseolus radiatus, Lin. — Lam. Enc. Bot. III (1789), 74. Ceylan, Chine. Cult. Jard. du Roi. — Desf. cat. 1804, 191.

Dolichos sinensis. Lin. — Lam. Enc. Bot. II (1786), 293. Chine. Cult. Jard. du Roi. — According to Aiton, Hort. Kew, first ed. 1789, III, 31, it was introduced (into the Royal Garden) by Thouin in 1776.

Gleditschia sinensis. Hort. Reg. Paris. — Lamarck, Enc. Bot. II (1786) 465. Cultivé depuis 9 ou 10 ans au Jardin du Rni. On le dit provenu de graines reçues de la Chine.

Gleditschia macracantha, Desf. catal. 1804, 182. Chine. Cult. Jard. d. Pl. — Desf., Arbres et arbrisseaux II (1809) 246. — Ind. Fl. Sin., I, 209: probably a variety of Gl. sinensis.

Cassia chinensis, Lam. Enc. Bot. I (1783), 644. Cultivé depuis peu au Jardin du Roi. On la dit originaire de la Chine. Nom chinois: *van dzian nam* seu *telo-dzin-zan* (names probably communicated by the collector who sent the seeds from China). — Jacquin, Collect. I (1786) p. 64 describes the same plant and figures it in his Icon. rar I, tab. 73. He says that it had been sent from Peking. He probably received seeds from his friend Lamarck. — Hemsley, Ind. Fl. Sin., I, 211 identifies it with *Cassia sophera.* L. The latter is communly cultivated at Peking under the Chinese name *Wang kiang nan.*

Cassia pumila, Lam., Enc. Bot. I, 651. Chine et Indes orientales. — This is not noticed in the Ind. Fl. Sin.

Adenanthera pavonina, Lin. — Lam. Enc. Bot. II (1786) 76. Chine et Molluques. — Ind. Fl. Sin., I, 214.

Amygdalus pumila Lin., or *Prunus sinensis* Pers, Desf. catal. 1804, 179. Chine, cult. Jard. d. Pl. — Ind. Fl. Sin., I, 219, *Prunus japonica,* Thbg.

Rosa semperflorens, Curtis. Bot. Mag. 284 (1794). Poiret Enc. Bot. VI (1804) 283. Chine. Cult. en France.

Rosa ternata, Poiret. l. c. 284. Chine. Cult. à Paris. — Ind. Fl. Sin., I, 250: *Rosa sinica,* Ait. or *R. laevigata,* Michx.

Cydonia sinensis. Thouin, Annales Mus. XIX (1812), 145, tab. 8, 9. Thouin says that it was introduced from Chine to the Jard. d. Pl. towards the end of the last century, and at the same time appeared in England and Holland. — *Pyrus sinensis,* Poiret. Enc. Bot. Suppl. IV, 452. — Hemsley in I. F. S., I, 256, proposes a new name: *Pyrus cathayensis.*

Pyrus spectabilis, Ait. Hort. Kew, first ed. (1789) II, 175. — Desf. cat. 1804, Chine cult. Jard. d. Pl. — Cultivated in England since 1780.

Mespilus japonica. Thbg. — Lam. Enc. Bot. IV (1796) 444. Chine, Japon. — Desf. catal. 1804, 174. Cult. Jard. d. Pl. — *Eriobotrya japonica,* Lindl.

Saxifraga sarmentosa, Lin. fil. — Poiret. Enc. Bot. VI (1804) 684, Chine, Japon. — Desf. catal. 1804, 163. Cult. Jard. d. Pl.

Hydrangea hortensis, Smith Icon. rar. t. 12 (1790). — Desf. catal.

1804. Cult. Jard. d. Pl. — It was introduced from China to England in 1789.

Kalanchoe spathulata, DC. — Desf. catal. 1804, 161. China, cult. Jard. d. Pl. — P. De Candolle published the above name only in 1810 in his "Plantes grasses", 65.

Terminalia vernix, Lam. Enc. Bot. I (1783), 350. Arbor Vernicis, Rumph. Amb. II, 259, tab. 86. — Lamarck, in stating, that this is the Chinese Varnish tree, was misled by Rumphius. As is well known the Chinese and Japanese Varnish is derived from Rhus vernicifera. — Terminalia vernix grows in the Molucca Islands.

Myrtus tomentosa, Ait. Hort. Kew, ed. 1, II, 159. Lam. Enc. Bot. IV, 411. Chine. Cultivé autrefois au Jardin des Plantes. — Desf. catal. 1804, 171. Cult. Jard. d. Pl. — Cultivated in England since 1776. — *Rhodomyrtus tomentosa*, Hassk. — I. F. S., I, 295.

Eugenia Michelii, Lam. Enc. Bot. III, (1789) 203, Indes orientales, Chine. Cultivé au Jardin du Roi. — According to DC, Prodr. III, 263 this is a plant of tropical America.

Lagerstroemia chinensis, Lam. Enc. Bot. III (1789), 375. "Chine, Japon, Molluques, Indes orientales. Cultivé au Jardin du Roi". — Same as *Lagerstroemia indica*, Lin.

Zanthoxylum trifoliatum, Lin., Lam. Enc. Bot. II (1786). Chine. Cultivé au Jardin du Roi. — *Acanthopanax aculeatum*, Seem. — Ind. Fl. Sin., I, 339.

Nauclea orientalis, Lam. Enc. Bot. IV (1796) 435. Indes, Chine. An Nauclea orientalis Lin? — Conf. DC. Prodr. IV, 345. Not mentioned in the Ind. Fl. Sin.

Hedyotis paniculata, Lam. Enc. Bot. III (1789) 79, Java (Sonnerat), peut-être aussi à la Chine. — DC. Prodr. IV, 427, refers this with (?) to *Oldenlandia alata*, Koenig. — Ind. Fl. Sin., I, 376.

Serissa foetida Willd., Poiret, Enc. Bot. VII (1806) 122. Indes orientales, Chine, Japon. Cult. Jard. d. Pl. — This is the *Lycium foetidum* of Linn. fil. (1781), who states that Thouin introduced it from China and sent it under the name of *Spermacoce fruticosa* (V. supra p. 83).

Aster chinensis. Lin. — Lam. Enc. Bot. I (1783) 308. "Reine Marguerite" des Jardins. Cette plante passe pour être originaire de la Chine. Cultivée au Jard. d. Pl. — Comp. supra p. 84. *Callistephus chinensis*.

Conyza chinensis, Lin. — Lam. Enc. Bot. II (1786) 83. Indes orient., Chine. Cult. Jard. du Roi. — *Blumea chinensis*, Hook. & Arn. V. supra p. 84.

Bidens chinensis, Lin. — Lam. Enc. Bot. I (1783) 413. Chine. Cult. Jard. du Roi.

Matricaria indica or *Chrysanthemum indicum,* Lin. — Lam. Enc. Bot. III (1789), 734. Originaire de la Chine, très commune dans les Indes orient. pour l'ornement des jardins. — Lamarck had not seen the living plant, but in the same year that he wrote the above note the Chrysanthe-mum indicum was first introduced into France, from China, by **Blancard,** a merchant of Marseilles. He then imported three different varieties of this beautiful Chinese garden flower, one with purple, one with white and one with violet flowers. He lost the two last, the first lived and was sent in 1790 to the Jardin du Roy at Paris. It was also cultivated in the Garden of Cels, who sent it to England. **Ramatuelle,** Canon at Aix, first described it in the Journ. d'Hist. nat. II (1792), 233 under the name of *Anthemis grandiflora.* Comp. Desfontaines, Hist. Arbres et Arbisseaux, 1809, I, 315.

Artemisia sinensis (chinensis), Lin. Moxa Chinensium. Desf. Catal. 1804, 96. Cult. Jard. d. Pl. Compare above p. 86.

Artemisia minima, Lin. — Lam. Enc. Bot. I (1783), 264. Chine. Cette plante a été cultivée au Jardin du Roi. Thouin introduced it into England, 1778. (Ait. Hort. Kew, 1789, III, 173). — *Myriogyne minuta,* Less. V. supra p. 86.

Artemisia capillifolia, Lam. Enc. Bot. I (1783) 267. "Cultivée au Jardin du Roi. On la croit originaire des Indes or. et de la Chine". — DC. Prodr. V, 176: L'Héritier has proved that the seeds from which this plant was raised in the Royal Garden, came from N. America, not from China.

Senecio Pseudochina Lin. — Poiret, Enc. Bot. VII, 79. Indes. — DC. Prodr. VI, 299: Thouin received it from Canton (L'Héritier mss).

Mogorium multiflorum, Lam. Enc. Bot. IV, 1796. "Chine, cote de Malabar". — DC. Prodr. VIII, 302, identifies this with *Jasminum pubes-cens,* Willd., which Roxburgh records from S. China (Fl. Indica ed. 1874 p. 31.)

Olea fragrans, Thunb. — Lam. Enc. Bot. IV (1796). "Chine, Japon".— Desf. Catal. 1804, 53, cult. Jard. d. Pl.

Convolvulus sinensis, Lam. Enc. Bot. III (1789) 557. V. s. in herb. Jussieu. — DC. Prodr. IX, 370. *Ipomoea sinensis,* Chois.

Cuscuta chinensis, Lam. Enc. Bot. II (1786) 229. Cette plante était entortillée autour d'un Basilic venu de la Chine. Cultivée en 1784 au Jardin du Roi.

Lycium chinense, Mill. Dict. n. 5., Lam. Enc. Bot. III (1789) 509. Chine. Cultivée depuis longtemps au Jardin du Roi.

Datura ferox, Lin. — Poiret, Enc. Bot. VII (1806) 460. Chine. Cult. Jard. d. Pl.

Justicia picta, Lin. — Lam. Enc. Bot. I (1783), 627. Molluques, Chine. — DC. Prodr. XI, 328, *Graptophyllum hortense,* Nees. — Not noticed for China by subsequent writers.

Vitex incisa, Lam. Enc. Bot. II (1786), 612. Cultivée au Jardin du Roi. On le dit originaire de la Chine.

Mentha perilloides, Lam. Enc. Bot. IV (1796), 112. Cette plante avait levé au Jard. d. Pl. dans des terres q'uon disait venir de Chine. — DC. Prodr. XII 163, *Perilla ocimoides,* L.

Hyssopus Lophanthus, Lin. — Lam. Enc. Bot. III (1789) 186. Chine septentrionale. Cult. au Jard. du Roi.

Chenopodium purpurascens, Hort. Reg., Jacq. Hort. Vindob. III (1776), 43, tab. 80. — Lam. Enc. Bot. I (1783), 196. Cultivé depuis longtemps au Jardin du Roi, où on le regarde comme originaire de la Chine. — In the Hort. Kew, first ed. (1789) this plant is identified with *Chenopodium atriplicis,* Lin. fil. (recorded from Siberia) and is stated there to have been introduced by Thouin from China in 1780. (I understand that Thouin sent it from Paris to England in 1780.)

Salsola canescens, Hort. Paris., Poiret, Enc. Bot. VII (1806) 287. "Cultivée au Jardin des Plantes. On la soupçonne originaire de la Chine".— Desf. Catal. 1804, 41. — In DC. Prodr. XIII, 2, p. 137 it is referred to *Echinopsilon diffusus,* which, however, has not been recorded from China.

Rheum palmatum, Lin. — Poiret, Enc. Bot. VI (1804) 193. Chine, Cult. Jard. d. Pl.

Daphne sinensis, Lam. Enc. Bot. III (1789) 438. An Daphne indica, Linn.? Chine. Cultivée au Jard. du Roi. — In DC. Prodr. XIV, 537 it is considered a distinct species. Hemsley in I. F. S., II, 395 reduces it to *Daphne odora,* Thunb.

Ficus pertusa, Lin. fil. Suppl. 442. — Lam. Enc. Bot. II (1786) 496. Martinique, Surinam, Isle de France. Il croît aussi à la Chine. —

Urtica nivea, Lin. — Lam. Enc. Bot. IV (1796), 643. Chine. Cult. Jard. d. Pl. — *Boehmeria nivea,* Hook. & Arn.

Thuya orientalis, Lin. — Poiret, Enc. Bot. VII (1806), 640. Chine. Depuis long temps cultivé en Europe.

Gingko biloba, Lin. — Lam. Enc. Bot. II (1786), 712. Japon, Chine. Cultivé depuis plusieurs années en France, en Angeterre etc.

Amomum Zingiber, Lin. — Lam. Enc. Bot. I (1783), 134. Malabar, Ceylan, Amboine et Chine.

Moraea (Ixia) chinensis, Thunb. — Lam. Enc. Bot. IV (1796), 274. Japon, Chine. Cult. Jard. d. Pl.

Convallaria japonica, Thunb. — Lam. Enc. Bot. IV (1796) 367. Chine, Japon. Cult. Jard. d. Pl. — Kunth, Enum. Pl. IV, 70, *Aneilema japonicum.*

Hemerocallis plantaginea, Lam. Enc. Bot. III (1789), 103. Cultivée depuis peu d'années au Jardin du Roi. Envoyée de la Chine par M. De Guignes. (Ch. L. Jos. De Guignes, French Consul at Macao, 1784—1801. V. infra Braam's Embassy). — In Schultes', Syst. Veg. VII, 421, this is identified with *Funkia subcordata*, Spreng.

Aletris chinensis, Lam. Enc. Bot. I (1783) 79; *Dracaena terminalis*, ibid. II, 324. Cult. Jard. d. Roi. Comp. supra p. 120.

Latania chinensis, Desf. Cat. 1804, 19. Cult. Jard. d. Pl. Comp. supra p. 119.

Cyperus Pangorei, Retz, Obs. Bot. IV, 10. — Poiret, Enc. Bot. VII, 261. Indes, Chine. V. s. herb. Lamarck. — Retz records it only from Tranquebar.

Polypodium dichotomum, Thbg. — Poiret, Enc. Bot. V (1804) 543. Indes, Chine, Japon. V. s. in herb. Lamarck. — *Gleichenia dichotoma* Hook. Sp. Filic. I, 12. — Benth. Fl. hongk. 442.

IX. Chinese Plants cultivated at Vienna in the second half of the 18th Century and described and illustrated by N. J. Jacquin.

Jacquin, Nicol. Joseph, Baron von. This distinguished botanist was born at Leyden, in 1727, studied medicine at Antwerp and Leyden, visited Paris and finally, in 1752, settled at Vienna. In 1755 the Emperor Francis I sent him, with the gardener R. van der Schott to the West Indies, to procure living plants for the newly (1753) established Imperial Garden at Schoenbrunn. They returned with a vast collection of rare tropical plants, in 1759. Subsequently Jacquin was appointed Professor of Chemistry and Botany at Schemnitz and in 1768 he took charge of the same professorships at the Vienna University and accordingly became Director of the Botanic Gardens of the University and afterwards of the Imperial

Gardens at Schoenbrunn. In the same year he described and figured a Chinese plant, the *Rosa chinensis*, in his Observationes Botanicae, III, p. 7, tab. 55 (black). He thought it was new, but subsequently reduced it to Linnaeus' *Rosa semperflorens* (v. infra). Jacquin introduced to the botanic garden many interesting exotic plants which still exist there, as for instance several beautiful specimens of the *Gingko biloba* and *Koelreuteria paniculata*, both Chinese trees, hardy in the climate of Vienna.

From 1770—76 Jacquin published the Hortus Botanicus Vindobonensis, in fol., 3 volumes, 300 coloured plates representing for the greater part exotic plants cultivated in the botanic garden, with descriptions. The following Chinese plants are illustrated in the Hort. Vindob:

Malva verticillata, Lin. I (1770) p. 15, tab. 40.

Urtica nivea, Lin. II (1772) p. 78, tab. 166.

Sinapis juncea, Lin. II, p. 80, tab. 171. Asia, China.

Hyssopus Lophanthus, II, 85, tab. 182. In China septentrionali sponte crescit.

Capsicum sinense, n. sp., III (1776) 38, tab. 67. A patria nomen stirpi indidi. In insula Martinica cultam vidi, fructusque in usum culinarem adhibitos. — Sonnerat brought the same from China. (V. supra p. 122).

Dolichos sinensis, Lin., III, 39, tab. 71.

Cucumis acutangulus, Lin., III, 40, tab. 73.

Chenopodium purpurascens, n. sp. III, 43, tab. 80. Jacquin received the seeds from Jussieu. In China sponte crescit. — Comp. supra p. 133.

Among the expeditions sent out subsequently by imperial order to various parts of America to complete the collection of living plants in the Garden of Schoenbrunn, I may mention that of the gardeners Francis **Boos** and **G. Scholl**. who on their way home, in 1786, visited the Cape of G. H. and made a rich harvest there. From thence Boos proceeded to Mauritius (Isle de France) and Bourbon, and returned to Vienna in 1788. Among the rare plants from these islands he brought also some plants of Chinese origin, cultivated there.

Jacquin kept up a correspondence with almost all the botanical authorities of his time in France, England etc, who used to send him seeds of rare exotic plants, especially Jussieu, and Lamarck, who knew him personally. Lamarck met him in Vienna, in 1782. See Lam. Enc. Bot. I, p. XXXIV.

From 1781—1793 Jacquin brought out the Icones Plantarum Rariorum, in 3 volumes, with 648 coloured plates. Descriptive and other details regarding the plants illustrated there, are found in Jacquin's Collectanea ad Botanicam Spectantia, 5, volumes, 1786—1796. In these works the following Chinese species are illustrated and described:

Cassia chinensis. Icon. rar. I (1786) tab. 73; Collect. I (1786) 64. Ex Peking transmissa. — Comp. supra p. 130. *C. chinensis*, Lam. — Jacquin, it seems, received the plant or the seeds, from Lamarck.

Saxifraga stolonifera, Jacq. Icon. rar. I, tab. 80; Jacq. Miscell. austr. II (1781) 327. — This is the same as *Saxifraga sarmentosa*, which Lin. fil. received from Japan. Jacquin received it from India. But it is a native of China.

Lychnis grandiflora, Jacq. Icon. rar. I, tab. 84; Collect. I (1786) p. 149. Habitat not mentioned. The plant is known to be a native of China and Japan. According to the Hort. Kew (1789) II, 117, this is the same as *Lychnis coronata*, published in 1784 in Thunberg's Flora jap. 187.

Dracaena terminalis. Jacq. Icon. rar. II (1790) t. 448; Collect. II (1788) 354. Jacquin seems to have received it from Mauritius, but it is of Chinese origin. Comp. Kunth, enum. pl. V, 23—25, *Cordyline Jacquini.*

Volkameria Kaempferi, Jacq. Icon. rar. III (1793) tab. 500; Collect. III (1789) 207. Cult. in hort. Schoenbrunn, ex insula Mauritii transvecta. — It is a native of China and Japan. *Clerodendron squamutum*, Vahl. Symb. II, 74. — Ind. Fl. Sin., II, 262.

Camellia japonica, Linn. — Jacq. Icon. rar. III, tab. 553; Collect I. (1786) 117. China. Japan.

Panax aculeatum, Ait. Hort. Kew III, 448, Zanthoxylum trifoliatum Linn. — Jacq. Icon. rar. III, tab. 634; Collect. IV (1790) 175. E China ortum. Cultum in horto Schoenbrunn. — Ind. Fl. Sin., I, 339, *Acanthopanax aculeatum*, Seem.

In 1791 Jacquin was entrusted by the Emperor Leopold II with the superintendence of the Imperial Garden at Schoenbrunn and from 1797—1804 he issued 4 volumes with 500 coloured plates entitled: Plantarum Rariorum Horti Caesarei Schoenbrunnensis Descriptiones et Icones. In this the following Chinese plants are depicted and described:

Amaryllis aurea, Ait. Kew, I 419. — Pl. rar. Schoenbr. I, p. 38, tab. 73. China.

Cookia punctata, Sonnerat iter II, p. 181, tab. 130. — Pl. rar. Schoenbr. I, p. 53, tab. 101.

Tetranthera laurifolia, Jacq. Pl. rar. Schoenbr. I p. 59, tab. 113. Sponte in China, colitur in insula Mauritii unde ad nos adlata fuit sub titulo "Cerisier de la Chine". — Comp. supra p. 119, Lamarck's *Litsea chinensis.*

Ulmus parvifolia, Jacq. l. c. III (1798) p. 6, tab. 262. Patria ignota. — Desfontaines, Tableau du Mus. etc, 1815, p. 242, identifies this with *Ulmus chinensis,* Persoon, Syn. I (1805) 67. According to DC. Prodr. XVII, 161, Ulmus parvifolia is a native of S. China and Japan.

Rosa semperflorens, Linn., Rosa chinensis, Jac. Obs. III, 7, tab. 55 (v. supra p. 135). Jacq. Pl. rar. Schoenbr. III (1798) p. 17, tab. 281. China.

Erigeron chinense, Jacq. Pl. rar. Schoenbr. III, p. 30, tab. 303. The seeds of this plant were sent to Jacquin from the Cape of G. H. with a notice that they had been received from China.

Rhapis flabelliformis, Ait. Kew, III, 473. — Jacq. l. c. III, p. 36, tab. 316. — Japan, China.

Volkameria japonica, Jacq. l. c. III, p. 48, t. 338. Sub hoc titulo in Germania colitur. Japonia. — Ait. Hort. Kew, IV (1812) 63, identifies this with *Clerodendron fragrans,* Ventenat, Malmais, (1803) tab. 70. Japan, China. — Ind. Fl. Sin., II, 260.

Daphne odora, Thbg. Fl. japon. 159, Ait. Hort. Kew, 1789, II, 26.— Jacq. l. c. III, p. 54, t, 351. Japan. — Ait. Hort. Kew records it from Japan and China. — V. supra p. 133. *Daphne sinensis,* Lam.

Sophora japonica, Lin. — Jacq. l. c. III, p. 55, t. 353. Japonia. Apud nos (Vienna) sub dio florens. — Sophora jap. was introduced into Europe from N. China. V. supra p. 48.

Vitex incisa, Lam. — Jacq. l. c. IV (1804) p. 14, tab. 427. Jacq. stated that this plant had been cultivated at Vienna for 40 years. He thinks it may perhaps be of N. American origin. But Jacquin is mistaken, V. incisa was introduced into Europe from China. V. supra p. 133.

We come now to the last of Jacquin's iconographical works representing interesting new plants, which is entitled: FRAGMENTA BOTANICA FIGURIS ILLUSTRATA, 1800—1809, 138 coloured plates in 6 fascicles. Dedicated to Sir Jos. Banks. In this, four new Chinese plants cultivated in the Schoenbrunn Garden are described and three of them illustrated.

Hypericum monogynum, Lin. — Jacq. l. c. p. 11. China. No figure. The plant was first figured by Miller. V. supra p. 108.

Latania chinensis, Jacq. l. c. p. 16, tab. 11, fig. 1 (only a part of the leaf represented). Jacq. states that this plant was brought ten years ago (1788) from Mauritius (by Boos v. supra p. 135) where it is known under the name of "Latanier chinos". — Comp. also supra p. 119, *Latania borbonica.*

Rosa bracteata, Wendl. — Ventenat, Hort. Cels, 1800, p. 28, tab.

28. — Jacq. l. c. p. 30, tab. 34, fig. 2. Crescit in China. (brought by Lord Macartney in 1794).

Bocconia cordata, Willd. Sp. pl. II, 841. — Jacq. l. c. p. 63, tab. 91, fig. 1. In China spontanea. — Introduced by Sir G. Staunton from China in 1795. See Bot. Mag. t. 1905.

At the age of 70, in 1797, Jacquin was pensioned, and in recognition of his important services to botany, the emperor made him a Baron in 1806. He died at Vienna, 24 Oct. 1817.

X. Chinese Plants cultivated in the Royal Kew Gardens in 1789.

The spot that now is occupied by the Royal Gardens of Kew, London, was originally private property with a residence called Kew House and in 1730 belonged to Mr. Molyneux, Secretary to King George II, when the son of the latter, the Prince of Wales, admiring the situation of Kew House, took a long lease of it. He then began to form the pleasure grounds and laid the foundation of the renowned Botanical Gardens. In 1759 Mr. **William Aiton,** born in 1731, a pupil of the celebrated Philip Miller of the Chelsea Physic Garden, was placed in charge of the Kew Gardens, which charge he retained till his death, in 1793. In about the year 1789 King George III purchased Kew House. (Comp. Kew Bulletin etc. 1891, 279 seqq., Historical Account of Kew to 1841).

Aiton was known for his knowledge of plants and his great skill in cultivating them. Under his direction the Kew Gardens soon became the grand repository of all the vegetable riches which could be accumulated with regal munificence by researches in every quarter of the globe. In 1789 he published the HORTUS KEWENSIS or Catalogue of the Plants cultivated in the Royal Gardens, Kew. This elaborate and important work, in the preparation of which Aiton was aided by the Swedish naturalists **Solander** and **Dryander,** appeared in 3 volumes, the plants according to the Linnaean arrangement, with 13 plates. It gives an account of the several foreign plants which had been introduced into the English Gardens at different times. A second edition of it was brought out by W. Aiton's son in 5 volumes, 1810—13. Of this we shall have to speak in another chapter.

The following list of Chinese plants is extracted from the first edition of the Hortus Kewensis with all the details regarding their introduction and description as given in that catalogue. My own observations are added in brackets.

II, 253, *Anona hexapetala*, Linn. Suppl. 270. Native of China and the East-Indies. Introduced 1758 by Hugh Duke of Northumberland. — Second. ed. III, 353. The same. Added: Willd. Sp. II, 1266. — (Comp. supra p. 65).

II, 402, *Brassica chinensis*, Lin. Sp. 932. China. Introduced 1770 by Mr. Richard (v. supra p. 66). — Second edition IV, 124. The same. Added: Willd. Sp. III, 550.

II, 403, *Sinapis chinensis*, Linn. Mant. 95. China. Introduced 1782 by Thouin (v. supra p. 67). — Second ed. IV, 126. The same. Added: Willd. Sp. III, 557, and Arduin Animadv. Bot. I (1759) 23, tab. 10. Sinapi chinensis, folio acanthi, siliquis glabris.

III, 404, *Sinapis juncea*, Lin. Sp. 934. China. Jacq. Vind. II, 80, t. 171. — Cult. 1731 by Philip Miller. Gard. Dict. ed. 1, n. 3. — H. K. 2d edition. IV, 126, adds: Willd. Sp. III, 557. — (Petiver, pl. raised at Chelsea). Cult 1710. Philos. Trans. n. 325, p. 49, n. 15. (not said that the seeds came from China).

II, 90, *Dianthus chinensis*, Lin. Syst. veg. 415. China. — Curt. Bot. Mag. 25 (1786). — Cult. before 1719. Philip Miller, Icon. 54. — H. K. 2d ed. III, 80, adds: Willd. Sp. II, 677. Cult. before 1713 by Thomas Fairchild*). Philos. Trans. n. 337, p. 62, n. 102 (Petiver, rare plants raised at Chelsea.)

II, 117, *Lychnis coronata*, Thbg., Fl. jap. (1784) 187. — *Lychnis grandiflora*, Jacq. Collect. I (1786) 149; Icon. rar. I, tab. 80. — L'Héritier, Stirp. nov. II, tab. 18. It seems that L. coronata is figured on L'Héritier's unpublished plates). Introduced in 1774 by Dr. John Fothergill. — China, Japan. — H. K. 2d edit. III, 133, adds: Willd. Sp. II, 808; — Curt. Botan. Mag. 223; — Agrostemma Banksia, Meerburg ic. (1775) 2, t. 28; — Hedona sinensis, Loureiro Cochin. 286.

---

*) Thomas **Fairchild**, 1667—1729, Gardener of the city gardens, Hoxton (J. Britten, Biogr. British Botanists).

III, 108. *Hypericum monogynum*, Lin. Sp. 1107. China. Introduced in 1753 by Hugh Duke of Northumberland.

H. K. 2d edit. IV, 421, adds: Willd. Sp. III, 1442. — Curtis Bot. Mag. 334 (1796).

II, 460. *Camellia japonica*, Linn. Sp. 982. — China, Japan. — Curt. Bot. Mag. 42 (1787). — Jacq. Icon. rar. III, tab. 553; Collect. I (1786) 117. — Cult. before 1742 by Rob. **James Lord Petre,** *). — Edwards, Birds, (1747) vol. 2, p. 67. — H. K. 2d edit. IV, 235, adds: Willd. Sp. III, 842; — Cult. before 1739 by Lord Petre, Coll. MSS; — 11 varieties enumerated.

II, 230. *Thea Bohea*, Lin. Sp. 734.

α, var. laxa, foliis elliptico-oblongis rugosis, — β, stricta, foliis lanceolatis, planis. — China, Japan. — Introduced 1768 by John Ellis.

H. K. 2d edit. III, 303, adds: Willd. Spec. II, 1180. — Var. α in first ed. is *Thea chinensis*, var. β in Bot. Mag. 998 (1807).

II, 446. *Alcaea rosea*, Lin. Sp. 966. — China. Gerard, the Herball (1597), 782, fig. 1, cult. in England (native country not mentioned). — J. Miller's Illustrations (1775): Alcea rosea.

H. K. 2d edit. IV, 209, name changed into *Althaea rosea*, Willd. Sp. III, 773. — China. Cult. in 1573, Th. Tusser, 500 points of Good Husbandry, fol. 40, n. 14 (nat. country not mentioned). — According to Bot. Mag. t. 3198 (1832) Althaea rosea was introduced into Europe from China, in 1753.

II, 449. *Malva verticillata*, Lin. Sp. 970. — China. — Jacq. Hort. Vindob. I (1770) p. 15, tab. 40. — Cult. in 1683 by James Sutherland, Hort. Edinb. 220, n. 1. (native country not mentioned).

H. K. 2d edit. IV, 217, adds: Willd. Sp. III, 785.

II, 452. *Urena lobata*, Lin. Sp. 974. — China. — Cult. in 1732 by James Sherard, M. D., Dillen. Hort. Elth. 430, t. 319, fig. 412.

H. K. 2d edit. IV, 222, adds: Willd. Sp. III, 800. — Cult. 1731 by

---

*) According to Britten, Biogr. British etc. Botanists, 1888—91, Lord Petre lived 1713—43, was Fellow of the Roy. Soc. He had a garden with large stoves under Ph. Miller's supervision, at Thorndon, Essex.

Philip Miller, R. S. n. 474 (Plants from Chelsea, presented to the Roy. Soc., enumerated in the Philos. Trans.)

II, 457. *Hibiscus Manihot,* Lin. Syst. veg. 630. — China, Japan. — Cult. in 1712 in the Chelsea garden, Philos. Trans. n. 333, p. 417, n. 64 (Petiver, rare plants grown in the Chelsea Garden).

H. K. 2d edit. IV, 229, adds: Willd. Spec. III, 825. — Ketmia folio Manihot serrato etc. Dillen. Elth. (1732), 189, t. 156, fig. 189.

III, 378. *Sterculia plantanifolia,* Lin. Suppl. 423. — Japan, China.— L'Héritier, Stirp. nov. II, tab. 93. (compare supra p. 126). — Cult. 1757 by Hugh, Duke of Northumberland. — Firmania, Marsili in Act. Patav. I (1786) 106, tab. 1. 2. — Outom chu, Lecomte, Mém. Chin I, 241. — Du Halde, Chine, II, 149, fig. in tab. ad pag. 154.

H. K. 2d edit. V, 339, adds: Willd. Sp. IV, 873.

III, 101. *Citrus sinensis,* Malus Arantia cortice dulci eduli, Bauhin Pin. 436. This is the var. β. of Linnaeus' Citrus aurantion, the bitter or Seville Orange, which Bauhin calls Malus Arantia major. — India. — Cult. 1629, Parkinson, Paradis, terr. 584.

H. K. 2d edit. IV, 420, adds: Willd. Spec. III, 1427. — Bishop Gibson, additions to Camden's Britannia, 1695, p. 165, states, that the first Orange trees were introduced to England by Francis Carew, before 1595 at Beddington in Surrey. Conf. Aubrey's Surrey, II, 160.

III, 443. *Ailantus glandulosa,* Desfont. in Act. Paris, 1786, 265, tab. 8. — China. — Rhus cacodendron, Erhart in Hannov. Magaz. 1783, p. 227. — Rhus sinensis foliis alatis etc., Ellis in Phil. Trans. XLIX, (1757) p. 870, t. 25, fig, 5 and vol. L, p. 446, t. 17. — Introduced about 1751 by d'Incarville, Ellis, l. c.

H. K. 2d edit. adds: Willd. Spec. IV, 974. — L'Héritier, Stirp. nov. I, 179, t. 84.

II, 7. *Koelreuteria paniculata,* Laxmann, Nov. Comm. Petrop. XVI (1772) p. 561, t. 18. — China. — Sapindus chinensis, Linn. Suppl. 228.— Koelreuteria paulinoides, L'Héritier, Sert. Angl. tab. 19. — Introduced about 1763 by George W. Earl of Coventry.

H. K. 2d edit. II, 351, adds: Willd. Sp. II, 330.

I, 366. *Rhus succedaneum,* Linn. Mant. 221. — China, Japan. — Introduced in 1773 by John Blake.

H. K. 2d edit. II, 163, adds: Willd. Spec. I, 1479. — Cult. by Ph. Miller in 1768. Mill. Dict. ed. 1768. Toxicodendron. n. 11.

III, 489. *Rhus semialatum,* Murray, Comm. Goett. 1784, p. 27, t. 3. — Native of Macao. David Nelson. Introduced in 1780 by Sir Joseph Banks.

H. K. 2d edit. II, 163, adds: Willd. Spec. I, 1479.

III, 54. *Robinia **Chamlagu,*** L'Héritier, Stirp. nov. (1784) p. 161, tab. 77. — China. — Introduced, 1773, by M. Richard (I understand Richard sent it from Paris to England. Comp. supra p. 129).

H. K. 2d edit. IV, 325, adds: Willd. Sp. III, 1138.

III, 31. *Dolichos sinensis,* Lin. Sp. 1018. — Jacq. hort. Vind. III, p. 39, t. 71. — Introduced, 1776, by Mr. Thouin (I understand the plant or the seeds were sent by Thouin from Paris to England). — Native of the East Indies.

H. K. edit. 2, IV, 292, adds: Willd. Sp. III, 1038.

II, 45. *Sophora japonica,* Lin. Syst. veg. 391. — Japan. — Introduced, 1753, by James Gordon.

H. K. 2d edit. III, 2, adds: Willd. Sp. II, 500. — Andrews, Repos. 585 (1809). — Japan, China.

III, 444. *Gleditschia horrida.* — North America. — Cult. 1700 by Bishop Compton*). Plukenet Almag. bot. Mantissa, 1.

H. K. 2d ed. V, 475: Willd. Sp. IV, 1098. — China. — Cult. 1774 by Messrs Kennedy & Lee.

(DC. Prodr. II, 479, same as *Gl. sinensis,* Lam. — Ind. Fl. Sin., I, 209.)

II, 203. *Rosa sinica,* Lin. Syst. veg. (ed. 14). 474. — China. — Cult., 1759, by Ph. Miller. — H. K. 2d ed. III, 261, the same.

(The R. sinica cultivated and described by Aiton is different from Linnaeus' R. sinica. See J. E. Smith in Rees' Cyclopedia. — Lindl. Ros.

---

*) **Compton,** Hon. Henry, 1632—1713, Bishop of Oxford 1674, of London, 1675, friend of Ray. He had a fine garden at Falham. (J. Britten, Biogr. Brit. Botanists).

(1821) p. 106, refers Linnaeus' plant to R. indica, and retains the name R. sinica for Aiton's plant. — I. F. S., I, 250, *R. laevigata,* Mchx.).

II, 175. *Pyrus spectabilis,* n. sp. — China. — Cult., 1780, by Dr. John Fothergill.

H. K. 2d ed. III, 208: Willd. Sp. II, 1018. — Curt. Bot. Mag. 267 (1794). — Schneevogt icon. pl. rar. (1793), 15.

II, 79. *Saxifraga sarmentosa,* Lin. Suppl. 240. — China, Japan. — Schreber, Monogr. Dioneae (?) p. 16, t. 2, 3. — Thunb. Fl. jap. 182. — Saxifraga stolonifera, Meerburg icon. (1775) 23. Jacq. Icon. rar. I, tab. 80; Miscell. II, 327. — Saxifraga ligulata, Murray, Comm. Goett. 1781, p. 26, t. 1. — Introduced, 1771, by Bemjamin Torin.

H. K. 2d ed. III, 67, adds: Willd. Sp. II, 646. — Curt. Bot. Mag. 92 (1789).

II, 159. *Myrtus tomentosa,* n. sp. — China. — Arbor sinensis Canellae folio minore, Pluken. Amalth. 21, t. 372, fig. 1. — Introduced about 1776 by Mrs Norman.

H. K. 2d ed. III, 189, adds: Willd. Sp. II, 968; — Curt. Bot. Mag. 250 (1794).

(This is *Rhodomyrtus tomentosa,* Hasskarl in Flora, 1842, Beibl. II, 35. — I. F. S., I, 295).

II, 230. *Lagerstroemia indica,* Lin. Sp. 734. — East Indies. — Introduced 1759 by Hugh, Duke of Northumberland.

H. K. 2d ed. III, 302 adds: Willd. Sp. II, 1178; — Curt. Bot. Mag. 405 (1798).

(Aiton gives India as the native country of L. indica, but Roxburgh, Fl. Ind. (ed. 1874) 404, expressly states, that it is an exotic from China, as does also Linnaeus).

III, 379. *Trichosanthes anguinea,* Lin. sp. 1432. — China. — Cult. 1755 by Philip Miller. — Miller icon. 21, tab. 32.

H. K. 2d ed. adds: Willd. Sp. IV, 598. — Bot. Mag. 722 (1804).

III, 383. *Cucumis acutangulus,* Lin. Syst. veg. 869. — India. — Cult. 1692 by Bishop Compton *). — Jacq. hort. Vind. III, 40, t. 73, 74.

---

*) v. supra p. 142, note.

H. K. 2d ed. V, 345 adds: Willd. Sp. IV, 612; — Pluken. Phyt. t. 172, fig. 1.

(Linnaeus states that this plant is a native of Tataria and China. Roxburg, Fl. Ind. 698, *Luffa acutangula,* says that it is commonly cultivated in India, but he never saw it in a wild state. In China also it seems to be met with only as a cultivated plant.)

I, 350, *Sium Sisarum,* Lin. Sp. 361. Native country? In the 2d ed. II, 144, Aiton says that it is a native of China. But this is an error. Comp. supra p. 82.

III, 448. *Panax aculeatum,* L'Héritier, Stirp. nov. II, tab. 99. — Zanthoxylum trifoliatum, Lin. Sp. 1455. — China. — Cult. 1773 by Dr. John Fothergill.

H. K. 2d ed. V, 482, adds: Willd. IV, 1125; — Jacq. Icon. rar. III, tab. 634.

(*Acanthopanax aculeatum,* Seem. — See I. F. S., ı, 339).

I, 293. *Gardenia florida,* Lin. Sp. 305. — Thbg. Gardenia (1780) n. 2.   α. flore simplici, β. flore pleno. — China, Japan, Cochinchina, South Sea Islands. — Introduced about 1754 by Captain Hutchinson of the "Godolphin" Indiaman, from the Cape of G. H. (See Martyn-Miller Gard. Dict.). — Philos. Trans. vol. LI (1759) p. 933, tab. 23, (article on Gardenia by J. Ellis).

H. K. 2d ed. I, 368, adds: Willd. Spec. I, 1225.

III, 184. *Conyza patula,* sp. nov. — Serratula foliis oblongo-ovatis obtuse dentatis etc. Ph. Miller, Icon. 165, tab. 247. — China.

H. K. 2d ed. adds: Willd. Spec. III, 1919. — Cult. before 1758 by Ph. Miller.

(*Vernonia chinensis,* Lessing. V. supra p. 109).

III, 208. *Aster chinensis,* Lin. sp. 1232. — China. — Introduced 1731 by Ph. Miller, Dict. ed. 8.

H. K. 2d ed. adds: Willd. Sp. III, 2038.

III, 343. *Xanthium orientale,* Lin. sp. 1400. — China, Japan, Ceylon. Cult. 1713 by Bishop Compton *), Philos. Trans. n. 337, p. 57, n. 84.

---

*) v. supra p. 142, note.

H. K. 2d ed. V, 268, adds: Willd. Sp. IV, 373. — Introduced 1685 by Sir Hans Sloane. — Ray, Hist. Pl. I (1686), 165, n. 2 (Lappa canadensis minori congener). (These references seem to be wrong).

III, 244. *Siegesbeckia orientalis,* Lin. Sp. 1269. — India, China. — Cult. 1730 by Ph. Miller, R. S. n. 424. (Plants raised at Chelsea and presented to the Royal Soc.).

H. K. 2d ed. V, 119, adds: Willd. Sp. III, 2219.

III, 173. *Artemisia minima,* Lin. Sp. 1190. — China. — Introduced 1778 (from Paris) by Thouin.

H. K. 2d ed. V, 103, the same plant named *Cotula minima.* — Artemisia minima, Burm. Ind. 177, t. 58, fig. 2. — Cult. 1768 by Philip Miller, Dict. ed. 8, Artemisia, 5.

(*Myriogyne minuta,* Lessing. V. supra p. 86).

I, 14. *Olea fragrans,* Thunberg, Fl. jap. 18, tab. 2. — Lin. Syst. veg. 57. — Quae fa, Osbeck iter, 250. — Cochinchina, China, Japan. — Introduced, 1771, by Benjamin Torin.

H. K. 2d ed. I, 22, adds: Willd. Sp. I, 46.

I, 257. *Lycium barbarum,* Lin. Var. β. *chinense.* — Lycium chinense, Ph. Millers's Dict. — Europa, Asia, Africa. — Cult. 1709 by the Dutchess of Beaufort. Br. Mus., Sloanian hortus siccus, 137, fol. 54.

H. K. 2d ed. II, 3. *Lycium chinense,* Willd. Sp. I, 1059. — Cult. 1696 in the Royal Gardens, St. James. Pluken. Almag. 317.

I, 238. *Datura ferox,* Lin. Sp. 255. — China. — Cult. 1731 by Ph. Miller, Dict. ed. 1, Strammonium, 3.

H. K. 2d edit. I, 387: Willd. Sp. I, 1107. — Strammonium ferox, Zanon. Hist. 212, t. 162.

I, 241. *Nicotiana fruticosa,* Lin. Syst. veg. 221. — China. — Cult. before 1699 by the Dutchess of Beaufort *). Br. Mus. Sloane MSS, 525, 3349.

---

*) This is, it seems, the Dowager **Dutchess of Beaufort**, mentioned in Jackson's Life of the great botanist W. Sherard (Journ. Bot. 1874, 131). Sherard, in 1700 was appointed tutor to her son Henry. Her garden then was the first in the kingdom. Comp. also J. Britten, Biogr. of British Botanists: Mary, Duchess of Beaufort, 1630—1714. Garden at Badminton.

H. K. 2d ed. I, 390: Willd. Sp. I, 1014.

(This is not a Chinese, but a South American plant. V supra p. 90).

II, 364. *Clerodendron fortunatum*, Lin. Syst. veg. 578. — East Indies. — Introduced 1784 by Messrs Kennedy & Lee. — Not noticed in the second ed. of H. K.

(Cl. fortunatum is a Chinese plant not met with in India. See Hooker's Flora of Brit. India, IV, 596).

II, 284. *Hyssopus Lophanthus*, Syst. veg. 529. — Siberia. — Cult. 1759 by Ph. Miller. Dict. ed. 7, n. 5. — Jacq. hort. Vind. II, 85, tab. 182.

H. K. 2d ed. III, 376, adds: Cult. 1752 by Miller. Plants raised in Chelsea Garden, presented to the Royal Soc. n. 1531. (V. supra p. 93).

I, 287. *Celosia argentea*, Lin. Sp. 296. — China. — Cult. 1714 by the Dutchess of Beaufort *). Br. Mus. Slonian hortus siccus, 131, fol. 70.

H. K. 2d ed. II, 58, adds: Willd. Spec. I, 1197.

I, 287. *Celosia coccinea*, Lin. Sp. 297. — China. — Cult. 1597. Gerard's Herball (Amarantus coccineus, figured, native country not noticed).

H. K. 2d ed. II, 58, adds: Willd. Sp. I, 1199; — Amarantus panicula speciosa cristata, Bauhin. Hist. Plant. II (1650) p. 969.

(C. coccinea, in DC. Prodr. XIII, 2, p. 242, is considered a variety of *Celosia cristata*, Lin.)

III, 348. *Amarantus tristis*, Linn. Sp. 1404. — China. — Cult. 1759 by Ph. Miller, Dict. ed. 7, n. 3.

H. K. 2d ed. V, 273, adds: Willd. Sp. IV, 385; — Willd. Hist. Amaranth. (1790), 21, tab. 5.

III, 350. *Amarantus cruentus*, Lin. Sp. 1406. — China. — Cult. 1728. Martyn Hist. Pl. rar. Dec. I, p. 6, tab. 6, (Amarantus sinensis etc.).

H. K. 2d ed. V, 277: Willd. Spec. IV, 392.

(I. F. S., ii, 320, = variety of *A. paniculatus*, Lin.

I, 311. *Chenopodium atriplicis*, Lin. Suppl. 171 (recorded by Pallas

---

*) V. supra p. 145, note.

from Siberia). — Chenopodium purpurascens, Jacq. hort. Vind. III, p. 43,
t. 80. — China. — Introduced 1780 by Thouin (from Paris into Kew
Gardens).

H. K. 2d ed. II, 98: Willd. Sp. I, 1300.

(Not noticed in the Ind. Fl. sin.).

II, 31. *Polygonum tinctorium*, n. sp. — China. — Introduced, 1776
by John Blake.

Hort. Kew 2d ed. II, 418: Willd. Sp. II, 445.

(In 1790 (resp. 1788) Loureiro Fl. cochin. 241, described the same
plant from Canton and gave it the same name).

II, 41. *Rheum undulatum*, Lin. Sp. 531. — Rheum Rhabarbarum,
Lin. Syst. veg. 385. — China, Siberia. — Cult. 1759 by Ph. Miller, Dict.
ed. 7, n. 2.

H. K. 2d ed. II, 430, adds: Introduced 1734. Ph. Miller icon.
II, p. 146.

II, 41. *Rheum palmatum*, Lin. Sp. 531. — China. — Cult. 1768 y
Ph. Miller, Dict. ed. 8.

H. K. 2d ed. II, 431. *Rh. palmatum*, Lin. Native of Bucharia. Introduced 1763 by John Hope, M. D., Philos. Trans. LV, 1765, p. 290,
t. 12, 13.

II, 42. *Rheum compactum*, Lin. Syst. veg. 385. — Tartary. — Introduced 1779 by Thunberg.

H. K. 2d ed. II, 431, adds: Cult. before 1758 by Ph. Miller. Icon.
II, p. 146.

I, 160. *Chloranthus inconspicuus*, Swartz in Philos. Trans. LXXVII
1787, p. 359, tab. 14. — L'Héritier, Sert. angl. 1788, t. 2. — Ni-
grina spicata, Thbg. japon. 65. — China. — Introduced 1781 by Dr. James Lind.

II. K. 2d ed. I, 270, adds: Willd. Sp. I, 688. Tea leaved Chloranthus, *chu lan*.

II, 26. *Daphne odora*, Icon. Kaempf. t. 16. — L'Héritier Stirp. nov. II,
t. 7, (probably one of the unpublished plates. V. supra p. 126). — Thunb.
Fl. japon. 159. — China, Japan. — Introduced 1771 by Benjamin Torin.

H. K. 2d ed. II, 411, adds: Willd. Spec. II, 421. — Smith, Exot. Bot. I (1804) 91, t. 47. — Jacq. Hort. Schoenbr. III, 54, t. 371.

III, 375. *Croton sebiferum*, Lin. Syst. veg. 863. — China. — Introduced 1755 by Hugh Duke of Northumberland.

H. K. 2d ed. V, 337, *Stillingia sebifera*, Willd. Sp. IV, 588. — Pluken. Amalth. 76, t. 390. (Arbor sinensis sebifera kieou-you, Amalth. 25, Phytogr. t. 390, fig. 2). — Cult. 1703 by Mr. Cole. Philos. Trans. n. 286, p. 1427, n. 90.

III, 342. *Morus alba*, Lin. Sp. 1398. — China. — Cult. 1596 by J. Gerard. Catal. horti Gerardi.

H. K. 2d ed. V, 266: Willd. Sp. IV, 368.

III, 371. *Thuya orientalis*, Lin. Syst. 861. — China. — Cult. 1752 by Ph. Miller, Dict. ed. 6, n. 3.

H. K. 2d ed. V, 322: Willd. Sp. IV, 509.

III, 475. *Cycas revoluta*, Thunb. Fl. jap. 229. — China, Japan. — Arbor calappoides sinensis, Rumph. Amb. I, 92, t. 24. — Tessio, Kaempf. Amoen. 897. — Introduced about 1758 by Richard Warner.

H. K. 2d ed. V, 409: Willd. Sp. IV, 844. — J. E. Smith in Linn. Soc. Trans. VI (1802) p. 312, t. 29. — Cult. before 1737 by John Blackburne Esq. *). Knowlton **). Jun. MSS.

(Miller-Martyn, Gard. Dict. 1797, records that Richard Warner Esq. of Woodford in Essex received the plant from Captain Hutchinson about 1758. His ship being attacked by the French, the head of the plant was shot off, but the stem being preserved, produced several heads, which being taken off, produced as many plants).

III, 302, cum tab. 12. *Limodorum Tankervilliae*, (L'Héritier Sertum angl. 1788, p. 28. In honour of Lady Tankerville). — China. — Introduced about 1778 by John Fothergill, M. D.

H. K. 2d ed. V, 205, *Bletia Tankervilliae*, Brown MSS. — Willd. Spec. IV, 122. — Andrew. Bot. Repos. 426 (1805) sub Limodorum. — Redouté Liliac (1802) 43. — Schneev., icon. rar. (1793) 5.

---

*) J. Blackburne, 1690—1786. He first built hothouses in the north of England. (J. Britten, l. c.).

**) Knowlton, Thomas 1702—82, formerly gardener to James Sherard at Eltham. He gave a variety of useful information (Aiton).

III, 499. *Epidendrum ensifolium,* Lin. Sp. 1352. — Limodorum ensatum, Thunb. Fl. jap. 29; Icon. Kaempf. t. 3. — China, Japan. — Cult. before 1780 by John Fothergill, M. D.

H. K. 2d ed. V, 213, *Cymbidium ensifolium,* (Swartz. Act. Upsal. VI, 77). Willd. Spec. IV, 110. — Smith. Spicil. bot. (1791) 22, tab. 24. — Andr. Bot. Repos. 344 (1803). — Epidendron sinense, Redouté, Liliac. 113.

I, 62. *Ixia chinensis,* Lin. Sp. 52. — China. — Cult. 1759 by Ph. Miller, Dict. ed. 7, Ixia, n. 1.

H. K. 2d ed. I, 123, *Pardanthus chinensis,* Ker (Koenig & Sims, Annals of Bot. I, 1805, 247). — Ixia chinensis. Bot. Mag. 171 (1791). — J. G. Müller, Spec. plant. ad vivum delineatae (1757). — *Belamcanda chinensis,* Redouté, Liliac. 121. — East Indies, China, Japan.

I, 419. *Amaryllis aurea,* Lin. fil. (an unpublished treatise on Liliaceous plants, which Linnaeus filius wrote in 1782, when he was in England). A. aurea was first published by L'Héritier in Sert. angl. 1788, 14, tab. 15 bis). — China. — Introduced, 1777 by John Fothergill, M. D.

H. K. 2d ed. II, 227: Willd. Sp. II, 85. — Curt. Bot. Mag. 409, (1798). — Redouté, Liliac. 61.

III, 402. *Smilax China,* Lin. Sp. 1459. — China, Japan. — Cult. 1759 by Ph. Miller, Dict. ed. 7, n. 5.

H. K. 2d ed. V, 388: Willd. Sp. IV, 778. — Smilax minus spinosa, China seu Sankiwa, Kaempf. Amoen. 781, t. 782.

I, 454. *Dracaena ferrea,* Lin. Syst. veg. 334. — China. — Introduced 1771 by Benjamin Torin.

H. K. 2d ed. II, 277: Willd. Sp. II, 157. — Dracaena terminalis, Jacquin. Icon. pl. rar. II, t. 448. — Redouté, Liliac. 91.

III, 473. *Rhapis flabelliformis,* L'Héritier, Stirp. nov. II, tab. 100 (inedita? (V. supra p. 126 note). — Chamaerops excelsa, Thunb. jap. 130 (only the var. β, Soo Tsiku is Rhapis flabellif.). — China, Japan. — Introduced about 1774 by J. Gordon.

H. K. 2d ed. V, 473: Willd. Sp. IV, 1093. — Bot. Mag. 1371 (1811).

I, 474. *Acorus gramineus*, n. sp. — China, cultivated. — Introduced 1786 by Allan Cooper, commander of the Atlas, Indiaman.

H. K. 2d ed. II, 305: Willd. Sp. II, 199. — J. E. Smith, Spicil. bot. (1791), 15, t. 17.

III, 459. *Pteris serrulata*, Lin. Suppl. 445. — Japan, China, Ceylon. — Filicula cheusanica s. Hemionitis multifido folio tenuissime serrato ad margines seminifero, Pluken. Amalth. 94, t. 407, fig. 2. — Kei son kusa, Kaempf. Amoen. 912. — Introduced about 1770 by James Gordon.

H. K. 2d ed. V, 520: Willd. Spec. V, 373. — Schkuhr, Cryptog. (1810), 85, t. 91.

To complete W. Aiton's statements on the introduction of Chinese plants, I now propose to give some biographical or other particulars, as far as obtainable, from other sources of information, regarding the persons mentioned in the Hortus Kewensis as having introduced, in the second half of the 18th century, plants from China into the gardens of England.

Northumberland. Hugh Percy, second Duke of, born 1742. According to Aiton he introduced, from 1753 to 1759, the following Chinese plants:

Hypericum monogynum, Sterculia platanifolia, Anona hexapetala, Croton sebiferum, Lagerstroemia indica. He was one of the richest magnates of his time. He had a splendid conservatory at Syon *). Ph. Miller dedicated the 8th edition of his Gardener's Dictionary, 1768, to Hugh Duke and Earl of Northumberland, Baron Warkworth of Warkworth castle, Vice Admiral of all America, His Majesty's Privy Counsel, Fellow of the Royal Society. He died 1817.

Coventry, George, William, Earl of. He introduced in 1763 *Koelreuteria paniculata* (Hort. Kew). In 1766 he received from China the *Calycanthus praecox*, and cultivated it in his conservatory at Croome. (Bot. Cabin. 617 (1822).

Gordon, James, an eminent nurseryman at Mile End near London, who introduced and cultivated with great skill and success many exotic plants. He was at first gardener with Lord R. Petre at Thorndon. Essex.

---

*) The Garden at Syon-house, Brentford, is known to botanical science for more than three centuries and a half. The historical *Mulberry tree*, planted there by an Earl of Northumberland in 1548, still exists. (See Gard. Chron. 1897, I, 50).

John Ellis named the genus *Gordonia* after him. Gordon was a correspondent of Linnaeus. The latter notices (Suppl. 240) that Gordon first cultivated *Saxifraga sarmentosa*. — According to Miller-Martyn, Gard. Dict., Gordon, in 1754, first cultivated successfully *Gardenia florida*. In the Hort. Kew, he is stated to have introduced, in 1753 *Sophora japonica*, in 1744 *Rhapis flabelliformis*, in 1777, *Pteris serrulata*, all Chinese plants. He died in 1780 (Rees' Cyclop., Britten, Biogr. Brit. Bot.).

Kennedy & Lee. They first cultivated, in 1774, *Gleditschia horrida* (Hort. Kew). According to Britten, 1. c., Kennedy, Lewis, fl. 1775—1818, nurseryman of the Vine yard, Hammersmith. He wrote much in the Botan. Repository. — Lee, James, 1715—95, gardener at Syon and Whatton, from 1745 at Hammersmith.

Torin, Benjamin, is stated in the Hort. Kew to have introduced, in 1771, the following Chinese plants: *Olea fragrans*, *Dracaena ferrea*, *Daphne odora*, *Saxifraga sarmentosa*, *Murraya exotica*. Besides this we know nothing of Torin. He is not noticed by Britten.

Ellis, John, 1710—76, a London merchant, a distinguished naturalist, in 1754 Fellow of the Royal Soc., and a Correspondent of Linnaeus. He wrote botanical articles in the Philos. Trans., in vol. XLIX (1757) p. 870 on the Varnish tree raised in the Chelsea garden (which subsequently proved to be *Ailantus glandulosa*), in vol. LI (1759) p. 933, tab. 23, article on *Gardenia florida*. V. supra pp. 48 and 144. — According to Hort. Kew he first introduced the Tea plant into England; B. Booth, Hist. of Camellia and Tea (1829) states that he raised the plant from seeds.

Hope, Dr. John. He first introduced *Rheum palmatum* into England and wrote an article on the plant in Phil. Trans. 1765, p. 290, t. 12. According to Britten, 1. c. he lived 1725—86, was a pupil of Bern. de Jussieu, 1761 Professor of Botany at the University of Edinburgh, and Fellow of the Roy. Soc.

Fothergill, John, M. D., 1712—1780, a much celebrated physician at London, and at the same time a distinguished naturalist, especially botanist. Fellow Roy. Soc. 1753. F. was very rich. He had, from 1752 a botanical Garden at Upton, and introduced many exotic plants. Aiton mentions the following Chinese plants as having been first introduced by Fothergill:

1773, *Panax aculeata.*

1774, *Lychnis coronata.*

1777, *Amaryllis aurea.* — According to the Botan. Repos. 95, 163, *A. Fothergilli* and *A. radiata*, were introduced in the same year by F. from China.

1778, *Bletia Tankervilleae.*

1780, *Pyrus spectabilis, Cymbidium ensifolium, Mazus rugosus.*

Lind, Dr. James. L'Héritier, 1788, (v. supra p. 127 note) states that Dr. Lind introduced *Chloranthus inconspicuus* from China. It appears that he himself brought the plant from that country. According to Britten, l. c. he lived 1753—1808. The Brit. Museum possesses plants collected by Lind at the Cape of G. H. and in the East Indies.

Blake, John Bradley, 1745—73. Aiton, in Hort. Kew states, that Blake introduced from China into the Kew Gardens, in 1773, *Rhus succedaneum,* in 1776 (it seems a mistake) *Polygonum tinctorium.* Britten, l. c. notices that Blake was Supercargo of the British E. I. Company at Canton where he died, 16 Nov. 1773, and that he sent many plants and seeds to Europe. The Index Fl. Sin., ɪ, 89, records *Sterculia nobilis* as having been gathered by Blake in China. In Miller-Martyn's Gard. Dict., sub *Smilax Pseudo-China*, a MS note regarding this plant by Blake is quoted, stating that it occurs in China (Canton) and that the Chinese call it *cum kong cunn.* It is frequently used by them instead of the true China root. A small quality of it, even in coldwater, tinges of a deep red; where as the true root yields a light yellow brown *). — Sub *Croton sebiferum*, the same work reproduces a MS account by Blake on the Chinese Vegetable Tallow.

In the Trans. Linn. Soc. Lond. I, 172, in an article on Begonia by J. Dryander (1789) mention is made of a volume of drawings representing Chinese plants, in Sir Joseph Bank's library. These drawings are stated to have been made by a Chinese at Canton, who had been instructed by the late Mr. Blake in the art of making botanical drawings.

Nelson, David. According to Aiton, in Hort. Kew, the *Rhus semi- alatum* was introduced in 1780 by Sir Joseph Banks. It (the seeds) had

---

*) I can make nothing of the above Chinese name. Smilax Pseudochina Lin. is an American plant. The S. Pseudochina of Loureiro, Fl. cochin. 765, *kim kang re* in Gochinchina, is according to Kunth, Enum., V, 262, *S. corbularia*; the S. Pseudochina of Thunb. Fl. japon. 152 is, according to Maximowicz, Mél. biol. VIII, 411, *Sm. herbacea*, Lin.

been brought from Macao by D. Nelson. The Rhus semialatum was first described and figured by Professor J. A. Murray in the Commentarii Soc. Reg. Scient. Goetting. VI (1784), p. 27, tab. 3. Murray had raised the plant from seeds received from Sir Jos. Banks, with a note stating that these seeds had been collected at Macao during James Cook's last voyage around the world. It is well known that after Captain Cook had been murdered, in Febr. 1779 at Owaihi (Hawaii), one of the Sandwich islands, his ships sailed to Kamtchatka, and on their way home visited Macao, where they staid from Dec. 4, 1779 to Jan. 12, 1780. From the Kew Bulletin, 1891, p. 297, we learn, that David Nelson was a gardener from the Kew Gardens, who accompanied Cook on his third voyage, 1776—80. When in 1787 H. M. S. "Bounty", Captain (subsequently Rear Admiral) Bligh (1753—1817) was despatched to Tahiti, to introduce the bread fruit into the West Indies, Nelson accompanied him. He was among those set adrift by the mutinous crew and died from the long exposure, after reaching Timor, in 1789. L'Héritier founded the famous genus *Eucalyptus* on *E. obliqua* figured in Sert. angl. 18, and which had been found by Nelson in Van Diemens Land. De Condolle, Syst. nat. veget. I, 7, states that Nelson gathered plants on the West coast of America. His herbarium is in the British Museum. The following Chinese plants are stated to have been brought home by Nelson:

Melastoma sanguinea, Bot. Mag. 2241 (1821), Macao, Nelson. Herb. Banks.

Evolvulus chinensis, DC, Prodr. IX, 447. Nelson, China. Herb. Brit. Mus.

Striga hirsuta, DC. Prodr. X, 503. Nelson, Macao.

Hypericum chinense, I. Fl. Sin., I, 73, China, Nelson.

Indigofera hirsuta, ibid. I, 157, Macao, Nelson.

Spermacoce hispida, ibid. I, 393. S. China, Nelson.

Gynura divaricata, ibid. I, 447, near Canton, Nelson, Brit. Mus.

Sideroxylon Wightianum, Hook. et Arn. Bot. Beech., 196. Ibid. II, 69.

Utricularia bifida, ibid. II, 222. China, Nelson.

Asparagus lucidus, Baker, Asparag. Journ. Linn. Soc. XIV, 605. Macao, Nelson.

Lindsaea flabellulata, Dryander, Linn. Trans. III (1797) 41. Nelson, Macao.

Bradley, Henry. He collected plants at Macao, about the year 1779, and sent them to Sir Joseph Banks. See Lindley, Collect. bot. (1821) 3,

Raphiolepis rubra, — Bot. Reg. 652 (1822) *Raphiolepis salicifolia*, — Masters, Conif. Japan, in Journ. Linn. Soc. XVIII, 502, *Cunninghamia sinensis*. — In 1790 Prof. Gaertner, Fruct. II, 127, t. 109, fig. 1, first described the *Bradleia sinica* (Euphorbiaceae). He states that he received the plant from Sir Joseph Banks, who proposed the genus name *Bradleia*. Martyn in Miller's Gard. Dict. 9th edit. suggests that the genus was named after Richard Bradley, the first Professor of Botany in Cambridge, -+- 1732. But I am inclined to think that Banks dedicated the name of the plant to Henry Bradley. Bentham Fl. hongk, 314, sub Glochidion, states that this plant is not known out of Hongkong, the adjacent continent and neighbouring islands. — The Ind. Fl. Sin. notices three Chinese plants in the herbarium of the Brit. Mus., gathered in 1779 by H. Bradley, viz: *Tephrosia villosa* (I, 158), *Acanthopanax aculeatum* (I, 339), *Tamarix chinensis* (I, 347).

Robertson. He collected plants at Whampoa, near Canton, apparently at the end of the last century. — Salisbury, Paradisus Lond. (1807) tab. 115, *Adina globifera*, states that this plant was gathered by Robertson, near Whampoa. The Ind. Fl. Sin. notices three plants viz: I, 147. *Rhus semialata*, I, 175, *Desmodium polycarpum*, II, 161, *Ipomoea philippinensis*, as collected by Robertson at Whampoa. We learn from Mr. J. Britten (Journ. Bot. 1882, 250) that Robertson's collection of Chinese plants still exists in the British Mus. Comp. also Dryander in Linn. Trans III, 41 (1794) *Lindsaya heterophylla*, Robertson.

Part III. THIRD PERIOD, FROM THE FIRST BOTANICAL EXPLORATION OF THE INTERIOR OF CHINA, IN 1793, TO THE FIRST WAR BETWEEN ENGLAND AND CHINA IN 1840.

Up to the period which we shall now proceed to sketch, botanical collections in China had been made only in Chusan, at Amoy (Cunningham, beginning of the 18th century), at Peking (d'Incarville, middle of the 18th cent.), at Canton and at Macao. During the 18th cent. and down to 1842 Canton was the only Chinese port open to European trade. Macao, where the Portuguese had settled about 1537 and which still belongs to them, was likewise accessible to Europeans. It seemed therefore that a new era of botanical exploration in China was inaugurated, when the Embassy of Lord Macartney to the Emperor of China, in 1793, had the opportunity of traversing a part of the country beyond the Great Wall, and was allowed to return from the Chinese capital to Canton through the interior of China, and when Sir George Staunton, Secretary to the embassy, brought home from this journey, a rare collection of Chinese plants gathered by him in regions where plants had never been collected by Europeans before. — Eleven months after Lord Macartney's return to Canton, a Dutch Embassy, headed by Titzing set out from Canton for the Court of Peking, and travelled overland traversing six provinces of the empire. Van Braam, the diarist of this journey occasionally gives notices of interesting plants, but no collections were made on the road. — In 1816, Lord Amherst, British Ambasandor to the Chinese Emperor, returned from Peking to Canton overland, and Dr. Clarke Abel, a naturalist who was in his suite, collected many plants during this interesting journey. But unfortunately this precious collection was lost, with the exception of a few plants, by shipwreck. After this, for nearly thirty years no European naturalist visited the interior of China, and nearly all the Chinese plants received in Western Europe during this period came from Canton, Macao and the adjacent islands.

I. Lord Macartney's Embassy to the Emperor K'ien lung 1793. — Sir George Staunton.

Staunton, Sir George Leonard. He was born in 1737 at Galway in Ireland, studied medicine at Montpellier and then, in 1762, began his career as a medical officer in the West Indies. There he acquired the friendship of Lord George **Macartney,** (1737—1806), Governor of the islands of Grenada and Tabago, and when, in 1789, the latter was appointed Governor of Madras, Staunton accompanied him in the capacity of Secretary and distinguished himself during the negotiations for peace with Tippo-Sahib. In 1784 he returned to England, in 1785 was made a Baronet of Ireland. In 1792 Lord Macartney was sent as British Ambassador to the Emperor of China and Staunton was appointed Secretary to this Embassy with the rank of Minister Plenipotentiary. Some years after the return of the embassy to England, Staunton published his work entitled:

AN AUTHENTIC ACCOUNT OF AN EMBASSY FROM THE KING OF GREAT BRITAIN TO THE EMPEROR OF CHINA . . . . . taken chiefly from the papers of His Excellency the Earl of Macartney etc; Sir Erasmus Gower, Commander of the Expedition, and other gentlemen in the several departments of the Embassy. By Sir George Staunton, Baronet. In two volumes with engravings (in 4); beside a folio volume of Plates. London 1797—98. — This book has been translated into French, Dutch and German. Sir George Staunton *) died at London, 1801. He was F. Linn. Soc., and F. Roy. Soc. His friend Lord Macartney, whose services were rewarded, in 1794, with the title of Earl, was appointed Governor-General of Cape Colony; but he returned in 1798, and died at London, in 1806.

In the following lines I propose to give a short account of Lord Macartney's embassy, extracted from Staunton's book and especially of the memorable journey overland, with special reference to the passages relating to Chinese plants.

---

*) His Son George Thomas **Staunton,** 1781—1858, accompanied his father to China. In 1799 he was appointed chief supercargo for the E. I. Company at Canton. Knighted 1810. In 1816 he accompanied the embassy of Lord Amherst to Peking. He was a distinguished sinologue, translated the Penal Code of China, 1810, and wrote several books on China.

Voyage undertaken on the occasion by H. M. S. "Lion" and the ship "Hindustan" of the E. I. Company's service, to the Yellow Sea and the Gulf of Peking.

Sir Erasmus Gower, Commander of the flag ship "Lion" and of the expedition.

Lord Macartney, Ambassador.

Sir George Leonard Staunton, Secretary of Legation.

Mr. Acheson Maxwell, Secretary of Lord Macartney.

Capt. Mackintosh, Commander of the "Hindustan".

Dr. Dinwiddie and Mr. J. Barrow *) both conversant in astronomy, mechanics and every other branch dependent on mathematics.

Dr. Gillan, Physician, — Dr. Scot, Surgeon.

Two botanic gardeners were provided, one at the public charge and one at the expense of a member of the embassy. (Probably J. Haxton, see farther on, who seems to have been engaged by Staunton).

On Sept. 26, 1792, the "Lion" and the "Hindustan" set sail from Portsmouth. They were accompanied by the brigs "Jackall," "Clarence" and "Endeavour". They passed by Madeira, Teneriffe, St. Jago, passage over the Line, and across the Atlantic to Rio Janeiro, Tristan d'Acunha, St. Pauls, Amsterdam, Sunda Straits, Batavia, Pulo Condore, and Cochinchina (Turan Bay).

On the 21 June 1793 they anchored at one of the Ladrones opposite 1793 Macao, and one of the brigs was despatched to Macao to confer with the Chinese authorities. These at first insisted upon the squadron's waiting at Canton for the Emperor's answer, but finally let them sail northwards. On June 23 they set sail, and, having passed the Straits of Formosa, on July 3, anchored near the island of Chusan. The brig "Clarence", with Mr. Barrow and others on board, proceeded to the city of Ting hae, in order to procure Chinese pilots. The pilots having been accorded, the squadron weighed anchor again. Having doubled the Shantung Cape, they anchored at Teng chou fu (about 50 miles N. W. of present Chefoo), and started again on July 23. On July 26 they had arrived at the bar near the mouth of the Pei ho river. On August 5 the Embassy went on board one of the brigs, passed the bar, and entered the river, where they were received by a Chinese yacht. Having passed Ta ku, they reached Tientsin on August 9 th.

Between Taku and Tientsin, Staunton observed *Barbadoes millet, Hol-*

---

*) J. Barrow published, in 1804, another account of Lord Macartney's embassy with the title: «Travels in China etc.». It contains little of special interest for botanists.

cus Sorghum, in Chinese *kao leang.* Groves of high and wide spreading *Pines (Pinus sinensis* Lamb.*). Panicum italiculum, P. crus galli, Sesam,* all cultivated near the banks of the Pei ho (II, 18, 43, 44).

After a short stay at Tientsin the Embassy proceeded in Chinese boats up the Pei ho river, and on August 16 reached Tung chow. Meanwhile the squadron had left the bar of Taku and sailed for Chusan.

From T'ung chow the Embassy went directly (passing, however through Peking), to Yuan ming yuan, one of the summer residences of the Emperor (about 7 miles N. W. of Peking), where a villa had been prepared for them between (the village of) Hai tien and Yüan ming yüan. The Emperor K'ien lung was then in Mongolia. Subsequently the Embassy was transferred to Peking, and while the Ambassador continued at Peking, some of the gentlemen often had occasion to pass from thence to the Imperial Palace in the country.

On Sept. 2, 1793, at the beginning of the cooler season, Lord Macartney with his followers set out for Jehol. He travelled in an English carriage, which he had brought with him. The Embassy passed by way of Ku pe k'ou.

Jehol, one of the summerresidences of the Manchu Emperors, lies beyond the Great Wall N. E. of Peking, about 41° N. lat. The route from Peking to Jehol crosses the Great Wall through the Gate of Ku pe k'ou. Jehol is a Mongol name. The Chinese call it Je ho, properly the name of a river on which the palace and the pleasure ground had been laid out in 1703. In the vicinity lies the city of Ch'eng te fu. The journey from Peking to Jehol takes from 6 to 8 days. (E. B.).

II, 170. "The Chinese extract an excellent oil from the kernels of the *Apricot* in lieu of olives (the wild Apricot tree abounds in the mountains of Narth China. E. B.); for more common purposes, however, from the seeds of *Sesamum,* of *Hemp,* of *Cotton* and of a species of *Mint (Perilla ocimoides,* E. B.). — The *Shepherds purse (Thlaspi bursa pastoris,* common in N. China, E. B.) is to be found in their salads. — The cup of the *Acorn (Quercus Bungeana)* serves them to dye black. —. The leaves of an *Ash* are used for rearing silkworms. (Staunton means probably *Ailantus glandulosa,* the "frêne puante" of d'Incarville, *ch'ou ch'un* in Chinese. V. supra p. 48, and Stan. Julien, Mûriers, vers à soie, 196. E. B.). — From the buds and the leaves of a species of *Colutea* a kind of green dye is obtained". (There seems to be some misapprehension. The genus *Colutea,* Bladder Senna, has not been recorded from Eastern Asia. I may observe, that the flowerbuds of *Sophora japonica* are used in North China for dyeing yellow and green, the leaves of *Koelreuteria paniculata*, which produces a bladder like fruit, for dyeing black. E. B.).

II, 174. "There is no traditional account of the introduction of the *Tobacco* plant into China. It seems to be natural". (Staunton is mistaken. There is evidence from Chinese records that Tobacco was introduced into China from the Philippines in the beginning of the 17th century. E. B.).

As the travellers advanced into Tartary, they saw *Oaks* of the two species called the English and the Russian. (There are three species of Quercus in those regions: *Q. mongolica, Q. dentata* and *Q. Bungeana.* E. B.). They saw also *Aspen (Populus alba) Elm trees (Ulmus pumila),* Hazel *(Corylus mandshurica* and *Cor. heterophylla* are common there), *Walnut trees.*

It is not stated when they arrived at Jehol.

II, 240. *Lien hwa,* Lotus flowers *(Nelumbium speciosum)* noticed in the ponds around Jehol.

After having been admitted to an audience by the Emperor K'ien lung, Lord Macartney returned to Peking, apparently towards the end of September.

On Oct. 7 the Embassy set out to return to Canton by the overland route. They performed the whole journey, with only two short interruptions, in boats.

They first proceeded down the Pei ho river to Tientsin, and then entered the Grand Canal*), and on Oct. 18 reached the border of the province of Shan tung, on Oct. 22, Lin sin choo (Lin ts'ing chou). "The Chinese cultivate the *Palma Christi (Ricinus communis),* largely. They eat the oil obtained from the seeds" (They eat it boiled, as boiling takes away its purgative effect. E. B.).

"On Oct. 25 the yachts arrived at the highest part of the Grand Canal being about ³/₅ of its entire length (at about the N. lat. of Yen chou fu). Here the river Luen (Wen on the Chinese maps), the largest by which the Canal is fed, falls into it with a rapid stream in a line which is perpendicular to the course of the Canal. A strong bulwark of stone supports the opposite western bank, and the waters of the Luen striking with force against it, part of them follow the northern, and part the southern course of the Canal. A circumstance which not being generally explained or understood, gave the appearance of wonder to an assertion, that if a bundle of sticks be thrown into that part of the river, they would soon separate and take opposite directions". (This peculiarity is already related by Marco

*) I may observe that the first Dutch Embassy to the Chinese Court, in 1656, passed by the Grand Canal (V. supra p. 23).

Polo. He states (Yule's 2d edition, II, 121) that near Sinjumatu (Tsi ning chou) the large river on which it is situated (the Grand Canal) has been divided by the people into two, making one branch of it flow towards Manzi (S. China) and the other towards Cathay (N. China). E. B.).

Nov. 2, arrival at the point of junction of the Grand Canal with the Yellow River.

Further on they crossed the Yang tze River (II, 425). "The land is chiefly cultivated with that peculiar species or variety of the *Cotton* shrub, that produces the cloth usually called *Nankeen* in Europe. The cotton wool is white in the common plant; but in that growing in Kiang nan it is yellow. It is asserted that the seeds of the Nankeen Cotton plant degenerate in other provinces".

II, 430. "The *Tallow* tree *(Stillingia sebifera)* grows near Soo chow foo. Candles are made of the tallow. The tree has been transplanted to Carolina, where it flourishes as well as in China".

On Nov. 9the Embassy arrived at Hang chow fu and was there informed that the "Lion" had mean while sailed, in October, for Canton, while the "Hindustan" was still at Chusan. The ambassador resolved to proceed with the greater part of his suite by the river way to Canton. The rest embarked on the "Hindustan".

II, 445. The Vale of Tombs near lake Sihoo, Hangchow foo. "Tombs surrounded by trees, such as different species of Cypress. . . . The Church yard yew did not, however, grow there, nor was it observed in any part of China. But a species of *Weeping Thuya* or Lignum vitae with long and pendant branches, unknown in Europe, overhung many of the graves. Plate XLI View of Lake Si hoo and Vale of tombs with weeping Cypress. Further on, II, 525 (list of plants coll. by Staunton) it is called *Cupressus pendula*". (This is the *Cupressus funebris*, Endlicher, Conif. 1847, p. 59. In the Gardener's Chron. 1849, 243 and in Paxton & Lindl. Flower Garden, I, 1850, 49, it is erroneously stated that the Vale of Tombs, where Staunton discovered this beautiful tree, lies near Jehol, in Tartary).

In the middle of November the Embassy left Hang chow foo and their boats sailed southward up the river Cheng tang chaung (Ts'ien t'ang kiang) *).

II, 457. *Tallow* trees, *Chest nut* trees, *Camphor* trees in abundance". Description of the mode of preparing Camphor.

II, 458. Great quantities of the *Arbor vitae* or *Thuya (Biota orien-*

---

*) The same river way was partly followed, in 1667, by a Dutch Embassy. V. supra p. 24.

talis) which grow to a prodigious height in the valley in which stands the city of Yen chow foo.

II, 460. Farther up the river they saw many *Sugar* plantations. *Orange* trees much cultivated. The oranges here present a great variety in size and colour. Some of them are smaller than the Portugal orange, and some as large as any produced in the Western Indies. But the sweetest and richest was a deep red orange, preferred to every other, and easily distinguishable by its pulp adhering to the rind only with a very few slight fibres. (Mandarin orange, *Citrus nobilis*).

II, 463. Staunton saw a kind of *Date* (the large fruit of *Zizyphus vulgaris* is called Chinese Date by Europeans in China) and *Lee chees (Nephelium Litchi)*, and describes the latter fruit.

"They have à fruit called *see chee* (shi tze). It is soft, reddish with a smooth skin containing a slight acid pulp with a kernel in the middle. It is of the size of a middling orange and looks as if flattened by weight, from its globular shape". (This is the Chinese Persimon, *Diospyrus Kaki, D. shi tze*, cultivated in many varieties. The fruit is generally seedless).

II, 467. The *Tea* shrub was for the first time seen growing. — In vol. I, 21 Staunton gives a notice of Tea. He says that Tea was not known in any part of Europe before the commencement of the last (17th) century. Some Dutch adventurers, seeking for such object as might fetch a price in China, introduced to the Chinese the herb Sage (Salvia officinalis), so much extolled by the Salernian school of physic as a powerful preservative of health. They accepted in return the Chinese Tea, which they brought to Europe. (Comp. supra p. 25). In England, about the middle of the last age infusions of tea were already sold in houses of public entertainment, and became an object of taxation to the legislature. The annual public sales of tea by the E. I. Comp. did not, however, in the beginning of the present (18th) century much exceed 50,000 pounds weight. The Company's annual sales now approach 20 millions of pounds. In the island of Corsica a small plantation of tea is said to be actually flourishing.

II, 467. "A plant very like the Tea plant flourished on the sides and at the very top of the mountains. The Chinese call it *cha wha* or flower of Tea, on account of the resemblance of the one to the other, and because its petals as well as the entire flowers of Arabian Jasmine are sometimes mixed among the teas, in order to increase their fragrance. This plant yields a nut, from whence is expressed an esculent oil, equal to the best which comes from Florence. It is cultivated on this account in vast abundance, and is particularly valuable from the facility of its culture in situations fit for little else." (This is the *Camellia Sesanqua* of botanists). A good drawing of the plant is given.

II, 468. At the town of Chan san shen (Ch'ang shan hien) the river ceased to be navigable. The Embassy then proceeded over a range of mountains and embarked again on a river, which brought them by the city of Kwang sin fu to the Po yang lake. Thence they ascended another river, the Kan kiang, up to the frontier of the Kwang tung province *).

II, 503. "Plantations of *Bamboo* abound in the province of Kiang si. — Large *Camphor* trees and trees of a still greater size, a species of the *Ficus*, in Chinese *yang shoo,* of which the branches spread horizontally of such an extent, that one tree may be almost sufficient to cover half an acre of land". *(Yung shu* is the Chinese name for *Ficus retusa,* the Chinese Banyan).

II, 505. *Wheat* cultivated.

City of Nan gan fu. Here the navigation ceased again. The Embassy were obliged to march over a mountain range, and then pursue their journey on the Pekiang river down to Canton. They embarked at Nan shoo foo (Nan hiung chou).

II, 511. City of Chau Chau fu (Shao chou fu). *Rice, Tobacco, Cotton.*

On the 19th of December 1793 the Embassy reached Canton. — Sojourn in this city and in Macao. — On March 17, 1794 the squadron left Macao. — St. Helena. — Arrival in England Sept. 6, 1794.

1794

Staunton, in his Account of Macartney's Embassy, gives four enumerations of plants, with their botanical names, gathered by him and others, during the journey of the Embassy in several provinces of China, in the whole about 400 species. It is not said who determined these plants, perhaps Dryander. The determinations are very superficially made; frequently only the genus name is given. It would seem from Mr. Hemsley's quotations in the "Ind. Fl. Sin.", that Lord Macartney likewise collected plants in China. Mr. Hemsley informs me, that some of the sheets belonging to this collection, in the British Museum, bear Staunton's name, some that of Macartney. He does not think that they represent different collections.

Staunton seems to have presented his collection of Chinese plants, among which there were probably a considerable number of duplicates, to Sir Joseph Banks and A. B. Lambert (of whom we shall have to speak farther on). I suppose that the Kew Garden also received a set of them. D. Don, at the end of Lambert's Pinus, 1828, in an account of Lambert's

---

*) From Nan ch'ang fu, S. W. of Po yang Lake, to Canton Lord Macartney's route coincides with that of the Dutch embassy, in 1656. V. supra p. 23.

herbarium, sub n. 20, states, that Sir G. Staunton gave to Lambert duplicates of all specimens collected during Lord Macartney's Embassy.

Alph. De Candolle, "Phytographie" (1880), 452, 458, says that
Staunton's plants were found in the collections of De Candolle (Aug. Pyramus, the father of Alphonse), Webb and Delessert.

Webb, Philipp Barker, 1773—1854, a wealthy Englishman of private
means, and a distinguished botanist, who used to live at Paris, was in possession of an immense herbarium of 80,000 species, which he bequeathed
to the Duke of Tuscany. This collection is now in the botanical Museum at
Florence (Phytogr. l. c., Britten, Biogr. Brit. Bot.).

Delessert, Baron Benjamin, 1773—1847, a rich French banker and
botanist, who had accumulated vast botanical collections, described, in 1845
by A. Lasègue, "Musée botanique de M. B. Delessert". In this book, p. 346,
we read that Delessert received from Lambert plants brought by Staunton
from China. Delessert undertook to publish a splendid iconographical work
on botany with the title: "Icones selectae Plantarum quas in Systemate
universali ex herbariis Parisiensibus praesertim ex Lessertiano descripsit
A. P. De Candolle". 5 vols in folio, 1820—1846, 508 plates, uncoloured.
After Delessert's death, his nieces presented his botanical collections to the
city of Geneva, where they are still kept.

Now Staunton's Chinese plants are found in the herbariums of the
British Museum and of Kew. Some of them have also found their way into
some private collections in France and Germany.

A. P. De Candolle, Syst. veg. I (1818) 513, named the genus *Stauntonia* in honour of the famous traveller and based it upon the *St. chinensis*
brought from China by Staunton, who has also been commemorated in several species names of Chinese plants.

Let us now turn to the four lists of Chinese plants, gathered by Staunton and others, as enumerated in his book. (My own observations are in
brackets. I have arranged the enumerations according to Benth. & Hook.
Gen. Pl.).

I, Plants collected by the the Gardener attached to the Embassy (probably J. Haxton) in the Province of Chili. II, 165—167. (It seems to
me that the greater part of these plants were gathered between Peking
and the Imperial summer palace Yüan ming yüan in August and September 1793).

Nymphaea Nelumbo. *(Nelumbium speciosum*. This beautiful plant covers all the lakes, ponds and ditches near the
summer palace).

Sisymbrium amphibium. *(Nasturtium amphibium* R. Br.).

Brassica. *(Brassica chinensis* is much culticated at Peking).

Lepidium latifolium. (Common in the environs of Peking).

Thlaspi. *(Thlaspi bursa pastoris,* ibid., common).

Cleome. (I suspect the plant thus named was *Gynandropsis viscida,* Bge.).

Arenaria rubra. (The Peking Flora has *A. juncea, A. formosa, A. lateriflora).*

Tamarix. (Probably *T. chinensis,* very common in the Peking plain).

Malva several species. *(M. mauritiana* and *M. pulchella).*

Hibiscus trionum. (Common near the summer palace).

Sida. *(Sida tiliaefolia* much cultivated at Peking).

Sterculia platanifolia. (Cultivated, Peking).

Tribulus terrestris. (Common in the plain).

Erodium ciconium. (Probably *Erodium Stephanianum,* common in the plain).

Impatiens Balsamine. (Gardenflower, Peking).

Ailantus glandulosa. (One of the most common trees at Peking).

Evonymus. *(E. Bungeana* in the plain, *E. alata* in the mountains).

Rhamnus. (The Peking Flora has 4 species of Rhamnus, of which *Rh. virgata* grows near the summer palace).

Trifolium melilotus. (Probably *Melilotus suaveolens).*

Astragalus. (The Peking Flora has 8 species of Astragalus).

Hedysarum striatum. *(Lespedeza striata,* Hook. & Arn., rare at Peking).

» several other species. (The Peking Flora has 3 species of Hedysarum and 10 of Lespedeza).

Dolichos hirsutus. *(Pachyrhizus Thunbergianus.* Common in the Peking mountains).

Sophora japonica. (A common tree in the plain).

Prunus armeniaca. (Much cultivated, also spontaneous in the mountains).

Potentilla. (The Peking Flora has 20 species of Potentilla).

Sempervivum tectorum. (Unknown for the Peking Flora. Perhaps *Umbilicus ramosissimus* is meant, which sometimes grows on roofs).

Cucurbita Citrullus. (Much cultivated).

Apium.

Lonicera caprifolium. (Unknown at Peking. *Lonicera japonica* much cult.).

Rubia cordata. (Common in the Peking plain).

Aster, two species.

Inula japonica. (Occurs in the Peking mountains. *I. chinensis* is more common in the plain).

Xanthium strumarium. (Common weed).

Eclipte erecta and prostrata. (Both = *E. alba,* not uncommon in marshes).

Bidens pilosa, and other species. (B. pilosa has not been found at Peking. The Peking Flora has *B. cernua, B. parviflora, B. bipinnata).*

Chrysanthemum. *(Chr. indicum.* Wild. in the plain).

Matricaria.

Artemisia capillaris. (Not met with at Peking).

    » integrifolia. *(A. vulgaris.* Common).

Prenanthes. (Perhaps *Lactuca denticulata* or. *L. versicolor).*

Sonchus oleraceus. (Rather *Sonchus arvensis).*

Statice limonium. (The Peking Flora has only *St. bicolor).*

Fraxinus. (Evidently *Fraxinus Bungeana).*

Nerium Oleander. (Cultivated at Peking).

Cynanchum. *(Cynanchum pubescens,* common).

Asclepias sibirica. *(Vincetoxicum sibiricum,* common in the plain).

Cuscuta. (The Peking Flora has two species: *Cuscuta chinensis,* yellow, in the plain; *C. japonica,* purple, in the mountains).

Solanum Melongena. (Cultivated).

Solanum other species. *(S. nigrum* and *S. septemlobum,* common).

Lycium chinense. (Common in the Peking plain).

Antirrhinum. (Linaria vulgaris, rare at Peking).

Incarvillea. *(Incarvillea sinensis,* common).

Sesamum orientale. (Cult.).

Vitex Negundo. *(Vitex incisa,* one of the most common shrubs in the mountains and in the plain).

Melissa.

Leonurus sibiricus. (Common).

Chenopodium aristatum. (Occurs in the Peking plain).

Chenopodium scoparium. *(Kochia scoparia,* Schrad. Common weed).

    » viride. (Variety of *Ch. album.* Common in the plain).

    » glaucum. (Common in the plain).

Atriplex. *(A. littoralis,* the only species met with at Peking).

Corispermum hyssopifolium and another species. (The Peking Flora has *C. tylocarpum* and *C. Stauntoni).*

Salsola altissima and other species. (The Peking Flora has *Suaeda Stauntoni* and 3 other species, and *Salsola collina).*

Blitum. (Perhaps *Amarantus Blitum,* which is common at Peking).

Amarantus caudatus. (Cultivated).

Polygonum tinctorium. (Cultivated).

    » lapathifolium. } (Common in the Peking plain).
    » aviculare.

Polygonum persicaria. (Common in the Peking plain).

Euphorbia cyparissias. *(E. Esula,* var. *cyparissioides).*

    »   Esula.

    »   tithymaloides. (Probably *E. lunulata,* common in the plain).

Acalypha. (The Peking Flora has only one species of this genus, *A. pau-ciflora).*

Cannabis sativa. (Common, wild and cultivated).

Morus. (Morus alba, many varieties).

Corylus avellana. *(Corylus heterophylla.* Peking mountains).

Salix. *(Salix babylonica,* common in the plain).

Pinus. *(P. sinensis* and *P. Bungeana.* Staunton, certainly, had met with both of these species in the environs of Peking. P. Bungeana is very conspicuous on account of its white bark).

Juniperus barbadensis. (The only Peking species is *J. chinensis).*

Hemerocallis japonica. (Cultivated).

Asparagus. (The Peking Flora has 4 species of Asparagus. That most commonly found in the plain is *A. gibbus).*

Typha latifolia. (This species is not common at Peking. *Typha angustifolia* is more frequently met with).

Cyperus odoratus. (An American plant).

    »   Iria. (Common in marshes).

Scirpus.   (7 Species are known from Peking).

Holcus.   (Probably *Sorghum vulgare,* much cultivated).

Panicum ciliare. (Not found at Peking).

    »   crus corvi. *(Panicum (Echinochloa) Crus galli.* Peking, wild and cultivated).

    »   glaucum. *(Setaria glauca,* common).

Rottboella.

Cenchrus racemosus. (Bunge records this for the Peking Flora).

Arundo phragmites. (Common).

Avena. *(A. sibirica* has been recorded from Peking).

Poa. (Four species are known from Peking).

Briza eragrostis. *(Eragrostis poaeoides.* A common grass in the Peking plain).

Cynosurus indicus. *(Eleusine indica.* Sometimes cultivated at Peking).

Lolium.

Andropogon ischaemum and other species. (Besides *A. ischaemum* three other species are known from Peking).

Equisetum. *(E. arvense* and *E. ramosissimum,* both not uncommon at Peking).

II. Plants collected between Peking and Jehol. II, 274—276.

Paeonia. (Two species, *P. albiflora* and *P. obovata*, are known from those regions).

Berberis. *(Berberis sinensis.* See DC. Syst. veg. II, 8).

Cistus. (This genus is unknown in Eastern Asia).

Dianthus. (Evidently *D. chinensis*, common in the mountains).

Tribulus terrestris. (V. supra p. 164).

Evonymus. (V. supra p. 164).

Rhamnus, 4 species. (These 4 species were, I suppose, *R. catharticus, R. argutus, R. virgatus, R. parvifolius).*

Vitis heterophylla, Thbg. (Common in the mountains north and west of Peking).

Sophora japonica. (V. supra p. 164).

Cassia procumbens, Thbg. *(Cassia mimosoides*, common in the mountains, north and west of Peking).

Sanguisorba officinalis. (Common in the same regions).

Crassula spinosa. *(Umbilicus (Cotyledon) spinosus*, DC. gathered by Kirilov, 1840, in the mountains near the Great Wall. Herb. hti Petrop.).

Crassula, other species.

Sedum. (The Peking Flora has 6 species of Sedum).

Bupleurum. (Probably *B. falcatum*, which is common in those regions).

Sambucus nigra.

    »   umbellata. (These two species do not occur in those regions. Staunton gathered evidently *S. racemosa*, which is common in the mountains north of Peking, and perhaps *S. Ebulus*, known from S. Manchuria).

Rubia cordata. (V. supra p. 164).

Valeriana. *(V. officinalis* is known from the Peking mountains and S. Manchuria).

Scabiosa leucantha and other species. (The only species known from the Peking mountains and S. Manchuria is *S. Fischeri).*

Aster and Matricaria.

Campanula. (Rather an *Adenophora).*

Lysimachia. (Three species are found in the mountains north of Peking of which *L. barystachys* is the most common).

Fraxinus. *(Fraxinus Bungeana*, v. supra p. 165).

Syringa vulgaris. (In the mountains traversed by Staunton there are *S. Emodi* and *S. villosa).*

Asclepias sibirica. (V. supra p. 165).

Swertia rotata. *(Ophelia diluta.* Common in the Peking mountains and
S. Manchuria).

Echium. (No Echium has been recorded from E. Asia).

Convolvulus, 2 species.

Solanum nigrum. (V. supra p. 165).

Capsicum. (Only cultivated).

Physalis Alkekengi. (Common in N. China and S. Manchuria).

Hyoscyamus niger. (Tatarinow gathered it at Kʻu pei kʻou).

»   other species. (Perhaps *H. physaloides (Physochlaina physal.*
Maxim. Ind. Fl. Peking).

Nicotiana. (Cultivated. V. supra p. 159).

Veronica, two species. (3 species are known from the Peking mountains, of
which *V. paniculata* is the most common).

Amethystea coerulea. (Common in the mountains).

Chenopodium, 3 species.

Polygonum Fagopyrum. (Cultivated and wild in N. China).

»   lapathifolium and another species. (V. supra p. 165).

Ulmus. (Two species in N. China, *U. pumila* and *U. macrocarpa.* The latter
in the mountains).

Morus. (V. supra p. 166). *Morus alba,* var. *mongolica* common in the mountains).

Quercus. (V. supra p. 159).

Salix. (V. supra p. 166).

Pinus.

Moraea chinensis. *(Pardanthus chinensis.* Only cultivated).

Convallaria multiflora. *(Polygonatum multiflorum).*

»   verticillata. *(Pol. verticillatum).*

Allium.

Asparagus.

Cyperus Iria and other species. (V. supra p. 166).

Panicum italicum. (Cultivated).

»   glaucum, viride, ciliare. (V. supra p. 166).

Aristida. *A. vulpioides* occurs in the Peking mountains).

Arundo.

Avena.

Poa.

Briza eragrostis. (V. supra p. 166).

Saccharum. *(Imperata,* of which 3 species grow in the Peking mountains).

III. Plants collected in Shan tung and Kiang nan *) October 1793.
II, 435—437.

Menispermum trilobum. Thbg.
Berberis cretica.
Nymphaea Nelumbo.
Sisymbrium amphibium.
     »   other species.
Cleome viscosa. (v. supra p. 164).
Cistus.
Viola.
Dianthus plumarius.
Stellaria.
Tamarix.
Thea.
Gossypium.
Geranium.
Oxalis corniculata.
Melia azedarach.
Ilex.
Evonymus.
Medicago lupulina.
     »   falcata.
Trifolium melilotus.
Astragalus, 2 species.
Aeschynomene.
Phaseolus.
Dolichos cultratus.
Sophora japonica.
Mimosa.
Rubus cordifolius.
Potentilla.
Agrimonia.
Rosa.
Crataegus.
Penthorum.
Myriophyllum spicatum.
Trapa.

Cucurbita Citrullus.
Inula japonica.
Chrysanthemum indicum.
Artemisia.
Prenanthes.
Fraxinus.
Convolvulus.
Solanum nigrum.
Nicotiana Tabacum.
Veronica Anagallis.
Antirrhinum.
Lindernia japonica.
Justicia procumbens.
Verbena officinalis.
Vitex Negundo.
Clerodendron.
Ocymum.
Mentha canadensis.
Lycopus europaeus.
Leonurus sibiricus.
Chenopodium aristatum.
     »   2 other species.
Celosia argentea.
Amarantus caudatus.
     »   tricolor.
Polygonum lapathifolium.
     »   dumetorum.
     »   amphibium.
     »   perfoliatum.
Laurus Camphora.
Viscum.
Croton sebiferum.
Agyneia impubes.
Phyllanthus.
Cannabis sativa.

*) General name for the provinces of Kiang su and An hui.

Morus alba.
»   nigra.
»   papyrifera. *(Broussonetia)*.
Ficus pumila.
Fagus Castanea.
Salix.
Thuya pensilis, new sp.
Valisneria spiralis.
Stratiotes.
Najas marina.
Cyperus difformis.
»   Iria.
»   odoratus.
»   2 other species.
Kyllingia monocephala.
Scirpus autumnalis.
»   miliaceus.
Leersia.
Oryza sativa.
Holcus.
Paspalum.
Panicum dactylon.
»   crus galli.
Rottboella.
Cenchrus.
Arundo phragmites.

Poa chinensis.
»   2 other species.
Cynosurus indicus.
Saccharum.
Anthistiria ciliaris.
Andropogon.
Schoenus aculeatus.
Chara.
Marchantia.
Hypnum.
Equisetum.
Polypodium hastatum, Thunb.
»   falcatum, Thunb.
»   species.
Pteris serrulata, Hort. Kew.
»   semipinnata.
»   caudata.
Asplenium.
Davallia chinensis, Smith.
Woodwardia.
Trichomanes chinense.
Azolla filiculoides.
Marsilea quadrifolia.
»   natans.
Lycopodium cernuum.
»   species.

IV. Plants collected in the provinces of Kiang si and Kuang tung.
II, 524—525.

Clematis.
Dianthus deltoides.
Thea.
Camellia Sesanqua.
Hibiscus syriacus.
Gossypium.
Sterculia platanifolia.
Xanthoxylon trifoliatum.
»   species.
Citrus trifoliata.

Ilex, species nova.
Indigofera tinctoria.
Arachis hypogaea.
Glycine.
Cassia obtusifolia.
Mimosa.
Rubus.
Rosa indica.
»   other species.
Crataegus bibas, Lour.

Crataegus glabra, Thunb.
Lagerstroemia indica.
Panax aculeata.
Jussiaea erecta.
Lonicera.
Nauclea orientalis.
Mussaenda frondosa.
Gardenia florida.
     »   asclepiadea.
     »   spec.
Psychotria asiatica.
     »   serpens.
Plectronia chinensis, Lour.
Rondeletia asiatica.
Dysoda fascicularis, Lour.
Serissa.
Elephantopus.
Chrysanthemum indicum.
Artemisia.
Sphenoclea zeylanica.
Azalea indica.
Plumbago zeylanica.
Bladhia japonica.
Myosotis scorpioides.
Convolvulus sericeus.
     »   Batatas.
     »   medius.
     »   obscurus.
Solanum diphyllum.
     »   verbascifolium.
     »   nigrum.
     »   species.
Capsicum.
Physalis.
Lycium japonicum.
     »   foetidum.
Datura.
Nicotiana Tabacum.
Justicia.

Illecebrum sessile.
Rumex.
Polygonum Fagopyrum.
Laurus Camphora.
Daphne indica.
Elaeagnus pungens.
Croton sebiferum.
Euphorbia.
Ficus indica.
Urtica nivea.
Quercus dentata.
Pinus Larix.
Juniperus barbadensis.
Cupressus pendula.
Burmannia, n. sp.
Convallaria japonica.
Smilax.
Allium.
Pontederia.
Tradescantia.
Juncus articulatus.
Sagittaria trifolia.
Kyllingia triceps.
Eriocaulon.
Panicum italicum.
Rottboella.
Cenchrus.
Arundo Bambu.
Triticum.
Saccharum.
Anthistiria ciliata.
Ischaemum aristatum.
Polypodium.
Pteris semipinnata.
Asplenium, 2 species.
Ophioglossum scandens.
Acrostichum siliquosum.
Lycopodium cernuum.

We shall now proceed to give a list of the plants gathered by Staunton and Macartney, which have since been determined and published by competent botanists, and notice the plants which those collectors introduced from China into English gardens. As can be judged from Staunton's enumerations, a considerable part of his specimens remain still unpublished, and, on the other hand, many of the published species are omitted in Staunton's lists. Thus De Candolle describes five species of Clematis brought by Staunton from China, but Staunton mentions only one species, in the fourth list.

Clematis chinensis, Retz. DC. Syst. I (1818) 137. Hab. in China et insulis Danes dictis (Danes' island, near Canton). Macartney, Herb. Banks.

Clematis terniflora, DC. Syst. I, 137. China, provincia Che kiang, G. Staunton. Herb. Banks. — I. F. S., 1, 7, = *Cl. recta*, Linn. var. *mandshurica*, Maxim.

Clematis longiloba, DC. Syst. I (1818) 136. China. G. Staunton. Commun. cl. Lambert. — I. F. S., 1, 5.

Clematis brevicaudata, DC. Syst. I (1818) 138. China, inter Peking et Jehol. Staunton. Herb. Banks. — I. F. S., 1, 3.

Clematis heracleaefolia, DC. l. c. 138. China, inter Peking et Jehol. Herb. Banks. — I. F. S., 1, 4.

Michelia fuscata, Blume. (*Magnolia fuscata*, Andr.) — I. F. S., 1, 24: Macartney, Staunton. Mus. Brit.

Stauntonia chinensis, DC. Syst. I (1818) 514. New genus. China, Staunton; herb. Banks et Lambert. — I. F. S., 1, 30.

Berberis sinensis, Desf. — DC. Syst. II (1821) 8: Cl. Lemmonier primus olim coluit hanc speciem ex seminibus chinensibus. Specimina spontanea a Cl. G. Staunton inter Peking et Jehol lecta vidi in herbariis Bank et Lambert. — I. F. S., 1, 31.

Nandina domestica, Thbg. — I. F. S., 1, 32: Che kiang, Staunton.

Euryale ferox, Salisb. — I. F. S., 1, 33: Hai nan, Staunton, Mus. Brit. Hort. Kew. — (I may observe that Staunton never visited Hai nan; probably Kiang nan is meant).

Bocconia cordata, Willd. — DC. Syst. II (1821) 91: China, Staunton. — Hort. Kew, 2d ed., III, 142: China, (Che kiang). John Haxton (a gardener, who accompanied Macartney's embassy; v. s. 163). Introduced 1795 by Staunton. — I. F. S., 1, 35.

Nasturtium indicum, DC. Syst. II (1821) 199. China, herb. Banks. — I. F. S., 1, 40: Macartney and Staunton, Mus. Brit.

Nasturtium microspermum, DC. l. c. Prov. Shan tung, Staunton, herb. Banks. — I. F. S., 1, 40.

Viola Patrinii, DC., var. *chinensis*, DC. Prod. I (1824) 293. China, Prov. Kiang nan. Staunton, herb. Banks. — I. F. S., 1, 53.

Silene Fortunei, Vis. — I. F. S., 1, 65. Che kiang, Staunton.

Tamarix chinensis, Lour. — I. F. S., 1, 347. China, Macartney.

Hypericum chinense, Lin. — I. F. S., 1, 73: China, Staunton, Macartney.

Eurya chinensis, R. Br. — I. F. S., 1, 76: China, Staunton.

Eurya Macartneyi, Champ. in Hook. Journ. Bot. III (1851) 307. — I. F. S., 1, 77. Lord Macartney, China.

Camellia japonica, Lin. — I. F. S., 1, 81: China, Macartney, Staunton.

Camellia Sesanqua, Thbg. — Staunton, Embassy, II, 466, figured; v. s. 162. — I. F. S., 1, 82: Kuang tung, Staunton.

Sida cordifolia, Lin. — I. F. S., 1, 85: China, Staunton.

Sida Stauntoniana, DC. Prod. I (1824) 460. China, Staunton. — I. F. S., 1, 84, sub *Sida acuta*, Burm.

Hibiscus Abelmoschus, Lin. — I. F. S., 1, 87: Kuang tung, Staunton.

Bombax malabaricum, DC. — I. F. S., 1, 89. China, Staunton.

Sterculia platanifolia, Lin. fil. — Debeaux, Flor. Tien tsin (1879) 14: Peking, Staunton.

Grewia Stauntoniana, G. Don, Gen. Syst. I (1831) 551. China, herb. Lambert. — I. F. S., 1, 92: *G. Microcos*, Linn.

Erodium Stephanianum, Willd. — I. F. S., 1, 99: Near Peking, Staunton.

Zanthoxylum alatum, Roxb. — I. F. S., 1, 105: Che kiang, Kiang nan, Staunton.

Zanthoxylum Avicennae, DC. — I. F. S., 1, 105: China, Staunton.

Zanthoxylum nitidum, DC. — I. F. S., 1, 106: Kuang tung: Danes' Island, Whampoa, Macartney.

Zanthoxylum schinifolium, Sieb. & Zucc. — I. F. S., 1, 107: Kiang su, Staunton.

Glycosmis pentaphylla, Correa. — I. F. S., 1, 109: China, Macartney, Staunton.

Murraya exotica, Lin. — I. F. S., 1, 109: China, Macartney.

Ailantus glandulosa, Desf. — I. F. S., 1, 112: Chi li Prov., Staunton.

Brucea sumatrana, Roxb. — I. F. S., 1, 112: China, Macartney, Staunton.

Ilex cornuta, Lindl. — I. F. S., 1, 115: Kiang si, Staunton.

Euonymus Bungeanus, Max. — I. F. S., ɪ, 118: Chi li, Shan tung, Staunton.

Euonymus chinensis, Lindl., Trans. Hort. Soc. VI (1826) 74. — I. F. S., ɪ, 119: China, Staunton.

Euonymus japonicus, Thbg. — I. F. S., ɪ, 120: Kiang nan, Staunton.

Euonymus Thunbergianus, Blume. — I. F. S., ɪ, 120: Che kiang, Staunton.

Celastrus articulatus, Thbg. — I. F. S., ɪ, 122: Kiang su, Staunton.

Paliurus ramosissimus, Poiret. — I. F. S., ɪ, 126: Kuang tung, Staunton.

Zizyphus vulgaris, Lam. — I. F. S., ɪ, 126: Chi li, Shan tung, Staunton.

Berchemia lineata, DC. — I. F. S., ɪ, 127: China, Macartney.

Sageretia theezans, Brogn. — I. F. S., ɪ, 131: China, Staunton.

Vitis japonica, Thunb. — I. F. S., ɪ, 134: Kiang su, Staunton.

Acer pictum, Thbg. — I. F. S., ɪ, 141: between Peking and Jehol, Staunton.

Rhus succedanea, Lin. — I. F. S., ɪ, 147: Kiang su, Staunton.

Crotalaria medicaginea, Lam. — I. F. S., ɪ, 152: China, Macartney.

Indigofera atropurpurea, Ham. — I. F. S., ɪ, 156: Kuang tung, Staunton.

Indigofera hirsuta, Lin. — I. F. S., ɪ, 157: China, Macartney, Staunton.

Indigofera virgata, var. *parvifolia,* DC. Prod. II (1824) 224. China, Staunton. — I. F. S., ɪ, 158.

Millettia reticulata, Benth. — I. F. S., ɪ, 159: Kwang tung, Staunton.

Caragana chamlagu, Lamarck. — I. F. S., ɪ, 163: Shantung, Staunton; spec. typ. Berberidis caraganaefoliae in Mus. Brit.

Astragalus complanatus, R. Br. — I. F. S., ɪ, 165: Chi li, Staunton. Mus., Brit., Herb. Kew.

Astragalus chinensis, Lin. fil. — Ait. Hort. Kew, 2d edit. IV, 362: Introduced into the Kew Gardens in 1795 by Sir G. Staunton. — I. F. S., ɪ, 165.

Aeschynomene indica, Lin. — I. F. S., ɪ, 170: Shan tung, Staunton.

Hedysarum latifolium, Roxb. — Bot. Reg. t. 355 (1819). The sample figured was sent by Mr. Lambert from Boyton House and formed part of a plant raised from seeds collected in China by Sir G. Staunton. — I. F. S., ɪ, 173. *Desmodium latifolium,* DC.

Alysicarpus vaginalis, DC. — I. F. S., ɪ, 179: China, Staunton.

Lespedeza bicolor, Turcz. — I. F. S., I, 179: Che kiang, Kiang su. Staunton.

Lespedeza juncea, Pers. — I. F. S., I, 181: Between Peking and Jehol, Staunton.

Lespedeza striata, Hook. & Arn. — I. F. S., I, 182; China, Staunton.

Lespedeza medicaginoides, Bunge. — I. F. S., I, 182: Staunton, near Peking.

Pueraria Thunbergiana, Benth. *(Pachyrhizus Thunb.* Sieb & Zucc.). — I. F. S., I, 191: Che kiang, Staunton.

Derris uliginosa, Benth. — I. F. S., I, 199: China, Macartney.

Sophora japonica, Lin. — I. F. S., I, 202: China, Macartney, Chili, Staunton.

Albizzia Millettii, Benth. — I. F. S., I, 216: China, Staunton.

Rubus rectangulifolius, O. Kuntze, Methodik, p. 60, 78: China, Staunton. In Herb. Vindob. — I. F. S., I, 236.

Rubus reflexus, Bot. Reg. t. 461 (1820): Samples of this species were gathered by Sir G. Staunton in the prov. of Canton and deposited in the herbariums of Sir J. Banks and Mr. Lambert. — I. F. S., I, 237.

Potentilla chinensis, Seringe in DC. Prodr. II (1825) 581. China. (ex herb. Lamb.) Lehmann, Nov. min. cogn. Stirp. pugillus III (1830) 22, P. chinensis. Petchely, Staunton. V. in herb. Banks et Lambert. — Earlier d'Incarville Peking. Supra p. 53.

Rosa bracteata, Wendland, Observ. (1798) 50. *Rosa Macartnea,* Dumont. Bot. cult. (1802). — Lindl. Ros. (1821) p. 10, var. β, *scabricaulis:* China, prov. Che kiang; Herb. Banks. Lindl. saw it also cultivated. — Bot. Mag. (1811) 1377. Rosa bracteata: Introduced by Lord Macartney from China. — Hort. Kew, 2d edit. III 267, says that Sir John (sic!) Staunton introduced it, in 1795 and that it was procured from China by John Haxton*). — I. F. S., I, 249.

Rosa indica, Lin. — Lindl. Ros. (1821) 106: China juxta Cantonem Sinarum; Staunton. Herbar. Banks. Lindl. saw it also cultivated. — I. F. S., I, 250.

Rosa hystrix, Lindl. Ros. (1821) 129: China, prov. Kiang si; Staunton. Herb. Banks et Lambert. — I. F. S., I, 250, *Rosa laevigata,* Michx.

Rosa microcarpa, Lindl. Ros. 130: China, prov. Canton, Staunton. Herb. Banks. — I. F. S., I, 251.

Rosa multiflora, Thbg. — Lindl. Ros. 119: China, Staunton. Herb. Lambert. — I. F. S., I, 253.

---

*) v. supra p. 163.

Photinia serrulata, Lindl. Trans. Linn. Soc. XIII (1822), 103. China, Staunton. Herb. Banks.

Penthorum chinense, Pursh Fl. Amer. sept. I (1814), 323. China, Staunton. Herb. Lambert. — I. F. S., I, 288, *Penthorum sedoides,* Linn.

Two species of *Eugenia,* collected by Macartney, are in the British Museum, not identified. — I. F. S., I, 298.

Sizygium odoratum, DC. Prodr. III (1828) 261. China, Staunton (Lambert). — I. F. S., I, 297, *Eugenia Millettiana,* Hemsl.

Melastoma malabathricum, Willd. — Hort. Kew, ed. 2d, III, 46: introduced into the Kew Gardens by Sir G. Staunton 1795, it seems from China. Comp. also Bot. Reg. (1822) 672. — Don in Mem. Wern. Soc. IV (1823) 288, reduces M. malab. to *M. macrocarpum,* Don. China. Ibidem: *M. candidum,* Don. China, Staunton, Herb. Lambert. — I. F. S., I, 299: All these species reduced to *M. candidum.*

Bupleurum chinense, DC. Prodr. IV (1830) 128. In China legit cl. Staunton. Comm. a cl. Lambert. — I. F. S., I, 327 sub *B. falcatum* Linn.: Chi li, Staunton.

Selinum Monnieri, Lin. — I. F. S., I, 333: China, Macartney.

Acanthopanax aculeatum, Seem. — I. F. S., I, 339: China, Staunton, Macartney.

Heptapleurum octaphyllum, Hance. — I. F. S., I, 342: Macao, Macartney.

Lonicera longiflora, DC. — I. F. S., I, 364. South China, Staunton.

Psychotria Reevesii, Wall. — DC. Prodr. IV (1830) 519. China, Staunton. Comm. a cl. Lambert. — I. F. S., I, 387 sub *Ps. elliptica.* China, Staunton.

Serissa Democritea, Baill. Kiang nan, Chekiang, Staunton. — I. F. S., I, 391.

Spermacoce hispida, Lin. — I. F. S., I, 392: Kuang tung, Staunton.

Calimeris integrifolia, Turcz. in DC. Prod. V, 259. Staunton, China. I. F. S., I, 412, *Aster holophyllus,* Hemsl.

Sphaeranthus africanus, Lin. — I. F. S., I, 423: China, Staunton.

Inula pseudocappa, DC. Prodr. V (1836) 469. China, Staunton. Comm. cl. Lambert. — I. F. S., I, 429. *Inula Cappa,* DC., South China, Staunton.

Bidens parviflora, Willd. — DC. Prodr. V (1836) 602. China, Staunton. — I. F. S., I, 435.

Chrysanthemum indicum, Lin. — I. F. S., I, 437: China, Staunton.

Myriogyne minuta, Lessing. — Debeaux, Flor. Tientsin (1879) 84. Shan tung, Staunton. — I. F. S., I, 440.

Artemisia lactiflora, Wall. — DC. Prod. VI (1837) 115. China, Staunton. — I. F. S., ɪ, 144.

Artemisia vulgaris, Lin. — Debeaux, l. c. 25. Peking, Staunton.

Artemisia lavandulaefolia, DC. Prodr. VI, 110. China, Staunton. — I. F. S., ɪ, 446, variety of *A. vulgaris*.

Artemisia scoparia, Waldst. — DC. Prodr. VI, 99. China, Staunton. — I. F. S., ɪ, 445.

Gynura divaricata, DC. — I. F. S., ɪ, 447: Canton, Staunton.

Emilia sonchifolia, DC. Prodr. VI, 302. China, Staunt. — I. F. S., ɪ, 449.

Senecio scandens, Ham. — I. F. S., ɪ, 457: China, Staunton.

Senecio Stauntonii, DC. Prodr. VI, 363. China, Staunton. Comm. cl. Lambert. — I. F. S., ɪ, 458.

Wahlenbergia homallanthina, DC. Prodr. VII (1839), 425. China borealis. Herb. Banks. — I. F. S., ɪɪ, 5: Staunton.

Adenophora sinensis, DC. Campanul. (1830), 354, t. 6. China, Staunton. — I. F. S., ɪɪ, 13.

Statice sinensis, Girard in Ann. Sc. nat., 1845, 329. China, Staunton. Herb. Webb. — DC. Prodr. XII, 642. Idem. Herb. De Cand., Webb. — I. F. S., ɪɪ, 35.

Ardisia japonica, Blume. — DC. Prodr. VIII (1844) 135. China, Staunton. Herb. Delessert. — I. F. S., ɪɪ, 65.

Ligustrum Stauntonii, DC. Prodr. VIII, 294. China, Staunton. Comm. cl. Lambert. — I. F. S., ɪɪ, 92. *Ligustrum sinense*, Lour.

Strophanthus dichotomus, β. *chinensis*, Ker in Bot. Reg. (1820) 469. Macao, Staunton. Herb. Banks. — I. F. S., ɪɪ, 97. *Strophanthus divergens*, Grah. (1827).

Cynanchum chinense, R. Brown in Mem. Wern. Soc. I (1808), 44. China, prov. Pechely; Staunton. — Earlier Incarville (supra p. 33). — DC. Prodr. VIII, 548. Idem. Herb. Webb. — I. F. S., ɪɪ, 105.

Metaplexis Stauntonii, R. Brown, l. c. p. 49; Roem. & Schult. Syst. VI, 111. Prov. Pechely; Staunton. — I. F. S., ɪɪ, 111.

Pentasachme glaucescens, Decaisne in DC. Prodr. VIII (1844) 627. China, Staunton. Herb. De Candolle et Maille, ex Herb. Lambert. — I. F. S., ɪɪ, 112.

Pentasachme Stauntonii, Decaisne, l. c. 627. China, Staunton. Herb. De Candolle et Maille, ex Herb. Lambert. — I. F. S., ɪɪ, 112.

Anchusa tenella, Hornem. Hort. Hafn. I (1813) 176. Introd. 1806. — *Bothriospermum tenellum*, Fisch. & Mey. — DC. Prodr. X, 116. China, Staunton. — I. F. S., ɪɪ, 151.

Argyreia, undetermined species *(A. mollis,* Choisy?) in the Brit. Mus., Che kiang, Staunton. — I. F. S., ii, 156.

Ipomoea Chryseides, Bot. Reg. 270. — DC. Prodr. IX (1845) 382, China, Staunton. Herb. Mus. Paris. — I. F. S., ii, 158.

Convolvulus chinensis, Ker in Bot. Reg. (1818) 322. Staunton in prov. Pechely. Herb. Banks. — I. F. S., ii, 165: *C. arvensis,* Linn.

Convolvulus calystegioides, DC. Prodr. IX, 413. China, Staunton. — I. F. S., ii, 167.

Solanum Osbeckii, DC. Prodr. XIII, 1 (1852), 179. China, Lord Macartney. Herb. Banks. — I. F. S., ii, 169: *S. biflorum,* Lour.

Solanum Dulcamara, Lin., var. *chinense,* DC. Prodr. l. c. 79. China, Staunton and Macartney; in Herb. Banks and Lambert. — I. F. S., ii, 169.

Solanum verbascifolium, Lin. — DC. Prodr. l. c. 115. China, Staunton in Herb. Lambert et De Candolle. — I. F. S., ii, 172.

Veronica paniculata, Lin. — DC. Prodr. X (1846) 465. China borealis, Staunton. — I. F. S., ii, 200, *V. spuria,* Lin.

Hygrophila assurgens, DC. Prodr. XI (1847) 90. China, Staunton. Herb. De Candolle ex Herb. Lambert. — I. F. S., ii, 237: *H. salicifolia,* Nees.

Lepidagathis hyalina, Nees., var. *dependens,* DC. Prodr. XI, 253. China, Staunton. Herb. De Cand. et Lambert. — I. F. S., ii, 244.

Rostellularia procumbens, Nees. DC. Prodr. XI, 372. China, Staunton. Herb. De Cand. — I. F. S., ii, 246: *Justicia procumbens,* Lin.

Callicarpa Reevesii, Wall. — DC. Prodr. XI, 641: China ad Canton, Staunton. — I. F. S., ii, 254.

Clerodendron fragrans, Vent., var. *pleniflora,* DC. Prodr. XI, 666. China, Staunton. — I. F. S., ii, 260.

Elsholtzia integrifolia, Benth. in DC. Prodr. XII (1848) 161. China, Staunton. Herb. Banks et De Cand. — I. F. S., ii, 277.

Elsholtzia Stauntonii, Benth. l. c. 160. Inter Peking et Jehol, Staunton. Herb. Banks. — I. F. S., ii, 278.

Lycopus lucidus, Turcz., DC. Prodr. XII, 179. China, Staunton? — I. F. S., ii, 282.

Salvia chinensis, Benth., DC. Prodr. XII, 355. China, prope Kia nang (mistake for province of Kiang nan), Staunton. — I. F. S., ii, 284, *S. japonica,* Thbg.

Nepeta tenuifolia, Benth., DC. Prodr. XII, 370. China, inter Peking et Jehol, Staunton. Herb. Banks. — I. F. S., ii, 290.

Scutellaria macrantha, Fisch., DC. Prod. XII, 424. China, inter Peking et Jehol, Staunton. — I. F. S., ii, 294: *Sc. baicalensis,* Georgi.

Scutellaria indica, Lin., DC. Prodr. XII, 417. China prope Macao, Staunton. — I. F. S., II, 295.

Scutellaria rivularis, Wall., DC. Prodr. XII, 426. China, in Kiang nan, Staunton. — I. F. S., II, 296.

Leucas molissima, Wall., var. *chinensis*, DC. Prodr. XII, 525. China, Staunton. — I. F. S., II, 304.

Amethystea coerulea, Lin. DC. Prodr. XII, 572. China, Staunton. — I. F. S., II, 310.

Ajuga decumbens, Thbg. — DC. Prodr. XII, 599. China, Staunton.— I. F. S., II, 315.

Celosia argentea, Lin. — DC. Prodr. XIII, II (1849), 243. China, Staunton. — I. F. S., II, 318.

Amarantus caudatus, Lin., var. *albiflora*, Mocq. — Debeaux Fl. Tche-fou (1877), 120: China, Staunton. — I. F. S., II, 319.

Amarantus paniculatus, Lin., var. *strictus*. Debeaux, l. c., China, Staunton. — I. F. S., II, 320.

Amarantus spinosus, Lin. — Debeaux, l. c. China, Staunton. — I. F. S., II, 320.

Achyranthes aspera, Lin., var. *indica*, DC. Prodr. XIII, II, 315. China, Kia nong (provincia Kiang nan), Staunton. Herb. De Cand., Fée, Delessert, Mus. Paris. — I. F. S., II, 322.

Alternanthera sessilis, R. Brown (1810). — DC. Prodr. l. c. 357., var. *Stauntonii*. China, Staunton. — I. F. S., II, 323.

Teloxys aristata, Moq. — DC. Prodr. l. c. 59. China, Staunton. Herb. Lambert. — I. F. S., II, 324, *Chenopodium aristatum*, Lin.

Corispermum Stauntonii, Moq. in DC. Prodr. l. c. 140. China, Staunton. — Debeaux, Fl. Tche fou, 118, Shan tung, Staunton. — I. F. S., II, 327.

Helicilla altissima, Moq. in DC. Prodr. l. c. 170. In China, Staunton. Salsola altissima, Dryander, mss in herb. Lambert, non Linn. (V. supra p. 165). — I. F. S., II, 328, *Suaeda glauca*, Bge, same as *Suaeda Stauntonii*, Moq. Chenop. enum. p. 31.

Polygonum amphibium, Lin. — DC. Prodr. XIV (1856) 115. China, Staunton. — I. F. S., II, 333.

Polygonum capitatum, Hamilt. (1802). — DC. Prodr. XIV, 129. China, Staunton. Herb. De Cand. — I. F. S., II, 335.

Polygonum chinense, Lin. — DC. Prodr. XIV, 130. China, Staunton. — Hort. Kew 2d ed. II, 420. Introduced into the Kew Gardens 1795 by Sir John (sic!) Staunton. Sent from China by Haxton. — I. F. S., II, 335.

Polygonum jucundum, Meisner, Monogr. Polyg. (1826) 71. China, Staunton. — I. F. S., II, 341.

Polygonum multiflorum, Thbg. — DC. Prodr. XIV, 136. China. J. G. Staunton. — I. F. S., II, 342.

Polygonum minus, Hudson. (1762). — DC. Prodr. XIV, 111. China, Staunton. — I. F. S., II, 342.

Polygonum orientale, Lin., var. *pilosum,* DC. Prodr. XIV, 123. China, Staunton. — I. F. S., II, 343.

Rumex chinensis, Campdera, Monogr. Rumex, (1819) 76. China, Staunton. Herb. De Cand. — I. F. S., II, 357, *R. maritimus,* Lin.

Cinnamomum Camphora, Nees. ab Es. — DC. Prodr. XV, I (1864) 24. China, Staunton. — I. F. S., II, 371.

Tetranthera polyantha, Wall. — DC. Prodr. XV, I, 182. China, Staunton. — I. F. S., II, 379, *Litsea citrata,* Blume.

Daphnidium strychnifolium, Sieb. — DC. l. c. 230. China, Staunton. — I. F. S., II, 392, *Lindera strychnifolia,* Villar.

Loranthus chinensis, A. P. De Candolle, Mém. Loranth. (1830) 28, t. 7. Staunton, China. Herb. Lambert. — I. F. S., II, 405.

Phyllanthus Emblica, Lin. — DC. Prodr. XV, II, 351. China, Staunton. — I. F. S., II, 421.

Phyllanthus puberus, Müller *(Agyneia impubes,* Lin.)., Debeaux, Fl. Tientsin, 127. Shan tung, Staunton. — I. F. S., II, 425.

Securinega obovata, Müller, DC. Prod. XV, II, 449. China, Staunton. Herb. DC. — *Xylophylla obovata,* Willd. Enum. pl. hort. Berol. I (1809) 329. Culta, Berol. Habitat unknown. — I. F. S., II, 427, = *Fluggea microcarpa,* Blume.

Mallotus ricinoides, Müller, l. c. 964. China, Staunton. Herb. DC. — I. F. S., II, 442.

Mallotus tiliaefolius, Müller, l. c. 969. China, Staunton. Earlier Ceylon. — I. F. S., II, 442.

Celtis sinensis, Persoon, Syn. Pl. I (1805) 286. China, ex hort. Celsii. — Planchon, Ulmac. (1848) 286. China. Inter Peking et Jehol, Staunton, Herb. Banks. — I. F. S., II, 450.

Sponia velutina, Planchon, l. c. 327. China, Staunton. Herb. Banks.— I. F. S., II, 451, *Trema amboinensis,* Blume. — According to Hook. fil Fl. Ind. V (1890) 484, same as *Celtis tomentosa,* Roxb.

Boehmeria densiflora, Hook. & Arn. Voy. Beech. 271. — DC. Prod. XVI, I, 215. China, Che kiang, Staunton.

Quercus mongolica, Fischer in Turcz. Cat. baic. dah. (1838) n. 1066.—

Carruthers, Oaks from N. China, in Journ. Linn. Soc. VI (1862) 31: between Peking and Jehol, Staunton.

Quercus obovata, Bge, Enum. Chin. (1831) 63 *(Q. dentata,* Thbg.). — Carruthers, l. c. 32: between Peking and Jehol, Staunton.

Quercus acutissima, Carruthers, l. c. 33 *(Q. serrata,* Thbg.?). Prov. Kiang si, Staunton.

Quercus aliena, Blume (1849). — DC. Prod. XVI, II. China, Staunton.

Cupressus funebris, Endlicher, Conif. (1847), 58. China, Staunton. In Herb. De Cand. — V. s. 160, Cupressus pendula, Staunton. — Lambert "Pinus", 2d edition (1828) II, 111, t. 50. Probably introduced to Kew by W. Kerr.

Glyptostrobus heterophyllus, Endlicher, Conif. (1847) 70. China, Staunton. — *Thuya pensilis,* Staunton, Embassy to China, II, 436 (v. s. 170). Lambert Pinus, 2d ed. 1828, II, 115, t. 51. — Benth. & Hook. Gen. Pl. III, 429, reduce Glyptostrobus to *Taxodium.*

Podocarpus chinensis, Wall. Catal. 6051. — Maxwell, Conifers of Japan, Linn. Soc. Journ. XVIII (1880) 502. China, Macartney. — Maximowicz, Mél. VII (1870) 562, = variety of *Pod. macrophylla,* Don.

Pinus lanceolata, Lambert "Pinus" (1803) 52, t. 34. China, Che kiang, Staunton. Banksian Herb. — Richard, Conif. (1826) 80, called the tree *Cunninghamia sinensis* (v. s. 43).

Pinus Massoniana, Lamb. Pin. (1803) t. 12; ed. 2, p. 16, t. 8, and ed. 3 (1832) 47, t. 29, *Pinus sinensis,* Lamb. — Maxwell, l. c. 505. China, Macartney.

Asparagus lucidus, Lindl. in Bot. Reg. 1844, Misc. n. 36. — Baker, Asparag. (1875) 605. Macao, Staunton.

Asparagus dauricus, Fischer *(A. gibbus,* Bge). — Baker, l. c. 599. China, borealis, Staunton.

Funkia subcordata, Sprengel, Syst. II, 41. — Baker Liliaceae (1871) 367. China bor., Staunton.

Bot. Mag. 1811, sub t. 1404. A *Dianella,* allied to *D. ensifolia.* China, Staunton.

Allium longistylum, Baker, Journ. Bot. 1874, 294. Inter Peking et Jehol, Staunton. Hb. Mus. Brit.

Allium exsertum, Baker, l. c. China, Kiang si, Staunton. Hb. Mus. Brit.

Monochoria vaginalis, Lin. — DC. Monogr. Phan. IV (1883) 525. China, Staunton.

Aneilema sinica, Bot. Reg. t. 659 (1822). Canton, Staunt. Herb. Banks.

Damasonium, several species brought from China by Staunton. Bot. Mag. sub t. 1201 (1809). Herb. Banks.

Cyperus difformis, Lin. — Debeaux, Fl. Tchéfou, 144. Shan tung, Kiang nan.

Cyperus rotundus, Lin. — Deb. Fl. Tientsin, 45. Province de Tchély, Staunton.

Cyperus sanguinolentus, Vahl. — Deb. Fl. Tchéfou, 142. Shan tung, Kiang nan, Staunton.

Killingia monocephala, Lin. — Deb. l. c. 146. Shan tung, Staunton.

Fimbristylis tomentosa, Vahl. — Deb. l. c. 150. Shan tung, Staunton. Herb. Staunt. sub Scirpo.

Fimbristylis miliacea, Vahl. — Deb. l. c. 151. Shan tung, Staunton.

Fimbristylis Stauntoniana, Deb. l. c. et Franchet, t. 3. Shan tung, Staunton.

Scirpus triqueter, Lin. — Dr. Hance in Journ. Bot. 1874, 330, Shan tung, Staunton.

Panicum crus galli, Lin. — Deb. Fl. Tientsin, 47. Cult. Peking. Macartney.

Panicum frumentaceum, Roxb. — Deb. Fl. Tchéfou, 154. Shang tung occid., cult. Staunton.

Gymnothrix japonica, Kunth. *(Panicum hordeiforme*, Thbg.). — Deb. l. c. 156. Kiang nan, Staunton.

Tragus tcheliensis, Deb. Fl. Tientsin, 48. Prov. de Tché ly, Staunton, sub *Cenchro racemoso*.

Ischaemum ciliare, Retz. — Deb. Fl. Tchéfou, 163. Kiang nan, Staunton.

Andropogon Ischaemum, Lin. — Deb. Fl. Tien tsin, 51. Prov. de Tchély, Staunton.

Sorghum halepense, Pers. — Deb. Fl. Tché fou, 164. Kiang nan, Staunton. Cultum.

Orypsis aculeata, Ait. *(Schoenus aculeatus*, Lin.) — Deb. l. c. 157. Shan tung, Staunton.

Cynodon dactylon, Pers. *(Panicum dactylon*, Lin.). — Deb. l. c. 159. Kiang nan, Staunton.

Chloris caudata, Trinius in Bunge, Enum. (1834) n. 404. — Deb. l. c. 158. Tchély, Shantung, Staunton, sub *Adropogone*.

Eleusine indica, Gaertn. *(Cynosurus indicus*, Lin.). — Deb. l. c. 159. Shan tung, Staunton.

Arundo Phragmites, Lin. — Deb. l. c. 157. Grand Canal, Staunton.

Phragmites longivalvis, Steudel. — Deb. Fl. Tien tsin, 50. Tchély, Staunton (sub *Arundo Phragmites).*

Eragrostis ferruginea, Beauv. *(Poa ferruginea,* Thbg.). — Deb. Fl. Tchéfou, 161. Shan tung, Kiang nan, Staunton.

Eragrostis tenella, Beauv. *(Poa tenella,* Lin.). — Deb. l. c. Shan tung, Kiang nan, Staunton.

Eragrostis pilosa, Beauv. *(Poa pilosa,* Lin.). — Deb. l. c. 159. Kiang nan, Kiang si, Staunton.

Eragrostis megastachya, Link. *(Briza eragrostis,* Lin.). — Deb. l. c. 160. Kiang nan, Staunton.

Bambusa flexuosa, Munro, Bambus. (1868) 101. Kuang tung, Staunton. Osbeck earlier.

Bambusa Beecheyana, Munro, l. c. 108. China, Staunton.

Phyllostachys Stauntoni, Munro, l. c. 37. China, Staunton. — Rivière, Bambous, 19.

Equisetum ramosum, Schleich. (1807). — Deb. Fl. Tientsin, 53; Fl. Tchéfou, 167. Tchély, Shan tung, Kiang nan, Staunton.

Selaginella Stauntoniana, Spring, Monogr. Selag. II (1842) 71. North China, Staunton. — Earlier d'Incarville.

Selaginella mongolica, Rupr., 1845. — Baker Syn. Selag. in Journ. Bot. 1883, 45: between Peking and Jehol, Staunton.

Asplenium incisum, Thbg. *(A. elegantulum,* Hooker). — M. Kuhn, Chin. Filices, in Journ. Bot. 1868, 268. Peking, Staunton.

Davallia elegans, Swartz, Filices (1806) 132. — J. E. Smith in Rees Cyclop. (1808) n. 13, says: brought by Staunton from China.

Lindsaea flabellulata, Dryander in Linn. Trans. III (1797) 41. Canton, Staunton.

II. A. E Van Braam Hoockgeest. 1794—95.

Eleven months after Lord Macartney had returned from Peking to Canton, a Dutch Embassy proceeded to Peking on a mission of salutation and respect to the Emperor K'ien lung, from the Government of Batavia, on the occasion of his reaching the sixtieth year of his reign. They appointed Isaac **Titsingh** of the Great Council of the D. E. I, who had resided many years in Japan, as Ambassador. Van Braam, the Dutch Consular Agent at Canton, occupied the second place; De Guignes, French Commissary in Macao accompanied the mission in the quality of interpreter. Other members of it were: R. Dozy, Secretary, — I. H. Blettermann, Physician,— A. van Braam, jun., nephew of A. E. van Braam.

Some years later an account of this embassy was published with the following title:

Voyage de l'Ambassade de la Compagnie des Indes orientales hollandaises, vers l'Empereur de la Chine, dans les années 1794 et 1795: où se trouve la Description de plusieurs parties de la Chine inconnues aux Européens et que cette Ambassade a donné l'occasion de traverser: Le tout tiré du journal d'A. E. van Braam Hoockgeest, chef de la Direction de la Comp. des Indes orient. holl. à la Chine et second dans cette Ambassade... orné de cartes et de gravures. — Publié en Français par M. L. E. Moreau de St. Méry à Philadelphie, 1797—1898, 2 vol. in 4. — Other edition, Paris 1798.

Van Braam, Andreas Everard, was born in Holland, in 1739. He went to China, in 1758, in the quality of supercargo of the D. E. I. Comp., and resided in Macao and Canton till 1773, when he returned to Holland. In 1783 he settled in the U. S. of N. America, but soon returned to Europe and was appointed Director of the D. E. I. Comp. at Canton. In 1794 he accompanied the Dutch Embassy to Peking, occupying the second place. In 1795 he left Canton, and arrived in 1796 at Philadelphia in the U. S. of N. America, where he spent the rest of his life. Van Braam left China with vast collections, especially drawings made by Chinese, amongst others coloured drawings of Plants and Animals.

De Guignes, Chrét., Louis, Joseph, son of the celebrated French orientalist (sinologist) Joseph De Guignes (+ 1800) — was, from 1784—1801, French commissary in Macao, accompanied in 1794 the Dutch Embassy to Peking, in the quality of secretary as he asserts. In 1796 he visited Manilla and Isle de France, and in 1801 returned to Europe. He published a book: "Voyage à Pékin, Manille et l'Isle de France", 1808, 3 volumes, with an atlas. I, p. 255 he speaks of drawings, which van Braam ordered to be made for him by Chinese painters. Soon after his arrival at Canton, De Guignes sent to the Jardin des Plantes at Paris the seeds of a Chinese plant, which was successfully cultivated there and described by Lamarck, Enc. bot. III (1789) 103, as *Hemerocallis plantaginea*.

The following brief account of the journey of the Dutch Embassy from Canton to Peking and back, is extracted from Braam's diary contained in the above mentioned book.

1794   The Embassy started from Canton in November 1794 and, as far as

Lake Po yang, followed the same riverway by which Lord Macartney had returned: Past Shao chou fu, Nan hiung chou, over the Mei ling mountain range to Nan an fu, then down the river to Kan chou fu, Kian fu, and Nan ch'ang fu*) near Lake Po yang, reaching at least Kiu kiang fu. Having crossed the Yang tze River, they continued their journey overland in sedan chairs and carts, evidently on account of the season of the year, passing through the cities of Lü chou fu, Feng yang fu (An hui province), Sü chou fu (in Kiang su). Then they entered the province of Shan tung and proceeded straight to Peking, passing through Yen chou fu, Tung p'ing chou, and Kao t'ang chou (all in Shan tung). In Chi li their road lay through Ho kien fu, Cho chou, Liang hiang hien, Lu kou k'iao bridge. They reached Peking on the 9th of January 1795. After a stay of 37 days in the Chinese capi- 1795 tal, the Embassy set out again on February 15th to return to Canton. They travelled by the same way they had come as far as Te chou in Shan tung, and then took a more easterly route passing through Ts'i ho hien, T'ai an fu, Sin t'ai hien, Meng yin hien (all in Shan tung), Su ts'ien hien (in Kiang su). At the marked town of Sin kan p'u they embarked on boats, and then followed on the Grand Canal the route taken by Lord Macartney, via Yang chou fu, Chen kiang fu, Su chou fu to Hang chou fu, the capital of Che kiang. Thence they continued by river, passing through Yen chou fu, and K'ü chou fu (both in Che kiang)**). At Ch'ang shan hien they left this river, proceeded 85 *li* by land, and at Yü shan hien (in Kiang si) again reached a navigable river on which they continued their journey in boats to Kuang sin fu, and Nan ch'ang fu. From the latter place they followed the same riverway by which they had come and arrived at Canton on the 10th of May 1795.

Let me now point out from the account of the Dutch Embassy some topics of botanical interest:

V. I, p. 43—47: Between Canton and Shao chou fu much *Buckwheat* cultivated, also *Wheat*. Vast plantations of *Sugar cane*.

I, 53—81: The slopes of the mountains near Shao chou fu presented regular plantations of a shrub covered with white flowers. This plant produces a nut from which the people express an oil much used for burning in lamps. The same shrub was seen in cultivation between Kan chou fu and Wan an hien (in Kiang si). (This is Camellia Sasankwa. Comp. supra p. 162).

I, 121: South of Lu chou fu, in the province of Kiang nan (An hui)

---

*) This route was followed 138 years earlier by a Dutch Embassy. V. supra p. 22.
**) Lord Macartney's route. V. supra p. 161.

the people cultivate excellent yellow *Carrots*, much superior to those sold at Canton.

I, 145, II, 41: In Shan tung and even in the northern part of Kiang nan, at Sü chou, *Millet (Setaria italica)* constitutes the ordinary food of the people. Millet is not cultivated in the south of the empire.

II, 33: Near Ts'i ho hien (in Shang tung) the travellers observed very large *Pears*, from 14 to 15 inches in circumference and of a fine golden yellow colour. The skin somewhat hard, but the pulp very juicy and of an agreeable flavour. (Comp. supra p. 2, Marco Polo on Shan tung pears). — At the same place they also saw wonderful *Turnips* (navets) of a crimson colour. (Perhaps the excellent red winter radish of North China).

II, 73: The reddish *Namking Cotton* they found growing in the south-western part of the province of Kiang nan (An hui) in the district of Feng yang fu, near the sea (?).

II, 123: Pat chak san, a place near Wu kiang hien (in Kiang su) near the border of Chekiang, is famed for the oil expressed from *Rape* seeds. This rape *(Brassica chinensis*, it seems) is cultivated there in great abundance.

Icones Plantarum sponte China nascentium, e Bibliotheca Braamiana excerptae.

London J. H. Boothe, foreign bookseller to His Majesty. 1821.

This is the title of a volume in folio, not a rare book, consisting of 30 unnumbered plates, on which are represented 32 plants in beautifully colour-ed drawings. Most of the plates are signed H. B. K. There is a Latin preface, in which it is stated that the plates now published have been selected from the collection of drawings of plants brought from China by Van Braam, now in the possession of William Cattley, Esq., the excellent patron of science. The editor, however, observes, that in publishing these drawings, he has in view rather to show the skill of the Chinese in the art of paint-ing, than to promote botanical science. He further remarks that among these plates there will be found representations of a new Bauhinia, of Rosa microcarpa and Rosa involucrata and an Orchid of very paradoxical ap-pearance. — This preface is dated London, Febr. 10th 1821, not signed.— Pritzel, Thesaur. Lit. Botan., 1872, n. 10779, states with respect to the Icones Braamianae: "In Catalogo bibliothecae Radcliffe Oxoniensis, liber H. B. Ker *) adscribitur". — It seems very probable that J. Lindley ar-

---

*) Ker, Charles Henry Bellenden, 1785—1871. See Britten & Boulger, Biogr. Brit. Bot.

ranged the plates for publication and wrote the preface. (See Seemann's Journ. Bot. III (1865) p. 385).

There is in the Kew library another copy of this volume which has a lithographed title page dated 1818 and an "Advertisement" signed W. Cattley. I owe to the kindness of Mr. Hemsley of Kew a copy of this title page with the advertisement, which read as follows:

Icones pictae Indo-Asiaticae Plantarum excerptae e codicibus. Dom. Cattley. — Scientiae et Artium fautori illustri Josepho Banks, Societatis Regiae Praesidi dignissimo summa reverentia dicantur. — Lithographice impressae a Moser. Londini MDCCCXVIII.

Advertisement. «The thirty figures which this work comprizes were from a collection of Chinese and Indian drawings — twenty four were printed on stone and were drawn by Mr. H. B. Ker, the remainder were etched on copper: after several of the figures had been executed, the assistance of Sir Joseph Banks was fortunately obtained in directing the selection of such plants as it would be most desirable to publish: it is intented at a future period to publish a further part which will consist of thirty figures. W. Cattley».

Mr. Hemsley has most graciously placed at my disposal the following identifications of the plants figured in the Icones Braamianae. Some identifications had been previously ventured by various botanists. The plates are quoted in the same order as arranged in the Kew copy.

Tab. 1. *Dicentra spectabilis*, Miq. (Hemsl.).

Tab. 2. Two figures. Fig. sin. *Cocculus Thunbergii*, DC. Prodr. I, 98. Fig. dext. *Clerodendron Siphonanthus*, R. B. (Hemsley).

Tab. 3. *Melastoma*, species dubia. (Hemsl.). Large rose-coloured flower.

Tab. 4. *Ruellia repens*, Lin. (lilac-coloured flowers). (Hemsl.).

Tab. 5. *Jasminum Sambac*, Lin., var. (Hemsl.).

Tab. 6. Tubers from which issue long-stalked heart shaped, 10—12 nerved, acuminate leaves; under side red coloured. This is, according to Mr. Hemsley a species of *Pogonia*, and the "curious Orchid" alluded to in the preface. — The little plant with small rose coloured flowers, represented on the plate as issuing from the same tubers, is a root parasite, Mr. Hemsley thinks, perhaps *Striga hirsuta*, Benth., figured in Hook. Exot. Flora t. 203, under the name of *Campuleia coccinea*.

Tab. 7. Plant with fingered leaves, large, white solitary flower. *Adansonia digitata*, Lin. (Hemsl.). This beautiful tree, a native of Africa, cultivated in India, is not met with in China.

Tab. 8. A leafy plant with beautiful blue flowers. *Daedalacanthus nervosus*, T. Anders. (Hemsl.).

Tab. 9. Two plants represented. Fig. sin. *Hyoscyamus niger?* (Hemsl.).

Fig. dext. *Kalanchoe laciniata,* DC. (Hemsl.). See DC. Prodr. III, 395, sub
K. ceratophylla.

Tab. 10. A. Rose, white flowers. In Walpers' Rep. II, 12, erroneously identified with Roxburgh's Rosa involucrata. Crépin, Monogr. Ros.,
refers the figure likewise to this plant, but according to Hemsley, Ind. Fl.
Sin., I, 249, the figure represents *Rosa bracteata,* Wendl.

Tab. 11. *Pancratium zeylanicum,* Lin. (Hemsl.).

Tab. 12. Cucurbitacea. Yellow flowers, hairy fruit. Described by Seringe from the drawing as a new plant, Momordica sycioides, but Hemsley,
Ind. Fl. Sin., I, 315, identifies it with *Momordica dioica,* Roxb.

Tab. 13. Cucurbitacea. White fringed flowers. *Trichosanthes chinensis,* Seringe in DC. Prodr. III, 315. Determined from the figure. According to Hemsl. l. c. I, 314, probably *Tr. cucumeroides,* Maxim.

Tab. 14. According to DC. Prodr. VII, 733, *Enkianthus reticulatus,*
Lindl. Same as *E. quinqueflorus,* Lour. See Ind. Fl. Sin., II, 18.

Tab. 15. *Viciae* spec., Walp. Repert. I, 718. — *Pisum arvense?*
(Hemsl.).

Tab. 16. Yellow flowers, hooked peduncle. *Artabotrys odoratissimus,*
R. Brown. DC. Prodr. I, 90.

Tab. 17. Bottle-gourd. *Lagenaria vulgaris,* var. *Cougourda,* Ser. in
DC. Prodr. III, 299.

Tab. 18. Water lily, large white flowers. *Nymphaea acutiloba,* DC.
Prodr. I, 116, described from the drawing.

Tab. 19. A prickly Rose. White flower, prickly fruit. Described from
the drawing as a new species, *Rosa amygdalifolia,* Ser. in DC. Prodr., II,
601. According to Crépin, Monogr. Ros., this is *Rosa laevigata,* Michx.
See Ind. Fl. S., I, 250.

Tab. 20. *Narcissus Tazetta.* Much cultivated in China.

Tab. 21. Amaryllidea with large flowers, white, tinged with purple.
Crinum latifolium, Lindl. (Hemsl.).

Tab. 22. *Trapa bicornis,* Linn. See DC. Prodr. III, 64. — It seems
rather *Tr. bispinosa,* Roxb.

Tab. 23. Leguminosa. Large red flowers and pod. Described from
the drawing as a new plant, *Butea Braamiana,* in DC. Prodr. II, 415. —
Mr. Hemsley means, perhaps, an *Erythrina.*

Tab. 24. Ovate, sometimes three lobed serrate leaves, white flowers,
5 petals. Described from the drawing as a new plant, *Hibiscus chinensis,*
in DC. Prodr. I, 455. Mr. Hemsley says the drawing represents, *Pavonia
odorata,* Willd.

Tab. 25. Aroidea. *Typhonium divaricatum*, Bl. (Hemsl.).

Tab. 26. Bauhinia. Identified by Maximowicz, Mél. biol. IX (1873) 74, with *Bauhinia chinensis*, Vogel. — *B. variegata*, Linn. See Ind. Fl. S., I, 213.

Tab. 27. *Rhododendron indicum*, Lin., var. (Hemsl.). See also DC. Prodr. VII, 726.

Tab. 28. A Rose with small, white single flowers. Described from the drawing as a new species, *Rosa fragariaeflora*, by Seringe in DC. Prodr. II, 601. Crépin, Monogr. Ros. takes it to be *R. microcarpa*, Lindl. But Hemsley, Ind. Fl. Sin., I, 249, declares it to be the wild state of *Rosa Banksiae*, R. Br.

Tab. 29. *Capparis acuminata*, Willd.? See DC. Prodr. I, 247. Mr. Hemsley identifies the drawing with *Capparis spinosa*, Linn.

Tab. 30. *Jasminum Sambac*, var. *trifoliata*, DC. Prodr. VIII, 302.

III. Kerr*), William. 1804—13.

Mr. John Smith, Curator of the Kew Gardens, writes in the Gard. Chron. 1881, II, 570, that W. Kerr was in 1803 appointed botanical collector for Kew at Canton in China and, after remaining there some time, visited Java and the Philippine islands, returned to Canton and forwarded collections of (living) plants to Kew. In 1812 he received the appointment of Superintendent of the Colonial Botanic Gardens at Colombo, Ceylon. In the following year he visited the mountain called Adam's Peak in this island, and on it discovered many new plants. He died in 1814.

The Indo-Chinese Gleaner for 1819, p. 122, publishes a letter by J. Livingstone (see further on) to the Rev. Morisson, dated Febr. 12th 1819, in which we read: "Mr. William Kerr was sent about 15 years ago, from the Royal Gardens, Kew, for the purpose of enriching that splendid collection with the stores of China. He came out from England on the same ship with me (Livingstone) and left China about 6 years ago".

In the Chinese Repository, published at Canton, III (1834), p. 86, it is stated that after his arrival at Canton, W. Kerr at first was very active; but after three or four years he became greatly changed. He was then unable to prosecute his work, in consequence of some evil habits he had

---

*) His name is sometimes erroneoulsy written Ker. He is not to be confounded with his contemporary John Bellenden Ker, a distinguished botanist, who in the beginning of this century wrote in the Botan. Magazine and afterwards in the Bot. Register.

contracted, as unfortunate as they were new to him. His salary was 100 pounds. All the plants sent by him to England were bought in the Flower Gardens Fate near Canton. — J. Livingstone, in a letter dated Macao, Febr. 5th 1819, published in the Trans. Hort. Soc. III, 421, writes that W. Kerr, sent out by the Royal Gardens to collect (living) Chinese plants, was not successful in transmitting them to England. Most of his plants arrived in an unhealthy condition.

A. P. De Candolle, Trans. Linn. Soc. XII (1818) dedicated to Kerr's memory the genus *Kerria*, founded upon Thunberg's *Corchorus japonicus*, a plant first introduced to Kew by Kerr.

Notwithstanding Livingstone's statement regarding Kerr's want of success in sending plants to England, a considerable number of new and interesting ornamental and other plants, transmitted by the latter from Canton, were cultivated in the Kew Gardens, as is testified in the second edition of the Hortus Kewensis, 1810—13.

The first consignment of living plants, selected by Kerr in Canton, was transmitted to Kew in 1804 by Captain Kirkpatrick. The Hortus Kewensis writes: "China, W. Kerr. Introduced by the Court of Directors of the East India Company in the 'Henry Addington', Captain Kirkpatrick", and notices the following plants:

Gardenia spinosa. (I, 370).	*Crataegus glabra.* (III, 202).
Gardenia radicans. (I, 368).	*Aster hispidus.* (V, 79).
Pittosporum Tobira. (II, 27).	*Sagittaria obtusifolia.* (V, 282).
Lilium japonicum. (II, 240).	*Begonia discolor.* (V, 284).
» *tigrinum.* (II, 241).	*Pinus lanceol.*(Cunghm. sin.). (V, 320).
Nandina domestica. (II, 314).	*Juniperus chinensis.* (V, 413).
Dianthus japonicus. (III, 79).	*Taxus macrophylla.* (V, 416).

Ibid: "Sent by W. Kerr from China, introduced in 1805 by the Court of Directors of the E. I. Comp. in the 'Winchelsea', Capt. Campbell:"

Mussaenda pubescens. (I, 372).   *Nymphaea pygmaea.* (III, 293).

Corchorus japonicus with double flowers. This is not noticed in the H. K. but the Bot. Mag. (1810) t. 1296, says that it was introduced into the Kew Gardens in 1805 by W. Kerr.

Ibid: "Sent by W. Kerr from China, introduced in 1806 by the Court of Directors of the E. I. C. in the 'Hope' E. I. man, Capt. Pendergrass, James:"

Gardenia micrantha. (I, 371).   *Lonicera flexuosa.* (I, 380).
Lonicera japonica. (I, 379).

Ibid: "Sent by W. Kerr from China, introduced by the Court of Directors of the E. I. Comp. in 1806 in the 'Wilmer Caske', Capt. Dodd.:"

Paederia foetida. (II, 64).

In 1807 W. Kerr sent *Rosa Banksiae.* (III, 258).

Ibid: "Sent by W. Kerr from China, introduced by the Court of Directors of the E. I. C. in the 'Cuffnels', Capt. Wellbank":

Camellia Sesanqua. (IV, 235).

Lambert, gen. Pinus 1828, II, 111, thinks that *Cupressus pendula,* was introduced from S. China by W. Kerr.

Hort. Kew, 2d ed. V, 206: *Bletia hyacinthina,* W. Kerr.

W. Kerr, sent also living plants and seeds from Canton to the Botan. Garden Calcutta. (See further on Roxburgh).

He presented also to the E. I. Comp. a collection of Chinese drawings of plants made from the life by a Chinese artist under the eye of W. Kerr (1806). These drawings are noticed in Murray, Crawford and Burnett, Account of China (1836) III, 338, as preserved in the Museum of the Company (India House). Comp. also Bot. Mag. (1814) sub t. 1649, Enkianthus quinqueflorus, and Sabine's articles on Chrysanthemum in Trans. Hort. Soc. IV (1821) 330 and on Paeonia Mouton, l. c. VI, 483.

IV. New Chinese Plants introduced into the Royal Gardens at Kew, 1789—1813.

Twenty one years after the appearance of William Aiton's Hortus Kewensis in 3 vols, 1789, his son W. T. **Aiton** began to bring out a new and enlarged edition of the Hortus Kewensis, which was completed in 5 vols, in 1813. He was assisted by Dr. **Dryander,** librarian to Sir Joseph Banks, and afterwards by Mr. R. **Brown.**

Aiton, William Townsend, born 1766, died 1849. After the death of his father, in 1793, he succeeded him in the Directorship of the Kew Gardens and held this post till 1841, when he retired. From the year 1797 he was a Fellow of the Linn. Soc. He was one of the founders of the Horticultural Society of London, in 1810.

The following new names of Chinese plants cultivated in the Kew Gardens, appear in the second edition of the Hortus Kewensis:

III, 315. *Paeonia Mouton,* Bot. Mag. (1808) t. 1154. — Paeonia suffruticosa, Andr. Bot. Repos. (1804) 373, ibid. (1807) 448, var. fl. *pur-pureo.* — China. Introduced 1787 by Sir Jos. Banks for the Kew Gardens.

III, 330. *Magnolia conspicua,* R. A. Salisbury, Paradisus Londonensis (1806), tab. 38 — China. Sinice: *Yu' lan.* — Introduced 1789, by Sir Jos. Banks.

III, 330. *Magnolia obovata,* Willd., Sp. II (1799) 1257. — M. purpurea, Bot. Mag. (1797), t. 390. — M. discolor, Ventenat, hort. Malmaison (1803) 24, tab. 24. — China. — Introduced, 1790, by William Henry Duke of Portland.

III, 330. *Magnolia tomentosa,* Willd., Sp. II (1799) 1257. — M. gracilis, R. A. Salisbury, Parad. Lond. (1806) 87. (Sent to Ch. Greville from Canton some years ago). — China, David Lance. — Introduced in 1804.

III, 330. *Magnolia pumila,* Sims, Bot. Mag. (1806), t. 977. — Andr., Bot. Repos. (1802) t. 266 (introduced 1793, by J. Slater). — China. — Introduced about 1786, by Lady Amelia Hume.

III, 330. *Magnolia fuscata,* Sims. Bot. Mag. (1807) t. 1008. — Andr., Bot. Repos. (1802) 229 (Introduced 1796, by T. Evans of Stepney). — M. anonaefolia, R. A. Salisbury, Parad. Lond. (1806) 5, (From China, Ch. Greville). — Introduced 1789, by Sir J. Banks.

III, 333. *Uvaria odorata,* Willd., Sp. II (1799) 1262. — Cananga, Rumph. Amb. II, 195, t. 65. — Java and China. — Introduced 1800, by Sir Jos. Banks.

II, 314. *Nandina domestica,* Willd. Sp. II (1799) 230. — Bot. Mag. (1808) t. 1109. (Thunb. Jap. 147). — China. — Will. Kerr. Introduced 1804. Kirkpatrick. (V. supra p. 190).

III, 293. *Nymphaea pygmaea.* — Castalia pygmaea, R. A. Salisbury, Parad. Lond. (1806) 68. — China. — W. Kerr. Introduced 1805, Capt. Campbell. (V. supra p. 190).

III, 142. *Bocconia cordata,* Willd. Sp. II (1799) 841. — China. — John Haxton. Introduced 1795, by Sir G. Staunton.

II, 27. *Pittosporum Tobira,* Willd., Sp. I (1797) 1130. — China, Will. Kerr. Introd. 1804. Kirkpatrick. (V. supra p. 190).

III, 79. *Dianthus japonicus,* Willd., Sp. II (1799) 673. — Thunb. Fl. jap. 183, t. 23. — China, W. Kerr. Introd. 1804, Kirkpatrick. (V. supra p. 190).

IV, 235. *Camellia Sesanqua,* Willd., Spec. III (1800) 842. Lady Banks' Camellia. — Thunb. Fl. jap. 273, t. 29. — Japan, China, W. Kerr. Introd. 1811, Wellbank. (V. supra p. 191).

III, 43. *Limonia trifoliata,* Willd., Sp. II (1799) 571. — Andr. Bot. Repos. (1801) 143. — China. — Introduced 1787, by Sir Jos. Banks.

III, 39. *Cookia punctata,* Willd., Sp. II, 558.— Jacq., Hort. Schoenbr., I, 53, t. 101. — China. — Introduced 1795, by Sir Jos. Banks.

IV, 420. *Citrus nobilis,* Loureiro, Cochin. II, 466. Mandaria Orange.— Andr. Bot. Repos. (1809) 608, (Imported from Canton, 1805). — China, Cochinchina. — Introduced 1805, by Sir Abraham Hume.

II, 354. *Dimocarpus Litchi,* Willd., Spec. II (1799) 346. — China, Tunquin, Cochinchina. — Litchi chinensis, Sonnerat, Iter Ind., II, p. 180, t. 129. — Introduced 1786, by Warren Hastings.

II, 354. *Dimocarpus Longan,* Longan, Du Halde, Chine I, 16. — China. — Introduced before 1786.

II, 162. *Rhus javanicum,* Willd., Sp. I (1797) 1478.— Java, China.— Introduced 1799, by Sir Joseph Banks.

IV, 362. *Astragalus chinensis,* Willd., Sp. III (1880) 1272. — Lin. fil. Dec. 5, t. 3. — China. — Introduced 1795, by Sir George Staunton.

IV, 368. *Astragalus sinicus,* Willd., Sp. III (1800) 1292. — Bot. Mag. (1811) t. 1350, (cultivated in the Kew Gardens, 1771). — Cultivated 1763, by Ph. Miller, Chelsea. R. S. n. 2059, (Chelsea plants presented to the Roy. Soc.). — Astragalus chinensis in Mill. Dict.

IV, 341. *Hedysarum tomentosum,* Willd., Sp. III, 1181. — Japan, China. — Introduced 1782. — Herb. Banks.

IV, 345. *Hedysarum lagopodioides,* Willd., Sp. III, 1203. — Burm. Ind. 168, t. 53, fig. 2. — China. — Introduced 1790, by Sir Jos. Banks.

IV, 291. *Phaseolus radiatus,* Willd., III, 1036.— China, Ceylon. Cult. 1732, by James Sherard, M. D. Dillen. Ht. Elth. 315, t. 235, fig. 304.

V, 457. *Desmanthus natans,* Willd., Sp. IV (1805) 1044. — Andr. Bot. Repos. (1810) 629. — East Indies China. — Introduced 1800, by Sir Jos. Banks. — Mimosa natans, Roxburg, pl. Coromand. II, 11, t. 119.

III, 314. *Corchorus japonicus,* Willd., II, 1218. — Andr. Bot. Rep. (1809) 587. (flore pleno). — Introduced 1804, by David Lance.

III, 258. *Rosa Banksiae,* R. Brown. Lady Banks' Rose. — China, Will. Kerr. — Introduced 1807.

III, 265. *Rosa multiflora,* Willd., Sp. II (1799) 1077. — Bot. Mag. (1807), t. 1059. (Thbg. Fl. jap. 214). — China. — Introduced about 1804, by Th. Evans (of the E. I. House).

III, 266. *Rosa semperflorens,* Willd., Sp. II, 1078, var. α. Smith, Exot. Bot. II (1805) p. 63, tab. 91, var. β. Bot. Mag. (1794) 284. Miss Lawr. Ros. t. 28. — Jacq. ht. Schoenbr. III, p. 17, t. 281. — Rosa diversifolia, Ventenat, Cels. 35. — China. — Introduced about 1789, by Gilbert Slater.

III, 267. *Rosa indica,* Willd., Sp. II, 1079. — Miss Lawr. Ros. t. 26. — China. — Introduced about 1789, by Sir Jos. Banks.

III, 267. *Rosa bracteata,* Willd., Sp. II, 1079. — Bot. Mag. (1811), t. 1377, (China. Brought by Lord Macartney). — Ventenat, Cels. 28. — Wendl. Hort. Herrenh. 4, p. 7, t. 23. — China. J. Haxton. — Introduced 1795, by Sir John (sic!) Staunton.

III, 202. *Crataegus glabra,* Willd., Sp. II (1799) 1004. — Japan, China, W. Kerr. Introduced 1804, Kirkpatrick. (V. supra p. 190).

III, 203. *Crataegus indica*, Willd., Sp. II, 1005. — India, China. — Introduced 1806, by James Drummond.

III, 63. *Hydrangea hortensis*, Willd., Sp. II, 633. — J. E. Smith, Icon. pict. pl. rar. (1790) 12. — Curt. Bot. Mag. (1799) 438. — China, culta. — Introduced 1788, by Sir Jos. Banks. — Schneevogt, Icon. pl. rar. 1793, 36.

III, 46. *Melastoma malabatrica*, Willd., Sp. II (1799) 592. — Bot. Mag. (1801), t. 529. — East-Indies, China. — Introduced 1795, by Sir George Staunton.

II, 347. *Grislea tomentosa*, Willd., Sp. II (1799) 321. — Roxburgh, Pl. Coromand. (1795) I, 29, t. 31. — Lythrum fruticosum, Andr. Bot. Rep. (1807) 467. — Woodfordia floribunda, R. A. Salisbury, Paradis. Lond. (1807) 42. — India, China. — Introduced 1804, by William Salisbury.

I, 267. *Trapa bicornis*, Willd., Sp. I (1797) 681. — Gaertner, Fruct. & Sem. II (1791), 84, t. 95. — China. — Introduced 1790, by Banks.

V, 284. *Begonia discolor*. — Begonia Evansiana. Andr. Bot. Repos. (1810) 627, (Evans introduced it 1808, from Pu lo Penang; Bot. Mag. (1812), t. 1473. — China, W. Kerr. Introd. 1804, by Kirkpatrick. (V. supra p. 190). — Sin: *Tsieou haï tang*, Mém. conc. Chin. III (1778) 443. — Grosier, la Chine I, 503. — *Tsou Hoy Tung*, Dryander in Trans. Linn. Soc. I (1791) 172, n. 3.

I, 379. *Lonicera japonica*, Willd., Sp. I (1797) 985. — Andr. Bot. Repos. (1809) 583. — China, W. Kerr. Introduced 1806, by Pendergrass. (V. supra p. 190).

I, 380. *Lonicera flexuosa*, Willd., Sp. I, 989. — China, W. Kerr. Introduced with the last.

I, 372. *Mussaenda pubescens*. — China, W. Kerr. Introduced, 1805, by Campbell. (V. supra p. 190).

I, 368. *Gardenia radicans*, Willd., Sp. I (1797) 1225. — Andr. Bot.

Repos. (1807) 491. — Thunberg, Gardenia (1780), n. 1, t. 1, fig. 1. — Cult. in Chinese gardens, W. Kerr. Introd. by Kirkpatrick, 1804. (V. supra p. 190).

I, 370. *Gardenia spinosa*, Willd., Sp. I, 1229. — Thunberg, l. c. n. 7, t. 2, fig. 5. — China, W. Kerr. Introduced 1806, by Kirkpatrick. (V. supra p. 190).

I, 371. *Gardenia micrantha,* Willd., Sp. I, 1230. — Thunberg, l. c. n. 8, t. 1, fig. 2. — China, W. Kerr. Introduced 1806, by Pendergrass. (V. supra p. 190).

II, 64. *Paederia foetida,* Willd., Sp. I, 1219. — Banks, Icon. Kaempf. tab. 9. — China, W. Kerr. Introduced 1806, by Dodd. (V. supra p. 191).

V, 59. *Aster hispidus*, Willd., Sp. III (1809) 2042. — Banks, Icon. Kaempf. tab. 29. — China, W. Kerr. Introduced 1804, by Kirkpatrick. (V. supra p. 190).

V, 27. *Conyza chinensis*, Willd., Sp. III, 1926. — Senecio amboinensis, Rumph. Amb. VI, 36, t. 14. — China, East Indies. — Cult. before 1796, by W. B. Coyte, M. D. Hort. Gippov. 107.

V, 27. *Conyza hirsuta,* Willd., Sp. III, 1926. — China. — Cult. 1767 by Phil. Miller, Gard. Dict. ed. 8, n. 18. — R. S. n. 2258. (Plants cult. Chelsea, presented to the Roy. Soc.)

V, 25. *Carpesium abrotanoides*, Linn. Sp. 1204. — China, Japan. — Cult. before 1768, by Phil. Miller, Gard. Dict. ed. 8, n. 2.

IV, 495. *Bidens chinensis,* Willd., Sp. III (1800) 1719. — Agrimonia moluccana, Rumph. Amb. VI, 38, t. 15. — China. — Introduced 1801, by Mr. John Hunnemann.

V, 95. *Chrysanthemum indicum,* Willd., Sp. III, 2147. — Bot. Mag. (1796), t. 327. — Anthemis grandiflora, Ramatuelle in Journ. d'Hist. nat. II (1792) 234. — Cult. 1764, by Ph. Miller. R. S. n. 2112. (Plants cult. in Chelsea Garden, presented to the Roy. Soc.). — Matricaria indica, Miller, Gard. Dict. ed. 8. — The Hort. Kew (1813) enumerates 8 varieties of Chrysanthemum indicum.

V, 38. *Senecio divaricatus,* Willd., Sp. III, 1977. — China. — Introduced 1801, by Sir Jos. Banks.

I, 318. *Azalea indica,* Willd., Sp. I (1797) 831.— Tsu tsu si, Kaempf. Amoen. exot. 845, t. 846. — China. — Introduced 1808, by the Court of Directors of the E. I. C. in the ship "Cuffnels", Capt. Wellbank.

I, 16. *Jasminum hirsutum,* Willd., I, 36. — China, East-Indies. — Cult. 1759, by Ph. Miller, Gard. Dict. ed. 7. Nyctanthes 2.

I, 24. *Syringa chinensis,* Willd., Sp. I, 48. — China. — Introduced before 1795, by Mons. Williams.

I, 19. *Ligustrum lucidum,* Wax tree. — Introduced from China about 1794, by Sir Joseph Banks.

II, 83. *Pergularia odoratissima,* J. E. Smith, Icon. pict. Pl. rar. (1790) 16. — Andr. Bot. Repos. (1801) 185. — China, cult. — Introduced about 1784, by Sir Joseph Banks.

II, 84. *Hoya carnosa,* R. Br. Mem. Wern. Soc. I, 27. — Bot. Mag. t. 788.— Smith, Exot. Bot. II, 21, t. 70. A native of Asia (China). Introd. 1802, by Sir Jos. Banks.

I, 276. *Cuscuta chinensis,* Willd., Sp. I (1797) 704. — China. — Introduced 1803, by Sir Jos. Banks.

IV, 53. *Mazus rugosus,* Loureiro, Fl. Cochin. 385. — Lindernia japonica, Willd., Sp. III (1800) 326. — China, Japan. — Cult. before 1780, by John Fothergill, M. D. — Herb. Banks.

IV, 34. *Bignonia grandiflora,* Willd., Spec. III (1800) 303. — Andr. Bot. Repos. (1807) 493. — R. A. Salisbury, Paradis. Lond. (1807) 61. — Bot. Mag. (1811) 1398. — China, Japan. — Introduced 1800, by the Right Hon. Charles Greville.

IV, 66. *Vitex ovata,* Willd., Sp. III, 390.— Japan, China. — Introduced 1796, by Sir Joseph Banks.

IV, 67. *Vitex incisa*, Willd., Sp. III, 392. — V. Negundo, Bot. Mag. (1797), t. 364. — China. — Cult. 1758, by Ph. Miller, V. chinensis. Miller, Icon. t. 275, fig. 1, 2.

IV, 63. *Clerodendron fragrans*, Ventenat Jard. Malmaison (1804) 70, floribus simplicibus. — Volkameria japonica, Jacquin, Ht. Schoenbr. III, 48, t. 338 (floribus plenis). — Japan and China. — Introduced about 1790, by G. Slater, Esq.

IV, 63. *Clerodendron squamatum*, Willd., Sp. III (1800) 385. — Volkameria Kaempferi, Jacq. Icon. pl. rar. III, t. 500. — China, Japan. — Introduced 1790, by Sir Jos. Banks.

II, 420. *Polygonum chinense*, Willd., Sp. II (1799) 453. — Burm. Ind. 90, t. 30, fig. 3. — China, John Haxton. — Introduced 1795, by Sir John (sic!) Staunton.

V, 420. *Nepenthes destillatoria*, Willd., Sp. IV (1805) 873. — Ceylon, China. — Introduced 1789, by Sir Jos. Banks.

II, 428. *Laurus glauca*, Willd., Sp. II (1798) 478. — Japan, China. — Introduced 1806, by James Drummond.

I, 271. *Elaeagnus latifolia*, Willd., Sp. I (1797) 689. — China, East Indies. — Cult. before 1712 in the Royal Gardens at Hampton Court. — Philos. Trans. n. 337, p. 221, n. 160 (not mentioned for China).

V, 328. *Croton aromaticum*, Willd., Sp. IV (1805) 549. — Halecus littorea, Rumph. Amb. III, 197, t. 126. — Ceylon, China. — Introduced 1793, in H. M. S. "Providence" by Rear-Admiral Bligh.

V, 486. *Ficus stipulata*, Willd., Sp. IV, 1139. — China, Japan. — Introduced about 1771.

V, 487. *Ficus pumila*, Willd. IV, 1140. — Kaempf. Amoen. exot. 803. — China, Japan. — Cult. 1759, by Ph. Miller. Dict. ed. 7, n. 8.

V, 416. *Taxus nucifera*, Willd. Sp. IV, 857. — Kaempf. Amoen. exot. 814. — Japan, China. — Cult. 1764, by Capt. Thomas Cornwall.

V, 416. *Taxus macrophylla*, Willd. IV, 857. — Banks, Icon. Kaempf. 24. — China, Will. Kerr. Introduced 1804, by Kirkpatrick. (V. supra p. 190).

V, 320. *Pinus lanceolata*, Lambert, Pinus (1803), 52, t. 34.— China, W. Kerr. Introduced 1804, by Kirkpatrick. (V. supra p. 190). *(Cunninghamia sinensis*, R. Br.).

V, 413. *Juniperus chinensis*, Willd., Sp. IV (1805) 751. — China, W. Kerr. Introduced 1804, by Kirkpatrick. (V. supra p. 190).

V, 206. *Bletia hyacinthina*, Brown, MSS. — Cymbidium hyacinthinum, J. E. Smith, Exot. Bot. I (1804), 117, t. 60. (Smith says, erroneously, I think, a native of Trinidad). — Bot. Mag. (1812), t. 1492 (China). — China, William Kerr. — Introduced about 1802, by Thomas Evans, Esq.

V, 214. *Cymbidium sinense*, Willd., Sp. IV (1805) 111. — Epidendron sinense, Andr. Bot. Repos. (1802) 216; Bot. Mag. (1805, t. 888. — China. — Introduced about 1793, by G. Slater, Esq.

V, 212. *Aërides odorata*, Willd., Sp. IV (1805) 131. — China. — Introduced 1800, by Sir Jos. Banks.

V, 425. *Musa coccinea*, Willd., Sp. IV, 895. — Andr. Bot. Repos. (1799) 47. — Bot. Mag. (1813), t. 1559. — China. — Introduced about 1792, by Thomas Evans.

II, 278. *Sanseviera carnea*, Andr. Bot. Rep. (1804) 361. — Sanseviera sessiliflora, Bot. Mag. (1804), t. 739. — China. — Introduced about 1792, by Thomas Evans, Esq.

I, 120. *Iris fimbriata*, Ventenat, ht. Cels (1800) 9.— Iris chinensis, Curt. Bot. Mag. (1797), t. 373, (introd. by Evans). — China. — Introduced 1792, by Sir Jos. Banks.

II, 240. *Lilium japonicum*, Willd., Sp. II (1799) 85. — China, W. Kerr. Introduced by Kirkpatrick. (V. supra p. 190). — This plant is not L. japonicum Thbg., although since his introduction in 1804, it has long been cultivated under this erroneous name. It is now known under the name of *Lilium Brownii*, which was first applied to it by Miellez, after a nursery man of Slough, in whose catalogue it was brought to notice, about 1838. See Elwes, Lilium, 1880. This beautiful Lily has in later times been found wild in Middle and West China.

II, 241. *Lilium concolor*, R. A. Salisbury, Paradis. Lond. (1806), t. 47. (Salisb. states that Ch. Greville cultivated it, native country unknown). — Bot. Mag. (1809), t. 1165. — China. — Introduced before 1806, by the Right Hon. Ch. Greville.

II, 241. *Lilium tigrinum*, Bot. Mag. (1809), t. 1237. — Lilium speciosum, Andr. Bot. Repos. (1809) 586 (nec aliorum). — China, Will. Kerr. Introduced 1804, by Kirkpatrick. (V. supra p. 190).

II, 247. *Uvularia chinensis*, Ker, in Bot. Mag. (1806) 916. — China. — Introduced 1801, by Sir Jos. Banks. — *(Disporum pullum,* Salisb.).

I, 9. *Philydrum lanuginosum*, Willd., Sp. I (1797) 17. — Bot. Mag. (1804), t. 783. — Garciana cochinchinensis, Lour. Fl. cochin. Loureiro sent specimens from Cochinchina to Europe in 1774. — China, Cochinchina, New South Wales. — Introduced 1801, by Sir Jos. Banks.

V, 282. *Sagittaria obtusifolia*, Willd. Sp. IV (1805) 408. — Rheede, Malab. XI, 93, t. 45, culituma. — East-Indies, China, Kerr. Introduced 1804, by Kirkpatrick. (V. supra p. 190).

V, 433. *Manisuris granularis*, Willd., Sp. IV, 945. — Roxburgh, Pl. Corom. II (1798) 11, t. 118. — East-Indies, China. — Introduced in 1784, by Sir Jos. Banks.

V, 432. *Ischaemum aristatum*, Willd., Sp. IV, 939. — China, East-Indies. — Introduced 1803, by A. B. Lambert.

V. English Promoters of Botanical Science, Eminent Botanists, Gardeners, Travellers etc. mentioned in the Second Edition of Hortus Kewensis and in the Early English Illustrated Botanical Periodicals, in Connection with the Introduction and Description of new Chinese Plants, from 1789 to about 1820.

The BOTANICAL MAGAZINE. This vast repertory of descriptions and drawings of plants of all quarters of the globe, a periodical frequently quoted in these pages, and which eight years ago celebrated the hundredth anniversary of its existence, was founded in 1786 by **Curtis**, William, born in 1746, at Alton, Hants. He was Superintendent of the Chelsea Gardens, 1772—77. The Botanical Magazine gives descriptions and coloured drawings of new exotic plants cultivated in English Gardens, and notices regarding their introduction and cultivation. After W. Curtis' death at Brompton, July 1799, the Bot. Mag. was continued by **Sims**, John (1749—1831) M. D., up to 1826. His coadjutor was **Gawler**, who afterwards, about 1805, changed his name into **Ker**, John Bellenden (1765—1842?). — **Edwards** Sydenham Teak (1769?—1819), botanical artist, illustrated the Bot. Mag. from 1799 to 1814.

Towards the close of the 18th century another English illustrated botanical periodical, having the same aims as the Bot. Magazine, was founded by **Andrews**, Henry C., of Knightsbridge, London, a botanical painter. It first appeared in 1797 with the title of the: BOTANIST'S REPOSITORY. After ten volumes of 664 plates had been published, the publication was discontinued in 1815, when Edwards started the "Botanical Register", of which we shall have to speak in another chapter.

Banks, Sir Joseph. This enlightened protector of natural science, whose honoured name is connected directly or indirectly with many botanical discoveries in foreign countries, and particularly in China, during the last quarter of the last and the first of the present century — was born at London, Febr. 13, 1743. His family originated from Sweden. His grandfather, who was physician in Lincolnshire, had amassed a considerable fortune and his father was in possession of vast landed property. Joseph received his primary education at Harrow and Eton. He early manifested a love of botany and carried the same taste to Oxford, where he was entered as a gentleman commoner of Christ's College. After taking the degree of M. A. Oxon. he left Oxford. His father died in 1761, and, as he was the only son, the whole fortune devolved to him. Thus he found himself possessed of ample means to satisfy his taste for scientific researches. In 1764 he made his first scientific expedition, to New Foundland and Labrador, and brought back a rich collection of plants and insects. In 1766 he became a Fellow of the Royal Society. When, in 1768, Captain Cook was sent out to observe the transit of Venus in the Pacific Ocean, Banks, through the influence of his friend Lord Sandwich, obtained leave to join the expedition and induced Dr. D. **Solander** (1736—1782), a distinguished pupil of Linnaeus, to accompany him. A rich harvest of discovery was made. The voyage occupied three years, and the expedition returned to England in 1771. The next year Banks made a voyage to Iceland in company with Solander. In 1778 he was elected President of the Royal Society, an honour which he held till his death. Banks was always a favourite with the King. In 1781 he was made Baronet, in 1795 he received the Order of the Bath, and in 1797 he was admitted to the Privy Council. In 1802 he was elected Member of the French Institute. During his long tenure of his office as President of the R. S. Sir Joseph did much to raise the state of science in Britain, and was at the same time most assiduous and successful in cultivating friendly relations with scientific men of all nations. His kindness and liberality were beyond praise. Not only was he most generous in affording to other naturalists the free use of his vast collections, but he was a munificent promoter of science and gave liberal aid to investigators. He was one of the founders of the Horticultural Society of London, in 1810.

Most of the English travellers of his time offered him their collections made in foreign countries. (Vide supra 152, 153, 162, Nelson, Bradley, Staunton). He bought also several important herbariums. Thus he amassed vast botanical collections with many rare exotic plants.

Sir Joseph died 19 June 1820 at Spring Grove, Isleworth, Middlesex. He bequeathed his valuable collections and library to the British Museum.

Although the high order of Sir Joseph Banks' scientific attainments and his learning were generally acknowledged by his contemporaries, he did not much care to perpetuate his name by writings. A short account of the Disease in Corn called the Blight, the Mildew and the Rust, 1805, a few botanical articles published in the first volume of the Transactions Hort. Soc. (On the time when Solanum tuberosum was first introduced, on Hill Wheat, on Zizania aquatica), an article on the Apple tree insect, ibidem, vol. II, an interesting MS memoir on Tea cultivation, written in 1788, and preserved in the British Museum. (See Journ. Soc. Arts, 1877, p. 200) — that is all posterity knows of Sir Jos. Banks' original scientific labours.

In 1781 he edited Houston's "Observations on American Plants", and in 1791 he published a part of Kaempfer's drawings of Japanese plants with he title: Icones selectae Plantarum quas in Japonia collegit et delineavit Engelbertus Kaempfer, ex archetypis in Museo Britannico asservatis. In folio, 59 plates.

Sir Joseph Banks' fame rests mainly upon his immense collections of objects of natural history, expecially plants, from various countries, his liberal contributions devoted to the aid of scientific research and his exertions to introduce and cultivate exotic plants. He had a splendid Garden at Spring Grove and another, with hothouses at Smallberry Green (Bot. Reg. (1817), tab. 220). The plants introduced by Banks were also cultivated in the Kew Gardens, which were much aided by his enlightened patronage, and to which he made innumerable donations of the rarest exotic plants. Since 1772 Sir Joseph was the virtual Director of the Kew Gardens, Aiton having the superintendence of them.

Jos. **Dryander,** (1748—1811, v. supra p. 191) was the keeper of Banks' herbarium and library. He published the Catalogus Bibliothecae historico-naturalis Josephi Banks, 1798—1800, 5 vols. which is illustrated with much curious information. In vol. II, 183, mention is made of 62 Chinese drawings of plants, in Banks' library, made at Canton. Comp. also supra p. 152 sub Blake, and Bot. Mag. (1813), tab. 1559, Musa coccinea, and tab. 2099 (1819), Mussaenda pubescens.

Linnaeus fil. in 1781 named in Sir J. Banks' honour the curious genus *Banksia* of which the latter had first brought several species from Australia, — Salisbury dedicated to him the Australian genus *Josephia* (Dryandra, R. Brown). The great botanist R. Brown named a beautiful thornless fragrant Chinese Rose *Rosa Banksiae* (v. supra p. 194) in compliment to the excellent Lady Banks. — Camellia sesanqua is called Lady *Banks' Camellia* by Aiton. V. supra p. 193. It first flowered in Banks' conservatory. Bot. Reg. t. 12.

Chinese plants introduced by Sir Joseph Banks:

1780. *Rhus semialatum,* Murr. V. supra p. 142.
1782. *Hedysarum tomentosum,* Thbg. V. supra p. 194.
1784. *Nelumbium speciosum,* Willd. — Bot. Mag. (1806), t. 903.
      Pergularia odoratissima, Smith. V. supra p. 197.
1787. *Limonia trifoliata,* Lin. V. supra p. 193.

1787. *Mespilus japonica*, Thbg. Bot. Reg. (1819), t. 365. — Trans. Hort. Soc. III (1822) 299. Introduced from Canton.

1788. *Hydrangea hortensis*, Smith. V. supra p. 195. — Kew Bull. 1891, p. 305: Sir J. Banks introduced the first Hydrangea hortensis to Kew about 1789 for the inspection of the curious. Its green petals were a puzzle to the botanists of the day.

1789. *Paeonia Moutan,* Sims. V. supra p. 191. Anderson in his Monogr. Paeonia, Trans. Linn. Soc. XII (1818), 248, says that Banks in 1789 introduced the variety β, *Banksia*, the first which appeared in Europe. We read in Loudon's Arboretum et Fruticetum Britan. I (1838) 252: "From Chinese drawings, and from the extravagant praises bestowed upon the Moutan by the Jesuit missionaries, an ardent desire was excited, in Sir J. Banks and others to import plants into England; and previously to 1786 the former engaged Mr. Duncan, a medical gentleman attached to the E. I. Company's service, to procure a plant for the Royal Gardens at Kew, where it was first received, through Mr. Duncan's exertions, in 1787. — One of the largest tree-paeonies within ten miles of London stood, till lately in the grounds at Spring Grove, where it was planted by Sir J. Banks. It was 6 or 8 feet high, and formed a bush 8 or 10 feet in diam., in 1825".
Magnolia conspicua, Salisb. V. supra p. 192.
Magnolia fuscata, Sims. V. supra p. 192.
Rosa indica, Lin. V. supra p. 194.

1790. *Clerodendron squamatum*, Vahl. V. s. p. 198.
Hedysarum lagopodioides, Lin. V. s. p. 194.
Trapa bicornis, Lin. V. s. p. 195.

1792. *Iris fimbriata*, Vent. (I. chinensis Curt.). V. s. p. 199.
Alpinia nutans, Rosc. V. Bot. Mag. t. 1903 (1817).

1794. *Ligustrum lucidum*, Ait. V. s. p. 197.

1795. *Cookia punctata*, Sonn. V. s. p. 193.

1796. *Vitex ovata*, Thbg. V. s. p. 197.

1796. *Rhus javanicum*, Lin. V. s. p. 193.

1801. *Uvularia chinensis*, Ker. V. s. p. 200.
Phylidrum lanuginosum, Banks. V. s. p. 200.
Senecio divaricatus, Lin. V. s. p. 197.

1802. *Hoya carnosa*. V. s. p. 197.

1803. *Cuscuta chinensis*, Lam. V. s. p. 197.

1805. *Paeonia albiflora,* Pall. var. *fragrans.* See Trans. Hort. Soc. II (1822) p. 278 (Sabine).

1807. *Lilium speciosum,* Andr. Bot. Repos. (1809) t. 586 *(Lilium tigrinum,* Gawl.).

1816. *Cacalia ovalis,* Bot. Reg. (1816), t. 101.

1817. *Aerides paniculatum,* Sir Jos. Banks' Aerides. Bot. Reg. (1817) t. 220.

Lambert, Aylmer Bourke, born at Bath, 2 Febr. 1761, died at Kew 10 Janr. 1842, was a private gentleman of property, who took a keen interest in botany and was well known in the scientific world. For some years he seems to have succeeded to the role of Sir Jos. Banks in the affairs of Kew. Educated at St. Mary's Hall, Oxford Univ. 1779. He was one of the founders of the Linnaean Society in 1788 and their Vice-President from 1796 till his death. From 1791 Fellow of the Royal Soc.

Lambert possessed a rich botanical library and a vast herbarium, at one time the richest in England in Siberian and S. American plants, which he placed at the disposal of men of science of all nations. Sir W. J. Hooker, in Lond. Journ. Bot. I (1842), 395, says that Lambert's Herbarium, which then was sold by auction, was one of the most extensive and valuable ever formed by a private individual. It comprised a great number of private herbariums from various countries, of which Lambert had gradually acquired possession. From China there were the collections made by Staunton and Capt. Beechey and 5 bundles of miscellaneous Chinese plants. According to the Gard. Chron. 1842, p. 439, Lambert's celebrated collection was sold for Liv. St. 1170. It is now in part in the British Museum. One portion of it was bought by Delessert. (V. s. p. 163).

Don, David (1799—1741), Librarian to Lambert and from 1822 to the Linnaean Soc., has given an account of Lambert's herbarium in the 2d edition of Lamberts Pinus, 1828. Seringe in DC. Prodr. II (1825), describes two new species of *Rubus* from China, communicated by Lambert, 567, *Rubus Lambertianus* and 557, *Rubus chinensis.*

Lambert possessed also drawings of Chinese plants made by Chinese artists. Lindley, Monogr. Ros. (1820) 406 sub Rosa xanthina refers to his collection. See also Bot. Repos. (1809), 583, sub Lonicera japonica, and Paxton's Mag. of Bot. III (1837) 51, Musa Cavendishii, Lamb.

Lambert had a garden and greenhouse at Boyton House, near Heytesbury, Wiltshire, where he cultivated many exotic plants and formed a vast museum.

Bot. Reg. (1819), 355. *Hedysarum latifolium,* cult. by Lambert, raised from seeds brought from China by Staunton.

Bot. Reg. (1820) 501. *Hovenia acerba* and Bot. Mag. (1822), 2360, *Hovenia dulcis,* both introduced by Lambert from China.

Lambert wrote many articles in the Transactions of Linn. Soc. and in the Transactions of Hort. Soc. He published a Description of the Genus Cinchona, 1797, and an Illustration of the Genus Cinchona. But his most important work is a DESCRIPTION OF THE GENUS PINUS, which appeared in 1803. In it we find, p. 52, tab. 34, the first scientific description and representation of a remarkable Chinese tree, the *Pinus lanceolata* or *Cunninghamia sinensis* of R. Brown, of which herbarium specimens had been brought by Staunton. A second, much enlarged edition of the Genus Pinus was published in 1828, in two volumes. The first volume, tab. 1—38, is devoted to the Genus Pinus and its then known species. In the second volume: continuation of Pinus, tab. 39—42. Other coniferous trees, tab. 43—63. Here are first described and figured:

Pinus Massoniana. I, p. 16, t. 8. China. The species represented in this plate is from the Banksian herbarium, having been brought by Mr. F. Masson from the Cape of G. H., where it was raised from seed sent from China.

Cupressus pendula. II, p. 111, tab. 50. Brought from China by Staunton. *(C. funebris,* Endlicher, Conif. (1847) p. 58).

Thuya pensilis. II, p. 115, tab. 51. Staunton, China. *(Glyptostrobus heterophyllus,* Endlicher, l. c. p. 70).

Thuya pendula. II, p. 115, tab. 52. Lambert states, that according to Loddiges, this is a native of Tartary. *(Biota orientalis,* var. *pendula.* Parlat. in DC. Prodr. XVI, 2, p. 462).

To this second edition is added an Appendix containing descriptions and figures of some other remarkable plants by Dav. Don (v. s. p. 205) and his Account of Lambert's Herbarium.

A third edition of the Genus Pinus, appeared in 1832. *Pinus sinensis,* p. 47, tab. 29. China. Lambert says: "all our knowledge of this species is derived from a Chinese drawing in the possession of the Hort. Soc., from which our figure was taken". (This is the same as P. Massoniana. See Parlat. in DC. Prodr. XVI, 2, p. 389).

The Australian genus *Lambertia* (Proteaceae) was dedicated to Lambert by his friend Dr. J. E. Smith. Trans. Linn. Soc. IV (1798), 214.

Aylmeria, Australian Genus, Martius. Caryophylleae.

Smith, Dr. Sir James Edward, born at Norwich 2 Dec. 1759; died at Norwich, 17 March 1828. M. D. Leyden 1786. F. R. S. 1785. He was the founder of the Linnean Society and their President from the beginning till his death.

THE LINNEAN SOCIETY of London, for the promotion of Botany and Zoology, was founded in 1788 to supplement the work of the Royal Society. The first volume of their Transactions appeared in 1791. In 1802 the Royal Charter was given to the Linn. Soc. by King George III of England. It is reproduced in the Transactions, VII (1804). J. E. Smith is there mentioned as the first President. Among the first Fellows we find the names of George Earl of Dartmouth, A. B. Lambert, R. A. Salisbury, Jonas Dryander. The Society having occupied for a long time Sir Joseph Banks' old house in Soho square, removed in 1857 to Burlington House and assumed the apartments it now occupies.

J. E. Smith, was a botanist of great ability. He was possessed of ample means and used to employ them most liberally for the advancement of botanical science. In 1784 he purchased in Sweden the Linnean collections, herbarium, library of books and manuscripts, which after his death came into the possession of the Linnean Soc., where they are still preserved. Smith himself relates regarding this transaction, in Rees' Cyclop., sub Linnaeus, as follows: "After the death of the younger Linnaeus (1783), the Linnaean Museum and Library reverted to his mother and sisters, and the former fixed her eyes on Sir Joseph Banks, as the most likely person to purchase the relics at the high price, as she thought it, of a thousand guineas. On his refusal, and by his kind recommendation and advice they came into the hands where they now are (they were brought to England by J. E. Smith. See Hooker, Journ. Bot. IV (1852), 218). The sale was precipitated before the return of the king of Sweden, then on his travels, lest he should oblige the heirs to dispose of the whole at a cheaper rate to the University of Upsala. This would actually have been the case, as appears from the exertions made by H. M. on his return, who sent a courier to the Sound, and a vessel by sea to intercept the ship that was bearing away the prize".

It appears that the purchaser was somewhat disappointed with respect to the value of the Linnaean herbarium, for the specimens in many cases were found to have been badly preserved, and a considerable number of species described by Linnaeus were wanting in the herbarium. But nevertheless Smith, and after him other botanists, have succeeded in elucidating some of the numerous doubts adhering to the Linnean species, for Linnaeus descriptions of plants are frequently riddles without the type specimens in his herbarium. Compare above p. 64 the chapter on Chinese plants known to Linnaeus.

From 1789—93. Smith published the Plantarum Icones hactenus ineditae, plerumque ad plantas in herbario Linnaei conservatas delineatae. Tab. 1—75. On tab. 36, Linnaeus' *Vatica chinensis* is figured. Smith says: "De nominis ratione nil memorat Linnaeus. Anne planta divinationis causa inter Chinenses superstitiosos plurimi habita?" Comp. supra p. 70.

From 1790—93 he brought out another iconographical work representing plants under the title: "Icones pictae plantarum rariorum, descriptionibus et observationibus illustratae". 18 coloured plates. Two Chinese plants for the first time depicted:

Hydrangea hortensis, tab. 12. Chinese Guelder Rose.

Pergularia odoratissima, tab. 16.

In his Tracts rel. to Nat. Hist., 1798, 247, he established the genus *Davallia* (Ferns) and described *D. chinensis*, (v. s. p. 104).

Smith published several articles in the Transactions of the Linnean Soc. Some of them refer to Eastern Asiatic plants:

III (1797), 330. *Salisburia adiantifolia.*

VI (1802), 312, tab. 29, 30. *Cycas revoluta.*

From 1804—5 appeared his Exotic Botany, in two volumes with coloured plates. On tab. 70 is figured *Asclepias carnosa*, Linn. V. supra p. 88.

Many very valuable botanical articles by Smith, referring also to Chinese plants, and frequently quoted by botanists, are found in Rees' Cyclopaedia. The complete title is: The Cyclopaedia or Universal Dictionary of Arts, Sciences, Literature by **Rees**, Abraham 39 volumes, all bearing the date 1819, and 6 volumes of plates As Messrs B. D. Jackson and H. Trimen have pointed out (Journ. Bot. 1877, 107, — 1880, 87), the Cyclopaedia was published from 1802 to 1820. The botanical articles were contributed by Dr. J. E. Smith from towards the conclusion of the letter C. (1809) to the end of the alphabet. The communications Smith sent amounted to 3348, besides 57 lives of botanists. — He wrote also in the Transactions of the Horticult. Society, of which he became a Hon. Member in 1819. (Trans. Hort. Soc. III, 338).

Smith was knighted in 1814. — Aiton in Hort. Kew, ed. first, III, 496, named the Genus *Smithia* after him. *(Sm. sensitiva*, India orientalis). His botanical collections are at the Linnean Soc.

Salisbury, Richard Anthony, *né* Markham, born at Leeds in 1761, died at London, March 1829, was a gardener and distinguished botanist. He had a garden at Chapel Allerton and M. Collinson's at Mill Hill. Salisbury was one of the founders of the Linnean Soc. in 1788, he was also a Fellow of the Royal Soc. (previous to 1805), and in 1810 First Secretary of the R. Hort. Soc.

From 1806—7 he published the PARADISUS LONDONENSIS, plants cultivated in the vicinity of the Metropolis, with coloured figures by W. Hooker. In it several new Chinese plants are depicted:

Tab. 5. *Magnolia anonaefolia.* A low shrub lately brought into the country from China and now in flower in the stove of the Right Hon. Charles Greville. — This is likewise figured in the Bot. Cab. t. 1072, (1825). According to DC. Prodr. I, 81, a variety of M. fuscata.

Tab. 33. *Hibiscus acerifolius.* Cultivated by the Chinese, distinct from H. syriacus. — Not mentioned in I. F. S.

Tab. 38. *Magnolia conspicua,* the *Yu lan* of the Chinese. Introduced some years ago by Sir Jos. Banks.

Tab. 47. *Lilium concolor.* Cultivated by the R. Hon. Ch. Greville. He does not know from whence he received it. — But in Bot. Mag. t. 1165 (1809) it is stated that it came from China.

Tab. 61. *Bignonia grandiflora.* B. chinensis Lam., Japan. — Hort. Kew, 2d edition, IV, 34, says: China and Japan. Introd. 1800, by Greville.

Tab. 68. *Castalia pygmaea.* Salisb. thinks that it is a native of China. Comp. above p. 192, Hort. Kew, *Nymphaea pygmaea.*

Tab. 88. *Magnolia gracilis.* Distinct from M. purpurea. Sent from China some years ago to the R. H. Ch. Greville. — According to DC. Prodr. I, 81, same as *M. Kobus.*

Tab. 115. *Adina globiflora.* This shrub came up in the earth of some plants sent from Canton to the R. H. Ch. Greville. — Robertson gathered it in a wild state near Wampu. (V. supra p. 154).

In 1807, Trans. Linn. Soc. VIII, 315, Salisbury proposed a new genus name *Belis (B. jaculifolia)* for a Chinese coniferous tree described by Lambert as Pinus lanceolata. But, as Belis resembles too much the Linnean genus name Bellis, R. Brown rejected it and called the new genus *Cunninghamia (sinensis).*

In Trans. Hort. Soc. II (1814), 156, Salisbury published an article on Loureiro's *Enkianthus quinqueflorus.*

Dr. J. E. Smith named the Eastern Asiatic genus *Salisburia (adiantifolia),* the *Gingko biloba* of Linnaeus, in honour of his friend R. A. Salisbury

(Trans. Linn. Soc. III (1797), 330) and acknowledges on this occasion his
acuteness and indefatigable zeal in the service of botany *).

Anderson, William, born at Edinburgh in 1766, was Gardener and
Curator at Chelsea, from 1814 till his death, Oct. 6, 1846. Fellow Linn.
Soc. 1815. He first cultivated *Azalea indica* in the Chelsea Garden, which
had been introduced (most probably from China) about 1810. See Loddiges in Bot. Cab. t. 275 (1818). Lindley, in Collect. Bot. (1821), tab. 3
states, that Anderson also first cultivated the *Raphiolepis rubra* (from
China). In 1817 Anderson introduced from China a new *Aloe*, which was
first described by Haworth, in Supplem. Plant. succul. (1819), p. 45, as
a Chinese variety of *A. barbadensis*. It was first figured, with an accurate
description, in 1877, in the Bot. Mag. t. 6301, under the name of *Aloe
chinensis*.

Anderson, George. Fellow Linn. Soc. 1800. Died 10 Janr. 1817.
After his death the Linnean Soc. in their Transactions, XII (1818) 248—
290, published his valuable MONOGRAPH ON THE GENUS PAEONIA, in which
are described the *Paeonia Moutan* and *Paeonia albiflora* with their beautiful garden varieties, introduced from China.

Greville, Charles Francis, the Right Honourable. According to Dr.
J. E. Smith, in Rees' Cyclopaedia, he was a Fellow of the Linn. Soc. and
Vicepresident of the Roy. Soc. He was one of the founders of the Hort.
Soc. of London, in 1804. In his garden and conservatory at Paddington the
rarest and most curious plants from various climates were cultivated with
peculiar success. He died in 1809 (Smith).

Rob. Brown, in dedicating to him the Australian genus *Grevillia*
(Trans. Linn. Soc. X, 1811, 168) speaks of him as being a gentleman
eminently distinguished by his acquirements in natural history, to whom
the botanists of this country are indebted for the introduction and successful cultivation of many interesting plants. From this it would appear
that in 1811, Greville was still alive. In the Trans. Hort. Soc. I, 276,
Febr. 1812, R. A. Salisbury speaks of the late Right Honour. Charles
Greville.

---

*) Not to be confounded with his contemporary **William Salisbury**, nurseryman of Chelsea,
partner and successor to Curtis in Brompton Garden, afterwards in Sloane Street, where he
had botanical classes. He flourished, 1797—1816. He first introduced, in 1804, *Grislea tomentosa*, from China or India. (v. supra p. 195).

Chinese plants introduced by Greville:

Paeonia Moutan, var. *rosea, semiplena*, 1794. Bot. Cab. t. 1035 (1825); Sabine, Paeonia Moutan in Trans. Hort. Soc. VI (1826), 476.

Bignonia grandiflora, 1800. V. supra p. 197.

Camellia japonica, var. flore rubro, pleno, 1801. Bot. Repos. t. 199, (1802).

Magnolia anonaefolia. Previous to 1806. V. s. p. 192.

Lilium concolor. Previous to 1806. V. s. p. 200.

Adina globifera. Previous to 1806. V. s. p. 209.

Magnolia gracilis. Previous to 1806. V. s. p. 192.

Camellia japonica, semi-double red variety., 1808. Booth, Camellia and Thea (1829) in Trans. Hort. Soc. VII, 549.

Amygdalus pumila (Prunus japonica). 1810. Bot. Reg. t. 27, (1815).— Bot. Mag. t. 2176, (1820).

Hume, Sir Abraham. Not mentioned in Britten & Boulger, Biogr. Ind. From the Dictionary of National Biography we learn that he lived 1749— 1838, and was a Fellow of the Royal Soc. It seems he was also a Fellow of the Hort. Soc. He had a garden at his estate Wormley-Bury in Hert- ford-shire, where he cultivated many rare exotic plants. (Trans. Hort. Soc. IV, 59). Florists especially were indebted to Sir Abraham for the first introduction of many kinds of Chrysanthemum and beautiful varieties of Chinese Paeonies, between 1798—1820.

J. Sabine, in his article on Chinese Chrysanthemums, 1821, Trans. Hort. Soc. IV, notices the following varieties introduced from China into Hume's Garden:

P. 344, 346, the *Rose* or *Pink Chrysanthemum*, and the *Buff* or *Orange*, called also the *Copper coloured*, were both brought from China by Capt. James Pendergrass in the "Hope" (v. s. p. 190), in 1798.

P. 341, 346. In 1802 Pendergrass brought the *Quilled Yellow* or *Straw-coloured Chrysanthemum*, and in 1806 the *Spanish Brown* variety.

P. 337. The *Quilled White* variety was brought from China for Hume in 1808, by Capt. George Welstead, in the "Alfred" Indiaman.

J. Sabine, in his treatise on Paeonia Moutan, Trans. Hort. Soc. VI, 1826 reports:

P. 469. The *Paeonia Moutan papaveracea* was imported 1802 for Sir A. Hume, by Capt. Pendergrass in the "Hope" E. I. man (v. s. p. 190).

In Loudon's Arbor. et Frut. Brit. I (1838) 254, we read: "North of London the largest plant (Moutan papaveracea) in the country is at the seat of Sir Abraham Hume, at Wormleybury, in Hertfordshire. It is 7 feet high, and forms a bush 14 feet in diam., after having been planted thirty years. It stands the winter, in general, very well. In 1835 this plant brought 320 flowers to perfection; but it has been known to bear three times that number".

P. 475. The *Paeonia Moutan Humei*, imported in 1817 by Capt. George Welstead in the "General Harris" E. I. man, and by him presented to Sir A. Hume. It was first described and figured in Bot. Reg. t. 379, (1819).

In Sabine's Account of seven Double Herbaceous Chinese Paeonies 1816, Trans. Hort. Soc. II, 279, we read: "*Paeonia albiflora Humei* was imported from China (Canton), in 1810, by Capt. Welstead (v. s. p. 211) for Sir A. Hume". This is the *Paeonia edulis*, var. *sinensis*, figured in Bot. Mag. t. 1768 (1815).

Other plants introduced from China into Sir A. Hume's Garden:

Trans. Hort. Soc. IV (1822), 59: Sir A. Hume communicates that in 1801, *Magnolia conspicua*, the *yu lan* of the Chinese, was planted in his garden in Wormley-Bury. This tree was in full blossom in 1816. Comp. also Bot. Cab. 1187 (1826): The same Magnolia tree mentioned in Hume's garden. It was then 20 feet high, produced 900 flowers in a season, large, white, appearing before the leaves, and very fragrant.

Bot. Reg. t. 456: *Viburnum odoratissimum,* from China, in Sir A. Hume's garden, flowered for the first time in 1820.

Bot. Reg. t. 529: *Cymbidium xiphiifolium* Lindley, flowered in 1820, in the hot house at Wormley-Bury, having been newly introduced by Sir A. Hume, as believed, from China. (Comp. Benth. Fl. hgk. 357).

Bot. Reg. t. 676: *Vanda teretifolia,* Lindl., a native of China, flowered in 1823, in Sir A. Hume's hothouse.

Hume, Lady Amelia, *née* Egerton, born 25 Nov. 1751, died at London 8 Aug. 1809. She married Sir A. Hume in 1771, was a pupil of Sir J. E. Smith, who named in compliment to her the Australian Genus *Humea,* Compositae, (Exotic Bot. I, 1804, p. 1, *Humea elegans).*

Dr. W. Roxburgh, likewise (about 1810), named a new genus, *Humea* (now reduced to *Brownlowia* Roxb., Tiliaceae) in honour of Lady Hume. In his Fl. Indica (ed. of 1874) 448, he writes: "I take the liberty of consecrating this genus to the memory of the late Lady Amelia Hume, that most amiable lady, by whose death Botany has lost one of its greatest admirers and best benefactors".

The following Chinese plants are recorded to have been introduced by Lady Hume's exertions:

About 1786, *Magnolia pumila*. V. s. p. 192.

In 1789, *Pergularia odoratissima*. Sir J. Smith in Rees' Cyclop. states: "The late Lady A. Hume received a fine specimen of this plant in 1789, which covered the stern of the ship with its fragrant green blossoms during a great part of the voyage, and has since been widely propagated in this country". Comp. also Smith, Icon. rar. pict. (1791), t. 16.

In Booth's treatise on Camellia and Thea (1829), Trans. Hort. Soc. VII, 539, the *Camellia japonica incarnata* or *Lady Hume's Blush*, is reported to have been imported in 1806 from China for Lady A. Hume. It was first figured in the Bot. Repos. t. 660 (1811).

Dr. Roxburgh sent *Jasminum hirsutum*, a plant he had received from China, to Lady A. Hume. Bot. Reg. t. 15, (1815).

Mean, James, a distinguished gardener, in the service of Sir A. Hume. He wrote horticultoral articles in the Trans. Hort. Soc. II (1813, 1814). He is mentioned by Sabine in Chin. Chrysanthemums, Trans. Hort. Soc. IV, 336.

Portland, Duke of, William Henry Cavendish Bentinck, born 1738, graduated, at Oxford, in 1757. Prime Minister of England. Died 1809. (Brit. Cyclop.). He had a garden at Bulstrode.

He introduced, in 1790, from China, *Magnolia obovata* (v. s. p. 192), same as *Magnolia purpurea*, said in the Bot. Mag. t. 390 (1797), to have first flowered in the collection of the Duke of Portland, at Bulstrode.

Preston, Sir Robert.

In the Bot. Repos. t. 199 (1802) it is stated that the *Camellia japonica*, var. *flore rubro pleno* was introduced about 1794 from China by R. Preston, Esq., Woodford, Essex. — The Bot. Cab. t. 397 (1819), referring to this importation, styles him Sir Robert Preston. — Booth, in his treatise on Camellia and Thea (1829), in Trans. Hort. Soc. VII, p. 531, says that the Double Red Camellia, was first introduced in 1794 by Sir R. Preston of Valleyfield, who then lived at Woodford in Essex.

Slater. Two persons of this family-name, Gilbert and John Slater, are frequently mentioned in the Hortus Kewensis, 2d edition, and the illustrated botanical periodicals of the beginning of this century in connection with the introduction of exotic and particulary of Chinese plants. As I have not been able to find any other information regarding these individuals, I must confine myself to quote the respective passages in the above mentioned works, frow which we can conclude that Gilbert Slater died between 1791 and 1794, and John previous to 1798. Perhaps both are identical.

Bot. Mag. t. 284 (1794): *Rosa semperflorens.* Introduced, in 1791, by the late Gilbert Slater of Knots-Green. He obtained this Rose from China. — In the Hort. Kew, 2d edit. (1812) III, 266, it is stated that Gilbert Slater introduced this Rose about 1789.

Bot. Cat. t. 1113 (1826): *Senecio speciosus,* Willd. A native of China, is said to have been introduced by Mr. Gilbert Slater, in 1789. — The Bot. Reg. t. 41 (1815), describing the same plant, says that it was introduced by George (probably a mistake) Slater, in 1789.

Hort. Kew, 2d edit. 1812, IV, 63: *Clerodendron fragrans,* Vent. Introduced about 1790, by G. Slater, Esq.

Bot. Cab. t. 796 (1823): *Clerodendron squamatum.* Introduced from China to this country, many years ago by the late Mr. Slater of Walthamstow, whom we have often had occasion to mention.

Bot. Cab. t. 513 (1821): *Hibiscus Rosa sinensis, flava.* First introduced from China by the late Gilbert Slater, Esq. of Walthamstow, a gentleman of whose zeal in introducing new plants we have already had occasion to speak.

Bot. Repos. t. 25 (1798): *Camellia japonica,* var. *flore albo pleno,* Double white Camellia. This delightful variety of the Camellia was first imported from China about the year 1793 by Capt. Connor of the "Carnatic" E. I. man, for the Gardens of the late J. Slater, Esq., a gentleman of most indefatigable spirit for the introduction of new plants to this kingdom; indeed it is to him we owe most of the plants received from China within these few years; he having procured a catalogue to be printed of all the described Chinese plants in that language, with the descriptions translated, and by various hands transmitted it to that country.

Bot. Repos. t. 91 (1800): *Camellia japon.,* var. *flore pleno variegato,* Double striped Camellia. This variety is of the same date (1793) in our gardens, as the double white, and was introduced through the same medium (J. Slater). It is very rare and found in but few collections. — B. Booth, Hist. and Descr. Camellia and Thea, 1829, Trans. Hort. Soc. VII, 531, reports with respect to the above two varieties: "Forty years ago none of the Double Camellias had been seen in a living state in the country. The Double White and the Double Striped varieties were the first that were introduced, having been brought from China, in 1792, by Capt. Connor of the 'Carnatic' for John Slater, Esq. of the India House".

Bot. Repos. t. 226 (1802): *Magnolia pumila*. China. We owe this plant to the late J. Slater, Esq. of Laytonstone. It was one of that rich cargo brought home in the "Carnatic" for him, by Capt. Connor, in 1793.— See also Bot. Mag. t. 977 (1806) Magnolia pumila.

Bot. Repos. t. 324 (1803): *Magnolia purpurea*. This fine species of Magnolia, a native of China and Japan, was introduced to us by the late Mr. Slater of Laytonstone, to whom we are so much indebted for the greatest number of beautiful plants which have been imported from thence by any individual.

Bot. Repos. t. 291 (1803): *Senecio Pseudo-China*. The roots of this plant were received at the same time with the Double Camellias, China Roses etc. from China, by the late J. Slater, Esq. of Laytonstone, Essex. Willdenow, Sp. Plant III, 1991, found that the figure in the Bot. Repos. did not represent the Linnean S. Pheudo-China, gave the plant figured by Andrews the name *Senecio speciosus*, the introduction of which in the Bot. Cab. t. 1113 (1826), however, is attributed to Gilbert Slater, in 1789. Mr. Brown (Gard. Chron. 1879, II, 615) has proved that the plant in question is from the Cape of G. H. not from China.

Bot. Repos. t. 216 (1802). *Epidendron sinense*. Introduced 1793 by the late J. Slater, Esq. of Laytonstone, at the same time with the two varieties of the Double Camellias, from China. But in the 2d edition of the Hort. Kew. (1813) V, 214, the introduction of this plant is attributed to G. Slater, 1793.

Bot. Mag. t. 438 (1799): *Hydrangea hortensis*, Smith. Introduced from China in 1790 by Sir J. Banks and about the same time by Mr. Slater.

Desvaux, Journ. de botan. I, 1808, 243, *Slateria japonica*, says that he dedicated the genus name to Slater, an English botanist and horticulturist, who first introduced from Japan the beautiful *Hydrangea hortensis*, which first flowered in his garden, in 1790.

Evans, Thomas, of Stepney, where he had a garden. This liberal patron of horticulture is frequently mentioned in the Hort. Kew, 2d edit. and in the illustrated botanical periodicals, at the end of the last and beginning of the present centuries. He was connected with the India House. He sent a botanical collector to Pu lo Pinang (Prince of Wales' Isl.), who discovered there several rare plants and in 1808 the beautiful *Begonia Evansiana*, which he introduced to Evans' garden. (Bot. Repos. t. 627,

1810). He died between 1810 and 1815, for in the Bot. Mag. t. 1783 (1815) we read sub *Rubus rosaefolius*, β *coronarius*: "This elegant Bramble was introduced some years since from the Princes of Wales' Isl. in the East India by the late Mr. Evans of Stepney, who devoted almost his whole income to the acquirement of new and rare plants, which he generously distributed among other collectors".

Evans is recorded to have introduced the following Chinese plants:

In 1792, *Sanseviera carnea*. Bot. Repos. t. 361 (1804). Same as *Sans. sessiliflora*, Bot. Mag. t. 739 (1804).

Musa coccinea, Bot. Repos. t. 47 (1799).

In 1796, *Magnolia fuscata*, Bot. Repos. t. 229 (1802).

Magnolia pumila, Bot. Mag. t. 977 (1806).

Iris chinensis, Bot. Mag. t. 373 (1797). *Evansia chinensis*, Salisb. in Trans. H. S. I, 303. Introd. by Th. Evans of the India House.

In 1802: *Bletia hyacinthina*, Hort. Kew (v. s. p. 199).

Chinese Chrysanthemum, the *Sulphur Yellow* variety. According to Sabine, Chin. Chrysanthemums, Trans. Hort. Soc. IV (1821), 341, this variety was brought over from China for Mr. Evans, in 1802, Sept., by Captain H. Wilson in the "Warley".

In 1804, *Rosa multiflora*, Hort. Kew, 2d ed. V. s. p. 194.

In 1810, *Ardisia lentiginosa*, Bot. Reg. t. 533 (1821).

Good, Peter. According to Britten & Boulger, Biogr. Ind. he was a Kew gardener, sent in 1796 to Calcutta. He died at Sydney, 1803. He is stated to have introduced, in 1796, *Holmskioldia sanguinea*, an elegant plant, which, according to Roxburgh, Fl. Ind. 480, had been previously introduced from China into the Botan. Gardens, Calcutta. Hort. Kew, 2d edit., IV, 65. Hemsl., I. F. S., II, 263, doubts whether it grows in China.

Raymond, Sir Charles, at Valentine House, Essex. He first cultivated, about 1793, *Camellia japonica, variegata*, one of the varieties of the double Camellias seen in England. Bot. Cab. t. 329 (1819).

Coyte, Rev. William Beeston, M. D., born, 1741, died at Ipswich, 1810. (Britten & Boulger. Biogr. Ind.). As he states in his Hortus Botan. Gippovicensis, 1796, 107, he cultivated *Conyza chinensis*. Hort. Kew 2d edit. V, 27.

Drummond, James. According to the Hort. Kew, 2d edit., J. Drummond introduced in 1806 into the Kew Gardens two Chinese plants, *Laurus glauca* and *Crataegus indica* (v. s. p. 195, 198).

I think this is the Drummond mentioned in the narrative of the Russian Admiral Krusenstern's circumnavigation of the globe (Reise um die Welt, II, 99). He calls him President of the English Factory, Canton, and acknowledges his kind assistance and hospitality during the stay of the Russian ships at Macao and Whampoa, from 20 Nov. 1805 to Febr. 9, 1806. Krusenstern had already made Drummond's acquaintance in Macao, in 1796, when the former was in the British service. Drummond used to live at Macao, where he had a beautiful garden, in which was the grotto of Camoens.

Haxton, John, Gardener, attached to Lord Macartney's Embassy to China. (Lasègue, Mus. bot. Delessert, p. 159). In the Hort. Kew. 2 d edition he is stated to have procured three Chinese plants for Sir G. Staunton, who presented them to the Kew Gardens, viz: *Bocconia cordata, Rosa bracteata* and *Polygonum chinense.* V. s. pp. 192, 194, 198.

In the Bot. Mag. t. 361 (1797) sub *Lycium japonicum,* it is noticed that Mr. Haxton, when in the suite of Lord Macartney's Embassy to China, found this plant cultivated by the Chinese.

In the Bot. Mag. t. 1649 (1814), we read, that Mr. Haxton sent to Sir Jos. Banks, from Macao, herbarium specimens of *Enkianthus quinqueflorus,* under the Chinese name *tsiao chung.* (Correct Chinese name *tiao chung,* suspended bells).

In 1831, Coley dedicated to Haxton the name of a new Australian genus, *Haxtonia,* now reduced to *Olearia,* Compositae.

Lance, David. According to Hort. Kew, 2 d edition, he introduced from China, 1804, *Magnolia tomentosa,* Willd. (same as *M. gracilis,* Salisb.) and *Corchorus japonicus (Kerria japonica).* V. s. pp. 192, 194.

Swainson, Isaac, ✝ 1806. He had a private bot. garden at Twickenham. Bot. Mag. 943 (1806): *Convolvulus bryoniaefolius* was raised by Swainson from seed received 1802 from China. — Mr. Hemsley, Ind. Fl. S., ii, 165, thinks, that this plant, a native of the Mediterranean region, is, through some error recorded from China, in the Bot. Mag.

Hunneman, John; died at London 1839. According to Hort. Kew, 2 d edition, he introduced, in 1804, *Bidens chinensis.*

Hibbert, George; born at Manchester, 1757, died 1837. F. L. Soc., 1793, F. R. S., 1811. He had a botanical garden at Clapham. Herbarium presented to Linn. Soc. — He is stated to have introduced the following Chinese plants:

In 1790, *Hemerocallis japonica,* Bot. Mag. t. 1433 (1812).

Hemerocallis coerulea, Bot. Repos. t. 6 (1797).

Hemerocallis alba, Bot. Repos. t. 194 (1801).

In 1795, *Paeonia Moutan rosea plena.* Bot. Cab. t. 1035 (1825).

Sabine, on Paeonia Moutan, 1826, Trans. Hort. Soc. VI, 477.

The Bot. Repos. t. 493 (1807), notices a fine specimen of *Bignonia grandiflora* in G. Hibbert's conservatory, where under the care of Mr. Knight, his botanic gardener, that valuable collection is in the most luxuriant state of cultivation.

Turner, Charles Hampden, of Rook's nest, Surrey, introduced the following varieties of Camellia japonica and other ornamental plants, through the medium of Capt. **Wellbank,** Robert, who brought them from China in the Hon. Court of Directors E. I. Comp. East-India man, the "Cuffnels", between 1810—20:

In 1810, *Camellia japonica paeoniaeflora rosea,* Bot. Cab. t. 238 (1818). — B. Booth, Camellia & Thea, 1829, Trans. Hort. Soc. VII, 547.

In 1816, *Camellia japonica variabilis,* Booth. l. c. 545.

Camellia japonica Pomponia, Bot. Cab. t. 596 (1821).— Booth. l. c. 546.

Glycine sinensis, Bot. Mag. t. 2083 (1819). — Bot. Reg. t. 650, (1822). — Bot. Cab. t. 773 (1823).

Sabine in Trans. Hort. Soc. VI (1826), 460.

In 1820, *Camellia japonica Welbankii,* Bot. Cab. t. 1198 (1826). — Booth, l. c., 552.

In the Hort. Kew, 2d edition, the same Capt. Wellbank is stated to have introduced in the "Cuffnels", for the Kew Garden, in 1808, *Azalea indica,* and in 1811, *Camellia sesanqua,* both from China. V. supra pp. 193, 197.

Colvill, James, (Sometimes quoted: Messrs Colvill), nurseryman, Chelsea, King's Road. Colvill introduced and cultivated many interesting exotic plants, between 1795 and 1825:

In 1795 the *Purple Chinese Chrysanthemum* first flowered in England in Colvill's garden. Bot. Mag. t. 327 (1796). — Sabine, Chin. Chrysanthemum, 1821, Trans. Hort. Soc. IV, 334.

In 1816, *Camellia japonica fimbriata*, the *Fringed White* variety, was first brought into notice by Mr. Colvill. Booth, Camellia & Thea, 1829, Trans. Hort. Soc. VII, 536. — Bot. Cab. t. 1103 (1826).

In 1818, *Chrysanthemum indicum*, δ, *superbum*, cult. by Messrs Colvill, Bot. Reg. (1820), t. 455. — Sabine, Chin. Chrysanth. 1821, Trans. Hort. Soc. IV, 338.

In 1820, *Strophanthus dichotomus*, Pers. var. *chinensis*, Bot. Reg. t. 469 (1820). First cultivated in Messrs. Colvill's nursery.

In 1821, *Osbeckia chinensis*, Lin., cultivated in the hothouse of Messrs Colvill. Bot. Reg. t. 542 (1821). — Bot. Mag. t. 4026 (1843). — But Bentham, Fl. Hongk. 115, observes that the figures given in these periodicals refer to some Indian species, not to O. chinensis.

Ardisia lentiginosa, Bot. Reg. t. 533 (1821). Messrs Colvill's nursery.

In 1822, *Amaryllis radiata*, L'Hérit. Bot. Reg. tab. 596, and *A. aurea*, L'Hérit., Bot. Reg. tab. 611 (1822). Both cultivated in Messrs Colvill's nursery.

Chrysanthemum, Pale Pink, variety, obtained from the Rose or Pink variety in the nursery of Mr. James Colvill, 1822.

Sabine, Chrys. 1826, Trans. Hort. Soc. VI, 325.

Clerodendron squamatum, Brown., Bot. Reg. t. 649 (1822). Cult. in the nursery of Messrs Colvill.

In 1823, *Mazus rugosus*, Lour., a native of China, raised at the nursery of Mr. Colvill from seeds received from the Bot. Garden at Berlin under the name of *Hornemannia bicolor*, Sweet, Flower Garden (1823) t. 36.

Jasminum paniculatum, first introduced, according to Roxburgh, from Canton, into the Bot. Gardens, Calcutta, flowered at the nursery of Mr. Colvill, Chelsea. Bot. Reg. t. 690 (1823). — I. F. S., II, 80.

Rosa involucrata, Roxb. cultivated in Mr. Colvill's Garden. Bot. Reg. t. 739 (1823).

In 1824, *Ixora crocata*, new species, introduced from China by Mr. Colvill. Bot. Reg. t. 782 (1824).

Sarcanthus rostratus, Lindl., from China, cult. in Mr. Colvill's nursery 1824. Bot. Reg. t. 981 (1826).

Rosa indica, Lindl., β, *odoratissima*, cult. in Messrs Colvill's nursery. Bot. Reg. t. 804 (1824).

In 1825, *Rosa microphylla*, Roxb. Flowered for the first time in Europe in Mr. Colvill's nursery. Bot. Reg. t. 919 (1825).

Camellia oleifera, Abel, flowered in Mr. Colvill's nursery. Bot. Reg. t. 942 (1825).

Knight, Joseph, born about 1781, died at Banbury, 27 July 1855. He was gardener to Hibbert (v. s. p. 218), afterwards nurseryman at Chelsea. (Britt. Boulg. Biogr. Ind.). He founded, previous to 1811, at Little Chelsea, King's Road, the Exotic Nursery, which in 1853 was bought by Mr. James Veitch (v. infra).

Chinese plants introduced or cultivated by J. Knight:

In 1811, *Paeonia Moutan,* var. *Rosea Plena,* cult. by J. Knight.

Sabine, Paeonia Moutan, 1826, Trans. Hort. Soc. VI, 477.

In 1814, *Enkianthus quinqueflorus,* Lour. first flowered in Mr. Knight's Exot. Nursery, Little Chelsea. Bot. Mag. t. 1649 (1814). — Bot. Reg. t. 884 (1825).

In 1816, *Camellia japonica,* var. *incarnata,* Blush Camellia, cult. by J. Knight, Chelsea. Bot. Reg. t. 112 (1816).

In 1825, *Camellia japonica,* var. ρ *Knight's* new *Warratah Camellia,* a variety produced by hybridisation. Bot. Mag. t. 2577 (1825).

In 1828, *Camellia japonica Knightii,* a crimson variety raised from seed by J. Knight. Bot. Cab. t. 1463 (1828).

In 1832, *Azalea indica variegata,* a celebrated plant brought home from China by Mr. M'Killigan, in 1832, now in possession of Mr. Knight, Chelsea. Bot. Reg. t. 1716 (1834).

Azalea indica lateritia, Brickred Chinese Azalea. Introduced with the preceding by M'Killigan and purchased by Mr. Knight. Bot. Reg. t. 1700 (1834).

The Trans. Hort. Soc. II, 400, report that J. Knight of Leecastle, in Sept. 1816, sent to the Hort. Soc. the fruit of *Dimocarpus Longan,* ripened by him. — The same statement is found in Bot. Mag. t. 4096 (1844): J. Knight at Lee Castle, Kidderminster.

Cypripedium purpuratum, Lindl. Bot. Reg. t. 1991 (1836). Introduced by J. Knight, who left no record of its origin. Afterwards it was found to be a Chinese plant = *C. sinicum,* Hance, Hongkong. See Reichb. fil. Bonplandia, 1855, 250. — Bot. Mag. t. 4901 (1856).

Lee and **Kennedy,** nursery of exotic plants at Hammersmith, an old firm, in the last and present century. In Britt. & Boulg. Biogr. Index we find:

Lee, James, of Hammersmith, 1715—95, gardener at Syon and Whatton, nurseryman, circa 1745, with Kennedy at Hammersmith.

Kennedy, Lewis, flourished 1775—1818, nurseryman of the Vine yard, Hammersmith. Father in law of H. C. Andrews (Botan. Repository).

Amygdalus pumila, Hort. Kew (Prunus japonica) was cultivated in 1774, in Lee's nursery. Bot. Mag. t. 2176 (1820).

In 1802 the *Changeable White Chinese Chrysanthemum* was first offered for sale by Messrs Lee & Kennedy in the Hammersmith nursery. Sabine, Chin. Chrys. 1821, Trans. Hort. Soc. IV, 336.

In 1813, Messrs Lee & Kennedy cultivated *Lilium concolor,* Bot. Repos. t. 663 (1815).

In 1815, Messrs Lee & Kennedy cultivated *Prunus japonica,* Thbg. Bot. Reg. t. 27 (1815).

In 1816, *Cacalia ovalis,* from China, first flowered with Messrs Lee & Kennedy. Bot. Reg. t. 101 (1816).

In 1818, *Camellia maliflora,* Lindl. first cultivated by Mr. Lee of the Hammersmith nursery. Booth, Camellia, Thea, 1829, Trans. Hort. Soc. VII, 526.

In 1820, *Rubus reflexus,* Canton Bramble, cultivated in Mr. Lee's nursery, Hammersmith. Bot. Reg. t. 401 (1820).

The charming *Primula cortusoides,* Lin., first described and figured in Gmelin's Flora sibirica IV, 85, t. 45, and discovered by Thunberg in Japan, was first cultivated in Europe by Lee & Kennedy, who raised the plant from seeds which Pallas sent from Siberia. See Andr. Bot. Repos. t. 7 (1797).

Davey, nurseryman and florist, Chelsea, King's Road, cultivated in 1809, *Camellia japonica semiduplex,* which he had received from China. Bot. Repos. t. 559 (1809).

Brookes, Samuel, nurseryman, at Ball's Pond, Newington Green, Northampton nursery, also Barr & Brookes, frequently mentioned between 1817—22 in connection with the introduction of exotic, particularly Chinese plants.

In 1818 or 19 Brookes sent out to China a gardener, Mr. Joseph **Poole** for the express purpose of collecting Chinese plants. He made the voyage in the "Lady Melville" Indiaman, Capt. John **Steward,** and brought home many curious plants. See Sabine, Chin. Chrysanthemum, 1821, Trans. Hort. Soc. IV, 333.

Chinese plants introduced or cultivated by Brookes:

In 1817, the *Superb White Chrysanthemum,* introduced for Barr & Brookes by Capt. Lockner, John Christopher, in the "Alfred" India man. Sabine, Chin. Chrys. 1821, Trans. Hort. Soc. IV, 338.

Psidium Cattleyanum, Lindl., stated by Cattley in Trans. Hort. Soc. IV, 315, to have been raised by Messrs Barr & Brookes from seed received from China, in 1818. — This tree is not indigenous to China.

In 1819, Mr. Brookes introduced through the medium of his collector Poole, the *Azalea indica alba* [Bot. Reg. t. 811 (1824)], the *Azalea indica purpurea plena* [Bot. Mag. t. 2509 (1814)], and *Azalea ledifolia* [Bot. Mag. t. 2901 (1829)].

In the same year and through the same medium were imported three varieties of Chinese Chrysanthemum (Sabine, Chrysant. 1821, Trans. Hort. Soc. IV, 333). One of them is the *Small Yellow Single*, the same which Linnaeus described as Chrys. indicum. (Sabine, Trans. H. S. V, 159; idem, Trans. Linn. Soc. XIV, 144).

In the same year, 1819, Sam. Brookes imported from China the *Prunus paniculata*, Thbg. Bot. Reg. t. 800 (1824).

At about the same time Brookes raised from seed received from China the *Melastoma sanguinea*, Bot. Mag. t. 2241 (1821).

In 1821, *Vanda teretifolia*, was introduced by him. Lindley, Collect. Bot. (1821), t. 6.

In 1822, he imported the *Prunus pseudocerasus* (Lindley in Trans. Hort. Soc. VI (1826), 90), and the *Prunus serrulata* (Lindley in Trans. H. S. VII (1830), 238).

In 1824, a new *Chinese Chrysanthemum*, the *Pale Buff*, was sent to Sam. Brookes by J. Reeves. Sabine, Chrysant. 1826. Trans. Hort. Soc. VI, 334.

Whitley, Reginard, nurseryman, King's Road, Fulham, 1808. Subsequently Whitley & C° or **Whitley, Brames & Milne,** 1819.

This firm introduced many exotic plants. The following were received from China:

In 1808, *Paeonia albiflora plena,* var. *Whitleyi*, introduced by Mr. Whitley, nurseryman, Fulham. Bot. Repos. t. 612 (1810). — G. Anderson, Paeonia, 1818, in Trans. Linn. Soc. XII, 256.

In 1817, *Hoya crassifolia*, Haworth, was sent from China, on board the "Wexford" E. I. man, to Mr. Reginard Whitley, who cultivates it in his nursery at Fulham. Sabine, in Trans. Hort. Soc. 1826, VII, 22.

Convolvulus chinensis, first discovered in China by Staunton, was raised in 1818 by Messrs Whitley & C°., Fulham, from seeds received from the Botan. Garden at Moscow. Bot. Reg. t. 322 (1818).

In the same year, the same raised from seed received from Russia, from Dr. Fischer, under the name of *Delphinium chinense*, the plant figured in Bot. Reg. t. 472 (1820) as *D. grandiflorum*, L. var. *chinense*.

Camellia axillaris, Bot. Mag. t. 2047 (1819). It has been several years in the Fulham nursery and is supposed to have been brought from China by Mr. Robarts, and presented by him to Messrs Whitley, Brames & Milne, with other Camellias.

About 1823, *Cydonia chinensis*, was cultivated in the nursery of Messrs Whitley & C°. Bot. Reg. t. 905 (1825).

Shepherd, John, born 1764?, died at Liverpool 27 Sept. 1836. He was for 35 year at the Botan. Gardens Liverpool (Britt. Boulg. Biogr. Ind.).

He cultivated in the Liverpool Botan. Gardens, in 1816, *Saxifraga cuscutaeformis*, a native of China and sent the plant to his friend Loddiges, who described it under the above name in the Bot. Cab. t. 186 (1818).

Callicarpa longifolia, Lam. — W. J. Hooker, Exot. Flora 133 (1825), states that the plant was sent to him by Mr. Shepherd, who had received it from China.

Justicia ventricosa, Hooker in Bot. Mag. t. 2766 (1827). Shepherd had received it in 1825 from Calcutta, whither it had been probably introduced from China.

Blandford, George, Marquis of, 4th Duke of **Marlborough,** also Lord **Spencer Churchill,** born 1766, died 5 March 1840. He used to live at White Knights, near Reading Berks, where he had a magnificent conservatory.

He cultivated in 1803, *Magnolia purpurea*, from China. It was then the first specimen of this species in Britain. Bot. Repos. t. 324 (1803).

He first cultivated in his magnificent aquarium at White Knights the wonderful waterplant *Anneslea spinosa (Euryale ferox,* Salisb.), a native of China and India, the seeds of which had been sent to him, from Calcutta by Roxburgh, in 1809. Bot. Repos. t. 618 (1810).

About 1815 he cultivated the *Gardenia amoena*, Bot. Mag. t. 1904 (1817), supposed to have been received from China or the East-Indies. — According to Hemsley, I. F. S., i, 383, an obscure plant.

Bagat, Lord. He had a garden at Blitfield, Staffordshire. He was F. Linn. Soc. and is reported to have cultivated, in 1813, *Mespilus (Eriobotrya) japonica*, Loquat, and to have presented ripe fruits of it to the Hort. Soc. — Trans. H. S. III (1818), 299; Bot. Reg. t. 365 (1819).

Long, Lady, daughter of Lady Hume (s. p. 212), married Sir Charles Long, who lived at Bromley Hill, Kent, where he had a garden and conservatory. Afterwards, between 1820—29, she became Lady **Farnborough.**

Jasminum revolutum, Bot. Mag. t. 1731 (1815), imported from China, bloomed the first time, in 1814 in the garden of Lady Long.

Camellia japonica, var. *involuta, Lady Long's Camellia,* Bot. Reg. t. 633 (1822); Trans. Hort. Soc. VII, 1829, 545. Introduced from China by Lady Long.

Vandes, Count de, introduced in 1805, from China, into his garden at Bayswater, *Lonicera japonica,* Bot. Repos. t. 583 (1809).

Vandes, Countess de, cultivated in 1820, at Bayswater, *Murraya exotica,* L., from China. Bot. Reg. t. 434 (1820).

She cultivated in 1820, *Chimonanthus fragrans,* Bot. Reg. t. 451 (1820).

In her garden flowered for the first time in England, in 1823, the *Holmskioldia sanguinea,* Retz. said by Roxburgh to be a native of China. Bot. Reg. t. 632 (1823). Mr. Hemsley, Ind. Fl. Sin., II, 263, doubts its Chinese origin.

Aylesford, Countess of, Finch, Louisa, *née* Thynne, born 1760, married in 1781, the 4th Earl of Aylesford, died 1832, at Coventry. (Britt. Boulg. Biogr. Ind.). This Lady took a great interest in botany.

She cultivated *Koelreuteria paniculata,* which flowered in her collection at Stanmore in 1818. Bot. Reg. t. 330 (1818).

She first cultivated *Jasminum undulatum,* Vahl., a native of China, which flowered in her collection at Stanmore, in 1820. Bot. Reg. t. 436.

Colebrooke, Sir Henry Thomas, born at London 1765. F. Linn. Soc. 1816; F. Rog. Soc. Sanscrit scholar. Chief Judge, Bengal. In India 1783—1815. Furnished oriental names for Roxburgh's Flora Indica. Collected Indian plants, wrote botanical articles in Trans. Linn. Soc., died at London, 1837. (Britt. Boulg. Biogr. Ind.). He was in possession of a collection of Chinese drawings of plants. Lindley, Roses, 1820, 9. Rosa microphylla, and Bot. Reg. t. 919 (1825). See also Journ. Bot. 1882, 60.

VI. Lord Amherst's Embassy to the Chinese Court, 1816—17. Dr. Clarke Abel.

In 1815 the British Government resolved upon despatching a second embassy to the Emperor of China. **Lord Amherst,** like Lord Macartney, Governor General of India, was appointed Ambassador to Peking, and Henry **Ellis** and Sir George Thomas **Staunton,** associated with him as second and third Commissioners; the latter a son of Sir George Leonard Staunton, who, as secretary, accompanied Lord Macartney to Peking, 1792—94. (V. supra p. 156, note).

The Rev. Rob. **Morrison,** one of the first English sinologists, since 1807, a resident in Macao, accompanied the embassy to Peking, as principal interpreter. — John Francis **Davis,** born 1795, was one of the members of this embassy. Afterwards he made himself known as an able sinologist and diplomatist, from 1844—48 Governor of Hong kong, in 1845, knighted.

At the suggestion of Sir Joseph Banks, Dr. Abel was appointed naturalist to the embassy and was fully instructed by this great and good man. His appointment at first was simply medical. **Abel,** Clarke, born 1780, M. D. After his return from China, in 1818, be became F. Linn. Soc. Subsequently appointed physician to Lord Amherst in India, he died at Cownpore, 14 Nov. 1826. He published an account of his travels in China:

NARRATIVE OF A JOURNEY IN THE INTERIOR OF CHINA AND OF A VOYAGE TO AND FROM THAT COUNTRY, IN 1816 AND 1817, by Clarke Abel, Chief Medical Officer and Naturalist to the Embassy of Lord Amherst to the Court of Peking. London, 1818. Engravings and Maps. — The latter are of little value. Generally they do not agree with the itinerary. — Comp. also: NARRATIVE OF A VOYAGE IN H. M. LATE SHIP. "ALCESTE" TO THE YELLOW SEA, ALONG THE COAST OF COREA, TO THE ILS. OF LEW CHEW, by **Mac Leod,** Surgeon, London 1817.

The Embassy left England in Febr. 1816 on board H. M. S. "Alceste", in company with H. M. S. "Lyra". — Rio Janeiro, Cape of G. H., Java.

On the 9th of July 1816, they anchored near the Chinese island of 1816 Hong kong with its high conical mountain. Dr. Abel succeeded in collecting some plants on the island. On the 13th they sailed from Hong kong for the Yellow Sea, on the 27th they entered the Gulf of Pe Cheli and on the 28th anchored 15 miles from the mouth of the Pei ho.

On the 12th of August Lord Amherst and his suite left the «Alceste» in his barge; they arrived at Ta koo and advanced by boats to Teen tsing (T'ien tsin) which place they reached the same day. Next morning they proceeded on their route up the Peiho.

Cultivated plants observed near the Peiho: Millet, Beans, *Sida tiliacfolia*, Fischer, one of the hemp plants of China, *Sesamum orientale*, from the seeds of which the Chinese extract an esculent oil. *Ricinus communis*, the Castor oil plant, continually occurred in patches or in fields. (86).

On August 16 they anchored before a village called Tsai tsun. Abel collected the following plants: *Polygonum lapathifolium*, *P. aviculare*, two species of *Chenopodium*, *Tribulus cistoides* (rather *terrestris*), *Hibiscus trionum*, *Statice limonium* (rather St. *bicolor*).*). — The *Holcus Sorghum* or *Kow leang*, tall corn of the Chinese, clothed the margin of the river. (87).

On the 19th, the Embassy reached Tung chow, where they remained more than a week. On the 21st Imperial Commissioners arrived to confer with the Ambassador about the question of the prostration. Lord Amherst declared his intention not to perform the ceremony. On the 27th the Embassy started for Yuen ming yuen, the Summer Palace of the Emperor (about 6. Engl. miles N. W. of Peking) Lord Amherst in an elegant barouche; four mules provided to draw it. At midnight they reached Peking, traversed the capital; at the dawn of the day the celebrated gardens of Yuen ming yuen were reached. Quarters for the embassy had been prepared in the village of Hai teen.

Fields of *Nelumbo, Lotus* (more correctly lakes and ponds covered with the Lotus plant) rearing high their glossy leaves and gorgeous flowers, edged by trees with the foliage of the Cassia *(Sophora japonica)*. (103). Ambassador urged to enter the imperial presence and to prostrate, refuses, is insulted, quits the palace. Departure of the Embassy from Yuen ming yuen, the same day. They pass only through the suburbs of Peking, proceed to Tung chow where they remain till 2d of Sept.

Plants cultivated near Tung chow: the elegant *Ipomoea quamoclit* trained on small frames of trellis work, *Begonia Evansiana*, *Lagerstroemia indica*, *Hemerocallis japonica*, *Punica granatum*, dwarfed, *Cassia Sophera*, *Nerium Oleander*, *Lychnis coronata*, *Tradescantia cristata* (I should think Abel saw *Commelyna communis*), a species of *Dianella* with purple flowers *(Ophiopogon spicatus?)*, *Hibiscus Plumbago*. "But cultivated and praised

---

*) In brackets my own observations.

above all others appeared the *Nelumbium speciosum,* the *leen wha* of the
Chinese, raised in capacious vases of water, containing gold and silver fish.
Its seeds, in size and form like a small acorn without its cup, are eaten
green or dried as nuts, and often preserved as sweet-meats; they have a
nutlike flavour. Its roots, sometimes as thick as the arm, of a pale green
without, and whitish within, in a raw state are eaten as fruit, being juicy
and of a sweetish refreshing flavour, and when boiled, are served as vegetables. Nelumbium we saw in all parts of the empire through which we
passed, but this plant seemed to flourish better in the north, than in the
southern provinces, and according to the missionaries it grows most luxuriantly beyond the great wall" *). (121).

Lycium chinense, is frequently met with in the neighborhood of Tung
chow, and is common in the province of Pe Che li. Abel was entirely unable
to observe any difference between this and Lycium barbarum, Linn., except
in the occasional size of the leaves. (123).

Tribulus terrestris, abundant on the banks of the Pe ho and the Eu ho
in Pe Cheli. The character "folia 6 jugata" given to this plant by its de-
scriber Linnaeus, does not at all apply to it as found by Abel in the north
of China. Like T. cistoides it has always folia 8 jugata and only differs
from that plant in the smallness of its corolla. (I found the leaves of the
Peking plant generally 7 jugate, sometimes 5 jugate E. B.).

Sophora japonica, common tree growing to a large size. "Holcus Sor-
ghum grows to 16 feet high. The large bunchy panicles were ripening
whilst we remained at Tung chow. A species of *Panicum* was also commonly
cultivated. — *Polygonum Fagopyrum,* cultivated as grain". (123).

"*Sesamum orientale* and *Ricinus communis,* much cultivated for the
esculent oils extracted from their seeds. The Chinese seem to use some
means of depriving the oil of the latter plant of its purgative properties.
(It seems to be unknown in Europe, that Castor oil when cooked loses its
purgative qualities. E. B.). The seeds (of Sesamum) are also eaten". (124).

Plants cultivated as vegetables: "*Solanum Melongena,* two species of
Capsicum, C. annuum and *C. chinensis;* several species of *Gourds* and *Cu-
cumbers,* several species of *Phaseolus,* and above all the vegetable which
they call *Pe tsae,* white vegetable, a kind of cabbage so named probably
because the Chinese blanch the plant, naturally green. (*Brassica chinensis.*
The leaves of this cabbage, near the root, are white). The quantity consumed of it all over the empire, and especially in Peking, is immense. This

*) It grows wild in the Ussuri river, about 46° N. lat. See Regel, Tent. Fl. ussur. 15.

vegetable may be considered, in relation to the Chinese, what the potatoe
is to the Irish. It is cultivated all over the empire». (124).

Fruits: Fine white *Grapes, Water melons* and other *Melons, Peaches,
Leen wha* (Nelumbium seeds), *Apples, Chestnuts,* esculent seeds of a *Pine*
said to come from Tatary *(Pinus koraensis).* Abel saw also seeds of *Taxus
nucifera.* (125).

"Among the plants raised for their fibres, the most conspicuous place
is occupied by the *Sida tiliaefolia, xing ma* (drawing p. 379), extremely
cultivated on the banks of the Pe ho, for the manufacture of cordage. This
plant seems to be confined to the northern provinces. — *Cannabis sativa,*
is also much cultivated, but the Chinese prefer the Sida». (125).

The Embassy remained at Tung chow from August 19th to Sept. 2d,
when they embarked for Teen tsing (T'ien tsin).

Plants collected: *Ulmus pumila,* a species of *Orobanche,* it seems *O.
coerulea, Viola tricolor.* (130).

On the 6th they reached T'ien tsin. After two days' stay there they
started, returning to Canton by the river way*). At first they proceeded
on the Grand Canal, reaching the borders of the province of Shan tung
on the 16th.

In the province of Pe Chely Abel had collected the following plants:
Salsola altissima (Suaeda glauca?), *Euphorbia tithymaloides, Lepidium latifolium, Hedysarum striatum* (Lespedeza striata), *Lonicera caprifolium* (evidently L. japon.), *Pontederia vaginata* (I think Monochoria Korsakowii, Rgl.),
Menyanthes nymphoides (130). — *Thuya orientalis,* in Shan tung. (145).

They passed *Tang chang foo* (Tung ch'ang fu) on the Grand Canal,
and on the 28th came to the mouth of a rapid river which feeds the Canal.
Staunton calls it Luen (supra p. 159) but Morrison said it is Wan. It is
reported to arise from 72 springs in the mountains called Tae shan, in the
Shan tung province. What Staunton states about this river and the dividing
of the waters is confirmed by Du Halde, I, 33. The temple where the
waters of the Grand Canal divide is called Fan shang (shuy) miao, or temple
of the division of waters. (147).

Farther on they passed, on the grand Canal, Hoai ngan foo, Kao yen
chow (Kao yu chow), Yang chow foo, Qua chow on the Yang tze kiang
and came to the large city of Tsing keang foo (Chin kiang fu) and the
Kin shan or Golden Hill (Abel means Silver Island in the Yang tze, down-
wards Chin kiang).

---

*) The route followed by the Dutch Embassy in 1656. V. s. 23.

Aquatic vegetables seen in the Grand Canal: — Besides the Nelumbium, the *Trapa bicornis* is extensively cultivated. Its fruits are sold in the markets as nuts are in Europe. The root of *Scirpus tuberosus*, is also sold and more highly estimated than the Trapa. It was not seen growing. (154).

The Embassy entered the Yang tze kiang on the 19th October and proceeded up the river to Nan king *).

In Nan king there was exposed for sale the cloth which bears the name "Nan king", but the raw *yellow Cotton*, from which it is supposed to be made was in vain looked for. (160).

On the walls of Nan king Abel met with *Rosa Banksiae, Cotyledon spinosa.* The expressed juice of the latter is said to be used by the Chinese women to dye their hair of a black colour and prevent boldness. *Hamamelis chinensis,* of Sir Joseph Banks' herbarium. *Ficus repens,* very abundant. It hides the walls by its profusion. (160).

Among the larger plants there were found: *Pinus Massoniana* of Lambert; the *Gingko* of Kaempfer, or *Salisburia adiantifolia* of Smith, *qua chow,* of the Chinese (probably *ya kio,* duck's foot). The fruit of the latter was exposed for sale in great quantities, whether as a fruit, a vegetable or a medicine could not be ascertained. (fruit used in medicine, kernel of the stone eaten). (160).

On the first of November the Embassy halted at the village (city) of Tung ling heen. Here Abel first saw the *Tallow tree, Croton sebiferum,* (Stillingea sebifera), here used as fire wood. (161).

On Nov. 3d they reached the small town of Ta tung, where they made a stay of four days, and Abel for the first time saw a *Tea* plantation. (164).

Several species of *Oaks* occurred in large plantations. The oaks seen here were remarkable for their beauty, but few for their size. The largest did not exceed 50 feet. They seemed to be used chiefly as pollards, considerable quantities of the branches being accumulated for fire wood or perhaps for charring. One of the largest and most interesting of these trees, which Abel has named *Quercus densifolia,* was an evergreen, closely allied to Q. glabra, Thbg., and resembled a laurel in the shining green of its foliage. It bears its branches and leaves in a thick head crowning a naked and straight stem. Its fruit grows in long, upright spikes, terminating the branches. Another species, which A. calls *Quercus chinensis,* grows to the height of 50 feet, and bears its fruits in long pendulous spikes. Of

---

*) Lord Amherst's route on the Yang tze, and across Po yang Lake to Nan chʻang coincides with that of the Dutch Embassy, in 1656. V. supra p. 23.

this he brought home a good specimen. The leaves of some of these oaks were distinguished by red prominent veins on the under surface; others by the size, and some by the hair-like processes which fringed their margin. Five distinct species were found in a short walk. (165).

Growing with the oaks there were some dwarf *Chestnuts*, the fruits of which were exposed for sale at the village, and were not larger than the common bon-nut of the country. (165)*).

From the shops of the same place Abel received several species of *Ferns*, which are used as vegetables, infused as tea and administred as medicine. The most general was the *Pteris piloselloides*.

Ginger was also much grown. (165).

The Embassy left Ta tung on Nov. 7th. On the 14th they left the Yang tze, entered Lake Po yang, reached the town of Ta koo tang (on the western shore of the lake, see Admiralty's map), where they were detained by boisterous weather till the 16th. Abel visited the neighbouring country and found it abounding in plantations of *Oaks* and *Firs*. Occasionally *Pinus lanceolata, (Cunninghamia sinensis)* was met, which was brought to Europe by the former Embassy. — A very distinct new genus was discovered here; Mr. R. Brown has called the plant *Abelia chinensis*, a straggling shrub with pendant branches. Its flowers were for the most part faded, but its permanent pink calyces, clustered into thick heads, gave it a beautiful appearance. — The *Tallow* tree also grew here in abundance, a most magnificent tree. (167).

The Embassy left Ta koo tang on the 16th and the same day reached Nan kang foo, situated at the foot of a lofty mountain covered with snow. Abel gathered the following plants in the neighbourhood: Among Ferns, *Adiantum flabellulatum, Asplenium lanceum, Aspidium varium, Blechnum orientale, Davallia*, very closely allied to D. canariensis, *Hydroglossi* spec., very common in the south of China, *Polypodium lineare, Pteris semipinnata, Pteris biaurita*, a new species of *Woodwardia*, allied to W. japonica, but differing from it in having acute lobes. (168). — *Ilex*, it seemed I. aquifolium. A species of *Mulberry*, it seems, with fruit of the colour and general appearance of a strawberry, but clustered around the branches, was also gathered in this part of the route. (169).

Having passed along the western shore of Lake Po yang, the boats quitted the lake at Woo ching chen, and proceeding on their route by a river arrived at Nan chang foo on the 23d.

*) *Castanea vulgaris*, Lam., var. *japonica*, A. DC. See Franchet Pl. David. I, 277.

The general cultivation on the Yang tze kiang and the Po yang Lake is *Rice*, where ever the nature of the land allows it. The *Arum esculentum*, the *Pe ts'ai* (Brassica chinensis), the *Arachis hypogaea*, are also exceedingly abundant and afford a vast quantity of vegetable food. — The *Pinus Massoniana*, and several species of *Oak*, mingled occasionally with the *Tallow* tree and *Camphor* trees, are the chief ornaments of the shores of the Yang tze and the Po yang Lake. The pine was often in considerable groves and appeared to grow to a great height. (171).

At Nan chang foo the Embassy re-entered upon the route pursued by Lord Macartney. They left Nan chang foo on the 27th and then entered a country higly ornamented with useful and beautiful plants. The most remarkable was a species of Camellia, which produces much of the oil consumed by the Chinese. Abel named this beautiful shrub *Camellia oleifera*. The Chinese call it *tcha yeou* or oil yielding tea plant. (A figure of the plant is given p. 174). The plant in general appearance, closely resembles the Tea plant. It was brought to England by the former Embassy and considered as the Camellia sasankwa, from which, however, it is very distinct. (Mr. Hemsley, I. F. S., 1, 82, reduces Abel's C. oleifera to. C. sasankwa). Abel sometimes found it of the magnitude of a moderate sized cherry tree, and always that of a large shrub from 6 to 8 feet high, and bearing a profusion of large, single, white blossoms. The places which it covered often looked in the distance, as if lightly clothed with snow, but on a nearer view, exhibited one immense garden. The Chinese cultivate it in large plantations, and procure from its seeds a pure esculent oil by a very easy process (which is detailed by Abel, 174).

The tree producing tallow, *Croton sebiferum* (Stillingia sebifera), was one of the largest, most beautiful, and most widely diffused of the plants found by Abel in China. He first met it a few miles south of Nan king and continued to observe it, in greater or less abundance, till his arrival at Canton. The tree has many Chinese names, among them *ya rieou*. (Abel quotes Du Halde, where the name reads *ya kieou*. The common name is *wu kiu*). High stem, spreading branches. The foliage has the green and the lustre of the laurel. Its small flowers of a yellow colour are borne at the ends of its terminal branches. Clusters of dark coloured seed vessels succeed them in autumin, and when matured burst asunder and disclose seeds of a delicate whiteness. To obtain the tallow the seeds are bruised, pounded, formed into a thick mass and heated with a small quantity of water in a large iron vessel, etc. (177).

The *Camphor* tree, *Laurus Camphora*, often adorned the banks of the

rivers during the Embassy's progress through the southern provinces of
China, Kiang si and Kuang tung. It grows to the size of our largest elms
or oaks. The largest that Abel measured was 20 feet in circumference and
about 50 feet high. Many of the branches were 9 feet in circumference.
The camphor obtained from this tree is less valued by the Chinese than
that which they procure from Borneo. The process pursued by the Chinese
to obtain Camphor has been described by Father d'Entrecolles in Lettres
édif. XII. Besides the Laurus Camphora, Abel found another species of
Laurel, highly impregnated with the pungency and flavour of the gum,
and which is probably used by the Chinese for its extraction. It so nearly
resembled its congener, when growing, as to require a close examination
to distinguish. It is, however, distinctly separated by the form of its leaves
and the distribution of the nerves. — The Camphor tree is one of the
principal timber trees of China, being used in building and in the fabrication of articles of furniture.

A species of *Ficus,* called *yung shoo* by the Chinese, much resembling
the Banyan in habit, grows very commonly in sandy soil on the banks of
rivers from Lake Po yang to the mountains of Mei ling. Its form gives a
singularly grotesque character to the scenery. Its trunk is made up of a
series of small stems always close together. Its branches are wide and
straggling, but scarcely overshadow its arching roots rising above the soil
and covering a considerable space of ground. (This is the *Ficus retusa,*
Lin. (180).

Although the land on both sides the river is favourable to the growth
of many useful plants, it is seldom very productive in corn. Between Lake
Po yang and the Meiling Mts the quantity of land cultivated in corn bears
no proportion to that which is entirely barren or covered with plantations
of Camellia oleifera.

The Embassy sailed up the Kan kiang River till December 6.th, then
passed the Se pa tan or 18 cataracts. On the 8th they anchored at See
chow. (On Staunton's map the 18 cataracts are marked south of Wan an
hien. See chow is not found on the Chinese maps).

The *Pinus Massoniana,* still continued to be the most general species
of fir, but was occasionally mingled with the *Pinus lanceolata (Cunninghamia sinensis).* Abel found the last tree seldom more than 8 or 10 feet high.
Abel also gathered the *Tea* plant, apparently in its native habitat, near no
plantation. It was a small shrub of what has generally been considered the
green variety.

The *Dryandra cordata,* Thbg., *(Elaeococca verrucosa,* A. Juss.), the *tong shoo,* of the Chinese, grew in the same place. From the seeds of the plant the Chinese extract an oil, which they use as a varnish for their boats and coarse articles of furniture. They often mix it with the more valuable varnish obtained from a species of *Rhus* and sell the compound as the superior article. Abel did not see the true varnish tree growing, but judging from specimens brought to him by his friends, thinks it to be an undescribed species of Rhus, not the Rhus vernix. (181).

In the remainder of his journey Abel found a new species of Eugenia, which, as it is probably the smallest of the genus, has been named *Eugenia microphylla.* It covered the declivities of almost every hill in the province of Kiang si. It is a very elegant plant strongly resembling a myrtle, grows to the height of 1 or 2 feet and bears thick terminal clusters of dark purple berries, which are eaten by the Chinese. (181).

On December 18th the Embassy reached the city of Nan gan (an) foo, situated at the northern base of the Mei ling Mountains. In its neighbourhood Abel collected *Eurya japonica (E. chinensis,* see farther on), which covered the hills in the greatest profusion. *Arachis hypogaea* much cultivated for the oil extracted from its seeds, and for the nourishment they afford to the common people as a fruit and a vegetable. Abel first met with the plant cultivated in fields on the banks of the Yang tze, and continued to observe it through the whole of the provinces of Kiang si and Canton. The seeds are roasted before they are eaten. The Arachis in China bears from 2 to 4 seeds in each capsule. (182).

The Embassy quitted their boats at Nan gan foo for a short land journey across the Mei ling Mountains.

Near the entrance of the pass on the Canton side, Abel saw a species of *Prunus* in full flower, in Chinese *mei hwa shoo;* hence the name of the mountain. (The *mei hwa* is *Prunus Mume,* S. & Z.). (184).

Having reached the city of Nan hiung foo they re-embarked. Large rafts of timber were floating down the river. It has been stated that the rafts of timber floated down to Canton are formed of *Pinus lanceolata,* but Abel found this to be an error. He never saw the latter in groves, but scattered among the *Pinus Massoniana,* and generally of a small size; and the rafts he saw were P. Massoniana. (191).

Farther on, down the river, a species of *Lycopodium,* was seen, resembling a tree in miniature. An exagerated figure of the plant is given in №238 of the botanical drawings in the India House. The head is umbrella-shaped and spreading,

On the 25th December Chao (Shao) chow foo was reached. Farther on they anchored at a narrow pass formed by rocks projecting from both banks of the river. Abel collected many native plants:

Myrtus tomentosus; Smilax China, famed for its sudorific properties and another species which seemed to be *Sm. lanceolata;* a *Begonia,* resembling B. grandis; a new *Camellia,* probably a variety of C. oleifera, but with narrow leaves and small flowers. The banks of some small streams were covered with a species of *Marchantia,* in full fruit, and one or two species of *Jungermannia.* Two *Rhexias.* — Plantations of the *Sugar cane,* were frequent in this part as well as in the southern part of Kiang si. (199).

From the time of crossing the Mei ling Mts till within two days' sail of Canton, they met with little else than a succession of sterile mountains. Then the river widened and deepened and the country became flatter. Groves of *Orange* trees, of *Bananas,* of the *Rose apple (Eugenia Jambos)* frequently relieved extensive *Rice* fields.

1817   On the first of January 1817, the Embassy reached Canton, where they found the Commanders of the "Alceste" and the "Lyra". These ships had meanwhile surveyed the Gulfs of Pe Tchely and Leoo tung, examined the S. W. coast of Corea, discovered its Archipelago, had visited the Lew chewan (Liu kiu) Islands, where a rich collection of plants had been made. On Nov. 2nd these ships had reached the coast of China. The "Lyra" was despatched to Macao, while the "Alceste" had anchored off the island of Lin tin, and subsequently had gone up the river and, notwithstanding the protest of the Chinese authorities, anchored beyond the Bocca Tigris.

Quarters had been prepared for the Embassy in the village of Ho nan, on the opposite side of the river to the British Factory, Canton.

Abel visited Canton. No *Opium* exposed for sale in the shops, probably because it is a contraband article, but it is used with *Tobacco* in all parts of the Empire. They have several methods of smoking Opium (described by Abel). Tobacco is sold every where, and cultivated in all provinces. Every body smokes. (214).

The Chinese consider the application of the *Moxa* as one of the most effectual remedies for local pain. The moxa is prepared by bruising the stems (the leaves) of a species of *Artemisia (A. vulgaris),* in a mortar and selecting the finest and most downy fibres. In this state it is applied in small conical masses upon the part effected, and these being set on fire, instantly consume without producing any severe pain. The fibre of the Artemisia is also used by the Chinese as tinder, after previously steeping it in a solution of nitre. (217).

Abel visited the famous nursery gardens at Fa tee (Fa ti), situated on the southern bank of the river, about 3 miles from Canton. He observed the following plants cultivated there:

The *Moutan* or *Paeony* tree. It is said never to survive more than three years at Canton or Macao. Chinese drawings represent a Yellow flowered Moutan, but Abel was unable to meet with any person at Canton who had seen it.

Camellia japonica, double flowered, many varieties. This is often confounded with the full flowered *Camellia oleifera.* The latter may be distinguished from C. japonica by its more silky calyx, and still more decidedly by its leaves being veinless beneath.

Azaleas, the most beautiful plants that Abel met in China.

A great variety of *Orange* plants were seen in full fruit. (220).

Vaccinium formosanum, is a sacred plant. Its flowers are gathered at the commencement of the Chinese New Year and placed in all the temples. (Abel evidently means *Enkianthus quinqueflorus).*

The *Lycopodium,* which Abel had seen growing in dry places in the Canton province, and which might perhaps be best compared to a fir tree in miniature, he found cultivated in the Fa ti gardens, but in pots, kept in a tub filled with water.

The *dwarf plants* seen in the gardens were *Elms* twisted into grotesque shapes. One of the principal methods of checking the growth and giving them the appearance of age, consists in taking up a young plant and putting it into a pot too small to allow the spreading of its roots. Afterwards they wound the bark in different places.

Abel examined the varieties of the *Tea* plant. He has little doubt that there are two species. (221).

On the 20th of January the Embassy on board the "Alceste" left Canton and anchored on the 23d off Macao. The country in the neighbourhood was found to be full of interesting plants. Abel discovered there *Nepenthes distillatoria.* The Chinese call it "pig basket plant". (236).

The "Alceste" left Macao on the 28th of January and shaped her course for Manilla, which they reached on the 3d of February, and left again after a short stay.

On the 17th of February, in the Gaspar (Banca) Straits, the "Alceste" struck on a coral reef and became a total wreck, but all people were saved. They embarked in the barge and cutter, which both reached Pulo Leat, a finely wooded island, near the spot where the catastrophe took place, and then sailed for Batavia. On the 12th of April the Embassy left Batavia on board the "Caesar", a ship chartered for their conveyance.

On May 27th, Cape of G. H. They re-embarked on the 11th of June. St. Helena, 27th June. Interview with Napoleon. Ascension, 7th July. On the 15th of August 1817, they landed at Spithead (Portsmouth).

All Abel's collections are buried in the Straits of Gaspar. It was by the kindness of Sir George Thomas Staunton, to whom he had presented a small collection of Chinese plants, that Dr. Abel obtained all the specimens of which he gave botanical notices in his work. Some of them were described in the Appendix to his book with the assistance of Mr. Robert Brown. The latter named the new genus *Abelia,* in compliment of the discoverer of this plant.

Abel's Chinese plants, of which botanical descriptions are given in the Appendix:

P. 363. *Quercus densifolia,* and *Quercus chinensis* (compare supra p. 229). Of the latter a figure is given p. 165.

P. 363. *Camellia oleifera,* (compare supra p. 231) and figure in Abel's narrative, p. 174.

P. 364. *Eugenia microphylla.* Compare supra p. 233. Mr. Hemsley, I. F. S., I, 298, states that he has not succeeded in finding the original of Abel's E. microphylla.

P. 374. *Hamamelis chinensis.* Botanical description and figure. R. Brown suggests that perhaps this plant may represent the type of a new genus for which he proposes the name *Loropetalum,* now generally adopted. — Gard. Chron. 1894, I, 342, figured. Introduced by J. Veitch, Chelsea.

P. 376. *Abelia chinensis,* with figure. Comp. supra p. 230.

P. 378, 379. *Eurya chinensis,* with figure. Kiang si and Kuang tung provinces. — This seems to be the E. japonica in Abel's narrative, v. supra p. 233. The latter, however, occurs also in S. China.

The drawing without name, p. 379, represents *Sida tiliaefolia,* Fischer *(Abutilon Avicennae,* Gaert.). V. s. p. 228.

Besides the above plants noticed in Abel's book, I may mention from his collection the following, which escaped the disaster and were subsequently described by various botanists.

Sir W. Hooker in Journ. Botan. & Kew Misc., IX, 1857, 339: *Lastrea opaca,* spec. nov. first discovered in China by Abel.

Hooker. Spec. Fil. III, 1860, 52, *Blechnum orientale,* Lin. — China, Abel.

W. Miquel in London Journ. Bot., VII (1848), 437: *Ficus pyriformis*, Hook. & Arn. China, Abel, herb. Hooker. Subsequently, in Ann. Mus. Lugd. Bat. III, 1867, 281, Miquel published it as a new species under the name of *Ficus Abelii*. But King, Ficus, 1888, 157, considers it to be *F. pyriformis*.

Schima Noronhae. This plant was first described by Reinwardt in Blume's Bijdr. (1825), 129, from specimens gathered in Java, but from the I. F. S., I, 80, we learn that Abel first discovered it in China.

According to Britt. Boulg. Biogr. Ind., the plants left from Abel's Chinese collection are now in the British Museum.

VII. Roxburgh and Wallich.

Roxburgh, William, M. D., born 1759, at Craigie, Ayrshire. Eminent botanist and successfull explorer of the Indian Flora. He entered the service of the E. I. Company, in 1766, as a medical officer of the Madras Army. He subsequently became Botanist of the Company, and in 1793, owing to his great merit, was removed to Calcutta to undertake the superintendence of the Company's Botanic Gardens there. Here he remained till 1814, when his declining state of health induced him to leave India. On his voyage home he visited the island of St. Helena, where he sojourned for some time. Roxburgh died at Edinburgh, May, 1815, not long, it seems, after his arrival in England. He was F. Linn. Soc., from 1799, also F. Roy. Soc. E.

Roxburgh's first work of importance was the Plants of the Coast of Coromandel, which was published at London in 3 folio volumes, with 300 coloured plates, by order of the E. I. Comp., under the direction of Sir Jos. Banks: I, 1795, II, 1798, III, 1819. — In 1814 there appeared his Hortus Bengalensis or a Catalogue of the Plants growing in the Hon. E. I. Comp.' Botanic Gardens at Calcutta. — His memorable work, the Flora Indica, remained in manuscript for some years after his death. In 1820 Drs Carey & Wallich undertook the publication of it. But this edition was not finished. It was only in 1832, that another complete edition of the Flora Indica appeared. Dr. Carey published the whole manuscript as left by the author, in 3 volumes. This is now a very rare work. Botanists are indebted to Mr. C. B. Clarke, Director of the Bot. Gardens, Calcutta, for the reprinting of Roxburgh's "Flora Indica" literatim from Carey's edition, in one volume, Calcutta, 1874.

Roxburgh's herbarium specimens are at Kew and at the British Mus.

Sir J. Banks, in Roxburgh's Coromandel Plants, 1795, named the genus *Roxburghia* (Roxburghiaceae) in honour of his friend.

In his Flora Indica, Roxburgh frequently mentions plants received from China (Canton) and cultivated in the Botan. Gardens, Calcutta. Some of them had been introduced previous to 1793, the date when he was appointed Director of the Gardens. Col. Robert Kyd is occasionally mentioned as having introduced Chinese plants into the Calcutta Gardens prior to 1793. Roxburgh received in the beginning of this century Canton plants forwarded by W. Kerr (v. s. p. 191). From his Flora Indica, 328, sub Scytalia Lichi, and 126, sub Ixora coccinea, we learn, that Roxburgh himself had visited China.

The following is a list of the Chinese plants mentioned in the Flora Indica. The figures refer to the pages of the last edition, 1874.

458. *Calycanthus praecox*, Willd. Introduced to the Bot. Gard. Calcutta from China.

452. *Magnolia pumila*, Bot. Repos. tab. 226. A native of China. In blossom all the year in the B. G. Calcutta, but never produced even half grown fruit. The genus is therefore doubtful.

453. *Magnolia obovata*, Willd. Introd. from China into the B. G. C. where it blossoms but never produces fruit.

453. *Magnolia fuscata*, Bot. Repos. tab. 229. Introduced from China into the B. G. C. It blossoms but never produces fruit.

456. *Uvaria odoratissima*, Roxb., (*Artabotrys odoratissimus*, R. Br. See Hook. Fl. Ind. I, 54). Introduced to B. G. C. from China.

733. *Menispermum hexagynum*, Roxb. A native of China, introduced to B. G. C.

301. *Nandina domestica*, Thbg. Introduced from Canton to B. G. C. by William Kerr.

450. *Nelumbium speciosum*, Willd. A variety with small flowers was brought from China to the B. G. C. — Another crimson variety is *hung lin* (hung lien) in Chinese.

498. *Sinapis brassicata*, Lin. Sinice *pak ts'ai*. (This name is generally applied to *Brassica chinensis*). Seeds procured from China, were sent to the B. G. C. by Mr. John Miller.

212. *Pittosporum Tobira*, Ait. Hort. Kew, edit. 2, II, 27. Introduced from China (by W. Kerr) to the B. G. C.

385. *Dianthus chinensis*, Lin. Native of China. Commonly cultivated in Bengal.

592. *Hypericum monogynum*, Lin. Introduced from China to the B. G. Calcutta. Sinice: *keem see jau taing (kin sze hai t̕ang*, Bridgm. Chin. Chrestom. 454).

518. *Althaea rosea*, Cav. In India only cultivated. The most beautiful sorts we have from China.

516. *Sida chinensis*, Retz. — *(Sida rhombifolia*, Lin. See Ind. Fl. Sin., I, 85).

518. *Sida Abutilon*, Lin. *(Sida tiliaefolia*, Fisch., *Abutilon Avicennae*, Gaert.). The seed has been received from Pekin into the B. G. C. under the name of *king ma*, and is said to be cultivated in that province as a substitute for hemp or flax.

525. *Hibiscus mutabilis*, Lin. A native of China. Cult. all over India.

529. *Hibiscus chinensis*, Roxb. Introd. from China into the B. G. C. (Wall. Cat., 1916, says: Roxburgh's plant not identical with H. chinensis in Braam's Icon. (v. s. p. 188). — Hook. Fl. Ind. I, 342, identifies it with Hibiscus Abelmoschus, Lin.).

520. *China Cotton*. It has been lately introduced into Bengal from China and its wool reckoned 25 per cent. better than that from Surat.

Ibid. sub *G. religiosum*, Willd.: *Nankeen*, a brown cotton, allied to *G. hirsutum*, introduced (from China) into Bengal, but the colour of the wool is very different from the colour of the cotton cloth called *nankeen*, which is no doubt dyed. — (According to Hooker, Fl. Ind. I, 347, *G. religiosum*, which yields Nankeen Cotton, is a variety of *G. herbaceum*).

506. *Sterculia Balanghas*, Lin., sinice *phim pho*. Cultivated in the B. G. C. — *(St. nobilis*, Smith. See I. F. S., I, 89, and Hook. Fl. Ind. I, 358). According to Hance and Parker, Canton plants, the name of *pin p̕o* is applied to *Sterculia lanceolata*, Cav.).

429. *Corchorus capsularis*, Lin. Cultivated in Bengal and China for its fibres.

360. *Gaertnera obtusifolia*, Roxb. Introd. from China to the B. G. C. prior to 1793. — *(Hiptage obtusifolia*, DC. See I. F. S., I, 96).

140. *Fagara nitida*, Roxb. *(Zanthoxylum nitidum*, DC. Prodr. I, 727). *Tcheum tsew* of the Chinese (Comp. Bridg. Chrest. 450, 64. Chin. Pepper, *chün tsiu*). Introduced into the B. G. C. from Canton by W. Kerr, in 1812.

362. *Murraya exotica*, Lin. Introduced from China many years ago.

365. *Cookia punctata*, Sonnerat. Sinice: *whung pi* (See Bridgm. Chrest. 446, 92). Introduced from China. Now common in Bengal.

589. *Citrus acida*, Roxb. var. 4th China-Gora-Neboo. Introduced 20 years ago from China. (Not identified).

590. *Citrus inermis,* Roxb. Sinice: *kumquat.* Introduced from China into the B. G. C. — (Citrus japonica, Thbg. See I. F. S., ɪ, 111).

322. *Amyris punctata,* Roxb. Chittagong. The Chinese gardeners say it grows also in Ohina. — *(Clausena excavata,* Burm. See Hook. Fl. Ind. ɪ, 505. Not mentioned in Ind. Fl. Sin.).

369. *Melia azedarach,* Lin. Sinice: *shum shu,* at Canton. [Bridgm. Chrest. 441, 44, Pride of India, *sham muk (shu)*]. Introduced from China into the B. G. C.

214. *Camunium chinense,* Rumph. *(Aglaia odorata,* Lour. — DC. Prodr. I, 537). Sinice: *sam yeip lan.* (Comp. Bridgm. Chrestom. 452, 3). Introduced from Canton into the B. G. Calcutta.

211. *Hovenia dulcis,* Thbg. A tree introduced by Dr. Buchanan into the B. G. C. He writes, 10th and 14th Nov. 1802, from Katumandu: "This tree was originally brought to Nepaul from China or some country subject to it".

205. *Zizyphus nitida,* Roxb. (Z. *vulgaris,* Lam. See I. F. S., ɪ, 126). Introduced from China into the B. G. C.

Ibid. *Zizyphus albens,* Roxb. Cultivated in Col. Palmer's Garden, near Calcutta, 1803. Originally brought from China. — *(Zizyphus oenoplia,* Miller. Gard. Dict. 3. See Hook. Fl. Ind. ɪ, 634).

328. *Scytalia Lichi,* Roxb. *(Nephelium Litchi,* Camb.). This very famous Chinese tree is now common in Bengal. Originally brought from China. I (Roxb.) have seen in China Lichi trees fully as large as a middling sized ash tree.

329. *Scytalia Longan,* Roxb. *(Nephelium Longana,* Camb.). Sinice: *longan.* Also a native of China, as well as of the mountains which form the eastern frontier of Bengal.

273. *Rhus succedanea,* Lin. Received from Dr. Berry at Madras, into the B. G. C., in 1801. It came originally from China.

551. *Crotalaria elliptica,* Roxb. Introduced from China into the B. G. Calcutta. — I. F. S., ɪ, 151.

555. *Phaseolus minimus,* Roxb. Sinice: *cham loe to.* Raised in the B. G. C. from seeds received from Canton. — Benth. Fl. Hgkg. 88.

560. *Dolichos obcordatus,* Roxb. Raised from seeds received from Canton. — *(Canavalia obtusifolia,* DC. Prodr. II, 404. See Hook. Fl. Ind. ɪɪ, 196; I. F. S., ɪ, 192).

564. *Dolichos phaseoloides,* Roxb. Introduced 1805, by W. Kerr, who sent seeds from China to the B. G. C. — *(Pueraria phaseoloides,* Benth., *Pachyrhizus trilobus,* DC. Prod. II, 402. — I. F. S., ɪ, 190).

572. *Flemingia* (new genus) *prostrata*, Roxb. Raised from seeds sent from China to the B. G. C., in 1805, by W. Kerr. — (I. F. S., i, 197).

358. *Caesalpinia chinensis*, Roxb. Introduced from China into the B. G. Calcutta. — *(C. Nuga,* Ait. Hort. Kew 2 edit. III, 32. See Benth. Fl. Hgk. 97).

422. *Gleditschia horrida*, Willd., a native of China and Cochinchina, most likely Loureiro's Mimosa fera. — *(Gled. sinensis,* Lam. See DC. Prodr. II, 479).

347. *Bauhinia corymbosa*, Roxb. Raised from seeds received from China. — (I. F. S., i, 212).

418. *Mimosa fruticosa,* Roxb. From China it has been introduced into the B. G. Calcutta under the Chinese name *tham yeaong ton* (not identified).

403. *Flat Peach*, from China. Fruit vertically compressed like a turnip. Cultiv. in B. G. C. — (Comp. Trans. Hort. Soc. IV (1822), 512, Flat Peaches of China, with a coloured drawing, cultivated in London).

Ibid. *Amygdalus cordifolius*, Roxb. A native of China, common in gardens about Calcutta. — (Unknown to Mr. Hemsley. I. F. S., i, 221).

Ibid. *Prunus triflora* (lapsu *trifolia),* Roxb. Sin.: *hong sum li.* Cultivated in Bengal, received from China. — (I. F. S., i, 122).

407. *Spiraea corymbosa*, Roxb. A native of China and India. — *(Sp. cantoniensis,* Lour. See Benth. Fl. Hgk. 105. Only known from S. China).

407. *Rosa chinensis,* Willd. Roxb. thinks it is Linnaeus' Rosa indica. A native of China. — (I. F. S., i, 250, *Rosa indica,* Lin.).

407. *Rosa glandulifera*, Roxb. Brought from China to the B. G. C.— (Hook. Fl. Ind. ii, 364, *Rosa alba,* Lin. Not recorded from China).

407. *Rosa semperflorens,* Willd. (Curt. Bot. Mag. tab. 284, 1794). Native of China, cultivated in Bengal. — (It is a variety of *R. indica.* See I. F. S., i, 250).

408. *Rosa diffusa,* Roxb. It was brought from Canton to the B. G. C. (Not identified).

408. *Rosa triphylla,* Roxb. Introduced from Canton into the B. G. C., previous to 1794. The Chinese gardeners call it *tsha te bay fa.* — (In Ind. Fl. Sin., i, 247, it is identified with *Rosa anemonaeflora,* Fortune ex Lindl. Journ. Hort. Soc. II, 316).

408. *Rosa microphylla*, Roxb. Sinice: *hoi tong hong* (hai tung hung. See Morrison Engl. Chin. Dict. sub Flowers, Reeves). Introduced from Canton into the B. G. C. — (I. F. S., i, 252).

408. *Rosa inermis*, Roxb. Native of China. Two varieties cultivated at Calcutta:

One with double white flowers, sin.: *po mou he wong (po mu hiang)*.

The other with double yellow flowers, sin.: *wong mou he wong* (huang mu hiang). — (Lindley, Roses (1820), 131, identifies the R. inermis with Rosa Banksiae, R. Brown in Hort. Kew. — I. F. S., ı, 248).

406. *Pyrus chinensis*, Roxb. Sin.: *cha li* (sha li = sand pear). The people about Calcutta call it *salli*. — *(Pyrus sinensis*, Lindley, in Trans. Hort. Soc. VI, 396, *sha le* or Chinese Sand pear. — I. F. S., ı, 257).

406. *Mespilus japonica*, Thbg. Introduced from China into Bengal, where it is much cultivated. Sin.: *loquat*. — (Eriobotrya japonica, Lindl.).

316. *Combretum chinense*, Roxb. Introd. from China into the Bot. G. C. — (Hook. Fl. Ind. ıı, 457. Assam, Tenasserim. Not recorded from China).

404. *Lagerstroemia indica*, Linn. Introduced from China.

143. *Trapa bicornis*, Roxb. A native of China. Sinice: *lin ko* (ling küe). — (The Chinese species is *Trapa bispinosa*, Roxb. l. c., but Roxb. records it only from Bengal).

702. *Bryonia tenella*, Roxb. Introd. from China into the B. G. C. — (Hook. Fl. Ind. II, 626, Melothria indica, Lour. — I. F. S., ı, 319).

395. *Cactus chinensis*, Roxb. Introduced from China to the B. G. C. According to Kew Index = *Opuntia Dillenii* (America australis).

236. *Gardenia florida*, Lin. Found in the gardens of Calcutta. A native of China.

180. *Vangueria spinosa*, Roxb. Bengal and China. Some plants introduced into the B. G. C. from China. — (Hook. Fl. Ind. ııı, 136, Bengal, Burma. Not in Ind. Fl. Sin.).

127. *Ixora coccinea*, Lin. Tanjore and China. Roxbg. saw it in China, in a wild state in great abundance. — (I. F. S., ı, 385).

127. *Ixora stricta*, Roxb. Moluccas. A variety of it with pale pink flowers was introduced from China, where it is called *hong mou tang*. — (Ind. Fl. Sin., ı, 385).

127. *Ixora alba*, Lin. Brought from China under the name of *ta mou tang*. Probably a variety of I. stricta. — (Hook Fl. Ind. ııı, 145).

195. *Serissa foetida*, Commers. Introduced from China into our gardens in India. — (I. F. S., ı, 391).

607. *Helianthus annuus*, Lin. Roxb., received several varieties of it from China and Persia.

604. *Tagetes erecta*, Lin. Originally from Mexico, but now considered as indigenous in Persia and China.

594. *Serratula cinerea,* Roxb. Very common in India. A variety with large flowers was received from China. — (DC. Prod. V, 24. *Vernonia cinerea,* Less. — I. F. S., I, 401).

170. *Lobelia radicans,* Thbg. Accidentally introduced from China into the B. G. C. — (I. F. S., II, 3).

412. *Diospyrus Kaki,* Lin. A native of China, introduced into the B. G. C., by the late Col. Kyd. The Chinese gardeners employed in the B. G. C., call it *chin* (?). — (I. F. S., II, 69).

33. *Jasminum paniculatum,* Roxb. Received from Canton into the B. G. C. Sin.: *sam yeip son hing* — (*sam ip su hing,* Bridgm. Chrestom. 454, 59. — *sam ip* = trifoliate). — (I. F. S., II, 80).

31. *Jasminum pubescens,* Lin. Brought from China into the B. B. C.— (I. F. S., II, 80).

50. *Fraxinus chinensis,* Roxb. A native of China, from thence introduced into the B. G. C. by the late Col. Robert Kyd, prior to 1793. — (I. F. S., II, 85. — Figured by D. Hanbury, Science Pap. 272, from a specimen in Herb. Kew).

35. *Olea fragrans,* Thbg. A native of China, cult. in the B. G. C.

34. *Phillyrea paniculata,* Roxb. Introduced from China into the B. G. C. before 1793. — *(Ligustrum lucidum,* Ait. — I. F. S., II, 92).

155. *Fagraea fragrans,* Roxb. Introduced from China into the Island of Pulo Pinang. — *(Cyrtophyllum fragrans,* DC. Prodr. IX, 31. — Mr. Hemsley, I. F. S., II, 121, thinks that this plant has been erroneously recorded as Chinese).

169. *Ipomoea pileata,* Roxb. Introd. from China into the B. G. C. — (I. F. S., II, 162).

163. *Convolvulus bilobatus,* Roxb. Raised in the B. G. C. from seed received from China. — *(Ipomoea Pes caprae,* Sweet., in DC. Prodr. IX, 349, which is reduced to *Ipomoea biloba,* Forsk. in I. F. S., II, 157).

190. *Solanum decemdentatum,* Roxb. Introduced from China into the B. G. C. — (Same as *Solanum biflorum,* Lour. See I. F. S., II, 169).

493. *Bignonia grandiflora,* Thbg. A native of China. Sinice: *tung vong fa* (?). — *(Tecoma grandiflora,* Lois. See I. F. S., II, 235).

42. *Justicia chinensis,* Lin. Introd. from China into the B. G. C. — *(Dicliptera chinensis,* Nees. See Benth. Fl. Hong. 266).

132. *Callicarpa purpurea,* Juss. Introd. from China into the B. G. C. in 1812. — (I. F. S., II, 254).

482. *Vitex incisa,* Lam. A native of China, whence Mr. W. Kerr sent seeds of it to the B. G. C. — (I. F. S., II, 257).

478. *Volkameria Kaempferi,* Jacq. Common in the gardens of Calcutta. Introduced from China. — (I. F. S., II, 262, Clerodendron squamatum, Vahl.).

408. *Hastingia coccinea,* Koenig. Brought originally from China, but also a native of Bengal.— (I. F. S., II, 263: *Holmskioldia sanguinea,* Willd. Mr. Hemsley thinks, that this plant has been erroneously reported as brought from China).

147. *Chloranthus inconspicuus,* Swartz. *Tcheu lan (chu lan,* Bridg. Chrest. 452, 17) of the Chinese. Received from Canton into the B. G. C. — (I. F. S., II, 368).

339. *Laurus dulcis,* Roxb. Introduced from China into the Calcutta Gardens. — In DC. Prodr. XV, I, 16, identified with *Cinnamomum Burmanni,* Blume. — (I. F. S., II, 371).

686. *Acalypha chinensis,* Roxb. Introduced from Canton into the B. G. C. — (In Benth. Fl. Hong. 303, identified with *A. indica,* Lin. — I. F. S., II, 438).

686. *Acalypha conferta,* Roxb. Introduced from China into the B. G. Calcutta. — In DC. Prodr. l. c. 870, identified with A. brachystachya, Hornem. — (I. F. S., II, 437).

691. *Sapium sebiferum,* Roxb., *Stillingia sebifera,* Mich. The tallow tree of China. Now very common about Calcutta.

263. *Ulmus virgata,* Roxb. Introd. from China into the B. G. C., by Sir John Royds. — (According to Planchon, Ann. Sc. nat. 1848, p. 280, this is the *Ulmus parvifolia,* Jacq. — I. F. S., II, 448).

658. *Morus atropurpurea,* Roxb., *M. rubra,* Loureiro. Introd. from China into the B. G. C. — (Bureau, in DC. Prodr. XVII, 244, a variety of *M. alba,* Linn.).

675. *Castanea pumila,* Willd. (American). Introduced from Canton into the B. G. C. Sinice: *fing lot* (Bridg. Chrest. 443, 7). — Perhaps the *C. pumila,* Blume, Bijdr. 525, or *C. mollissima,* Blume, Mus. Lugd. Bat. I, 286. DC. Prodr. XVI, II, 116).

678. *Thuja orientalis,* Lin. Introduced from China into the gardens of India. Sin.: *piem fa (pien po,* or *pin pak,* Bridg. Chrest. 440, 20).

741. *Juniperus aquatica,* Roxb. Shrubby. Leaves single, distichous, linear, not mucronate. Sin.: *theng tsong* (?). Common in the vicinity of Canton and from thence introduced into the B. G. C., by Mr. W. Kerr. — (Obscure plant. Parlatore in DC. Prod. XVI, II, 492, thinks, a Glyptostrobus or Taxodium. — Dr. Hance, China Review, II, 133, identifies it with *Glyptostrobus heterophyllus,* Endl.).

741. *Juniperus chinensis*, Lin. Introd. from China into the B. G. C.

741. *Juniperus chinensis*, Roxb. Introduced from China into the B. G. C. — (According to Parlatore in DC. Prod. XVI, II, 516, this is *Podocarpus chinensis*, Wall.).

741. *Juniperus communis*, Lin. Sin.: *tien tsong*.

741. *Juniperus cernua*, Roxb. Sin.: *ying loe* (?).

741. *Juniperus dimorpha*, Roxb. Sin.: *kong nam tsong* (?). (Parlatore, l. c. 488, refers both to *J. chinensis*, Lin.).

613. *Cymbidium tessalatum*, *Epidendron tesselatum*, Roxb. Corom. Plts I, n. 42. When in flower, a very beautiful plant; suspended in a room or elsewhere, it will continue to grow for several months, though I believe it will not flower. In this manner it, or *Epidendron flos aeris*, has been brought from China to the coast under the name of the *Air* or *Cameleon* plant, and represented as one of the most wonderful productions of nature, because it will only thrive when so suspended. (Comp. s. pp. 12, 199. — Hook. Fl. Ind. VI, 52, *Vanda Roxburghii*, Bot. Reg. t. 506, India, Ceylon.

23. *Alpinia calcarata*, Rosc., *Renealmia calcarata*, Andr. A native of China and from thence introduced by Capt. J. Garnauld, into the B. G. C. in 1799.

223. *Musa coccinea*, Andr. Bot. Repos. I, 47. Sinice: *ouang chok chee* (?). Introduced from China into the B. G. C.

57. *Iris chinensis*, Bot. Mag. tab. 373. Introduced from China into the B. G. C.

57. *Moraea chinensis*, Lin., *Pardanthus chinensis*, Ker. A native of India and China.

Crinum sinicum, Roxb., noticed in Bot. Mag. 1221, p. 7, (1819). Bulbs of this plant were brought from China, in 1809, to the Bot. G. C.— Kew Index, = *Crinum asiaticum*, Lin.

287. *Amaryllis radiata*, L'Hérit., sin.: *yuk lan* (?) and *Amaryllis aurea*, L'Hérit. Both introd. from China into the B. G. C.

296. *Hemerocallis fulva*, Lin. Introduced probably from China.

296. *Hemerocallis cordata*, Thbg. Introduced from China into the B. G. C. by W. Kerr. *(Funkia subcordata*, Spr.).

292. *Dracaena ferrea*, Lin. A native of China, cultivated in Bengal.

627. *Arum divaricatum*, Lin. Sent from Canton to the B. G. C. by W. Kerr. — (Benth. Fl. Hgk. 342).

70. *Scirpus tuberosus*, Roxb. Plants of Coromandel, tab. 231. The Water Chestnut of the Chinese. Sinice: *pi tsi, maa tai, pu tsai* (comp. Bridg.

Chrest. 451, 91). It was transmitted from Canton by Mr. Duncan at the desire of the Governor General, for the B. G. C. — *(Eleocharis tuberosa,* Schult.).

112. *Poa cylindrica,* Roxb. Raised in the B. G. C. from seeds received from Canton. According to Kew Ind. = *Eragrostis ciliata,* Nees. India orientalis.

98. *Panicum lineare,* Lin. A native of China, accidentally introduced into the B. G. C.

97. *Panicum filiforme,* Lin. Brought from China to the B. G. C.

80. *Saccharum sinense,* Roxb. Introduced from China into the B. G. C. in 1796. "This seems to be the sort employed over China for making their sugars". Roxb. gives an interesting account of the Chinese sugar canes, their mode of culture and manufacturing the sugar in China, extract from a letter from Mr. A. Duncan, Surgeon to the Factory at Canton to Richard Hall, Esq. President of the Select Committee, dated Canton, 26 Oct. 1796. — Kew Index, = *S. officinarum,* Lin.

107. *Agrostis maxima,* Roxb. A large species, native of Bengal. It seems that the Chinese use the leaves to pack up various articles. — *(Thysanolaena acarifera,* Nees. Pl. Meyen. 181. Benth. Fl. Hgk. 417).

306. *Bambusa nana,* Roxb. Sinice: *keu fa* (?). A native of China, cultivated in Bengal. — (Found also in Japan. Franch. Savat. Enum. Jap. II, 606).

746. *Lycopodium aristatum,* Roxb. Introduced to the B. G. C. from China. — Baker, Selaginella, Journ. Bot. 1884, p. 88: = *Selaginella exigua,* Spring, Lycop. II, 238. Assam.). Not recorded from China.

Wallich, Nathaniel, originally Nathan Wolff, a distinguished botanist, who did much to further the botanical exploration of India. Born 1786 at Copenhagen, where he studied medicine, and took his degree of M. D. In 1807 he went in the quality of surgeon to Serampore, near Calcutta, then belonging to the Danes. Subsequently in 1813, he entered the service of the British Bengal army as a Medical Officer. Superintendent of the Calcutta Garden, 1815—1846. Wallich made botanical collections in various parts of India, and also received many Indian plants from other collectors. In 1828 he arrived in England with an enormous number of specimens, nearly 7000 species.

From 1824—26 W. published at Calcutta the TENTAMEN FLORAE NEPALENSIS, with 50 plates. — In 1828 his MS CATALOGUE OF DRIED

SPECIMENS OF PLANTS IN THE E. I. COMPANY'S MUSEUM was lithographed at London. — From 1830—32 he edited the PLANTAE ASIATICAE RARIORES, in 3 vols, with 300 col. plates.

Wallich died at London, 28 Apr. 1854. He was F. Linn. Soc. from 1818, F. R. Soc. from 1829.

The genus *Wallichia* (Palms) was dedicated to the memory of Wallich by Roxburgh, Corom. Plts, III, t. 295. The genus *Wallichia*. DC. (Byttneriaceae) Prodr. I, 501, is now reduced to *Eriolaena*. (Gen. Plant. I, 221).

Wallich, when superintendent of the Calcutta Garden, like his predecessor Roxburgh, received living plants and herbarium specimens from Canton. His principal correspondent and collector there was John Reeves. (See farther on). Let me record some of these Chinese plants received by Wallich:

In the first edition of Roxb. Fl. Ind. II (1824), 104, Wallich describes *Psychotria Reevesii* and states that this plant was introduced from Canton into the B. G. C. in 1821 by his worthy friend John Reeves, to whose knowledge and zeal that institution owes many valuable additions of Chinese plants. — (I. F. S., I, 387).

Wall. Cat. n. 3168. *Eupatorium Reevesii*. Sent by Reeves. — (I. F. S., I, 405).

Wall. Cat. n. 1830. *Callicarpa Reevesii*. China, J. Reeves. — (I. F. S., II, 254).

Ibid. n. 1804. *Clerodendron canescens,* Wall. Introduced Calc. Garden by J. Reeves.

Ibid. n. 1207. *Zanthoxylum nitidum,* Wall. (non DC.). Received from China. — (I. F. S., I, 108, *Toddalia aculeata,* Pers.).

Gymnema parviflorum, Wallich Cat. 8184. Introduced into Bot. Gard. Calcutta, from China. — Fl. Hongk. 227, = *G. affine,* Dcne., in DC. Prod. VIII, 622. Peculiar to China.

Wall. Plant. As. Rar. I, 80, tab. 93, *Justicia ventricosa*. Introd. from China into the B. G. C. by Mr. Reeves. Bot. Mag., tab. 2766 (1827). — (I. F. S., II, 247).

Sir Jos. Hooker in Bot. Mag. t. 6784 (1884): *Dendrobium aduncum,* described 1842 by Lindl. in Bot. Mag. Misc. p. 58. He had received the plant from Wallich, Calcutta. It is not an Indian but a Chinese plant, rediscovered lately by Mr. Ch. Ford in South China.

Roumea chinensis, Wall. mss. Introduced from China by Reeves into the Calcutta Gardens. DC. Prodr. XIV, 538. *Daphne Roumea.* — (I. F. S., II, 395).

VIII. Ventenat, — Jacquin Jun.

Ventenat, Etienne, Pierre, born 1757, Professor of Botany at Paris, Member of the French Institute. He died in 1808. He published several iconographical works on exotic plants cultivated in two renowned gardens at Paris in the beginning of the present century. I may notice the Chinese plants described and depicted by Ventenat. He first published:

DESCRIPTIONS DES PLANTES NOUVELLES ET PEU CONNUES, CULTIVÉES DANS LE JARDIN DE J. M. CELS. Paris, an VIII (1800). 100 uncoloured drawings. The volume is dedicated to Mr. Cels "cultivateur de l'Institut national de France".

Cels, Jacques Martin, born 1743, a pupil of Bern. de Jussieu cultivated exotic plants for sale. He died in 1806. (Dictionnaire de biographie). His garden was situated in the plain of Mont rouge, at a distance of $2\frac{1}{2}$ kilomètres from Paris. It was then one of the richest gardens in France.

Tab. 9. *Iris fimbriata*. Introduced by Cels from China in 1797. — Maximow. Mél. biol. X, 736: same as *I japonica*, Thbg. and *I chinensis*, Curtis in Bot. Mag. tab. 373 (1797).

Tab. 23. *Agyneia impubes*, Lin. (v. supra p. 98). This, a native of China, was raised 4 years ago in Cels' garden from seed received from the Bot. Garden in Isle de France (Mauritius).

Tab. 28. *Rosa bracteata*, Wendl. This plant was brought from China by Lord Macartney (v. supra p. 175) and was introduced into Cels' Garden in 1798.

Tab. 35. *Rosa diversifolia*. A native from China and Bengal. According to Lindley, Roses, 108, this is the *Rosa semperflorens*, Curtis, Bot. Mag. 284 (v. supra pp. 135, 194).

In 1803. Ventenat published another volume with 60 uncoloured drawings of plants, under the title: Choix des plantes dont la plupart sont cultivées dans le jardin de M. Cels. No Chinese plants figured.

I may further notice that, after Blancard had first introduced from China the *Chrysanthemum indicum*, in 1789, it was cultivated in Cels' Garden (v. s. p. 132).

Persoon, Synops. Plant. I (1805), 291, reports that *Ulmus chinensis*, Pers., is cultivated in Cels' Garden. According to Desfont. Tabl. Mus. Catal. hort. Paris, 1815, p. 242, this is the same as *Ulmus parvifolia*, Jacq. (v. supra p. 137). — Persoon, l. c. 292, describes his *Celtis sinensis*, likewise from a specimen cultivated in Cels' Garden.

Ventenat, having been invited by Madame Bonaparte to describe the plants cultivated in her gardens at Malmaison, published 1803—4, a work in two volumes under the title of JARDIN DE LA MALMAISON, with 120 coloured drawings. The following Chinese plants are depicted in it:

Tab. 18. *Hemerocallis coerulea,* Andr. Bot. Repos. t. 6 (1797). China.

Tab. 24. *Magnolia discolor,* Japan, China. This is, according to DC. Prodr. I, 81, a variety of *Magnolia obovata,* Thbg.

Tab. 37. *Magnolia pumila,* Audr. Bot. Repos. t. 226 (1802). China.

Tab. 44. *Indigofera macrostachya,* sp. nov. Received from China. — Comp. I. F. S., ɪ, 157. Northern provinces of China, Peking.

Tab. 19. *Mespilus japonicus,* Thbg. Introd. into France from Canton in 1784. Cultivated also at Isle de France.

Tab. 70. *Clerodendron fragrans,* sp. nov. or *Volkameria fragrans,* cultivated for several years in Paris. Received from Java. — (But it is a Chinese plant, first discovered by Staunton, 1793. See I. F. S., ɪɪ, 260).

On p. 27, note, Ventenat describes a new Chinese *Ionidium, I. hetero-phyllum,* from Jussieu's herbarium (sent by d'Incarville from South China, v. s. p. 54). Not mentioned in Ind. Fl. Sin.

Jacquin, Joseph Franz, Baron von, son of N. J. Jacquin (v. s. p. 134), born at Schemnitz, 7 Febr. 1766. He was like his father, Professor of Botany and Chemistry at the Vienna University. He died at Vienna, 4 th Dec. 1839. Among his botanical works we may notice the following:

ECLOGAE PLANTARUM RARIORUM AUT MINUS COGNITARUM QUAS AD VIVUM DESCRIPSIT ET ICONIBUS COLORATIS ILLUSTRAVIT J. F. Jacquin. In 2 vols. Vol. I, tab. 1—100, was published, 1811—16, — Vol. II, tab. 101—169, was edited in 1844, after the author's death, by his daughter and Prof. Fenzl.

P. 5. *Schollia crassifolia,* Jacq. — Asclepias carnosa, Lin.; Bot. Mag. t. 788 (1804). — Hoya carnosa in Hort. Kew (v. s. p. 197). — A native of China and the Sunda Islands. Cultivated in the Schoenbrunn Gardens. Sir Joseph Banks, in 1804, had presented to the Gardens a living specimen. The plant was first described by Linnaeus from an herbarium specimen brought by Bladh from China. (V. supra p. 88).

P. 54. *Sida tiliaefolia,* Fischer. Jacquin states (in about 1812) that he received seeds of this plant many years ago with a Chinese name on the label. In 1808 Dr. Fischer sent the same seeds from the Gorenki Garden (near Moscow), marked: *Sida tiliaefolia,* ex Tibet. — This is *Abutilon Avicennae,* Gaert. It is much cultivated near Peking.

P. 133. *Ipomoea dasysperma,* sp. nov. A native of China. The seeds, from which the plant was raised at Schoenbrunn, had been received in 1814 from England. — I. F. S., II, 159, 161, cultivated in S. China.

J. F. Jacquin published yet another iconographical work on plants with the title: ECLOGAE GRAMINUM RARIORUM AUT MINUS COGNITORUM QUAE AD VIVUM DESCRIPSIT ET ICONIBUS COLORATIS ILLUSTRAVIT J. F. Jacquin, 49 plates. The publication of this work commenced in 1813; it was completed, after the author's death, in 1844, by Prof. Fenzl, in accordance with the MS left.

Tab. 40. Represents *Coix exaltata,* Jacquin, a new species. — Link, Enum. hti Berol., II (1822) 377, considers it a variety of *C. La-chryma,* Lin.

IX. The Royal Horticultural Society of London and its Exertions to introduce Chinese Plants.

THE HORTICULTURAL SOCIETY OF LONDON originated in 1804. The founders of it were:

Forsyth, William (1737—1804), a pupil of Ph. Miller and Superintendent of the Roy. Gardens at St. James and Kensington. F. L. S.

Knight, Thomas Andrew (1759—1838), the celebrated vegetable physiologist. F. R. S.; F. L. S.

Wedgewood, John, son of the inventor of the Wedgewood pottery.

Banks, Sir Joseph. (V. supra p. 201).

Aiton, William Townsend. (V. supra p. 191).

A Royal Charter was granted to the Society by George III in 1810, and in 1812 it began to issue its TRANSACTIONS, which have been of the greatest service to horticulture.

The first President was **George, Earl of Dartmouth.** After him T. A. **Knight,** was President from 1811—1838.

Greville, Ch. F. (v. s. p. 210) was Counsellor and the first Treasurer.

Salisbury, R. A. (v. s. p. 209) was the first Secretary. He was soon succeeded by Jos. **Sabine.**

This useful institution possessed a magnificent garden, in which many interesting exotic plants were cultivated. It was situated in Chiswick not far from the Kew Gardens. The garden at Chiswick was obtained in 1822 on a lease from the Duke of Devonshire, an enthusiastic Supporter of the Society, who subsequently became its President.

The first volume of the Transactions of the Hort. Soc. appeared in 1812. It contains articles dated 1805. The other volumes were published: II, 1813, 1814, — III, 1822, — IV, 1822, — V, 1824, — VI, 1826, — VII, 1830. — A second series, 1835—48, was issued in 3 volumes.

In 1846 the Society began to publish the Journal of the Hort. Soc. of which from 1846 to 1855, 9 volumes appeared. A new series was commenced in 1869.

Under the wise and enlightened guidance of its worthy Secretary Jos. Sabine, the Hort. Soc. soon made rapid advances to unexampled eminence. Four years before the Chiswick Garden was started, the Society extended its field of operations by adopting the Kew plan of sending out collectors to foreign countries. It first began to direct its attention to the botanical riches of China, in which country (at Canton) the Society then had a zealous member and correspondent, John **Reeves,** to whom it owes the introduction of many beautiful ornamental plants.

Dr. Lindley writes in his Collect. Bot. (1821), sub tab. 7, Primula sinensis: "One of the many objects which occupy the Hort. Society, is the introduction of ornamental plants to the gardens of the country and the free distribution of them, when procured. As it is difficult to form a very correct idea of the beauty of the plants from the appearance they assume when dried, in which state only, a great proportion of tropical vegetables is known to residents in Europe, it was determined by the Society that a person should be employed in making drawings of plants in the country where they grow. For several reasons China was selected for a beginning, and particularly as being the residence of John Reeves, Esq., a correspondent and very active member of the Society, under whose immediate superintendence the draughtsman could be placed. By the direction of this gentleman, a considerable number of drawings have already been sent to England and many of the plants they represented introduced".

This collection of beautifully executed drawings of plants, which still exists in the Hort. Soc.' library, is frequently quoted by Lindley, Sabine and other botanists, who describe ornamental plants from China*).

Sabine, Joseph, born 1770, an able botanist, originally a barrister. He became Fellow of the Linn. Soc. in 1798, and was also F. of the Roy.

---

*) A similar set of Chinese drawings of ornamental plants executed at Canton, in 1806, was in the library of the E. I. Company. It had been sent by W. Kerr (v. s. p. 191). — In Bot. Mag. sub tab. 2997—98 (1830). Sir W. Hooker speaks of a collection of Chinese drawings of Plants made at Canton, which had been presented to him by Mrs. Halket.

Soc.; Secretary of the Hort. Soc., 1810—1830, when he resigned, after having rendered great services to the Society. He died at London, 24 Jan. 1837.

Aug. Pyr. De Candolle dedicated to him the name of a new genus, *Sabinea* (Leguminosae, American), in Ann. Soc. nat. 1825, 92. — Regarding *Chrysanthemum Sabini,* Lindley, v. supra p. 86.

Sabine wrote several important articles on Chinese ornamental plants:

In 1821, in Trans. Hort. Soc. IV, 326—354: Account and Description of the Varieties of Chinese Chrysanthemum, which at present are cultivated in England. With coloured plates. 12 varieties detailed.

In 1822, in Trans. Hort. Soc. V, 149—162: Further Account of Chinese Chrysanthemums. Coloured Plates. 8 new varieties described, which had recently flowered in the Soc.'s garden.

In 1823, in Trans. Linn. Soc., XIV (1825), p. 144, seqq.: Generic and Specific Characters of the Chrysanthemum indicum of Linnaeus and the Plants called Chinese Chrysanthemums. Coloured plates. — Sabine separates from the *Chrysanthemum indicum* (small yellow flowers), the cultivated Chinese Chrysanthemums with large double flowers of various colours, which he names *Chrysanthemum sinense,* and of which the original single flowered form has also been met with in a wild state in China and Japan.

In 1824, in Trans. Hort. Soc. V, 412—428: Account and Description of five new Chinese Chrysanthemums. Coloured plates. 27 varieties then were known to Sabine.

In 1826, in Trans. Hort. Soc., VI, 322—359: Account of several new Chinese and Indian Chrysanthemums with additional observations on the Species and Varieties. Coloured plates. 21 new varieties described.

In 1826, in Trans. Hort. Soc. VI, p. 465—488: On the Paeonia Moutan or Tree Paeony. Coloured plates. 9 varieties described.

Lindley, Dr. John. This experienced horticulturist and celebrated botanist was born on Febr. 5, 1799, at Catton, near Norwich, where his father owned a nursery garden, and was educated in the Norwich Grammair School. He early manifested a taste for botanical studies, in which he afterwards gained reputation. When about 16, he went for a time to Belgium to a seed merchant. After his return he remained with his father a few years and devoted himself to botanical, horticular and entomological pursuits. At Norwich he made the acquaintance of Mr. (afterwards Sir) William Hooker, was subsequently introduced to Sir Joseph Banks and proceeded to London, where in 1819 he was employed by the latter as assistant librarian.

In 1820, there appeared Lindley's first botanical work, the Mono-
graphia Rosarum, based principally upon the material found in the Banksian herbarium. 16 Chinese Roses are described in it. Coloured drawings.

In the same year he read before the Linnaean Society, of which he
had become a Fellow, an interesting Memoir on Pomaceae. This paper,
beautifully illustrated, was printed in vol. XIII (1822), 88—106, of the
Trans. of the Linn. Soc. *Eriobotrya japonica, Photinia serrulata, Raphiolepis indica, Chaenomeles japonica,* all new Eastern-Asiatic genera, proposed by Lindley.

In 1821, he published, at the expense of Mr. W. Cattley (see farther
on) the Collectanea Botanica, or Figures and Botanical Illustrations
of Rare and Curious Exotic Plants. Coloured drawings, 41 plates. The
following Chinese plants (all from Canton it seems) are described and figured there:

Tab. 3. *Raphiolepis rubra,* Crataegus rubra, Lour. Banks' herbarium.
Macao, Bradley. V. s. p. 153.

Ibid. *R. phaeostemon* and *R. salicifolia,* two new Chinese species introduced by the Hort. Soc.

Tab. 6. *Vanda teretifolia,* native of China, from Cattley's collection,
where it flowered.

Tab. 7. *Primula sinensis,* Sabine. Introduced from China by Thom.
Palmer.

Tab. 16. *Psidium Cattleyanum,* Sabine. Cultivated by Cattley. — (It
was for some time erroneously believed to be a native of China).

Tab. 17. *Chloranthus monostachys,* Brown, Bot. Mag. tab. 2190. Introduced by the Hort. Soc. (J. Reeves).

Tab. 30. *Spiranthes pudica.* Introd. from China, 1821, by the Hort.
Soc. Roots of this pretty plant were found among the earth of some plants
received from China in 1821. Trans. Hort. Soc. VI, 85.

Tab. 38. *Vanda multiflora.* Introduced from China by Cattley.

Tab. 39. B. *Sarcanthus rostratus.* Imported from China by the Hort.
Soc. Flowered 1822 in Cattley's Collection.

About the same time Lindley is believed to have edited the Icones
Braamianae (v. s. p. 186).

In 1822, Lindley became Assistant Garden Secretary (Chiswick) to
the Hort. Soc., of which Sabine then was Honorary Secretary. In 1826 he
was appointed sole Assist. Secretary, having duties to perform, both in
London, Regent Street, where the meetings were held, and at the Chiswick
Gardens. From this time he may be said to have become the mainspring

of the Society. Exotic plants, especially from China, were most successfully cultivated in the Society's Gardens. Lindley used to determine and describe the new plants, which had flowered there, in the "Trans. Hort. Soc.": Vol. VI, 1826, 62—100; 261—299. New and rare plants which have flowered in the Chiswick Gardens, from the first formation of the Hort. Soc. to 1824. Continued in Vol. VII, 1830, 46—76; 224—253. Some of them were published in the Botanical Register which Lindley conducted from 1828, but in which he had written some years earlier.

In 1829 Lindley was appointed to the Chair of Botany at University College, London. He also lectured on Botany from 1831 at the Royal Institution, and from 1835 at the Botan. Gardens Chelsea of the Society of Apothecaries. He was Praefectus Horti Chelsea 1836—1853. Before Mr. Bentham's resignation, 1841, Lindley took the designation of Vice-Secretary of the Hort. Society, and continued in this office till 1858, when he resigned and became Secretary to the Society and Member of the Council.

In 1841 Lindley had established the GARDENER's CHRONICLE. After his death in 1865, Dr. Maxwell T. Masters succeeded him as editor of this periodical and still conducts its. From 1846—55 L. edited the Journal of the Hort. Soc. — In conjunction with Joseph Paxton he published the FLOWER GARDEN OF NEW OR REMARKABLE PLANTS, of which illustrated periodical 3 vols appeared, London, 1850—53. In these periodicals likewise, many of the novelties from the Chiswick Garden and other gardens were described by Lindley. He determined and described almost all the plants introduced for the Hort. Soc. from China by J. Reeves, J. Potts, J. D. Parks and R. Fortune, from 1818 to 1860.

Among his numerous botanical writings I may further notice his important works on Orchids:

GENERA AND SPECIES OF ORCHIDACEOUS PLANTS, 1830—1840.

SERTUM ORCHIDACEUM, a Wreath of the most beautiful Orchidaceous Flowers, 1837—42. With beautifully coloured drawings.

FOLIA ORCHIDACEA. An Enumeration of the known species of Orchids, 1852—59. Incomplete.

In 1863 Lindley was compelled by the bad state of his health to resign the Secretaryship of the Hort. Soc. He died Nov. 1st 1865 at Turnham Green, Middlessex.

In 1828 Lindley became F. Roy. Soc. In 1832 he received the degree of Ph. D. from the University of Munich, in 1835 he was elected Corresp. Member of the French Institute.

Humboldt and Kunth in 1823 named a new American genus, *Lindleya* (Rosaceae) in honour of the great botanist. Many specific names of new plants have been dedicated to him. I may notice some Chinese species which bear his name.

Amygdalopsis Lindleyi, Carrière in Rev. Hort. 1862, 91. — The same plant was called *Prunopsis Lindleyi*, by André, Rev. Hort. 1883, 336. — I. F. S., I, 222. *Prunus triloba*, Lindl.

Eupatorium Lindleyanum, DC. Prodr. V, 1836, 180.

Buddleia Lindleyana, Fortune, Bot. Reg. 1844, Misc. 25.

Clerodendron Lindleyi, Decaisne in Fl. des Serres IX, 1853, 17.

Lindley left a vast herbarium which was purchased by the University of Cambridge, except the Orchideae, which were purchased for the Kew herbarium.

Cattley, William, of Barnet, where he had a garden and a conservatory. C. was a rich merchant and liberal patron of horticulture in the first quarter of the present century, and one of the most ardent collectors of rare plants of his time. He was J. Lindley's first patron and paid him a regular salary for drawing and describing new plants. Mr. Hemsley kindly informs me that a letter from Cattley, preserved in Lindley's correspondence, relates to this connection. Cattley financed the edition, by Lindley in 1818, of the "Icones Braamianae" (v. supra p. 186) and the "Collectanea Botanica", in 1821, in which Lindley described and figured three new Chinese orchids from Cattley's collection: *Vanda teretifolia, V. multiflora, Sarcanthus rostratus* (v. supra p. 253) and also the *Psidium Cattleyanum*, tab. 16. The first account and a figure of the fruit of this beautiful plant was given by Cattley in 1820, in the Trans. Hort. Soc. IV, 315. He then cultivated it successfully in his conservatory, and states that it was originally raised from seed received from China, in 1818, by Messrs Barr & Brookes. But these seeds could hardly have come from China, where this plant is unknown. See I. F. S., I, 295. In the Collect. Bot. tab. 33, 37. Lindley established the new genus *Cattleya* (American Orchids) in honour of his patron and friend.

In the Bot. Reg. tab. 1314 (1830) Lindley described and figured the *Dendrobium moniliforme*, Sw., cultivated by Cattley. The plant had been introduced from China several years before by the Hort. Soc. According to Veitch, Orchids, this was not the plant described by Linnaeus (Epidendron moniliforme) and Swartz, but a new species which Reichenbach named *D. Lianawianum*.

Veitch, Orchids, II (1887) 4, sub Cattleya, states that Cattley possessed the finest collection of Orchids then known. He died in 1832. At his
death the collection passed into the hands of Mr. Knight, Veitch's predecessor in the Roy. Exot. Nursery, Chelsea. — Mr. Hemsley writes me,
that Sir J. D. Hooker told him that there is a son of W. Cattley still living.

Reeves, John (Senior). The following particulars regarding the life
of this distinguished naturalist and collector of natural objects in South
China are principally derived from the Chinese Repository and the obituary
notices of him as found in the Proc. Linn. Soc. 1856. p. XCIII, and in the
Gardener's Chron. 1856, 212.

John Reeves born May 1st 1774, was the youngest son of the Rev.
Jonathan Reeves of Westham, near London. Left an orphan at an early
age, he was educated at Christ's Hospital, and afterwards entered the
counting house of a tea broker, where he acquired so thorough a knowledge
of teas, as to recommend him in 1808, to the office of Inspector of Tea in
England in the service of the Hon. E. I. Comp. In 1812 he proceeded to
China as Assistant and subsequently became Chief Inspector of Tea in the
Comp.'s establishment at Canton. He generally resided at Macao, but during the tea season lived at Canton. During his long employment in China,
from 1812—1831, he twice visited England. From a letter by John Livingstone to the Hort. Soc. of London dated Jan. 1818 (see farther on) we
learn that Reeves, in 1817, returned to China from England, and Sabine
in Trans. Hort. Soc. VI, 324, states that Reeves in 1824 returned from
China to England on board the «Warren Hastings». He seems to have gone
back to China early in 1826. In the Chin. Repository II (1833) p. 226,
we read: John Reeves, Esq. of the Hon. Comp.'s establishment (Canton)
left China about two years ago.

J. Reeves became a Fellow of the Linn. Soc. in 1817 and of the Roy.
Soc. in the same year. He was also a Fellow of the Hort. Soc. and of the
Zoological Societies. From the time of his return to England, in 1831, he
resided at Clapham, where he died, 22nd March 1856. The Gardener's
Chron. 1856, in giving notice of his death says: "John Reeves, a name
dear to all who are familiar with England and her gardens, one of the
Nestors of Horticulture, died".

J. Reeves went to China when little was known of China's natural
productions and of Chinese gardens. From the time of his arrival in that
country he devoted his leisure to investigating the resources of the country
and to the pursuit of various branches of science, making it his principal

object to procure specimens of the natural productions and transmit them to England to such individuals or Societies as appeared most likely to turn them to account. His principal correspondent for some years after his arrival in China was Sir Joseph Banks. During the whole period of his residence in China, 1812—31, he contributed largely to English horticulture and to the Horticural Society of London in particular: not only by his own direct shipments but also by collecting plants during the spring and summer, establishing them well in pots, previous to the shipping season, and then commending them to the care of the captains of the Comp.'s ships, to whom he was also always able to recommend the most desirable plants for transportation to England, and to whom he succeeded in communicating the enthusiasm which animated himself. Reeves employed himself with indefatigable zeal in sending home all that he found most rare and beautiful among living plants in the Canton gardens. He was either the immediate or indirect source from which we derived the Chinese Azaleas, Camellias, Moutans, Chrysanthemums, Roses and numberless other treasures, which have been for so many years the glory of English collections. Not a Company's ship at that time sailed for Europe without her decks being decorated with the little portable greenhouses which preceded the present Wardian cases.

It was in the early part of his stay in China that R. made the fine collection of *Fishes,* which, together with his drawings, furnished the ground work of Sir John Richardson's valuable "Report of the Ichthyology of the Seas of China and Japan", published in the Reports of the British Association for 1845. Richardson says: "Mr. John Reeves was long resident at Macao. With an enlightened munificence he caused beautiful coloured drawings to be made, mostly of the natural size, of no fewer than 340 specimens of fish, which are brought to the markets of Canton. A copy of these drawings left by Mr. Reeves at Macao with Mr. Beale (v. infra) formed the basis of the enumeration of Chinese Fishes in Bridgman's "Chinese Chrestomaty", 479—489.

The Hort. Soc. of London is indebted to J. Reeves for a fine collection of coloured drawings of Chinese ornamental plants, executed in his own house under his superintendence by Chinese draughtsmen. Such drawings first brought us to a knowledge of the *Chinese Prime rose* (v. infra p. 258) *Dendrobium nobile,* many of the finest *Camellias, Chrysanthemums, Azaleas, Moutans,* and above all of the *Glycine (Wistaria) chinensis,* which plants were subsequently introduced into English gardens. In this way was formed that collection of authentic drawings of Chinese plants, by far the most extensive in Europe, which now forms part of the library of the Horticular Society.

A similar collection is now in the British Museum. Mr. Carruthers, Report Bot. Dep. Brit. Mus. for 1877, states that 654 Chinese drawings of plants, executed under the superintendence of the late John Reeves, were presented by Miss Reeves (his daughter or perhaps grand daughter).

Sir W. J. Hooker, in Kew Journ. Bot. II (1850), 250, and IV (1852) 50, notices a faithful Chinese drawing of the then little known Chinese *Rice paper* plant, in the possession of J. Reeves, Clapham.

J. Reeves' exertions were not confined to the collection of living plants. If we look into the private herbaria, there likewise we find marks of his industry and liberality; at the British Museum and in the Museum at Kew the name of Reeves occurs whereever Chinese productions are met with. The name of *Reevesia thyrsoides* given by Lindley to a beautiful Chinese plant discovered by Reeves, is the simple epitaph inscribed to his memory in the annals of science.

Living Chinese plants introduced through the medium of John Reeves, as mentioned in the English Botanical Periodicals:

Bot. Reg. tab. 496 (1820), and Trans. Hort. Soc. VII (1830), 246. *Rubus parvifolius,* Lin. Introduced in 1818, from China by the Hort. Soc. (most probably sent by Reeves).

Bot. Mag. tab. 2190 (1820) *Chloranthus monostachys.* Introduced from China by the Hort. Soc., through their correspondent John Reeves of Canton.

Sabine, Chinese Chrysanthemums, 1821, Trans. Hort. Soc. IV, 334: In June 1820 Capt. C. O. Mayne of the Atlas brought a box containing 12 varieties of *Chrysanthemum,* which was entrusted to his charge by our active correspondent Mr. John Reeves, Canton. It arrived safe, with only the loss of a single plant.

Sabine, Trans. Hort. Soc. VI (1826), 324, 334: Mr. Reeves, who returned to England, on board the "Warren Hastings", in 1824 brought with him a few Chrysanthemums, which he gave to Mr. S. Brookes, who presented plants of them to the Hort. Soc. Two of these were new. One of them was named the *Pale Buff.*

Bot. Mag. tab. 2497 (1824) *Laurus aggregata.* The plant from which the drawing was made flowered in the Hort. Soc.'s garden last February. It had been sent from China by John Reeves in the "Orwell", Capt. Lindsay, in 1821. (Obscure plant, not mentioned in Ind. Fl. Sin.).

Lindley, Collect. Botan. 1821, tab. 7, *Primula sinensis,* Sabine MS.

Among the drawings sent by Reeves from Canton to the Hort. Soc., in 1819, there was one of what appeared to be a very handsome gigantic species of Primula, accompanied by dried specimens which confirmed the general accuracy of the figure. It was immediately ordered to be sent home. Seeds and a living plant were accordingly procured at Canton by Mr. Reeves, but the latter unfortunately perished during its passage and the seeds, of which there was an abundance, did not vegetate. Captain Reeves (mistake for Rawes) has, however, been subsequently more fortunate, having succeeded in bringing over a plant which he presented to his relation Thomas Palmer, Esq. of Bromley, Kent.

Trans. Hort. Soc. VI (1826) 90. Lindley describes the *Prunus pseudocerasus* or Bastard Cherry, which had been presented to the Hort. Soc. in 1822 by S. Brookes. It had also been sent to the Hort. Soc. from China by Mr. Reeves under the name of *ying to*.

Trans. Hort. Soc. VII, 238. *Prunus serrulata*, Lindley. Sent to the Hort. Soc. from China by Mr. Reeves, in 1822, under the name of *yung to* (ying tao). (A variety of P. pseudocerasus. I. F. S., I, 221).

Ibidem p. 239. *Prunus salicina*, Lindley, the Chinese Plum. Sent with the preceding by Mr. Reeves. Chin. name *tsing chok li*. (See Bridgm. Chin. Chrest. 446, 77, Green Plum). — An obscure plant.

Bot. Mag. tab. 2659 (1826). *Astranthus cochinchinensis*, Lour. Sent to the Hort. Soc. from China by John Reeves. It flowered in the Chiswick Garden in June 1824. — (I. F. S., I, 312: *Homagium fagifolium*, Benth.).

Bot. Reg. t. 1308 (1830), *Blackwellia padiflora*, Lindley. Sent from China to the Hort. Soc. by John Reeves. — (I. F. S., I, 312, same as the preceding).

Glycine sinensis, Sims, Bot. Mag. tab. 2083 (1819), *Wistaria chinensis*, DC, Prod. II (1825), 390, is said to have been first sent to the Hort. Society, London by J. Reeves from Canton, in 1818. This specimen exists still in the Chiswick Garden. Loudon in "Hort. Brit." (1830) 315, named it *Wistaria Consequana*, and Paxton in "Magazine of Bot." VII (1840) 127 gave a coloured figure of the plant under this name. Fortune, "Tea Countries", 335, says that it was introduced into England from a garden near Canton, belonging to a Chinese merchant named Consequa. In the Bot. Reg. 1840, Misc. Notes, p. 41, we read: "The Wistaria sinensis in the Garden of the Hort. Soc. is a magnificent specimen 180 feet long, covering about 1800 square feet of wall. In the spring of 1839 it produced 675,000 flowers. The latter are lilac-coloured and fragrant". — Gard. Chron. 1855, 299: "The magnificent Wistaria sinensis of the first living plant, now an aged tree, in the Hort. Society's Garden, arrived in 1818". — Pharmac. Journal, 1896, II, 437: "The original plant of Wistaria chinensis, introduced by John Reeves, is still growing at Chiswick." — Reeves was, however, not the first to introduce this beautiful plant into England. The Bot. Cab. 1823, sub tab. 773 reports, that in 1816 Capt. Wellbank (v. infra) brought home a living specimen for Turner of Rook's Nest. But Reeves' plant is the finest specimen ever seen in Europe.

Bot. Cab. t. 1765 (1831), *Lagerstroemia indica*, Lin. Loddiges says: "We received it from our very kind friend, Mr. Reeves, in 1825, with some other varieties of this elegant plant."

Bot. Reg. tab. 1320 (1830), *Capparis acuminata*, Lindley says: "This beautiful plant was sent from China by John Reeves to the Hort. Soc. in whose garden it blossomed in Sept. 1828". (V. s. p. 189. Plantae Braam.).

Booth, Camellia and Thea, 1829, in Trans. Hort. Soc., VII, 531, says, sub *Camellia japonica*, varieties: "Of these very ornamental plants the Hort. Soc. has formed an extensive collection. Many of them are the result of importations made at different times from China of plants that were procured through the friendly agency of John Reeves, Esq.".

Bot. Reg. t. 1501 (1832). Lindley describes the *Camellia japonica*, var. *Reevesiana*, Reeves' Crimson Camellia. "One of the most striking varieties hitherto imported. It flowered in the nursery of Mr. Tate. Sent probably by Mr. Reeves (perhaps by Mr. Beale). We have named it in compliment to Mr. Reeves, to whom this country is under the greatest obligations for the zeal and liberality, with which he devoted himself, during a long residence in China, to the collection and transmission to England of all that is rare, beautiful and useful in the Flora of the Celestial Empire".

Bot. Cab. tab. 1648 (1830). *Andromeda chinensis*, Lodd. Loddiges writes: "After striving for more than twenty years in vain to obtain it, we received this elegant plant in June 1829 from our valuable friend Mr. Reeves of Canton. It flowered in August 1830". (I. F. S., ii, 14, *Vaccinium bracteatum*, Thbg.).

Mr. John Smith, Historia Filicum, 1875, 256, *Cibotium Barometz*, J. Sm., Polypodium Barometz, Lin. A native of China. The plants cultivated at Kew were introduced some time before 1834, by John Reeves, who was for many years "tea taster" at Canton for the E. I. Comp. He learned that this fern was the origin of the fabulous story of the "Tartar Lamb".

John Reeves also sent living plants from Canton to the Calcutta Bot. Gardens then superintended by Wallich: *Psychotria Reevesii*, *Roumea chinensis*. (v. s. p. 247).

There can be no doubt that all the living plants sent by John Reeves to England and to Calcutta had been procured at Macao or Canton. In the neighbourhood of the same places there had also been gathered the greater part of his herbarium specimens now found in the herbariums of the British Museum, Kew and the Calcutta Gardens. A few of his plants are stated on the labels to have been collected in the Foo keen province, namely in the tea district of An ki hien (N. W. of Ts'üan chou fu). Whether Reeves himself visited this province, or a Chinese collector had been sent thither, remains uncertain.

Chinese plants first discovered by John Reeves, as recorded by various botanists from his herbarium specimens:

Reevesia thyrsoidea, Lindl. in Brande's Journ., n. s. II (1827), 112. Lindley described the plant from a dried specimen sent to the Hort. Soc. from China by J. Reeves and a Chinese drawing of the half-ripe fruit forwarded by the same indefatigable correspondent of the Society. Two years later Lindley figured the same plant in Bot. Reg. t. 1236, from a specimen which flowered at Chiswick, and had been introduced by J. D. Parks (see farther on). — I. F. S., I, 90.

Helicteres spicata, Colebr., MS, ex G. Don Syst. I (1831) 507. India or. — I. F. S., I, 91. Canton, Reeves.

Elaeocarpus decipiens, Hemsl., I. F. S., I, 94. First sent by Reeves from Kuang tung.

Ilex Reevesiana, named by Fortune in Gard. Chron. 1851, 5, but, it seems, not sent by Reeves. — I. F. S., I, 116: perhaps identical with *I. Fortunei,* Lindl.

Euonymus gracillimus, Hemsl., I. F. S., I, 119. China, Reeves, Herb. Kew.

Ormosia pachycarpa, Champ., Benth. Fl. Hong. 96. First gathered by Reeves near Canton. — I. F. S., I, 204.

Photinia prunifolia, Lindl., Bot. Reg. sub. tab. 1956 (1836). — I. F. S., I, 262. Canton, Reeves.

Viburnum sempervirens, K. Koch, Hort. Dendr. (1853) 300. — I. F. S., I, 356: Canton, Reeves.

Abelia uniflora, R. Brown, in Wallich, Pl. As. Rar., I (1830) 15. Only the name. Sent from China by J. Reeves. It was first described by Lindley Bot. Reg. 1846, sub tab. 8, and in Lindl. & Paxt. Flow. Gard., II (1851) 145, fig. 208: received in 1824. — I. F. S., I, 359. Reeves gathered the plant in Fu kien, in the tea district.

Psychotria Reevesii, Wall. in Roxburgh. Fl. Ind. II, 104. — I. F. S., I, 387. Earlier Staunton. V. s. p. 176.

Eupatorium Reevesii, Wallich Catal. (1828) n. 3168. Sent by Reeves from China to Calcutta. — I. F. S., I, 405.

Azalea squamata, Lindl., Journ. Hort. Soc., I (1846) 152. Reeves sent dried specimens and Chinese drawing from China. — I. F. S., II, 23. *Rhododendron Farerae.*

Choripetalum obovatum, Benth. — I. F. S., II, 62, *Embelia obovata,* Hemsl., China, Reeves.

Strychnos paniculata, Champ., Benth. Fl. Hongk., 232. — Ind. Fl. Sin., II, 121. Canton, Reeves.

Erycibe obtusifolia, Benth. l. c. 236. — Ind. Fl. Sin., II, 156. South China, Reeves.

Callicarpa Reevesii, Wall. catal. n. 1830. Reeves sent it from China to Calcutta. — Staunton earlier. — I. F. S., II, 254.

Clerodendron canescens, Wall. cat. n. 1804. V. s. p. 247. — I. F. S., II, 259.

In Benth. Labiatae (1832—36) mention is made of several plants of this order first gathered in China by Reeves, viz:

p. 309. *Salvia plebeja,* R. Brown, Prod. Fl. Nov. Holl. (1810) 501. China, prov. Fu kien.

p. 543. *Stachys aspera,* Michx., var. *glabrata.* Prov. of Fu kien.

p. 438. *Scutellaria rivularis,* Wall., Fu kien. — Earlier Staunton.

p. 45. *Plectranthus amethystoides,* China.

Daphniphyllum calycinum, Benth. Fl. Hongk., 316. Canton, Reeves. — I. F. S., II, 429.

Microdesmis caseariaefolia, Planch. in Hook. Icon. Pl. ad tab. 758 (1848). — I. F. S., II, 433. Reeves, S. China.

Croton lachnocarpum, Benth. Fl. Hongk. 308. — I. F. S., II, 435. Reeves, S. China.

Lindley, in Gen. et Spec. Orchid. (1830—40), notices the following Orchids, of which Reeves had sent herbarium specimens:

p. 22. *Dienia congesta,* Lindl. Malaxis latifolia, Smith in Rees Cycl. Nepal. Figured in the coll. of Chin. draw. Hort. Soc.

p. 26. *Liparis nervosa,* Lindl., Ophrys nervosa, Thbg. Figured l. c.

p. 49. *Bolbophyllum bicolor,* Lindl. Sp. nov. Figured l. c.

p. 83. *Dendrobium auriferum,* Lindl. Sp. nov. Figured l. c.

p. 84. *Dendrobium catenatum,* Lindl. Japan and China. Figured l. c.

p. 216. *Vanda multiflora*, Lindl. Coll. Bot. (1821) 38, Sp. nov. — = *Acampe multiflora*, in Lindl. Fol. Orchid. I (1853).

In Hook. Lond. Journ. Bot. II (1843) 207, Berkeley gives an account with a figure of a curious fungus from China, produced on a caterpillar, and which the Chinese call *hia ts'ao tung chung*, a plant in summer and in winter a worm. (See my Earl. Europ. Res. Flora Chin. p. 30). Berkeley found specimens of it sent over by Reeves in the collection of the British Museum. The same had been sent, in 1723, from Peking to Paris. Réaumur described and figured it in Mém. Acad. 1726, p. 302, t. 17. Berkeley named it *Sphaeria sinensis*.

In Morrison's English Chinese Dictionary, 1822, p. 174, under the article Flowers, we find a list of Chinese and scientific names of Canton plants drawn up by Reeves, Names of 148 plants which flower in each month of the year.

There is in the Transactions of the Medico-Botanical Soc. of London, N. 1, June 1828, an ACCOUNT OF SOME OF THE ARTICLES OF THE MATERIA MEDICA EMPLOYED BY THE CHINESE, from a letter addressed to the Director by John Reeves, F. R. S. and Corresp. Member of the Med. Botan. Soc. This letter is dated, Canton, March 7, 1826. Reeves had arrived not long before from England and had not yet been in Macao. The article is of little importance.

Besides plants, Reeves collected also zoological objects, which he likewise forwarded to Europe. After his return to Europe, he also wrote several articles in zoological periodicals, on Chinese Mammalia, Birds, Frogs, Tortoises, Fishes. A list of these publications is given in Cordier's Bibl. Sin. 173.

Reeves, John Russell, son of John Reeves, like his father a naturalist and collector of plants and zoological objects. From the short notices of his life, as given in the Gard. Chron. 1877, I, 604, in the Journ. Bot. 1877, 192, and in Britt. & Boulg. Biogr. Ind., we learn that he was born in 1804, was resident in Canton for thirty years of his life, in the service of the E. I. Comp., and during this time paid attention to all branches of Natural History, collected a herbarium and was the means of introducing into Engl. gardens several Chinese plants, some of which, such as *Reevesia thyrsoides* and *Spiraea Reevesia* bear testimony by their names to the good service he rendered to botany and horticulture. J. R. Reeves died at

Wimbledon, Surrey 1 May 1877. He became a Fellow of the Roy. Soc. in 1832, and was a few years before his death a Member of the Council of the Roy. Hort. Society. He was also a Fellow of the Linnaean Soc.

In the above quoted obituary notice of John Reeves, in the Proc. Linn. Soc. 1856, it is said: "His son John Russell Reeves Esq., also a valuable Fellow of our Soc., has likewise presented various fish procured at Macao".

We have not been able to ascertain when J. R. Reeves went to China and when he returned to Europe. In the Chin. Repository, V, 1838, p. 428, we find the name of John R. Reeves noticed among the residents of Lintin, an island in the estuary of the Canton River, where in former times European ships trading with Canton used to anchor. This may be Reeves jun., for Reeves sen. had left China in 1831. In the next list of European residents in China, l. c. X, 1841, Reeves' name does not appear.

The correspondents of the Gard. Chron. and the Journ. Bot. are undoubtedly mistaken in stating that the genus name *Reevesia* was dedicated to Reeves jun., for, as we have seen above, Lindley, who in 1827 established this genus, says that the plant had been sent from China by John Reeves. But, as to *Spiraea Reevesii*, first described by Lindley, in 1844, Bot. Reg. t. 10, as a Chinese plant introduced by Mr. Reeves, those correspondents seem to be right in ascribing its introduction to Reeves jun. The name appears first in 1841, Bot. Reg. Miscel. Notes, p. 45, where Lindley notices the *Spiraea lanceolata*, Poir., received from China and then in cultivation in London, stating that in gardens it is known as *Sp. Reevesiana*. Afterwards he found that it was different from Sp. lanceolata. Herbarium specimens of the Sp. lanceolata were first brought to France by Commerson from Isle de France (the Mauritius) where the plant had probably been introduced from China. (V. supra p. 119). Poiret first described it in the Encycl. Botan. VII, 1816, 354. Cambess, Monogr. Spiraea, Ann. Sc. nat., 1824, I, 366, tab. 25, figured it and proved that it was identical with Loureiro's *Sp. cantoniensis*. Loudon, Arboret. et Frut. Brit. II, 1838, p. 732, writes that Sp. lanceolata, a native of the Mauritius and China, had then not yet been introduced into European gardens. The statement in the Flore des Serres, t. 1097 (1856), by Planchon, that Sp. Reevesiana, at first considered to be Sp. lanceolata, was introduced from China into English Gardens by Reeves, in 1824, seems to be an unfounded assertion, and from the above accounts we may conclude that Sp. Reevesiana was introduced by J. R. Reeves junior, between 1838 and 1841. Mr. Hemsley, I. F. S., i, 224, considers Sp. Reevesii identical with Sp. lanceolata and *Sp. cantoniensis*, the latter being the oldest name of the plant.

The two Reeves, father and son, who both have introduced into England living Chinese plants and both sent herbarium specimens from China,
are not always distinguished by the botanists who published their plants,
their personal names and dates being frequently omitted in these records.
Mr. Hemsley, who has kindly taken the trouble to look up for me a number
of Reevesian plants in the Kew Herbarium, writes that they are all
labelled: Mr. John Reeves or Mr. Reeves, ex herb. Brown (thus not original labels); localities: some Canton, some China; no dates.

As Reeves sen. left China in 1831, we ascribe to Reeves jun. all the
plants recorded in the English botanical periodicals as having been introduced by Reeves posterior to that year.

In the Trans. Hort. Soc., New Series, II (1835—42) p. 417—19,
the following five plants, cultivated in the garden of the Society, are stated
to have been sent from China by Reeves:

Deutzia scabra, Thbg., imported in 1833. See also Bot. Reg. tab.
1718 (1834).

Acer palmatum, Thbg., 1832.

Kerria japonica, DC., single flowered form, 1836. See also Bot. Reg.
tab. 1873 (1836). Only the double flowered form was previously known in
European Gardens. V. supra p. 190.

Prunus japonica, Thbg., single flowered form, 1835. See also Bot.
Reg. t. 1801 (1835). Previous to that time only the double flowered form
was known in Europe. V. supra p. 211.

Dolichos Soja, Linn., 1838.

The *Canna Reevesii,* first described in Bot. Reg. tab. 2004 (1837),
is said there to have been raised in the Garden of the Hort. Soc. from
seeds sent by Mr. Reeves from China.

The beautiful orchid, *Dendrobium nobile,* according to Lindley, in
Gen. & Spec. Orchid. p. 79, was introduced by Mr. Reeves, who brought a
live plant from China, which first flowered with Messrs Loddiges in 1837.
In Veitch, Orchids, it is stated that this plant was introduced by John
Russell Reeves, who had purchased it in the market of Macao. Sir J. D.
Hooker, however (Bot. Mag. t. 5003) says, that J. Reeves, sen. introduced
the plant.

In the Journ. Hort. Soc. VIII, 1853, p. 58, Lindley notices *Hibiscus
syriacus,* L. var. *chinensis,* raised from seed presented to the Society by
Mr. J. Reeves, in June 1844, under the name of *koorkun vellory.* (not a
Chinese name).

I may finally observe that some of the herbarium specimens from

China mentioned above under Reeves sen., may have been gathered by his son. The Ind. Fl. Sin., does not distinguish between the two Reeves.

Livingstone, John. Surgeon of the E. I. Comp's medical service in China. In a letter to the Rev. Mr. Morrison, dated Macao, 1819 (v. supra p. 189) Livingstone says that he came out to China in 1803. But this may have been his second voyage to that country for from another letter (see farther on) it appears that he first arrived in China, in about 1793. He lived in Macao. The Chinese Missionary Recorder, VII, 176, states that Dr. Livingstone, in 1820, had a dispensary for the Chinese at Macao. In 1817 he was elected Corresp. Member of the Hort. Soc. In a letter dated Macao, 15 Jan. 1818, published in the Trans. Hort. Soc., III, 183, he writes to the Society:

"By the hands of my friend Mr. Reeves, on his return to China from England, I had the honour to receive a copy of the Charter and Bye-Laws of the Hort. Soc. of London, together with the information contained in your letter to him, that the Soc. has pleased to appoint me one of its corresponding members. In return for this distinguished mark of attention, I shall take much pleasure in contributing my best endeavours to forward the views of the Soc. About 25 years pretty close attention to the Gardens of China, having enabled me to become familiar with their horticulture, I intended to lay before the Soc. a general outline of the subject". — Then follows an article entitled: ACCOUNT OF A METHOD OF RIPENING SEEDS IN A WET SEASON WITH NOTICES OF THE CULTIVATION OF CERTAIN VEGETABLES AND PLANTS IN CHINA, p. 183—186.

In the same volume of the Transactions, p. 421—429, is an article by Livingstone, dated Macao, 5 Febr. 1819: "On the Difficulties which have existed in the Transportation of Plants from China to England". On p. 422 he writes: "The Fa te or Flower Gardens are situated on the banks of the river at a short distance from Canton. To these Chinese nursery gardens strangers used to have access at all times, but for the two or three last years these visits have been restricted to two or three days in a month, say the 8th, 18th and 28th, and they must besides pay 8 dollars for the permission to go thither. In these gardens may be seen all the plants for which a demand exists among the Chinese themselves, but they will be found to consist of a very small variety comprehending only showy or odoriferous plants, shrubs and trees, and such fruit trees as are commonly cultivated in the gardens. To these may be added abundance of dwarf trees, which the Chinese greatly admire and pay a high price for. The soil of these gardens, and indeed of the banks of the river, consists of strong alluvial clay. The plants are either kept in the ground, or they are placed in pots invariably filled with the same clay. It is from these nurseries that Europeans are generally supplied with the plants which they send or carry home".

In vol. IV of the Transactions Hort. Soc., p. 224—231, we find an article by Livingstone, dated Macao, 1820: ON THE METHOD OF DWARFING TREES AND SHRUBS, AS PRACTISED BY THE CHINESE, INCLUDING THEIR PLAN OF PROPAGATION FROM BRANCHES.

Lastly, in 1821, an interesting paper by Livingstone: ON THE STATE OF CHINESE HORTICULTURE AND AGRICULTURE, WITH AN ACCOUNT OF SEVERAL ESCULENT VEGETABLES USED IN CHINA, was read before the Hort. Soc. and published in vol. V of the Transactions, p. 49—56: Chinese agricultural processes, implements etc.; manure. — *Convolvulus reptans, Basella nigra, Amaranthus polygamus* and *A. tristis, Sinapis pekinensis,* a new species of *Brassica,* procumbent, *Fluted bitter Cucumber.* Of all these plants L. sent seeds to the Society.

The following Chinese plants are recorded to have been introduced into England by Livingstone:

The Bot. Repos. t. 612 (1810), *Paeonia albiflora,* flore *pleno,* states that Mr. Livingstone, in 1808, brought seeds of this plant from China, calling it a Yellow Paeony. It was raised in the nursery of Mr. Whitley. V. supra p. 222.

Bot. Mag. t. 2908 (1829), *Crinum plicatum,* Livingstone MSS. Professor W. J. Hooker writes: "About 5 years since (1824), my valued friend Dr. Livingstone obligingly communicated to our Glasgow Botanic Gardens a bulb of this singular plant from China and he sent me a drawing of the natural size made in that country. Our plant flowered in the spring of last year. Livingstone cultivated it at Macao".

Bot. Mag. tab. 2802 (1828), *Baeckia frutescens,* Lin. Professor W. J. Hooker writes: "This plant, first discovered by Osbeck, was sent to the Glasgow Bot. Garden, together with a great number of other botanical rarities from China, in 1827, by Dr. Livingstone".

Prof. Hooker further states, in Bot. Miscell. I (1830) 88, that Dr. Livingstone, in about 1805, first brought from China to Europe the substance known under the name of Chinese *Rice paper* (pith of *Aralia papyrifera).*

Trans. Hort. Soc. VII, 226 (1827): Livingstone, on the GRAFTING OF ROSA BANKSIAE IN CHINA.

Loudon, Arbor. et Frut. Brit. I, 1838, p. 177, reports that Living-stone introduced the (American) *Magnolia grandiflora,* to Macao. (Informa-tion given, it seems, by Reeves).

The Kew Herbarium possesses the following Chinese plants, which Livingstone seems to have first discovered, or first sent specimens of them to Europe.

Illicium religiosum, Sieb. & Zucc., Fl. Japonica, I, 1835, p. 5. Sie-bold says that this plant is a native of China and was introduced into Japan. Livingstone discovered it in China (Macao). Sent also by Millett from Macao. See Ind. Fl. Sin., ɪ, 23.

Photinia pustulata, Lindley, Bot. Reg. adnot. t. 1956 (1836). Parks, Canton. — I. F. S., ɪ, 263, Livingstone, Macao.

Photinia serrulata, Lindley, var. *obovata,* Hooker & Arn. Bot. Beech. 185. The late Dr. Livingstone, China.

Eugenia operculata, Roxb. Ind. Fl. Sin., ɪ, 297. Canton, Livingstone,. Millett.

Cupia mollissima, Hook. & Arn., Bot. Beechely, 192; received from the late Dr. Livingstone. — *Webera mollissima,* Benth. — Ind. Fl. Sin., ɪ, 381.

From the above statements and dates we can conclude that Living-stone died posterior to 1827 and before Hook. & Arn. prepared the "Botany of Beechey's Voyage, China, 1833—38".

Drummond, Henry Andrews, Captain of the E. I. man "Castle Huntley". He introduced in 1819, for the Hort. Soc. of London two new Chinese varieties of *Chrysanthemum,* the *Quilled Flamed Yellow* and the *Quilled Pink,* which for the first time flowered in the Soc.'s garden in the autumn of 1821. See Sabine, Chrysanthemum, in Trans. Hort. Soc. IV, 334, 348.

The same imported in 1823 for the Hort. Soc. from China the *Double White* variety of *Camellia Sasanqua,* and which had been so long sought after by collectors. Lindley, Bot. Reg. tab. 1091 (1827).

Mayne, Charles Otway, Captain of the E. I. man "Atlas". Sabine, in Trans. Hort. Soc. IV, Chin. Chrysanthemums, 334, writes: "In June of 1820, from the skilful management of Capt. Ch. O. Mayne of the "Atlas", a box containing 12 varieties of *Chinese Chrysanthemum,* which was en-trusted to his charge by Mr. J. Reeves, arrived safe, with the loss only of a single plant". After these varieties had successively flowered in the So-ciety's Garden, Sabine described and named them in Trans. Hort. Soc. V,

151—158, 412, 413. One of them, the *Semi-double Quilled Orange*, is represented there in a coloured drawing.

In 1824 Capt. Mayne brought another collection of Chinese Chrysanthemums, which he presented to the late Duchess of Dorset, having previously given cuttings of them to W. Wells of Redleaf, from whom the Hort. Soc. subsequently received the plants. Several of them turned out to be old kinds, but one proved to be entirely new, the *Early* or *New Blush*. Sabine, Trans. Hort. Soc. VI, 324, 326.

Nesbitt, Captain of the Hon. E. I. Comp.'s ship, "Essex", imported for the Hort. Soc., in 1820, from China the *Camellia oleifera,* first noticed, a few years earlier in Abel's Journey (v. s. p. 231). Bot. Reg. t. 942 (1825).

Mayer, Captain of the Hon. E. I. Comp.'s ship "Atlas", imported for the Hort. Soc. from China, in 1820, a new variety of *Chrysanthemum,* named the *Expanded Light Purple* by Sabine, Trans. Hort. Soc. V, 153.— Bot. Mag. tab. 2556 (1824).

Wilson, John Peter, Captain of the ship "Cornwall", introduced for the Hort. Soc., from China, in 1820, the *Sha le* or Chinese Sand Pear, which Lindley named *Pyrus sinensis.* Trans. Hort. Soc. VI, 396 (1824). Bot. Reg. tab. 1248 (1829).

Jamieson, Captain of the "Earl of Balcarras" E. I. man, introduced for the Hort. Soc. from China, in 1821, the *Glycosmis citrifolia.* Lindley in Trans. Hort. Soc. VI, 72 (1824). — I. F. S., I, 109.

Lindsay, Captain of the "Orwell" E. I. man, brought from China for the Hort. Soc., in 1821, the *Laurus aggregata,* which had been sent by John Reeves (comp. supra p. 258). Bot. Mag. tab. 2497 (1824).

Potts, John, Gardener to the Hort. Soc. of London, in 1821 was sent to China and Bengal to procure living plants for the Society. He performed the voyage on board the "General Kyd", Captain Nairne, visited Canton, Macao and Calcutta and, returning in the same ship, reached London in August 1822. Trans. Hort. Soc. V, 427, VII, 25.

Potts brought home valuable collections of living plants and herbarium specimens, especially from China. This most meritorious collector unfortunately died soon after returning from the East. L. c. VI, 66.

Sabine, Trans. Hort. Soc. V, 427, writes, that Potts had made (in Canton) an assemblage of 40 varieties of Chinese *Chrysanthemums*, corresponding to the drawings then in the possession of the Soc., which he sent to England in 1822, but the whole was lost in consequence of an accident which befell the ship in which they were embarked, in her voyage home. — But he was successful with the rest of the living plants procured by him in China and Calcutta. From China he is recorded to have introduced the following:

Paeonia albiflora, var. *Potsii*, Bot Reg. tab. 1436 (1831). Lindley states: "This splendid plant originated in China, from which country it was brought in 1822 to the Hort. Soc. by the late Mr. J. Potts, after whom Mr. Sabine named it. It is by far the handsomest of the varieties of the P. albiflora".

Guatteria rufa, DC. Syst. I, 504 (India, Java). — Bot. Reg. tab. 836 (1824). Lindley says, a native of China, sent in 1822 to the Hort. Soc. by J. Potts. — I. F. S., i, 26, *Uvaria microcarpa*, Champ.

Camellia euryoides, Lindl. Bot. Reg. tab. 983 (1826). This is one of the stocks on which the Chinese graft their varieties of Camellia japonica. The grafted portion of a Camellia brought from China for the Hort. Soc. by J. Potts, in 1822, having died, the stock sprang up and produced this plant, which flowered for the first time in England, in March 1826, in the Chiswick Garden. — I. F. S., i, 81.

Camellia japonica, var. *Paeoniaeflora alba*, White Waratah. Booth, Camellia, Thea, Trans. Hort. Soc. VII, 548. Introduced by Potts, in 1822, but known at Kew in 1810.

Sterculea lanceolata, Cavan. — Bot. Reg. ta. 1256 (1829). — Introduced in 1822 from China by Potts. — I. F. S., i, 89.

Grewia affinis, Lindl., Trans. Hort. Soc. VI (1826), 265. Brought from China by Potts, 1822. — I. F. S., i, 92.

Zanthoxylum nitidum, DC. Prod. I, 724. — Bot. Mag. t. 2558 (1825). Brought from China by Potts, 1822. — I. F. S., i, 106.

Cyminosma pedunculata, DC. Prod. I, 722. — Lindley in Trans. Hort. Hort. Soc. Brought from China, in 1822 by Potts. — I. F. S., i, 109, *Acronychia laurifolia*, Blume.

Tephrosia (?) chinensis, Lindl. Trans. Hort. Soc. VII (1826) 58. Living plant brought by Potts from China, 1822. Not found in Ind. Fl. Sin.

Nauclea Adina, Smith in Rees' Encycl., *A. globiflora*, Salisb. — Bot. Reg. t. 895 (1825), Trans. Hort. Soc. VI, 264. Introduced 1822 by Potts from China. — I. F. S., i, 370.

Ardisia punctata, Lindl. Bot. Reg. tab. 827 (1824), Trans. Hort. Soc. VI, 262. Introduced 1822 from China by Potts. — I. F. S., ii, 64, *Ard. crenata*, Sims.

Diospyros vaccinioides, Lindley in Hooker's Exot. Flora I (1825) t. 139; Trans. Hort. Soc. VI, 261. Introd. from China by Potts, 1820. — I. F. S., ii, 71.

Diplolepis ovata, Lindley, Trans. Hort. Soc. VI (1826) 268. Introd. from China, 1822, by Potts. — Omitted in Ind. Fl. Sin.

Hoya Pottsii, Traill, Trans. Hort. Soc. VII, 16 (1826). The late Mr. J. Potts, on his return from China, August 1822, and shortly before his death, gave Mr. Sabine a single leaf of this Hoya, gathered in one of his excursions near Macao. It was carefully planted and sent forth a shoot from its base in the spring of 1824. In the next year it produced perfect flowers. — I. F. S., ii, 116.

Hoya angustifolia, Traill. l. c. 29. Introduced by Potts from China. — I. F. S., ii, 116.

Callicarpa rubella, Lindley, Bot. Reg. t. 883 (1825) and Trans. Hort. Soc. VI, 263. Introduced by Potts from China. — I. F. S., i, 255.

Callicarpa longifolia, Lamarck, Enc. Bot. I, 563. (Malacca, Sonnerat). — Trans. Hort. Soc. VI (1826) 263: Sent by Potts from China. — I. F. S., ii, 253.

Tetranthera laurifolia, Jacq. — Bot. Reg. t. 893 (1825). Introduced by Potts from China. — I. F. S., ii, 385. *Litsea sebifera*, Pers.

New plants, herbarium specimens, brought by Potts from China:

Itea chinensis, Hook. & Arn. Bot. Beech. 188. — I. F. S., i, 278. Potts, South China.

Doellingeria trichocarpa, sp. n. DC. Prod. V, 263. In China legit Potts, Herb. Lindley. — I. F. S., i, 416, *Aster striatus*, Champ.

Pottsia (new gen. Apocyn.) *cantonensis*, Hook. & Arn. Bot. Beech. 199, t. 43. The plant is evidently named after Potts, but it is not said there, that he gathered it. Only Millett is noticed as collector. — I. F. S., ii, 96.

Tylophora hispida, Decaisne, DC. Prodr. VIII, 610 (Callery). — I. F. S., ii, 113. Potts, China.

Fimbristyles decora, Nees and Meyen in Wight, Contr. Bot. India (1834) 101. China, Potts.

Fuirena Rottboellii, Nees, l. c. 94. China, Potts.

Parks, John, Damper. A most successful botanical collector for the Hort. Soc. of London, in China (Canton, Macao) and Java, in 1823. He left

London on board the "Lowther Castle" E. I. man, Captain Thomas Baker, and returned in the same ship in May 1824. He was instructed to collect in China, amongst other rarities, as many good varieties as possible of Chrysanthemums. Part of the collection was sent home by him and arrived in the spring of 1824, on board the "General Kyd", Captain Nairne, the same gentleman, who in 1821 had shown so much care and attention to Potts, when the latter went with him to Bengal and China. The remainder of Parks collection were brought by himself in the "Lowther Castle", in May 1824. The number of distinct living *Chrysanthemums* received by these two consignements was 20, of which 16 proved to be new and were described by Sabine, all varieties of Sabine's large flowered *Chrysanthemum sinense,* with the exception of two, which belonged to Linnaeus' *Chr. indicum,* small flowering, *double yellow* and *double white.* (Trans. Hort. Soc. V, 427, VI, 323—347).

Besides Chrysanthemums, Parks introduced from China several species and varieties of *Camellia:*

Camellia japonica Parksii, Parks' Striped Rose Cam.

Cam. japon. imbricata, Crimson Shell Camellia.

Cam. japon. Sabiniana, Sabine's White Camellia.

(Booth, Camellia and Thea in Trans. Hort. Soc. VII (1829) 555—57).

Camellia reticulata, Lindl. Bot. Reg. t. 1078 (1827), first imported from China, in 1820, by Captain Rawes (see farther on), was also brought by Parks in 1824.

Camellia euryoides, Lindl., first imported by Potts (v. supra p. 270) was likewise brought by Parks.

The white fragrant *Rosa Banksiae,* had been introduced from China in 1807. Roxburgh in his Hort. Bengal., 1814, p. 38, note, had indicated the existence, in China, of a double yellow variety of this Rose. Lindley, accordingly, instructed Parks, when he was sent to China, to lose no opportunity of securing this valuable variety, in which he was so fortunate as to succeed, having brought several plants home with him, upon his return in 1824. See Bot. Reg. t. 1105 (1827), *Rosa Banksiae lutea.*

Parks imported also another interesting Chinese Rose described by Lindley in the Trans. Hort. Soc. VI, 286, as *Rosa indica,* var. *orhroleuca,* long known before in Europe by Chinese drawings.

Other living plants brought from China by Parks and cultivated at Chiswick in the garden of the Hort. Soc:

Dianthus arbuscula, Lindl., Bot. Reg. t. 1086 (1827). — According to Mr. Hemsley, I. F. S., ɪ, 63, a semi-double of form *D. Caryophyllus,* L.

Eurya chinensis, Abel (v. supra p. 236). Trans. Hort. Soc. VI, 271.

Reevesia thyrsoidea (v. s. p. 261). Lindl. Bot. Reg. tab. 1236 (1829).

Grewia affinis, brought by Potts (v. supra p. 270) and also by Parks. Trans. Hort. Soc. VI, 265.

Blackwellia fagifolia, Lindl. Trans. Hort. Soc. VI, 269. Imported by J. Reeves (v. supra p. 259) and Parks.

Canthium dubium, Lindley, Bot. Reg. tab. 1026 (1826). — I. F. S., ɪ, 384, *Diplospora viridiflora,* DC.

Hoya trinervis, Traill in Trans. Hort. Soc. VII, 26. — I. F. S., ɪɪ, 116, same as *H. Pottsii* (v. s. p. 271).

Callicarpa longifolia, Lam. — Trans. Hort. Soc. VI, 263. Imported by Potts and Parks from China. (V. supra p. 271). — I. F. S., ɪɪ, 253.

Clerodendron lividum, Lindl. Bot. Reg. t. 945 (1826); Trans. Hort. Soc. VI, 267. — I. F. S., ɪɪ, 260: *Cl. fragrans,* Vent.

Eria rosea, Lindl. Bot. Reg. t. 978 (1826).

Coelogyne fimbriata, Lindl. Bot. Reg. tab. 868 (1825); Trans. Hort. Soc. VII, 69.

Sarcanthus succisus, Lindl. Bot. Reg. tab. 1014 (1826).

Glossula tenticulata, Lindl. Bot. Reg. tab. 862 (1824). — *Glossaspis tenticulata,* Lindl. Gen. Spec. Orchid. 284.

Hellenia abnormis, Lindl. Trans. Hort. Soc. VII, 60. — Ind. Kew. = *Alpinia chinensis,* (?).

Aspidistra punctata, Lindl. Bot. Reg. tab. 977 (1826); Trans. Hort. Soc. VII, 66.

Barnardia scilloides, Lindl. Bot. Reg. t. 1029 (1826).

Parks also brought herbarium specimens from China. In the Index Fl. Sin. his name appears frequently but it is there always written *Parkes.* Mr. Hemsley, on my inquiry, kindly informed me that the name is misspelt in the Kew Herbarium. Parks' plants formed part of Bentham's herbarium, received by him from the Hort. Soc. and are now at Kew. Mr. H. thinks that the mistake may have originated with the person who wrote the labels in the Hort. Soc. Parkes is a common name in England, — Parks uncommon. In DC. Prodr. his name is sometimes erroneously written *Parker,* as in vol. XI, 147, sub Ruellia chinensis, XIV, 613, sub Elaeagnus gonyanthes.

The following Chinese plants have been first discovered by Parks:

Evodia meliaefolia, Benth. Fl. Hgk. 58. — I. F. S., ɪ, 104.

Indigofera venulosa, Champ. — Fl. Hgk. 77. — I. F. S., ɪ, 158.

Desmodium reticulatum, Champ. — Fl. Hgk. 84. — I. F. S., ɪ, 176.

Lespedeza viatorum, Champ. — Fl. Hgk. 86. — I. F. S., ɪ, 183.

Eriosema chinense, Vogel, Plant. Meyen, 31. — I. F. S., ɪ, 197.

Photinia pustulata, Lindl. V. supra p. 268.

Eupatorium Reevesii, Wall. (V. supra p. 262). DC. Prod. V, 179. China, Reeves. — Canton, Parkes, herb. Lindley.

Hemigraphis (Ruellia) chinensis, T. Anders. — I. F. S., ɪɪ, 238.

Plectranthus amethystoides, Benth. (v. supra p. 262).

Elaeagnus gonyanthes, Benth. in Kew Journ. Bot. V (1853), 196. Parkes, commun. by Lindley.

Goodyera procera, Hook. Exot. Fl. t. 39, (1823). Nepal, Wallich. — Maxim. Mél. biol. XII (1888), 926. Parkes, China.

Leblanc, Captain Thomas, is reported by Booth, Camellia and Thea, Trans. Hort. Soc., 1829, VII, 553, to have imported from China, in 1821, for the Hort. Soc. a new variety of *Camellia japonica*, which was named *Leblanc's Red Camellia*.

X. Information regarding Chinese Plants found in Systematic Works on Botany, published after Linnaeus.

After the great reformer of Botanical Science, and founder of the binary nomenclature, Linnaeus, had disclosed his principles for the systematic arrangement of plants, at first in the "Systema Naturae", 1735, and more fully in the "Species Plantarum", 1753 and 1763 (v. supra p. 57), his systematic works on Botany went through numerous editions published and enlarged by eminent botanists in the last quarter of the 18th and the first of the present century. The works most frequently quoted by botanists are the 13th and 14th editions of the Systema Vegetabilium, both brought out, in 1774 and 1784 by J. A. **Murray,** Professor of Botany in Goettingen. These editions contain all the new plants discovered after the publication of Linnaeus' Species Plantarum, but, besides the names, only short, imperfect diagnoses of the species are given, and the native countries not noticed.

Willdenòw, Carl, Ludwig, born 1765 at Berlin, studied medicine at Halle. In 1798, appointed Professor of Natural Hist. at the Med. Chirurg. College, Berlin; in 1801, Botanist of the Academy of Sciences and Director of the Bot. Garden, Berlin, Professor of Botany. He died in 1812.

W. published the 5th edition of Linnaeus Species Plantarum, vol. I of which appeared in 1797, — vol. II, in 1799, — vol. III, 1800, — vol. IV, 1805. — These four volumes give descriptions of all the *Phanerogamous Plants* at that time known to botanists. — The first part of vol. V, 1810, contains the *Filices*. The second part was prepared after Willdenow's death, but issued only in 1830: *Musci frondosi* by F. **Schwaegerichen.** — Vol. VI, was published by H. F. **Link,** 1824—1830: *Cryptogamia, Hyphomycetes* and *Gymnomycetes*.

Although W. states in the Preface, that he possessed a vast herbarium, it does not seem that he had seen many Chinese species. The following Chinese plants, or plants believed to be of Chinese origin, were first described by him:

Syringa chinensis, Willd. Sp. Pl. I (1797) 48. This plant was first noticed in the first edition of Willdenow's "Berliner Baumzucht" (1796), 378, where he described it from a living specimen received from Holland. It was believed there to have been sent from China. W. thinks that it may be a hybrid. It was introduced into England before 1795 (v. supra p. 197). — See also I. F. S., ii, 83.

Bocconia cordata, Willd. Sp. Pl. II (1799), 841. China? W. does not say from whence this plant was sent to him, but as we have seen (supra p. 172), it had been introduced from China by Sir G. Staunton, in 1795.— I. F. S., i, 35.

Canna chinensis, was described by Willd. in Magaz. Gesell. Nat. Fr. Berlin, II, 1808, 170, as a new plant raised by him from seeds received from Canton. — Ind. Kew = Canna orientalis, Roscoe (1828), India.

Celtis sinensis, Willd. Berlin. Baumz., 2d edition (1811), 81, described from a specimen cultivated in the Botan. Garden, Berlin, which had not then flowered. China. — Schult. Syst. Veget. VI, 1820, 306, named it *C. Willdenowiana,* for he thought that it was not identical with Persoon's C. sinensis. But Hemsley, I. F. S., ii, 450, takes them to be the same.

Persoon, Christian Hendrick, born at Cape Town, in 1755. He studied medicine and natural sciences at Leyden and Goettingen, became a medical practitioner at Paris, where he died in 1837. P. was one of the most distinguished mycologists of his time. He published the following systematic works on botany:

Linnaeus' Systema Vegetabilium, 15th edition 1797.

Synopsis methodica Fungorum, 2 vols, 1801.

Synopsis Plantarum, seu Enchiridium Botanicum, complectens enumerationem systematicam specierum hucusque cognitarum. 2 vols, Paris, 1805—1807.

Species Plantarum, seu Enchiridium Botanicum. 2 vols. Petropoli, Acad. Scient. 1817—21.

In the above mentioned works we find short diagnoses of the species then known to botanists, with general indications of their native countries. Amongst the new plants first described by Persoon, only one is recorded from China.

Celtis sinensis, Pers. Syn. Pl. I (1805) 292. China. Ex horto Celsii. (V. supra p. 248).

Ulmus sinensis, Pers. l. c. I, 291, Ulmus chinensis, Ulmus pumila quorumdam (Thouin), "Thé de l'Abbé Gallois" per ironiam (Cels). Colitur in hortis et durante hyeme in frigidariis. — This is the *Ulmus parvifolia*, described by Jacquin in 1798. (V. supra p. 137). — I. F. S., ii, 448.

Vahl, Martin, born in 1749 at Bergen, in Sweden, a pupil of Linnaeus, in 1779 lecturer at the Bot. Garden, Copenhagen; from 1785 till his death, in 1804, Professor of Botany, Copenhagen. He possessed a rich herbarium.

From 1790—94 Vahl published the Symbolae Botanicae in 3 volumes. In this work he first described two new plants, both gathered by Dr. Dahl in the island of Hainan, viz: *Ardisia humilis* (III, 40) and *Croton laevigatum* (II, 97). — I. F. S., ii, 65, 435.

In 1804 he began to publish an Enumeratio Plantarum, descriptions of many new exotic plants; but he died before the work was finished. In this were first described two plants from Father d'Incarville's Chinese collection, which he had examined in Jussieu's herbarium, viz: *Syringa villosa*, I, 38, and *Ficus hirta*, II, 201. (V. supra p. 48).

Swartz, Olof, born in 1760, at Norrköping, Sweden, visited the West Indies, 1783—1787. Professor of Botany at Stockholm. Died in 1818. He wrote several papers on the order of Orchideae. In Act. Holm. 1800, p. 207, he described a new Chinese Orchid: *Platanthera dentata*, Sw.

In his Synopsis Filicum, published in 1806, he described a few new Ferns, received from China:

Meniscium triphyllum, Sw. Syn. Fil. p. 19, 206. East India, China.— Syn. Filic. 391.

Pteris crenata, Sw. l. c. p. 96, 290. India, China. — Syn. Filic. 155, = *Pteris ensiformis*, Burm.

Polypodium quercifolium, known to Linnaeus from India, is recorded from China by Swartz, l. c. 32. — Syn. Filic. 367.

Adiantum pallens, Sw. l. c. 125, 323. Mauritius? Chusan. — *Ochropteris pallens,* J. Sm. — Hook. Sp. Filicum II, 55, states that Swartz gives Chusan, in China, as native country, apparently upon the authority of a most unsatisfactory figure in Plukenet's Amalth: "Filix Adianto nigro officinar. similis", t. 403, fig. 2 (v. supra p. 44). — Syn. Filic. 127.

Hornemann, Jens Wilken, born in 1770, from 1801 Professor of Botany at Copenhagen, where he died in 1841. He cultivated many rare exotic plants in the Bot. Garden at Copenhagen.

From 1813—15 he published the HORTUS REGIUS BOTANICUS HAFNIENSIS; SUPPLEMENTUM, 1819. In this he described as new 6 Chinese plants introduced into the Botan. Garden in 1806 (most probably received from Canton).

Eleusine tenerrima, I, p. 79. (Known before from India, Burman's *Poa chinensis.* V. supra p. 103). — *Leptochloa chinensis,* Nees (1829).

Anchusa tenella, I, p. 176. — *Bothriospermum tenellum,* Fisch. & Mey. Ind. sem. hti Petrop. 1834, 23. Cult. St. Petersb. — First discovered by Staunton in China. Introd. 1806. — I. F. S., II, 152.

Acalypha pauciflora, II, p. 909. — Ind. Fl. Sin., II, 437, = *A. australis,* Lin.

Acalypha brachystachya, II, p. 909. — I. F. S., II, 437.

Phyllanthus cantoniensis, II, p. 910. — I. F. S., II, 423, = *Ph. urinaria,* Lin.

Sisymbrium atrovirens, Suppl. p. 72. — I. F. S., I, 40: reduced to *Nasturtium indicum,* DC. Collected by Staunton in China.

Roemer, Joh. Jac. (1763—1819, Professor of Botany at Zurich), and **Schultes,** Jos. Aug. (1773—1831, Prof. of Botany at Vienna, subsequently at Innsbruck and Landshut) undertook conjointly the publication of a new edition of Linnaeus' Systema Vegetabilium, which appeared in 8 volumes, and Mantissae, 3 vols, from 1817—1830; unfinished. Detailed descriptions of the species. Botanical literature is given with great completeness. It does not seem, however, that many new plants were described by the authors of this systematic work; at least we have not met with any new Chinese plant first recorded by Roemer and Schultes.

Sprengel, Curt (1766—1833, Professor of Botany and Director of the

Botanic Garden at Halle) published the 16th EDITION OF LINNAEUS' SYSTEMA VEGETABILIUM, 1825—28, in 5 volumes, with a SUPPLEMENT in the 5th vol. This is much inferior to Roem. & Schult.'s edition. It gives only short diagnoses of the species, and general indications of the native countries. No attention is paid to botanical literature.

The only Chinese plant described there as new, is the *Galium chinense*, Spr. Suppl. p. 7. "Mongolia Chinae contermina. Ante aliquot annos culta in Horto Hallensi." — I suspect the seeds of this plant had been sent to Halle from Russia, for at that time no Europeans, besides Russians had access to Mongolia and China from the north. It is however not mentioned, either in Maximowicz's "Index Florae Mongolicae" or in his "Enumeration of East. Asiat. Species of Galium" (Mél. biol. IX, 259). Hemsley, I. F. S., I, 394, thinks that it is the same as *Galium asprellum*, Michx. Fl. Amer. bor. (1803) or *Galium dahuricum*, Turczan.

De Candolle, Augustin Pyramus, born 4 Febr. 1778 at Genéva. He studied botany at Paris, from 1804 gave lectures on botany at the Collège de France. From 1810—15 Professor of Botany at Monpellier. From 1816 till his death Professor of Botany at Geneva. He died there, Sept. 9, 1841.

In 1818 this eminent botanist undertook the publication of a much extended systematic work on botany, with the title REGNI VEGETABILIS SYSTEMA NATURALE. Whilst the previous publications of this kind were all arranged according to the Linnaean system, De Candolle first adopted for his work a natural system established by himself. After two volumes of it had been issued, 1818 and 1821, it was relinquished by the author who, in 1824 commenced the publication of the PRODROMUS SYSTEMATIS NATURALIS REGNI VEGETABILIS. These important works left far behind all previous productions in descriptive and systematic botany. De Candolle possessed a vast herbarium, containing many rare collections of exotic plants, and besides this he had seen and examined the rich herbariums preserved at Paris and London. Accordingly descriptions of a considerable number of new Chinese plants appear in the Syst. Nat. and in the Prodromus with interesting statements regarding collectors, localities and the herbariums in which the described plants are preserved. Seven volumes of the Prodromus, concluding with Monotropeae, were published by Aug. Pyr. De Candolle and after his death, in 1841, this gigantic work was steadily carried on under the editorship of his no less illustrious son:

De Candolle, Alphonse, born at Paris, 27 Oct. 1806, was educated at Geneva and destined by his father for a legal career. Accordingly he studied

law at the Academy of Geneva, and took his degree of Dr. of Law in
1829. He, however, never practised this profession, but devoted himself
most ardently to the study of botany. In 1831 he became Honorary Prof.
of Botany at the Academy of Geneva, and assisted his father in the admini-
stration of the Botan. Garden, and in 1835, when the latter retired,
succeeded him in the professorship, which post he held till 1850, when,
from political motives, he felt obliged to resign this chair. The remaining
part of his life was spent at Geneva in complete devotion to his cherished
pursuits, and to the preparation and publishing of those important botanical
works, upon which his high reputation rests as one of the most accomplished systematic botanists of our time and the great authority on the history
of cultivated plants. He retained his health and his eminent mental vigour
unimpaired till his death, which carried him away on April 4th 1893.
A. De Candolle was one of the few "associés étrangers" of the French Institute. Of the Prodromus, which was the great object of his life, D.
published ten (properly thirteen) volumes, from VIII, 1844, to XVII, 1873.
He was, however aided in this arduous undertaking by more than thirty of
the most distinguished botanists, who wrote more than one half of the Prodromus. But this great work remains unfinished, comprising only the
Dicotyledons of Phanerogams and even these incomplete by the omission of
the large family of Artocarpeae.

In 1878 Alph. De Candolle, in conjunction with his son Casimir, commenced the publication of a work: Monographiae Phanerogamarum, of
which he wrote the first part, "Smilaceae". This was followed by a series
of monographs by various eminent botanists of which hitherto seven volumes
have appeared.

Kunth, Carl Sigismund, 1788—1850, Professor of Botany at Berlin.
He published an Enumeratio Plantarum omnium hucusque cognitarum. Only
5 volumes of this important systematic work appeared, from 1833—1850,
comprising the greater part of the orders of Monocotyledons. Detailed descriptions of the species, exhaustive quotations of authors and works.

Smilax Gaudichaudiana, Kunth, Enum. V, 252, seems to be the only
Chinese plant first described by Kunth. It had been gathered by Gaudichaud near Canton.

In 1836 there appeared at Edinburgh a work in 3 volumes with the
title: An Historical and Descriptive Account of China by H. **Murray**,
J. **Crawfurd**, P. **Gordon**, Capt. Th. **Lynn**, W. **Wallace** and G. **Burnett**.

Burnett, Gilbert Thomas, 1800—1835, Professor of Botany at King's College, London, from 1831—35, F. L. S., wrote the botanical part of the work in the third volume. He is stated to have enjoyed in composing these chapters an unreserved communication with Mr. Reeves, (I suppose Reeves jun.) a resident for upwards of 17 years at Canton, and to have had access to the materials in possession of the Hon. E. I. Company.

Chap. VI, General view of Chinese Botany, pp. 282—305.

Chap. VII, Fragments towards a Flora of China, being a Catalogue of Chinese Plants, arranged according to the System of Linnaeus, pp. 305—331.

Chap. VIII, Conspectus of the Flora of China, arranged according to the Natural Orders, etc., pp. 331—387. This is nothing but a list of botanical names of plants without quotations of authors or works.

G. v. Martens, Tange der Preuss. Exped. Ost Asien, 1866, frequently quotes Burnett in connection with Chinese Algae.

XI. Introduction of Chinese Plants into English Gardens, between about 1820 and 1840, as recorded in English Illustrated Botanical Periodicals.

The Botanical Magazine. V. supra p. 201. We have seen that this valuable periodical illustrative of exotic botany, founded by W. **Curtis,** in 1786, after the death of the latter, in 1799, was continued by J. **Sims** up to 1826 or volume LIII.

With vol. LIV (resp. I) a new series begins. The Bot. Mag. conducted by **Curtis,** Samuel, 1779—1860, a son-in-law of W. **Curtis.** The descriptions of plants are by **Hooker,** William Jackson, Professor of Botany at Glasgow, who was knighted in 1836 and in 184 made Director of the Roy. Kew Gardens.

The Botanical Register, consisting of coloured figures of exotic plants cultivated in British Gardens, with their history and mode of treatment. This periodical was started under the above title in 1815, and continued till 1847, altogether 2702 coloured plates, in 33 volumes.

Vol. I—XXIII contain tab. 1—2014.

Vol. XXIV—XXXIII contain 688 plates, numbered according to the volumes. According to Pritzel, all the descriptions of Plants in the Bot. Reg. from 1815 to 1824, are by **Ker,** John Bellenden, olim **Gawler,** although his articles are not signed. — From vol. XIV (1828) tab. 1190, the Bot. Reg. was conducted by J. **Lindley,** but many of the articles in the preceding volumes are already signed by him.

Edwards, Sydenham Teak, botanical artist, illustrated the Bot. Reg. from 1815—19, when he died. From 1820 to 1832 we find on the plates the name of M. Hart, after 1832 that of Miss Drake.

The BOTANICAL CABINET, consisting of coloured delineations of plants from all countries, with a short account of each, directions for management etc. by Conrad **Loddiges** & Sons, the Plates by George Cooke. — 20 volumes, London 1818—1833.

Loddiges, Conrad, a German by birth, established a Nursery and Botanic Garden at Hackney, which under his able superintendence and after his death, in 1820, under that of his two sons, attained the highest reputation both at home and abroad.

Loddiges, George, born at Hackney, in 1784, continued after his father's death the publication of the Bot. Cab. He was a Fellow of the Linn. Soc. (1821), also a Fellow of the Hort. Soc., of which for some years he was one of the Vice-Presidents. He died in 1846. See Proc. Linn. Soc. I (1847), 334.

The following Chinese plants were introduced by the Loddiges:

Camellia japonica, var. *atrorubens,* intr. 1809. Bot. Cab. t. 170 (1818).

Gardenia amoena, Bot. Mag. tab. 1904 (1817), Bot. Reg. tab. 735 (1823), cultivated by Messrs Loddiges. Said to have been introduced from China by the Marquis of Blandford. — I. F. S., I, 383.

Ardisia crenulata, Bot. Cab. tab. 2 (1818). Lately introduced from China. — Bot. Mag. tab. 1950 (1817) *Ardisia crenata.* Communicated by Messrs Loddiges, 1816. — I. F. S., II, 63.

Azalea sinensis, Bot. Cat. t. 885 (1824). Loddiges received this plant from China, in 1823. — I. F. S., II, 30.

Cirrhopetalum chinense, Lindley in Bot. Reg. 1843, tab. 49. Imported by Loddiges from China.

Dendrobium aduncum, Wallich MSS ex Lindley in Bot. Reg. 1842, Misc. p. 58, et 1846, tab. 15. Wallich had sent the plant from Calcutta to Loddiges. Its native country was unknown, when Mr. Charles Ford, in 1883, rediscovered it in South China. See Bot. Mag. t. 6784 (1884).

Bambusa striata, Lindley in Penny Cyclop. III (1835), 357. Cultivated by Loddiges, who received the plant from China. — Bot. Mag. tab. 6079 (1874).

Bambusa glauca, Loddiges, Collection of Plants, 1823. China. Lindley, l. c. — According to Munro, Bambus, 89, same as *B. nana,* Roxb.

Bambusa nigra, Lodd. l. c. 1823, Lindl. l. c. *Phyllostachys nig.,* Munro.

Sweet, Robert, 1783—1835, nurseryman, had a nursery at Stockwell, 1810, and afterwards at Parson's Green. Fellow Linn. Soc. 1812. He died at Chelsea. He wrote:

The British Flower Gárden, containing coloured figures and descriptions of the most ornamental and curious herbaceous plants. Drawings by E. D. Smith, 3 volumes, 1823—29, with 300 tab. col. Second series, 4 volumes, 1831—38, with 412 tab. col. Many Chinese plants are figured and described in this work.

Bot. Mag. t. 1726 (1815): Sweet cultivated in the Stockwell nursery *Crataegus indica,* a native of India and China.

The Bot. Mag. tab. 2043 (1819) reports that *Ilex chinensis* was imported from China, in 1814, by Messrs Malcolm and Sweet.

Allnut, nurseryman. In his conservatory at Clapham Common there flowered in Febr. 1813, the variety of *Camellia japonica,* called *Waratah* or *Anemone*-flowered Camellia. Bot. Repos. tab. 662 (1815). — This variety is first noticed in Hort. Kew, 2d edit. IV (1812) 235.

Bot. Cab. tab. 1237 (1827): *Camellia japonica,* coccinea. Raised from seeds by Mr. Allnut of Clapham.

Palmer, Thomas Carey, of Bromley in Kent. His relation Captain **Rawes,** Richard, of the ship "Warren Hastings" introduced for him from China, the following plants:

Iu 1816, *Tasseled White Chrysanthemum.* Sabine, Chrysant. in Trans. Hort. Soc. IV, 339. It first flowered in 1818.

Glycine sinensis. See Sabine in Trans. Hort. Soc. VI, 460.

Wellbank's White Camellia (supra p. 218); Booth, Camellia in Trans. Hort. Soc. VII, 552.

Camellia maliflora, Lindl. Booth, l. c. 526.

In 1820, *Camellia reticulata,* Capt. *Rawes'* Camellia, Lindl. Bot. Reg. tab. 1078 (1827).

Camellia sasanqua, var. β. *stricta,* Mrs *Palmers'* Camellia, Bot. Reg. tab. 547 (1821).

Paeonia Moutan, var. *Rawesii,* Sabine, P. Moutan, in Trans. Hort. Soc. VI, 479.

Primula praenitens, Ker, in Bot. Reg. tab. 539 (1821). Same as *Pr. sinensis,* Sabine, in Lindl. Collect. Bot. (1821) t. 7. — I. F. S., II, 42.

In 1824, *Camellia japonica speciosa, Rawes'* variegated Waratah. Booth, Camellia, in Trans. Hort. Soc. VII, 554.

Kent, William, Gardener at Clapton. Fellow Linn. Soc. 1813. Accompanied Reinwardt to the Indian Archipelago, about 1815, died before 1828 (Brit. Boulg. Biogr. Index). He had a fine collection of living plants at Clapton. Capt. Rawes (v. supra. p. 282), introduced for him from China, in 1820:

Camellia japonica crassinervis, called also *Kent's Hexangular Camellia,* Bot. Cab. t. 1475 (1828); Booth, Cam. in Trans. Hort. Soc. VII, 542.

Bot. Reg. tab. 461 (1820), *Rubus reflexus,* cultivated by Mr. Kent at Clapton, in 1820.

Chandler, Alfred, nurseryman and floral artist at Vauxhall.— Chandler and Buckingham's nursery noticed in the Bot. Cab. and Bot. Reg.

In 1819 they raised from seed *Camellia japonica, althaeiflora,* Bot. Cab. tab. 1794 (1831).

In the same year they produced a variety of *Camellia japonica,* a hybrid, which was named Cam. jap. *Chandleri.* Bot. Mag. tab. 2571 (1825); Bot. Cab. tab. 1897 (1832).

In 1825 they obtained another hybrid, *Camellia japonica anemonaeflora variegata,* or *Chandler's* Waratah Camellia. Bot. Reg. t. 887.

Camellia japonica corallina, a variety raised from seeds by Chandler in 1829. Bot. Cab. tab. 1586.

Chandler and Buckingham published, in 1825, the "Camellia Britanica" with 8 coloured drawings representing new varieties of Camellia, which they raised from the Waratah Camellia (supra p. 282), fertilized by pollen of many other sorts.

Chandler published, in 1831, "Illustrations and Descriptions of Camellias cultivated in England". 40 vol. plates.

Clifford, Dowager Lady. Garden and hothouse at Paddington. She cultivated in 1820 *Artabotrys odoratissima,* a Chinese plant, which for the first time in Europe produced fruit.

Griffin, W. In his conservatory at South Lambeth he cultivated, in 1822, *Neottia australis,* R. Br. var. β. *chinensis,* Lindl. Bot. Reg. t. 602. —

It is not, as stated there, the same as the *Spiranthes pudica,* Lindl. Collect. Bot. tab. 30 (supra p. 253). See Trans. Hort. Soc. VI, 85.

Braddick, J. London, cultivated, in 1822, the *Flat Peaches* of China. He had received the plant from Java, where it had been brought from China. At Canton esteemed a good fruit. Trans. Hort. Soc. IV (1822) 512. Coloured drawing, flowers and fruit. The latter is flat, green, downy, tinged with red. — Comp. also: Rev. hort. 1870, p. 111, c. tabula and Decaisne, Jard. Fruit. Mus. 1874, p. 42, *Persica platycarpa.*

Basington, of the Kingsland Nursery.

He introduced from China, in 1823, *Symplocos sinica,* Lindl. Bot. Reg. t. 710 (1823), and *Camellia japonica,* var. *luteo albicans, Basington's* new Camellia. Bot. Reg. tab. 708. But, with respect to the latter, Booth, Camellia, in Trans. Hort. Soc. VII, 552, says that it is erroneously stated to have been introduced by Basington, the plant being the same as *Camellia Welbankii* (supra p. 218), and Basington received it from Turner.

Previous to 1832 B. intr. *Thea viridis,* var. *latifolia,* Bot. Cab. t. 1828.

Tate, nurseryman, London, Sloane Square.

He introduced in 1823, from China, *Lonicera flexuosa.* Bot. Reg. tab. 712 (1823).

In 1829 he imported *Rhododendron sinense,* var. *flavescens.* Sweet Fl. Gard. t. 290. Rhod. (Azalea) sinense had been first introduced in 1823 (supra p. 281).

Rhododendron Farrerae, Tate in Sweet Fl. Gard. 2 d ser., t. 95 (1831). This beautiful species was introduced 1829, from China, by Captain Farrer of the "Orwell" E. I man, for Tate. — I. F. S., II, 23. Earlier Reeves.

Rhododendron indicum, L. var. *ignescens.* Sweet, l. c. t. 128 (1832). This splendid flame-coloured variety was likewise imported by Capt. Farrer for Tate.

Herbert, William, Reverend and Honourable, of Spofforth, Dean of Manchester, 1778—1847. MONOGRAPH OF AMARYLLIDACEAE 1837. See Britt. Boulg. Biogr. Ind.

He introduced the following Chinese plants:

Before 1818, *Ipomoea chrysoides,* n. sp. Bot. Reg. tab. 270 (1818). Raised by H., from seed which came from China under the vernacular name of *quoll fa.* — I. F. S., II, 158.

Before 1844. *Caloscordum nerinefolium,* Herbert in Bot. Reg. 1844, Misc. n. 64; ibidem, 1847, tab. 5. This plant was sent to Spofforth by J. Trevor Alcock, Esq. who received it from Chusan, when that island was occupied by our troops. — Baker, in Journ. Bot 1874, 290, named it *Allium nerinifolium.*

Wells, William, of Redleaf, nurseryman. He is stated to have cultivated the following Chinese plants:

In 1823 he received from China *Azalea pontica,* var. *sinensis* (same as *A. sinensis,* Bot. Cab. tab. 885. V. supra p. 281). Bot. Reg. tab. 1253 (1829).

New Chinese Chrysanthemums brought in 1824 by Captain Mayne (supra p. 269) and given to Wells.

In 1825 he cultivated *Enkianthus reticulatus,* Lindley. Bot. Reg. tab. 885.

From about 1828 Wells cultivated the *Bauhinia corymbosa,* Roxb. It flowered first in 1838. Bot. Reg. tab. 47 (1839).

In 1842 the *Double Red Azalea indica* first flowered with Wells. Bot. Reg. 1842, tab. 56.

Goodhall, Henry. He cultivated in 1824 the *Small Yellow Chrysanthemum.* Sabine, Trans. Hort. Soc. V, 415. Col. drawing.

Books. He is reported by Sabine, Trans. Hort. Soc. V, 414, to have first cultivated the *Quilled Salmon Coloured Chrysanthemum.*

Dorset, Duchess of. She died before 1826, had a garden at Knowle. Her gardener Mr. Ashworth.

She cultivated, in 1824, new varieties of Chrysanthemum presented to her by Captaine Mayne. V. supra p. 269.

Barclay, Robert, born 1757. He devoted great attention to horticulture. His taste for gardening first displayed itself in 1781, when he went to reside at Clapham. In 1805 he removed to Buryhill, Surrey, where he established a garden in the midst of a country remarkable for its natural beauties. He died 22 Oct. 1830. F. Linn. Soc. He was acquainted with Dr. J. E. Smith, Sir Jos. Banks and W. Hooker. The latter dedicated to him vol. LIV (1827) of the Botanical Magazine.

In 1816 B. cultivated *Phyllanthus turbinatus,* Sims, Bot. Mag. tab.

1862 (1816); Bot. Cab. t. 731 (1823). He had raised it from seed received from China. — I. F. S., II, 427, = *Breynia fruticosa*, Hook. f.

Before 1821 he introduced from China *Anthemis apiifolia*, Brown MS. in Bot. Reg. tab. 527 (1821). It had come also to the Hort. Soc.'s garden, under the name of *Pyrethrum chrysanthemifolium*. See DC. Prod. VI, 7, *Anthemis parthenioides*.

In 1824 B. cultivated the *Small Yellow Chrysanthemum*. Sabine, in Trans. Hort. Soc. V, 415.

In Paxton's Mag. of Bot. III (1837), 51, we read that *Musa Cavendishii*, Lambert (supra p. 205), a Chinese plant, was sent to Barclay, from the Mauritius, in 1829, by the late Charles Telfair *), who stated, in a letter, that he had obtained the species two or three years previously from China. It has a very valuable fruit. — Same as *M. chinensis*, Sweet.

Fairbairn, gardener to His Highness, Prince Leopold at Claremont. He first, in 1827, caused *Renanthera coccinea*, Lour. to blossom. Bot. Reg. tab. 1131 (1828).

Messrs **Malcolm, Gray** & C°. Nursery at Kensington. They cultivated, in 1827, *Primula praenitens (sinensis)*, var. *albiflora*. Sweet Flow. Gard. tab. 196 (1827).

Parmentier, nurseryman? He is stated in the Bot. Cab. tab. 1751 (1831), to have first cultivated *Liparis priochilus*, which he had received from China.

Press, George, Gardener to Edward Gray, Esq., Fellow Hort. Soc. Harringay House, Hornsey. Press cultivated several new varieties of Camellia japonica.

In 1824, *Camellia japon. punctata, Gray's* invincible Camellia. Hybrid. Bot. Reg. tab. 1267 (1829).

1830. *Cam. japonica variegata, simplex*, raised from seed. Bot. Cab. tab. 1694 (1830).

1831. *Cam. jap. Pressii.* Raised from seed. Bot. Cab. tab. 1745.

1831. *Camella japonica, Rosa mundi*. Raised from seed. Bot. Cab. tab. 1866 (1832).

---

*) **Telfair,** Charles, born 1777?, Belfast. Surgeon. Founded the Bot. Gardens at the Mauritius and Réunion. Died at Port Louis, Mauritius, 1833. New genus *Telfairia*, Hook. *Lonicera Telfairii*, Hook. & Arn. Bot. Beech. 190. (Britt. Boulg. Biogr. Ind.).

Fox-Strangways, the Hon. William, afterwards 4th Earl of Ilchester, 1795—1865, F. L. S., F. R. S. He was an ardent horticulturist, whose garden at Abbotsbury in Dorsetshire was famous for its collection of rare and interesting plants. Before 1842, he introduced *Jasminum floridum,* Bunge, from China. Bot. Mag. tab. 6719 (1883). Lindley described it first as *Jasm. subulatum,* in Bot. Reg. 1842, app. n. 58.

Annesley, George, Earl of **Mountnorris,** more generally known by the title of his youth Lord Valentia, born at Arley Castle, Straffordshire, 7 Dec. 1770, educated at Oxford. Succeeded to the earldom, on the death of his father, in 1816, died 23 July 1844. Fellow Roy. Soc. and Linn. Soc. (Proc. Linn. Soc. I, 245). — He was a worthy patron of botanical science and horticulture. Whilst travelling in India, in the beginning of this century, he discovered a wonderful waterplant, to which Roxburgh gave the name *Anneslea spinosa (Euryale ferox,* Salisb.), common also in China and recorded much earlier by the Jesuit Missionaries.

Sweet, Flow. Gard. t. 238 (1834) states that the Earl of Mountnorris' successful culture of the Tree Paeony of China has been rewarded by the production of several splendid varieties, far excelling any of those imported from China. Sweet, l. c., figures *Paeonia Moutan* var. *variegata,* a hybrid.

Lindley, Bot. Reg. t. 1456 (1831), and t. 1678 (1834), reports that the Earl of Mountnorris raised in his garden at Arley Hall, in Worcestershire, the *Semi-double Tree Paeony,* a hybrid, and the *Double-White* variety.

Northumberland, Hugh Percy, third Duke of, born 1785. (Comp. supra p. 150). He was educated at Eton and afterwards at St. John's College, Cambridge and took his degree of M. A. in 1805, and that of L. L. D. in 1809. In 1817 he succeeded his father in the dukedom. He afterwards filled several dignified and important offices in the state. His leisure hours were much employed in the study of botany, astronomy and mechanics. On his garden at Syon (supra p. 150) his Grace annually expended large sums, erected extensive conservatories, successfully cultivated tropical fruit trees, introduced horticultural novelties, and many rare and interesting plants flowered and produced for the first time at Syon. He died at Alnwick Castle, 11 Febr. 1847. From 1833 Fellow Linn. Soc., also Fell. Roy. Soc.

Bot. Reg. tab. 1729, *Euphoria Longan,* a Chinese fruit tree, cultivated in 1833 at Syon.

Bot. Reg. 1844, Miscell. n. 36. *Asparagus lucidus,* Lindl. Cultivated at Syon. Received from Macao.

Smith, nurseryman at Islington, is reported in Sweet's Flow. Gard. tab. 320 (1836) to have imported from China, in 1832, *Daphne odora,* Thb., var. *rubra.*

Young, Messrs of Epsom, nurserymen, are stated in Paxton's Mag. of Botany III (1837), 51 to have purchased at Barclay's sale (supra p. 285). *Musa Cavendishii.*

Ibid. V (1838), 3, *Aconitum chinense,* Paxt. Cultivated 1837 in Messrs Young's nursery.

XII. Botanical Collectors in the South of China (Macao, Canton), between about 1825—1840.

VISIT OF H. M. SHIP "BLOSSOM", CAPTAIN F. W. BEECHEY, TO MACAO, APRIL 1827.

Beechey, Frederick William. This distinguished British naval officer and navigator was born at London in 1796. In 1818 and 1819 he took part in the arctic expeditions of Franklin and Parry, and in 1825 he was appointed to the Sloop "Blossom", which was intended to explore Behring's Strait and the N. W. coast of Arctic America, in concert with Franklin, who advanced by land. He returned, however, from Kotzebue Sound without having met with Franklin. He discovered several islands in the Pacific. After his return to England, Beechey wrote:

NARRATIVE OF A VOYAGE TO THE PACIFIC AND BERING'S STRAIT, PERFORMED IN H. M. SHIP "BLOSSOM" UNDER THE COMMAND OF CAPTAIN F. W. BEECHEY, IN THE YEARS 1825—28. 2 vols. 1831.

In 1854 he was made Rear Admiral, and in the following year he was elected President of the R. Geogr. Society. He died at London, 29 Nov. 1856.

From Beechey's narrative we learn that the "Blossom" departed from 1825 Spithead 19th May 1825. Rio Janeiro, passage around Cape Horn, Valparaiso, Salaz y Gomez Isl., Pitcairn Isl., Otaheite, Sandwich Isls, Kamtchatka, Petropavlovsk, Bering's Strait, Kotzebue Sound, St. Paul's Isl., Aleutean Isls, St. Francisco, Sandwich Isls, Assumption (Ladrones Isls), 1827 Bashee Isls (S. E. of Formosa). — On April 10th 1827 the "Blossom" anchored in the Typa or outer harbour of Macao, where they stayed till April 30th. The naturalists of the expedition visited Macao and its neighborhood, but did not proceed to Canton.

After leaving Macao, the "Blossom" made sail for Manilla, then directed her course to the Loo choo (Liu kiu) Ils and put into the harbour of
Napa kang situated at the southern end of the largest island of that group,
where she lay from the 16th till the 25th of May. Then steering eastward
they came to the Bonin Islands (Bonin Sima in Japanese) and on June
7th entered the port of an island not visited before by Europeans. Beechey
named it Peel Isl. and the harbour Port Lloyd. He declared this island,
the largest of the Bonin group, the property of the British Govt. After a
stay of a week, they again got under sail and proceeded for a second time
to Petropavlovsk and Bering's Strait. On their voyage home they touched
at Monterey, S. of San Francisco, Mazatlan, San Blas, Acapulco, Valparaiso, Rio Janeiro, and on October 12th 1828 landed at Spithead.   1828

Two naturalists were attached to Captain Beechey's expeditions,
Messrs G. T. **Lay** and A. **Collie,** the latter of whom was surgeon to the
"Blossom" (see farther on). The plants they had collected during the voyage
in various parts of the world were determined by two eminent botanists Sir
W. J. **Hooker** and G. A. **Walker-Arnott,** who published conjointly, from
1830—41, London, the BOTANY OF CAPTAIN BEECHEY'S VOYAGE, comprising an Account of the Plants collected by Messrs Lay and Collie and other
officers of the expedition during the Voyage to the Pacific and Bering's
Strait, etc., in the years 1825—28. The plants are arranged according to
the countries where they had been gathered. China occupies 92 pages,
166—258, Loo choo and Bonin, p. 258—275. These were the first botanical collections from these remote islands, which reached Europe *). According to Mr. B. D. Jackson, Journ. Bot. 1893, 298, the publication of
the book began in 1830 and finished in 1841. The part containing the
Chinese plants was issued between 1833—38.

The authors state: "The Chinese plants (460 species) were chiefly
collected about Macao. Through the kindness of Mr. Millett, we are in
possession of many plants gathered by himself at Macao and in the adjacent
islands. The collection includes also many from the herbarium of the Revd.
G. H. Vatchell, Chaplain of the Factory at Canton".

Capt. Beechey's name has been commemorated by the authors, p. 270,
in the *Ficus Beecheyana* discovered in Loo choo. — Munro, in his Monogr.
Bambus. (1868), named a bamboo from Macao *Bambusa Beecheyana*.

Collie, Alexander, Surgeon R. N., Fellow Linn. Soc. 1825. On board

---

*) The plants gathered in 1816 in Loochoo (H. B. M. S. «Alceste») were lost (v. supra
p. 234, 236).

of the Blossom, 1825—28. Collected plants with Lay, during the voyage, especially in California 1827. He died in Dec. 1835, in King George's Sound. Britt. Boulg. Biogr. Index.

Lay, George Tradescant. Naturalist attached to Capt. Beechey's expedition, 1825—28. He collected plants near Macao, in Loochoo, Bonin Isls, in 1827, and subsequently at Canton. Those from Beechey's expedition were described by Hooker & Walker-Arnott in Bot. Beechey, where the new genus *Laya* (v. infra) is dedicated to him.

In 1836 Lay again went to China, as Agent of the British and Foreign Bible Society. In the same year an American expedition to China and the Malayan Archipelago was set on foot by several rich American Houses (Oliphant, Talbot, King, Morss.). They purchased a small vessel, the "Himmaleh", and sent her out from New York to be employed in voyage of exploration through the Archipelago and about the China and Japan Seas, principally for the purpose of aiding missionaries in circumlating religious books on the coasts of China and the neighbouring countries. The "Himmaleh" arrived at Macao in August 1836. Lay then accompanied the expedition, which left Macao on Dec. 3d 1836, visited Singapore, Macassar, the islands of Ternate, Tidore, Mindanao, Borneo, and returned to China in July 1837. An "Account of the Voyage", which was a total failure in respect to missionary work, was written by Lay and published in New York. (Dr. William's Middle Kingdom, 1883, II, 330).

After returning from this cruise, Lay resided at Macao and became a learned sinologist distinguished by great abilities. Subsequently he entered the British Consular Service and served as interpreter of the Special British Mission to China under Sir Henry Pottinger, 1840—42. In 1843 he was appointed British Consul at Canton. (Chin. Repos. XI, 1842, 114, XII, 392). In June 1844 Lay was despatched to Fu chow as the first British Consul at that place. In 1845 he seems to have been transferred to Amoy. See Fortune's Wanderings, 378. The Chin. Repos. XVI, 1847, p. 75, reports the death of Mr. Lay, British Consul at Amoy, from pernicious fever, in that year.

Lay seems to have sent to England plants gathered near Canton. Britt. Boulg. Biogr. Ind. state that his Chinese plants are preserved in the British Museum. — In the Proc. Linn. Soc., I, 180 (Dec. 1843) we read: Mr. T. Lay, H. M. Consul at Canton, presented a box of specimens of the *keih-seen-me* (?), a species of *Alga*, related to Nostoc and eaten as a delicacy among the Chinese.

Lay wrote numerous articles on China and the Chinese, most of them published in the Chinese Repository. I may notice the following, which are of botanical interest:

EUPHORBIACEOUS PLANTS IN CHINA, Chin. Repository V, 1836, 437—443: *Stillingia sebifera* or *Tallow tree, Acalypha indica, Jatropha curcas* (he rather seems to mean *Elaeococca verrucosa*).

NATURAL HISTORY OF MACASSAR. Chin. Repos. VI, 1838, 449, seqq.

REVIEW OF BLANCO's FLORA DE FILIPINAS. L. c. VII, 1838, 422—37.

A CALENDAR FOR FOUR MONTHS (July, August, September, October 1844) OF THE WEATHER, NATURAL HISTORY AND COUNTRY OPERATIONS AT FOO CHOW FOO, by G. T. Lay, H. M. Acting Consul at that place.

This valuable paper was published in the Trans. Hort. Soc. London, 2d series III (1843—48), 237—245. The continuation appeared in the Journ. Hort. Soc. I, 1846, 119—126, and comprises the months of November, December 1844, January 1845. Lay's Calendar gives Meteorological Observations, Observations on Husbandry, Fruits, Flowers and Vegetables in Season, Animal Kingdom, Events, General Remarks.

List of plants gathered near Macao, in 1827 by the collectors of the Beechey Expedition, chiefly by Lay and recorded as novelties by Hooker & Arnott in Bot. Beech., and by other botanists.

Cocculus diantherus, H. & A. Bot. Beech. 167. Beechey, Millett, Vachell. — I. F. S., I, 28, = *C. Thunbergii*, DC.

Capparis pumila, Champ. Kew Journ. Bot. III (1851), 260. — I. F. S., I, 51. Millett, Beechey (1827).

Pittosporum pauciflorum, H. & A. l. c. 168, tab. 32. — I. F. S., I, 58. — Icon. Pl. tab. 1579 (1887).

Calophyllum spectabile, Willd. in Bot. Beech. 174, is, according to Gard. & Champ. Kew Journ. Bot. I (1849), 303, a new species, *C. membranaceum.* — I. F. S., I, 75.

Emmenanthus (gen. nov.) *chinensis*, H. & A. l. c. 217. — *Ixonanthes chinensis*, Benth. Kew Journ. Bot. III (1851), 308. — Hook. & Benth. Gen. Pl. I, 245. — Champion, Linn. Trans. XXI, tab. 13. — I. F. S., I, 96.

Prinos asprellus, H. & A. l. c. 176, t. 36. — *Ilex asprella*, Champ. Kew Journ. Bot. IV (1852), 329. — I. F. S., I, 115.

Ilex pubescens, H. & A. l. c. 176. — I. F. S., I, 117.

Cissus cantoniensis, H. & A. l. c. 175. — *Vitis cantoniensis*, Seem. Bot. Herald (1857) 370. — I. F. S., I, 131. — Also Vachell.

Acer trifidum, Thbg.? H. & A. l. c. 174. According to Maximowicz, Mél. biol. X (1880) 603, Thunberg's A. trifidum in *Lindera triloba,* Bl., and the plant noticed as A. trifidum by H. & A. is a new species of Acer.— I. F. S., ɪ, 142. Earlier Cunningham. V. supra p. 37.

Connarus microphylla, H. & A. l. c. 179. — *Rourea microphylla,* Planchon in Linnaea XXIII (1850), 421. — I. F. S., ɪ, 149. Also Millett.

Connarus (?) juglandifolius, H. & A. l. c. 179, is, according to Benth. Fl. Hongk. 69, *Rhus succedaneum,* Lin.

Bowringia (new gen.) *callicarpa,* Champion in Kew Journ. Bot. IV (1852), 75. — According to I. F. S., ɪ, 201, first discovered by Beechey expedition (1827).

Laya (new gen.). H. & A. l. c. 182, t. 38. Lay, also Millett. *Ormosia emarginata,* Benth. Kew Journ. Bot. IV (1852), 77. — I. F. S., ɪ, 204.

Gleditschia australis, Hemsley, I. F. S., ɪ, 208, Beechey.

Inga demidiata, H. & A. l. c. 181. Lay. — Named *Pithecolobium Clypearia,* by Benth. in London. Journ. Bot. III (1844), 209. — I. F. S., ɪ, 217. Also Millett.

Inga bigemina, Willd., noticed in Bot. Beech. 182, is a new plant: *Pithecolobium lucidum,* Benth., Lond. Journ. Bot. III (1844), 207. — I. F. S., ɪ, 217.

Itea chinensis, H. & A. l. c. 189, t. 39. — I. F. S., ɪ, 278. Earlier discovered by Potts.

Syzygium buxifolium, H. & A. l. c. 187. — Named *Eugenia chinensis,* by Hemsley, I. F. S., ɪ, 298.

Memecylon scutellatum, H. & A. l. c. 186. — Same as *M. ligustrifolium,* Champ. Kew Journ. Bot. IV (1852), 117. — Fl. Hongk. 117. — I. F. S., ɪ, 302. Also Vachell.

Memecylon nigrescens, H. & A. l. c. 186. — I. F. S., ɪ, 302.

Viburnum nervosum, H. & A. l. c. 190. — As this name was preoccupied by a species of Don's, Bentham in Fl. Hgk. 142, named it *Vib. venulosum.* — I. F. S., ɪ, 356. Gathered also by Reeves.

Hedyotis macrostemon, H. & A. l. c. 192. Lay, Millett. — I. F. S., ɪ, 374. Also Vachell.

Cupia (Webera) corymbosa, Willd., noticed in Bot. Beech. 192, is, according to Hook. fil., Fl. Ind. III (1882), 104, and I. F. S., ɪ, 380, a new species, *Webera attenuata,* Hook. fil. l. c., Hongkong, perhaps also Khasia Mts.

Psychotria scandens, H. & A. l. c. 193. — In Benth. Fl. Hgk. 161, reduced to *P. serpens,* Linn. — I. F. S., ɪ, 387.

Verbesina prostrata, H. & A. l. c. 195. Also Vachell, Millett. — I. F. S., I, 434, *Wedelia prostrata,* Hemsl.

Myrsine ardisioides, H. & A. l. c. 197. — DC. Prod. VIII, 104, obscure plant.

Styrax suberifolium, H. & A. l. c. 196, tab. 40. — I. F. S., II, 77.

Tabernaemontana mollis, H. & A. 199. — I. F. S., II, 96.

Holarrhena affinis, H. & A. l. c. 198. — Benth. in Fl. Hgk. 221, identifies it with *Aganosma laevis,* Champ. Kew Journ. Bot. IV (1852), 335. — According to I. F. S., II, 96, an obscure plant.

Ecdysanthera rosea, H. & A. l. c. 198. — I. F. S., II, 98. Also Vachell, Millett.

Ficus setosa, H. & A. l. c. 216, tab. 49.

Fimbristylis podocarpa, Nees ab Esenb. in Wight Contrib. Bot. Ind. 1834, 98. — Bot. Beech. 225.

Lomaria longifolia, H. & A. Bot. Beech. 257. — Not found in Hook. Spec. Filicum.

New plants discovered by Beech. Exped. in Loo choo and Bonin Islands:

Pittosporum pauciflorum, var. β. H. & A. l. c. 259. Loochoo, Bonin Isl. — I. F. S., I, 58, perhaps a new species.

Elaeocarpus photiniaefolius, H. & A. l. c. 295, tab. 53. — Bonin. — I. F. S., I, 95.

Zanthoxylon Arnottianum, Maxim. Mél. biol. VIII (1871), 372. Noticed as Z. piperitum, DC. in Bot. Beech. 261. Bonin Isl. — Gathered in the same island, a year later, by Dr. Mertens (Russian expedition).

Lespedeza striata, H. & A. l. c. 262. Bonin Isl. — I. F. S., I, 182. Earlier Staunton, Peking. Supra p. 175.

Raphiolepis integerrima, H. & A. l. c. 263. Bonin Isl. — I. F. S., I, 264: reduced to *R. japonica,* Sieb. & Zucc. Fl. Jap. I, 162.

Sedum uniflorum, H. & A. l. c. 263. Loo choo. — I. F. S., I, 288.

Distylium racemosum, Sieb. & Zucc. Fl. Jap. I (1835), 178. — According to I. F. S., I, 289, discovered by Beech. expedition (1827) in Loo choo.

Lonicera affinis, H. & A. l. c. 254. Loo choo. — I. F. S., I, 259.

Hedyotis Grayi, Hook. fil. in Benth. & Hook. Gen. Pl. II, 57, and Maxim. Mél. biol. XI (1883), 779. H. multiflora H. & A. l. c. 264 (nec Cavan.). Bonin Isl.

Mussaenda glabra, Vahl, in Bot. Beech. 264. Loo choo. — According to I. F. S., I, 379, a new species.

Cupia corymbosa, Willd., noticed in Bot. Beech. 264, Bonin Isl., is, according to I. F. S., i, 381, not the plant mentioned p. 192 under the same name for Macao (supra p. 292) but *Stylocoryne subsessilis*, A. Gray, Bot. Jap. (1859), 394.

Cirsium brevicaule, A. Gray, l. c. 396. Cirsium japonicum H. & A. (non DC.). Bot. Beech. 266. Loo choo. — Maxim. Mél. IX (1874), 324, = *Cnicus japonicus*, var. *brevicaule*.

Lysimachia lineariloba, H. & A. l. c. 268. Loo choo. — I. F. S., ii, 53.

Sideroxylon ferrugineum, H. & A. l. c. 266, tab. 55. Bonin Isl. — I. F. S., ii, 68.

Buddleia curviflora, H. & A. l. c. 267. Loo choo. — I. F. S., i, 119.

Callicarpa subpubescens, H. & A. l. c. 305. Bonin Isl. — Maxim. Mél. XII, 1886, 507.

Croton Cumingii, Müll. Arg. gathered about 1835 by Cuming in the Philippines, was, according to Ind. Fl. Sin., ii, 434, earlier discovered by the Beech. exped. in Loo choo. L. c. 270, *Cr. polystachyus*, H. & A. (non Spreng).

Rottlera aurantiaca, H. & A. l. c. 270. Loo choo. — Identified with *R. tinctoria*, Roxb. in Benth. Fl. Hongk. 307. — I. F. S., ii, 440, sub *Mallotus philippinensis*, Müll. Arg.

Ficus Becheyana, H. & A. l. c. 271. Loo choo. — Fl. Hongk. 329. — According to King, Ficus, 1888, 141, = *F. erecta*, Thbg.

Ficus pumila, Lin., in Bot. Beech. 271 is, according to Miquel, Syn. Ficus, Lond. Journ. Bot. VII (1848), 435, a new species, which he names *F. insularis*. — Maxim. Ficus, in Mél. XI, 1881, 332.

Juniperus taxifolia, H. & A. l. c. 271, Bonin.

Carex Boottiana, H. & A. l. c. 273. Bonin Isl. — Not to be confounded with *C. Boottiana*, Benth. (1845). Comp. Franchet, Enum. Jap. II, 561.

Spodiopogon aureus, H. & A. l. c. 273. Loo choo. — *Ischaemum aureum*, Hackel, Andropog., in DC. Monogr. VI, 1889, 224.

Vilfa elongata, Nees ab Esenb. in Bot. Beech., 248, 274. Beechey exped. Loo choo. Vachell and Millett, Macao.

Woodwordia prolifera, H. & A. l. c. 275, tab. 57. — Hook. Spec. Fil. III (1860) 68, identifies it with *W. orientalis*, Swartz, Syn. Fil. 117.

Dictyata spinulosa (Alga) H. & A. l. c. 275. Loo choo.

Vachell, Rev. George Harvey. According to Britt. Boulg. Biogr. Index, he flourished, 1808—36. In 1821 B. A. Cambridge Univ. Subsequently

Chaplain to the Factory of the E. I. C. at Macao, where he made botanical collections, now preserved in Herb. Kew and Fielding (Cambridge).

Some of Vachell's labels in Lindley's herbarium, as quoted by Nees v. Esenbeck, bear the dates 1825 to 1830, Macao. Meyen (see further on) made Vachell's acquaintance at Macao, in 1831. — In the Chin. Repository, III (1834), 143, we read: "On July 29 the Rev. Vachell arrived from England to assume the duties of Chaplain to the Factory of Macao. He came out in the same ship with Lord Napier, Superintendent of British Trade, Captain Elliot etc." Vachell had evidently been on leave. In Chin. Rep. V, 429, the Rev. Vachell is mentioned among the residents of Macao in 1836. His name does not appear in the next list given for 1841.

Vachell seems to have given his Chinese botanical collections to the Rev. John Stevens Henslow (1796—1861), from 1825 Professor of Botany, Cambridge. Sir W. J. Hooker, in Bot. Beech., 166, 195, states that Vachell's Chinese plants were communicated to him by Professor Henslow, Cambridge, and Lindley, in his Gen. et Spec. Orchid. repeatedly notices that he received Vachell's plants through the medium of Prof. Henslow.

The greater part of Vachell's botanical collections, from the neighbourhood of Macao and the adjacent islands, were described in the Botany Beech. by Hooker & Arnott. Vachell's name occurs also in Sir W. J. Hooker's Species Filicum, 1846—64, and in the same author's enumeration and description of Chinese Ferns in Kew Journ. Bot. IX 1857 (Bentham's Florula Hongk.). Lindley published Vachell's Orchids in his Gen. et Spec. Orchid. 1830—40. Nees v. Esenbeck (1776—1858), from 1831 Professor at Breslau, described in Bot. Beech., in Linnaea, 1834, in Wight's Contrib. Bot. India, 1834, and in Plantae Meyenianae, 1843, the new Cyperaceae and Gramineae from Vachell's collection, placed at his disposal by Lindley. (See Linnaea IX, 1834, 273).

Vachell rediscovered many of the obscure plants named and imperfectly described by Loureiro, in 1788, in his Flora Cochinchinensis.

The following localities appear on the labels affixed to Vachell's herbarium specimens, as given by the above mentioned botanists.

Macao. Some times Hooker writes Oman. Ao men is the Chinese name for Macao.

Lappa or Lappas Islands. I may observe that there is only one island which bears the name of Lappa. It is pretty large and hilly and is situated beyond the inner harbour of Macao, west of the city. On English maps it is generally marked Priests Island.

Whanum Isl. It lies S. of Macao, is also called Montanha.

Putoy Isl. Situated S. E. of Macao, belongs to the Ladrones Group*).

Chicow. This is perhaps Chukchow, one of the Ladrones Islands.

Lama. Probably Lamma Island, S. W. of Hongkong.

Tynon Bay, stated by Vachell to be 45 miles east of Macao. (Plantae Meyen p. 61, sub Cyperus canescens, Vahl). This name is not found on modern maps. Perhaps it is misspelt for Tytam Bay, on the southern coast of Hongkong.

Danes' Island, in the Canton River, S. E. from Canton, with the Whampoa anchorage and British Consulate.

Wight and Arnott in Prodr. Fl. Ind. (1834), 272, established the genus *Vachellia* in honour of the Rev. G. H. Vachell "who has lately", as they say, "contributed largely by means of specimens to make the botany of China better known to Europe". This genus was founded upon *V. farnesiana* (Acacia farnesiana, Willd.). Bentham & Hooker do not even retain it as a section of the genus Acacia.

New plants discovered by Vachell, near Macao etc.:

Crotalaria Vachellii, Hooker & Arnott in Bot. Beech. 180. — Ind. Fl. Sin., I, 151, = *C. elliptica*, Roxb. Fl. Ind. III, 279. — Also Millett.

Hedyotis Vachellii, H. & A. l. c. 194. — I. F. S., I, 376.

Toxocarpus Wightianus, H. & A. l. c. 200. Vachell and Millett. — I. F. S., II, 101.

Lithospermum chinense, H. & A. l. c. 202. Vachell. — I. F. S., II, 154.

Vandelia oblonga, Bentham, Scrophul. Ind. (1835), 35. Vachell. Bot. Beech. 202. Vachell. — I. F. S., II, 188, sub *Torenia oblonga*, Hance.

Siphonostegia chinensis (new genus), Benth. Scroph. Ind. (1835), 51, and Bot. Beech. 203, tab. 44. Vachell, Macao and adjacent islands. This plant was discovered at about the same time (1831) by Bunge in North China. — I. F. S., II, 202.

Callicarpa nudiflora, H. & A. l. c. 206, tab. 46. Vachell, Millett. — reduced, in Fl. Hgk. 270, to *C. Reevesii*, Wall. — I. F. S., II, 254.

Leucas Benthamiana, H. & A. l. c. 204. Vachell, Millett. — DC. Prodr. XII, 525, = *Leucas mollissima*, Wall. Pl. Asiat. rar. I, 62, var. *chinensis*. Earlier Staunton. — I. F. S., II, 304.

Chenopodium Vachellii, H. & A. l. c. 269. Loo choo. Same plant recorded in Bot. Beech. 207, under the name of *Ch. acutifolium*, Kit. Vachell, Macao. — Bentham in Fl. Hgk. 282, reduces it to *Ch. acuminatum*, Willd. — I. F. S., II, 323.

---

*) Another island of the same name is marked on English maps south-east of Hongkong.

Euphorbia bifida, H. & A. l. c. 213. Vachell. Peninsula of China (evidently the peninsula on which Macao is situated). — I. F. S., II, 412.

Euphorbia Vachellii, H. & A. l. c. 213. Vachell, Macao. — DC. Prod. XV, II, 25, = *E. serrulata*, Reinw. in Blume Bigdr. 635 (previous to 1822). — I. F. S., II, 417.

Apaturia chinensis, Lindley, Gen. & Spec. Orchid. 131. Vachell, Macao. Comm. Rev. Henslow.

Arundina chinensis, Blume. Lindley, l. c. 125, Vachell, Macao. Comm. Rev. Henslow.

Eriocaulon cantoniense, H. & A. Bot. Beech. 219. — Vachell, Millett.

Cyperus Meyeni, Nees v. Esenb. Pl. Meyen, 59. Vachell, Millett. S. China. — Meyen, Manilla.

Kyllinga nana, Nees v. Es. in Linnaea, IX (1834), 286. Only name. Bot. Beech. 224, Vachell.

Fimbristylis sub-bispicata, Nees v. Es. and Meyen in Pl. Meyen, 75. Vachell.

Rhynchospora chinensis, Nees v. Esenb. in Wight Contrib. Bot. Ind. (1834), 115. China, Meyen. — Bot. Beech. 226, Vachell. — Hook. fil. Fl. Ind. VI, 671, reduces this to *R. glauca*, Vahl, Enum. II, 233. Khasia, Ceylon.

Lepidosperma chinense, Nees v. Es. and Meyen, Pl. Meyen, 117. — Bot. Beech. 228. Vachell 1825, herb. Lindley. — Fl. Hgk. 398.

Cladium chinense, Nees v. Es. Linnaea IX (1834), 301. Only name. — Nees in Pl. Meyen. 116, Vachell, Macao. Herb. Lindley. — Bot. Beech. 228, Vachell, Millett. — Fl. Hgk. 397, = *Cladium Mariscus*, Brown, Prodr. Fl. Nov. Holl. 1810.

Gahnia tristis, Nees v. Esenb. in Bot. Beech. 228. Vachell, Millett. Fl. Hgk. 398.

Scleria ciliaris, Nees v. Es. in Wight Contr. Bot. Ind. (1834), 117. Macao, Vachell, Millett. — Kunth, Enum. II, 357, named it *S. chinensis*. — Fl. Hgk. 400.

Scleria Neesiana, H. & A. in Bot. Beech. 229. Vachell, Millett.

Carex valida, Nees v. Es. in Wight, l. c. 123 and Bot. Beech. 230. Vachell.

Paspalum chinense, Nees in Bot. Beech. 231. Macao, Vachell. Hook. & Arnott identify it with *Panicum filiforme*, Lin., which Roxburgh received from China. (Supra p. 246).

Panicum caesium, Nees in Bot. Beech. 235. Vachell.

Arundinella glabra, H. & A. l. c. 237. Vachell.

Pogonatherum refractum, Nees in Pl. Meyen, 182. Macao, Vachell, May 1829. Herb. Lindley. — Bot. Beech. 239, Vachell, Millett. — Hackel Andropogon in DC. Monogr. VI, 193 identifies it with *P. saccharoideum*, Beauv. var. *monandrum*. (Andropogon monandrum, Roxb.).

Meoschium lodiculare, Nees in Bot. Beech. 246, Pl. Meyen, 195. Macao, Vachell, 1830. — Fl. Hgk. 426 = *Ischaemum barbatum*, Retz. — Hackel, l. c. 205 = *Ischaemum aristatum*, Lin.

Andropogon hamatulus, Hook. & Arnott. Bot. Beech. 244. Danes' Isl., Vachell. — Hackel, Andropog. l. c. 606 = variety of *A. nardus*, Lin.

Andropogon Vachellii, Nees v. Es. Pl. Meyen, 188. Macao, Vachell, Sept. 1829. Herb. Lindley. — Bot. Beech. 243. — Fl. Hgk. 423.

Homoeatherum (new genus) *chinense*, Nees v. Es. in Bot. Beech. 239. Macao, Vachell, July 1829. Herb. Lindley. — Hackel, Andropog. l. c. 457, = variety of *Andropogon apricus*, Trin.

Anthistiria caudata, Nees in Bot. Beech. 245. Macao, Vachell. Herb. Lindley.

Arundo Henslowiana, H. & A. l. c. 248. Vachell, Danes' Isl. Herb. Henslow.

Nothochlaena pilosa, H. & A. l. c. 255. Macao, Vachell. — W. J. Hooker, Spec. Filic. V (1864), 116 = *N. hirsuta*, Desv. *(Pteris hirsuta*, Poir. 1804).

Osmunda Vachellii, Hooker, Icon. Pl. I (1837), tab. 15, Lappas Isl., Vachell. — Hook. Kew Journ. Bot. IX (1857), 360, reduced to *O. javanica*, Blume. Enum. Pl. Jav. (1830), 254. — Bot. Beech. 255, also Lay.

Polypodium subtriphyllum, Hook. & Arnott, l. c. 256, tab. 50. Macao, Vachell. — Hook. Spec. Filic. IV (1862), 52, named *Aspidium subtriphyllum*.

Millett, Charles. He lived at Macao and Canton (during the tea season) at about the same time as Vachell, collected plants in the neighbourhood of Macao, Lappa Isl., Canton, on Pak wan Hills which he used to sent to his friend Prof. W. J. Hooker, Glasgow, who mentions him first as being in China in 1827. Millett's name does not appear in the lists of foreign residents in China for 1836 and 1841, as given in the corresponding volumes of the Chinese Repository. According to Britt. Boulg. Biogr. Index, Millett was also in Ceylon and Malabar. His plants are now in Herb. Kew. The same authors state that he was M. D., it seems on the authority of Wight and W. Arnott, who in their Prodr. Fl. Ind. (1834), 263 say: "We have named the new genus *Millettia* after Dr. Millett, Canton". This genus, of the order Leguminosae, is represented in China by 8 species. (See Ind. Fl. Sin., I, 159).

In the Bot. Mag. 1832, tab. 3148, Thea viridis, Prof. W. J. Hooker writes: "To request information on green Tea, I wrote to my valued friend Charles Millett, Esq. of Canton, who holds a high official situation in the Company's Factory there. In a letter dated Canton, 12 Dec. 1827, Mr. Millett communicated the following particulars on the subject.

'The Tea plant is almost as scarce in this neighbourhood, as it is in England. The Tea country is at a great distance from hence, and the Teas brought to Canton are several months on their route by inland navigation. Of the plants there are two kinds, of which one has a leaf of a much darker green than the other. This difference may partly arise from cultivation: but it is to the various modes of preparation, that the "green" and the "black Teas", as they are called in England, of the shops are due. In proof of this, we sent home, last year, green Tea from the black Tea plant. You may, therefore, conclude that, though there are two plants, differing as much in appearance and growth as any two varieties of the Camellia japonica, each, by proper management, will produce black or green Tea, indifferently. The varieties of Teas from the several provinces, arise from soil, culture, mode of preparation, and above all, from the part of the shrub whence the leaves are pulled. From the same individual plant, indeed, there are three crops or gatherings annually; the first affords the finer Teas, of which the Pour chong is the produce of the larger leaves of the young shoots. The extreme shoots, with the opening leafbuds constitute the Peko. This is in England commonly supposed to be the flowers: but an examination after infusion will clearly show its origin. The first picking takes place in June, the second in July, and the third in August'".

Prof. W. J. Hooker notices two Chinese plants introduced by Millett in 1830 and 1831 into the Bot. Gardens, Glasgow, viz:

Bot. Mag. t. 2997, 2998, *Renanthera coccinea,* Lour. and Bot. Mag. tab. 3043, *Urena lobata,* Linn.

Millett's dried specimens, like those collected by Vachell, were chiefly worked up by Hooker & Arnott in Bot. Beech. (v. supra p. 289). Hooker described the ferns in his Gen. et Spec. Filicum. Nees v. Es. determined the Glumaceae.

New plants discovered by Millett:

Kadsura chinensis, Hance in Benth. Fl. Hongk. 9. — I. F. S., I, 25. Earlier Millett.

Cleyera Millettii, H. & A. Bot. Beech. 171, tab. 33, fig. 1. Hooker states: "Millett's collection contains a very fine plant. We think the species well deserving of bearing the name of its discoverer, who has rendered so much service to Botany during his long residence in China". — Benth. & Hook. Gen. Plant. I, 183, *Adinandra Millettii.* — I. F. S., 1, 76.

Connarus Roxburghii, H. & A. l. c. 179. Millett, Macao. — Planchon, Prodr. Connar. in Linnaea XXIII (1850), 420, named it *Rourea Millettii.*— I. F. S., 1, 150.

Tephrosia oraria, Hance, Journ. Bot. 1886, 17. Ford. — Ind. Fl. Sin., 1, 158, 489. Earlier, Millett.

Millettia nitida, Benth. in Lond. Journ. Bot. I (1842), 484. Fine flowering specimen from Millett in herb. Hooker. — I. F. S., 1, 159.

Millettia speciosa, Champ., Kew Journ. Bot. IV (1852), 73. — I. F. S., 1, 159. Millett.

Lespedeza chinensis, G. Don, Gen. Syst. II (1831), 307. China, Japan. — I. F. S., 1, 180. Millett.

Dalbergia Millettii, Benth. Syn. Dalberg, in Journ. Linn. Soc. IV, Suppl. (1860), 34. Millett.

Dahlbergia monosperma, Dalzell, Bot. Ind. (1850), 36, Benth. l. c. 48, Millett. — I. F. S., 1, 198.

Caesalpinia Millettii, H. & A. Bot. Beech. 182. Millett, also Lay. — I. F. S., 1, 205.

Pterolobium subvestitum, Hance, Journ. Bot. 1884, 365. Faber. — I. F. S., 1, 207. Earlier Millett.

Bauhinia Championi, Benth. Fl. Hongk. 99, Champion. — I. F. S., 1, 212. Earlier Millett.

Albizzia Millettii, Bentham in Lond. Journ. Bot. III (1844), Notes on Mimos., 89. Millett. — I. F. S., 1, 216. Earlier Loureiro, Staunton.

Syzygium odoratum, H. & A. Bot. Beech. 187. — I. F. S., 1, 297, *Eugenia Millettiana,* Hemsley. Millett, Vachell. Earlier Loureiro, Staunton.

Eugenia minutiflora, Hance, Journ. Bot. 1871, 5. Sampson & Hance. — I. F. S., 1, 297. Earlier Millett.

Hedyotis uncinella, H. & A. l. c. 192. Millett. — I. F. S., 1, 376.

Melodinus suaveolens, Champ., Kew Journ. Bot. IV (1852), 333. Champion, Hance. — I. F. S., II, 94. Earlier Millett.

Pottsia cantonensis, H. & A. Bot. Beech. 198. (V. supra p. 271).

Gymnema sylvestre, R. Brown, 1808. Bot. Beech. 200. Bentham in Kew Journ. Bot. V (1853), 54 — *G. sylvestre,* var. *chinense.* Fl. Hgk. 227 — *G. affine,* Decaisne (v. supra p. 247, Wallich). — I. F. S., II, 112. Millett.

Cuscuta Millettii, H. & A. Bot. Beech. 201. Grammica aphylla, Lour.? Millett. — I. F. S., ɪɪ, 168. According to Engelmann = the widely spread *C. obtusiflora,* H. B. K.

Buchnera densiflora, H. & A. l. c. 203. Millett. — I. F. S., ɪɪ, 201, reduced to *B. cruciata,* Hamilton, in D. Don, Prodr. Fl. Nepal. (1825), 91.

Callicarpa integerrima, Champ. K. J. B. V (1853), 135. Champion. Fl. Hgk. 270. — I. F. S., ɪɪ, 253. Millett.

Viburnum chinense, Bot. Beechey, 190, note. Millett. DC. Prodr. XI, 632 = *Premna serratifolia,* Linn. — I. F. S., ɪɪ, 256 = *P. integrifolia,* Lin.

Clerodendron castaneifolium, H. & A. Bot. Beech. 205. Millett. — I. F. S., ɪɪ, 261 = *Cl. fortunatum,* Linn.

Piper chinense, Miquel in Lond. J. B. IV (1845), 439. Millett, herb. Hooker. — I. F. S., ɪɪ, 365.

Glochidion molle, H. & A. Bot. Beech. 210. Millett. — Müller, Arg. in DC. Prodr. XV, ɪɪ, 279, named it *Phyllanthus Arnottianus.* — I. F. S., ɪɪ, 424, *Glochidion Arnottianum.*

Ficus pyriformis, H. & A. l. c. 216. — This is *Ficus Millettii,* Miquel in Lond. J. B. VII (1848), 438. — Comp. Fl. Hongk. 328.

Urtica Millettii, H. & A. l. c. 214, Millett. — DC. Prod. XVI, ɪ, 63, considered a dubious plant.

Cryptomeria japonica, Don (1841), described from Japanese specimens. W. J. Hooker, Icon. Plant. VII, 1844, tab. 668, notices that Millett first sent specimens from China (Macao).

Habenaria Linguella, Lindl. Gen. et Spec. Orchid. (1834), 325. Herb. Hooker. Millett.

Pontederia ovata, H. & A. l. c. 218. Millett. — Not mentioned in Solms' Monogr. Ponteder. 1883.

Perotis longiflora, Nees v. Esenb. in Bot. Beech. 247. Millett. Herb. Lindley. — Fl. Hgk. 418 = variety of *P. latifolia,* Ait.

Eragrostis Millettii, Bot. Beech. 252. Millett, Vachell, Meyen.

Bambusa verticillata, Willd. Bot. Beechey, 254. Millett. — Munro, Monogr. Bamb. (1868), 105 = new species: *B. Beecheyana.* — Earlier Staunton.

Bambusa tuldoides, Munro, l. c. 93. Millett, Canton. Bentham in Fl. Hongk. 434, *B. Tulda,* Roxb.

Lindsaya variabilis, H. & A. l. c. 257, tab. 52, Millett. — Fl. Hgk. 446 = *L. heterophylla,* Dryander (1794).

Meyen, Franz Julius Ferdinand, born at Tilsit in 1804. M. D. 1826.
In the course of a voyage of circumnavigation undertaken by him,
1830—32, he visited South China, where he collected plants. After his
return to Europe he published a work in two volumes: "Reise um die Erde,
ausgeführt auf dem Königl. Preussischen Seehandelsschiff "Prinzess Louise",
Capt. W. Wendt, in den Jahren 1830—32, von Dr. F. J. F. Meyen 1834,
1835. From his narrative we learn, that the "Princess Louise", a ship
belonging to the Prussian Govt. was despatched to South-America and
China for the purpose of enquiring about the trade of these countries.
Meyen accompanied the expedition in the capacity of physician and naturalist. The "P. L." set sail from Hamburg on the 7th Sept. 1830, called
at Rio Janeiro, doubled Cape Horn, arrived at Valparaiso, visited the ports
of Coquimbo and Copiapo in Chile, and Arica in Peru. Then they sailed to
China across the Pacific, visiting Honolulu (Sandwich Isl.) on their way.

On August 14th 1831 the "P. L." reached the estuary of the Canton
River and anchored in the Port of Cap Syng moon near the north point of
Lantao Isl. *).

Meyen botanized in Lantao, visited Macao, where he experienced
a kind reception from Mr. Lindsay, Secretary General of the E. I. Comp.,
and the Rev. Mr. Vachell. Meyen notices the splendid gardens of Mr. Beale
(see further on) at Macao, in which beautiful plants were cultivated and
rare Chinese beasts and birds kept. On September 2d the "P. L." left the
Chinese coast and sailed for Manilla, from which excursion she returned on
Nov. 11th. The ship anchored at the island of Lin tin, N. W. of Lantao
and nearer to Canton, and there she remained for about four weeks. Meyen
went to Macao, visited Canton and its environs. He saw the celebrated Fa
ti or Chinese flower gardens there. On Dec. 12th 1831, the "P. L." left
the waters of China for the homeward voyage. St. Helena 13 Febr. 1832,
Cuxhaven, near Hamburg 19 April 1832.

In 1835 Meyen published, in the Verhandl. d. Vereins z. Beförd. d.
Gartenbaues in Preussen, XI, 258, an interesting article (in German) on a
PECULIAR CHINESE COTTON PLANT, from which the cotton cloth known in
Europe under the name of *Nanking,* is derived, with a figure of the plant.
Meyen saw living specimens of this interesting plant, which yields a yellow
wool in Mr. Beale's garden in Macao. He brought seeds of it with him
to Berlin, where it was successfully cultivated, and produced flowers and

---

*) Cap Syng moon is not a promontory as Nees v. Es., Vogel, and other botanists be-
lieved. The name is Chinese, more correctly Kap suy moon (swift-water passage). This passage
is formed between the north point of Lantao and the mainlands NW. of Hongkong.

ripened capsules containing the characteristic yellow wool. Meyen considered it to be a distinct species which he named *Gossypium Nanking*.

After his return to Europe, Meyen was appointed Professor of Botany at Berlin and set himself about to work up his botanical materials. He died on Sept. 2d 1840.

The botanical collections made by Meyen in S. America, Honolulu, Manilla and China, about 1350 species, are now preserved in the Museum of the Bot. Garden, Berlin. Only a small part of these plants were described by Meyen himself. The bulk was worked up by a number of German botanists and published in a separate volume of the Nova Acta Acad. Caes. Leop. Carol. Naturae curiosorum, XVI, Supplem., with the title: F. J. F. Meyen, OBERVATIONES BOTANICAE IN ITINERE CIRCUM TERRAM INSTITUTAE, 1843. Opus posthumus, 495 pages *). 244 Chinese plants noticed.

Nees v. Esenb. named the acanthaceous genus *Meyenia* in honour of Meyen. Wallich, Pl. As. rar. III (1832), 78. Walpers, Nees, Schauer dedicated several new species names to his memory.

New Chinese plants discovered by Meyen:

Clematis Meyeniana, Walp. Observ. 297. Meyen, Cap Syng moon. — I. F. S., i, 5: also Vachell.

Cissampelos hypoglauca, Schauer, Observ. 479. Cap Syng moon. — Omitted in Ind. Fl. Sin.

Sida fallax, Walp. l. c. 306. Meyen. — I. F. S., i, 85. Earlier Bot. Beechey, 79, Sandwich Isls, under the erroneous name of *Sida rotundifolia*, Cav.

Brucea amarissima, Walp. l. c. 322. Cap Syng moon. — Not mentioned, either in Planchon's Simarubeae, Lond. J. B. V (1846) or in Ind. Fl. Sin. Perhaps *Br. sumatrana*, Roxb. See I. F. S., i, 112.

Cissus diversifolia, Walp. l. c. 314. Cap Syng moon. — Fl. Hongk. 54 = *Cissus cantoniensis*, Bot. Beech. 175. (Supra p. 291).

Crotalaria leiocarpos, Vogel, Obs. 8. Promont. Syng moon. — I. F. S., i, 151.

---

*) G. Walpers described the greatest part of the collection, 38 orders.

J. C. Schauer: Myrtaceae, Apocyneae, Asclepiadeae, Bignoniaceae, part of Amarantaceae, Orchideae, Irideae, Commelinaceae, Eriocaulaceae.

C. G. Nees v. Esenb.: Compositae, Solaneae, Acanthaceae, Cyperaceae, Gramineae.

Th. Vogel: Leguminosae, Scrophularineae (partim).

H. Grisebach: Gentianeae.

I. F. Klotzsch: Euphorbiaceae.

I. G. Goldmann: Filices.

J. Meyen & Jul. de Flotow: Lichenes.

F. A. G. Miquel, K. M. Gottsche, J. B. Lindenberg: Hepaticae, etc.

Crotalaria splendens, Vogel, Obs. 8. China, Meyen. — I. F. S., ɪ, 151 = *C. elliptica,* Roxb. Fl. Ind. = *C. Vachellii,* Bot. Beech. 180.

Indigofera chinensis, Vogel, Obs. 14. Macao. — I. F. S., ɪ, 156.

Tephrosia vestita, Vogel, Obs. 15. Promont. Syng moon. — I. F. S., ɪ, 158. Also Vachell.

Marquartia tomentosa, Vogel, Obs. 35, tab. 1, 2. — Bentham, Lond. J. B. I (1842), 484, determined the same plant, gathered by Millett, earlier and named it *Millettia nitida* (supra p. 300). — Fl. Hgk. 78. — I. F. S., ɪ, 159.

Wistaria dubia, Walp. Obs. 324. China. — I. F. S., ɪ, 162.

Desmodium formosum, Vogel, Obs. 29. Macao. — I. F. S., ɪ, 172.

Desmodium viride, Vogel, Obs. 29. Macao. — I. F. S., ɪ, 177.

Eriosema chinense, Vogel, Obs. 31. China. — I. F. S., ɪ, 197. Earlier Parks. (Supra p. 274).

Bauhinia chinensis, Vogel, Obs. 42. Canton, culta. — I. F. S., ɪ, 213 = variety of *B. variegata,* Linn.

Syllysium buxifolium, Meyen & Schauer, Obs. 334. China. — I. F. S., ɪ, 298. Same as *Syzygium buxifolium,* Bot. Beech. 187 = *Eugenia sinensis,* Hemsl. (supra p. 292).

Ferula marathrophylla, Walp. Obs. 347. Cap Syng moon. — I. F. S., ɪ, 335.

Vernonia Gomphrena, Walp. Obs. 253. Lintin. — I. F. S., ɪ, 402.

Vernonia eriosematoides, Walp. Obs. 254. Cap Syng moon. — I. F. S., ɪ, 429 = *Inula Cappa,* DC. Prodr. V, 469, known long before from India.

Aster panduratus, Nees v. Es. Obs. 258. Lintin. — I. F. S., ɪ, 415.

Aster Walpersianus, Nees v. Es. Obs. 259. Lintin. — I. F. S., ɪ, 417.

Eurybia rhodotricha, Nees v. Es. Obs. 259. Lin tin. — This name is not found elsewhere, not even in the Kew Index.

Conyza syringaefolia, Meyen & Walp. Obs. 263. Cap Syng moon. — I. F. S., ɪ, 420.

Bidens Meyeniana, Walp. Obs. 271. Cap Syng moon. — I. F. S. ɪ, 435.

Schistocodon Meyenii (new genus), Schauer, Obs. 363. — Promont. Syng moon. — Fl. Hgk. 224 = *Toxocarpus Wightianus,* Bot. Beech. 200 (supra p. 296). — I. F. S., ɪɪ, 101.

Mitrasachme chinensis, Griseb. Obs. 51. Cap Syng moon. — Fl. Hongk. 230 = *M. nudicaulis,* Reinw. in Blume Bijdr. (1825), 849. — I. F. S., ɪɪ, 117.

Vandellia limosa, Walp. Obs. 394. Lintin. — I. F. S., ɪɪ, 190.

Scoparia gypsophiloides, Walp. Obs. 394. — I. F. S., ɪɪ, 193.

Polygonum aviculare, Lin., noticed by Walp. Obs. 408, Cap Syng moon, is, according to Koch, Linnaea XXII (1849), 205, a new species which he names *Polygonum Meyeni.* Hemsl. in I. F. S., ɪɪ, 347, reduces it to *P. plebejens,* R. Brown, Prod. Fl. N. Holl. (1810), 420.

Cymbidium Meyeni, Schauer, Obs. 433, Macao. — Reichenb. fil., in Linnaea XXV (1852) 227, named it *Arundina Meyeni.*

Cymbidium micans, Schauer, Obs. 433. Macao. — According to Reichenb. fil., Hong kong Orchids, in Bonplandia, 1855, p. 250 = *C. ensifolium,* Swartz.

Chaeradoplectron Spiranthes, Schauer, Obs. 436, t. 13, C. Promont. Syng moon. — According to Benth. & Hooker Gen. Pl. III, 626 = *Coeloglossum lacertiferum,* Lindl., known before from India.

Centrochilus gracilis, Schauer, Obs. 435, t. 13. Prom. Syng moon. — According to Gen. Pl. III, 626 = *Habenaria tipuloides,* Lindl. *(Orchis tipuloides,* Linn.).

Commelina ochreata, Schauer, Obs. 447. Lin tin, Macao. — Fl. Hgk. 376 = *Commelyna salicifolia,* Roxb.

Cyperus Meyenii, Nees v. Es. Obs. 57. Meyen, Manilla, Vachell and Millett, China (supra p. 297).

Cyperus radians, Nees, Linnaea IX (1834), Gen. Cyp. 9, (only name). — Obs. 63. Meyen, Promont. Syng moon.

Haplostylis Meyeni, Nees v. Es. in Wight Contrib. Ind. (1834), 115; Bot. Beech. 227. Meyen, Vachell, Millett. — *Sphaeroschoenus Wallichii,* Nees, Obs. 97. Promont. Syng moon, Macao, Meyen. — Kunth Enum. II (1837), 289 = *Rhynchospora Wallichiana.* — Fl. Hgk. 396. — Hooker Fl. Ind. VI 668. Wallich, Cat. 3422. Known earlier from India.

Panicum heteranthum, Nees, Obs. 174. China. — Fl. Hgk. 410 = *P. barbatum,* Kth Enum. I, 84. Known earlier from India.

Miquelia barbulata, Nees, Obs. 178. Prom. Syng moon. — Fl. Hgk. 417, reduced with? to *Garnotia stricta,* Brongniard in Duperrey's Voy. (1822—25). Also in Ceylon.

Thysanolaena (nov. gen.) *acarifera,* Nees, Obs. 181. Promont. Syng moon. — Fl. Hongk. 417. Known earlier from India, = *Melica latifolia,* Roxb.

Spodiopogon obliquivalvis, Nees, Obs. 185. Macao, Promont. Syng moon. — Fl. Hgk. 426. Common in India.

Meoschium Meyenianum, Nees, Obs. 197. Cap Syng moon. — Fl.

Hgk. 426 = *Ischaemum barbatum*, Retz, and same as *M. lodiculare*, Nees, (supra p. 298). — Hackel Andropog. l. c. 205 = variety of *Ischaemum aristatum*, Lin.

Eragrostis geniculata, Nees, and Meyen, Obs. 203. Cap Syng moon. — Fl. Hongk. 433. Not known out of S. China.

Lycopodium amentigerum, Goldmann, l. c. 468. China. — Fl. Hongk. 436 = *L. cernuum*, Lin.

Pteris ensiformis, Goldmann, Obs. 457. Meyen, China? — An Pteris ensiformis Burm. in Hook. & Bak. Syn. Fil. 155?

Ramalina digitata, Meyen & Flotow, Obs. 212. Canton, ad ramos Theae chinensis.

Fusarium Caries, Nees, Obs. 478. In spicis Meoschii lodicularis. Cap Lin tin.

Beale, Thomas, a rich English merchant long resident in Macao, who took a lively interest in Chinese ornamental plants and used to send them to England. According to the Chinese Repository, XI, 59, 60, he was born about 1775, came to China in 1792 and died in Macao, Dec. 1842. He had in his garden one of the richest collections of Chinese flowers, 2500 pots, in the cultivation of which he spent much time.

The late Dr. Wells Williams, in his "Recollections of China", in Journ. N. China Br. Asiat. Soc., VIII, 8, records that Beale had collected, at his residence in Macao, a fine garden and splendid aviary, which was deservedly a great celebrity. When Dr. W. first saw it (in 1833) there were 200 birds in it, about 20 of them being large and magnificent pheasants. Mr. Beale was the first to send to England the "Reeves" Pheasant, which he had procured from the interior at great expense. He had also a number of "Medallion Pheasants" and several other rare kinds, which he was the first to collect. From the interior of the aviary rose two large Longan trees *(Nephelium Longan)*, among the branches of which the birds might disport themselves, while in the centre of it was a pond where the various kinds of ducks could indulge in their specific propensities. It was altogether a most interesting collection. — Comp. also Fortune's Tea Countries, 6.

From a notice by Lindley, Bot. Reg. tab. 1501 (1832), sub Camellia japonica, var. Reevesiana, we may conclude, that Beale used to send living plants from China to England. He was a friend of John Reeves, and like this gentleman one of the Company's chief officers. (See Yule's article on Indian Tea cultivation in Journ. Soc. of Arts, 1877, 207). In a letter by J. Reeves, dated Clapham, June 3 d, 1835, and published in Loudon's Gard.

Magaz. XI, 1835, 437, Reeves states that Beale in 1830 had in his garden at Macao a tree of *Magnolia grandiflora*, 20 feet high. When Reeves quitted China, he left to Beale a copy of his beautiful drawings of Chinese Fishes. (V. supra p. 257).

Dorwand. In the Ind. Fl. Sin., ɪ, 3, sub Clematis chinensis, Dorwand is mentioned as having collected the plant in China. On my enquiry, Mr. Hemsley kindly informed me, that in Wight's herbarium, there are a few plants labelled: Dorwand, China. Dr. R. Wight was in India, 1819—1853.

Gaudichaud Beaupré, Charles, a distinguished French botanist, 1789— 1864, Professor of Pharmacy, from 1810 in the service of the French Navy. In 1837 Member of the French Academy, in the same year Foreign Member of Linnean Soc. G. took part, in the capacity of naturalist, in two French circumnavigations of the globe: the first, 1817—1820, on board the corvette "L'Uranie", commanded by Capt. Freycinet. In Dec. 1835 he embarked at Toulon for the second circumnavigation, on board the corvette "La Bonite", Capt. Vaillant, despatched by the French Government for the purpose of conveying several consuls and consular agents to their respective posts in various parts of the globe. The "Bonite", having successively visited Rio Janeiro, Montevideo, Patagonia, Valparaiso, the Galopagos Isls, the Sandwich Isls, Manilla, arrived on Jan. 1st 1837 at the roads of Macao, and, 1837 after a stay of several weeks there, set sail for the homeward voyage: Touron in Cochinchina, Singapore, Pulo Penang, Malacca, Calcutta, Pondichery, Isle de France, St. Helena. Reached Brest, Dec. 1837. — See Gaudichaud, BOTANIQUE DU VOYAGE AUTOUR DU MONDE, EXÉCUTÉ EN 1836 ET 1837 SUR LA CORVETTE "LA BONITE". 5 vols, 156 planches 1844—1866.

During the stay of the "Bonite" at Macao, G. amassed considerable botanical materials. He was assisted in his explorations by M. Callery, a distinguished sinologist and able botanist, then resident at Macao, who accompanied the French botanist to search for plants in the adjacent islands. G. visited also Canton.

The botanical collections brought home by G. are kept in the Mus. d'Hist. nat., Paris. Mr. A. Franchet kindly informed me that there are among them about 500 species from Macao. Gaudichaud's name appears frequently in DC. Prodr. in connection with Chinese plants. A set of his plants exists also at Kew. In the Ind. Fl. Sin. he is frequently quoted.

New Chinese plants discovered by Gaudichaud:

Maesa sinensis, DC. Prod. VIII, 82. China merid. Gaudichaud, also Callery. — I. F. S., ii, 60.

Solanum Macaonense, DC. Prod. XIII, i, 265, Gaudichaud, Macao; also Callery. — I. F. S., ii, 172 = *S. torvum*, Swartz, Prod. Fl. Ind. Occid. (1788) 47.

Smilax Gaudichaudiana, Kunth, Enum. V (1850) 252, Gaudichaud, S. China. — Fl. Hongk. 370.

Polypodium chinense, Mettenius, in Kuhn, Adnot. Fil. Chinae, in Journ. Bot. 1868, 270. Gaudichaud, Canton. — Not found in Hook. & Bak. Syn. Fil.

Geoffroy, Captain of a French vessel, is noticed in Vilmorin's "Bon Jardinier" for 1841, as having introd. from China *Sinapis brassicata*, Lin., the Chinese Cabbage *Paï tsaï (Brassica chinensis*, Lin.), *Basella rubra.*

Blume, Karl Ludwig, a German, born at Brunswick, 1796. He studied medicine, and took the degree of M. D. In 1817 he went to Java, entered the Dutch service as Director of the Board of Health. His inquiries into the native medicines led him to the study of botany. He made a large botanical collection and was subsequently (1822?) appointed Director of the Bot. Gardens, Buitenzorg. He returned to Europe in 1826 and was made Superintendent of the Government herbarium at Leyden, where he died in 1862.

In his Bijdragen tot de Flora of Nederlandsch Indie, Batavia, 1825—26, and in his Museum botanicum Lugdono-Batavum, Leyden, 1849—56, Blume described a number of Chinese plants, which he believed new, all from specimens cultivated in Java.

Magnolia parviflora, Bl. Bijdr. 9. In hortis Javae colitur. Verosimiliter c China allata. — I. F. S., i, 24. "Obscure plant".

Viola inconspicua, Bl. Bijdr. 58. Verosimiliter e China in Javam introducta. — Ind. Fl. Sin. i, 55. It inhabits the mountains of Java and not China.

Prunus (Cerasus) chinensis, Bl. Bijdr. 1104. China. — I. F. S., i, 219: *P. japonica*, Thbg.?

Mespilus spiralis and *M. sinensis*, Bl. Bijdr. 1102. E regno chinensi introductae. — I. F. S., i, 259: Both these species names are obscure.

Azalea mollis, Bl. Bijdr. 853. E China introducta in Javam. — Maximowicz Rhod. As. or. 1870, 28 = *Rhododendron sinense*, Sweet.

Azalea mucronata, Bl. l. c. In Javam e China introducta. — Maxim.
l. c. 35 = *Rhododendron ledifolium*, Don.

Ardisia quinquegona, Bl. Bijdr. 689. E China introducta. — I. F.
S., ii, 66 = *A. pentagona*, A. DC., Trans. Linn. Soc. XVII (1836) 124. —
Cultivated specimen of Blume's plant in herb. Kew.

Diospyros chinensis, Blume, Catal. Hort. Buitenz. 1823, 110; Bijdr.
670. Only name. — I. F. S., ii, 69 = *D. Kaki*, Lin.

Cinnamomum Cassia, Bl. Bijdr. 570. China. Colitur in Java. — This
is the tree which yields the true Chinese Cassia bark. It was first described
by Blume. — I. F. S., ii, 371.

Cinnamom. chinense, Bl. Bijdr. 569. In Javam e China introducta. —
Meissner, in DC. Prod. XV, i, 17 = variety of *C. Burmanni*, Bl. l. c.
According to Blume C. Burmanni wild in Java. It occurs also in a wild
state in S. China and yields also Cassia bark. — I. F. S., ii, 371.

Litsea chinensis, Bl. Bijdr. 565. E China allata. Same as *Iozoste chi-
nensis*, Bl. Mus. Bot. Lugd. Bat. I (1850) 364, where Bl., referring to
Bijdr., writes erroneously Laurus (for Litsea) chinensis. — I. F. S., ii, 379.
Common in Hongkong.

Daphnidium sinense, Bl. Mus. Bot. I (1850) 352. In Sina. — I. F. S.,
ii, 392. Hemsley calls it *Lindera sinensis*, Hemsl. and quotes only Blume.

Melanthesa chinensis, Bl. Bijdr. 592. In hort. bot. Bogor. culta et
forte e China introducta. — Fl. Hgk. 313. — I. F. S., ii, 427: *Breynia
fruticosa*, Hook. fil. Fl. Ind. V, 331 *(Andrachne fruticosa*, Lin.).

Castanea pumila, Bl. Bijdr. 525. In hortis Javae; e Sina introducta.
In Mus. Bot. I (1850) 286, Blume calls it *C. mollissima*. According to Ind.
Kew., Blume's C. pumila is = *Quercus ferox*, Roxb., *Castanea tribuloides*,
Lindl. in Wall. Pl. As. rar., *Castanopsis tribuloides*, ADC.

Arundina chinensis, Bl. Bijdr. 402. Java, forte e China introducta. —
Lindl. Gen. et Spec. Orchid. 125. — Fl. Hgk. 355.

XIII. First Russian Botanical Explorations in Eastern Asia.

During the first half of the 17th century the conquest of the whole of
Siberia from the Ural Mountains to the Sea of Okhotsk was accomplished
for Russia by the warlike and enterprising Cossacks. Cities and fortresses
were built in the newly-acquired vast country. Russia now became for long
stretches of country the boundary of the Chinese Empire towards the north,

although the frontier line for long distances was undefined owing to the large tracts of sand and mountains separating the two Empires.

From 1832 to 1756 the Russians established a fortified line against the Kirghiz or Kirghiz-Kasaks, the nomadic inhabitants of the so-called Kirghiz Steppes extending from the southern Volga, the. Caspian and Aral Lake to the Irtish and Tarbagatai. This line of fortresses and fortifications at first followed the Ural River upwards (Uralsk, Orenburg, Orsk, Verkhne-uralsk), then turned eastward (Troitsk, Zverinogolovskaya on the Tobol, Petropavlovsk, Omsk). From Omsk it continued south-eastward along the Irtish River to Semipalatinsk and Ust-kamenogorsk. These Kirghiz, some hordes of whom were tributary to the Chinese, gradually surrendered to Russia. Farther east the Siberian boundary was formed by the Chinese territories situated north of the T'ien shan Mountains and termed T'ien shan pe lu by the Chinese, with the Lakes Balkash, Ala kul, Zaisan and Issik kul and the Ili River. They were occupied in the first half of the 18th century by the Eleuths, a Mongol race. These were for a long time at war with China and finally succumbed to the Emperor K'ien lung, in 1759. Nearly all of them were exterminated. After having re-established their authority in the T'ien shan pe lu, the Chinese divided it into three commanderies: Ili, on the west, Tarbagatai on the north and Kurkara usu in the east. The government was under the control of Manchu officers residing at Ili (Kuldja, founded 1764 on the Ili River). The inhabitants of this country are composed of the remnants of the Eleuths, for a great part of Kirghiz; Chinese are in a minority. As the Eleuths called themselves Dsungar, our geographers are accustomed to term these regions Dsungaria, as far east as the Altai. — From Tarbagatai to the junction of the Argun and Shilka, which when united form the Amur, the old frontier was the same as shown on our modern maps and bordered Mongolia and Manchuria. The country to the south-east of Lake Baikal, inhabited by the Buriat-Mongols had already in 1644 been annexed by Russia, and in it, in 1666, the fortress of Selenginsk was built. Farther east the Tungus tribes, called Daur, had likewise surrendered. The city of Nerchinsk had been founded in their territory. The country of the Buriats and Dauria now from the province of Transbaikalia. Already in 1644 some Russian adventurers had ventured to sail down the Shilka and the Amur, and in 1651 a Russian colony was founded on the latter river. But their fortified city there, which they called Albazin, existed only 37 years. In 1688 the Russian settlers, after a long siege by the Chinese army, were driven out and partly made prisoners. Having crossed the Amur at its formation, the old undefined boundary stretched northward and then turned eastward, following the Yablonoi Mountain Range to the Sea of Okhotsk.

Messerschmidt, Daniel Gottlieb, a German, was the first naturalist who explored the natural riches of Siberia. According to Pallas, Nord. Beitr., II (1781) 98, he was born in 1685 and studied at Halle, where he received the degree of Doctor in 1707. H. G. Bongard, "Esquisse historique des trav. bot. en Russie" (1834), reports that Peter the Great, when paying a visit, in 1716, to the botanist Breynius in Danzig requested him to recommend an able naturalist who could explore the natural productions of Russia. Breynius proposed Messerschmidt, who in 1719 went to St. Petersburg and from 1720 to 1727 travelled in Siberia, especially in its eastern part and made vast collections. The latter, however, do not exist now, but the botanical specimens have been worked up by Amman, Gmelin and Pallas. M. left an Account of his travels and collections in MS, 4 vols, which is preserved in the Library of the Acad. of Sc., St. Petersb. From Russian Dauria, he made, in 1724, an expedition into the steppes of Chinese Mongolia, visited the great lake Dalai nor, and collected plants and other natural objects there. It was in these regions that he observed a new wild equine animal, which he called: Mulus dauricus foecundus, subsequently described by Pallas as *Equus hemionus,* and here he discovered the plant which Linnaeus named *Messerschmidia Arguzia (Tournefortia Arg.,* Roem. & Sch.). — After returning to Moscow, M. married and died in 1730. His widow married G. W. Steller (1709—1746), Member of the Russian Academy.

The exploration of Siberia was continued in the last century by several distinguished German naturalists in the Russian service, viz:

Gmelin, Joh. Georg, born in 1709 in Tubingen, came to Russia in 1728, became a Member of the Academy, travelled in Siberia from 1733—1743, then returned to Tubingen, where he was appointed Professor of Botany, and died in 1755. — From 1747—69 his FLORA SIBIRICA was published in 4 vols.

Pallas, Peter Simon, born in 1741 at Berlin, studied natural sciences. In 1767 he was called by the Empress Catherine II, to be a Member of the Academy and Professor of Natural Sciences at St. Petersburg. In the next year he commenced his extensive travels of exploration in the Russian Empire, which lasted till 1774. From 1771—73 he was in Siberia. He finally spent 15 years at Simferopol, and then returned to Berlin, where he died in 1811. — Pallas, REISE DURCH VERSCHIEDENE PROVINZEN DES RUSSISCHEN REICHES, in 3 vols., 1771—1776.

In Siberia Pallas was for a time assisted in his explorations by J. G. **Georgi,** Professor at the Russian Academy and J. P. **Falck,** Professor of Botany at the Apothecaries' (now Botanical) Garden, St. Petersb.

None of these travellers extended their explorations beyond the Siberian frontier. But Gmelin procured some interesting plants from China, which he received through the medium of a physician, named **Heuke** or **Heucke,** who, as he relates in his Flora Sibirica, brought them from Peking.

Flora Sib. IV, 68, n. 90, Fumaria caule recto ramoso duobus petalis hemicyclicis supra caudatis. Stupendae pulchritudinis planta, quae e Sinis a chirurgo Heuke adportata fuit etc. — Heuke brought a living plant of *Dicentra spectabilis,* Miq. V. supra p. 66.

Fl. Sib. III, 172, n. 3, sub Prunus inermis floribus sessilibus etc.: Ramum hujus fruticis in campis apricis sinensibus, per quos ex Sibiria per Mongolarum regiones ad Sinas itur, mihi attulit chirurgus Heucke, qui comitatui sinico interfuit. — Schlechtendal, in Abh. Halle, II (1854) 22, described this plant from Gmelin's specimen under the name of *Amygdalus Heuckeana.* See Maxim., Mél. XI (1883) 671.

Fl. Sib. I, 182, n. 31, Thuya strobilis laevibus, squamis obtusis, Lin. hort. Cliff, 449. A chirurgo allata est, qui comitatui mercium causa ad Sinas proficiscenti interfuit. Dicit se in rupibus hinc inde legisse, quae occurrerunt a Kiachta Pekinum versus iter facienti. — Gmelin, as appears from the reference to Linnaeus, took it to be Thuja occidentalis, but Ledebour, Fl. Ross. III, 680, says that Gmelin's specimen is *Thuja orientalis,* a common tree in North China.

Brassica violacea, was raised by Linnaeus, and *Hyssopus Lophanthus,* by Haller in Goettingen, from Chinese seeds sent by Gmelin. V. supra p. 66, 93.

Gmel. Fl. Sib. III, 254, n. 11, *Lepidium* foliis inferioribus lanceolato-serratis, superioribus linearibus integerrimis, floribus diandris tetrapetalis etc. Chirurgus, qui comitatui sinico interfuit, in Krasnagora, intra murum sinicum collegit. (Krasnagora in Russian = Red mountain).

Martini, A. Ph. who accompanied Gmelin in Siberia, from 1740, brought home specimens of a curious fruit, which he distributed among botanists. Gaertner, de Fruct. et Sem. II (1791) 291, tab. 142, described and figured the fruit and named the unknown plant *Pugionium cornutum.* J. Mayer, to whom Martini had given specimens, states, in Abh. d. Böhm. Ges. 1786, 246, that they had been gathered by the physician Heucke in Mongolia. Comp. Ledebour, Pugionium cornutum, in Abh. Münch. Acad.

IV, 1846, Abth. 3, and Maximowicz, Mél. biol. X (1880) 575. — It is well known that Prczewalski, in 1871 brought complete specimens of the plant from the Ordos.

Gmelin, Fl. Sib. III, 265, n. 27: Eruca foliis subtriangularibus, ex sinuato-dentatis glabris. Detailed description; figured tab. 61. "E semine sinico in horto academico enata est. An Brassica violacea Linnaei?" — Ledebour, l. c. identifies it with Bunge's *Orychophragmus sonchifolius* of Peking.

Sievers, Johannes, a Germann apothecary and botanist, who accompanied from 1890—95, an expedition sent out by Imperial order to explore the mountains on the southern boundary of Siberia from the Ural to Dauria, and search for good medicinal Rhubarb and for localities where true Chinese Rhubarb could be successfully cultivated. The letters he wrote on his travels, were published by Pallas in his Neueste Nord. Beitr., III (1796) 143—370. In 1794 Sievers botanized on Chinese territory. He was the first botanist to visit the Tarbagatai Mountains. He passed near Chuguchak, the capital of Tarbagatai, advanced towards the Alakul Lake, but did not reach it, saw the Irtish River near where it flows out from the Zaisan Lake. Pallas, in Nov. Act. Acad. sc. Petrop. X, 1797, 369—381, published: "Plantae Novae ex herbario et schedis defuncti botanici J. Sievers praesertim in Songaria, inter fl. Irtish et lacum Alakul, vel etiam in aliis Sibiriae regionibus lectae describuntur et in tabulis repraesentantur". 10 species. — *Pyrus Sieversiana,* Ledeb. Fl. Altaica, II, 222. Sievers, ad fl. Uldjar ad radicem m. Tarbagatai. Sievers also first discovered the *Abies Schrenckiana.*

ADMIRAL KRUSENSTERN'S CIRCUMNAVIGATION OF THE GLOBE.

In 1803 Emperor Alexander I of Russia gave orders to fit out two ships, the "Nadedja" (Hope) and the "Neva", for a circumnavigation of the globe. **Krusenstern,** Admiral Adam Johann von, was appointed to the command of this expedition. He was born in 1770, in the Russian province of Esthonia, was educated in a government naval school, and entered the navy. Having been appointed by his government to serve in the English fleet, for several years, 1793—99, he visited America, India, China, and then returned to Russia.

The principal object of the voyage was to visit the Russian possessions on the N. W. coast of America and to open commercial relations with Japan.

But great attention was also paid to the cause of science. Three naturalists were appointed to join the expedition:

Langsdorff, Georg, Heinrich von, born in Hessen, 1774.

Tilesius von Titenau, Wilhelm, born in Thüringen, 1769.

Horner, Dr. Jean Gaspard, a Swiss, born at Zürich 1774. He conducted the astronomical part of the proceedings.

The "Nadejda" and the "Neva", with all the members of the expedition on board, took their departure from the roads of Kronstadt on Aug. 7th 1803. They visited the Canary Isls., Sta Catharina (Brazil), went around Cape Horn, and then directed their course to Petropavlovsk (Kamtchatka), touching on their way at Nukahiva (Marquesas Isls.).

After a short stay at Petropavlovsk, the Admiral with his two ships, on Aug. 30th 1804, sailed to Japan, kept along its eastern coast, passed van Diemens Strait and on October 5th reached Nagasaki, where the admiral remained till the 16th of April 1805. On his way back he sailed along the west coast of Japan, visited the Aniva Bay of Sakhalin and reached Petropavlovsk on June 4th.

Here Langsdorff quitted the expedition and performed alone a voyage to the Russian possessions on the N. W. coast of America, visited San Francisco, and returned to Petropavlovsk. On May 13, 1807, he embarked for Okhotsk, crossed Siberia from East to West and arrived at St. Petersburg, Febr. 16, 1808. He brought with him rich botanical collections from Japan, Kamtchatka, America and Siberia, which are now kept in the Museum of the Academy, St. Petersburg. In 1813 he published an account of his voyages and travels. In 1808 Langsdorff was elected Member of the Academy. Subsequently the Russian Government appointed him Consul General at Rio Janeiro, where he continued to collect plants for the Academy. L. retired from service 1831 and died in 1852, at Freiburg.

After this digression, let us return to the Admiral whom we left with his two ships at Petropavlovsk. The expedition quitted this place again on June 30th 1805.

They sailed between the Kurile Isls. to Sakhalin and the eastern coast of the island was surveyed up to the northern point. On Aug. 28th the ships returned to Petropavlovsk, and on Oct. 4th the expedition started for the homeward voyage. They first directed their course to China, and after a rather long passage got sight of Formosa, sailed through the channel between this island and the Bashee Island and on Nov. 20th 1805 the two ships cast anchor in the roads of Macao. At Macao Krusenstern and his officers were treated with the utmost kindness and liberality by Mr. J.

Drummond, then President of the E. I. Comp.'s Factory. Krusenstern knew him from his first visit to China (about 1796). The Russian Admiral, his officers, Tilesius and Horner, were all lodged in his magnificent house at Macao. In the adjoining beautiful garden was the grotto of Camoens. The Admiral was permitted by the Chinese authorities to sail up the Canton River with the Neva to Whampoa. It was not till the 9th Febr. 1806 that the Russian expedition left China. Passing by Gaspar Strait, Java, Christmas Isl., Cape of G. H., St. Helena, the "Neva" and the "Nadejda" reached Kronstadt Aug. 19, 1806.   1806

Krusenstern published an account of his voyage in German: REISE UM DIE WELT, 1803—6, in 3 volumes, 1810—12, with an Atlas. He died in 1846 at his estate in Esthonia.

Tilesius after his return from the circumnavigation was elected Member of the Russian Academy. He returned to Germany in 1817, died at Mülhausen, 1857.

Dr. Horner, likewise Member of the Academy, left St. Perersburg in 1808, died at Zürich 1834.

The naturalists of the expedition collected during the voyage natural objects, especially plants, in various parts of the world, also in Japan and China.

Dr. Horner directed his attention particularly to Marine Algae, and transmitted his collection to the celebrated British algologist Dawson Turner (1775—1858), who described and figured several of Horner's Algae in his "Fuci", or coloured figures and descriptions, etc. 1808—19.

Fucus microceratus, Turn. l. c. II (1809) 151. In mari Coreano, Dr. Horner.

Fucus Horneri, Turn. l. c. I (1808) 34. Straits of Corea. Dr. Horner.

Fucus Thunbergii, Roth. Turn, l. c. II, 159. In mari Coreano, Horner.

Horner's name appears frequently in G. v. Martens' "Tange, Preuss. Expedition nach Ost Asien 1866. Nord-Chines. u. Japan. Meer".

As to the botanical collections formed by Langsdorff and Tilesius, during the circumnavigation, the former had given his plants to Fischer, and Tilesius' plants are found in the herbarium of Ledebour (from 1811 to 51, Professor of Botany at Dorpat) and in that of Fischer. These are now in the possession of the Botan. Garden, St. Petersb. to which, after the death of these botanists, their botanical collections devolved.

Viola Langsdorfii, Fischer, in DC. Prod. I, 296, is from Unalashka, one of the Aleutian Islands.

Viola japonica, Langsdorff, in herb. Fischer, DC. Prod. I, 295, was discovered by Langsdorff near Nagasaki, in 1805.

Pedicularis Langsdorfii, Fischer, in Steven, Monogr. Pedic. 1823, from Unalashka.

Calamagrostis Langsdorfii, Trinius de Gram. unifl. 1824, 224. Sibiria.

Artemisia sachaliensis, Tilesius in herb. Fischer. DC. Prodr. VI, 96. It was discovered by Tilesius on the eastern coast of the island of Sakhalin in 1805 (v. supra p. 314). He thus was the earliest botanical collector in that island.

Artemisia Tilesii, Ledeb. in Mem. Acad. St. Petersb. V (1812) 568. Kamtchatka. — A variety of *A. vulgaris,* Linn. See Ledeb. Fl. Ross. II, 586.

Serrulata Tilesii, Ledeb. Mem. Acad. l. c. 562. Kamtchatka.

Sisymbrium Tilesii, Ledeb. l. c. 548. Kamtchatka.

Pedicularis villosa, Ledeb. in Spreng. Syst. II (1825) 780, and Ledeb. Fl. Ross. III, 289. Siberia, Tilesius. — I may observe that Tilesius never visited Siberia. — According to Maximowicz, Mél. X (1877) 115, no locality given by Tilesius.

Maximowicz, Rhododendr. (1870) 40, found in the herbarium of the Bot. Garden a Rhododendron gathered by Tilesius in 1806 in South China (Macao). It was the *Azalea obtusa,* first described by Lindley in 1846 from specimens sent by Fortune from Shanghai.

Tilesius also collected Marine Algae. Many of his Algae gathered at Nagasaki and in the Corean Strait are noticed by G. v. Martens in the above mentioned work. B. Seemann, "Botany of the Herald", Introduction, states, that Tilesius has done much towards advancing our knowledge of the Algae of the China Seas.

DR. FISCHER AND THE IMPERIAL BOTANICAL GARDENS, ST. PETERSBURG.

Fischer, Dr. Friedr. Ernst Ludwig von, born 1789, in Halberstadt, received the degree of Dr. from a German University, in 1804. He commenced his career in Russia as Superintendent of Count Razumovski's garden at Gorenki. Count Razumovski, Alexei Kirillovich (son of Kirill Grigoriewich R., 1728—1803, Field-Marshal, the able President, from 1746—98, of the Russian Academy) was, in the reign of Alexander I, first Curator of the Moscow University and finally Minister of Public Instruction. He was also President of the Soc. of Naturalists at Moscow. In 1798 he had laid out a magnificent garden with hothouses at Gorenki, near Moscow *).

---

*) There was in the second half of the last century at Moscow a splendid Botanic Gard., famed for its rare plants, founded in 1756 by Prokop Akinfievich Demidov. It was destroyed in 1812 when the Russians set the city on fire to force the French away.

It was first superintended by J. Redovski, a Russian botanist, and when the latter left for a scientific expedition to Eastern Siberia, in 1804, Fischer accepted the offer made to him to take Redovski's place.

In 1808 F. published the "Catalogue du Jardin des Plantes du comte Alexis de Razoumovsky à Gorenki près de Moscou". It is nothing but a bare list of names of plants; native countries not mentioned. A second edition was issued in 1812.

When count Razumovski died, in 1822, this splendid garden, in which besides interesting Siberian plants, many rare exotic plants were cultivated, was abandoned, and in the next year, Fischer was appointed Director of the Imperial Botanic Garden at St. Petersburg.

It may be not out of place to say here a few words regarding the history of this celebrated institution, which now possesses one of the richest collections of living exotic plants, from northern countries as well as from the tropics, and a most extensive and valuable herbarium. As far as Eastern Asiatic and Central Asiatic plants are concerned, this herbarium is, no doubt, the richest in Europe. The late Professor E. R. von Trautvetter, Director of the Bot. Garden, 1866—1875, † 1889, published in Act. hti Petropol. II, 1873, p. 149—259, a History of the Imp. Bot. Gardens, St. Petersburg (in Russian) from which the following particulars are extracted:

In 1714 Peter the Great ordered a garden to be laid out in one of the islands of the Neva River, for the purpose of cultivating medicinal plants. The new garden, therefore, was called the Apothecaries' Garden, and the island in which it was situated, now within the precinct of the city, still bears the name of Apothecaries' Island. This garden, although it occupied a vast area and although botanical celebrities superintended it and lectured there *), was unserviceable for scientific pursuits.

What was in St. Petersburg called the Botanic Garden, during the last and in the beginning of the present century was a garden belonging to the Academy of which the site has been changed several times. According to F. J. Ruprecht, "Beiträge zur Geschichte der Kais. Akademie", Mél. biol. V, 1865, 73, at the time of the botanist J. **Amman**, Member of the Acad. 1733—41, this Garden was situated on the island Vassili Ostrov, and now forms the garden of the Rom. Cath. Eccles. Academy, Vass. Ostr.,

---

*) **Buxbaum,** Johann Christ., 1727, Member of the Academy, † 1730.
Siegesbeck, Joh. Georg, 1742, Memb. Acad.
Falk, Joh. Peter, † 1774.
Rudolphi, Joh. Heinr., End of 18th cent.
Stephan, Christ. Friedr. † 1817 as Professor in Moscou.

1st Line. But towards the end of the last cent., as can be proved from documents, the Bot. Garden of the Imp. Academy, was situated in the southern part of the city, near the place where the Technological Institution now stands. The Academy sold this enclosure in 1815.

As to the Apothecaries' Garden, it is only since 1823 that it has acquired importance as a scientific institution. In this year the Emperor Alexander I ordered it to be placed upon a new footing. Considerable sums were assigned to acquire rare exotic plants for the conservatory and to form a herbarium, for the Apothec. Garden had no herbarium, all the botanical collections hitherto made by Russian botanists being kept in the Botan. Museum of the Academy. The name of the transformed Apothec. Garden was changed into that of Imperial Botanic Garden and Fischer was appointed its Director. A library was established.

In 1824 Fischer was sent to England, France and Germany to purchase living exotic plants. He returned with a magnificent collection of 2320 species.

Extensive collections of living plants for the Botan. Garden were made in Brazil by **Langsdorff**, Russian Consul General in that country (v. supra p. 314), and L. **Riedel,** the botanist commissioned by the Russian Government to explore Brazil. Between 1822—35, Riedel made immense collections of Brazilian specimens for the herbarium.

The old Apothecaries' Garden belonged to the Medico-chirurgical Academy. In 1830 the new Botan. Garden was placed under the control of the Ministry of the Imp. Court.

Fischer is justly regarded as the founder of the Botan. Garden at St. Petersburg. It is, however, to be regretted that he never kept a register of the numerous botanical collections sent to the Garden from various quarters, and that he used to incorporate the most interesting collections, especially those received from China, with his private herbarium *). Happily Fischer's herbarium was, after his death, purchased for the Garden.

In 1850 Fischer was compelled, owing to some irregularities dis-

---

*) It is now a rule that none of the botanists composing the Staff of the Imp. Bot. Gard., are allowed to possess a private herbarium. All plants sent to their address have to be delivered to the Garden's herbarium. — An inventory of the herbarium of the garden was first drawn up in Sept. 1856. This list gives very little information regarding the various collections. Generally only the name of the collector and the country are noticed, not the date. The original labels have disappeared. All the plants of these old collections have now been distributed in the general herbarium, arranged according to the system of Endlicher. The plants from Eastern Asia (Japan, China, Mongolia etc.) including Przewalski's and Potanin's collections, form a separate herbarium. The plants from Russian Turkestan are likewise kept apart.

covered in the accounts of the Garden, to relinquish the post of Director, which he had held for 27 years. He died at St. Petersburg, 5 June 1854.

Fischer's herbarium, about 60,000 species, many exotic plants, was bought from his widow, by Imperial order, for 1000 roubles, for the Botanic Garden. The labels are all in Fischer's handwriting. As to the Chinese specimens found in Fischer's herbarium, many of them are simply labelled China, without any information regarding the collector and the date. Some of these plants were evidently received in the first quarter of this century and most probably from Peking *). Some other Chinese plants described by Fischer or by other botanists, to whom Fischer had sent seeds, are not found in his herbarium.

Chinese plants first mentioned by Fischer:

Delphinium chinense, Fischer. Under this name Fischer distributed seeds before 1816. Plants raised in England. See Bot. Cab. 1818, t. 71.— DC. Syst. I (1818) 351: *D. grandiflorum,* Linn. var. *chinense,* Fischer in litt. — Bot. Reg. 1820, tab. 472.

Sida tiliaefolia, Fischer. See above p. 228, 249. Most probably received from Peking.

Phyllolobium chinénse, Fischer. Of this plant Fischer had sent a description and seeds, before 1818, to the Botanic Garden, Halle, where it was cultivated. K. Sprengel, Novi preventus hort. Halensis et Berolinensis (1818) n. 72, named it *Sphaerophysa chinensis.* There is a specimen of Ph. chinense in Fischer's herbarium, but this was brought in 1840 by Kirilov from Peking. The plant is common in the Peking mountains. — Bunge, Astragal. (1867) 1, proved that Fischer's Ph. chinense was identical

---

*) Since the beginning of the last century there has been in Peking a Russian Ecclesiastical Mission. As Russian caravans from Siberia not unfrequently visited the Chinese capital in former times, there is reason to believe that the missionaries occasionally sent the seeds of interesting plants to Russia, or they brought them themselves when they returned from China. It would seem that in this way *Xanthoceras sorbifolia,* a beautiful tree, peculiar to Peking, was introduced into the Crimea about 70 years ago. When in 1866 this tree was introduced from Peking into the Jardin des Plantes, by Father David, and flowered there a few years later, it was believed that it was a plant quite new to European gardens. But in 1876, Mr. Schoene, assistant gardener to the Bot. Garden at Nikita, situated on the southern coast of the Crimea, reported that at Karasson, 20 versts east of Nikita, there was growing a fine specimen of this tree, at least 50 years old, and another specimen of the same about 20 years old, was to be seen in the Nikita Garden. These trees there flower abundantly every year and produce ripe fruit. Having seen this interesting notice in the Gartenflora, 1876, 346, I wrote to Dr. Zabel, then Director of the Nikita Garden, to inquire about the introduction of the tree. He sent me leaves and flowers from his specimen, but stated, that nothing is known regarding the introduction.

with *Astragalus complanatus*, R. Brown, gathered previously by Staunton in China. (V. supra p. 174).

Galium chinense, Sprengel. V. supra p. 278.

Ipomoea sinensis, Fischer, Hort. Gorenki, 2d edition, 1812, 24; only the name. Choisy in DC. Prod. IX, 390, after stating that Fischer's Ipomoea sinensis is unknown to botanists, gives the same name to *Convolvulus sinensis*, Descr. in Lam. Enc. III (1789) 557. Herb. Jussieu.

I may observe, that Ker in Bot. Reg. tab. 322 (1818) figures a new plant which he calls *Convolvulus çhinensis*, stating that it was raised by Messrs Whitley & C°., from seeds received through Berlin from the Bot. Garden, Moscow. The drawing looks like the common Peking *C. arvensis*, Linn. var. *sagittaefolius*, Turcz., mentioned by Bunge, same as *C. sagittaefolius*, Fischer, Hort. Gorenki, 1812, 28. — I. F. S., II, 165.

Nicotiana chinensis, Fischer in litt. in Lehmann, Genus Nicotiana (1818) 18. China. Lehmann thinks that it is perhaps Loureiro's N. fruticosa. — Dunal in DC. Prod. XIII, I, 559. — I have little doubt that Fischer's specimens came from Peking. I have in vain tried to rediscover this plant among the sorts of Tobacco cultivated in the environs of Peking. I met there only with N. Tabacum Lin. and N. rustica, Lin. (My specimens were determined by Maximowicz). Fischer's specimen is not found in the Herb. hti Petrop.

Lophanthus rugosus, Fischer & Meyer, Ind. Semin. Hti Petrop. 1834, 31. Described from specimens raised from seeds received from China, where it it said to be cultivated. — There is a specimen of the plant in Hb. Mus. Acad., gathered by Tatarinov, 1847 north of Peking, not said whether wild or cultivated. I only saw it there cultivated. — Ind. Fl. Sin., II, 288.

Polygonum tinctorium, Chinese Indigo, was first described under this name by Aiton, in Hort. Kew. 1789. The plant had been introduced from Canton into the Kew Gardens by Blake (supra p. 152). At about the same time Loureiro described it under the same name as a Canton plant, in his Flora Cochin. 297. P. tinctorium does not seem to have then been cultivated in Europe as a tinctorial plant and was subsequently lost. Fischer reintroduced it into Europe about fifty years later. I suspect he had received seeds from Peking, where it is much cultivated as a tinctorial plant. Fischer in 1833 sent seeds of it to Professor A. R. Delile, Montpellier, and from this time P. tinctorium was much cultivated in France and afterwards also in Belgium. (Vilmorin, Bon Jardinier, 1841, Diction. d'Hist. nat. p. Arago etc. X (1847) 758.

Bot. Mag. tab. 233 (1822). *Iris Pallasii,* Fischer, β. *chinensis.* Communicated to Hooker by Mr. Anderson from the Chelsea Garden, in May 1820, where it was raised from seeds sent by Dr. Fischer from Gorenki; who informed H., on his late visit, that the plant is a native of Chinese Mongolia.

Avena chinensis, Fischer in Hort. Gor. (1808); only the name. — Roemer & Schult. Syst. Veg. II (1817) 669 = Variety of *Avena nuda,* L. Omnium fertilissimam esse. Fischer. — This plant is cultivated in the Peking mountains and Mongolia. When ripe it drops the grains from the husk.

Plantago mongolica, Decaisne, in DC. Prod. XIII, i, (1852) 707. Decaisne had received this plant from Dr. Fischer, St. Petersburg, with a note, that it came from Mongolia. The plant does not seem to exist in the Herbarium Hti Petrop.

Franchet, in Pl. David. I, 246, records Pl. mongolica, Decaisne, as having been gathered by David in the Urat, in South Mongolia, in July 1866.

RUSSIAN CIRCUMNAVIGATIONS OF THE GLOBE 1815—26.

Nine years had elapsed since the termination of the first Russian circumnavigation of the globe, when a liberal patron of science, Count N. P. Rumiantsov, fitted out, at his own expense, a ship, the "Rurik" for a new circumnavigation. Otto von **Kotzebue,** Captain-lieutenant of the Russian Navy, who had accompanied Krusenstern, 1803—6, was appointed to the command of the ship. To the expedition there were attached three naturalists: Adelb. von **Chamisso,** a German naturalist and poet, Professor J. F. **Eschscholtz** of the Dorpat University, and **Choris.**

This voyage lasted from 1815—18 and was followed, in 1823, by another naval expedition under the command of O. v. Kotzebue, ordered by the Emperor Alexander. The ship destined for this purpose was the "Predpriatie" (Enterprise). Eschscholtz again joined the expedition as naturalist. As both these voyages were particularly devoted to the exploration of the Pacific Ocean, neither the China nor the Japan Seas were visited by Kotzebue.

In the same year, when the "Predpriatie" returned to Kronstadt, in 1826, a new naval expedition was fitted out, by order of the Emperor Nicolas, to complete Kotzebue's explorations. The following summary of this voyage is extracted from a book which bears the title:

VOYAGE AUTOUR DU MONDE EXÉCUTÉ PAR ORDRE DE SA MAJESTÉ,

L'Empereur Nicolas sur la Corvette "Séniavine", 1826—1829, par Fr. **Luetke,** Captaine de vaisseau. Traduit du russe par F. Boyé. 1835. 3 volumes avec un Atlas lithographié d'après les dessins originaux d'Alexandre Postels et du Baron Kittlitz.

Dr. Ch. H. **Mertens** and Baron **Kittlitz** accompanied the expedition as naturalists, A. **Postels** as painter.

The Corvet "Seniavin" (named in honour of the Russian Admiral D. N. Seniavin (1763—1830), commanded by Capt. Lieutenant Fr. P. **Luetke** (born 1797), left the roads of Kronstadt, 1st Sept. 1826. She was accompanied by the Corvet "Moller", Capt. Staniukovich. The ships touched at Portsmouth, Teneriffa, arrived at Rio Janeiro Janr., 7, 1827. Then they pursued their voyage around Cape Horn to Valparaiso. Having quitted this port on April 15th, they directed their course to the Russian possessions in N. W. America. They sailed first to New Arkhangelsk (Sitka Isl.), then visited the Aliaska Peninsula, Petropavlovsk (Kamtchatka). From the latter place they sailed, end of 1827, to the Caroline Islands, landed at the island of Ualan, Jan. 5, 1828. After this they made sail to the north-westward and visited the Bonin Group. From 29th April to 15th May they remained at Port Lloyd on Peel Island, the largest of the Bonin Islands. This island had been discovered a year before by Capt. Beechey (v. s. p. 289). Then they returned to Petropavlovsk, and visited Bering Strait. On Nov. 10 they quitted Petropavlovsk for the homeward voyage, visited Manilla, passed Gaspar Strait, went around the Cape of G. H., arrived at Kronstadt, 6th Sept. 1829.

Luetke subsequently became Governor of the Grand-duke Constantin. In 1835 he was promoted to the rank of Rear-admiral; in 1855 Admiral. In the same year he was created a Count. He was one of the founders of the Imp. Geogr. Society, St. Petersburg, 1845. From 1864 President of the Academy of Sciences. He died 1882.

Mertens, Dr. Ch. H., born at Bremen in 1796, studied natural history and medicine at Goettingen, Halle; M. D. 1820; went to Russia, appointed naturalist to Luetke's expedition. After the return of the "Seniavin", Mertens was elected a Member of the Russian Academy, but a short time after died, 17th Sept. 1830. He had collected during the voyage 2500 herbarium specimens, amongst which there were many Ferns and Algae. This collection he handed over to the Academy. Dr. Siebold in Japan seems to have received duplicates. The ferns were sent for determination to Prof. G. Kunze, Leipzig. He enumerated or described the ferns gathered by

Mertens in Bonin sima (Peel Isl.), in his Pteridographia japonica, in Bot. Zeitung, 1848. The following are recorded as new:

524. *Asplenium trigonopterum*, Kze. — Figured in Mettenius, Gen. Asplenium, (1857) 107, tab. 5. — Hook. Spec. Fil. III (1860) 172, reduces it to *Aplenium nitidum*, Swartz.

525. *Asplenium Mertensianum*, Kze. — Hook. Sp. Fil. III, 211.

541. *Lindsaya repanda*, Kze.

585. *Alsophila Mertensiana*, Kze. Filix arborea in silva ins. Bonin sima a Kittlitzio delineata in Luetke, Voy. Atlas, tab. 40.

Some other new plants gathered by Mertens in Bonin sima, were described by Siebold and Maximowicz:

Ilex Mertensii, Maximowicz de Ilice, 42.

Raphiolepis Mertensii, Sieb. & Zucc. Flor. Jap. I (1841) 164.

Hedyotis cordata, Sieb. & Zucc. Fam. nat. Fl. Jap. (1846) n. 599.

Webera subsessilis, Maxim. Mél. XI (1883) 789.

Piper Postelsianum, Maxim. Mél. XII (1886) 531. Earlier described by Miquel, Piperac. in Lond. Journ. Bot. IV (1845) 431 as *Pothomorphe subpeltatum*, Ceylon, Walker, — Bonin Sima, Dr. Mertens. Herb. Petrop.

Gahnia Boninsima, Maxim. Mél. XII (1886) 559.

Ficus variolosa, Lindl. in Lond. Journ. Bot. I (1842) 492. Hongkong Hinds. — According to Maxim. Mél. XI (1881) 336, first discovered by Mertens in Bonin Sima.

Comp. also, **Bongard,** "Mémoire sur la végétation des isles de Bonin sima". Bull. scient. Acad. St. Pétersb. II, 1837, 371.

Postels, Alexander, the painter attached to the expedition, collected Algae. See: "Postels and Ruprecht, ILLUSTRATIONES ALGARUM in itinere circa orbem auspiciis navarchi Fr. Luetke, annis 1826—29, executo etc. collectarum. 1840".

EXPLORATION OF THE NORTHERN CHINESE AND MONGOLIAN FLORA BY RUSSIAN BOTANISTS.

Bunge, Dr. Alexander von. This eminent Russian botanist (he was of German extraction) was born at Kiev in 1803. He studied medicine at Dorpat University and in 1825 took his degree of M. D. He had early commenced to take interest in botany and was encouraged and guided in his botanical studies by C. F. v. Ledebour, from 1811 Professor of Botany at Dorpat, author of the Flora Rossica and many other works on Russian plants. Bunge soon became a botanist of considerable ability. In 1826 he was appointed District-Physician at Barnaul (S. Siberia), and accompanied

in the same and the next year Ledebour and C. A. Meyer on their travels
in the Altai Mountains. After this he lived some time in Kolyvan, and was
finally transferred to Zmeinogorsk, where he remained till 1830, and found
ample opportunity to study the flora of this part of Siberia. On the recommendation of Alex. von Humboldt, whose acquaintance B. had made when
in 1829 this celebrated naturalist visited the Altai, the young botanist
obtained a scientific mission from the Academy of St. Petersburg. The Russian Government sent a new Ecclesiastical Mission (the 11th) to Peking to
relieve the 10th, which had been there since 1820. Colonel **Ladijenski** was
appointed pristav (conductor) of this mission. Dr. P. **Kirilov** accompanied it
in the capacity of physician. Three scientific men were selected by the
Academy to proceed with the mission to Peking: Dr. Bunge, as botanist,
G. **Fuss** as astronomer and meteorologist, and the engineer Kovanko as
mineralogist. Bunge joined the mission on their road through Siberia. Having traversed this vast country, they passed the Chinese frontier at Kiakhta,
end of August (old style) 1830, reached Urga in the middle of September,
crossed the Gobi desert and arrived at Peking 17th Nov. 1830.

Portions of Bunge's diary, written in Mongolia, were published in the
Dorpater Jahrbücher, IV, 1835, pp. 251, 341. In Berghaus' Annalen der
Erd- und Völkerkunde, IX, 1834, a letter by Bunge, dated Urga, 19 Sept.
1830, is published, and in the same vol. p. 452, seqq. we find an interesting article written by the same author, entitled: BAROMETRISCHES NIVEL-
LEMENT UND NATURGEMÄLDE DER CHINESISCHEN MONGOLEI. It had been
communicated by Alexander v. Humboldt. Bunge and the other naturalists
spent the winter 18 $^{30}/_{31}$ in Peking. In the month of March 1831 Bunge
performed a journey to the ruins of Tsagan balgasun, situated beyond
the Great Wall (Kalgan) in the grassy plain of S. Mongolia, where the
Russian escort of the mission and the beasts of burden had been left behind.
After his return from Mongolia B. botanized during the first half of the
summer 1831 in the Peking plain and among the mountains west, south-
west, north, and north-east of the capital. The labels in his Chinese collection are dated: April, May and June 1831. It must, however, be remarked that while staying at the Buddhist monastery Ta pei sz' in the
Western mountains, in the month of May, he incurred the displeasure of
the Chinese authorities, and during the month of June, till his departure,
was not allowed to walk beyond the walls of Peking. (See his article on
Siphonostegia, 273).

On the 6th of July, Bunge and his scientific colleagues left Peking,
together with the 10th Eccles. Mission, which was returning to Russia. In

traversing Mongolia, they followed between Kalgan and Urga a new and more westerly route. Kiakhta was reached in the beginning of September 1831. Bunge's notices of the journey back through Mongolia, are found in Linnaea XVIII (1843) 1.

Bunge brought home from this expedition about 420 species of plants gathered in North China (envirous of Peking, road from Peking to the Great Wall), and also an interesting collection made in Mongolia. He had formed, besides, in these Regions a collection of Coleopters whis was subsequently described by Faldermann in .Mem. Acad. Sc. St. Petersb. II, 1835: "Coleopterum ab illustr. Bungio in China boreali, Mongolia et montibus Altaicis collectorum descriptio".

Having passed the winter in Irkutsk, Bunge, in 1832, proceeded to explore the eastern part of the Russian Altai Mountains, and then went to St. Petersburg. In 1833 he was elected a Corresponding Member of the Imp. Academy, St. Petersb., and appointed Professor of Botany at the Kazan University. From this place, in 1835, he explored botanically the steppes situated east of the lower Volga. When, in 1836, Ledebour retired Bunge succeeded him as Professor of Botany at Dorpat, which post he held till 1867, when he resigned. His quiet scientific occupations at Dorpat were interrupted for two years, 1858—59, when he accompanied a diplomatical and scientific expedition through Persia to Herat, headed by N. Khanikov. After his retirement from public service, B. spent the remaining years of his life at Dorpat in complete devotion to botanical pursuits. In this period he published many important botanical treatises in the Memoirs of the Academy, of which, in 1875 he had become an Honorary Member. Bunge died on July 6 th 1890.

As has been noticed above, Bunge, after returning from China, in Sept. 1831, remained in Irkutsk till the spring of 1832, and immediately set himself to determine and describe the botanical collection made by him in Northern China. Unfortunately he could not spend much time on this work and, besides, had only a small botanical library at his service. In 1832 he sent to the Academy a memoir with the title:

Enumeratio Plantarum quas in China boreali collegit A. Bunge, anno 1831.

It was read March 7, 1832, and published in the Mémoires des savants étrangers of the Acad., vol. II, 1835, pp. 75—147.

Among the 420 species noticed in this paper, Bunge describes many new and interesting plants of the Peking Flora. C. B. von Trinius, member of the Academy, then a great authority for Gramineae, described for Bunge

the specimens belonging to this order. Some of the novelties, which Bunge
had not been able to determine in Irkutsk, were described a few years
later by him or by other botanists.

In 1835, when Bunge was professor in Kazan, he published there, in
the Memoirs of that University, IV, p. 154, seqq. an article entitled: New
Genera and Species of Chinese and Mongolian Plants. First Decade
(only one decade has appeared) with 3 plates. Title and Introduction in
Russian, description of plants in Latin. Bunge proposes first a new genus
of Leguminosae, *Campylotropis*, now reduced to *Lespedeza*, and then describes 4 new species of *Lespedeza* of the Peking Flora, 4 new species of
Patrinia and finally proposes a new genus of Verbenaceae: *Caryopteris*.

Other papers written by Bunge on Chinese and Mongolian plants will
be noticed further on, in the list of novelties.

A complete set of Bunge's botanical collections made in China and
Mongolia, is preserved in the Botan. Museum of the Academy. Bunge also
distributed duplicates of his Eastern Asiatic plants to the great botanical
institutions in Europe and to distinguished botanists. A. De Candolle seems
to have possessed a great many of Bunge's plants, as can be judged from
the Prodromus.

Localities mentioned in Bunge's Enum. Chin. bor.:

The Chinese names of the localities, where Bunge collected, are frequently incorrectly rendered by him. It may, therefore, not be out of place
to give here a list of these names, adding the correct spelling according to
the English mode of transliterating the Chinese sounds.

Palitschuan = Pa li chuang, a large village about 3 miles west
of Peking.

Flumen Chunj che = Hun ho River. It flows west of Peking, 8 miles
distant at the nearest point.

Kantai, in humidis prope Kantai. — K'ang t'ai is a small village, in
a sandy plain, about 4 miles S. W. of Peking. But Bunge means, I have
no doubt, Feng t'ai, which is a general name for a number of villages
situated between K'ang t'ai and the capital. The inhabitants of these villages are almost all engaged in gardening. There are many swamps, ditches
and ponds about, in which water plants are cultivated.

Zing ho = Ts'ing ho, a small river and a large village, 5 miles
N. W. of Peking.

Si schan = Si shan (Western Mountains), a general name for the
mountains west of Peking.

Montes Zui wei schan = Ts'ui wei shan Mts, a part of the Si shan, about 10 miles west of Peking, with picturesque Buddhist monasteries (Pa ta ch'u), where the foreign Legations now live during the hot season.

Templum Da bei ssy. Ta pei sze is one of the monasteries of the Ts'ui wei shan, where Bunge had established his headquarters for April and May 1831.

Templum Tan dsche ssy. The large monastery of T'an che sze lies west of the Hun ho, 18 miles from Peking.

Ssi jui ssu = Si yü sze is a monastery in the mountains S. W. of Peking, 37 miles distant. Is lies 13 miles S. W. of the district city of Fang shan.

Pan schan. The nearly isolated mountain P'an shan (granite, 2500 ft.), lies N. E. of Peking, 53 miles distant.

Thermae Tan schan = T'ang shan, a hill in the plain, hot springs, Imperial Palace, gardens, 18 miles N. of Peking.

Lun züan ssy. The monastery of Lung ts'üan sze lies at the skirt of the northern Peking mountains, 22½ miles from the capital.

Tschan pchin shou = Department city of Ch'ang p'ing chou. N. W. of Peking, near the northern mountain chain.

Nanj kou = Nan k'ou, name of a village at the southern entrance of the Kuan kou Pass (Bunge writes Guan gou) in the northern chain. Road to Kalgan. Europeans call it Nan k'ou Pass.

Zui jun guan, = Kü yung kuan, a fortress in the Nan k'ou Pass.

Ba da lin, = Pa ta ling. The most elevated part of the Nan k'ou Pass, where the Inner Great Wall passes.

Tscha dao, = Ch'a tao, a village at the northern entrance of the Nan k'ou Pass.

Jü ling, = Yü lin, a post station on the road to Kalgan, west of Ch'a tao.

Montes Zsimin shan, Urbs Zsimin. There are on the same road, towards Süan hua fu, a hill and a post station, both called Ki min.

Dschan dsia kou = Chang kia k'ou. This is the Chinese name for Kalgan, where the Outer Great Wall passes.

As to the localities in Mongolia, where Bunge collected plants I beg the reader to refer to my Map of Mongolia, appended to Palladius' Travels in Mongolia, published in Russian, in 1892 by the Russian Geograph. Society. My Introduction to the book as well as the map have been translated into French by Mr. Boyer in the Journ. Asiat. 1893. On the map Bunge's itineraries are marked.

Novelties in Bunge's collections of Chinese and Mongolian plants *).

Bunge enumerates 420 species of Northern Chinese plants, and among these he proposed 11 new genera viz: *Orychophragmus, Xanthoceras, Oresitrophe, Thladiantha, Calysphyrum, Myripnois, Urostelma, Dorcoceras, Bothriospermum, Ceratostigma, Anemarrhena.*

To these must be added 6 more new genera from Bunge's Chinese and Mongolian plants, which he afterwards described, viz: *Sarcozygium, Campylotropis, Hemistepta, Pycnostelma, Phtheirospermum, Caryopteris.*

Of Bunge's new genera and species a few have been subsequently reduced by other botanists to genera and species already described. On the other hand, some plants of Bunge's collection, which he believed to be known, were proved by other botanists to be new species: *Corydalis Bungeana,* Turcz., *Xanthoxylum Bungei,* Planch., *Indigofera Bungeana,* Walpers, *Catalpa Bungei,* C. A. Meyer, *Fraxinus Bungeana,* DC., *Jasminum nudiflorum,* Lindley, *Syringa oblata,* Lindl., *Polygonum Bungeanum,* Turcz., *Celtis Bungeana,* Blume.

Clematis intricata, Bge, Enum. n. 3. China borealis. — Maximowicz, Mél. biolog. IX (1876) 583, = *Cl. orientalis,* Linn. var. *intricata.* — Ind. Fl. Sin., I, 7.

Thalictrum foeniculaceum, Bge, Enum. n. 4. Lun züan ssy. — Figured in Regel, Monogr. Thalict. 1861, t. 3. — I. F. S., I, 8.

Pulsatilla (Anemone) *chinensis,* Bge, Enum. n. 6. Zui wei schan, Ssi jü ssy, Lun züan ssy. — Earlier d'Incarville (supra p. 52). — Figured in Regel, Fl. Ussur. tab. 2. — I. F. S., I, 10.

Ranunculus chinensis, Bge, Enum. n. 10. Ssi jü ssy. — I. F. S., I, 14: *R. pensylvanicus,* Lin.

Ranunculus hydrophilus, Gaudich. — Bge, Enum. n. 7. Ssi jü ssy, Dsü jun guan. — Steudel considered it a new species, *R. Bungei.* — I. F. S., I, 13: *R. aquatilis,* Linn. — V. supra p. 52. d'Incarville.

Ranunculus oryzetorum, Bge, Enum. n. 9. Lun züan ssy. — I. F. S., I, 16: *R. sceleratus,* Lin.

Trollius chinensis, Bge, Enum. n. 11. China borealis. — I. F. S., I, 17: *T. asiaticus,* Lin.

Corydalis racemosa, Persoon. Bge, Enum. n. 26. Prope Pekinum. — Turczaninov. Bull. Mosc. XIII (1840) 6, new spec. = *C. Bungeana.* — Earlier Incarville (supra p. 52). — I. F. S., I, 36.

---

*) As has been related above (v. s. p. 52) a part of Bunge's novelties from Peking, 28 species, had been discovered there more than 80 years earlier by d'Incarville.

Cheiranthus aurantiacus, Bge, Enum. n. 27. China borealis, montes.—
I. F. S., I, 39. — Maxim. Pl. Potan. chin. (1889) n. 129: *Erysimum
aurantiacum.*

Cardamine lyrata, Bunge, Enum. n. 30. Ssi jü ssy. — Ind. Fl.
S., I, 43.

Dontostemon crassifolius, Bge, in Maxim. Amur, 46, 480. — Maxim.
Fl. Mongol. (1889) 57: Bunge between Kiakhta and Kalgan, 1830.

Andreoskia dentata, Bge, Enum. n. 33. Pan schan. — Earlier d'Incarville (v. supra p. 52). — Ledeb., Fl. Ross. I, 175, calls it *Dontostemon
dentatus.* — I. F. S., I, 45.

Erysimum macilentum, Bge, Enum. n. 36. Prope Pekinum frequens.—
I. F. S., I, 46.

Orychophragmus sonchifolius, Bge, Enum. n. 40. Prope Pekinum frequens. — Gmelin, d'Incarville earlier. (V. supra pp. 52, 313). — Hook. fil.
in Bot. Mag. tab. 6243 (1876) reduces Bunge's new genus to *Moricandia,
M. sonchifolia.* — I. F. S., I, 47.

Gynandropsis viscida, Bge, Enum. n. 42. Zing ho. — I. F. S., I, 50:
same as *G. pentaphylla,* DC. Prod. I, 238 (Cleome pentaphylla, Linn.).

Viola prionantha, Bge, Enum. n. 44. Prope Pekinum frequens. —
Maxim. Mél. IX (1876) 722 = *V. Patrinii,* DC. var. *chinensis,* Ging. in
Prodr. I, 293. — Earlier Incarville, Staunton. (V. supra pp. 52, 173). —
I. F. S., I, 53.

Lychnis Bungeana, Fischer, Bot. Reg. tab. 1864 (1836). Bot. Mag.
tab. 3594 (1837). Cultivated by Fischer, St. Petersb. before 1835. It was
raised from seeds brought by Bunge from Peking, where the plant is cultivated. — I. F. S. I, 65.

Stellaria gypsophiloides, Fenzl. in Ledebour, Flora Ross. I (1842)
380. Mongolia, Bunge. Herb. Acad. Petrop. — Maximowicz, Fl. Mongol.
1889, 100.

Tamarix juniperina, Bge, Enum. n. 170, et Tentamen Generis Tama-
ricum species def. 1852, 45. Kantai. — I. F. S., I, 347.

Sterculia pyriformis, Bunge, Enum. n. 57. Peking, culta. — I. F.
S., I, 90: *St. plantanifolia,* Linn.

Grewia parviflora, Bge, Enum. n. 57. Pan chan, Zui wei schan, Lun
züan ssy. — I. F. S., I, 93.

Sarcozygium (new genus) *Xanthoxylon,* Bge, in Linnaea, XVII, 1843,
Neue Gattung aus d. Familie d. Zygophyllaceae, p. 8, tab. 1. Bunge, Mongolia: Ude et inter Schilin Chuduk et Boroldschi. — Maxim. Ind. Mongol.
in Prim Amur. 480. *Sarcoz. tripteris,* Bge, in shed. — I. F. S., I, 97. —

Baillon, Hist. Pl. IV, 417 reduces Bunge's new genus to *Zygophyllum*, *Z. Xanthoxylon*. — Maxim. Fl. Mong. 1889, 124.

Oxalis fontana, Bunge, Enum. n. 74. Pan chan. — I. F. S., I, 99: *O. stricta*, Linn.

Peganum Nigellastrum, Bge, Enum. n. 78. Inter Tscha dao et Dschang dsia kou, Mongolia. — I. F. S., I, 103.

Zanthoxylon nitidum, DC. Bunge, Enum. 77. Colitur in hortis. — Planchon in Ann. Sc. nat. 1853, 82, new species: *Z. Bungeanum*. — I. F. S., I, 105.

Citrus microcarpa, Bge, Enum. 60. In calidariis Pekinensibus. — I. F. S., I, 111: *Citrus japonica*, Thbg.

Rhus ailanthoides, Bge, Enum. n. 86. Zui wei schan. — Planchon in Lond. Journ. Bot. V, 1846, 373, refers it to *Picrasma: P. ailantoides*. — I. F. S., I, 112 = *P. quassiodes*, Benn. Pl. Jav. rar. (1844) p. 198, note. First discovered, about 1802, by Hamilton in Nepal.

Euonymus micranthus, Don. Bunge, Enum. n. 79. Lun züan ssy. — Maxim. Amur. p. 470 (Index Fl. Pekin.), in a note, describes this as a new species: *E. Bungeanus*. — I. F. S., I, 118: earlier Staunton.

Zizyphus vulgaris, Lam. var. *spinosa*, Bunge, Enum. 81. Prope Pekinum. Frequentissima et molestissima. — About 18 years ago I sent from Peking complete herbarium specimens of this shrub to Prof. Decaisne, Paris. The same plant had also been sent to him by Father David. He declared that he was not able to distinguish it from the *Z. Lotus*, Lin. of N. Africa. But Maximowicz (Fl. Mong. 1889, 136) does not agree with this view, and maintains the name given to the plant by Bunge. — I. F. S., I, 126.

Rhamnus globosus, Bge, Enum. n. 83. Prope Pekinum. — Maxim. Rhamn. (1866) 13, fig. 24—35, and Flora Mong. (1889) 137, refers it to *Rh. virgata*, Roxb. (India). But Hemsley, I. F. S., I, 129 identifies it with *Rh. tinctoria*, Waldst. et Kit. of Hungary.

Rhamnus parvifolius, Bge, Enum. n. 82. Frequens prope Pekinum. — Figured in Maxim. Rhamn. fig. 36—47. — I. F. S., I, 129.

Ampelopsis humulifolia, Bge, Enum. n. 69. Montes Zui wei schan. — Figured in Regel, Fl. Ussur. t. 3. — I. F. S., I, 133 = *Vitis heterophylla*, Thbg. Fl. Jap. 103.

Ampelopsis serianaefolia, Bge, Enum. n. 70. Lun züan ssy. — *Vitis serianaefolia*, Maxim. Mél. IX (1873) 149. — I. F. S., I, 136.

Ampelopsis aconitifolia, Bge, Enum. n. 71. Lun züan ssy. — I. F. S., I, 136: same as preceding.

Vitis bryoniaefolia, Bge, Enum. n. 67. Ssi jü ssy, Tan dsche ssy. —
I. F. S., ɪ, 136: *Vitis vinifera,* Linn.

Vitis ficifolia, Bunge, Enum. n. 68. Pan schan, Ssi schan. — I. F.
S., ɪ, 134: *V. labrusca,* Lin.

Aesculus chinensis, Bge, Enum. n. 63. In sylvaticis montosis circa
Pekinum. — I. F. S., ɪ, 139.

Xanthoceras (Gen. nov.) *sorbifolia,* Bge, Enum. n. 65. In montosis
Chinae bor. rarior. — I. F. S., ɪ, 140.

Acer truncatum, Bge, Enum. n. 62. In montosis circa Pekinum. —
I. F. S., ɪ, 142.

Pistacia chinensis, Bunge, Enum. n. 84. Zui wei schan. — I. F.
S., ɪ, 148.

Melilotus graveolens, Bge, Enum. n. 94. Kantai. — I. F. S., ɪ, 155 =
M. suaveolens, Ledeb. (1824, Dahuria).

Indigofera micrantha, Desv., Bunge, Enum. n. 95. Lun züan ssy. —
Walpers in Linnaea, XIII (1838) 525, new spec.: *I. Bungeana.* — Earlier
Incarville (v. supra p. 53). — I. F. S., ɪ, 156.

Caragana Chamlagu, L'Hérit., Bunge, Enum. n. 98, is according to
Turczaninov a new species, *C. chinensis* (unpublished). See Maxim. Ind.
Pekin. 470. But I. F. S., ɪ, 163 reduces it to C. Chamlagu.

Gueldenstaedtia multiflora, Bge, Enum. n. 106. Prope Pekinum frequens. — Earlier Incarville (v. supra p. 53). — I. F. S., ɪ, 164.

Gueldenstaedtia stenophylla, Bge, Enum. n. 107. Prope Pekinum frequens. — I. F. S., ɪ, 164.

Astragalus scaberrimus, Bunge, Enum. n. 105. Prope Pekinum frequens. — I. F. S., ɪ, 166.

Oxytropis bicolor, Bge, Enum. n. 102. Zui wei schan, Palitschuan. —
Earlier d'Incarville (supra p. 53). — I. F. S., ɪ, 167.

Oxytropis hirta, Bunge, Enum. n. 101. Zui wei schan. — Ind. Fl.
Sin., ɪ, 167.

Hedysarum brachypterum, Bge, Enum. n. 108. Zsi min i, Suan hua
fu. — I. F. S., ɪ, 169.

Lespedeza Caraganae, Bunge, Nova gen. et spec. Chinae, Mongoliae
Decas., 1835, 162. Prope Pekinum. — I. F. S., ɪ, 179.

Lespedeza floribunda, Bunge, l. c. 164. Prope Pekinum. — I. F.
S., ɪ, 181.

Lespedeza macrophylla, Bunge, l. c. 161. In montosis prope Pekinum. — Ind. Fl. Sin., ɪ, 183 = *L. villosa,* Persoon, Syn. II (1807) 318
(Amer. bor.).

Lespedeza medicaginoides, Bunge, l. c. 166. Prope Pekinum, Tscha dao. — Earlier Staunton, Peking (supra p. 175). — I. F. S., ɪ, 182.

Lespedeza macrocarpa, Bunge, Enum. n. 109. Zui wei schan, Nanj kou. — *Campylotropis* (nov. gen.) *chinensis,* Bunge, Nova gen. et spec. Chinae, Mongoliae, Decas., 1835, 158. — I. F. S., ɪ, 182: *Lespedeza macrocarpa.*

Vicia gigantea, Bge, Enum. n. 112. Zui wei schan. — I. F. S., ɪ, 184.

Vicia tridentata, Bunge, Enum. n. 114. Prope Pekinum. — I. F. S., ɪ, 185.

Phaseolus anguinus, Bge, Enum. n. 117. — Prope Pekinum cultus. — I. F. S., ɪ, 192.

Gleditschia heterophylla, Bunge, Enum. n. 123. Lun züan ssy, Ssi jü ssy. — I. F. S., ɪ, 209.

Cercis chinensis, Bge, Enum. n. 124. In hortis Pekinensibus culta. — I. F. S., ɪ, 213.

Acacia macrophylla, Bunge, Enum. n. 120. Pan schan. — I. F. S., ɪ, 216 = *Acacia Lebbek,* Willd. (Egypt).

Amygdalus pedunculata, Pallas. Bge, Enum. 126. In hortis Pekinensibus colitur. Bunge notices 3 varieties, all cultivated. — Maxim. Mél. XI (1883) 665, 666, proves that they have nothing to do with the plant described by Pallas. Bunge's varieties α and γ are *Prunus Petzoldi,* C. Koch Dendrol. I (1869) 92. China, Cultiv. Europe, and var. β, floribus maximis is *Prunus triboba,* Lindley, Gard. Chron. 1857, 268. — Ind. Fl. Sin., ɪ, 222.

Amygdalus communis, Linn., Bge, Enum. n. 125. In hortis Pekinensibus colitur. — Maxim. l. c. 667 = *Persica Davidiana,* Carrière, Rev. hort. 1872. — I. F. S., ɪ, 220.

Prunus humilis, Bunge, Enum. 133. Zui wei schan, Si jü ssy. — I. F. S., ɪ, 218.

Prunus pauciflora, Bge, Enum. n. 132. Zui wei schan. — Maxim. l. c. 694. — I. F. S., ɪ, 220.

Prunus trichocarpa, Bge, Enum. n. 131. Spontanea in montosis prope Pekinum; nec non in hortis culta. — According to Sieb. & Zuccar. Fl. Japon. I, 222, this is *Prunus tomentosa,* Thunb. Fl. Jap. 203. — I. F. S., ɪ, 222.

Spiraea dasyantha, Bge, Enum. n. 136. In montosis prope Pekinum.— I. F. S., ɪ, 224.

Spiraea sorbifolia, Lin., Bunge, Enum. n. 139. Bada lin ad murum magnum. — Regel & Tiling Fl. Ajan. (1858) 81, in adnot describe Bunge's

Peking plant as a new species, *Sp. Kirilovii*. — Maxim. Spiraeac. 225. —
I. F. S., I, 227, sub *Sp. sorbifolia*.

Rubus crataegifolius, Bge, Enum. n. 140. Pan schan. — Figured
in Rgl. Fl. Ussur. tab. 5. — I. F. S., I, 230.

Rubus purpureus, Bunge, Enum. n. 139. Pan schan. — Maxim. Mél.
VIII (1871) 392 = *R. parvifolius*, Lin. — I. F. S., I, 235.

Potentilla ancistrifolia, Bge, Enum. n. 145. Si jü ssy. — Lehmann,
Revisio Potent. (1856) 43, tab. 18. — I. F. S., I, 240.

Potentilla discolor, Bunge, Enum. n. 149. Zui wei schan. — I. F.
S., I, 241.

Potentilla exaltata, Bge, Enum. n. 142. Prope Pekinum. — Same as
P. chinensis, Ser. Staunton, Peking (supra p. 175). — I. F. S., I, 241.

Potentilla multicaulis, Bunge, Enum. n. 147. Prope Pekinum frequens. — I. F. S., I, 244 = *P. sericea*, Lin.

Potentilla songarica, var. β. *chinensis*, Bge, Enum. n. 146. Ssi jü ssy.
Pot. sischanensis, Bge, MSS in Lehmann, nov. recens. Potent., 1851, 3. —
Lehm. Revisio Potent. 1856, 33, tab. 9. — I. F. S., I, 244.

Agrimonia viscidula, Bge, Enum. n. 152. Prope Pekinum. — I. F.
S., I, 246 = *A. Eupatoria*, Lin.

Rosa pimpinellifolia, Lin. Bunge, Enum. n. 155. Colitur in hortis
varietas floribus majusculis sulphureis. — According to Hemsley, I. F.
S., I, 253, probably *Rosa xanthina*, Lindley Roses (1820) 132, described
and figured by Lindley from a Chinese drawing.

Pyrus betulaefolia, Bunge, Enum. n. 161. Si jü ssy. Tan schan. —
I. F. S., I, 256.

Crataegus pinnatifida, Bunge, Enum. n. 157. Tan dsche ssy. — I. F.
S., I, 259.

Oresitrophe (new genus) *rupifraga*, Bge, Enum. n. 187. Lun züan ssy,
Si jü ssy. — I. F. S., I, 271.

Deutzia grandiflora, Bge, Enum. n. 184. China borealis. — Earlier
d'Incarville. — Figured in Max. Hydrang. 1867, t. 3. — I. F. S., I, 276.

Deutzia parviflora, Bge, Enum. n. 185. Pan schan. — Earlier d'Incarville. — Figured in Maxim. Hydrang. tab. 3; in Rgl. Fl. Ussur. t. 5. —
I. F. S., I, 276.

Sedum sarmentosum, Bunge, Enum. n. 183. Si jü ssy. — Ind. Fl.
Sin., I, 286.

Thladiantha (nov. gen.) *dubia*, Bge, Enum. 173. Prope Pekinum. —
I. F. S., I, 316.

Sanicula chinensis, Bge, Enum. n. 189. Pan schan. — I. F. S., I, 326.

Bupleurum octoradiatum, Bunge, Enum. n. 188. Zui wei schan. — I. F. S., ɪ, 327.

Peucedanum rigidum, Bunge, Enum. n. 190. Ad. fluv. Chunj ho. — I. F. S., ɪ, 335.

Viburnum fragrans, Bge, Enum. n. 194. Colitur in hortis Pekinensibus. — Earlier d'Incarville (supra p. 53). — I. F. S., ɪ, 352.

Calysphyrum floridum, Bge, Enum. n. 196. Colitur in hortis Pekinensibus. — Sieb. & Zuccar. in Fl. Japon. I (1839) 75: *Diervillea florida.* — I. F. S., ɪ, 368.

Leptodermis oblonga, Bge, Enum. n. 197. Lun züan ssy, Zui wei schan. — I. F. S., ɪ, 390.

Galium gracile, Wall., Bge, Enum. n. 198. Lun züan ssy. — *Galium Bungei,* Steudel, Nomencl. bot. — I. F. S., ɪ, 394, sub *G. gracile,* Bge.

Galium pauciflorum, Bge, Enum. n. 199. Si jü ssy. — Maxim. Mél. IX (1873) 259 = *G. Aparine,* Lin. — I. F. S., ɪ, 393.

Patrinia heterophylla, Bunge, Enum. n. 201, and Bge, Nova Gen. et Spec. Chin. Mongol. Decas. 1835, p. 174. China borealis. — Ind. Fl. Sin., ɪ, 396.

Patrinia hispida, Bge, Nov. Gen. etc. p. 176, c. icone. Prope Pekinum. — I. F. S., ɪ, 397 = *P. scabiosaefolia,* Fischer, 1826. (Dahuria).

Patrinia ovata, Bge, l. c. p. 174, c. icone. Prope Pekinum. — I. F. S., ɪ, 398 = *P. villosa,* Jussieu, 1823 (Japan).

Patrinia scabra, Bge, l. c. p. 171, c. icone. In montosis prope Pekinum. — I. F. S., ɪ, 398.

Inula ammophila, Bunge, in DC. Prod. V (1836) 470. In sabulosis Chinae borealis. — I. F. S., ɪ, 428.

Inula salicina, Pallas; Bge, Enum. n. 21. In herbidis prope Kantai.— Ruprecht, in Maxim. Amur. 149, considers it a new plant, *I. chinensis.* — I. F. S., ɪ, 428 = *I. Britannica,* Lin.

Eclipta thermalis, Bge, Enum. n. 224. Thermae Tan schan. — I. F. S., ɪ, 433 = *Eclipta alba,* Hassk. Pl. Jav. rar. (E. erecta, Lin.).

Artemisia eriopoda, Bge, Enum. n. 211. Pan schan, Guan gou. — I. F. S., ɪ, 442.

Artemisia monostachya, Bunge, fide sched. in herb. Acad. Mongolia, Bunge. Maxim. Ind. Mongol, in Prim. Amur. p. 482.

Cacalia aconitifolia, Bge, Enum. n. 208, Zui wei schan. — *Syneilesis* (nov. gen.) *aconitifolia,* Maxim. Amur. 165 et 473, c. icone, tab. 8. — I. F. S., ɪ, 449, *Senecio aconitifolius,* Turcz. in Pl. Kiril. Chinae borealis, 1837, 155.

Cineraria subdentata, Bunge, Enum. n. 218. Tan schan. — Maxim. Mél. VIII (1871) 15 = *Senecio campestris*, DC. Prod. VI, 361. (Cineraria campestris Retz Observ.) Europe, Siberia. — I. F. S., I, 450.

Acarna chinensis, Bge, Enum. n. 204. Zui wei schan, Guan gou. — *Atractylis chinensis*, DC. Prod. VI, 549. — I. F. S., I, 459 = *Atractylis ovata* et *lancea*, Thunb. Fl. Jap. 306.

Cirsium segetum, Bge, Enum. n. 202. Prope Pekinum. — Maxim., Mél. biol. IX (1874) 333, names it *Cnicus segetum*. — I. F. S., I, 462.

Cirsium lyratum, Bge, Enum. n. 203. Lun züan ssy, Tan schan. — In Dorpater Jahrb. I, 1833, 221, Bunge establishes upon this plant a new genus, and calls it *Hemistepta lyrata*. It was cultivated in the Bot. Garden, St. Petersb. See Fischer & Meyer Semina hti Petrop. II, 1835, 38. — DC. in Prod. VI, 539 calls it *Aplotaxis Bungei*. — Benth. & Hook. Gen. Pl. II, 472, 1236, are inclined to reduce Hemistepta to *Saussurea*. Franchet, Enum. Jap. calls the plant, therefore *S. Bungei*. — I. F. S., I, 463 = *S. affinis*, Spreng in DC. Prod. VI, 540. (India orient.).

Saussurea eriolepis, Bunge in litt. in DC. Prod. VI, 535. China borealis. — I. F. S., I, 464.

Saussurea pectinata, Bge in litt., 1833, in DC. Prod. VI, 538. China borealis. — I. F. S., I, 467.

Myripnois (gen. nov.) *dioica*, Bge, Enum. n. 213. Zui wei schan. — Earlier d'Incarville. — I. F. S., I, 472.

Prenanthes sonchifolia, Bge, Enum. n. 226 (non Willd.). Prope Pekinum. — *Youngia sonchifolia*, Max. Amur. 180. — *Lactura denticulata*, var. *sonchifolia*, Maxim. Mél. IX (1874) 360. — I. F. S., I, 480.

Sonchus lactucoides, Bge, Enum. n. 228. Kan tai. — Max. Ind. Fl. Pekin. in Prim. Amur. 474 = *Mulgedium tataricum*, DC. (Sonchus tataricus, Linn.). — I. F. S., I, 484.

Scorzonera albicaulis, Bunge, Enum. n. 230. Lun züan ssy. — I. F. S., I, 488.

Azalea macrantha, Bge, Enum. n. 235. Colitur in frigidariis Pekinensibus. — Maxim. Rhodod. (1870) 37, *Rh. indicum*, L. var. *macrantha*. — I. F. S., II, 25.

Rhododendron leucanthum, Bunge, Enum. n. 234. Colitur rarius in frigidariis Pekinensibus. — Maxim. l. c. 35 = *Rh. ledifolium*, G. Don, III (1834) 846, var. *leucanthum*. — I. F. S., II, 27.

Statice bicolor, Bge, Enum. n. 303. Tschadào, Kalgan, S. Mongolia. Bunge notices two varieties α *latiflora* and β *densiflora*. — Fisch. & Meyer Sertum Petropolitanum, 1846, tab. 17, col. drawing (var. α).

Statice Bungeana, Boissier in DC. Prod. XII, 642. This is var. β of St. bicolor. Bunge, Mongolia meridionalis. — Ind. Fl. Sin., ɪɪ, 35, sub St. bicolor.

Ceratostigma (new gen.) *plumbaginoides,* Bunge, Enum. n. 302. In umbrosis prope Pekinum unicum tantum specimen florens Bunge legit. — Boissier in DC. Prod. XII (1848) 695, named it *Valoradia plumbaginoides.* — Bot. Mag. tab. 4487 (1850). Same as *Plumbago Larpentae,* Gard. Chron. 1847, 732 (Fortune). — I. F. S., ɪɪ, 36.

Androsace saxifragaefolia, Bunge, Enum. n. 297. Prope Pekinum. — d'Incarville earlier (supra p. 53). — I. F. S., ɪɪ, 45.

Lysimachia barystachys, Bunge, Enum. n. 298. In montosis prope Pekinum. — d'Incarville earlier (supra p. 53). — Figured in Regel Fl. Ussur. tab. 9. — Carrière in Rev. hort. 1881, 90, with col. drawing. Cultivated, Paris. Dr. Regel brought the living plant, in 1878, from Bot. Garden St. Petersb. to the Paris exhibition. — I. F. S., ɪɪ, 47.

Lysimachia pentapetala, Bunge, Enum. n. 299. Kantai, Guan gou. Duby in DC. Prod. VIII (1844) 67 founded upon this plant a new genus: *Apochoris pentapetala.* — In Ind. Fl. Sin., ɪɪ, 55. Bunge's old name is restored.

Diospyros Schitse, Bge, Enum. n. 237. Saepe culta, fere spontanea ad radices montium Pekin. — I. F. S., ɪɪ, 69. *D. Kaki,* L.

Jasminum floridum, Bge, Enum. n. 239. Quasi spontaneum prope Kantai (ubi colitur). — Bot. Mag. t. 6719 (1883). — I. F. S., ɪɪ, 78.

Jasminum angulare, Vahl, (Cape of G. H.); Bunge, Enum. n. 238. Frequens colitur ob florum multitudinem. — Lindley in Journ. Hort. Soc. 1846, 153, proved that this is a new species: *J. nudiflorum.* — Bot. Reg. 1846, tab. 48. — I. F. S., ɪɪ, 79.

Syringa chinensis, Willd. Bge, Enum. n. 241. Frequens in hortis. — In Gard. Chron., 1859, 868, this Peking plant was described as a new species, *S. oblata,* by Lindley. — I. F. S., ɪɪ, 83.

Fraxinus floribunda, Wallich. Bunge, Enum. n. 343. Frutex. Montes prope Lun züan ssy et Si jü ssy. — DC. Prod. VIII (1844) 275, found that this was not the Indian species, but a new plant which he called: *Fr. Bungeana.* — I. F. S., ɪɪ, 84.

Periploca sepium, Bge, Enum. n. 244. In montibus prope Pekinum.— Earlier d'Incarville (supra p. 53). — I. F. S., ɪɪ, 101.

Asclepias paniculata, Bge, Enum. n. 245. Lun züan ssy. In DC. Prod VIII (1844) 512, Bunge calls the same plant *Pycnostelma chinense,* (new genus), — I. F. S., ɪɪ, 102.

Cynanchum atratum, Bge, Enum. n. 251. Si jü ssy, Tan schan. — I. F. S., ɪɪ, 104.

Cynanchum versicolor, Bunge, Enum. n. 250. Lun züan, Zui wei shan. — I. F. S., ɪɪ, 109.

Cynanchum pubescens, Bge, Enum. n. 248. China borealis. — D'Incarville earlier (supra p. 52). — Hemsley, I. F. S., ɪɪ, 105, finds it identical with *C. chinense*, R. Brown in Mem. Wern. Soc. I (1808) 44. (Staunton).

Asclepias hastata, Bunge, Enum. n. 246. Zui wei schan. — Decaisne in Prod. VIII (1844) 549 = *Cynanchum Bungei*. — Turczaninow in Bull. Mosc. XXI, 1848, 255 = *Symphyoglossum* (new gen.) *hastatum*. — I. F. S., ɪɪ, 105. *Cyn. Bungei*.

Urostelma (nov. gen.) *chinense*, Bge, Enum. n. 247. Kantai, Dsü jung guan. — Decaisne in Prod. VIII, 511, found it identical with *Metaplexis Stauntoni*, R. Brown (1808). (Staunton, v. s. p. 177). — I. F. S., ɪɪ, 110.

Ophelia chinensis, Bunge, MSS. in DC. Prod. IX (1845) 126. China borealis. — I. F. S., ɪɪ, 139. *Swertia chinensis*.

Bothriospermum (nov. gen.) *chinense*, Bunge, Enum. n. 266. Prope Pekinum. — Earlier Incarville (v. s. p. 53). — Cultivated in Bot. Garden St. Petersb. See Ind. Sem. Petrop. 1834, 23, sub *B. bicarunculatum*. — I. F. S., ɪɪ, 151.

Cuscuta fimbriata, Bge, MSS. Peking. Turcz. Plantae Kirilovi 1837, n. 148. — Engelmann, Cuscuta, 1860 = *Cuscuta chinensis*, Lamarck. — I. F. S., ɪɪ, 168.

Solanum septemlobum, Bge, Enum. n. 272. Prope Pekinum nec non usque ad fines mongolicos. — Earlier d'Incarville. — I. F. S., ɪɪ, 172.

Mimulus tenellus, Bge, Enum. n. 278. Pan schan. — I. F. S., ɪɪ, 181.

Tittmannia obovata, Bunge, Enum. n. 279. Si jü ssy. — *Vandellia obovata*, Walp. Rep. III, 249. — Maxim. Mél. IX (1874) 402, found that Bunge's plant was *Mazus rugosus*, Loureiro. — I. F. S., ɪɪ, 183.

Gerardia glutinosa, Lin.; Bunge, Enum. n. 280. Frequens prope Pekinum. — Bunge was mistaken. When the plant was cultivated in the Bot. Garden St. Petersb., I understand raised from seeds brought by Bunge, it turned out to be *Rehmannia glutinosa*, Liboschitz *). See Fischer & Meyer.

---

*) The plant in question had been cultivated 65 year earlier at St. Petersburg by Gaertner, who named it *Digitalis glutinosa* (v. supra p. 115). Joseph **Liboschitz,** according to Pritzel was a physician who lived in St. Petersburg about 1811, and died at Vienna in 1824. In his herbarium was found the plant upon which he had founded the new genus *Rehmannia*, but not published. The genus name refers most probably to Dr. **Rehmann,** who in 1805, accompanied the Russian Embassy, headed by Count Golovkin, to China. But the Embassy was obliged to return from Urga. See my Botanicon Sinicum I, 193.

Ind. Sem. hti Petrop. 1835, 36, *Rehmannia chinensis,* Fischer, where it is stated that the plant grows in N. China and Mongolia. I do not think that it is found in Mongolia. Rehmannia glutinosa (or chinensis) was figured in Bot. Reg. tab. 1960, (1837) and in Bot. Mag. tab. 3653, (1838), from plants sent by Fischer. — I. F. S., ii, 193.

Siphonostegia chinensis, Bentham, Scroph. Ind. 1835, 51. The genus Siphonostegia was established by Benth. upon incomplete specimens gathered by Vachell near Macao. Time not given, 1825—31 (v. supra p. 296). Bunge in Bull. scient. Acad. St. Petersb. VII (1840) 273, on the Genus Siphonostegia, reports that he saw the same plant, in July 1831, in the Guan gou defile, north of Peking, and received afterwards from a Peking friend complete herbarium specimens. Bunge gives a detailed description of it. — I. F. S., ii, 202.

Phtheirospermum (new genus) *chinense,* Bge in Fischer & Meyer, Ind. Sem. hti Petrop., 1834, 37. China borealis. The plant was raised in the Bot. Garden from seeds brought by Bunge. — I. F. S., ii, 204.

Orobanche canescens, Bge, Enum. n. 301. Lun züan ssy. — As this specific name was preoccupied by a species of Presl's (1820), Steudel, Nomencl. Bot., proposed the name *O. albo-lanata.* — I. F. S., ii, 221.

Dorcoceras hygrometrica, (new genus), Bge, Enum. n. 301. Zui wei schan, Guan gou. — R. Brown, in Plant. Jav. rar. (1840) 120, named the plant *Boea hygrometrica.* — Figured in Bot. Mag. tab. 6468 (1879). — I. F. S., ii, 234.

Catalpa syringaefolia, Sims., Bge, Enum. n. 254. Frequens in hortis Pekinensibus. — C. A. Meyer, Dissert. Catalpa, in Bull. scient. de l'Acad. St. Petersburg, II (1837) 49, describes it as a new species, *C. Bungei.* — I. F. S., ii, 234.

Clerodendron foetidum, Bge, Enum. n. 296. Colitur. As this generic name was preoccupied by a species of D. Don's, Steudel, Nomencl. Bot., proposed the name *Cl. Bungeanum.* — I. F. S., ii, 259.

Caryopteris (gen. nov.) *mongolica,* Bunge, Nov. Gen. et Spec. China, Mongolia, 1835, 178. Mongolia. — Coloured figure in Rev. hort., 1872, 451. It is there stated, that the plant was cultivated in the Garden of the Museum in 1845, but subsequently disappeared. — I. F. S., ii, 264.

Salvia milthiorrhiza, Bge, Enum. n. 284. Zui wei schan, Si jü ssy. — D'Incarville earlier. — I. F. S., ii, 286.

Salvia minutiflora, Bge, Enum. n. 285. In montosis prope Pekinum.— According to Benth. in DC. Prod. XII, 355 = *S. plebeja,* R. Brown, Prod. Nov. Holl., 1810, 501. — I. F. S., ii, 287.

Scutellaria viscidula, Bge, Enum. n. 294. China borealis, Dsi min i, Jü lin. — I. F. S., ɪɪ, 298.

Stachys chinensis, Bge in Bentham Labiatae (c. 1835) 544. China borealis. DC. Prod. XII, 471. — Maxim. Fragm. 1879, 45 = *St. aspera*, Michx. Fl. Amer. bor. var. *chinensis*. — I. F. S. ɪɪ, 300.

Stachys affinis, Bge, Enum. n. 289, Kantai. — Maxim. Fragm. 1879, 46, same as *St. Sieboldi*, Miq. Prol. Jap. (1868) 64. Siebold discovered it earlier than Bunge. — I. F. S., ɪɪ, 301, prefers the name *St. Sieboldi*, because the name St. affinis is already occupied, since 1833, for an Arabian plant.

Lagochilus ilicifolius, Bge in Bentham, Labiatae (c. 1835) 641, DC. Prod. XII, 515. Mongolia australis, Bge.

Ajuga ciliata, Bge, Enum. n. 287. Si jü ssy. — I. F. S., ɪɪ, 314.

Ajuga multiflora, Bge, Enum. n. 286. Prope Tschan pin dschou. — Bentham in DC. Prod. XII, 596, reduces it to *A. geneviensis*, Lin. — I. F. S., ɪɪ, 315.

Agriophyllum gobicum, Bge in Enum. Salsolacearum omnium in Mongolia hucusque collectarum, Mél. biol. X (1879) 284. Bunge, Mongolia media et australis.

Kochia mollis, Bunge, l. c. 283. In medio deserto Gobi. Bunge, 1830, 1831.

Kalidium gracile, Fenzl. in Ledeb. Fl. Ross. III, 769, note. — Bunge l. c. 287: in deserto Gobi prope Chailassutu, Boroldschi.

Schoberia glauca, Bge, Enum. n. 310. Kantai. — *Suaeda glauca*, in Bge, Salsol. Mongol. 293, and in Bunge, Salsolaceae Chinae, Japoniae et Mandshuriae in Herbar. Petropol., in Acta Hti Petrop. XIII, 1893, 21. — I. F. S., ɪɪ, 328.

Halogeton arachnoides, Moquin in DC. Prod. XIII, ɪɪ, 205; Bunge Sals. Mongol. 305. Bunge, Mongolia australis, 1831.

Calligonum gobicum, Bge, MSS. See Meisner in DC. Prod. XIV, 29 = variety of *C. mongolicum*, Turcz. 1832, in Decad. tres etc.

Polygonum pensylvanicum, Lin.; Bunge, Enum. n. 321. Kantai. — According to Turczaninov, Bull. Mosc. 1840, 77, Bunge's plant is a new species, *P. Bungeanum*. — I. F. S., ɪɪ, 335.

Polygonum interruptum, Bunge, Enum. n. 323. Zui wei schan. — I. F. S., ɪɪ, 341.

Aristolochia contorta, Bunge, Enum. n. 328. Zui wei schan, Dsü jun guan. — I. F. S., ɪɪ, 361.

Passerina Chamaedaphne, Bunge, Enum. n. 326. China borealis. —

Meisner, in DC. Prod. XIV, 547, named it *Wikstroemia Chamaedaphne.*—
I. F. S., II, 397.

Euphorbia lunulata, Bge, Enum. n. 330. Montes Pekinenses. — D'Incarville earlier (v. supra p. 54). — I. F. S., II, 415.

Andrachne chinensis, Bge, Enum. n. 332. Frequens in montosis circa
Pekinum. — D'Incarville earlier (supra p. 54). — I. F. S., II, 420.

Croton tuberculatus, Bge, Enum. u. 334. Lun züan ssy, Si jü ssy. —
Müller, Arg. in DC. Prod. XV, II, 734, calls it *Argyrothamnia tubercu-
lata.* — *Speranskia tuberculata,* Baillon, Étude Gén. Euphorb. — Ind. Fl.
Sin., II, 436.

Celtis chinensis, Pers.; Bunge, Enum. n. 345. In montosis circa Pekinum, non rara. — Blume, Mus. Lugd. Bat. II, 1852, 71, described it
as a new species, *C. Bungeana.* — I. F. S., II, 449.

Morus constantinopolitana, Poiret; Bunge, Enum. n. 341. Zui wei
schan. — Bureau in DC. Prod. XVII, 241 = *Morus alba,* Lin., var. *Bungeana.* — I. F. S., II, 455.

Quercus chinensis, Bunge, Enum. n. 347. Zui wei schan. DC. Prod.
XVI, II, 50. — As this specific name was already occupied for one of Abel's
oaks (supra p. 229), Mr. F. B. Forbes, in Journ. Bot. 1884, proposed the
name *Q. Bungeana.*

Quercus obovata, Bge, Enum. n. 348. In montosis prope Pekinum. —
DC. Prodr. XVI, II, 13. — According to Hance, Adversaria (1864) 243,
same as *Q. dentata,* Thbg. Japan.

Pinus Bungeana, Zuccarini, in Endlicher, Syn. Conif., 1847, 166.
Zuccarini had received herbarium specimens of this brought by Bunge from
Peking. Bunge, in his Enumeratio does not mention this splendid tree, the
white-barked Pine, so common near Buddhist monasteries. — Some years
ago Dr. A. Henry discovered it, in its wild state, among the mountains
of Hupeh.

Iris oxypetala, Bge, Enum. n. 358. Frequens in montosis et pratensi-
bus. — D'Incarville earlier. — Maxim. Mél. X (1880) 699 = *Iris ensata,*
Thunb. var. *chinensis.*

Iris Bungei, Maxim. l. c. 695. Mongolia australis, Bunge.

Asparagus Gibbus, Bunge, Enum. n. 370. In cultis prope Pekinum,
frequens. — Baker, Asparagaceae, in Journ. Linn. Soc. XIV (1875) 599 =
variety of = *A. dauricus,* Fischer.

Asparagus trichophyllus, Bge, Enum. n. 369. Lun züan ssy. — D'Incarville earlier. — Baker, l. c. 600.

Polygonatum macropodum, Turcz., Pl. Kiril. Chin. 1837, n. 195. — Maxim. Mél. XI (1883) 848. Bunge, cemet. Dyn. Ming.

Anemarrhena (nov. gen.) *asphodeloides,* Bge, Enum. n. 373. Montes prope Lun züan ssy.

Allium macrostemon, Bge, Enum. n. 372. Prope Pekinum. — Omitted in Baker's Alliums of India, China, Japan, Journ. Bot. 1874.

Arum macrourum, Bunge, Enum. n. 379. Zui wei schan, Lun züan ssy. — H. G. Schott, Prod. Aroid (1860) 20 = *Arum ternatum,* Thunb. Jap. = *Pinellia tuberifera,* Tenore, Sem. hti Neap. 1830.

Sagittaria macrophylla, Bge, Enum. n. 355. In stagnis prope Pekinum (culta). — DC. Monogr. III, 1881, Micheli, Alismac. = *S. sagittaefolia,* Lin., var. *diversifolia,* Micheli.

Carex leucochlora, Bge, Enum. n. 386. Zui wei schan.

Carex heterostachya, Bge, Enum. n. 387. Prope Pekinum. — As this name was already occupied by Torrey & Desvaux, Debeaux, Fl. Tientsin, 45, proposes the name *C. Bungeana.* — According to Maximowicz, Fragmenta, 1879, 66 = *Carex nutans,* Host., Gramin. (1801) fide Boot.

Carex heterolepis, Bge, Enum. n. 388. Pan schan.

Lappago racemosa, Willd.; Bge, Enum. n. 401. Nanj kou. — According to Debeaux, Fl. Tientsin, 48, a new plant, *Tragus tcheliensis.*

Hierochloee borealis, Roem. & Schult.; Bunge, Enum. n. 394. Pan schan. — *H. Bungeana,* Trinius, in Acad. Petrop. sciences nat. 6 série, III, 1840, Trin. Phalarid. 82. Tan schan, Bunge. This plant was unknown to Maximowicz. No specimen in the herbariums of St. Petersb.

Stipa Bungeana, Trinius, in Bge, En. n. 402. Zui wei schan, Si jü ssy.

Chloris caudata, Trinius, in Bge, Enum. n. 404. Prope Pekinum.

Eragrostis orientalis, Trinius, in Bge, Enum. n. 410. Ubique frequens.

Melica scabrosa, Trinius, in Bge, Enum. n. 411. Zui wei schan, in ruderatis prope Pekinum. — Incarville earlier (v. supra p. 54).

Poa linearis, Trinius, in Bge, Enum. n. 406. Zui wei schan.

Poa spondylodes, Trinius, in Bge, Enum. n. 407. Pan schan, Si jü ssy.

Triticum ciliare, Trinius in Bge, Enum. n. 415. Kan tai.

Triticum chinense, Trinius in Bge, Enum. n. 416. Kan tai.

Selaginella mongolica, Ruprecht, Distrib. Cryptogam. Ross. in Beitr. z. Kenntn. Pflanzenk. Russl., 1845, 32. Bunge? Earlier d'Incarville (v. supra p. 54).

Adiantum Capillus Junonis, Ruprecht l. c. 49. China borealis. Bge.— Hooker & Baker, Syn. Filicum. 114.

Athyrium sinense, Ruprecht, l. c. 41. China borealis, Bunge. — Not noticed in Synopsis Filic.

List of Bge's botanical papers, in which East. Asiat. plants are dealt with:
Enumeratio Plantarum Chinae borealis, 1832. (v. s. p. 325).
Hemistepta lyrata, 1833. (v. s. p. 335).
Phtheirospermum, 1834. (v. s. p. 338).
New Genera and Species of Chin. and Mong. Plants, 1835. (v. s. p. 326).
Siphonostegia, 1840. (v. s. p. 338).
Sarcozygium, 1843. (v. s. p. 329).
Tamarix, 1852. (v. s. p. 329).
Generis Astragali Species Gerontogeae. Mém. Acad. St. Pétersb. 7série, XI (1868) n. 16, and XV (1869) n. 1.
Astragali Turkestanici. Acta Hti Petrop. III (1874) 101—107.
Species Generis Oxytropis. Mém. Acad. St. Pétersb., 7série, XXII (1874) n. 1.
Astragaleae Turkestaniae, written in 1876, published in 1880 in Fedchenko's Travels, fasc. 15, p. 160—316. — A Supplement appeared in Acta Hti Petrop. VII (1880) 363—380.
Salsolacearum novarum Turkestaniae Descriptiones. Acta Hti Petrop. V (1878) 642— 46.
Enumeratio Salsolacearum Mongolicarum, 187. (v. s. p. 339).
Plantagineae Centralasiaticae. Acta Hti Petr. VI (1880) 392—94.
Enumeratio Salsolacearum Centralasiaticarum. Ibid. 403—459.
Salsolaceae Herbarii Petropolitani, in China, Japonia et Mandshuria collectae. Acta Hti Petrop. XIII (1893) 15—22.

Ladijenski, Colonel Michael Vassilievich, pristav (supra p. 324) of the Eccles. Mission, with which Bunge went to China, in 1830. He collected some plants in the neighbourhood of Peking and in Mongolia, which are found in the herbariums of the Academy and the Bot. Garden, St. Petersburg. His plants are occasionally noticed by Maximowicz. We owe to L. the first map of the Environs of Peking, published in 1848 by the Russian General Staff and of which an English edition was published at London in 1860. L. returned in 1831 with the 10th Eccles. Mission and Bunge to Russia, was subsequently entrusted by the government with important commissions, in 1858 was made Lieutenant-General; and in 1865 retired from the service.

Kuznetsov, Ilia*), a young Cossack who used to accompany Turczaninov (see further on) on his travels in Cis- and Trans-Baicalia, and whom he
sent with the Eccles. Mission in 1830 to Peking, to collect plants for him
(Turcz. Fl. baical. dahur. Preface, p. 19).

The new plants gathered by Kuznetsov in 1831 between Kalgan and
Kiakhta were described by Turczaninov in 1832 in the Bull. Soc. Natur.
Mosc. V, p. 180—206: Decades tres plantarum novarum Chinae
borealis et Mongoliae Chinensi incolarum.

Turczaninov, here, does not say that these plants were collected by
Kuznetsov, but Bunge, who described several plants of this collection, positively states that it was made by Kuznetsov. The latter does not seem to
have been in Peking, having probably been left with the other Cossacks,
forming the escort, at Tsagan balgassu (v. supra p. 324). But he visited Kalgan.

New plants gathered by Kuznetsov:

Clematis aethusaefolia, Turcz. Bull. Mosc. V, 1832. Decades tres
(v. supra) n. 2. Mongolia chinensis ad limites Chinae. — Maxim. Mél. IX
(1876) 586, Fl. Mongol. (1889) 5. — I. F. S., ι, 1.

Clematis fruticosa, Turcz. l. c. n. 1. Mongolia chinensis. — Maxim.
Mél. l. c. 582, Fl. Mongol. 2.

Atragena macrosepala, Ledeb. Fl. Altaica, II, 376, in nota, and Ledebour Icon. Pl. Fl. Ross. tab. 11. — In Turcz. Fl. Baic. dahur. I, 26 we
read: "Ledebour in Fl. Alt. descripsit A. macrosepalum, in horto Dorpa-
tensi cultam, quam in Dahuria, prope Nerczinsk provenire dicit. Hujus
specimina e China boreali a Kuznetsowio lecta accepi, in Dahuria vero
nullibi vidi". — I. F. S., ι, 5.

Hesperis trichosepala, Turcz. 1832, l. c. n. 3. Prope oppidum Kalgan,
Junio m. (1831), China borealis. — In Ind. Fl. Sin., ι, 45, this is reduced
to *H. aprica,* Poiret, but Maxim. Fl. Mongol. 53, maintains it as a distinct species.

Viola micrantha, Turcz. 1832, l. c. n. 4. Kalgan, Majo m. 1831,
China borealis. — Maxim. Mél. IX (1876) 746, same as *V. acuminata,*

---

*) There are several Russian botanical collectors and botanists of this name. I may
mention the K. to whom Reichenbach, Illustr. Spec. Aconit. (1823) n. 38, tab. 21, dedicated the
name *A. Kusnetzowii,* and who had brought the plant from Yakutsk. We will have to speak,
further on, of D. Kuznetsov, a naval officer, who collected, in 1854—55, plants on the Manchurian coast and in Japan. A young intelligent botanist of the same family—name, likewise
interested in Asiatic plants, was for some years Assistant to the Herbarium of the Botan.
Gardens, St. Petersburg. In 1896 he was appointed Professor of Botany at Dorpat University.
See farther on.

Ledeb. = *V. canina*, Lin., var. β *acuminata*, Rgl. known also from Siberia, but it seems Kuznetsov discovered the plant first in China. — Turczaninov's specific name micrantha was preoccupied by a species of Presl's (Sicilia). I. F. S., 1, 52. — Maxim. Fl. Mongol. 80.

Dianthus foliosus, Turcz. l. c. n. 5. Mongolia chinensis. — Maxim. Fl. Mongol. 84, = variety of *D. chinensis*, Lin.

Phaca macrostachys, Turcz. Bull. Mosc. XIII (1840) 66. Mongolia chinensis. — Bge, Monogr. Astragalus, II, 1869, in Mém. Acad. Petrop. 7sér. XV, 30, named this plant: *Astragalus mongholicus*, Kuznetsov, Mongolia australis.

Astragalus tenuis, Turcz. Catal. baic., 1838, 354. Only the name. — Bge, Monogr. Astrag. II, 21. Kuznetsov, Mongolia chinensis. — Ind. Fl. Sin., 1, 167.

Phaca brachycarpa, Turczan. Bull. Mosc. V, 1832, n. 6. Mongolia chinensis, July (1831). — Bge, Astrag. II, 27, names the plant *Astragalus zacharensis* *). Kuznetsov, Zagan balgassu, Mongoliae australis.

Oxytropis ciliata, Turcz. l. c. 1832, n. 7. Tsagan balgassu, Mongolia chinensis australis, Majo m. (1831). — Bunge, Monogr. Oxytropis, in Mém. Acad. Petrop. 7sér. XII, 1874, 130. Zagan balgassu, Kuznetsov.

Oxytropis ochrantha, Turcz. l. c. 1832, n. 9. Prope Tsagan balgassu, Mongoliae chinensis, australis, Majo m. (1831). — Bunge, Oxytropis 149. Zagan balgassu, Kuznetsov.

Oxytropis racemosa, Turcz. l. c. 1832, n. 8. Mongolia chinensis prope Chadatu, Augusto m. (1831). — Bge, Oxytrop. 161. Chadatu, Kuznetsov.

Amygdalus pilosa, Turcz. l. c. 1832, n. 10. Prope Tsagan balgassu, Majo m. (1831). — Maxim. Mél. XI (1883) 664, *Prunus pilosa*.

Spiraea pubescens, Turcz. l. c. 1832, n. 11. Kalgan, Majo m. (1831).— I. F. S., 1, 227.

Cotoneaster acutifolia, Turcz. l. c. 1832, n. 12. Mongolia chinensis australis, Junio, Julio (1831). — I. F. S., 1, 260, reduced to *C. integerrima*, Med. of W. Europe (Mespilus Cotoneaster, Lin.).

Ribes pulchellum, Turcz. l. c. 1832, n. 13. Mongolia chinensis ad limitem Chinae et prope oppid. Urga. — Maxim. Mél. IX (1873) 241.

Peucedanum Falcaria, Turcz. l. c. 1832, n. 14. — Max. Ind. Mongol. 482. I have not found this name anywhere else.

Aster alyssoides, Turcz. l. c. 1832, n. 21. Mongolia chinensis, prope Chapchaitu, Augusto m. (1831). — I. F. S., 1, 409.

---

*) This specific name is evidently taken from Chakhar, as the grassy steppes of S. Mongolia are called, where the Chakhar Mongols live.

Conyza salsoloides, Turcz. l. c. 1832, n. 20. Mongolia chinensis, circa Chadatu, Augusto m. (1831). — DC. Prod. V, 470 = *Inula ammophila,* Bge, β. *salsoloides.*

Artemisia achilleoides, Turcz. l. c. 1832, n. 18. Mongolia chinensis, circa Mogoitu, Augusto m. (1831). — DC. Prod. VI, 130 = *Tanacetum achilleoides.* — Maxim. Mél. XI, 1872, 520. Dubious plant.

Artemisia trifida, Turcz. l. c. 1832, n. 19. Cum priore. — DC. Prod. l. c. = *Tanacetum trifidum,* Maxim. l. c. Dubious plant.

Cineraria mongolica, Turcz. l. c. 1832, n. 22. Mongolia australis, circa Boroldschi, Julio m. (1831). — DC. Prod. VI, 315, *Ligularia mongolica.* — Schultz, Bip., Flora (1845) 50, *Senecio mongolicus.* — Maxim. Mél. XI (1881) 242. — I. F. S., ɪ, 155.

Echinops Gmelini, Turcz. l. c. 1832, n. 17. Mongolia chinensis media et australis. — Turczan. believed that the plant he described was identical with a species recorded by Gmelin from Siberia. — Ledebour found that he was mistaken, and named the present Mongolian plant *E. Turczaninowii* (Ledeb. Fl. Ross. II, 1884, 657). — *E. Gmelini,* Ledeb. Fl. Altaica IV, 1833, 45, is quite different.

Carduus leucophyllus, Turcz. l. c. 1832, n. 16. Mongolia chinensis australis, prope stationem Taltun. — DC. Prodr. VI, 623. — Maxim. Mél. IX (1874) 303.

Saussurea intermedia, Turcz. l. c. 1832, n. 15. Mongolia chinensis.— DC. Prod. VI, 537.

Scorzonera divaricata, Turcz. l. c. 1832, n. 23. Mongolia chinensis.— DC. Prod. VII, 125. — Maxim. Mél. XII (1888) 737.

Statice tenella, Turcz. l. c. 1832, n. 26. Mongolia chinensis. — DC. Prod. XII, 641.

Androsace longifolia, Turcz. l. c. 1832, n. 25. Mongolia chinensis australis. — DC. Prod. VIII, 49. — Maxim. Mél. XII (1888) 749.

Bothriospermum Kusnetzowii, Bunge, in Delectus seminum horti bot. Dorpatensis, 1840, p. VII, note. "Kusnetzow, Mongolia chinensis ad fines chinenses". B. chinense Fischer & Meyer, (non Bunge) in Ind. Sem. hti Petropol., 1835, 23. It was cultivated in Dorpat and St. Petersburg. — DC. Prod. X, 116. — Maximovicz, Mél. VIII, 1872, 560. — Ind. Fl. Sin., ɪɪ, 151.

Convolvulus tragacanthoides, Turcz. l. c. 1832, n. 24. In ruinis oppidi mongolici Tsagan balgassu, Majo m. (1831). — DC. Prod. IX, 400.

Calligonum mongolicum, Turcz. l. c. 1832, n. 27. Mongolia chinensis. — DC. Prod. XIV, 29.

Diarthron (gen. nov. Thymel.) *linifolium,* Turcz. l. c. 1832, n. 28. In montosis Mongoliae chinensis borealis, inter oppidum Urga et Kiachta, Augusto m. (1831). — D'Incarville (v. supra p. 54) discovered this plant more than 80 years earlier in the Peking mountains. It has also been gathered by David near Jehol and by Fauvel at Chefoo.— I. F. S., ii, 401.

Polygonatum macropodum, Turcz. l. c. 1832, n. 29. Prope oppidum Kalgan, Junio m. (1831) lectum. — Maxim. Mél. XI (1838) 849.

Allium condensatum, Turcz. Bull. Soc. nat. Mosc. XXVII (1854) II, 121; Baicalia. — Maxim. Amur, 281: Mongolia, Turcz. Pl. exsicc. — In Herb. Hti Petrop. is a specimen of this from Mongolia, 1831. Probably Kuznetsov. — Regel, Allium, 1875, n. 70.

Woodsia subcordata, Turcz. l. c. 1832, n. 30. China borealis, Junio m. (1831). — Not found in Hook. & Baker's Syn. Filic.

As Turczaninov reports (Fl. Baic. Dahur. I, 20), Kuznetsov died (in about 1835) on his way home from a jurney to Okhotsk.

Rozov, Grigori, a young man who went to Peking with the 11th Russian Eccles. Mission, to learn the Chinese, Manchu and Mongolian languages there. He was, in 1830 Bunge's fellow-traveller through Mongolia to Peking, and when Rozov ten years latter returned from China, he transmitted to Bunge a small collection of plants made by him near Peking and in Mongolia during the homeward journey in 1841. See Bunge in Linnaea XVII, 1843, p. 5. Bunge calls him missionary, but Professor V. P. Vasiliev (with the Mission in Peking, 1840—50) told me that he was not a missionary but a student, attached to the Mission.

Bunge notices from Rozov's collection two novelties.

Zygophyllum Rosowii, Bge in Linnaea XVII, 1843, 5. Mongolia. — Maxim. Fl. Mongol. 125.

Oxytropis gracillima, Bge, Linnaea, l. c. and Bunge, Oxytrop. 1874, 160. Mongolia media.

Kirilov *), Porphyri Yevdokimovich. Born 1801. He studied medicine at the Medico-Chirurg. Academy, St. Petersb., where he took his degree of M. D. He was appointed Physician to the 11th Russian Eccles. Mission, and in company with Bunge in 1830 proceeded with this mission (v. supra

---

*) He is not to be confounded with Kirilov, Ivan (Jean), his contemporary, who in 1836 collected for Turczaninov in the regions near Lake Baikal (Flora Baik.-Dahur. I, 20) and who is, I suspect the same J. Kirilov, who in 1841 published with G. Karelin: «Enumeratio Plant. in desertis Soongoriae orientalis et in jugo Alatau 1841 collectarum».

p. 324) through Mongolia to Peking. During his long residence of more than ten years in the Chinese capital he devoted himself to the investigation of the flora of the Peking plain and the adjacent mountains. He was the first botanist to visit, in about 1835, the celebrated Mount Po hua shan, about 60 miles west of Peking. He published an account of his journey in the "Strekoza", a Russian newspaper then appearing at Kiakhta, as the late Archimandrite Palladius, who knew K. personally, informed me.

Dr. Kirilov left Peking in May 1841 and returned with the Eccles. Mission to Russia. Having traversed Mongolia, they reached Kiakhta towards the end of the summer. In 1842 he went to St. Petersburg and was appointed Interpreter for Chinese to the Asiatic Department of the Foreign Office. He died in 1864 (?).

Kirilov's botanical collections were made in the neighbourhood of Peking, 1831—40, and in Mongolia in 1830, and more especially in 1841 on his way home. His first collection of Chinese plants, 200 species, K. forwarded to Turczaninov, who then was living in Eastern Siberia. Turczaninov described and enumerated them in "Bulletin de la Société des naturalists de Moscou", X (1837) II, p. 148—158: ENUMERATIO PLANTARUM QUAS IN CHINA BOREALI COLLEGIT ET MECUM, COMMUNICAVIT CL. MEDICUS MISSIONIS ROSSICAE PORPHYRIUS KIRILOV.

Kirilov seems also to have sent, from Peking, plants to Dr. Fischer, Director of the Bot. Gard. St. Petersb., or perhaps he handed these plants over to him, in 1842, after his return from China. In Fischer's herbarium, now in the herbarium of the Bot. Garden, there are many of Kirilov's Chinese and Mongolian specimens. The original labels have not been preserved, and all the labels are in Fischer's handwriting: "Kirilov, China borealis or Mongolia Chinensis" *). Very seldom particulars. The dates are frequently erroneously given, as e. g. 1831, 1843 for Mongolian plants. The same errors are occasionally met with in Maximowicz's Flora Mongol., for Maxim. copies Fischer's labels. Kirilov's plants are likewise found in the Bot. Museums of the Acad. of Sc. and the Med. Chirurg. Academy.

In 1842 Kirilov presented to the apothecary G. Gauger, St. Petersb. an interesting collection of 54 Chinese drugs, which the latter described and figured in Repertorium für Pharmacie etc. in Russland, vol. VII (1848), p. 565, seqq.

---

*) By this appellation Mongolia proper subject to China is meant, extending from the Great Wall to Kiakhta, in opposition to Mongolia Rossica or Transbaicalia, likewise inhabited by Mongols.

New Chinese and Mongolian Plants discovered by Dr. P. Kirilov, and described by C. A. Meyer, Ruprecht, Regel, Maximowicz.

As to the Mongolian plants, some of them may have been discovered at the same time by Bunge.

Clematis Kirilowii, Maxim. Mél. IX (1876) 583. Near Peking.

Clematis tubulosa, Turcz., Pl. Kiril. 1837, n. 3. China borealis. The plant was cultivated in 1846 in Chelsea; Bot. Mag. tab. 4269. I suppose it had been received through the medium of Fischer. — Maxim. Mél. IX (1876) 589. — I. F. S., ɪ, 4, reduced to *Cl. heracleaefolia,* DC. Syst. I, 138 (Staunton).

Anemone barbulata, Turcz. Pl. Kiril. Chin. borealis, 1837, n. 6. — I. F. S., ɪ, 10.

Kadsura chinensis, Turcz. l. c. n. 14. Chin. bor. — *Maximowiczia chinensis,* Rupr. & Maack in Mél. biol. II (1857) 439. — I. F. S., ɪ, 25, *Schizandra chinensis,* Baillon.

Myricaria brevifolia, Turcz., Decad. quatuor Pl. nov. in Bull. Soc. nat. Mosc., vol. XIII (1840) n. 21. Mongolia chinensis. — Probably received from Kirilov. — Maxim. Fl. Mong. 113: probably a variety of *M. germanica,* Lin. var. *alopecuroides,* Max., Fl. Tangut.

Tilia pekinensis, Ruprecht, Mél. biol. II, 1856, 413 and 1857, 519. Kirilov, Peking. — Maxim. Mél. X (1880) 586 = *Tilia mandshurica,* Rupr. l. c. (Maximowicz, Amur). — I. F. S., ɪ, 94.

Caragana rosea, Turcz.; Max. Ind. Pekin. 470. — This is *C. frutescens,* DC., var. floribus roseis, Turcz. Pl. Kiril. Chin. bor. 1837, n. 55.

Astragalus monophyllus, Bge, in Max. Mél. X, 1880, 642. Mongolia, ad tractum mercatorium prope Zagan tugurik, Kirilov.

Hedysarum mongolicum, Turcz., Fl. Baical-dahur. 337. Mongolia chinensis. — Maxim. Mél. XI (1881) 211. Kirilov, ad tractum mercatorium orientalem. 1830.

Spiraea Kirilowii, Regel & Tiling, Fl. Ajan. (1858) 81, in nota. Colitur in hort. Pekin. — Maxim. Spiraeac., 1879, 225, sub *Sorbaria Kirilowii.* — I. F. S., ɪ, 227 = *Spiraea sorbifolia,* Lin.

Sorbus discolor, Maxim. Amur, 103, in nota. China borealis. — In herb. Mus. Acad. Petrop.: Kirilov, Peking. — Max. Mél. IX (1873) 170 = variety of *Sorbus Aucuparia,* Lin. — I. F. S., ɪ, 255.

Saxifraga pekinensis, Maxim. Amur, 120, in nota; Peking. — Herb. Mus. Acad., Kirilov. — I. F. S., ɪ, 268.

Hydrangea vestita, Wall. Tent. Nepal., var. *pubescens,* Max. Hydrang., 1867, 10. Peking, Kirilov, Tatarinov. — I. F. S., ɪ, 274.

Philadelphus pekinensis, Ruprecht, Mél. biol. II, 542. — Maximov. Hydrang., 42 = variety of *Ph. coronarius,* Lin. China borealis. Kirilov.

Cotyledon fimbriata, Turczaninov, Catal. Baical., 1837, n. 469. Only name. Kirilov, Mongolia. — Maximovicz, Mél. XI (1883) 727. — Ind. Fl. Sin., ɪ, 281.

Sedum Kirilowii, Regel & Tiling, Fl. Ajan. (1858) 92, in nota. — Maxim. l. c. 733. Kirilov, Po hua shan. — I. F. S., ɪ, 285.

Penthorum intermedium, Turcz. Pl. Kiril. Chin. bor. 1837, 82. — Maxim. Mél. l. c. 774 = *Penthorum sedoides,* Lin. var. β, *chinense. (P. chinense,* Pursh. 1816, Staunton). — I. F. S., ɪ, 288.

Trichosanthes Kirilowii, Maxim. Amur, 482, in nota. Mongolia chinensis australior, ubi Kirilov e China rediens legit, 1841. — I. F. S., ɪ, 313.

Panax Ginseng, C. A. Meyer, in Gauger's Repert. für die Pharmacie Russland's, I (1842) p. 516—528, cum icone. An abstract of this original article appeared in Bull. Cl. phys. mathem. Acad. St. Petersb. I, 1843, 338. Meyer, describes the true Manchurian Ginseng plant, so highly valued by the Chinese, from a specimen, root leaves and fruits, kept in Fischer's herbarium (now in Herb. Hti Petrop.). This had been procured by Dr. Kirilov, who received it from a Chinese Mandarin, his friend. This Chinese officer had been at the head of one of the expeditions sent every year by the Chinese Emperor to Manchuria, to collect Ginseng root in the forests there. This is still the only herbarium specimen of this famous plant, in its wild state, found in European herbariums. See also Regel Tent. Fl. Ussur. (1861) p. 73, *Panax quinquefolium,* Lin., var. *Ginseng,* and Regel's Gartenflora, 1862, 314, tab. 375. Mr. Fr. Schmidt, Member of the Academy, when, in 1861, travelling in Russian Manchuria, was shown, near the river Sui fun, by a Chinese, a wild Ginseng plant, but it was sterile (Schmidt, Reisen, Amur, Sachalin, 1868, p. 5). — I. F. S. ɪ, 338, sub *Aralia.*

Eleutherococcus senticosus, Maxim. Amur, 132. There is in Hb. Mus. Acad. Petrop. a specimen labelled Kirilov, Peking. — Regel Gartenflora, 1863, tab. 393. — I. F. S., ɪ, 342. — Figured in the Atlas appended to the Exploration of the Amur 1859.

Abelia biflora, Turcz. Pl. Kiril. Chin. bor. 1837, n. 93. Maxim. Mél. XII (1886) 477. — I. F. S., ɪ, 358.

Lonicera chrysantha, Turcz. l. c. n. 91. — Maxim. Mél. X (1877) 68. Peking, Kirilov. — I. F. S., ɪ, 364.

Lonicera Tatarinowii, Max. Amur, 138 in nota. — Maxim. Mél. l. c. p. 61. First discovered by Kirilov, near Peking. — I. F. S., ɪ, 367.

Galium linearifolium, Turcz. Pl. Kiril. China bor., 1837, n. 96. —
Maxim. Mél. IX (1873) 265. — I. F. S., ɪ, 394.

Valeriana heterophylla, Turcz. Catal. Baical. — Dahur. (1837) n. 574,
only name, and Turcz. Pl. Kiril. l. c. 1837, n. 99. — Maxim. Ind. Pekin.
473, thinks that this is perhaps *V. dubia*, Bunge, in Ledebour, Fl. Alt. I
(1829) 52.

Eupatorium Kirilowii, Turcz. Pl. Kiril. Chin. bor., 1837, n. 108. Fi-
gured in Gartenflora 1875, 354, tab. 850. Introduced from the Ussuri. —
I. F. S., ɪ, 404 = *E. Lindleyanum*, DC. Prod. V (1836) 180 (herbarium
Lindley, China).

Calimeris integrifolia, Turcz. l. c. (1837) n. 110. The plant was first
described in DC. Prod. V (1836) 259. Turczaninov had it sent from Dahuria. — Earlier Staunton. — I. F. S., ɪ, 412. Hemsley proposes a new
name, *Aster holophyllus*.

Aster ageratoides, Turcz. l. c. n. 109. — According to Maximowicz
(in schedula, see Franchet, Enum. Jap. I, 222) = *A. trinervius*, Roxb. —
I. F. S., ɪ, 416.

Inula linearifolia, Turcz. l. c. 114. — Regel, Fl. Ussur. 85 = variety of *Inula Britannica*, Lin. — I. F. S., ɪ, 429.

Inula repanda, Turcz. l. c. n. 113. — Maxim. Amur, 150. — I. F.
S., ɪ, 429 = *I. Britannica*, Lin.

Pyrethrum lavandulaefolium, Fischer. Kirilov, Peking. — Max. Mél.
VIII (1872) 517 = variety of *Pyrethrum (Chrysanthemum) indicum*. Kirilov discovered this wild form of P. indicum near Peking.

Pyrethrum (Chrysanthemum) sinense, Sabine, who knew only the cultivated form. — Kirilov discovered it wild in S. Mongolia. See Maxim. l. c.
518. Lindley possessed a wild growing specimen from Macao.

Artemisia Kirilowii, Turcz. Pl. Kiril. Chin. bor. 1837, n. 118. —
Maxim. Ind. Pekin. 473 = *A. desertorum*, Spr. var. *Kirilowii*. — Maxim.
Mél. VIII (1872) 526, sub *A. japonica*, Thbg.

Senecio Kirilowii, Turcz. in litt. 1837, in DC. Prod. VI, 361. China
borealis. — Maxim. Mél. VIII (1871) 15 = *S. campestris*, DC. (Cineraria
campestris, Retz). — I. F. S., ɪ, 450.

Saussurea nivea, Turcz. Pl. Kirilov, Chin. bor. 1837, n. 103. —
Maxim. Ind. Pekin. 473 = *Sauss. eriolepis*, Bge, (v. supra p. 335). —
I. F. S., ɪ, 464.

Rhododendron mucronulatum, Turcz. l. c. n. 125. — Maximovicz,
Rhododendron (1870) 43 = *Rh. dauricum*, Lin., var. *mucronulatum*. —
I. F. S., ɪɪ, 22.

Rhododendron micranthum, Turcz. l. c. n. 126. — Earlier d'Incarville (supra p. 53). Maxim. Rhodod. 18, tab. 4. — I. F. S., II, 27.

Syringa pubescens, Turcz. Dec. quatuor Sibir. etc. Bull. Soc. nat. Mosc. 1840, n. 30. China borealis. Collector not mentioned, but most probably Dr. Kirilov had sent the plant. There are in Herb. Hti Petrop. specimens gathered by Kirilov, Peking. — I. F. S., II, 83. Mr. Hemsley, following Decaisne, reduces this, erroneously, to *S. villosa*, Vahl. (Comp. supra p. 48).

Syringa (Ligustrina) Pekinensis, Ruprecht, Mél. biol. II (1857) 371, in nota. — Maxim. Amur, 194 = *S. amurensis*, Rupr., l. c. var. *pekinensis*. — Herb. hti Petrop. Kirilov, Peking. — D'Incarville earlier (supra p. 53). — I. F. S., II, 82.

Gentiana diluta, Turcz. Pl. Kiril. Chin. bor., 1837, n. 141; Catal. Baical-Dahur., 1837, n. 771 (only name). — DC. Prodr. IX, 126 = *Ophelia chinensis*, Bunge, var. *daurica*, Bunge, MSS. — I. F. S., II, 139, *Swertia chinensis*.

Convolvulus acetosaefolius, Turczaninov, Dec. quatuor Sibir. etc., 1840, n. 31. China borealis. — Specimen in Herb. Hti Petropolitanum. Probably Kirilov. — Ind. Fl. Sin., II, 164 = *Calystegia hederacea*, Wallich (124, Nepal).

Tittmannia stachydifolia, Turcz. Pl. Kirilov, China borealis, 1837, n. 152. — Maxim. Mél. IX (1874) 404, named it *Mazus stachydifolius*. — I. F. S., II, 183.

Scutellaria pekinensis, Maxim. Amur, 476, in nota. China borealis. In Herb. Hti Petrop. a specimen gathered by Dr. Kirilov. — Max. Fragm. 1879, 42 = *S. indica*, Lin., var. *pekinensis*. — I. F. S., II, 295.

Phlomis umbrosa, Turcz. Decad. quatuor etc., 1840, n. 35. China borealis (probably Kirilov). — I. F. S., II, 306.

Phlomis mongolica, Turcz., Fl. Baic. Dahur. II (1856) 434, in nota. Mongolia chinensis. Probably Kirilov.

Halimocnemis microphylla, Turcz. herb., unpublished. According to Bunge, Salsol. Mongol. 1879, 299, this is = *Salsola gemmascens*, Pallas (Siberia). Kirilov and Bunge, 1830, Mongolia.

Polygonum volubile, Turcz. Dec. quatuor, 1840, n. 37. Enatum e seminibus in China bor. collectis. (Kirilov?) — Meisn. in DC. Prod. XIV, 144 = *P. triangulare*, Wallich. Catal. n. 1689. Nepal.

Thesium chinense, Turcz., Pl. Kirilov. China bor. 1837, n. 175. — I. F. S., II, 408.

Euphorbia pekinensis, Ruprecht in Maximov. Amur, 1859, 239. —

Euphorbia pekinensis, Boissier, Cent. Euphorb. (1860) 31. — DC. Prod. XV, II, 121. — In Herb. Hti Petrop. are specimens gathered by Kirilov. — I. F. S., II, 445.

Ulmus macrocarpa, Hance, Journ. Bot. 1868, 333 (David, Jehol). — Maxim. Mél. IX (1872) 22. Kirilov, Peking.

Carpinus Turczianowii, Hance in Journ. Linn. Soc. X, 1869, 203 (Williams 1865, Peking). Hance states: "This is doubtless the plant enumerated by Maximowicz in his Index Fl. Pekin. (477) without any trivial name or description, from Turczaninov's collection." — Maxim. Mél. XI (1881) 315. Kirilov, Peking, 1831, misit Turczaninov.

Dioscorea polystachya, Turcz., Pl. Kiril. China bor. 1837, n. 198. — One specimen in Herb. Hti Petrop.

Asparagus brachyphyllus, Turcz., Decad. quatuor etc. 1840, n. 39. China borealis (probably Kirilov). — Baker, Asparag. in Journ. Linn. Soc. XIV, 1875, 600.

Polygonatum officinale, All. var. *Maximowiczii,* Max. Mél. XI (1883) 847. (Fr. Schmidt, Fl. Sachalin. 185, spec. prop.). — This is the Pol. spec. indeterm. in Max. Amur, 275. Kirilov, Peking.

Turczaninov, Nikolai Stepanovich, a distinguished Russian botanist who by his collections and writings has done much to throw light on the Flora of the northern part of Eastern Asia. Although he had never visited China, his name is closely connected with the early Russian exploration of the Flora of North China.

Turczaninov, who had studied at the Kharkov University, early commenced to take interest in botany and finally acquired a thorough knowledge of this science. He is well known for his extensive investigations into the Flora of Cis- and Transbaicalia, which he executed with the consent of the Russian Government for the Imperial Botan. Gardens and for the Academy, while holding an official position in Irkutsk. From 1828 to 1836 he explored the mountainous regions surrounding the Baikal Lake and the tracts of land east and south of this lake, as far as the Chinese frontier, to the Argun River and to the Chikoi, what is now the Russian province of Transbaicalia. The eastern part of it, inhabited by the Dahurs, who belong to the Tungus stock, is also called Russian Dahuria, whilst the western and southern part of the province is peopled by the Buriats, who are Mongols. Turczaninov terms this latter tract also "Mongolia Rossica", to distinguish it from the Mongolia Chinensis or Mongolia proper, the vast land which extends between the Russian frontier and the Great Wall of China.

In 1833 T. descended the Shilka in a boat, entered the Amur and sailed down this river as far as the old fortress of Albazin. He was the first botanist who collected on the Amur, and discovered there a new plant, the *Lespedeza bicolor*, described by him in 1840, in the Decad. quatuor Pl. Sibir. etc., n. 20, and figured in the Atlas appended to the Exploration of the Amur, 1859. See also DC. Prod. VI, 534, *Saussurea amurensis*, Turcz.

In 1831 he began to receive plants from Chinese Mongolia and Northern China, collected, as we have seen, by Kuznetsov and Dr. P. Kirilov. He described them in three papers published in the Bull. Soc. nat. Moscou:

1. DECADES TRES PLANTARUM CHINAE BOREALI ET MONGOLIAE CHINENSI INCOLARUM (Kuznetsov. 1831), in Bullet. Soc. nat. Moscou, tome V, 1832, p. 180—206. V. supra p. 343.

2. ENUMERATIO PLANTARUM QUAS IN CHINA BOREALI COLLEGIT ET MECUM COMMUNICAVIT CL. MEDICUS MISSIONIS ROSSICAE PORPHYRIUS KIRILOV. (200 species). In Bull. Soc. natur. Moscou, X, 1837, II, p. 148—158. V. supra p. 347.

3. DECADES QUATUOR PLANTARUM HUCUSQUE NON DESCRIPTARUM SIBIRIAE MAXIME ORIENTALIS ET REGIONUM CONFINIUM INCOLARUM. In Bull. Soc. natur. Moscou, XIII, 1840.

In 1831 Turczaninov had been elected Corresp. Member of the Academy, St. Petersburg. His botanical exploration of Transbaicalia seems to have lasted till 1836. He then set himself about to determine and describe the vast collections formed in these regions.

In 1837 there appeared, in the Bull. Soc. nat. Moscou, his CATALOGUS PLANTARUM IN REGIONIBUS BAICALENSIBUS ET IN DAHURIA SPONTE NASCENTIUM. It gives only names of plants. His great work, FLORA BAICALENSI-DAHURICA, seu descriptio plantarum in regionibus cis- et transbaicalensibus atque in Dahuria sponte nascentium was likewise published in the Bull. Soc. nat. Moscou, 1842—56. Turczaninov's Preface (in French) is dated, Krasnoyarsk, 11 Janvier 1841. Krasnoyarsk is the capital of the Siberian province of Yeniseisk, where T. then was Acting Governor. The Flora Baic.-Dahur. appeared also separately, in two large volumes, but this edition was soon out of print and is now an extremely rare book.

Turczaninov sent complete sets of his botanical collections to the Bot. Garden and the Academy, St. Petersburg. He used also to communicate interesting new plants to Alph. De Candolle, who published them in the Prodromus, and commemorated T.'s name in the new genus *Turczaninowia*, Prod. V, 257, now reduced to *Aster*.

There are in T.'s herbarium some apparently new plants from Mongolia, named, but never described by him. I may quote from Maxim. Ind. Mongol: *Potentilla hypoleuca* (481), *Saussurea crepidifolia* (483), *Axyris rosmarinifolia* (484).

Turczaninov finally settled at Taganrog, where he continued working up his materials. He was a wealthy man and purchased several exotic collections, amongst others Mexican and African plants, which he partly described in the Bull. Mosc. In his later days, at the time of the Crimean war, however he lost all his wealth through connections with a merchant to whom he had entrusted his fortune. Thus he died in rather reduced circumstances at Kharkov in Dec. 1863. His rich herbarium, including all his Eastern Asiatic species, he had made over (or sold?) to the Kharkov University.

Part IV. PERIOD FROM THE FIRST WAR BETWEEN ENGLAND AND CHINA, 1840, TO THE SECOND WAR, 1860.

As has been recorded in a previous chapter (supra p. 29), in 1684, a Factory was established at Canton, by the British E. I. Company, whose agents had previously traded at Amoy and in Formosa. From this date the import of Tea into England grew rapidly and until 1834, when the Company's monopoly expired, the Factory at Canton was celebrated throughout the world for the vastness of its trade. Besides Macao, which belonged to the Portuguese, Canton was the only port in China where European ships were admitted. But foreigners were subjected to intolerable oppression by the native authorities. Most of the foreign merchants engaged in the tea trade used to live in Macao, the greater part of the year. They went to Canton only for the tea season. When in 1834 the privileges of the E. I. Comp. had ceased, the hardships on the English merchants at Canton had become so unbearable that the British Government determined to send out a Minister to superintend and protect the English trade of this port. **Lord Napier,** William John, was selected for this office. He arrived with a squadron at Macao, July 15th 1834, where were associated with him in the commission, John F. **Davis,** and Sir G. B. **Robinson,** former servants of the Company. But when Lord Napier arrived before Canton with two men of war, to negotiate with the Chinese Governor-General, the latter did not admit him and ordered the foreign trade to be stopped. Lord Napier died after but a few month's residence in China, in October 1834. Sir G. B. Robinson took his place, but neither he nor his successor Captain **Elliot,** (who

had come out to China with Lord Napier) succeeded in making arrangements with the Chinese authorities for the re-establishment of a regular
trade. The chief cause of complaint adduced by the Chinese was the smuggl-
ing trade in Opium carried on by English merchants. Opium brought to
China from India was the most lucrative merchandize for the British trade,
but its importation was prohibited by the Chinese Government.

In 1838 the Chinese Emperor sent the Commissioner Lin to Canton
to put down the Opium trade. Lin stopped the foreign trade at Canton,
imprisoned the foreigners in the factories and demanded all the British
owned Opium in the store ships to be given up, promising to re-open the
trade as soon as the Opium had been surrendered. Then Capt. Elliot agreed,
in April 1839, that all the Opium in the hands of British subjects should
be delivered to the Chinese, although he had no authority to make this
concession and exceeded his powers. Accordingly 30,000 chests of Opium
were handed over to the Chinese authorities. All the Opium was destroyed
by them. But this step taken by Elliot neither induced the Chinese to
re-open a regular trade — they manifested hostile intentions and even
attempted to attack the British fleet anchored at Hongkong — nor was it
approved by the British Government. The proceedings of the Chinese were
considered a "casus belli". In 1840, accordingly, war was declared. This
Anglo-Chinese war is known in history as the "Opium War".

On June 28, 1840 the Plenipotentiary, Admiral George **Elliot,** arrived
with the British fleet and blocked the port of Canton. One part of the
British forces, 5 men of war, 3 steamers and 21 transports under Commodore Sir Gordon **Bremer** made sail for the Chusan Island, and on July 4,
reached the Ting hai Harbour. The city of Ting hai, the capital of Chusan, was captured after a short resistance. Two days later the British
Plenipotentiaries, Admiral Elliot and Captain Elliot arrived at Ting hai.
After arranging the government of the island, the stations of the troops and
the blockading of Amoy, Ningpo and the mouth of the Min (Fu chou) and
Yang tze Rivers, the Plenipotentiaries left Ting hai and proceeded to the
Gulf of Chili. They anchored off the Pei ho River, August 11. Capt. Elliot
went ashore to confer with the Governor General of Chili, who was at Ta
ku, and deliver him Lord Palmerston's letter to the Emperor. While await-
ing the reply, the ships visited the coast of Liao tung to procure provisions.
When they returned to the Pei ho, it was arranged with the Governor
General, that the latter should meet the English Plenipotentiaries at Canton.
On Sept. 15th the squadron returned to Chusan and in the middle of November sailed for Canton. Negotiations with the Chinese authorities were

resumed, but without any result, the latter refusing to grant full indemnity for the Opium destroyed. Commissioner Lin was recalled by the Emperor. Commodore Bremer, who after the recall of Admiral Elliot had taken the command of the fleet, on January 7th 1841 attacked and took the forts at 1841 Chuen pi and Tai kok tau, at the Bocca Tigris (entrance to the Canton River). After this the suspended negotiatións were resumed, and on January 20th a preliminary arrangement upon the following points was concluded: the cession of the island of Hong kong to the British crown, an indemnity of 6 millions of dollars to be paid by the Chinese Government, direct official intercourse upon an equal footing, and immediate resumption of English trade at Canton. By these arrangements Chusan and Chuen pi were to be immediately restored to he Chinese, and the English allowed to occupy Hong kong. The latter two stipulations were then carried out into effect, Chusan was evacuated by the British trops, end of February 1841. For the rest both parties had to wait for the ratification of the treaty. But this was rejected by the Chinese Emperor as well as by Queen Victoria, and the hostilities were resumed. On Febr. 26 the British took all the forts in the Bocca Tigris and moved towards Canton. The British troops occupied the factories of Canton, just two years after the Commissioner Lin had imprisoned the foreigners there. On March 20th the Chinese proposed a suspension of hostilities, by which trade was allowed to proceed, and it was agreed upon. Meanwhile Major General Sir **Hugh Gough** arrived from India to take command of the British land forces. For about six weeks British trade at Canton was carried on uninterruptedly. But the Chinese had profit-ed by the truce and gathered considerable forces around Canton. When their hostile intentions became manifest, Hir Hugh Gough arrived with all the British land and naval forces at Canton, and on May 25th gained a victory over the Chinese troops. Then the Chinese again proposed a truce, which was agreed to after they had paid a ransom and withdrawn their troops from Canton.

On the 9th of August 1841, there arrived at Macao, direct from England, Sir Henry **Pottinger,** a Major General in the E. I. Comp.'s service and Admiral Sir William **Parker,** the former to supersede Capt. Elliot, as Superintendent of Trade and Plenipotentiary of the Queen, the latter to replace Commodore Bremer as Commander of the Fleet. Sir H. Gough continued to command the British land forces.

On August 21 an expedition moved northward under the joint com-mand of Sir H. Gough and Admiral Parker, consisting of 9 ships of war. The force reached Amoy, attacked and took the city on Aug. 27. The

island of Ku lang su was garrisoned by a detachment of 550 troops.
3 ships were left to protect them.

On Sept. 29 the British fleet again entered the harbour of Tinghai
in Chusan. The fortification was attacked and taken and the whole island
occupied for a second time.

On Oct. 9 the fleet proceeded to Chin hai, a city with a citadel
at the mouth of the Yung River upon which Ningpo is situated. The
place was taken by assault, and on the 13th Ning po surrendered without
resistance.

1842   In Febr. 1842 Sir H. Pottinger returned to Hong kong, Sir Hugh
and the Admiral remaining in the north. On the 18th of May the British
forces attacked and captured Chapu, an important emporium on the Hang
chou Bay. After this the expedition proceeded northward to the mouth of
the Yang tze kiang, and on June 16 reached the embouchure of the Wu
sung River on which, higher up, Shanghai is situated. The fortifications of
Wu sung were captured and on the 19th Shang hai surrendered. Sir
H. Pottinger, accompanied by Lord Saltoun, now rejoined the expedition,
with large reinforments. Preparations were made for proceeding up the
Yang tze, to interrupt the communication by the Grand Canal across the
river. The Chinese had concentrated their troops at the city of Chin kiang
fu, situated near the spot where the Grand Canal crosses the Yang tze. The
British fleet sailed up the Great River and attacked the city, which was
bravely defended by Man chu and Mongol soldiers, but notwithstanding
their energetic resistance was taken on July 21st.

On the 9th of August, 1842 the fleet reached Nan king, when the
Chinese Government proposed terms of peace. After much discussion, Sir
H. Pottinger, on August 26, concluded a Treaty with the Imperial Commissioners, by which, besides Canton, four additional Chinese ports, Amoy,
Fu chou, Ning po and Shang hai were declared open to British trade
and residence. The island of Hong kong was ceded to England. Six millions of dollars to be paid as the value of the Opium destroyed in 1839,
twelve millions for the expenses incurred in the expedition sent out, three
millions due to British merchants. Chusan was rendered to the Chinese in
1846, after the 21 millions had been paid by the Imperial Government.

After this brief account of the memorable "Opium War", let me now
proceed to show what important botanical discoveries in China resulted
from this expedition, the ensuing cession of Hong kong and the opening of
four new Chinese ports to European access.

I. Botanical Collectors in Chusan during the British Occupation of the Island, 1840—1846.

Cantor, Theodore, M. D., of Bengal Med. Service. He was Assistant Surgeon to H. M. 26th Regiment, the Cameronian, which during the first occupation of Chusan, in 1840, was cantoned near Ting hai. Cantor employed himself in making zoological and botanical collections in the island for the Museum of the Court of Directors of the E. I. C., and published some interesting papers on Chusan and its natural productions:

COLLECTIONS MADE BY DR. CANTOR ON THE EXPEDITION TO CHINA, in Calcutta Journ. of Nat. Hist. V, 1841. Reproduced with corrections in the Chinese Repository, X (1841) 434—38.

Cantor, GENERAL FEATURES OF CHUSAN WITH REMARKS ON THE FLORA AND FAUNA OF THAT ISLAND. Annals of Nat. Hist., 1842, p. 265—278, 361—371, 481—494. Journ. Asiat. Soc. Bengal XXIV, 1855.

From these papers we learn, that Cantor, in 1840, when at Calcutta, was ordered to assume the medical charge of a detachment of H. M. 26th Regiment, the Cameronian, with which he embarked for China. During the whole month of June they were detained in the island of Lantao (west of Hong kong). In July they sailed to Chusan, where Cantor remained till the end of November, when, having become the victim of a violent cerebral fever, he was ordered to sea by the medical board at Chusan. It seems he was then sent back to India.

W. **Griffith** (1810—45), Superintendent of the Bot. Garden, Calcutta, to whom Cantor transmitted his botanical collections, wrote an ACCOUNT OF THE BOTANICAL COLLECTION BROUGHT FROM THE EASTWARD BY DR. CANTOR: Plants from the Straits of Malacca, from Lantao Island, Chusan, and a few from Peking, the bulk of the Chinese plants being from Chusan. This paper was published, after Griffith's death, in the Journ. Asiat. Soc. of Bengal, XXIII, 1854, and is found reprinted in Trübner's Oriental Series, II (1886) 257—72. We find there enumerated 133 Chusan plants. Generally only the genus name in given. One new plant, representing a new genus of Cucurbitaceae, is described and figured: *Actinostemma tenerum*, Griff. There are also mentioned 4 plants from Teng chou, Pekin, and 20 from Toki, Pekin.

By Toki, perhaps Taku, near the mouth of the Pei ho Riv., is meant, and by Teng chow, the city of T'ung chou, east of Peking. There is a city called Teng chou, on the northern coast of Shan tung. It is, however, quite certain, that Cantor never visited these places. He was in Chusan, in

August 1840, when Admiral Elliot and Captain Elliot sailed to the mouth
of the Peiho and the latter landed at Ta ku (v. s. p. 356). It is also beyond
doubt that from 1816 (Lord Amherst's Embassy) till 1860, no Englishman
had been admitted to the Chinese capital. I am unable to explain the
existence of Peking plants in Cantor's collection.

Mr. Baker, Journ. Bot. 1874, 290—92, notices three species of
Allium from Cantor's plants, all labelled Peking, viz: *Allium tuberosum,*
Roxb., *A. tenuissimum,* Lin., and *A. nerinifolium,* Baker. The latter plant
has not been met with by other collectors in the vicinity of Peking. It was
first described under the name of *Caloscordium nerinifolium* by Herbert in
Bot. Reg. 1844, Misc. n. 64, and figured in Bot. Reg. XXXIII (1847)
tab. 5. The specimen had been received from Chusan.

Clerodendron cyrtophyllum, first described by Turczaninov in Bull.
Soc. Nat. Mosc., 1863, II, 222, from Fortune's collection made in Chusan,
about 1844, had been discovered four years earlier by Cantor, in the
same island. Ind. Fl. Sin., II, 259. Cantor's name occurs frequently in the
latter work.

Cantor sent from Chusan an empty cone, which was believed to belong
to an unknown coniferous tree. A few years later, Fortune gathered complete specimens of the tree, which Lindley described as *Fortunea chinensis*
in Journ. Hort. Soc., I, 1846, 150.

General W. Munro, in his Monogr. Bambus., Trans. Linn. Soc. XXVI
(1868) 96 and 111, described two new species of Bambusa, gathered by
Cantor in Lin tao (Lan tao Isl., west of Hong kong), viz: *Bambusa brevifolia* and *B. Cantori.*

Alexander, William T., Surgeon of the British Navy. He is frequently
mentioned in Sir W. J. Hooker's papers and systematic works on Filices, in
connection with Ferns gathered by Alexander in China. His name appears
first in Kew Journ. Bot. V, 1853: *Aspidium podophyllum,* Hooker. Chusan,
Dr. Alexander. But generally Hooker calls him simply T. Alexander.

Alexander collected chiefly in Chusan, but there are in his collection
also plants from various places on the Chinese coast.

In Hooker's Journ. Bot., VII, 1848, 273—78, W. Wilson published
46 species of Mosses gathered by T. Anderson, surgeon to H. M. S.
"Plover", on the Chinese coast from Chusan to Hong kong, from December
1845 to March 1846. As the localities noticed there for Anderson's mosses
are the same as those given on the labels of Alexander's ferns, I was
inclined to believe that the latter belonged also to the "Plover" and that

his ferns were likewise gathered in 1845 and 46. My supposition was not only confirmed by Mr. J. Britten, the learned Editor of the Journ. of Botany, but this gentleman, moreover, ascertained that the species of Mosses described by Wilson in the above mentioned paper under Anderson's name, in Wilson's herbarium are written up as from Alexander. Thus the names of Alexander and Anderson seem to represent only one person. To clear up the matter Mr. Britten made enquiries at the Admiralty and was informed by Lord Walter Kerr that the medical officer (assistant surgeon) of H. M. S. "Plover" — the surveying vessel in the East Indies and China, in 1845—6, was William T. Alexander *).

Besides Loo choo Islands, Chusan, Fu chou fu, Hong kong, the following names of localities on the Chinese coast occur on the labels of Alexander's collections: **)

Woo sung. (Name of a place near Shang hai, which latter lies on the Wu sung River).

Sa mou, Bay. (Most probably San mun Bay, 29° N. Lat. is meant).

Bullock Bay. (27°, 46' N. L. near Wen chou Bay).

Pihqwan, mountainous island. (Near Namqwan Port, about 27°, N. L. the Admiralty Chart marks Pihqwan Harb. between two islands).

Sam sah Bay. (Samlah Bay, Sam lan Bay (Hooker) probably misprints for Sam sah B. 26°, 30').

Koo lung, Koo lung lu Island. (Island Koo lung su, opposite Amoy).

Tung zan inlet, old fort. (Tung tan, Tung lan, in Hooker, denote probably the same locality: Tung san Harbour and Island, 23°, 45').

Cow loon. (Locality opposite Hong kong, on the mainland).

New Ferns discovered by Alexander:

Alsophila podophylla, Hooker, Kew Journ. Botan. IX (1857) 334. Chusan. — Hook. 2d Cent. of Ferns (1860) tab. 66. — Syn. Fil. 43.

Cheilanthes Chusana, Hook. Spec. Fil. II (1858) 95, t. 110. Chusan. Syn. Fil. 135, reduced to *C. mysurensis*, Wall., India.

Asplenium elegantulum, Hook. Spec. Fil. III (1860) 190. — 2d Cent. of Ferns, 1861, tab. 28. — Syn. Fil. 217, reduced to *A. incisum*, Thunberg, Japan.

---

*) It appears from the list of charts of the Chinese coast, as given in Cordier's Bibliotheca sinica, pp. 87—92, that the «Plover» was in the waters of China from 1840—49. Her able commander, Capt. R. Collinson, then surveyed the Chinese coast between Chusan and Hongkong. The ship was subsequently employed in the relief expedition to search for Sir John Franklin. B. Seemann, in Botany Herald, reports, that they met with the «Plover» at Chamisso Island on July 15, 1849, and then sailed together to the Polar See.

**) See my Map of China.

Aspidium podophyllum, Hook. Journ. Bot. V, 1853, 236, tab. 1. Chusan, Fu chou fu. — Syn. Fil. 261, *Nephrodium podophyllum.*

Nephrodium decipiens, Hook. Spec. Fil. IV, 1862, 86. Fuchou fu. — Syn. Fil. 260.

Home, Sir Everard. He collected plants in Chusan and at Ningpo, previous to 1844. Sir W. J. Hooker, in his Icon. Plant., VII, 1844, t. 668, reports that Sir Everard Home had sent him specimens of *Cryptomeria japonica,* with cones, from Chusan. This is probably the Home whose name occasionally appears in the Ind. Fl. Sin.:

I, 65. *Silene Fortunei,* Vis. Home, China, — Fortune, Chusan. — Home gathered the plant probably earlier than Fortune.

I, 115. *Ilex cornuta,* Lindl. Fortune, Shanghai, — Home, Ning po.

Fortune, Robert, botanized in Chusan, in 1844 and 45, and again in 1850, 53. He introduced into England a number of rare plants from that island. An account of his travels and collections will be given in another chapter.

II. Botanical Exploration of the Island of Hong kong.

As has been recorded in a previous page (supra p. 357) the Chinese authorities at Canton and the Commander of the British forces, on 20 Jan. 1841, came to an agreement according to which Chusan was to be restored to China and the island of Hongkong ceded to England. According to Sir Edw. **Belcher,** in his «Voyage of the Sulphur», the official act of taking position of the island of Hong kong by the British government, took place on Jan. 25th 1841. The first Governor was Captain Charles **Elliot.** (V. supra p. 355).

H. M. S. "Sulphur" in 1836 was sent out on a surveying voyage to the Pacific Ocean and finally to the western shores of America, under the command of Capt. Beechey (v. supra p. 288), who, however, was invalided home and succeeded by Captain Belcher, Edward, born 1799. Both of these distinguished officers were well known for the interest they took in natural history. The Royal Kew Gardens sent out, to accompany the expedition a collector **Barclay,** George. Two of the officers of the "Sulphur", R. B. **Hinds,** Surgeon and A. **Sinclair,** Assistant Surgeon, also distinguished themselves by their ardour in pursuit of plants during this interesting voyage. When the "Sulphur", on her way home, arrived at Singapore, Oct. 1840, Capt. Belcher found there despatches from their Lordships, directing him to join the commander in chief in China. The "Sulphur" reached Macao on Dec. 14, 1840, and took part in the Anglo-Chinese war. She quitted the China waters on Nov. 21st, 1841. Ceylon, Sechelles, Madagascar, Cape of G. H. Arrived in England, July 19, 1842. — Belcher was knighted. In 1843 he published: NARRATIVE OF A VOYAGE ROUND THE WORLD OF H. M. S. "SULPHUR", 1836—42, two volumes. In the same year B. was appointed to H. M. S. "Samarang" and from 1843—1846 surveyed the islands of the Eastern Archipelago, Loo chou Isls, Corean Archipelago etc. In 1852 he was directed with 5 ships to proceed to the Arctic regions, to search for Sir John Franklin. B. died 1877 at London.

Hinds, Richard Brinsley, Surgeon to the "Sulphur", an ardent botanist. He made the first collection of Hong kong plants, which reached England. Previously, in 1816, Dr. C. Abel had collected some plants there (supra p. 225), but they were lost. Hinds' stay round the island was only for a few weeks, in January and February 1841, but he was enabled, on his return to England, to place in the hands of G. Bentham, Hong kong specimens of nearly 140 species, the enumeration of which Bentham published in Hooker's London Journ. of Bot. I (1842) pp. 482—494, at the end of Hinds' REMARKS ON THE PHYSICAL ASPECT, CLIMATE AND VEGETATION OF HONG KONG. — In 1844 Hinds edited the BOTANY OF THE VOYAGE OF H. M. S. "SULPHUR". Botan. descriptions by G. Bentham. According to Biogr. Bot., Hinds died before 1861.

Among the 140 Hongkong plants gathered by Hinds, G. Bentham, l. c., describes 21 as entirely new. A few other novelties from this collection have been described by other botanists.

Viola tenuis, Benth. Lond. Journ. Bot. I (1842) 481. But in Fl. hgk. 20, Bentham reduces this to *Viola diffusa,* Ging. in DC. Prod. I, 298. — I. F. S., I, 52.

Atalantia monophylla, DC. (India)., Benth. Lond. J. B. l. c. 483, Hinds, Hongkong. But Champion in Kew J. B. III (1851) 328, considers it a new species, *Sclerostylis Hindsii,* which Oliver named *Atalantia Hindsii.* — Fl. Hgk. 51. — I. F. S., I, 110.

Cansjera lanceolata, Benth. l. c. 491. — Fl. Hgk. 296, reduced to Gmelin's *C. Rheedii.* — I. F. S., I, 115.

Euonymus nitidus, Benth. L. Journ. Bot., l. c. — Fl. Hongk. 62. —

I. F. S., I, 119 = *Euonymus chinensis*, Lindl. (1825), gathered earlier by Staunton.

Catha monosperma, Benth. l. c. 483. — *Celastrus Hindsii*, Benth. Kew J. B. III (1851) 334. — I. F. S., I, 123.

Melastoma calycinum, Benth. l. c. 485. — I. F. S., I, 299 = *M. candidum*, Don (1823).

Allomorphia pauciflora, Benth. l. c. 485 *(Oxyspora* in Fl. Hgk. 116).— *Blastus Hindsii*, Hance, Add. Fl. Hgk. (1871) 103. — I. F. S., I, 301.

Ammannia subspicata, Benth. l. c. 484. — In Fl. Hgk. 111, reduced to *A. rotundifolia*, Roxb. Fl. Ind. I, 485. — I. F. S., I, 303.

Hedyotis acutangula, Champ. K. J. B. IV (1852) 171. — I. F. S., I, 372. Hinds, Hong kong.

Hedyotis recurva, Benth. l. c. 486. — In I. F. S., I, 374, reduced to *H. macrostemon*, Bot. Beech. 192 (supra p. 292).

Vernonia solanifolia, Benth. l. c. 486. — I. F. S., I, 402.

Vernonia congesta, Benth. l. c. 487. In Fl. Hgk. Bentham reduces it to *Inula Cappa*, DC., long known from India. — I. F. S., I, 429.

Amphirhapis leiocarpa, Benth. l. c. 480. — In Fl. Hgk., 179, reduced to the common *Solidago Virgaurea*, Lin. — I. F. S., I, 406.

Diplopappus baccharoides, Benth. l. c. 487. — *Aster baccharoides*, Steetz in Seem. Bot. Herald (1857). — I. F. S., I, 409.

Diplopappus laxus, Benth. l. c. 487. — In Fl. Hgk. 175, reduced to *Aster trinervius*, Roxb. Fl. Ind. III, 433. — I. F. S., I, 416.

Gnaphalium confertum, Benth. l. c. 488. — In Fl. Hgk. 187, reduced to *G. hypoleucum*, DC. (1834) India. — I. F. S., I, 426.

Senecio Hindsii, Benth. l. c. 488. — In Fl. Hgk. 190, reduced to *S. chinensis*, DC. (Loureiro, Canton), in I. F. S., I, 457, to *S. scandens*, Hamilt. (1825, Nepal.).

Brachyramphus ramosissimus, Benth. l. c. 489. — Fl. Hgk. 193 = *Ixeris ramosissima*, A. Gray. — Ind. Fl. Sin., I, 480, *Lactuca denticulata*, Maxim. — Thunberg first gathered the plant in Japan.

Barkhausia tenella, Benth. l. c. 488. — In Fl. Hongk., reduced to *Ixeris versicolor*, DC. — I. F. S., I, 485 = *Lactuca versicolor*, Maxim. — Long known from Japan (Prenanthes chinensis, Thbg.).

Enkianthus uniflorus, Benth. l. c. 480. — I. F. S., II, 18. This name falls as it was founded on a mixture of two different plants.

Choripetalum obovatum, Benth. l. c. 490. — *Samara obovata*, Benth. Kew J. B. IV (1852) 205. — Hemsley in I. F. S., II, 62, names it *Embelia obovata*.

Henslovia frutescens, Champ., Kew J. B. V (1853) 194. — I. F. S., II, 409, Hinds, Hongkong.

Glochidion macrophyllum, Benth. l. c. 491. — Müller Arg. in DC. Prod. XV, II, 282, named the plant *Phyllanthus Benthamianus*. — I. F. S., II, 425.

Ficus variolosa, Lindl. (MS). Benth. l. c. 492. — Fl. Hgk. 328. — Earlier discovered by Mertens in Bonin sima. (v. supra p. 323).

Broughtonia chinensis, Lindl. (MS). Benth. l. c. 493. In Kew J. B. VII, (1855) 38, Lindley names it *Laeliopsis chinensis*. — According to Benth. & Hook. Gen. Plant. III, 532, this plant is an *Epidendron* from Central America, erroneously believed by Hinds to be a Chinese plant.

Arundinaria Hindsii, Munro, Bambus (1868) 31. Hong kong, Hinds, 1841, in herbarium Bentham. — Gard. Chron. 1894, I, 239. Cultivated at Kew.

Meniscum simplex, Hook. Lond. J. B. 1842. tab. 11, p. 295. Hong kong, Hinds. — Hook. Filic. exot. 1857, tab. 83.

The genus *Hindsia* (Cinchonaceae) was founded by Bentham upon an American plant.

Hance, Dr. Henry Fletcher. Reserving for the next period a full account of the life and the labours of this distinguished botanist and investigator of the Chinese Flora, we for the present, confine ourselves to notice his early exploration of the Flora of Hong kong. Hance arrived at Hong kong on September 1st 1844, when he was 17 years of age, and entered the civil service of Hong kong and successively filled various offices in the colony. He seems to have resided in Hong kong about 12 years, when he was attached to the Canton Consulate, but he soon returned to Hongkong where he again spent 3 or 4 years. In 1861 he was appointed British Vice-Consul at Whampoa, near Canton. Hance zealously applied himself to the study of the Flora of the island. He remitted a few descriptions of species from Hongkong, which he believed to be new, to Sir W. J. Hooker, who published them in his Lond. Journ. Bot. VII (1848) and Kew Journ. Bot. I, 1849. He sent also diagnoses of more than 50 new Hongkong plants to Dr. Walpers, who inserted them in his "Annales Bot. Syst.", vol. II and III (1851—53). — H. G. Reichenbach, fil. described in the "Bonplandia" 1855, 249, 250, some new Orchids gathered by Hance in Hong kong. — Other novelties, discovered successively by Hance and his friends in Hong kong were published subsequently by him in Bentham's Flora Hongk. and in various botanical periodicals.

List of Hong kong plants, gathered by Hance and described by him or other botanists as new:

Clematis oreophila, Hce, Walpers Ann. Bot. II (1851) 1. — Seemann, Bot. Herald, 361 = *Cl. Meyeniana*, Walp. (supra p. 303). — I. F. S., ɪ, 6.

Ranunculus holophyllus, Hce, Symbolae and Fl. Sin. (1861) 220. — I. F. S., ɪ, 16 = *R. sceleratus*, Lin.

Leontoglossum scabrum, Hce, Walp. Ann. II, 18, III (1853) 812. — Seem. Bot. Her. 361; Fl. Hgk. 7 = *Delima sarmentosa*, Lin. — I. F. S., ɪ, 22, sub *Tetracera sarmentosa*, Vahl.

Kadsura chinensis, Hce, in Fl. Hgk. 9. — I. F. S., ɪ, 25. Earlier Millett.

Uvaria badiflora, Hce, Walp. Ann. II, 19. — I. F. S., ɪ, 25 = *U. microcarpa*, Champ. Earlier Reeves.

Uvaria rhodantha, Hce, l. c. 19. — I. F. S., ɪ, 26 = *U. purpurea*, Blume (1825).

Uvaria calamistrata, Hce, Journ. Bot. 1882, 77. Gathered, in fruit, in Hongkong, 1861. — I. F. S., ɪ, 25.

Artabotrys hongkongensis, Hce, Journ. Bot. 1870, 71. Hance gathered the plant in 1853. Champion earlier. — I. F. S., ɪ, 26.

Capparis sciaphila, Hance, Adversaria (1864) 206. — Gathered in 1861. — I. F. S., ɪ, 51 = *C. pumila*, Champ. Earlier Beechey.

Phoberus saevus, Hance, Walp. Ann. III, 825. — Seem. Bot. Her. 363 = *Phob. chinensis*, Lour. — I. F. S., ɪ, 57, *Scolopia chinensis*, Clos.

Xylosma senticosum, Hance, Journ. Bot. 1868, 328. Gathered in 1861. — I. F. S., ɪ, 57.

Cerastium petiolare, Hce, Kew J. B. I, 143. — I. F. S., ɪ, 67 = *Stellaria aquatica*, Scop. (Cerastium aquaticum, Lin.).

Stellaria fecunda, Hce, Walp. Ann. II, 95. — I. F. S., ɪ, 67, same as the preceding.

Stellaria leptophylla, Hce, l. c. 95. — I. F. S., ɪ, 69 = *Stellaria uliginosa*, Murr. (1770).

Hypericum nervatum, Hce, l. c. 188. — I. F. S., ɪ, 73 = *H. japonicum*, Thbg.

Elodea chinensis, Hce, London J. B. VII (1848) 472. — I. F. S., ɪ, 74 = *Cratoxylon polyanthum*, Korth. Earlier Bladh, supra p. 113.

Malvastrum ruderale, Hce, Walp. Ann. III, 830. — I. F. S., ɪ, 84 = *Malvastrum tricuspidatum*, As. Gray. Earlier known from America.

Abutilon cysticarpum, Hce, Walp. Ann. II, 157. — I. F. S., ɪ, 86 = *Abutilon indicum*, G. Don (Sida indica, Lin.).

Megabotrya meliaefolia, Hce, l. c. 259. — I. F. S., ɪ, 104 = *Evodia meliaefolia*, Benth. Fl. Hgk. 58. Earlier Parks.

Hedera hypoglauca, Hce, l. c. 724. — I. F. S., ɪ, 131 = *Vitis cantoniensis*, Seem. Earlier Beechey.

Vitis succisa, Hce, l. c. 231. — Ind. Fl. Sin., ɪ, 132 = *V. flexuosa*, Thbg., Japan.

Averrhoa sinica, Hce, l. c. 241. — I. F. S., ɪ, 149 = *Rourea microphylla*, Planch. Earlier Beechey.

Desmodium acrocarpum, Hce, London J. B. VII, 473. — I. F. S., ɪ, 176 = *Desmodium triquetrum*, DC., long known from India.

Mucuna macrobotrya, Hce, Walp. Ann. II, 422. — I. F. S., ɪ, 190.

Dahlbergia Hancei, Benth. Journ. Linn. Soc. IV, Suppl. (1860) 44. — I. F. S., ɪ, 198.

Derris chinensis, Benth. l. c. 104. — I. F. S., ɪ, 199.

Rubus leucanthus, Hce, Walp. Ann. II, 468. — I. F. S., ɪ, 234.

Kalanchoe macrosepala, Hce, Journ. Bot. 1870, 5. — I. F. S., ɪ, 281.

Calyptranthes mangiferifolia, Hce, Walp. Ann. II, 629. — I. F. S., ɪ, 297 = *Eugenia operculata*, Roxb. Fl. Ind. II, 486.

Hydrocotyle lurida, Hce, Walp. Ann. II, 690. — I. F. S., ɪ, 324 = *H. asiatica*, Lin., var. *crispata*, Maxim. (Japan).

Hydrocotyle perexigua, Hce, l. c. 691. — I. F. S., ɪ, 325 = *H. rotundifolia*, Roxb.

Gardenia daphnoides, Hce, l. c. 796. — I. F. S., ɪ, 383 = *Diplospora viridiflora*, DC. Earlier Parks.

Galium sororium, Hce, l. c. 734. — I. F. S., ɪ, 393 = *G. Aparine*, Lin.

Artemisia apiacea, Hce, l. c. 895. — I. F. S., ɪ, 441. Earlier Thunberg, Japan. *A. Thunbergiana*, Maxim.

Cirsium oreithales, Hce, l. c. 944. — I. F. S., ɪ, 461 = *Cirsium chinense*, Gard. & Champ. (1849). Earlier Fortune.

Gerbera amabilis, Hance, l. c. 947. — I. F. S., ɪ, 473 = *Gerbera piloselloides*, Cass. (Arnica piloselloides, Lin.).

Dubyaea ramosissima, Hce, l. c. 1028. — I. F. S., ɪ, 480 = *Lactuca denticulata*, Maxim. Long known from Japan. Thunberg.

Stylidium sinicum, Hce, l. c. 1030. — I. F. S., ɪɪ, 1 = *St. uliginosum*, Swartz, 1807.

Scaevola lativaga, Hce, l. c. 1055. — I. F. S., ɪɪ, 2 = *S. Koenigii*, Vahl, 1794.

Choripetalum Benthamianum, Hance, Walp. Ann. III, 10. — Same as *Ch. obovatum*, Benth. (1842). Hinds. V. supra p. 364.

Olea Walpersiana, Hce, l. c. 17, and *O. consanguinea*, l. c. 18. — I. F. S., ii, 92: Both = *Ligustrum sinense*, Loureiro.

Lycimnia suaveolens, Hce, l. c. 31. — I. F. S., ii, 94 = *Melodinus suaveolens*, Champ. Earlier Millett.

Exacum bellum, Hce, Lond. J. B. VII, 472. — I. F. S., ii, 122 = *E. tetraganum*, Roxburgh, India.

Ipomoea fulvicoma, Hance, Kew Journ. Bot. I (1849) 176. — Seem. Botan. Her. 401 = *Ipomoea cymosa*, Roem. & Schult. India. — Ind. Fl. Sin., ii, 159.

Convolvulus ianthinus, Hance, Walp. Ann. III, 113. — I. F. S., ii, 166 = *C. parviflorus*, Vahl, 1794.

Evolvulus pudicus, Hce, l. c. 115. — I. F. S., ii, 166 = *Evolvulus alsinoides*, Lin.

Solanum immane, Hance, l. c. 165. — I. F. S., ii, 170 = *Solanum ferox*, Lin.

Mazus vandellioides, Hce, l. c. 193. — I. F. S., ii, 183 = *Mazus rugosus*, Loureiro.

Anisocalyx limnanthiflorus, Hance, l. c. 195. — I. F. S., ii, 186 = *Herpestis Monniera*, Humb. Bonpl. (1817).

Utricularia extensa, Hce, l. c. 13. — I. F. S., ii, 223 = *Utricul. flexuosa*, Vahl, 1794.

Gutzlaffia (nov. gen.) *aprica*, Hce, Kew J. B. I, 143. — In Fl. Hgk. it is named *Strobilanthes apricus*. — I. F. S., ii, 239.

Callicarpa brevipes, Hce, Advers. 232. — I. F. S., ii, 252.

Clerodendron haematocalyx, Hce, Walp. Ann. III, 238. — I. F. S. ii, 259 = *C. canescens*, Wall. Cat. (1828).

Clerodendron pentagonum, Hce, l. c. — I. F. S., ii, 260 = *Clerod. fortunatum*, Lin.

Mentha reticulosa, Hance, l. c. 247. — I. F. S., ii, 279 = *Perilla nankinensis*, Dcne (1852), *Dentitia nankinensis*, Loureiro.

Teucrium fulvum, Hce, l. c. 270. — I. F. S., ii, 314 = *Teucrium quadrifarium*, Hamilt. (1825), Nepal.

Chavica puberula, Benth. in Fl. Hongk. 335. — Maximovicz named it *Piper puberulum*. — I. F. S., ii, 365.

Dichelactina (nov. gen.) *nodicaulis*, Hance, Walp. Ann. III, 376. — I. F. S., ii, 421 = *Phyllanthus Emblica*, Lin.

Phyllanthus hongkongensis, Muell. Arg. in Flora, 1865, 370. — I. F. S., ii, 424, sub *Glochidion hongkongensis*, Muell. Arg.

Croton Hancei, Benth in Fl. Hgk. 308. — I. F. S., ii, 434.

Hancea (new gen.) *Hookeriana,* Seem. Bot. Her. 409, tab. 96. —
I. F. S., ii, 440, sub *Mallotus Hookerianus,* Müll. Arg.

Hancea muricata, Bentham in Fl. Hgk. 306. — I. F. S., ii, 439, =
Mallotus Furetianus, Muell. Arg. *(Rottlera Furetiana,* Baillon). Furet
gathered the plant in about 1854, Hance probably earlier.

Artocarpus hypargyrea, Hance in Fl. Hgk., 325.

Engelhardtia chrysolepis, Hce, Symb. 227. Hongkong 1861. — DC.
Prod. XVI, ii, 142 = *E. Wallichii,* Lindl. var. *chrysolepis.*

Quercus thalassica, Hance, Kew J. B. I (1849) 176. — DC. Prodr.
XVI, ii, 84.

Quercus Hancei, Bentham in Fl. Hgk. 322. — DC. Prodr. l. c. 96.

Quercus bambusifolia, Hance, MSS, figured in Seem. Bot. Her. t. 91.
See also Journ. Bot. 1875, p. 364.

Cleisostoma virginale, Hance, Journ. Bot. 1877, 38. Gathered in
Hongkong, 1861.

Tropidia grandis, Hce, Add. Fl. Hgk. (1871) 128. Gathered in 1866
in Hongkong.

Cypripedium sinicum, Hance, Walp. Ann. III, 602. — Reichb. fil.
in Bonpl. 1855, 120 = *Cypriped. purpuratum,* Lindl. (1837). V. supra
p. 220.

Alpinia stachyoides, Hce, Add. Fl. Hgk., 126. Gathered in 1857.

Thysanotus chinensis, Benth. in Fl. Hgk. 372. Hance, Hongkong. —
Hance Advers. 245.

Phoenix Hanceana, Naudin, Man. Acclim. (1887) 407. Cultivated in
Southern France. — Comp. *Ph. farinifera,* Roxb. in Hance's Add. Fl. Hgk.
129, Adversaria (1864) 246, and Hance in J. B. 1879, 174. According
to Kew Ind. = *Ph. humilis,* Royle.

Eriocaulon heteranthum, Bentham in Fl. Hgk. 382.

Scirpus chinensis, Munro in Seem. Bot. Her. 422.

Scleria radula, Hance, Manip. (1863) 232.

Garnotia patula, Munro in Fl. Hgk. 416.

Garnotia drymeia, Hance, Manip. 233. Gathered in 1862.

Zoysia sinica, Hce, Journ. Bot. 1869, 168.

Ischaemum leersoides, Munro, Proc. Amer. Acad. IV, 363. — Fl.
Hongk. 425.

Ischaemum ophiuroïdes, Munro, l. c. — Fl. Hgk. 425.

Arundinaria sinica, Hce, Manip. 225.

Arundinaria longiramea, Munro, Bamb. (1868) 19.

Bambusa dumetorum, Hce, Walp. Ann. III, 791. — In Munro, Bamb. 136 = *Schizostachium dumetorum.*

Dendrocalamus latifolius, Munro, l. c. 152.

Asplenium Hancei, Baker, Syn. Fil. (1868) 208. — Hance, Add. Fl. Hongk. 139. Discovered, 1857.

Aspidium latipinna, Hce, Add. Fl. Hgk. 141. — Syn. Fil. 292.

Pteris insignis, Mettenius; Hance, Hong kong. See Journ. Botan. 1868, 269.

There are in the above list of Hong kong plants, 66 described by Hance and proposed as new species. The Kew botanists, however, admit only 20 of them to be really novelties.

During his residence in Hong kong, and previous to 1851, Hance seems to have received a small collection of plants from Pratas Island, situated about 200 miles S. E. from Hongkong. In Walp. Ann. II, he described two plants from that spot, believing them to be new, viz:

Portulacca psammotropha (l. c. 659), described from living specimens dug up there. It proved afterwards to be identical with *P. australis,* Endl. from the Gulf of Carpentaria. See Hance in J. B. 1871, 201.

Psammanthe marina (l. c. 660). It was found afterwards to be a variety of *Sesuvium portulacastrum,* Lin. Hance, l. c. 202.

Morris. Dr. Hance has commemorated the name of this botanical collector, regarding whom he gives no particulars, in two plants:

Dianthus Morrisii, Hce, in Lond. J. B. VII (1848). Description. In arenis insulae Lin tin leg. clar. Morris. — Lin tin lies in the estuary of the Canton River, N. W. of Hongkong. — Hance calls the same plant *Tunica Morrisii,* in Walp. Ann. II (1852) 101. — I. F. S., ı, 63. Seemann says it is *Dianthus caryophyllus,* Linn.

Diospyrus Morrisiana, Hce, Walp. Ann. III (1852) 14. Gathered, in Hongkong, where it has also been found by other collectors. Met with also near Canton. Hance in J. B. 1880, 299. — I. F. S., ıı, 70.

Gützlaff, Rev. Dr. Karl, a distinguished sinologist, who has written many important works and articles on China. He was a German, born in 1803, in Pomerania, protestant missionary, went in 1826 to Batavia, proceeded in 1828 to Bangkok, from which place he performed a voyage to Tien tsin, in a Chinese junk. In 1831 he arrived at Macao. Engaged as Chinese interpreter on Opium ships he had opportunity to do missionary work. He thus travelled up and down the Chinese coast for several years. In 1832 the "Lord Amherst" was fitted out at the expense of the E. I. C. to visit the ports of the Chinese coast. This experimental commercial expedition was conducted by Mr. Lindsay and the Rev. Gützlaff. Starting from Canton they went to Amoy, Ning po and the mouth of the Yang tze, but the local Chinese authorities refused all propositions for the establishment of trade. In 1834 G. published: "Journal of Three Voyages along the Coast of China, in 1831, 32, 33".

In 1835 G. took service with the British Govt. and acted as Interpreter during the Opium War. In 1842, when he island of Chusan was under British control, G. occupied the office of Chinese Magistrate there. After the war he lived in Hongkong, occupying the post of Chinese Secretary. In 1849 he went on leave to Europe. Soon after his return to Hong kong, in 1851, he died there.

G. collected in Hongkong some plants which he presented to Hance. The latter dedicated to the memory of his friend a new genus name. Kew J. B., I (1849) 141: Hance, Description of a new genus of Acanthaceous plants of China. "*Gutzlaffia*, dicatum sinologo indefesso et eruditissimo Carolo Gutzlaff, Theol. Doct., scientiarum artiumque in imperio coelesti conditionis facile studiorum principi. *G. aprica*, Hongkong". — Hance's Gutzlaffia is now reduced to the genus *Strobilanthes*, Bl., but the species discovered by Gützlaff was new.

Harland, William Aurelius, Medicinae et Chirurgiae Doctor. He was in China before 1847, occupied the post of Government Surgeon at Hong kong, where he died in 1857, Sept. 12. (Mayers & Dennys, Treaty Ports of China and Japan, 80). Dr. J. Legge writes me, in 1894: "I knew well Dr. Harland. He was in charge of one of the hospitals in Hongkong, thought much of for his medical skill, and esteemed for his general intelligence. He had acquired considerable knowledge of the Chinese written characters".

I 1847 H. published in Trans. China Br. R. As. Soc. No I, a Treatise "on Chinese Anatomy and Physiology", — in the same Trans., 1850, an article on "Chinese Manufacture of Magnetic Needles and Vermilion", — ibidem 1855, VII: "On the Topography and Natural History of Turon in Cochinchina" and in the same, 1853, "Notes on Chinese Medical Jurisprudence".

H. was one of the early botanical explorers of Hongkong and intimately associated with Hance (Advers. 248), who described most of his plants; and from Hance's papers it appears (Advers. 230) that Harland, in 1853, botanized also near Shang hai and on Turon Bay, Cochinchina (J. B. 1866, 171). — Bentham, in Fl. Hgk. Pref. 10, writes: "The late Dr. W. A. Harland, brought to this country, in 1857, a very valuable set of Hongkong plants, including many that had escaped the notice of previous collectors. He allowed me to select specimens of all that appeared new or interesting". I suppose that Harland's botanical collections finally found their way to Kew.

Plants from Harland's Hongkong collection described as new by Hance and other botanists (Some of Harland's plants labelled S. China. One is from Cochinchina):

Gymnosporia Harlandi, Hance, J. B. 1866, 171. Ad sinum Turon, Cochinchina, 1855, beatus Dr. Harland.

Vitis corniculata, Benth. in Fl. Hgk. 54. — I. F. S., i, 132.

Harlandia bryonioides, Hance, Walp. Ann. II, 648. — I. F. S., i, 319 = *Zehneria umbellata,* Thwaites, Enum. Ceyl., 125. Earlier Loureiro.

Aucuba chinensis, Bentham, Fl. Hongk. 138. Also Hance. — I. F. S., i, 346.

Hedyotis consanguinea, Hce, Manip. (1863) 221. Harland, S. China, Hance, Whampoa. — I. F. S., i, 373. Perhaps same as *H. fruticosa,* Lin.

Samara longifolia, Benth. in Fl. Hgk. 205. Hong kong; Harland, Hance. — I. F. S., ii, 62. Hemsley names it *Embelia longifolia.*

Reptonia laurina, Benth. in Fl. Hgk. 208. Harland. — I. F. S., ii, 68. It is named *Sarcosperma laurina,* in Hook. & Benth. Gen. Plt. II, 655.

Utricularia Harlandi, Oliver, MS. in Fl. Hongk. 257. — Ind. Fl. Sin., ii, 223.

Litsea rotundifolia, Hemsley, I. F. S., ii, 385. Harland, S. China.

Balanophora Harlandi, J. D. Hooker, Trans. Linn. Soc. XXII (1859) 426. — I. F. S., ii, 410.

Buxus Harlandi, Hance, Add. Fl. Hgk. 123. Harland and Hance. — I. F. S., ii, 418 = *B. sempervirens,* Lin. *).

Ficus Harlandi, Benth. in Fl. Hgk. 330. — Maximowicz, Ficus, in Mél. biol. XI (1881) 330.

Pellionia scabra, Benth. Fl. Hongk. 330. Harland, Hance. — DC. Prod. XVI, i, 166.

Quercus Harlandi, Hce, Walp. Ann. III (1853) 382. — Seem. Bot. Her. t. 89. — DC. Prod. XVI, ii, 96.

---

*) I may observe here, that the name of *Buxus chinensis* was erroneously applied by Link in Enum. Ht. Berol. (1822) 386 to *Simmondsia californica,* Nutt. See DC. Prod. XVI, I, 23.

Habenaria heptoloba, Benth. in Fl. Hgk. 362. Also Hance.

Pandanus urophyllus, Hance, J. B. 1875, 67. Harland, 1855, Hong kong. — See also Hce, Add. Fl. Hgk. 129.

Arisaema penicillatum, N. E. Brown, Journ. Linn. Society, XVIII (1881) 248 *(A. laminatum,* Benth. Fl. Hongk. 342, nec Blume). Harland, Hongkong.

Scleria Harlandi, Hce, Advers. (1864) 248.

Carex Harlandi, Boott, Illust. Car. (1859) 87, tab. 255.

Carex ligata, Boott in Fl. Hgk. 402. Harland, Hance.

Carex manca, Boott, l. c. 402, Harland.

Carex nexa, Boott, l. c. Harland, Hance.

Carex tenebrosa, Boott, Ill. Car., 88, t. 256. Harland, Champion.

Aristida chinensis, Munro, Proc. Amer. Acad. IV, 363. — Harl., Hance.

Woodwordia Harlandi, Hooker, Fil. exot. (1857) tab. 7.

Aspidium controversum, Hance, Manip. (1863) 235. Subsequently Hance found (Advers. 259, Journ. Bot. 1878, 113), that it was *Aspidium amabile,* Blume. Dr. Harland, 1856, prope pagum Sung tong, adversus ins. Hongkong.

Acrostichum Harlandii, Hook., Spec. Fil. V (1864) 274. Same as *Gymnopteris decurrens,* Hook., Fil. exot. (1857) tab. 94, *Chrysodium Harlandi,* Hce. in Add. Fl. Hgk. 138. — Syn. Fil. 418.

Osmunda bipinnata, Hooker (non Linn.), Fil. exot. tab. 9. — Syn. Fil. 426.

Dill, M. D., mentioned in Benth. Fl. Hgk. Pref. 11. He was one of the earlier botanical collectors in Hongkong. Dr. J. Legge wrote me last year that he knew Dr. Dill, who came to Hongkong in 1844, but died in a few years. Sir W. J. Hooker quoted his name in connection with Hong kong Ferns transmitted to him by Dill, who seems to have first discovered the *Asplenium davallioides,* Hooker, K. J. B. IX (1857) 343; Spec. Fil. III (1860) 212; — 2d Cent. Ferns, tab. 40. — Benth. Fl. Hongk. 451. — Syn. Fil. 222.

Philippi, Th. — Reichenbach fil. in Linnaea, XXV (1852) 227, described a new Orchid, *Arundina Philippi,* discovered by Th. Philippi in China. According to Dr. Eichler, Jahrb. d. Kgl. Bot. Gartens und Museums, Berlin, I, 1881, 121, 148, the latter was Keeper of the Herbarium of the Museum from 1841 to 48, and from 1844—46 collected plants for it in India and China. From information kindly given by Prof. Dr. J. Urban of the aforesaid Museum, we learn that in 1844 Ph. was despatched by the Prussian Govt. to Mergui, Tenasserim, to negotiate the acquisition of some landed property there. His Indian plants were collected at this spot. He also brought a few plants from China, it is unknown from what part of China. From the fact, however, that Arundina Philippi (reduced in Fl. Hongk. 355, to *A. chinensis*, Bl.) is a common Hongkong plant, we may suppose that he paid a visit to this island, which a few years earlier had been occupied by the British Govt., and then was but little explored. — In 1849 Ph. emigrated to Chile, and became Professor of Natural history at Concepcion, where he died in 1852.

Champion, John George, an ardent explorer of the Flora of Hongkong. He was born at Edinburgh, in 1815. In 1831 he was gazetted Ensign in the 95th Regiment. During his residence in Ceylon, 1838—47, Captain Champion formed extensive collections of the plants of that island, in connection with Dr. G. Gardner, Superintendent of the Botan. Gardens at Peradenia, who, in 1846, named after him the new genus *Championia* (Gesneraceae). Having been removed with his regiment to Hongkong, he remained there from 1847 to-50, and during his leisure time employed himself sedulously in investigating the Flora of the island. The herbarium which he formed there was of the highest scientific importance and abounded in novelties. Champion early transmitted to his friend Dr. Gardner, several entirely new species, descriptions of which that botanist published in Sir W. Hooker's Kew Journ. of Bot., Vol. I (1849) 240, 308, 321. In 1851 Champion returned to England and brought with him a fine collection of between 500 and 600 species of phaenogamous plants and ferns from Hongkong, of which a complete set was placed in Bentham's hands, who published them, together with Champion's valuable notes, in the Kew Journal of Bot., vols I—VII, 1851—57. (See further on sub Bentham). In the Linn. Trans. XXI (1855) 111 there is a Treatise by Champion on the Ternstroemiaceae of Hongkong, published after the author's death. After leaving China, Ch. was at the Cape of G. H. — In 1854 Champion, then Major, proceeded to the Crimea with the 95th Regiment. He was present at the victory of the Alma, Sept. 20, to the gaining of which Champion and his glorious regiment greatly contributed. In the heroic battle of Inkerman, Nov. 5, Major Champion and his 95th were again foremost and saved the allied forces from destruction. It was here he fell at the head of his men, dangerously wounded. He was then promoted to the rank of Lieutenant-Colonel. Champion died in hospital at Scutari on Nov. 30th, 1854.

New plants recorded from Champion's Hongkong collection:

Clematis parviloba, Gardener & Champion in Kew Journ. Bot. I (1849) 221. — I. F. S., I, 6.

Clematis uncinata, Champ., K. J. B. III (1851) 255. — I. F. S., I, 7.

Magnolia Championi, Benth. in Fl. Hongk. 8 (Champion, l. c. 255, noticed it first as Thalauma pumila Bl.). — I. F. S., I, 24 = *Magnolia pumila,* Andr. (1802).

Artabotrys hongkongensis, Hance (v. supra p. 366).

Cyclea deltoidea, Miers in K. J. B., l. c. 258. — I. F. S., I, 29.

Capparis membranacea, Gard. & Champ., K. J. B. I (1849) 241. Also Hance. — I. F. S., I, 50.

Viola confuca, Champ., Kew Journ. Bot. III (1851) 260. — I. F. S., I, 55 = *Viola serpens,* Wall. in Roxb. Fl. Ind., ed. Carey, II (1824) 449. N. India.

Polygala hongkongensis, Hemsley in I. F. S., I, 60, c. icone.

Garcinia multiflora, Champ., K. J. B. III, 310. — I. F. S., I, 75.

Garcinia longifolia, Champ., l. c. 311. — I. F. S., l. c.

Pentaphylax (gen. nov.) *euryoides,* Gard. & Champ., in Kew J. B., I (1849) 245. — Champ. in Trans. Linn. Soc. XXI (1855) 114, tab. 12. — I. F. S., I, 77.

Heptaca (?) *latifolia,* Gard. & Champ., K. J. B. l. c. 243. — *Actinidia Championi,* Benth. in Fl. Hgk. 26. — I. F. S., I, 78.

Camellia assimilis, Champ., K. J. B. III (1851) 309, et Trans. Linn. Soc. l. c. 112. — Seem. Bot. Her. 367, tab. 77. — I. F. S., I, 80.

Camellia salicifolia, Champ., l. c. 309, and Trans. Linn. Soc. l. c. 112. — I. F. S., I, 82.

Pterospermum heterophyllum, Hance, Journ. Bot. 1868, 112. — According to I. F. S., I, 91, noticed by Benth. in Fl. Hongk. 39, as *Pterosp. acerifolium,* Willd.

Friesia chinensis, Gard. & Champ., K. J. B. I, 243. — Hook. fil. in Fl. Hgk. 43 named it *Elaeocarpus chinensis.* — I. F. S., I, 94.

Zanthoxylum pteleaefolium, Champ., K. J. B. III, 330. — I. F. S., I, 104 = *Evodia triphylla,* DC. Prod. I (1824) 724.

Zanthoxylum cuspidatum, Champ., l. c. 329. — I. F. S., I, 106.

Schoepfia chinensis, Gard. & Champ., Kew Journ. Botan. I, 308. — I. F. S., I, 115.

Ilex cinerea, Champ., K. J. B. III (1851) 327. — I. F. S. I, 115.

Ilex graciliflora, Champ., l. c. 328. — I. F. S., I, 116.

Ilex memecylifolia, Champ., l. c. — I. F. S., I, 117.

Ilex viridis, Champ., l. c. — I. F. S., ɪ, 118.

Euonymus laxiflorus, Champ., l. c. 333. — I. F. S., ɪ, 120.

Euonymus longifolius, Champ., l. c. 332. — I. F. S., l. c.

Evonymus hederaceus, Champ., l. c. 333. — I. F. S., l. c.

Catha Benthami, Gard. & Champ., K. J. B. I, 310. — In K. J. B. III, 334, Bentham names it *Celastrus Championi.* — I. F. S., ɪ, 122.

Acer reticulatum, Champ., K. J. B. III, 312. — I. F. S., ɪ, 141.

Crotalaria brevipes, Champ., K. J. B. IV (1852) 44. — I. F. S., ɪ, 152 = *C. sessiliflora,* Lin.

Millettia Championi, Champ., K. J. B. IV, 74. — I. F. S., ɪ, 159.

Mucuna Championi, Benth. in K. J. B. IV, 49. — I. F. S., ɪ, 189.

Caesalpinia vernalis, Champ., K. J. B. IV, 77. — I. F. S., ɪ, 206.

Bauhinia Championi, Benth. V. supra p. 300, sub Millett.

Albizzia Championi, Benth. in K. J. B. IV, 79, is the same as *Pithecolobium lucidum,* Benth. in L. J. B. III (1844) 207. Earlier Beechey. V. supra p. 292. — I. F. S., ɪ, 217.

Eriobotrya fragrans, Champ., Kew Journ. Bot. IV, 80. — Ind. Fl. Sin., ɪ, 261.

Eustigma oblongifolium, Gard. & Champ. in K. J. B., I (1849) 312.— I. F. S., ɪ, 291.

Rhodoleia (new genus) *Championi,* Hook. Bot. Mag. t. 4509 (1850). Champion, who discovered it in Hongkong, in Febr. 1849, writes that it is the handsomest of Hongkong flowering trees. Champion and Braine sent seeds of it to Kew, where it was cultivated. Hance, in Notes & Quer. Ch. Jap. 1870, 28, states that this tree is famed for the extreme beauty of its flowers, and its rarity, there being but two trees known, growing together, on the island and the plant having not yet been met with anywhere on the mainland.

Liquidambar chinensis, Champ., Kew J. Bot. IV (1852) 164. — In Bentham & Hooker Gen. Pl. I (1862) 669 named *Altingia chinensis.* — I. F. S., ɪ, 291.

Acmena Championi, Benth., K. J. B. IV, 118. — Hemsley in I. F. S., ɪ, 296, names it *Eugenia Championi.*

Hedera parviflora, Champ., Kew J. B. IV, 122. — Bentham in Fl. Hgk. 137, names it *Dendropanax parviflorum.* — I. F. S., ɪ, 343.

Hedera protea, Champ., l. c. — *Dendropanax proteum,* Benth. in Fl. Hgk. 136. — I. F. S., ɪ, 343.

Benthamia japonica, var. *sinensis,* Benth., K. J. B. IV (1852) 165.— Hemsley, I. F. S., ɪ, 345, names it *Cornus hongkongensis.*

Lonicera reticulata, Champ., K. J. B. IV, 167. — I. F. S., ı, 366.
Also Hance.

Thysanospermum diffusum, Champ., l. c. 168. — I. F. S., ı, 371.

Ophiorrhiza pumila, Champ., l. c. 169, — I. F. S., ı, 378.

Mussaenda erosa, Champ., l. c. 193. — I. F. S., ı, 378.

Randia canthioides, Champ., l. c. 194. — I. F. S., ı, 381.

Randia leucocarpa, Cham., l. c. — I. F. S., ı, 382.

Guettardella (gen. nov.) *chinensis,* Champ., l. c. 197. — Benth. &
Hook., Gen. Pl. II (1873) 100, reduce the genus to *Antirrhoea; A. chinensis.* — I. F. S., ı, 384.

Mephitidia chinensis, Champ., l. c. 196. — Bentham in Fl. Hgk. 160,
named it *Lasianthus chinensis.* — I. F. S. ı, 388.

Aster brevipes, Benth., Fl. Hgk. 175. — I. F. S., ı, 409.

Ainsliaea fragrans, Champ., Kew Journ. Bot. IV (1852) 236. —
I. F. S., ı, 470.

Lactuca brevirostris, Champ., l. c. 237. — I. F. S., ı, 479.

Rhododendron Championae, Hook. Bot. Mag. t. 4609 (1851). It was
discovered by Captain and Mrs Champion at Fort Victoria in Hongkong,
Apr. 18, 1849. Champion sent also a drawing and seeds. Sir W. Hooker
writes: We have named this species in compliment to his amiable and accomplished lady, whose partiality for plants equals that of her husband
and who accompanied him on many of his botanizing excursions. — Ind.
Fl. Sin., ıı, 21.

Lysimachia alpestris, Champ., l. c. 299. — I. F. S., ıı, 47.

Ardisia chinensis, Benth., Fl. Hgk. 207. — I. F. S., ıı, 63.

Ardisia primulaefolia, Gard. & Champ., K. J. B. I (1849) 324. —
I. F. S., ıı, 66.

Diospyros eriantha, Champ., Kew Journ. Bot. IV (1852) 302. —
I. F. S., ıı, 69.

Symplocos crassifolia, Benth., Fl. Hgk. 212. — I. F. S., ıı, 72.

Styrax odoratissimum, Champ., l. c. 304. — I. F. S., ıı, 76.

Olea marginata, Champ., l. c. 330. — In Gen. Plant. II (1876) 677,
named *Osmanthus marginatus.* — I. F. S., ıı, 88.

Melodinus fusiformis, Champ., l. c. 332. — I. F. S., ıı, 93.

Alyxia sinensis, Champ., l. c. 493. — I. F. S., ıı, 95.

Holostemma pictum, Champ., Kew J. B., V (1853) 53. — In Gen.
Plant. II (1876) 760, named *Graphistemma pictum.* — I. F. S., ıı, 103.

Pentasachme Championi, Benth. in K. J. B., V, 54. — I. F. S., ıı, 112.

Stephanotis chinensis, Champ., l. c. 53. — I. F. S., ıı, 114.

Dischidia chinensis, Champ., l. c. 55. — I. F. S., II, 116.

Strychnos angustiflora, Benth., Journ. Linn. Soc. I (1857) 102. — I. F. S., II, 121.

Ehretia longiflora, Champ., l. c. 58. — I. F. S., II, 145.

Argyreia Championi, Benth. in Fl. Hgk. 236. — In Gen. Plant. II (1876) 869, named *Lettsomia Championi.* — I. F. S., II, 156.

Aeschynanthus chinensis, Gard. & Champ., K. J. B., I (1849) 320. — I. F. S., II, 224 = *Aeschynan. acuminatus,* Wall. Cat. 6397. — Earlier from India.

Strobilanthes Championi, T. Anders. in Fl. Hgk. 261. — I. F. S., II, 239 = *Strob. flaccidifolius,* Nees, in DC. Prod. XI (1847) 194. Earlier known from India, Assam, Burma.

Ruellia tetrasperma, Champ., K. J. B., V (1853) 132. — T. Anders. in Fl. Hgk. 262, named it *Strobilanthes radicans.* — I. F. S., II, 242. Also Harland.

Adhatoda chinensis, Benth. in K. J. B., V, 134. — T. Anders. in Fl. Hgk. 264, named it *Justicia Championi.* — I. F. S., II, 244.

Rungia chinensis, Benth. in Fl. Hgk. 266. — I. F. S., II, 247.

Aristolochia longifolia, Champ., Kew Journ. Bot. VI (1854) 116. — I. F. S., II, 362.

Chavica sinensis, Champ., l. c. 116. — In DC. Prod. XVI, I, 361, it is named *Piper sinense.* — I. F. S., II, 366. Also Harland.

Alseodaphne chinensis, Champ., Kew J. B., V, 198. — Hemsley in I. F. S., II, 374, names it *Machilus chinensis.*

Machilus velutina, Champ., l. c. 198. — I. F. S., II, 378.

Daphne Championi, Benth. in Fl. Hgk. 296. — I. F. S., II, 395.

Wikstroemia nutans, Champ., l. c. 195. — I. F. S., II, 400. — Fortune earlier.

Glochidion eriocarpum, Champ., Kew Journ. Bot. VI (1854) 6. — I. F. S., II, 424.

Aporosa leptostachya, Benth., Fl. Hgk. 317. — Described as *Scepa chinensis,* by Champ., l. c. 72. — DC. Prod. XV, II (1866) 472. Eealier Callery. — I. F. S., II, 429.

Endospermum chinense, Benth., Fl. Hgk. 304. Also Hance, Harland. — I. F. S., II, 444.

Gironniera (?) *nitida,* Benth., Fl. Hgk. 325. Only male specimens. — I. F. S., II, 452.

Ficus Championi, Benth., K. J. B. VI, 76. — Maximowicz, Ficus, in Mél. biol. XI (1881) 336.

Ficus impressa, Champ., K. J. B. VI, 76. — Maxim. l. c. 340. — King, Ficus, 1888, 133, reduces it to *F. foveolata,* Wall.

Ficus hibiscifolia, Champ., l. c. 77. — Fl. Hgk. 329 = *Ficus hirta,* Vahl, Enum. II (1806) 201. — King, l. c. 149.

Ficus chlorocarpa, Benth., Fl. Hgk. 330. — Maxim. l. c. 330.

Quercus fissa, Champ., l. c. 114. — Seem. Bot. Her. 415, t. 92. — DC. Prod. XVI, II (1864) 104.

Quercus Championi, Benth., K. J. B. VI, 113. — DC. Prod. l. c. 94.

Castanea concinna, Champ., K. J. B. VI, 115. — Also Hance. — In DC. Prod. l. c. 110 named *Castanopsis concinna,* Alph. DC.

Cottonia Championi, Lindl., Kew J. B. VII (1855) 35. — Fl. Hgk. 357. — Hooker fil. & Thomson found it also in the Khasia Mountains, and Sir J. Hooker, in Flora Ind. VI (1893) 26, proposed for it a new name *Diploprosa* (gen. nov.) *Championi.*

Appendicula bifara, Lindley, K. J. B. VII, 35.

Platanthera stenostachya, Lindl., l. c. 37. — Benth. in Fl. Hgk. 362 named it *Habenaria stenostachya.* — Also Hance. Perhaps earlier from Moulmein, Lobb. 1847.

Habenaria Miersiana, Champ., K. J. B. VII, 37. — Fl. Hgk. 363.

Platanthera Championi, Lindl., K. J. B. VII, 38. — Reichenb. fil. in Linnaea, XXV (1852) 226, had named it *Pl. Galeandra.* — Fl. Hgk. 363, *Habenaria Galeandra,* Benth. It seems Fortune earlier.

Smilax hypoglauca, Benth., Fl. Hgk. 369. Champion, Hance.

Carex tenebrosa, Boott, Illust. Caric., 1860, 88, t. 256. — Champion, Harland. — Fl. Hgk. 402.

Aspidium Championi, Benth., Fl. Hgk. 456. — Hook. Spec. Fil. IV (1862) 30.

Champion also collected Coleopters in Hongkong, as Dr. J. Mc. Lelland stated in Journ. As. Soc. Bengal XVII (1848) 206—9.

Eyre, J., Lieut. Colonel, Roy. Artillery, was stationed in Hongkong, from 1849 to 51, and made botanical collections there. B. Seemann, who in December 1850, spent three weeks in Hongkong, writes in his Voyage of the Herald, p. 20: "There are at present in Hongkong, two gentlemen, Dr. H. F. Hance, and Lieut. Col. Eyre, who take great interest in botany. (It seems Champion then had already left). They made several excursions with me to the most profitable localities and pointed out some of the rarest productions of the flora. Dr. Hance was unfortunately suffering from intermittent fever, which has shaken him so much during the last four months, that he will be compelled to return to England before the commencement of the rainy season. He was, therefore, unable to accompany me frequently. Lieut. Col. Eyre makes almost daily excursions. He possesses, besides a considerable herbarium, a beautiful set of coloured drawings of Hongkong plants chiefly executed by himself. Many of the figures represent species new to science; there is one called by Captain Champion *Camellia euryoides*" *). Before 1861 Col. Eyre was promoted to the rank of General. See Fl. Hgk. Pref. 11.

New plants discovered in Hongkong by Col. Eyre:

Camellia hongkongensis, Seemann in Linn. Trans. XXII (1859) 342, tab. 60. First discovered by Colonel Eyre, Hongkong, 1849. — Ind. Fl. Sin., I, 81.

Eyrea (gen. nov.) *vernalis*, Champ., K. J. B. III (1851) 331. Hooker writes: "The genus is named by Champion in honour of his friend Col. Eyre, who lately returned from Hongkong with a considerable herbarium collected there. He has also brought over a valuable collection of Chinese seeds, which he presented to the Roy. Gardens, Kew, and a large and interesting set of rice-paper drawings of plants". — The genus Eyrea has been now reduced to *Turpinia*. The plant in question had been received long ago from China by the Hort. Soc. It flowered in their garden and disappeared. Lindley described and figured it in Bot. Reg. tab. 1819 (1835). It was discovered again by Fortune near Amoy, in 1843, but Fortune's specimens for more than forty years lay unexamined. Meanwhile Champion had found the same plant in Hongkong and Gardner in K. J. B. I (1849) 301, described it as *Staphylea simplicifolia*, and Champion, in 1851, as *Eyrea vernalis*. — Seemann in Bot. Her. 371, changed Lindley's original name *Ochranthe arguta*, into *Turpinia arguta*, under which later name it figures in I. F. S., I, 143.

Ophiorrhiza Eyrii, Champ., K. J. B. IV (1852) 170. — Now reduced to *O. japonica*, Blume, Bijdr. (1825) 978. — I. F. S., I, 378.

Azalea myrtifolia, Champ. in Bot. Mag. sub t. 4609. Discovered by Col. Eyre, March 1849, Hongkong. — Maximowicz, Rhodod., 1870, 45, reduces this to *Azalea ovata*, Lindl. (1846) Fortune. — I. F. S., II, 28.

Gacrtnera hongkongensis, Seem. Bot. Her. 384. Eyre, Champion. — I. F. S., II, 121.

Quercus Eyrei, Benth., Kew J. B. VI (1854) 114. — DC. Prod. XVI, II (1864) 105.

*) *C. euryoides* was published by Lindley in 1830. V. s. p. 270.

Bowring, Dr. Sir John, and his eldest son John C. Bowring, both residents in Hongkong in the beginning of the second half of this century. Both keenly interested in botany, to which science they rendered many services.

Bowring, Sen. was born Oct. 17, 1792 at Exeter. He commenced his career as a merchant. By cultivating his talents he succeeded in acquiring considerable scientific attainments and became especially skilled in languages. 1829 L. L. D., Groningen. In about 1847 he was appointed British Consul at Canton. From April 1852 to Febr. 1853 he was in charge of the office of Plenipotentiary and Governor of Hongkong, in the absence of Sir George Bonham. But on the return of the latter Bowring applied for leave of absence for a year, and visited Java on his way home. After his return to Hong kong, he assumed, on April 13th 1854, the duties of Governor of the Colony and was knighted. He made a voyage to Siam, to conclude a treaty of commerce, and returned thence on May 11th 1855. In 1858—59 he visited the Philippines. He published accounts of these voyages. In May 1859 Sir John resigned his office and left the colony. See Treaty Ports of China and Japan sub Hongkong. He died in Exeter, Nov. 23, 1872. He was F. R. S. and F. Linn. Soc.

Bowring, John C. **Jun.** was a merchant in Hong kong. As my friend Mr. Th. Sampson wrote me, some years ago, Bowring Jun. is now living at a place near Windsor in England. English botanists have frequently confounded the two Bowrings, father and son.

In his Kew Journ. Bot. V (1853) 236, Sir W. J. Hooker proposed the genus name *Bowringia**) for a beautiful arborescent Fern of Hong kong, *B. insignis,* (figured l. c. VI, tab. 2). He writes as follows: "We are indebted to the Messrs Bowring for living plants of this curious Fern, which we named in compliment to Dr. Bowring and his son J. C. Bowring, Esq., members of a family no less distinguished for their love of literature than for their patronage of science, when ever opportunity presents itself: and the Messrs Bowring have contributed largely to our knowledge of the Natural History of Hongkong, and have further been a means of obtaining a correct knowledge of the famous Chinese Rice-paper".

The history of the discovery in China (Formosa) of the Rice-paper plant, *Aralia (Fatsia) papyrifera,* Hook., its introduction into the Kew

---

*) It had evidently escaped Sir W. Hooker's memory that in the same Kew Journ. Bot. IV (1852) 75, Champion had already proposed a new genus name *Bowringia* for a leguminous plant. Mr. J. Smith changed therefore, in 1856 the name proposed by Hooker into *Brainea.* V. infra sub Braine.

Gardens by the exertions of the Bowrings etc. were related at length by Sir W. Hooker in several papers. See Bot. Miscell., I (1830) 88. — Kew Journ. Bot. II (1850) 27, 250; IV (1852) 25, 50, 348; V (1853) 79; VII (1855) 92. Two living plants of Aralia papyrifera arrived in a healthy state at Kew, in 1852.

In the same year J. C. Bowring published a valuable article on the Rice-paper. Plant of China in Trans. China Br. Roy. Asiat. Soc., Hong kong, Part III (1852) p. 37—43.

In Kew Journ. Bot. VIII (1856) 317 Hooker reports that he received from China bulbs of the Californian Soap plant, sent by his excellent friend Sir John Bowring, who informed him that the Chinese use the bulbs as soap without any artificial preparation. The Chinese brought it probably from California. It is *Scilla pomeridiana*.

It does not seem that Sir John Bowring himself made botanical collections in Hongkong. The Hongkong specimens noticed by Hooker, Hance, and Champion under Bowring's name were all gathered by J. C. Bowring.— Hance in Add. Fl. Hongk., 144 states: "Mr. J. C. Bowring seems to be the only person who has paid any attention to the *Moss*-flora of Hongkong, and who has formed a collection of these plants".

Mr. Sampson informs me that J. C. Bowring was a great fancier of beetles, and that he was said to have paid 1000 dollars for one specimen from South-America. — R. Swinhoe, in Sclater's Ibis, III (1861) 326, said that J. C. Bowring was then the best entomologist perhaps, this side the Cape.

Plants from J. C. Bowring's Hongkong collection recorded as new by Hance, Champion, Hooker:

Bowringia (new gen.) *callicarpa*, Champ., Kew J. B. IV (1852) 75. Bentham adds: "Genus named by Champion in honour of his friend John C. Bowring Esq. who has been for some time investigating the Flora of Hongkong with much zeal and who has formed large collections there". — I. F. S., I, 201. Beechey earlier.

Pygeum phaeostictum, Hce, J. B. 1870, 72. J. C. Bowring, Hong kong, many years ago. — Maxim., Mél. biol. XI (1883) 708, named it *Prunus phaeostictum*. — I. F. S., I, 221, where the plant is referred to *Prunus punctata*, Hooker, f. Fl. Ind. II (1879) 317. Earlier Griffith, Khasia.

Begonia Bowringiana, Champ., K. J. B. IV (1852) 120. Discovered, it seems, by Champion. Seem. Bot. Her. 379, named it *Doratometra Bowringiana*. — Bot. Mag. 5182 (1860). — I. F. S., I, 322 = *Begonia laciniata*, Roxb., long known from India.

Parechites Bowringii, Hance, J. B. 1868, 299. Hance states: This plant was given to me more than ten years ago (before 1858) by Mr. J. C. Bowring, who had formerly gathered it in Hong kong. — Hemsley, in I. F. S., II, 99, names it *Trachelospermum Bowringii.*

J. C. Bowring's name is frequently met with in Sir W. Hooker's various works on Ferns. Bowring had sent him the Ferns collected by him in Hongkong and on the mainland opposite the island, 15 miles from the latter, at a place the name of which Hooker writes, Sung tong, Sung long or Lung tong, unknown to Mr. Charles Ford, whom I asked about the matter. Compare above 373, sub Harland.

W. Mitten, Musci Ind. Orient., in Journ. Proc. Linn. Soc. vol. I, suppl. 1859, 26, described a new Moss from Hongkong, *Leucobryum Bowringii.*

Braine, C. J., a British merchant in Hongkong (Messrs Dent & C°). Dr. J. Legge wrote me that he remembers seeing him there in 1844. Braine made a collection of Hongkong ferns of which he sent a set to Sir W. Hooker, who noticed them in the Kew Journ. Bot. IX (1857) 333—63: Ferns of Hongkong. He also sent some ferns from Chusan. Seem., in Bot. Her. 368, mentions Braine's Garden in Hongkong, 1850. Compare also Fortune, Tea Countries, 5, Description of Messrs Dent & C°.'s Garden at 'Green Bank', under the care of Mr. Braine (1848).

Bot. Mag. sub tab. 4509 (1850): Braine presented to Kew seeds of *Rhodoleia Championi.*

Hooker in Kew Journ. Bot. II, 1850, 250: "C. J. Braine, Esq. a gentleman who has recently returned from Hongkong, brought a rich collection of living plants for the Kew Gardens and many curious vegetable productions for the Kew Museum, together with a thin volume of well executed drawings by a Chinese artist, on Rice-paper, — drawings exhibiting the several conditions of the Rice-paper, from the preparation of the seed to the packing of the material for exportation. We cannot conjecture to what family the plant belongs".

Seem. Journ. Bot. IV (1866) p. 15: Mr. John Smith, Curator of the Kew Gardens writes: "In 1851 (rather 1850) Mr. C. J. Braine, on his return from Hongkong brought with him a collection of living plants which he presented to Kew. Amongst these were several epiphytal Orchids artificially attached to stems of Tree Ferns about one foot in length and about a foot in circumference. They were considered to be dead, and placed with the Orchids in the hothouse. About two years after I was much surprised to find that two of them had pushed out lateral buds, which in due time were transferred into pots, and ultimately became fine plants. About the same time the late Sir W. Hooker had received specimens of the same Fern from Sir John Bowring, and finding it to be the type of a new genus, he dedicated it to this gentleman (v. supra p. 381), giving him, instead of Mr. Braine, the credit of having introduced the living plant to Kew. Some time after I found that Mr. Bentham had previously applied the name Bowringia to a Leguminose plant. Bringing these facts to the notice of Sir W. Hooker, I proposed to rename the plant *Brainea* (1856)". — *Brainea insignis,* Hook., figured in Filices Exot. (1857) tab. 38.

Seemann, Dr. Berthold, German, born Febr. 28, 1825, at Hanover, educated at the Lyceum there. Taking a lively interest in botany and full of a desire to travel in foreign countries, he came to Kew and worked in the garden under the then curator, Mr. John Smith. On the recommendation of Sir W. Hooker he was in 1846 appointed, by the Admiralty, naturalist to H. M. S. "Herald", Captain H. Kellett, C. B. The Herald had been employed, since June 1845, on a surveying expedition in the Pacific. From Plymouth, June 26, to Rio Janeiro, around Cape Horn to Valparaiso, Chile, Gallopagos Is., Ecuador, Panama, Vancouver's I., San Francisco. The post to which Seemann was appointed had become vacant by the accidental death of Mr. Thomas Edmonston, a young promising botanist, at Sua, on the coast of Ecuador, Jan. 24, 1846. Seemann left England in August and reached Panama in September. As the "Herald" had not returned from Vancouver's I., Seemann profited by the delay to explore the Isthmus. On Jan. 17th 1847. S. joined the "Herald", which had returned from the North, and remained with her until the completion of her voyage round the world. The "Herald", in 1848, received order to proceed to the Polar Sea in order to cooperate with the vessels composing the relief expedition to search Sir J. Franklin. Honolulu, Petropavlovsk, where they found the schooner Nancy Dawson, Chamisso I. in Kotzebue Sound. After waiting in vain for the arrival of H. M. S. "Plover", the "Herald" returned to Panama. In 1849 the "Herald" proceeded again to the Arctic Regions, passed Behring's Strait, reached Chamisso I. in July 15, and there met the Plover. Then the three vessels "Herald", "Plover" and "Nancy Dawson", left Kotzebue Sound, sailed to the North, and attained 72° 53'. Returned to Kotzebue Sound Sept. 2d. Leaving the "Plover" to winter in Kotzebue Sound, the "Herald" departed for the South, and reached Mazatlan on the west coast of Mexico in Nov. In 1850 the third and last voyage to the Arctic Ocean was accomplished. The "Herald" reached Chamisso I., in May, and started with the "Plover" for the North. In October the "Herald" returned to Honolulu, leaving the "Plover" in the North. Seemann had the opportunity of exploring nearly the whole west coast of America, frequently making long journeys inland, Peru, Ecuador, Mexico. He also collected materials for the Flora of the extreme north-west of America. After a stay of a fortnight at Honolulu, the "Herald" began her homeward course. Nov. 19th in sight of Assumption, passed Formosa and the Bashee Group. Hongkong was reached on December 1st 1850. The "Herald" remained there till December 22. Seemann gathered some plants in the island and near Canton. See above p. 379. Voyage continued, Singapore, Cape of G. H., St. Helena, Ascension. Arrived at Spithead June 6th 1851.

The Admiralty requested Mr. Seemann to publish the results of this voyage, and he accordingly produced, in 1853, the Narrative of the Voyage of H. M. S. "Herald", being a Circumnavigation of the Globe and three Cruises to the Arctic Regions in Search of Sir John Franklin, from 1845 to 51. 2 vols. The animals collected during the voyage were described by the late Sir John Richardson, and, in the years 1852—57, the botanical results appeared in Seemann's Botany of the Voyage of H. M. S. "Herald".

The plants in this work, which contains 100 plates, are arranged according to the countries visited: 1, Flora of Western Esquimaux-land, — 2, Flora of the Isthmus of Panama, — 3, Flora of North-Western Mexico, and 4, Flora of the Island of Hongkong, p. 347—432.

In the preparation of this book the author had the advantage of the assistance of Sir William and Dr. J. D. Hooker. Besides these, J. Steetz determined the Compositae, — Col. W. Munro the Grasses, — J. Smith the Filices.

As to the Flora of the Island of Hongkong it contains an enumeration of 773 phaenogamous plants and ferns. Seemann himself made some collections there and at Canton, but the enumeration of Hongkong plants is chiefly based upon Dr. Hance's collections, for this gentleman, in 1851, being on a visit to England, had entrusted the whole of his Hongkong herbarium to Seemann.

There is no new plant in Seemann's Chinese collection. The *Pothos Seemanni*, Schott, in Bonplandia, V (1857) 44, and Seem. Bot. Her. 416, is according to Bentham, Fl. Hgk. 344, *Pothos scandens*, Lin.

S. founded and edited a botanical journal, in German, under the title "Bonplandia" 1853—63, Hanover and London.

In 1860 B. explored the Fiji Islands in the South Pacific, and in 1865 commenced the publication of the Flora Vitiensis.

In 1863 he started, the JOURNAL OF BOTANY, London, and conducted it to the end of 1869, when he was obliged to give it up.

In 1864 S. visited Venezuela, and in 1866 explored Nicaragua for the Central American Association. In 1871 he visited the same country again. There he was seized with fever and died near the Javali Goldmine, Oct. 10, 1871. The degree of Ph. Dr. had been conferred on Seemann by the Goettingen University, after his return from the circumnavigation of the globe. Comp. his biography in Journ. Bot. 1872, p. 1.

AMERICAN EXPEDITIONS TO EASTERN ASIA. DR. W. WIL-LIAMS. CH. WRIGHT.

In 1852 the Government of the United States of N. America resolved upon sending two naval expeditions to Eastern Asia. The first of them, which left in Nov. 1852, was commanded by Commodore Perry, who had received instructions to endeavour to open commercial connection with Japan. After the return of the expedition there appeared:

Narrative of the Expedition of an American Squadron to the China Seas and Japan, performed in the years 1852—1854, under the command of Commodore M. C. Perry, United States Navy, by order of the United States, by Francis L. Hawks. D. D., L. L. D., with numerous illustrations. Washington, 1856—60, in 3 vls. The first vol. contains the narrative of the voyage, from which we extract the following summary:

The squadron consisted of the "Mississipi", flagship of Comm. **Perry,** the steamers "Powhatan", "Alleghany" and "Vermont", the sloops of war "Vandalian", "Macedonian". The steamships "Susquehanna and the sloops of war "Saratoga" and "Plymouth" were already on the E. I. stations and were to form part of the expedition.

On Nov. 24, 1852, the "Mississipi" took her departure from Norfolk. The other ships followed somewhat later. On Dec. 12, 1852, the "Missis-1853 sipi" anchored at Funchal, Madeira — St. Helena January 10, 1853, — Capetown, Jan. 24, — Mauritius, Febr. 18, — Point de Galle, Ceylon, — Singapore, March 25, — Ladrones I. near Macao, April 6. Visits Macao, Hong kong, anchors at Whampoa. Visit to Canton. The other ships of the squadron arrive.

On the evening of April 28, the flagship was again under weigh, and sailed for Shanghai leaving the Saratoga at Macao to await the arrival of Dr. S. W. **Williams** from Canton, who had been appointed Interpreter to the Expedition, and instructed to join the flagship at Lew chew. Leaving Shanghai, the "Mississipi" arrived on May 25, at Napha, the chief town and port of Great Lew chew, where they found the "Saratoga" with Dr. Williams, who then took up his quarters permanently in the "Susquehanna", which became the flagship. Inland exploration of the island. Notices of plants. Cocoa palms.

On June 9th, the "Susquehanna", the flagship, got under way for the Bonin I! having the "Saratoga" in tow. The "Mississipi" was left at Napha. On June 14th the flagship arrived at Port Lloyd, on Peel I., one of the Bonin group. A coal depôt for steamers was established there. June 18, the "Susquehanna" with the "Saratoga" returned to Napha, which they reached June 23.

On July 2d, 1853, the commodore on the flagship departed for Japan. He was followed by the "Mississipi", "Saratoga" and "Plymouth". On the 7th the squadron anchored off the city of Uraga, on the western side of the Bay of Yedo. The vice-governor of that place came on board the "Susquehanna". It was explained to him, that the Americans had come on a friendly mission, and that the commodore bore a letter from the President of the U. St. to the Emperor. The Commodore refused to go to Nagasaki in order to wait for the Imperial Commissioners. Visit of the Governor, who proposes to refer the matter to Yedo. Perry assents. Survey of the Bay of Yedo. — The reply from Yedo received July 12th. Agreement of the Emperor to receive through a commissioner the President's letter. Landing of Perry and his suite, their reception, delivery of the letter. The squadron proceeds further up the Bay towards Yedo. Perry transfers his Pennant from the "Susquehanna" to the "Mississipi". He decides not to wait for the reply of the Emperor, and leaves Yedo Bay, declaring his intention to return in the ensuing spring. On July 17 the squadron steams out of the Bay. Arrival at Napha, July 25, where a coal depôt was built. Departure of the Commodore for Hongkong, August 1st. Capt. Kelly, Commander of the Plymouth, was left at Napha with instructions to visit the Bonin I! about the beginning of October, which order he executed. The Commodore remained at Macao, ordering the squadron to rendezvous at Cap sing mun (v. supra p. 302).

It had been originally Perry's intention to wait until the spring had set in, before going to the north, but the suspicion of the movements of the French and Russians induced him to alter his plans. The expected storeship "Lexington", having fortunately arrived, the Commodore sailed from Hongkong, in the "Susquehanna", on the 14th of January 1854, for Lew 1854 chew, in company with the "Powhatan", "Mississipi" and the store-ships "Lexington" and "Southampton". Previous to leaving Napha for Japan Perry had received a communication from the Governor General of the Dutch Indies, conveying information, at the request of the Japanese authorities at Nagasaki, of the death of the Emperor of Japan. The "Macedonian" sailed on Febr. 1st, in company with the "Vandalian", "Lexington", and "Southampton". The flagship with the Commodore followed on the 7th, in company with the "Powhatan" and the "Mississipi". On Febr. 11th the squadron entered the outer bay of Yedo, and anchored. Japanese officials came on board. They desired the commodore to go with his ships to Urago, but finally proposed Yokohama, about 8 miles from Yedo, as the proper place for negotiations. The Commodore assented. After lengthy negotiations the Japanese finally agreed to open for the trade of the Americans the ports of Simoda and Hakodate. The treaty was signed at Kanagawa, March 31, 1854.

After H. A. Adams, Commander of the Saratoga, on April 4, had left for Washington with the treaty, the Commodore departed for Simoda, April 19, and on May 9, left for Hakodate, where he arrived May 16. Visits and rambels of the Americans on shore. — On June 3, departure for Simoda.

On June 28, 1854 the whole squadron got under weigh from Simoda and arrived at Napha 1st of July. The "Macedonian" and the "Supply" were ordered to Formosa to search for coal. They visit Ke lung, proceed to Manilla and Hongkong. The flagship "Mississipi" directed her course from Napha directly to Hong kong, where Commodore Perry, with the assent of his Govt., delivered the command of the squadron to Capt. Abbot and proceeded, Sept. 14, by the English mailsteamer "Hindustan" and by the overland route from India, to Europe. At New York he arrived Jan. 12, 1855. The "Mississipi" reached Brocklyn April 23, 1855.

Vol. II of the work, 1857, contains a series of various observations and scientific papers by different gentlemen belonging to the expedition.

On p. 217 Perry states that Mr. W. **Heine,** who accompanied the expedition as an artist attributed chiefly to the procurement of birds. Botanical specimens were gathered by the chief interpreter Dr. Wells **Williams** and by Doctors **Green, Fahs,** and **Morrow.**

On p. 305—332 Professor **Asa Gray,** published an

ACCOUNT OF THE BOTANICAL SPECIMENS-LIST OF DRIED PLANTS COLLECTED IN JAPAN (1854) BY DR. WELLS WILLIAMS, AND DR. JAMES MORROW. These collections were chiefly made at Simoda, Yokohama, Hakodate. The Carices were determined by Dr. Boott, — the Filices by Daniel C. Eaton, Esq., — the Mosses by W. S. Sullivant, Esq., — the Algae by Dr. W. Harvey, Dublin.

Though the Flora of Japan does not come strictly within the province of our investigations, we thought it might not be out of place to give here a list of the novelties discovered in that country (in conjunction with Dr. Morrow) by Dr. Wells Williams, the great sinologist, whose merits with respect to the Botany of China we shall have to record in another chapter. The figures refer to the pages of A. Gray's paper.

306. *Clematis Williamsii,* A. Gray. Simoda, April 20th 1854. — Maxim. Mél. IX (1876) 601.

308. *Viola Grypoceras,* A. Gr. Yokohama, March. — Maxim. Mél. l. c. 743 = variety of *V. sylvestris,* Kit.

308. *Viola laciniosa,* A. Gr. Hakodate. — Maxim. l. c. 746 = *V. canina,* L., var. *acuminata,* Regel. Earlier known from Siberia.

311. *Rubus hydrastifolius,* A. Gray. Simoda. — Maxim. Mél. VIII (1871) 383 = *R. trifidus,* Thbg.

311. *Rubus coptophyllus,* A. Gr. Yokohama. — Maxim. l. c. 384 = *R. palmatus,* Thbg.

313. *Lonicera Morrowi,* A. Gr. Hakodata. — Maxim. Mél. X (1877) 70. — Cultivated at Kew. V. Kew Bull. 1897, App. I, 37.

315. *Azalea serpyllifolium,* A. Gr. Simoda. — Maximow., Rhodod. (1870) 42, cum icone.

316. *Lithospermum japonicum,* A. Gr. Simoda. — Maxim. Mél. VIII (1872) 542.

318. *Tricercandra* (gen. nov.) *quadrifolia,* A. Gray, Hakodate, Yokohama. — Solms in DC. Prodr. XVI, i (1869) 476.

Euphorbia Guilelmi, A. Gr., in Wright's Japan Plants 406. Williams and Morrow, Simoda Hakodate, Wright, Yokohama. — Maxim. Mél. XI (1883) = *E. Sieboldiana,* Morr. Decne (1836).

Quercus phyllireoides, A. Gray, in Wright's Japan Plants, 406. Williams and Morrow, Simoda, — Wright, Tanegasima.

Salix padiflora, Anderson in A. Gray, l. c. 451. Simoda, Williams and Morrow.

319. *Cephalanthera japonica,* A. Gr. Simoda. — Miquel, Prol. Jap. 141 = *C. falcata,* Lindl. Orchid., 412.

321. *Smilacina japonica,* Asa Gray, Hakodate. — Maxim. Mél. XI (1883) 857.

321. *Disporum smilacinum,* A. Gray, Simoda, Hakodate. — Maxim. l. c. 858.

323. *Carex monadelpha,* Boott. Simoda. — Franchet, Enum. Jap. II, 135 = *C. tristachya,* Thbg, teste Maximowicz.

324. *Carex puberula*, Boott. Simoda. — Franch. l. c. 136.

324. *Carex transversa*, Boott. Yokohama. — Franch. l. c. 149.

325. *Carex dispalatha*, Boot. Hakodate. — Franch. l. c. 151.

324. *Carex pisiformis*, Boot. Simoda. — Franch. l. c. 142.

325. *Carex conica*, Boot. Simoda. — Franch. l. c. 143.

326. *Carex excisa*, Boot. Yokohama. — Franch. l. c. 143.

326. *Carex lanceolata*, Boott. Hakodate. — Franch. l. c. 134.

326. *Carex Morrowii*, Boott. Simoda. — Franch. l. c. 145.

327. *Carex anomala*, Boott. Simoda. — Franch. l. c. 132 = *C. Gibba*, Wahlenb. (1803).

327. *Carex incisa*, Boott. Hakodate. — Franch. l. c. 128.

327. *Carex villosa*, Boott. Simoda. — Franch. l. c. 142.

328. *Alopecurus malacostachyus*, Asa Gray, Simoda, Yokohama. — Miq. Prol. 165 = *Alopecurus japonicus*, Steud. Glum. (1854) 147. — Franch. l. c. 158.

330. *Aspidium erythrosorum*, Eaton. Simoda. — Franch. l. c. 239. — Syn. Fil. 273.

330. *Hypnum japonicum*, Sullivant.

331. *Rytiphloea latiuscula*, Harvey, Hakodate.

331. *Polysiphora Morrowii*, Harvey, Hakodate.

331. *Pol. japonica*, Harvey.

331. *Lomentaria catenata*, Harv. — Simoda.

332. *Gymnogoryrus flabelliformis*, Harv. — Simoda.

332. *Gym. pinnulatus*, Harv. — Hakodate.

332. *Cystoclonium* (?) *armatum*, Harv. — Hakodate.

332. *Nemastoma livida*, Harv. — Simoda.

The second American Expedition to Eastern Asia departed seven months after Comm. Perry's squadron had left America. Its chief object was the exploration of the Pacific Ocean. Besides, the commander of the expedition was instructed to join Perry's squadron and reinforce it during the negotiations with the Court of Yedo. For the latter destination, however, he arrived too late. This second expedition is known under the name of: United States North Pacific Surveying Expedition, 1853—56. With this title A. W. Habersham, Lieut. U. S. N. in 1857, published a much confused account of it, in which no dates are given with the exception of the date of departure from and that of return to America. A quite satisfactory narrative of these cruises, compiled from official sources was published in German by W. **Heine** (the artist who had been attached to Perry's expedition), 1858—59, with the title: *Die Expeditionen in die Seen von China, Japan und Ochotsk, unter Commando von Commodore C. Ringgold und Commodore John Rodgers*, im Auftrage der Regierung der Ver. Staaten unternommen, 1853—56. 3 volumes.

The squadron of this expedition, placed under the command of Commodore Cadwallader **Ringgold,** consisted of 5 ships:

the sloop of war "Vincennes", the flag ship — the screw-steamer "Hancock", — the brig "Porpoise" — the schooner "Fenimore Cooper", — the transport "John P. Kennedy".

Two naturalists were on board the "Vincennes": Mr. **Stimpson,** zoologist, and Mr. Charles **Wright,** botanist. Mr. **Small** *), Wright's assistant.

On June 20, 1853 the squadron put to sea from Norfolk and on Sept. 20, arrived in Simon's Bay, Cape of G. H., where they remained for 7 weeks. On Nov. 9th the "Hancock", "Cooper" and "Kennedy" departed directly for China, via Batavia and Gaspar (Banca) Strait, whilst the "Vincennes" and "Porpoise" shaped their course to Australia, and reached Sidney December 26. Departed from Sidney Jan. 8, 1854. — Vanicora I., where 1854 in 1788 Lapérouse perished, — Marianne I°. — On March 17th Macao was reached, on the 19th sailed to Hongkong, where the other ships of the squadron were met. Comm. Ringgold returns to the United States in bad health. Lieutenant commanding, John **Rodgers,** takes the command of the expedition.

On Sept. 12, 1854 the "Vincennes", followed by the "Porpoise", left Hong kong and sailed for the Lew chew I°, whilst the "Hancock" and the "Cooper" proceeded to Fu chou and Shanghai. Between these places they experienced very stormy weather and were obliged to put into Bullock Harb. (v. supra p. 361). Arrived Oct. 7, at Shanghai, where they found the steamer "Powhatan" of Perry's squadron with Mc Lean, the Commissioner for the U. S. on board. It was resolved upon proceeding to Peking. The "Powhatan", with the "Hancock" and "Cooper" in tow, steamed northward to the Gulf of Chili. They reached the mouth of the Pei ho, but the Chinese authorities refused to negotiate with the Americans. Then the "Powhatan" returned to Hongkong, leaving the "Hancock" and the "Cooper" in the Gulf of Chili. These ships were intended to survey the coast from the Peiho to the Great Wall, but having been shattered with violent storms, they were obliged to return to Shanghai, where they arrived on Nov. 28, and had to stay there for two months, to have the vessels repaired.

---

*) *Vaccinium Smallii,* A. Gray. Kamtchatka.

In February 1855, they visited Wen chou and the port of Ki lung in Formosa, and on Febr. 13th were again at Hong kong, where the flag ship had arrived a fortnight earlier.

When the "Vincennes" and the "Porpoise" had left Hongkong, Sept. 12, 1854, they took their course to the Bonin I. In the Bashee Channel, south of Formosa they experienced a violent Typhoon, during which the "Porpoise" was lost. The "Vincennes" reached Port Lloyd on Peel I., of the Bonin Group, and stayed there till Nov. 6, when she left for Lew chew. Napha was reached Nov. 17. The surveying operations commenced by Perry were continued. On Dec. 13, the "Vincennes" left Napha and made sail for the I. of Kiu siu (Japan). On the 28th of Dec. she reached Kagosima Bay. Stay of 9 days, the Bay surveyed. The naturalist went fre-
1855 quently on shore. — Left the Bay on Jan. 6, 1855, landed on the 9th at the southern end of the island of Tanega sima. — Janr. 11, Alcmene and Pacific I. surveyed. (Not found on maps). — Janr. 19, landed on the island of Kika sima. — Januar, 22, landed on the eastern coast of Ou sima (Oshima, North Lew chew) left Jan. 24. — Jan. 29, reached Hong kong.

In the second half of March the "Hancock" was despatched to search after the "Porpoise" and survey the S. W. and E. coast of Formosa. The "Porpoise" not found. Vain attempts to land on the island. On April 9th the "Hancock" anchored in the port of Napha. Soon afterwards the "Vincennes" and the "Cooper" arrived. After a short stay the "Vincennes" steered northeastward to Northern Lew chew. — The strait between Ou sima and Katana sima surveyed, April 29. — Fo kow Bay (it seems on the N. W. coast of Oshima) visited, May 3. Wright goes on shore. — Cleopatra I. (N. of Oshima), May 7, Toukara, Sabine. — Yakuno sima (further N. E.) visited, May 8. — Then the "Vincennes" passed up along the eastern coast of Tanega sima, followed by the other ships, and on May 13th they anchored off Simoda. This Japanese city had been recently (Dec. 23) destroyed by an earth quake, during which the Russian Frigate "Diana" was wrecked. Visit to the inlet of Heda, where the crew of the "Diana" had retired, to conceal themselves from the French and English ships. On May 28, the American ships left Simoda. The "Vincennes" steered directly to Hakodate. The "Hancock" followed. The "Cooper" was ordered to proceed between Kiu siu and Nippon and survey the west coast of the latter.

After a stay of some weeks in Hakodate, the "Vincennes" made sail for Kamtchatka, June 26, 1855, leaving instructions for the other two ships to perform the homeward voyages by different ways. Avatcha Bay, July 8, — Bering I., 16th, — passed Bering Str., entered the Arctic

Ocean, attained lat. 70°, 12′, Aug. 13. — The coast of the land of the Tchukches surveyed — St. Matthew I., Sept. 19. — Between the Aleutian I⁬. to S. Francisco, which was reached in the middle of October. — Otaheite, Sandwich I⁬. — New York, July 1856.

The "Hancock" left the Japan Sea in July 1855. — From Hakodate reached Matsumai (S. W. end of Yezo), July 5, then along the western coast of Yezo, Stroganov Bay, Cape Romantsov, Cape Soya on Lapérouse Strait, July 15. — Through the Sea of Okhotsk to Kamtchatka, Bolsheretsk Harb. July 25. — Along the west coast of Kamtchatka to Penjinsk Bay. Returned along the Siberian coast, Tausk Bay, Ayan, Aug. 31. — Shantar I⁬ visited, mouth of the Amur, around the northern point of Sakhalin, Sept. 17. — Crossed the Kurile Group, arrived at S. Francisco October 20, 1855.

The "Fenimore Cooper" sailed from Hakodate across the Kurile to the Aleutian I⁬ and then directly to San Francisco.

The store-ship "J. Kennedy" having become unseaworthy, had been left in China.

Wright, Charles. The following brief notices regarding this able American botanist and botanical collector, have been extracted from an interesting review of the life and the labours of Ch. Wright, which we owe to the kindness of Mr. C. S. Sargent, Director of the Arnold Arboretum, who is, I understand, the author of it.

Ch. Wright was born Oct. 29, 1811, at Wethersfield in Connecticut. In 1831 he entered Yale College, graduating in 1835. His fondness for botany was developed while he was in college. In 1837 he emigrated to Texas, then an independent republic. Here he occupied himself with land-surveying, explored the country, collected plants, became a deer-hunter, learned to dress deer-skins, made mocassins and leggins. For some time he was also a teacher. In the summer of 1847 and 1848 he had an opportunity of carrying his botanical exploration farther south and west. He botanized on the Rio Grande, on the Mexican frontier and returned with interesting collections to Connecticut. In 1849 he went to Texas again and opened to our knowledge the Flora of Western Texas, then annexed to the U. S. In 1852 his collections were partly published in vol. III of the Smithsonian Contributions, as Plantae Wrightianae. In 1851 W. was attached to Col. Graham's party, surveying the U. S. and Mexican boundary. He returned next year with extensive collections, which were the basis of the second part of Plantae Wrightianae, published 1853. Many new species and genera. A new Acanthaceous genus from Texas was named *Carlo-Wrightea* by Asa Gray in 1878.

Wright's next expedition was a long one round the world, as botanist to the N. Pacific Exploring Expedition of which we have given a brief account in the preceding pages. W. joined it in the spring of 1853 and performed the whole voyage in the flagship "Vincennes". A stay of 7 weeks at Simon's Bay (Cape of G. H.) resulted in an important collection of about 800 species within a small area. The voyage was thence to Sidney and Hong kong, which latter was reached March 19, 1854. The collection of over 500 species made there, was in part the basis of Bentham's Flora Hongkongensis. — In the autumn of 1854 W. made interesting collections on the Bonin and Loo choo I²., and later upon the islands between the latter and Japan. Still more extensive and important were the collections made in Japan, especially those on the island of Yezo, near Hakodate, although the stay there was brief. Also those made in Bering's Strait, mainly on Kiene or Arakamchechen I. (S. W. of Bering's Strait, near the coast of the Tchuktches), where the scientific members of the expedition passed the month of August. Reaching San Francisco in October, Wright was detached from the expedition and came home by way of Nicaragua.

In the autumn of 1856, W. began his prolific exploration of the Flora of Cuba, which extended to 1865. After this he passed a large portion of several years at Cambridge, taking a part of the work of the Gray Herbarium. The latter years of his life he spent at his native place Wethersfield, where, on Aug. 11, 1886, he suddenly died. — Mr. Sargent concludes: "Ch. Wright accomplished a great amount of useful and excellent work for botany in the pure and simple love of it; and his memory is held in honorable and grateful remembrance by his surviving associates".

Sets of Wright's botanical collections in America and other parts of the world are found in Cambridge and in the principal herbaria in Europe. The St. Petersburg Botan. Garden possesses the plants collected in Hong kong, Lew chew, Bonin I²., Japan etc. and also his American collections.

As to the Hongkong collection, G. Bentham writes in his Flora Hongkongensis, Pref. 10: "Mr. Charles Wright, of the U. S., so well known for the beauty and excellence of the specimens distributed from his various botanical expeditions, was naturalist on board the U. S. S. the "Vincennes" of the North Pacific Exploring Expedition. During this cruise he staid at Hongkong from March to September 1854, and from January to April 1855, and has proved himself as zealous and active on this as on other occasions, for he brought away specimens of above 500 species, several of them of great interest, and not received from any other source. An almost complete set has been remitted to me for publication by Dr. Asa Gray".

Hongkong plants gathered by Wright and recorded as new:

Grewia glabrescens, Bentham in Fl. Hongk. 42. — I. F. S., I, 92. Also Hance.

Berchemia (?) *sessiliflora,* Benth. in Fl. Hgk. 68. — I. F. S., I, 128.

There is among Ch. Wright's specimens in the Herb. Hti Petrop. one marked on the label, with which it was sent by A. Gray: *Sedum Formosae,* sp nov. Insula Soong kong. Maximowicz, Mél. XI (1883) 768, found that Wright's plant was identical with *Sedum Alfredi,* Hance, gathered, 1869 near Canton and understood that Wright gathered it on or near Formosa. But the latter never visited Formosa. There is a small island Sung kong, S. E. of Hongkong. The I. F. S., I, 283, sub Sedum Alfredi, correctly states that Wright's Soong kong is an island in the Kuang tung province. See also l. c. 313, sub Passiflora ligulifolia *).

Passiflora ligulifolia, Mast. Trans. Linn. Soc. XXVII (1871) 632. — Soong kong I. — I. F. S., I, 312.

Hedyotis loganioides, Benth. Fl. Hgk. 149. — I. F. S., I, 374.

Marsdenia lachnostoma, Benth. l. c. 226. — I. F. S., II, 113.

Solanum Wrightii, Benth. l. c. 243. — I. F. S., II, 172. Not indigenous in Hongkong, but a native of S. America.

Gmelina chinensis, Benth. l. c. 272. — I. F. S., II, 257.

Phyllanthus leptoclados, Benth. l. c. 312. — I. F. S., II, 422.

Glochidion Wrightii, Benth. l. c. 313. — I. F. S., II, 426.

Rottlera cordifolia, Benth. l. c. 307. — I. F. S., II, 441 = *Mallotus repandus,* Müll. Arg., variety. Earlier known from Timor, Java.

Ficus Wrightii, Benth. l. c. 329. — Maxim. Ficus, in Mél. XI (1881) 341. — King, Ficus (1888) 133, reduces it to *F. foveolata,* Wall.

Fimbristyles leptoclada, Benth. l. c. 393.

Arthrostyles chinensis, Benth. l. c. 397.

Apocopis Wrightii, Munro, Proc. Amer. Acad. IV, 363. — Benth. l. c. 421.

Selaginella xipholepis, Baker, Journ. Bot. 1885, 155.

Selaginella heterostachys, Baker, l. c. 177.

Trichomanes latemarginale, D. Eaton, Fil. Chin. Jap. 3. — Syn. Fil. 79.

---

*) Wright visited also other islands near Hongkong, viz: Pu toy, S. W. of Sung kong, not to be confounded with Pu tu I., of the Chusan Group. Comp. Maxim. Mél. IX (1874) 405, sub Pterostigma grandiflorum, — Lemna I. to the S. W. of Hongkong. Benth. Fl. Hgk. 155, sub Randia sinensis.

Wright and Small: Loo choo Archip., especially Great Loo choo, Oo sima. New Plants:

Gymnosporia diversifolia, Max. Mél. XI (1881) 204. Small, Usima. — I. F. S., ɪ, 123. Hemsley names it *Celastrus diversifolius.*

Rubus abortivus, O. Kuntze, Meth. (1879) 68. Loo choo Archip. — I. F. S., ɪ, 229.

Rubus Grayanus, Maxim. Mél. VIII (1871) 382. Ins. Katona sima et Yakuno sima. — I. F. S., ɪ, 231.

Photinia Maximowiczii, Decne, Pomac. (1874) 143. Bonin. — Ph. arbutifolia, Asa Gray, Bot. Japan, 388 (non Lindley). — Maxim. Mél. IX (1873) 180; ibid. XII (1888) 726, *Ph. Wrightiana.*

Vaccinium Wrightii, A. Gray, Bot. Jap. 398. Ou sima, Katona sima.

Statice Wrightii, Hance, Advers. (1864) 236. Loo choo Iˢ. — Ind. Fl. Sin., ɪɪ, 35.

Androsace patens, Wright in A. Gray, l. c. 401. Ou sima.

Stimpsonia (nov. gen. Primul.) *chamaedryoides,* Wright in Mém. Amer. Acad. VI (1859) 401. Katona sima, Loo choo Iˢ. — I. F. S., ɪɪ, 46.

Fraxinus insularis, Hemsl. I. F. S. ɪɪ, 86. Loo choo Archip.

Premna glabra, Asa Gray, in shed. Maxim. Mél. XII (1886) 512. Loo choo Archip. — I. F. S., ɪɪ, 255.

Philoxerus Wrightii, Hook. & Benth. Gen. Pl. III (1880) 40, (nomen tantum). Maxim. Mél. XII (1886) 528. Loo choo. — I. F. S., ɪɪ, 323.

Cleidion ulmifolium, Müller Arg. in Flora (1864) 481. Loo choo Archip. — I. F. S., ɪɪ, 442.

Nanocnide lobata, Wedd. in DC. Prod. XVI, ɪ (1869) 69. Loo choo.

Potamogeton Wrightii, Morong in Bull. Torrey Bot. Club. XIII (1886) 158. C. Wright, Loo choo Iˢ.

Carex parciflora, Boott, in A. Gr., Bot. Jap. 418. Oo sima, Hakodate.

Carex Ringoldiana, Boott, l. c. 419. Oo sima.

Carex discoidea, Boott, l. c. 419. Loo choo.

Carex sociata, Boott, l. c. 420.

Selaginella boninensis, Baker, Journ. Bot. 1884, 178. Bonin.

Athyrium cysteroptoides, Eaton, Filices Japan, 1859, 110. Wright, Oo sima, Katona sima, Loo choo. — Hook. Spec. Fil. III, 1860, 220, sub *Asplenium.* — Syn. Fil. 225.

Nephrodium Eatoni, Baker, Syn. Fil. 276. Wright, Loo choo.

Gymnogramme Wrightii, Hook. Spec. Fil. V (1864) 160, tab. 303. Wright, Loo choo. — Syn. Fil. 388.

Wright, Bonin. New Plants:

Photinia Wrightiana, Maxim. Mél. XII (1888) 726.

Webera subsessilis, Maxim. Mél. XI (1883) 789. Bonin sima. Postels, flor. (v. supra p. 323). Wright, fructif.

Psychotria hamalospermum, Asa Gray, Bot. Jap. 393.

Ixeris linguaefolia, A. Gr. l. c. 398.

Wright's Japanese Plants. Some of them collected by Small.

A small number was gathered by Wright, from December 29th 1854, to Jan. 3rd 1855, on the shore of Kago sima Bay at the southern end of Kiu siu and a few were picked up on Tanega I., Jan. 9th. The principal collection was made at Simoda in May, and at Hakodate in June 1855, where also previously plants had been gathered by Drs Williams and Morrow (v. supra p. 389).

The greater part of Wright's plants collected in Japan were published by **Asa Gray** with the title:

Diagnostic Characters of New Species of Phaenogamous Plants collected in Japan by Mr. Charles Wright, botanist to the U. S. Surveying and Exploring Expedition, Commander C. Ringgold and' afterwards J. Rodgers. With Observations upon the Relations of the Japanese Flora to that of North America. Memoirs of the Amer. Academy of Art and Sciences, vol. VI, 1859, 377—452. The Carices by Boott. — Appendix: A. J. Anderson, Salices e Japonia.

Daniel C. Eaton published Wright's Ferns, some of them were also described by Sir W. Hooker. — Some of Wright's Japanese plants were also described by Miquel, in his Prol. Florae Japon., and by Maximowicz in the Mél. biolog. The following were reported as new:

Diphylleia cymosa, Mich., in A. Gray, Jap. Pl. 380, is according to Fr. Schmidt, Fl. Sachal. 109, a new species, for which he proposes the name *D. Grayi*. — Franchet, Enum. Jap. I, 24.

Actinidia platyphylla, A. Gray, herb. in Miq. Prol. 203, Yezo, Small (June 1855). — Maxim. Mél. XII (1886) 425 = *A. Kolomikta*, Max. — Evidently first gathered by Small.

Prunus virginiana, Mich. in Asa Gray, l. c. 386, Japan, is according to Maxim. Mél. XI (1883) 704, a new species, *Prunus Grayana*.

Rubus Wrightii, Asa Gray, l. c. 387, Simoda, Hakodate, is according to Maximowicz, Mél. VIII (1871) 383 = *Rubus crataegifolius*, Bunge, (1830).

Potentilla reptans, var. *trifoliata*, A. Gray, l. c. 387, Small, Hakodate, is according to Max. Mél. IX (1873) 156, a new species, *P. centigrana.* — I. F. S., ɪ, 241.

Pourthiaea coreana, Decaisne, Pomac. (1874) 148. Corea, Tsu sima. Wright, Wilford. — Hemsley in I. F. S., ɪ, 263, unites Decaisne's *Pourthiaea villosa*, *P. Calleryana*, *P. lucida*, *P. coreana*, *P. Oldhami* et *P. Thunbergii*, l. c. 147—149 to one species for which he proposes the name *Photinia variabilis*, Hemsl.

Rodgersia (new gen.) *podophylla*, Asa Gray, l. c. 389. Hakodate. — I. F. S., ɪ, 266.

Chrysosplenium Grayanum, Maxim. Mél. IX (1876) 769. — A. Gray, l. c. 389, *Chr. ovalifolium*, M. Bieb. Hakodate.

Panax repens, Maxim. Mél. VI (1867) 254. — Asa Gray, l. c. 391, sub *P. quinquefolium.*

Viburnum Wrightii, Miq. Prol. Jap. (1866) 155. — Asa Gray, l. c. 393. *V. erosum*, Thunb. Small, Hakodate.

Erythrochaete dentata, A. Gray, l. c. 395. Nippon, Yezo. — Unknown to other botanists.

Cirsium pectinellum, A. Gr. l. c. 395. Nippon, Yezo. — Maxim. Mél. IX (1874) 308, sub *Cnicus.*

Lampsana parviflora, A. Gr. l. c. 396. Simoda, Hakodate.

Ixeris stolonifera, Asa Gray, l. c. 396. Simoda, Hakodate, Kago sima Bay — *Lactuca* in Benth. & Hook. Gen. Pl. II, 526.

Ixeris albiflora, A. Gr. l. c. 397. — Cape Siriki saki, Nippon. *Lactuca* in Gen. Pl. l. c.

Leucothoe Grayana, Maxim. Mél. VIII (1872) 613. — Asa Gray, l. c. 399. *L. chlorantha*, DC.

Primula japonica, A. Gr., l. c. 400. Hakodate. Franch. l. c. I, 299.

Eritrichium Guilelmi, A. Gr: l. c. 403. Hakod. — Franch. l. c. I, 336.

Scrophularia alata, A. Gr. l. c. 401. Hakodate. — Franch. l. c. I, 342.

Ajuga pygmaea, A. Gr. l. c. 402. Simoda. — Maxim. Mél. XI (1883) 812, cum icone.

Daphne pseudomezereum, Asa Gray, l. c. 404. Simoda. — Franch. l. c. I. 403.

Quercus dentata, Thbg., var. *Wrightii*, DC. Prod. XVI, ɪɪ, 13.

Juniperus conferta, Parlat. in DC. Prod. l. c. 481. Wright, Japonia, in herb. Hooker. — Maxim. in Mél. VI (1867) 375, had named it *J. littorale*, (a year earlier).

Arethusa japonica, Asa Gray, l. c. 409. Hakodate, Small. — Franch. l. c. II, 35.

Iris gracilipes, A. Gr. l. c. 412. Hakodate. — Max. Mél. X (1880) 735.

Smilax stenopetala, A. Gr. l. c. 412. Kago sima Bay, Hakodate. — Maxim. Mél. VIII (1871) 405.

Asparagus Wrightii, A. Gr. l. c. 413. Hakodate. — Franch. l. c. II, 58 = *Asparagus Schoberiodes,* Kth., Enum. V (1850) 70. Earlier Zollinger, Siebold.

Polygonatum falcatum, Asa Gray, l. c. 414. Simoda. — Maximowicz, Mél. XI (1883) 851 = *Polygonatum giganteum,* Dietr. ex Kunth Enum. V, 136, var. *falcatum.*

Polygonatum lasianthum, Maxim. l. c. 850. Small, Hakodate. — A. Gr. l. c. notices it as *P. multiflorum,* All.

Lilium medeoloides, A. Gr. l. c. 415. Hakodate. — Franch. l. c. II, 63.

Heloniopsis (gen. nov.) *pauciflora,* A. Gr. l. c. 416. Cape Rumantsev ins. Yezo. — Maxim. Mél. VI (1867) 211.

Trillium Smallii, Maxim. Mél. XI (1883) 862. — A. Gr. l. c. 413, notices it as *Tr. erectum,* Lin., var. *japonica.* Small, Hakodate.

Arctiodracon (new gen.) *japonicum,* A. Gr. l. c. 408. Hakodate. — Schott named it *Lysichiton japonicum.* He had described the new genus *Lysichiton* (same as Arctiodracon) two years earlier, 1857, founded upon *L. camtchaticum.* — Franch. l. c. II, 9.

Eleocharis pileata, Asa Gray, l. c. 417. — Franch. l. c. II, 110, sub *Scirpus.*

Carex nana, Boott, in A. Gr. l. c. 418. Hakodate.

Carex confertiflora, Boott, l. c. 418. Hakodate.

Carex papulosa, Boott, l. c. 418. Hakodate.

Carex ringens, Boott, l. c. 419. Hakodate.

Carex picta, Boott, l. c. 418. Hakodate.

Carex parciflora, Boott, l. c. 418. Hakodate.

Carex micans, Boott, l. c. 419. Simoda.

Daniel C. **Eaton,** Characters of some New Filices from Japan and Adjacent Countries, collected by Mr. Charles Wright in the Northern Pacific Exploring Expedition under Captain J. Rodgers. American Academy Proc. IV, 1857—60, p. 110—111.

The following recorded as new plants gathered by Wright in Japan:

Woodsia polystachyoides, Eaton in Wright's herb. — Hook. 2 d Cent. Ferns, 1861, tab. 2. Hakodate. — Syn. Fil. 48.

Hymenophyllum Wrightii, van den Bosch, Neederl. Kruidk. Archiv. IV (1858) 391. Hakodate. — Syn. Fil. 58. Perhaps, H. rarum, R. Br.

Adiantum monochlamys, Eaton, l. c. 110. Simoda. — Hook. 2d Cent. Ferns, 1861, tab. 50. — Syn. Fil. 125.

Asplenium Wrightii, Eaton MS. Hook. Spec. Fil. III (1860) 112, tab. 182. Japan, Tukono sima. — Syn. Fil. 204.

Wilford, Charles, collector for the R. Gardens, Kew, whose botanical explorations in various parts of Eastern Asia will be detailed in another chapter, remained in Hongkong from Nov. 1857, to June 1858. Bentham, Fl. Hgk., Pref. 11, states: "Mr. Ch. Wilford remitted to this country above 400 species from Hongkong. This collection has been of considerable use to me, the specimens being good, usually in several duplicates, and often accompanied by memoranda of their stations, with occasionally a few other notes".

Among Wilford's Hongkong plants described in Bentham's Fl. Hgk., there are only four novelties and three of them hitherto known only from Wilford's herbarium:

Clematis crassifolia, Benth. l. c. 7. — I. F. S., ı, 3.

Tetrathyrium (nov. gen.) *subcordatum,* Benth. l. c. 133. — Oliver in Hook. Icon. Pl. (1884) tab. 1417, describes and figures it from specimens gathered in Hongkong by Mr. Ch. Ford, under the name of *Loropetalum subcordatum.* — I. F. S., ı, 291.

Mesona chinensis, Benth. l. c. 274. — I. F. S., ıı, 267.

Pellionia brevifolia, Benth. l. c. 330. — DC. Prod. XVI, ı (1869) 167.

Bowman, J. C., mentioned in Benth. Fl. Hgk., Pref. 11. His name appears frequently in Sir W. J. Hooker's enumeration of Hongkong Ferns, in K. J. B. IX (1857) 333—363.

Barthe, J., M. D. French, Dr. Hance's friend. They botanized together in Hongkong, in 1855, 1856.

In Benth. Fl. Hgk., 115, and in Symb. (1861) 223, Hance described the *Dissochaeta Barthei:* "in praeruptis montis Victoriae, ins. Hongkong, primo legi cum amicissimo J. Barthe, Franco-gallo, D. M. chirurgo navali, mense januario a. 1856, cui sacratam volui". — In Benth. & Hook. Gen. Pl. I (1862) 751, this plant is considered to be the prototype of a new genus, and named *Barthea chinensis.* — I. F. S., ı, 300. — The plant was earlier discovered by Fortune.

In Advers. (1864) 205, Hance described *Caltha palustris,* Lin., var. δ *Barthei:* ad sinum Jonquières, ins. Sakhalin, a. 1855 coll. amic. Dr.

J. Barthe. Franch., Adonis (1894) 89, described *Adonis Barthei,* discovered by B. near Nikolayevsk on the Amur.

Lorraine is mentioned by Sir W. J. Hooker in his Spec. Fil. as having collected Ferns in Hong kong. — Hance, Advers. (1864) 254, sub *Davallia Lorrainii,* writes: "in ins. Penang legit C. W. B. Lorrain (sic!) M. D., strenuus, dum viveret, filicum penangianarum messor".

Furet, L. French Missionary of the Missions Étrangères. I learn from Mr. A. Franchet of the Mus. d'Hist. nat. Paris, that the Abbé Furet, in 1856, presented to the Museum a collection of 350 species of plants, gathered in Hongkong. In the Revue de l'Orient et de l'Algérie, XVI (1854) 399—401, and nouv. sér. III (1856) 23—28, — 127—132, M. Léon de Rosny published letters written to him by L. Furet from Hongkong on the voyages of the latter to Japan and the Loo chou I⁵ Furet signs: "Missionnaire apostolique, supérieur de la Mission au Japon et aux isles Lou tchou".

The *Rottlera Furetiana* described by Baillon in his Étude gen. Euph. (1858) 426 *(Mallotus Furetianus,* Müll. Arg.), gathered by Furet in Hong kong, in about 1854, was probably earlier discovered by Hance. V. supra p. 369. — I. F. S., II, 439.

Urquhart, Colonel, made botanical collections in Hong kong, apparently towards the end of the fifties. His name is frequently mentioned in Hemsl. Ind. Fl. Sin. Bentham in Fl. Hgk. Pref. 11 states that in Sir W. J. Hooker's herbarium there is a fine set of Hongkong ferns transmitted to him by Col. U. Urquhart's name appears frequently in Hooker's Spec. Fil. IV (1862). One of his ferns, Nephrodium spinulosum Desv., l. c. 126 is labelled: Urquhart, N. China, Ta lien whan. From this we may conclude that U. took part in the North China campaign, 1860. It was in Ta lien Bay that the British forces assembled, before proceeding to Chili.

Bentham, George. This illustrious botanist, the author of the Flora Hongkongensis, was born near Portsmouth Sept. 22, 1800. His father, General Samuel Bentham (subsequently Sir Samuel) was sent, in 1805, by the British Admiralty to St. Petersburg, where the family resided until 1807. Here G. Bentham is said to have acquired his knowledge of the Russian language. When the war broke out between England and Russia, General B. returned to England. After 1815 the Bentham's removed to France, the General finally bought an estate near Montpellier and George became his father's farm-manager. At the same time he examined the wild plants of southern France and, having acquired some knowledge of botany, made researches into the Flora of the Pyrenees, and published in 1826 a Catalogue of the Plants of the Pyrenees, in French. In the same year the estate was relinquished, and the Bentham family returned to England. George decided to embark in some profession and began to study jurisprudence, and in 1832 took his degree, but he did not neglect botany. Soon after his return to his native country he was elected Fellow of the London Linnean Society. In 1829 he undertook the secretaryship of the Roy. Horticultural Society, which post he held till 1840, and entered into friendship with his predecessor Jos. Sabine, then Hon. Secretary and J. Lindley (v. supra pp. 251, 252). After his father's death, in 1832, he devoted himself exclusively to botany. Since that time his name became connected with the exploration of the Flora of China and in his numerous botanical articles and works he described a great number of new plants received from various collectors in China. In 1832 he began to publish his first important work: LABIATARUM GENERA ET SPECIES, finished in 1836.

As has been recorded on a previous page (374), Capt. Champion, who from 1847 to 1850 made large botanical collections in the island of Hong kong, first sent his Chinese plants for determination to Dr. G. Gardner, Peradenia, but the latter died, in 1849, and Champion, after his return to England, in 1851, placed a complete set of his Hongkong plants in the hands of G. Bentham, who from 1851—55, in the Kew Journ. of Botany, vol. III to IX, published the

FLORULA HONGKONGENSIS, an Enumeration of the Plants collected in the Island of Hongkong by Captain J. G. Champion, the determinations revised and the new species described by G. Bentham. — The Orchids (VII, 1855, 33—39) by J. Lindley; the Filices (IX, 1857, 333—44, 353—363), by Sir W. J. Hooker.

In 1861 appeared G. Bentham's FLORA HONGKONGENSIS, a Description of the Flowering Plants and Ferns of the Island of Hongkong, with a map of the Island. Bentham had made use of all the botanical materials then known from Hongkong. In the determination of the plants he was aided by several distinguished botanists: Dr. J. Lindley, Sir W. J. Hooker, Dr. J. D. Hooker, Colonel Munro, Prof. D. Oliver, Dr. Boott and others. Total number of species described 1056. This remarkable book exhibits on every page the vast botanical knowledge of the author and serves as a model for accurate characteristic and at the same time popular descriptions of plants. — Two years later, in 1863, B., assisted by Baron F. Mueller, began the publication of the Flora Australiensis, the 7th and last volume of which was issued in 1878. But the main work of Bentham's life, done in conjunction with his friend Dr. Sir J. D. Hooker, is the Genera Plantarum, which the celebrated authors began in 1862 and brought to a conclusion in 1883.

In 1861 G. Bentham was elected President of the Linnean Society and held this post till 1874. In 1862 he became F. R. S. In 1878 he was made Companion of the Order of St. Michael and St. George. He died at London, Sept. 10, 1884.

In his last will Bentham bequeathed considerable sums to the Royal and Linnean Societies and left a part of his estate to be held upon trust to apply the same in preparing and publishing botanical works, as his trustees may consider best for the promotion of botanical science. — The trustees now continue the edition of Hooker's Icones Plantarum.

Bentham accumulated a vast herbarium, which in 1854, he made over to the Kew Gardens. This collection and that of his intimate friend and associate Sir W. J. Hooker form the basis of the unrivalled Kew Herbarium.

Lindley in 1832 dedicated to Bentham's memory a new genus *Benthamia* founded upon an Indian plant discovered by Wallich, *B. fragifera*, subsequently discovered also in Central China. Another species was found in Hongkong. But now this genus is reduced to *Cornus*. See I. F. S., i, 345. Among the numerous species of plants named after Bentham I may notice the following from China:

Clematis Benthamiana, Hemsl. I. F. S., i, 2.
Indigofera Benthamiana, Hance. See l. c. I, 156.
Photinia Benthamiana, Hance. See l. c. I, 262.
Anodendron Benthamianum, Hemsl. l. c. II, 98.
Oreocharis Benthami, Clarke. L. c. II, 226.
Chelonopsis Benthamiana, Hemsl. l. c. II, 298.

III. Robert Fortune.

The travels and explorations of R. Fortune in China, beginning with the year 1843, inaugurate a new era in the history of botanical discoveries in that country. Previous to that time the Chinese plants known to our botanists in Europe and introduced into our gardens, came from Canton or Macao. In 1701 J. Cunningham (v. supra p. 31) had sent to England an interesting collection, chiefly made in Chusan, and in about the middle of

the 18th century a small collection of Peking plants made by Father d'Incarville was received in Paris (v. supra p. 47). But these collections had
long fallen into oblivion, and were only in later years rediscovered and
examined. We have seen (p. 162) that in 1793 G. Staunton, during Lord
Macartney's journey overland (i. e by the Grand Canal and the rivers) from
Peking to Canton had gathered plants and brought them to England. These
and a few plants from about the same regions collected in 1816 by Dr.
C. Abel (supra p. 236), which had happened to be saved from shipwreck,
were up to the time of Fortune's explorations the only botanical specimens
from the interior of China found in European herbariums. The Flora of
Peking and Mongolia had been explored in 1830 and 1831 by Dr. A. Bunge
(v. supra p. 324).

Fortune, during 18 years, in which he paid repeated visits to China
succeeded in exploring botanically not only the neighbourhood of the Chinese
ports then opened to European trade, but visited also some interesting spots
of the interior in the provinces of Fu kien, Che kiang and An hui, not seen
before by Europeans, from which parts he brought home rich collections of
living plants, seeds and herbarium specimens. The numerous beautiful ornamental plants, most of them new, introduced by him from China and Japan
into the gardens of England, are still objects of admiration with florists.
Fortune was an excellent gardener and botanist, and an acute observer. All
his notes regarding Chinese plants, Chinese gardening and husbandry possess even now a high interest. I thought it therefore useful to give in this
chapter, besides a summary of all his travels, an account of his interesting
notes regarding Chinese plants as found scattered over his works, and also
to add a list of the Chinese plants introduced by him, and the new
herbarium specimens gathered by him, as far as these have been hitherto
described.

R. Fortune was born in Berwickshire, Scotland, on the 16th of Sept.,
1812, and was educated in the parish school of Edrom. He showed an
early preference for gardening. After having served out his apprenticeship
under experienced gardeners, he entered the Botanic Gardens, Edinburgh,
under the elder Mac Nab. Here he remained between two and three years,
and, in 1842, went to London, being appointed Superintendent of the hothouse department of the Roy. Hort. Soc., at Chiswick. Under the auspices
of that Society he proceeded, in the next year to China.

Fortune visited China four times, 1843—45, 1848—51, 1853—56,
1861, and accordingly published four interesting accounts of his travels
in that country.

1. R. Fortune. THREE YEARS WANDERINGS IN THE NORTHERN PRO-
VINCES OF CHINA*), INCLUDING A VISIT TO THE TEA, SILK, AND COTTON
COUNTRIES, WITH AN ACCOUNT OF THE AGRICULTURE AND HORTICUL-
TURE OF THE CHINESE, NEW PLANTS, etc. With numerous illustrations.
London, 1847.

The occupation of Hongkong and the island of Chusan by British
forces and the opening of new ports in the Chinese Empire after the treaty
of Nan king, in 1842, appeared to present so favourable an opportunity
of acquiring valuable plants, that the Council of the Horticultural Society
of London deemed it advisable to send a collector to that country, which
had for so many years been the richest of all fields from the horticultural
point of view. In February 1843, R. Fortune, having offered himself for
the service, was commissioned by the Society to proceed to China and to
spend there two or three years in exploring such districts as were accessible to Europeans, for the purpose of introducing new plants to the country.
He sailed on Febr. 26, 1843, in the ship "Emu", and on the 6th of July 1843
Hong kong was reached. Here Fortune remained nearly seven weeks,
botanizing in the island. On August 23, he left Hongkong in a British
vessel and sailed for Amoy. About halfway between Hong kong and Amoy
the ship touched at the island of Namoa and sailing up the coast reached
Amoy. During his stay here, about a month, F. was continually travelling
in the interior of the islands of Amoy and Ku lang su, and over the adjacent mainland, going sometimes a considerable distance up the rivers,
prosecuting his botanical researches. Towards the end of September he left
Amoy and sailed for Chusan. The ship experienced very stormy weather
and was at last obliged to put into the Bay of Chinchew, where after a
stay of two days, Fortune embarked in another vessel to proceed again on
his voyage towards the Formosa Channel. But the ship had hardly left
Chin chew Bay when she met a dreadful gale and, nearly wrecked, was
driven southward. After having been three days in the storm, they succeeded in finding shelter in Chimoo Bay. Fortune profited by the stay
here, to visit the hills on the adjacent mainland, where he gathered several
interesting plants. The little vessel being sufficiently repaired, they continued their voyage. The Chusan group of islands was reached in ten
days, (second half of October). Fortune landed at Ting hai the principal

---

*) I may observe here that Fortune by Northern Provinces of China, North China,
does not mean, as an unprejudiced reader might presume, the provinces of Chili, Shan tung
etc., — but Che kiang and Kiang su and An hui. The northern-most spot visited by him during
his wanderings, was Su chou fu, in lat. 31° 19'.

town of the large island of Chusan, then in the hands of the English. His
first visit to Chusan occupied only a short time but during the next two
years he had at all seasons of the year frequent opportunities of visiting
this beautiful island of which he gives an interesting account, particularly
of its rich flora.

After leaving Chusan, F. visited Ning po, in the autumn of 1843.
He inspected the gardens of some Mandarins. Notice of the dwarfed trees
of the Chinese and Japanese.

At the end of 1843 F. paid a visit to Shang hai and then proceeded
with his collections to Chusan, which he made his head-quarters in the
north of China, and then sailed for Hongkong, where he seems to have
1844 arrived in January 1844. The various collections made in the north were
shipped to England. Afterwards he visited Canton and Macao. Near the
former place the celebrated "Fa tee" Gardens (v. s. p. 266) attracted his attention. In company with Mr. T. Lay he visited the mountains near Canton.

At the end of March 1844, F. sailed again for the northern provinces
and spent the summer partly in Chusan, partly in the neighbourhood of
Ning po and in Shang hai. In the latter place he passed three weeks, in
the second half of April. About the beginning of May he set out upon an
excursion to visit the green tea district near Ning po, more than 20 miles
to the south west of that city. About 13 miles of the journey was performed by water on a canal ending at the foot of the hills. In the centre of the
tea district there is a large and celebrated temple Tein tung (T'ien t'ung),
where Fortune lodged during his stay in this part of the country. In the
middle of June he went again to Shang hai and visited the hills 30 miles
west of the city. He then also paid a visit to Soo chow foo, which place
he reached by boat, via Ca ting (Kia ting hien). Besides Chusan, F. also
explored the flora of some of the other islands of the Chusan Archipelago,
namely Poo to and Kin tang. In October F. returned from Chusan to
Shang hai, where he was laid up with a severe attack of fever. Having
recovered, he sailed for Hongkong, reached this place in November and
shipped his collections to England.

1845   In the beginning of January 1845, F. sailed for Manilla and spent
about three weeks in the interior of Luzon. From this district he sent to
the Hort. Soc. a living specimen of an extremely rare Orchid, the beautiful
Phalaenopsis amabilis, Bl. On the 14th of March he returned to his old
station in the north of China (Chusan?). Having completed his botanical
researches in the district of Ning po, F. started on a journey overland to
Shang hai. On May 18, he crossed from Chin hai, at the mouth of the

Ning po River, the Bay of Hang chow, in a junk, to Cha poo. Here he hired a boat to bring him on the various canals across the country to Shang hai. They passed via Ping hoo through an extensive silk district, where the mulberry tree was the principal object of cultivation, reached Sung kiang foo and finally Shang hai.

When F. had finished his business in Shang hai he left that city and sailed for Foo chow foo. Here he inspected the gardens and nurseries, visited a black tea district in the mountains near Foo chow, and then took passage in a Chinese junk for Chusan. Having reached the harbour of Chusan, F. immediatly went on board an English vessel, which took him to Shang hai, where he had left the greater part of his collections.

On the 10th of October F. left the north of China for Hong kong. On December 22, he embarked with the rest of his collections at Canton for London in the ship "John Cooper". After a long but favourable voyage they anchored in the Thames, on May 6, 1846. The plants arrived in excellent order and were immediatly conveyed to the garden of the Hort. Soc. at Chiswick.

Shortly after his return to England, F. was appointed, in 1846, on the recommendation of Lindley, to the Curatorship of the Botanical Garden of the Soc. of Apothecaries at Chelsea. In 1848 he was obliged to resign this appointment, on beeing commissioned by the E. I. Company to proceed to China again. Of this, his second expedition to China, he gave a detailed account in a book with the following title:

2. R. Fortune. A Journey to the Tea Countries of China; including Sung-lo and the Bohea Hills; with a short notice of the East India Company's Tea Plantations in the Himalaya Mountains. With map and illustrations. London 1852.

Before writing this book, F. had sent, when still travelling in China, pretty detailed accounts of the progress of his journey, which lasted from 1848 to-51, and his botanical discoveries. The Gardeners Chronicle published these letters, addressed probably to Lindley, 16 in number, as they were received, with the exception of the first, under the title Notes of a Traveller, viz:

№ ii, to vi, in Gard. Chr. 1849, resp. 133, 196, 214, 277, 284.
№ vii, » xiv, »   »   » 1850, » 70, 84, 116, 212, 228, 372, 757, 821.
№ xv, » xvi, »   »   » 1851, » 5, 340.

These notes contain many details not recorded in the author's book.

Having been deputed by the Hon. Court of Directors of the E. I. Company to proceed to China for the purpose of obtaining the finest varieties of the Tea plant, as also native manufacturers and implements for the Government Tea plantations in the Himalayas, — Fortune left Southampton on June 20, 1848, in one of the P. & O. Co's steamers and landed in Hongkong on August 14. Here he visited several of the gardens laid out by European residents since the formation of the colony. He gives an interesting description of Messrs Dent & Co.'s fine garden at «Green Bank», then under the fostering care of Mr. Braine (supra 383). Then he took the earliest opportunity of going northwards to Shang hai, which he reached in September. He visited the gardens of the foreign residents. The most beautiful ornamental gardens then were those of Mr. Beale and Messrs Mackenzie. The former was the son of Thomas Beale of Macao, mentioned in a previous chapter (p. 306)*). F. found in his garden nearly all the new plants which he had sent from China to the Hort. Soc. from 1843 to 1846.

F.'s next object was to obtain seeds and plants of the tea shrub for the E. I. Company's plantations in the North-West Provinces of India, and to procure them from those districts in China where the best teas were produced. Accordingly he determined to start for the **Hwuy chow Green-Tea District** in Southern An hui upwards of 200 miles inland from either Shang hai or Ning po. The whole country around Shang hai is intersected with rivers and canals, so that the traveller can visit by boat almost all the towns and cities in that part of China. Some of the canals lead to the large cities of Sung kiang foo, Soo chow foo, Nan king, and onward by the Grand Canal to the capital itself. Others again, running to the west and south-west, from the highways to the Tartar city of Cha poo, Hang chow foo, and to numerous other cities and towns, which are studded over this large and important plain. F., travelling by boat, proceeded in a south-westerly direction. Mao Lake, about 30 miles from Shang hai; then Kea hing foo was reached, which lies in the great Hang chow silk district. The mulberry was observed in great abundance (v. supra p. 407). By the Grand Canal, via Seh mun yuen (Shi men hien) Tan see (T'ang si) to Hang chow foo, where he arrived October 23. The country around this

---

*) **Beale**, Thomas Chay, is mentioned in the Chin. Repository, XIX, XX, 1850—51, among the European residents in Shang hai. He was then chief of the firm Dent. Beale & C⁰, and Portuguese Consul. B. rendered great services to Fortune, who named a new Berberis and an Azalea in compliment to his friend. B. died before 1860. Comp. T. 15—17, 314—15, 330, 341, — R. 2, 3, 195, — Y. 170.

city may well be called "the Garden of China". The Grand Canal, which terminates here, with its numerous branches, waters it. Hang chow is not a sea port and does not communicate by water with the large river Tsien tang kiang which near Hang chow falls into the Bay of the same name. F. had therefore to proceed overland, 2 miles to the city of Kan du, situated on the left bank of the river, where is the sea port of Hang-chow. The river Tsien tang kiang has its sources far away amongst the mountains to the westward. One of its branches rises amongst the green-tea hills of Hwuy chow, another near the town of Chang shan, on the borders of Kiang si, and a third on the northern side of the Bohea mountains. All the green and black tea comes down the Tsien tang River on its way to Shang hai, and at Hang chow is transshipped from the river-boats into those, which ply upon the Grand Canal.

From Kandu F. proceeded by boat up the Tsien tang River, passed the towns of Fu yang and Tung yu (lü), and reached Yen chow foo, a large town in a rich and fertile country. Here the northern branch of the river unites with that coming from the southward, taking its rise partly on the borders of Kiang si, and partly on the northern sides of the Great Bohea Mountains. F. went up the northern branch. On the 29th and 30th October he passed Tsa yuan (Ch'a yüan chen on the Chin. map), Tsa sa pu, Kang ku and Shang i yuen, all places of considerable note, the last must contain at least 100,000 inhabitants (perhaps Shun an hien is meant). On 31 Oct. Wae ping was reached, a city of considerable size on the borders of the Hwuy chow district. Farther on, two days up the river, was the town of Tun che (k'i), an important trading place. It forms the port of Hwuy chow foo *), from which it is distant about 20 miles. All the large Hang chow and Yen chow boats are moored and loaded here, the river beeing too shallow to allow of their procceding higher up. Nearly all the green teas which are sent down the river to Hang chow foo, and thence onward to Shang hai, are shipped at this place. The green teas destined for Canton are carried across a range of hills to the westward, where there is a river which flows in the direction of Lake Po yang. Nearly all the low land in the extensive and beautiful valley in which Tun che is situated, is under tea cultivation, the soil is rich and fertile and the bushes grow most luxuriantly. At Tung che F. left the boat, and having hired a chair, took the road for Sung lo and Hieu ning. At about 10 miles from Tung che F. reached the house of a farmer, within 2 miles of the foot of the Sung lo

*) Rather of Hiu ning hien (tea district).

shan, where he took up his quarters and spent a week in examining the vegetation of the hills, collecting seeds and obtaining information regarding the cultivation and manufacture of tea. The far-famed Sung lo hill, where green tea is said to have been first discovered, is situated in the district of Hieu ning (in southern An hui) and appears to be between 2 and 3000 feet above the level of the plains. It is very barren and produces but little tea now. The low lands of the Hieu ning district and those of Moo (Wu) yuen, situated a few miles further south, produce the greater part of the fine green teas of commerce. Hence the distinction between hill-tea and garden-tea, the latter name applied to the teas cultivated in the plains.

On Nov. 20, F. left this country, went back to Tun che, where he embarked and proceeded down the river. In three days he arrived at Yen chow foo and on the sixth day after leaving Tung che reached Ne chow, a small but busy town higher up the river than Hang chow, and on its right bank. It stands on the main road between Hwuy chow and Ning po. Having left the Hwuy chow boat, F. with his baggage proceeded through the town to the terminus of a small canal, where another boat was engaged to take him to the city of Shao hing foo and farther on to a town called Tsaou o (Ts'ao wo, Chin. map), a place not very far from the source of the Ning po River. At the latter town F. left the Ne chow boat, and walked about a mile across the country to another small town named Pak-wan (Pai kuan, Chin. map.). It stands on the banks of a river which falls into the Bay of Hang chow. The journey was continued on another canal, two embarkments passed, and the waters of the Ning po River entered. City of Yu-eou (Yü yao), Ning po.

On his arrival at Ning po, F. engaged a Chinese boat to take him to Kin tang or Silver Island in the Chusan Archipelago, where the green-tea shrub is cultivated very extensively. He examined the vegetation of this beautiful island, gathered tea seeds, and then left for Shang hai via Chapoo 1849 (v. s. p. 407). He arrived at Shang hai in the middle of January 1849, visited the flower-shops there, and finally left for Hong kong to ship his tea-plants to India. His spare time in Hong kong was spent in rambling about the hills. He was frequently accompanied by Captain Champion (v. supra p. 374).

After this F. sailed for Foo chow foo and paid a visit to the celebrated Koo shan mountain, a few miles to the eastward of the city. It is about 3000 feet high; and about 1000 feet lower than the summit is a famed Buddhist temple. A few days later F. determined to proceed up the Min River to the Bohea Mountains. But when he had ascended the river by boat as far as the town of Suiy kow (Shui k'ou), about 80 miles from Foo chow, he found that he had not brought money enough with him and returned to Foo chow, instructing his servants to proceed to the Bohea Hills and to Hwuy chow in order to purchase tea plants there and bring them to Ning po, which commission they executed. F. himself sailed in a Portuguese lorcha to Ning po. He then, in the spring, again visited the temple of Tein tung, explored the mountains west of it, and the Tung hoo Lakes (south of Ning po?), collecting objects of natural history.

On May 15, 1849 F. set out for a long journey, to visit the Black-Tea Districts of Woo-e-shan (Wu yi Mountain, Bohea in the local dialect). He proceeded first via Shao hing foo and Pa kwan to Ne chow (supra p. 410), and engaged a boat to take him up the Tsien tang River. At Yen chow foo, already visited in the previous autumn, where two rivers unite (supra 409), F. went the larger branch, which has its sources near the town of Chang shan, where the three provinces of Che kiang, An hoi and Kiang si meet. He passed Ta yang, Nan che or Lan chee (Lan ki hien), Long yeou, arrived on June 1st at the important city of Chu choo foo (Kü chou fu). About a mile above the city a river coming from the S. W. and having its sources on the northern side of the Fo kien mountains, falls into the river from Chang shan ascended by F. At the latter place the river was no longer navigable and F. had to cross the water-shed between the Tsien tang River and Lake Po yang in a chair. At Yuk shan (Yü shan hien) a boat was again engaged to take the traveller as far as Quan sin foo (Kuang sin fu). From this place a smaller boat brought him to Ho kou (k'ou)*), an important inland town and great emporium of the black-tea trade. From this point he had to journey overland in a southerly direction to the Bohea Mountain range, across which he had to go on his way to the Woo-i-shan. The road he pursued was a beautiful highland road with passes of considerable steepness. He arrived at a small town named Yuen shan, by which a mountain stream passes and winds round the hills on its way to Ho kow, where it falls into the Kin kiang. After leaving the town a large tea-growing country was reached. Farther on Chu chu**), a small town near the (northern) foot of the Bohea Mountains, properly so called.

Now the far-famed Bohea Ranges lay before the traveller in all

---

*) From the mouth of the Ts'ien t'ang River to Ho k'ou Fortune's route coincides with that followed in 1793 by Lord Macartney's Embassy.

**) According to Yule (Marco Polo, II, 212). Fortune's Chu chu, not found on my Chinese map, is the Cugiu in Polo's route between Kiang si and Fu kien.

their grandeur, with their tops piercing through the lover clouds, and showing themselves far above them. F. means that the highest peaks here may
be 6 or 8000 feet above the level of the sea. Leaving this spot, F. soon
arrived at the foot of the central and highest range, and began the ascent
towards the mountain pass on a paved road, which connects the province of
Fu kien with that of Kiang si and is a busy thoroughfare. He arrived at
last at the celebrated "Gates" or huge doors which divide the two provinces.
The pillars of these gates have been formed by nature. In these mountains
F. discovered many new plants, which he dug up and took with him. He
got them carried several hundred miles in safety and at least deposited
them in the garden of his friend Mr. Beale, at Shang hai. Subsequently
they were sent to Europe.

The streams which flowed from the sides of the hills now ran to the
southward. After travelling about 13 miles along one of the many sources
of the river Min, F. came to a small town named Ching hu. He was now
on the outskirts of the great Black-Tea country of Fu kien. Large quantities of tea-plants under cultivation were observed. Farther on Tsong-
gan- (Chung an) hien was reached, a large town in the midst of the black-
tea country, where nearly all the teas of this district are packed and
prepared for exportation. — It is situated in a plain of no great extent,
surrounded by hills on all sides. From this place the traveller took the road
to Woo-e-shan which was about 15 miles farther on (according to the Chin.
map, west of Chung an).

The farfamed Woo-e-shan (Bohea Hill) stands in the midst of the
plain above noticed and is a collection of little hills, none of which appear
to be more than a thousand feet high: They abound in temples. F. went to
one of the principal temples, near the top, the next day he visited another
temple. A stream, "the stream of nine windings" divides the Woo-e-shan
Hills into two districts, the north and south: the north range is said to
produce the best teas, i. e. the finest "Souchongs" and "Pekoes" *).

F. spent three days on the Woo-e Hills and saw a great part of them
and their productions. Then he went down to Tsin tsun (probably Sing
ts'un of the Chin. map) a small town built on the banks of one of the
branches of the Min River (it seems the same as the stream of nine windings) and one of the chief marts for black tea. In order to see more new
ground, F. determined neither to go down the river Min to Foo chow, nor

---

*) It would seem that Fortune applies the name of Woo-e-(Bohea) shan also to the high
range forming the boundary between Fu kien and Kiang si. But the Woo-e «par excellence»
is situated at the southern foot of the main range.

to return by the way he came, but to take another route which led eastward to the town of Pou ching hien, then across the Bohea Range and
down its northern side into Che kiang. Accordingly he took his course in
an easterly direction. On the morning of the third day after leaving the
Woo-e Hills the small town of She mun was reached at the foot of a very
high range of mountains. Having crossed this mountain, F. arrived on the
same evening at the little town of She pa ke and then entered a wide and
beautiful valley, in the centre of which appeared the city of Pou ching
hien, situated on a pretty river, one of the tributaries of the Min. A
considerable trade in tea is carried on here, Pou ching being a tea district.
From his place F. travelled in a northerly direction. After about 10 miles
he passed a large town the name of which had escaped his memory. The
way led through an undulating country to the Bohea Mountains amongst
which there was a small place, named Tsong so. Having crossed several
passes in these mountains, F. reached the borders between Fu kien and Che
kiang, at the border-town Ching che on the banks of a small mountain-
stream which flows to the westward. On the same day the city of Er-she-
pa tu was reached. On the next day he passed over a mountain on the
top of which stands the celebrated temple of Shan-te-maou. Farther on
the city of Sha co, built on both sides of a river near its sources was
reached, and on the next day he arrived at Ching hoo (Ts'ing hu, Chin.
map) a small bustling town and a place of considerable importance, being
at the head of one of the branches of the Tsien tang River (supra p. 409).
A boat was engaged here to take the traveller down the river, via Kiang
shan, Kü chow, Yen chow, to Ne chow (supra p. 410)*).

After having been absent on this long journey nearly three months,
F. arrived at Shang hai in due time (in August). The tea-plants procured
in Woo-e-shan reached Shang hai in good order. F. then remained at the
latter place for several months under Mr. Beale's hospitable roof. In October
and November he procured a large supply of tea-seeds and young plants
from Hwuy chow, and from various parts of the province of Che kiang.
These were all brought to Shang hai in order to be prepared and packed
for the long voyage to India, and were finally taken down to Hongkong
under F's own care and sent to Calcutta by four different vessels.

F. spent the winter in Hong kong, prosecuting the botanical exploration of the island, sometimes in company with Captain Champion. (Comp.

---

*) I may observe that, in 1667 a Dutch Embassy, proceeding from Fu chou to Peking
(supra p. 24), had followed from P'u ch'eng to the mouth of the Ts'ien t'ang River the same
route as Fortune.

1850 Hook. Kew Journ. Bot. III, 260, sub Capparis pumila). In April 1850 he went again to Shang hai and visited the various Chinese gardens and nurseries in the vicinity of the city, from which he procured many interesting plants for the Horticultural Society. Then he left Shang hai for the tea-districts of Ning po, passing a part of the summer there and having his head-quarters in the old monastery of Tein tung. In the end of June he proceeded to the island of Chu san, which then was again in the possession of the Chinese. Visit to Poo too or "Worshipping Island" (supra p. 406). In the beginning of September he left again for the mainland and took his quarters in the districts near Ning po. In the end of December he sailed for Shang hai. In the meanwhile, owing to the care of Mr. Beale, some good Chinese tea-manufacturers from the interior had been procured, to be sent to India, with a large assortment of implements for the manufacture of tea.

1851 On 16 Febr. 1851, all collections having been packed, F. left Shang hai for Hong kong and from the latter place at once went onwards in a British steamer to Calcutta, where he arrived on March 15, with nearly 2000 young tea-plants and 17,000 germinated seeds. With these he proceeded to the north-western provinces of India and thus served to lay the foundation of what is now a lucrative industry. In September 1851 he left Calcutta for England.

3. R. Fortune. A RESIDENCE AMONG THE CHINESE: INLAND, ON THE COAST, AND AT SEA. BEING A NARRATIVE OF SCENES AND ADVENTURES DURING A THIRD VISIT TO CHINA, FROM 1853 TO 1856. With illustrations. London, 1857.

In the end of 1852 Fortune was deputed a second time by the E. I. Company to proceed to China for the purpose of adding to the collections of tea-plants, seeds etc. already formed, and particularly of procuring some first rate black-tea makers for the experimental tea-farms in India. He seems to have left England in the beginning of 1853. The narrative in his
1853 book begins with the 14th of March 1853, when he proceeded from Hong kong to the north of China. During this third series of his travels in China he supplied the columns of the Gardener's Chronicle with particulars of his journey and articles on Chinese plants under the title of: LEAVES FROM MY CHINESE NOTE-BOOK, viz:

№ I to III in Gard. Chron. 1853, resp. 631, 741, 822.
№ IV » VII » » » 1854, » 54, 217, 533, 758.
№ VIII » X » » » 1855, » 242, 318, 502.

The accounts given in these papers are not always repeated in the Residence among the Chinese.

Fortune arrived at Shang hai in the second half of March. Mr. Beale's large and interesting garden was of the greatest value for him, as he could store his various collections there until an opportunity occurred for having them shipped to their destination. After a short stay in Shang hai he sailed for the town of Ning po and, on his arrival at this port started immediately for the tea-districts in the interior. He proceeded by boat on a canal, the end of which was at the foot of the hills to which he was bound, about 10 or 12 miles from Ning po. In May he visited the celebrated temple of A-yu-wang or, as he generally calls it, Ayuka's temple, in these regions and made it for several days his head-quarters. Having completed his investigations there, he returned to Ning po, and then visited the old city of Tse kee (Tz'e k'i hien), about 10 or 12 miles northwest from Ning po and near one of the branches of the river which flows past that town. The scenery around it is of the most romantic and beautiful description. The city stands on a flat plain, and is surrounded by hills varying in height from 200 to 1000 feet above the level of the plain. F. was staying near Tse kee for some time, making excursions amongst the hills and collecting plants and insects.

In the month of July he again took up his quarters in the old Buddhist temple of Tein tung, repeatedly visited during his former travels. It is situated amongst the mountains some 20 miles south-east (mistake for south-west, v. supra p. 406) from Ning po and in the midst of an extensive tea-country. — In the end of August he visited the island of Chusan, and in the beginning of September returned to Shang hai. Having finished his work there, he took his departure, in the beginning of October, for the tea-districts in Che kiang in order to make collections of seeds and plants for the government plantations in the Himalaya. He established himself in his old quarters in the temple of Tein tung. While making purchases of tea-seeds in this district, he dispatched two Chinese to the districts of Moo (Wu) yuen and Ping shui, producing fine teas. The first is in the Hwuy chow country, the other 9 miles south of Shao hing foo. Both his messengers returned in due time with rich collections of seeds.

During his travels in the province of Chekiang, F. also visited the celebrated waterfalls near a place named Seue-tow-sze, or the "Snowy Valley Temple", which is situated amongst the mountains some 40 or 50 miles to the south-west of Ning po, and examined the natural productions of these hills. As far as the foot of the mountains the journey was performed

by boat up the river, which flows through the fertile Ning po plain. Then
a mountain pass was ascended which led to the entrance of the Snowy
Valley, estimated at about 2000 feet above the level of the sea. This
valley is surrounded on all sides by mountains and perpendicular rocks. A
beautiful waterfall shoots down over a precipice.

After this F. proceeded with his various collections to Shang hai and
deposed them safely in Beale's garden. Subsequently they were transshipped
to Hongkong, accompanied by F. himself, and finally sent on to Calcutta in
four different ships. The whole arrived at their destination in excellent condition. No fewer than 23, 892 tea-plants, upwards of 300 chesnut-trees,
and a large quatity of other things of great value in India, now growing in
the Himalayas, were the results of this year's (1853) labour.

When the various consignments had been despatched, F. went up to
Canton for a few days, in order to get some reliable information as to the
mode of scenting tea by the Chinese for the foreign market. Before leaving
this place, he visited Howqua's Garden, situated near the well known Fa
tee nurseries (v. supra p. 266).

1854   In the beginning of March 1854, F. sailed for Foo-chow-foo. After
a short stay there, he took passage in the American steamer "Confucius"
chartered by the Chinese government to convey money to the island of
Formosa. Having crossed the channel in the direction of the north-west end
of Formosa, the steamer entered a river, which leads up to the important
town of Tam shuy and anchored there, April 20. As the steamer had to
remain a day in this port, F. went on shore and met there with the famous
"Rice paper plant", *Aralia papyrifera*. The "Confucius" then steamed
to Shang hai.

On arriving at Shang hai F. lost no time in returning again to the tea-
districts in the interior of Che kiang. During the summer and autumn of
1854, he had many opportunities of visiting new districts in the interior.

In October he engaged a boat at Ning po to take him up to one of
the sources of the river, which flows past the walls of that city, through the
fertile plain of Ning po. The town of Ning-kang-jou was reached, beyond
which the river is not navigable for boats of any size. This city is situated
in the hills on the way to the Snowy Valley visited in the previous year.
F. spent several days there and made excursions into the surrounding
country. Subsequently he proceeded westward among the mountains to the
temple of Tsan tsing, nearly 2000 feet above the level of the sea, one or
one and a half day's journey from Ning po. The surrounding mountains
were apparently from 3 to 4000 feet high. Farther on he reached the

summit of a steep pass, and entered a valley, in which was the village of
Poo-in-chee. Another high mountain-pass had now to be got over. This
was the highest range in this part of the country. He at last descended into
another valley, and reached the temple of Quan-ting for which he was
bound, fully 3000 feet above the sea. F. went to the head of the valley
and then came to the top of a range about 4000 feet above the sea. Having
returned to Ning po, he proceeded, with all his collections, by steamer, to
Shang hai. During the succeeding winter and spring months F. was engaged
in packing and dispatching to India and Europe the numerous collections
of plants, seeds and other objects of natural history formed in the summer
and autumn.

In the month of April 1855 F. paid another visit to the old city of 1855
Tse-kee near Ning po, and then determined to proceed to the great silk
country of Hoo chow, and to the hills on the western side of the plain of
the Yang tze kiang. He, accordingly, left Ning po for Shang hai, by the
inland route. Having engaged boats, he went, in company with some friends,
up the northern branch of the river to the ancient city of Yu yao, then left
the mainstream, and turned into a canal, which brought them to the bay,
crossed the latter and arrived at the ancient city of Kan poo, where canal
boats were engaged to continue the way through the silk districts to Shang
hai. This place was reached on the 3d of June, and on the 8th F. took his
departure for the great silk district for which the province is famed all
over the world. His boat proceeded up the Soo chow branch of the river to
the old city of Cading (Kia ting hien) where he remained for several days,
inspecting the natural productions of the country, which is exceedingly fertile. Leaving Cading, F. pursued his journey to the westward in the direction of Tsing poo. Lakes extended in all directions for many miles. P'ing
wang, a small bustling town on the edge of the lakes, near the eastern
borders of the great silk country. A broad and beautiful canal with mulberry trees on its banks stretched far away to the westward and led to the
great silk-towns of Nan tsin (Nan sin) and Hoo-chow-foo. Nan tsin, where
he spent several days, lies in the centre of the great silk country. The
country resembles a vast mulberry garden. Sailing onwards to the west on
the wide and beautiful canal he came within view of the mountain ranges,
which form the western boundary to the great plain of the Yang tze kiang,
through which he had been passing for several days. On the 7th of June
the city of Hoo chow foo was reached on the same canal. F. spent a week
in its vicinity, and visited some hills and temples, employing himself in
examining the natural productions and in making entomological collections.

The great trade in Hoo-chow, as at Nan tsin, is in raw silk. The T'ai hoo Lake with the Tung ting shan Island are seen to the north. On the west the prospect is bounded by a long range of mountains to which F. intended to proceeded. Following the Hoo chow Canal westward he found a wide and deep river which takes its rise amongst the hills in the far west. It is called Lun ke (Lung k'i) by the natives, and probably one of its most distant sources is near the celebrated T'ien mu shan — a mountain said to be the highest in that part of China *).

In sailing up this river he observed that the plantations of *Mulberry* still formed the staple crop of the country. The large monastery of Hoo shan, situated in the midst of rich and luxuriant vegetation not very far from the banks of the river, was visited. Then, continuing up the river, F. arrived at a place called Kin hwa, where he remained for two days and employed himself in making entomological collections and examining the productions of the district. He then went onwards to a small town called Mei che (k'i) which was as far as the river was navigable for boats, and from 36 to 40 miles west from Hoo chow foo. Mei che appears to be almost the western boundary of the great silk country. The town is surrounded by hills on all sides. Considerable quantities of tea are produced in the mountains to the west. — After a short stay at Mei che, F. returned to Hoo chow foo. He spent the next few days partly in the country to the northward bordering on the T'ai hoo Lake, and partly near the town of Nan tsin.

Having now finished his inspection of the silk districts, F. commenced his journey eastwards and reached Shang hai (in July). On August 10, he embarked with the Chinese tea manufacterers engaged for the tea farms in India and sailed for Hong kong, where the Chinese were shipped for Calcutta. Then F. returned once more to the north. On reaching Ning po he lost no time in proceeding onward to the interior of the country. About the end of October he found himself once more in front of the old temple of Tsan tsing (supra p. 416). With the object of obtaining various kinds of seeds, more particularly those of the "golden pine", *Abies Kaempferi*, he went again to the high mountains further west, visited in the last year, reached the "Valley of the Nine Stones", about 4000 feet high and then returned to Ning po.

Fortune had now brought his work in China to a successful termination. Many thousands of tea-plants, obtained in the finest districts, had

---

*) The Tien mu shan was visited in Nov. 1854 by the Rev. W. H. Medhurst. See Shanghai Almanac 1855.

reached their destination in the Himalayas in good condition; abundant supplies of implements used in these districts had also been sent on, and two sets of first-rate black-tea manufacturers from Fu kien and Kiang si had been engaged, and were now on their way to the north-west provinces of India. In accordance with the instructions received from the government of India, F. had also introduced many of the useful and ornamental productions of China, such for example as timber and fruit trees, oil-yielding plants, dyes, etc. These things were sent partly to the Government Gardens and partly to the Agricultural and Horticultural Society.

F. now sailed for Hong kong and Canton and then to Calcutta, where he arrived February 10, 1856. He proceeded once more to inspect the 1856 tea-plantations in the North-Western Provinces and the Punjab. On November 9th he left India and reached Southampton December 20th 1856.

4. R. Fortune. YEDO AND PEKING. A. NARRATIVE OF A JOURNEY TO THE CAPITALS OF JAPAN AND CHINA. With notices of the natural productions, agriculture, and trade of those countries etc. With maps and illustrations. London, 1863.

In the summer of 1860 Fortune took his departure for the "far East", 1860 with the view of making collections of objects of natural history and works of art. On October 12th 1860 he arrived from China at Nagasaki, left this place and its beautiful scenery on October 19th and proceeded on his voyage to the ports of Kana gawa and Yukuhama, near Yedo. He started for Yedo on November 13, and remained there for 15 days. During his stay there he made many excursions into the surrounding country and visited the gardens. On December 17th he left with his collections for Shang hai, where he arrived January 2nd 1861, and was kindly received by Mr. 1861 Webb, the worthy successor of his old friend the late Mr. Beale.

In the spring of 1861, F. returned to Japan, his object being to inspect the natural productions of the country during the spring and summer months. The steam-ship in which he took his passage, was bound for Kanagawa, but called at Nagasaki on her way. After a short stay at the latter place, the steamer continued her voyage to Yukuhama, where F. then took up his residence and made excursions in the neighbourhood of this place and of Kanagawa. On May 20, he proceeded to Yedo and remained there for a few days. On July 4th he went, in company with Dr. Dickson from China and others, to the towns of Kanasawa and Kamakura, a few miles south from Yukuhama. On July 29 he left Japan for Shang hai. During his travels in Japan he wrote NOTES ON THE BOTANY of Japan, which were published in the Gard. Chron. 1861, 145, 312, 385, 456, 576, 773.

When F. had finished his work in Japan, the Chinese war had been brought to a successful termination, and he was enabled to visit the new ports of Che foo and Tien tsin, on the Gulf of Chili, and also the capital Peking, itself. Accordingly, on August 11, he sailed from Shang hai for Che foo and reached that port on the 16th. Left Che foo on the 1st of September in a French steamer for Tien .tsin. F. visited some Chinese gardens in the neighbourhood of Tien tsin. On September 17 started for Peking. After rambling over this great city, he examined the Chinese nursery gardens two or three miles south of Peking, and then paid a visit to some celebrated Buddhist temples in the western mountains, known under the name of Pa ta choo. Here he first met with full grown specimens of the beautiful white-barked pine, *Pinus Bungeana*.

Having finished his work in Peking and packed up the collections formed there, he left that city on September 28. Shang hai was reached October 20. F. employed the next fortnight in preparing his Japanese and Chinese collections for their long voyage around the Cape of G. H. The cases reached England, and nearly every plant of importance had been introduced alive. Long shelves filled with these rare and valuable trees and shrubs of Japan were exhibited by Mr. Standish at different botanical and horticultural exhibitions in London. A part of the collection, however, was brought home by the overland route under Fortune's own care. On the 2d of January 1862, F. landed at Southampton.

Fortune formed one of the Committees of the International Exhibition of 1866, but from that time gradually retired from horticultural pursuits, and betook himself to farming in Scotland. Nevertheless he occasionally came among his old friends and companions. He died on the 13th of April 1880, at Brompton. (See Gard. Chron. 1880, April).

As had been stated an a previous page, one of the chief objects of Fortune's missions to China was the introduction of valuable living plants into England and India, in which he admirably succeeded. After his return from the first expedition, the Journ. of the Hort. Soc. published in Vol. I, 1846, 215 and 221 two lists of the living plants sent home by him from China in the beginning of 1844, and in the autumn of 1844, and on p. 223 he enumerates himself the plants brought with him in 1846. In 1860 he published in the Gard. Chron., 169, NOTES ON SOME CHINESE PLANTS RECENTLY INTRODUCED INTO ENGLAND. A short time before his death he wrote a paper on the PLANTS INTRODUCED BY HIM AT DIFFERENT TIMES FROM CHINA AND JAPAN, FOR THE HORTICULT. SOCIETY OF LONDON AND ON HIS OWN ACCOUNT, which was published in the Gard. Chron. 1880, I, 72, 179, 234.

As to the herbarium specimens gathered during his first journey, 1843—45, in various parts of China, he transmitted them likewise to the Hort. Society. This collection, which may be estimated at 450 species, was distributed by the latter, in 1847, to the great botanical institutions in Europe and to several distinguished botanists of various countries. There exist also at Kew a few dried specimens of plants discovered and introduced by F. during his other journeys, but the bulk of his herbarium specimens date from his first journey. From the statements of various botanists who have described F.'s plants, and especially from Mr. Hemsley's Index Florae Sinensis, we can derive some information regarding this interesting collection. The greater part of the specimens are numbered on the labels (highest number hitherto quoted 183). Dates appear only exceptionally. As to the localities more than 150 labels have: China, sometimes South China or North China. In other cases the following localities are noticed on the labels: Hong kong, Canton, Macao, Chin chew (Ts'üan chou fu), Fu chou fu, Chusan, Che kiang, Ning po, Shang hai. One part of Fortune's herbarium specimens has its own numeration (highest number published, 161) and is marked A. on the labels. The Horticult. Society, in distributing these plants, has erroneously put on these labels: China borealis. Mr. Hemsley, who notices about a hundred of these plants marked A. considers A. to mean Amoy and he is, no doubt right. We have seen (v. supra p. 405) that in 1843 F. was staying for a month at Amoy (August, September) exploring the island and the adjacent mainland. But in general the information given on the labels, (by the Hort. Soc., I suppose) is not reliable at all, and frequently does not agree with F.'s books, especially with respect to the localities. For further particulars on the subject see my article: "On some Old Collections of Chinese Plants", in Journ. of Bot. 1894, 295.

BOTANICAL RESULTS OF FORTUNE'S TRAVELS IN CHINA AND JAPAN.

We now propose to put together into a connected form and in systematical sequence all the valuable notes of the traveller on Chinese plants scattered over his books on China *) and found in various papers he wrote on the subject. At the same time we shall give detailed notice of all the living plants introduced by him into England from China and Japan, and of those of his herbarium specimens which have been recorded as novelties by various botanists.

Clematis lanuginosa, Lindl. in Paxt. & Lindl. Flow. Gard. III (1852) p. 107, short diagnosis, tab. 94, coloured. This magnificent plant flowered last spring in the nursery of Messrs Standish and Noble of Bagshot, who received it from Fortune. We have a wild specimen from that enterprising traveller, marked: Hills of Che kiang, July 1850, and he has favoured us with the following memorandum:

"This pretty species was discovered at a place called Tein tung (v. s. pp. 406, 414) near Ning po. It is there wild on the hill sides. Its fine, star-shaped azure blossoms are there seen from a considerable distance, rearing themselves proudly above the shrubs to which it clings for support. It is very attractive". — The flowers of this species are much larger and more hairy than those of the Japanese Cl. azurea grandiflora, to which it bears some resemblance. The flowers are about 5 inches in diam. Red stamens. Leaves coriaceous.

Fine coloured drawings of this plant are also found in Fl. d. serres, VIII (1853) tab. 411, and XI, (1855) tab. 1176, — in Illust. hort. 1854, tab. 14; — in Lemaire, Jard. fleur. IV (1854) tab. 363.

Comp. also Maximowicz, Mél. IX (1876) 600, — and I. F. S., ı, 5.

Clematis Fortunei, Moore in Gard. Chron. 1863, 460, 676, c. fig. A fine plant with large double sweet-scented white flowers, introduced from Japan for Mr. Standish in 1862 by Fortune. — Fl. d. serres, XV, (1863) tab. 1553, — Belg. hort. XIV (1864) 33.

According to Maxim. Mél. IX (1876) 600, this is a variety of *Cl. patens,* Morr. Dcne, or *Cl. coerulea,* Lindley, introduced from Japan by Siebold previous to 1837.

Thalictrum Fortunei, S. Moore in Journ. Botan. 1878, 130. Herb. spec., Fort. China, n. 28 (probably Ning po).

---

*) Abbreviations used with respect to Fortune's works:
W. = Wanderings in China, 1843 to 46.
T. = Tea Countries, 1848 to 51.
R. = Residence among the Chinese, 1853 to 56.
Y. = Yedo and Peking, 1860 to 61.

Maxim. Fl. As. Or. Fragm. 1878, 3 = *Th. baicalense*, Turcz. Cat. Baical. (1839) 5, var.? *minor*. — I. F. S., ɪ, 8.

Anemone japonica, S. & Z. Fl. Jap. I, 15, tab. 5, (1835); Bot. Reg. 1845, tab. 66. — Fortune found this plant (in 1843) near Shang hai amongst the graves of the natives. It was in full flower in November, when other flowers have gone by, and is a most appropriate ornament to the last resting places of the dead. (W. 333, 409).

Fortune sent living specimens of this beautiful plant to the Hort. Society. They were received June 20, 1844, and in October 1846, F. saw them in full bloom in the garden of the Society at Chiswick. Lindley, in Journ. Hort. Soc. I, 1846, 61, states that the flowers are nearly 3 inches in diam. and consist of a considerable number of bright purple leaves (sic!) of a somewhat obovate form and about $\frac{1}{2}$ inch wide. — Bot. Mag. (1847) tab. 4341. — I. F. S., ɪ, 11.

Aconitum autumnale, Lindley, Journ. Hort. Soc. II (1847) 77. Short diagnosis. Fortune sent a living plant from Chusan, which was received 1846, April 6. Also cultivated in the gardens of Ning po. It flowers in winter or late in autumn. Most nearly related to Aconitum japonicum, Thbg, from which it differs in its pubescence. Flowers lilac and white, in a simple spike, of an unpleasant, heavy small. See also Lindl. & Paxt. Fl. Gard. I (1850) 187, c. fig.

A. autumnale as well as A. sinense S. & Z. (1846) are now reduced to the widely spread *A. Fischeri*, Reichenb. Illustr. Acon. t. 22 (1825). N. E. Asia, North America. — I. F. S., ɪ, 20.

Aconitum chinense, Paxton, Mag. Bot. V (1838) tab. 3, col. plate. Introd. by Dr. v. Siebold from Japan, about 1833. — Bot. Mag. tab. 3852 (1841). Probably cultivated in China. — According to Mr. Hemsley, I. Fl. S., ɪ, 20, this is the same as *A. sinense*, Lindley (non Sieb.) in Paxt. Fl. Gard. I (1850) p. 187, fig. 116, said there to be a native of Japan. Deep violet flowers which appear in the autumn. As there is in the Kew herbarium a specimen of this plant labelled: Fortune, China, Hemsley named it *Aconitum Fortunei*.

Paeonia Moutan, Sims, Tree Paeony. This Chinese plant was first introduced into England, in 1786, from Canton, owing to the exertions of Sir Joseph Banks (v. s. p. 204). It was a garden variety and succeeded perfectly in the open air. Fortune during his travels in China introduced about 40 new varieties of this beautiful plant to which he paid a special attention in his records. It was in the winter of 1843 that, when paying a visit to Shang hai, he first saw tree paeonies in their native country. He writes:

"Those varieties of this flower, which are yearly brought from the northern provinces to Canton, and which are now common in Europe, have blossoms, which are either rose-coloured or white: but it was always asserted, although not believed, that in some part of China purple, blue and yellow varieties were produced". (In his instructions from the Hort. Society he was told to look out for these varieties). F. secured from the vicinity of Shang hai some most striking and beautiful kinds. Amongst them were *lilacs* and *purples;* some nearly *black;* and one which the Chinese called the *yellow,* which, however, was only white with a slight tinge of yellow near the centre of the petals. There were altogether 12 or 13 fine new varieties which F. sent home in the beginning of 1844. (W. 124, 125; — Fortune, in Journ. Hort. Soc. I (1846) 215; in Gard. Chron. 1880, I, 179, Chinese Tree Paeony).

In January 1844, F. saw Moutans in the celebrated Fa-tee Gardens near Canton. They are not natives in the south of China, but are brought down in large quantities every year, about the month of January, from the northern provinces. They flower soon after they arrive, and are rapidly bought up by the Chinese to ornament their houses; after which they are thrown away, as they will not flower a second season. They are sold according to the number of flower buds they may have upon them. The varieties seen at Canton are mostly different from those met at Shang hai. (W. 143, Gard. Chron. l. c.).

F. again visited Shang hai about the middle of April 1844, when the tree paeonies there were in bloom, and purchased another collection of them, which he forwarded to England in the autumn. The flowers were very large and fine, and the colours were *dark purples, lilacs,* and *deep reds.* (W. 245, 246, Journ. Hort. Soc. I (1846) 221, 223).

When, in the middle of January 1849, F. was in Shang hai, he was struck with the facility with which the Moutan Paeony had been brought into full bloom by the Chinese gardeners, notwithstanding the cold season. Their blooms were tied up, to keep them from expanding too rapidly. (T. 121).

In April of 1850 F. inspected again the Moutan Gardens near Shang hai. They are situated about 5 or 6 miles west of Shang hai, near the village of Ta who, in the midst of an extensive Cotton country. On the road he met a number of coolies, each carrying two baskets filled with tree-paeonies in full flower, which were being taken to the markets for sale. When he reached the gardens he found many of the plants in full bloom and certainly extremely handsome. The *purple* and *lilac*-coloured kinds were particularly striking. One, a very dwarf kind, and apparently a distinct species, had finely cut leaves, and flowers of a *dark velvety purple*, like the Tuscany Rose of our gardens. This the Chinese call the *black* Moutan. It is probably the same which Dr. Lindley has described in Journ. Hort. Soc. and named *Paeonia atrosanguinea*. Another kind, called the *tse* (tze) or *purple*, has double flowers of a large size. The third is called the *lan* or *blue*; this is a lilac variety, with flowers of the colour of Wistaria si-nensis. — The double *whites* are also numerous and handsome. The largest of these Dr. Lindley, has named *P. globosa*, but there are four or five others nearly as large and double. Some of them have a slight lilac tinge. The most expensive is one called *wang* (huang) or yellow by the Chinese: it is a straw-coloured variety, rather pretty, but not so handsome as some of the others. — The reds, *hong,* are also numerous. There are about half-a-dozen new varieties of reds in these gardens: one of them, called *van yang hony* was the finest flower F. ever saw. The flowers are of a clear red colour, unlike any of the others, perfectly double, and each measures 10 inches across. Altogether he numbered about thirty distinct varieties in these gardens. — Nearly all these fine varieties of the Moutan are quite un-known in Canton, because the Canton gardens are supplied with moutans by another district, which lies much further to the west than Shang hai. — The propagation and management of the moutan seem to be much better understood at Shang hai than in England. In the beginning of October large quantities of a herbaceous paeony are seen heaped up in sheds, and are intended to be used as stocks for the moutan. The bundle of tubers which form the root is pulled to pieces, and each of the finger-like rootlets forms a stock upon which the moutan is grafted. F. describes in detail the mode of grafting used by the Chinese. Budding is never practised in the country. — In the gardens of the mandarins the tree-paeony frequently at-tains a great size and produces between 300 and 400 blooms every year.

The tree-paeony is found wild *) on the mountains of the central pro-vinces of China, and is cultivated as a garden plant in all parts of the

---

*) The province of Shen si and especially the northern part of it seems to be the native country of the moutan tree. The Shen si t'ung chi or Chinese description of that province, pub-lished about 200 years ago, reports that in the district of Han ch'eng, situated on the western bank of the Yellow River, between 35° and 36° N. lat., there is, 60 *li* N. W. of the district city,

Empire. It is not indigenous to South China. The Canton gardeners carry on a large trade with the moutan growers, who bring the plants yearly in boats from the provinces of Hoo nan and the western parts of Kiang nan, a distance of a least one thousand miles. This takes place in the winter months when the plants are leafless and in a state of rest. The roots are packed in baskets, and have scarcely any soil adhering to them; in this simple manner they are distributed all over the Empire without suffering any injury. On their arrival in the south they are immediately potted by the purchasers, and, owing to the difference in the temperature, soon come into bloom. As soon as the flower-buds are fairly formed, the plants are eagerly bought up by the natives to ornament their balconies, halls and gardens. The moutan, when brought down into the hot climate of Canton, will not thrive for any length of time. It blooms well the first year, but, being deprived of its natural period of rest, — that is a cold winter — it gets out of health. The southern Chinese, therefore, never attempt to preserve it after it has once bloomed. (T. 320 seqq.).

The following new varieties of Paeonia Moutan introduced by Fortune from China have been named or described by Lindley, some of them figured in Paxt. Lindl. Flower Garden:

In the Journ. Hort. Soc., II (1847) 308, Lindley notices the *P. Moutan picta*, a very handsome flesh-coloured variety, which Fortune had sent from Canton in 1844, and which had flowered in the Hort. Society's Garden in 1847.

In the same Journal, III (1848) 236, six more of Fortune's Moutans are mentioned which flowered in 1848, viz:

P. Moutan atropurpurea.
» » *Banksia* (v. s. p. 204).
» » *parviflora.*
» » *lilacina* (v. supra Fortune 425).
» » *globosa* (v. s. Fortune 425).
» » *salmonea.* — This is figured in Paxt. Lindl. Flower Gard. II (1850) p. 9, plate 20 (colour). The Society received it in April 1846.

---

a hill called Mou tan shan, where the mou tan tree grows in great profusion, in a wild state. In spring, when these trees are in blossom, the whole hill appears tinged with red, and the air round about for a distance of ten *li* is filled with fragrance. The people in this district as well, as in the prefecture of Yen chou (which lies farther north) use the tree for fuel. No European botanist has ever visited these regions. Mr. Hemsley, I. F. S., 1, 22, states that in the Kew Herbarium there are only cultivated specimens of Paeonia Moutan. It has been observed in a wild state in the mountains of Japan (Miquel Prol. Jap. 197).

Very large flowers. The outer petals, when fully blown, are of a pale salmon colour, the inner have a deep rich tint of the same.

Paeonia Moutan versicolor, Lindley in Journ. Hort. Soc. IV (1849) 224. Received from Fortune, in April 1846, from North China. It flowered in the Soc.'s Garden, 1849. Large semi-double flowers. Petals deep purple near the base, fading to a rose-lilac near the outside.

P. Moutan atrosanguinea, l. c. 225. Fortune sent it from Hongkong and Shanghai as "very dark, nearly black" (v. s. 425). It is deep blood-coloured, the darkest variety in colour of all the tree-paeonies yet culti-vated. Paxt. Lindl. Flow. Gard. I, 1850, p. 161, plate 31 (colour).

In his article on the Chinese Tree Paeony, Gard. Chron. 1880, I, 179, Fortune says that then most of the beautiful varieties of this plant he had formerly introduced, were lost to English gardens.

Chimonanthus fragrans, Lindley. This plant has been known for a long time from Japan. Fortune saw it much cultivated in the gardens of Shang hai. The Chinese call it *la mei.* Chinese ladies use it for decorating their hair. (T. 16, 320).

F. found this plant in a wild state in the mountains above Yen chou fu in Che kiang. It was also quite common on the hill sides near Hwuy chow fu, in southern An hui, 1849. (T. 57, 79).

Aquebia quinata, Decne, Rajania quinata, Thbg. — Sieb. & Zucc. Fl. Japon. I (1836) 143, tab. 77. This plant has been known for a long time from Japan. Fortune first discovered it in China and introduced it into England.

Lindley figured the plant in Bot. Reg., 1847, t. 28. The drawing was made from one of the plants sent home by Fortune. The latter had informed Lindley that it was one of the wild plants of Chusan. He found it growing on the lower sides of the hills, in hedges, where it was climbing on other trees and hanging down in graceful festoons from the ends of their branches. The colour of its flowers in China is of a dark brown, not unlike the Mag-nolia fuscata, and they are very sweet-scented; a delightful fragrance. — In the garden of the Hort. Soc., where it flowered for the first time in England, the flowers are much lighter in colour and nearly scentless.

In the Journ. Hort. Soc., II (1847) 160, Lindley states that A. quinata, received from Fortune in 1845, flowered in the Soc.'s garden, but is not very ornamental. — I. F. S., i, 30.

Berberis Fortunei, Lindley, Journ. Hort. Soc. I (1846) 231, 300. Short diagnosis, uncoloured drawing. Fortune, who brought the living plant with him, when he returned from China in May 1846 (l. c. 223), furnished to Lindley the following memorandum: "Evergreen bush with pretty pinnated and serrated dark green leaves. It grows 2—4 feet high, flowers in the autumn months. Short spikes of yellow flowers, generally 6—7 together. Found in a nursery garden near Shang hai (1845). The Chinese call it *che wang chok* or "blue and yellow bamboo", so named from the peculiar tint of bluish green which the leaves have and from the yellow flowers". —
The first syllable in the Chinese name is evidently wrong. — Ind. Fl. Sin., ɪ, 31.

End of October 1848, when proceeding to the Sung lo tea district, F. discovered, before reaching Tun che, at the border of the Hwuy chow district in An hui, in a garden, a fine evergreen *Berberis,* belonging to the section of Mahonias, and having of course pinnated leaves. Each leaflet was as large as the leaf of an English holly, spiny and of a fine dark shining green colour. The shrub was about 8 feet high, much branched, and far surpassed in beauty all the other known species of Mahonia. When staying at Sung lo shan he procured three good plants of this Berberis, which arrived safely in England. It was the *Berberis japonica,* R. Br. (T. 81, 99, 100, 150). At first F. supposed that it was a new plant and named it *Berberis Bealii,* in compliment to his friend Beale of Shang hai. It flowers in winter. Flowers in terminal spikes, 6—9 inches long, yellow. Fruit of a glaucous blue. Chinese name *shae ta kong la.* (Fortune's Notes of a Traveller, Gard. Chron. 1850, 212. — Fortune's Chin. pinnated Berberries in Journ. Hort. Soc. VII (1852, 225).
Lindley found that it was identical with *B. (Mahonia) japonica,* R. Br., known for a long time from Japan. Lindley in Paxt. Fl. Gard. I (1850) p. 11, figures a leaflet of the plant received from Standish and Noble to whom F. had given his B. Bealii. The plant is figured under the latter name in Bot. Mag. t. 4852 (a. 1855). Ibidem t. 4846 is figured *B. Bealii,* var. *planifolia,* introduced to Standish and Noble by Fortune from a district considerably to the north of Shang hai (?). — In I. F. S., ɪ, 31, B. japonica and B. Bealii are reduced to *B. nepalensis,* Spreng.

In May 1849 F. observed near Long yeou in Che kiang a species of *Berberis,* apparently a variety of the English one, which was extensively cultivated there, probably for medicine or as a dye.

Berberis consanguinea, Fortune, Chin. Pinnated Berberries, in Journ. Hort. Soc. VII (1852) 225. This was found by F.'s servant in the same district as B. Bealii, in 1848. It is easily distinguished from the latter by the colour of its stem and leaves, which are of a light green. The leaflets are also more lanceolate in form and have a tendency to vary, that is, some are lanceolate and some ovate. This is scarcely inferior.

Berberis trifurca, Lindl. in Paxt. Fl. Gard. III (1852) p. 57, fig. 258, black, a leaf figured. A curious species of pinnated berberry, allied to B. nepalensis, an evergreen shrub. Long leaflets which have a few coarse toothing near the base, then a long toothless interval and at the point three stout teeth. The terminal leaf is sessile. China, Fortune. — Fortune, l. c. 225, states, that he found this plant (in 1849 or 50) in the province of Che kiang and near the coast. It is there cultivated extensively on account of the dye which it furnishes. He bought it in a cottage garden. It may be easily known from the other species by its dark green lanceolate leaflets. The young stems are covered with scales of a clear reddish purple, colour, which give it a marked appearance.

This and the preceding F. gave to Standish & Noble at Bagshot. They had not yet produced fruit.

Nandina domestica, Thbg. Fortune, in T. 122 (Jan. 1849), states: "In the winter season at Shang hai a plant with red berries is seen in the gardens, which takes the place of our English Holly. It is the N. domestica. The Chinese call it *tein chok* (t'ien chu) or "Sacred (Heavenly) Bamboo". Large quantities of its branches are brought at this time from the country and hawked about the streets. Each of these branches is crowned with a large bunch of red berries, not very unlike those of the common holly, and, when contrasted with the dark, shining leaves, are singularly ornamental. It is used chiefly in the decoration of altars, not only in the temples, but also in private dwellings and in boats — for here every house and boat has its altar. The Nandina is found in English gardens, but from these specimens no idea can be formed of its beauty. It does not appear to produce its fruit so freely in England as it does in China". — Fortune observed this plant also in the gardens near Hwuy chou foo (T. 79) and at Ning po (R. 80), but not in a wild state. — I. F. S., I, 32.

Euryale ferox, Salisb. When Fortune visited, in 1855, the country between Shang hai and the T'ai hu Lake, he here and there came upon the broad prickly leaves of this plant, covering the surface of the water. — I. F. S., I, 32.

Nelumbium speciosum, Willd., *Lotus*. This elegant water plant was frequently met with by F. during his travels in China and Japan.

"On the sides of the Canton River, both below and above the city of Whampoa, large quantities of the water lily, or Lotus are grown, which are enclosed by embankments in the same manner as the rice fields. This plant is cultivated both as an ornament, and for the root, which is brought in large quantities to the markets, and of which the Chinese are remarkably fond. In the summer and autumn months, when in flower, the lotus fields have a gay and striking appearance, but at other seasons the decayed leaves and flowers, and the stagnant and dirty water, are not at all ornamental to the houses which they surround". (W. 137).

Poo to Island and temples, July 1844. "In front there was a large artificial pond, filled with the broad green leaves and noble red and white flowers of the Nelumbium speciosum, a plant in high favour with the Chinese. Every body who went to Poo-to admired these beautiful water-lilies". (W. 170).

In the southern provinces of China large quantities of Nelumbium are grown for its roots, which are much esteemed. (W. 307).

Woo-e-shan (Bohea Mountains), June 1849: "A small lake was seen glistening through the trees, and covered with the famous *lien wha* or Nelumbium, the noble leaves of which were rising above its surface. This plant is held in high esteem and veneration by the Chinese and always met with in the vicinity of Buddhist temples". (T. 227).

She pa ky, N. E. of the Bohea Mountains, Che kiang, July 1849: "This place is situated in a fine, fertile valley. Rice is the staple production, but large quantities of Nelumbium are also cultivated in the low irrigated lands. The rhizoma, or underground stem, of this plant is largely used by the Chinese as an article of food, and at the proper season of the year is exposed for sale in all the markets. It is cut into small pieces and boiled, and, like the young shoots of the bamboo, is served up in one of the small dishes, which crowd a Chinese dinner-table. An excellent kind of arrowroot is also made from the same part of this useful plant". (T. 251).

Visit to Poo to Island, July 1850: "In the front of the temples, there is a pretty lake filled with the Nelumbium, which then was in full bloom. No flower could be more beautiful or more majestic than the Nelumbium was at this season. The eye rested on thousands of these flowers, some of which were white, others red, and all were rising out of the water and standing above the beautiful clear green foliage. The leaves themselves, as they lay upon the smooth surface of the lake, or stood erect upon long footstalks, were scarcely less beautiful than the flowers, and both harmonized well together. Gold, silver and other kinds of fishes were seen swimming swiftly to and fro, enjoying themselves under the shade of the broad leaves". — In the garden of a mandarin at Ning po F. once observed a very beautiful variety of the Nelumbium, the flowers of which were finely striped. He proposes to call it *N. vittatum*. It is extremely rare, he could not succeed in procuring a plant to sent to England. — The Nelumbium plant is fitted by nature to endure a very low degree of temperature in winter. It is abundant in all parts of the province of Kiang nan (An hui and Kiang su), at Shang hai, Soo chow, and Nan king, where the winters are very severe. The ponds and lakes are often frozen up. During the spring and summer months the plants form and perfect their leaves, flowers, and fruit; in autumn, all these parts which are visible above water gradually decay, and nothing is left in a living state exept the large roots, which remain buried deep in the mud, and they continue in a dormant state until the warmth of spring again calls vegetable life into action.

The Nelumbium or *lien wha* is cultivated very extensively in China for the sake of its roots, which are esteemed an excellent vegetable, and are much used by all classes of the community. The roots attain their largest size at the period when the leaves die off; and are dug up and brought to market during the winter months in the north of China. The seeds are also held in high estimation; they are commonly roasted before being served up to table. (T. 348—52).

Dicentra (Dielytra) spectabilis, Miq. This beautiful plant was known to Gmelin, who, when travelling in Siberia, 1733—1743, had received a living specimen from Peking (v. s. p. 312). Fortune introduced living plants from China, which he brought with him for the Hort. Society, when he returned to England in May 1846.

Lindley, Journ. Hort. Soc., I (1846) 233, New Plants from the Hort. Soc.'s Garden: *Dielytra spectabilis*, DC., Fumaria spectabilis, Lin. Brought home by Fortune from gardens in North China. Beyond all comparison the handsomest plant of the order of Fumariaceae. The stems $1\frac{1}{2}$ feet high, have from 3 to 4 axillary racemes of beautiful flowers, each a full inch long, and nearly $\frac{3}{4}$ wide with the two saccate petals of a delicate rose colour, and the intervening projecting, narrow ones white with a purple tip.

Fortune's note on D. spectabilis in Journ. Hort. Soc., II (1847) 178, tab. 3 (coloured): "One of those plants of which the Chinese Mandarins in the North are so fond, and which they cultivate with much pride in their little fairy gardens". He found it first in the Grotto Garden in the island of Chusan. The Chinese call it *hong pak moutan* or "red and white Moutan" flower, for the flowers are red and white and the leaves not unlike those of Paeonia Moutan. F. never met with it in the southern part of China. It flowered for the first time in England in the garden of the Hort. Soc. at Chiswick.

Flore des Serres III (1847) t. 258. — Bot. Mag. t. 4458 (1849). — Paxton Mag. Bot. XV (1849) 127. Colour. drawing.

Fortune T. 318: Dielytra spectabilis in the gardens of Shang hai (1850, April). Its large purse-like blooms of a clear red colour, tipped with white, and hanging down gracefully from a curved spike, and its moutan-like leaves, render it a most interesting plant, and one which will become a great favourite in English gardens. — I. F. S., i, 35.

Corydalis Sheareri, S. Moore, Journ. Bot. 1875, 225. This plant collected by Dr. Shearer near Kiu kiang, in 1873, was earlier discovered by Fortune. Specimen in the Kew Herb. labelled: Fortune, China, 10. — Ind. Fl. Sin., i, 38.

Brassica chinensis, the oil plant is extensively grown in the northern provinces of China, both in the province of Che kiang (including Chusan Island) and also in Kiang soo, and there is a great demand for the oil, which is pressed from its seeds. It is a species of cabbage producing flower stems 3 or 4 feet high, with yellow flowers, and long pods of seed like all the cabbage tribe. In April, when the fields are in bloom, the whole country seems tinged with gold, and the fragrance which fills the air, particularly after an April shower, is delightful. It is planted out in the fields in autumn, and the seeds are ripe in April and May, in time to be removed from the land before the rice crops. (W. 54, 90, 309, 349).

The cabbage oil plant is also much cultivated in Japan. Fortune saw it near Nagasaki, Kanagawa etc. In that country it is one of the staple winter productions. Early in the month of April the hill-sides are yellow with its flowers. In May the seed-pods are swelling and coming fast to maturity. They ripen near Yedo about the end of the month, and the oil harvest begins. The plant is not cut like corn, but is pulled up by the root, and laid on the field where it has been growing. When it has lain for a few days to dry, the labourers (women chiefly) take the stalks, and tread out the seeds upon mats. The stalks then are burned, and the ashes used for the summer crops, which are now being sown to take the place of the rape. (Y., 172, 182, 184, 270).

Lindley, New Plants cultivated in the Hort. Soc. Garden, in Journ. Hort. Soc. I (1846) 67: *Brassica chinensis,* received from Fortune, Nov. 23, 1844. Description of the plant.

I. F. S., i, 46: *Brassica chinensis,* Lin., reduced to *Brassica campestris,* Lin.

White Cabbage. Several varieties of the cabbage tribe, which seem indigenous to China, are grown extensively in the winter season both in the south and the north of China. These never produce a solid heart like our cabbages, and are of no value when imported to England. But the celebrated *pak (pai) tsae* or *White Cabbage* of Shang tung and Peking, is a very different plant; it is never grown in the south of China, but is produced in the summer months in the north. Large quantities of this delicious vegetable are brought south every autumn, in the junks which sail at the commencement of the N. E. monsoon in October. — At Peking the Shan tung cabbage is very plentiful. (W. 309, — Y. 362).

When returning from China, in May 1846, F. brought some seeds of the true Shan tung cabbage for the Hort. Society. See Journ. Hort. Soc. I (1846) 223.

Isatis indigotica, Fortune. After passing, in June 1844, the cotton district (west of Shang hai), F. came into a tract of country in which a cruciferous plant seemed to be principally cultivated. From this plant a kind of Indigo or blue dye is prepared; it is called *tein-ching (tien tsʻing =* Indigo) by the Chinese. Very large quantities of this substance are brought to Shang hai and all the other towns in the north of China, where it is used in dyeing the blue cotton cloth, which forms the principal article of dress of the poorer classes. F. brought home (in 1846) living specimens of the plant which produces this dye. It flowered in the gardens of the Hort. Society and proved to be a new *Isatis.* Fortune proposed the above specific name. This plant supplies in the north of China the true Indigo plant, Indigofera tinctoria. Its leaves are prepared in the same manner as the common one. The colour of the liquid at first is a kind of greenish blue, but after being well stirred up and exposed to the air, it becomes much darker. It is probably thickened afterwards by evaporation. Fortune was inclined to believe that this is the dye used to colour the green teas. (W. 251, 307).

Lindley in Journ. Hort. Soc. I (1846) 269, gives a short description of the *Isatis indigotica,* accompanied with a black and white drawing, and Fortune there furnishes additional details regarding the process used by the Chinese in the preparation of this kind of indigo.

In 1855, when F. again travelled in Kiang su he found the Shang hai indigo largely cultivated in the *Ke-wang-meou* (?) district, a few miles south of Ca ding (Kia ting hien). (R. 333).

Hemsley in I. F. S., i, 49, believes that Fortune's I. indigotica is only a variety of the common *Isatis tinctoria,* Lin.

Fortune, when visiting Peking, in Sept. 1861, observed there the *red turnip-radish,* which, as he states, is sent south every winter and made to flower in pots or flat saucers amongst pebbles and water at the time of the Chinese new year. (Y. 363).

It seems to me that F. means the red Chinese summer radish, which Regel received from Peking, and which he figured and described in his Gartenflora XI (1862) 407.

Scolopia chinensis, Clos in Ann. Sc. nat. 4 sér. VIII (1857) 249. Fortune, China, n. 134. — In I. F. S., i, 57, reduced to *S. crenata,* Clos. Same as *Phoberos chinensis,* Lour. Fl. Cochin. 318.

Flacourtia chinensis, Clos. l. c. 219. Fortune, China, Herb. Delessert. — I. F. S., i, 57, reduced to *Xylosma racemosum,* Miq., known before Fortune from Japan.

Pittosporum glabratum, Lindley, New Plants from the Hort. Society's Garden, in Journ. Hort. Soc. I (1846) 230. Short diagnosis. Fortune sent a living plant from Hongkong. Received May 1st 1845. It was found growing on the top of the hills. A dwarf shrub with deep green rather blistering convex leaves, which shine as if varnished when young and are somewhat glaucous underneath. The flowers appear early in spring in terminal, sessile umbels. They are small, of a pale-greenish white, very sweet-scented.

I. F. S., i, 58: Fortune, Amoy, A , 126. The same plant was described by Turczaninov in Bull. Soc. Imp. Mosc. 1863, 562, as *P. Fortunei.* — I. F. S., i, 58.

Silene Fortunei, Vis., Ind: Hort. Patav. 1847, ex Linnaea, XXIV, 181. Culta in hto Patavino sub nomine Lychnidis Fortunei. — I. F. S., ɪ, 65: China, Fortune, 36, Chusan, Fortune.

Cleyera Fortunei, Hook. fil. in Gard. Chron. 1895, I, 10. This plant was cultivated in England for upwards of thirty years, but has only recently flowered. It is first alluded to as a new plant introduced by Fortune to Standish, from Japan. It was since that known as *Eurya latifolia* variegata or *Cleyera japonica* variegata. Sir Joseph Hooker figured it in the Botan. Mag. tab. 7434 (1895).

Actinidia chinensis, Planchon in Hook. Lond. Journ. Bot. VI (1847) 303. Only a short diagnosis. Fortune, China. — I. F. S., ɪ, 78: Fortune, 1846, n. 39. — Figured in Hook. Icon. Pl. tab. 1593 (1887).

Camellia japonica, Lin. This plant was found to grow spontaneously in the woods of the island Poo to. There are specimens from 20 to 30 feet in height and with stems thick in proportion. The variety, however, was only the well known single red. (W. 173).

The district round Foo chow foo seems to be the great Camellia garden of China, and in no other part of the country did F. ever see these plants in such perfect health, or so beautifully cultivated. (W. 383).

When in April 1854, F. visited Shang hai he saw some good Camellias in bloom there. They are generally grown in pots under such shelter as mat-sheds and other buildings of a like kind can afford. Two of these varieties are particularly striking. Their flowers are of the most perfect form, and they have striped and self-coloured blossoms upon the same plant. Fortune sent them to Mr. Glendinning's nursery at Chiswick. (R. 242).

In Journ. Hort. Society, I (1846) 223, Fortune states that in May 1846, when he returned from China he brought with him living plants of *Camellia hexangularis* and a variety which he calls the *Star.* The C. hexangularis, remarkable for the regular disposition of its petals and the peculiar elegance of its flowers, till then was known only from a collection of Chinese drawings of plants in the possession of the Hort. Society. — In the Journ. Hort. Soc. III (1848) 237, Lindley writes: "Five of the varieties of Camellia received from Fortune, have now flowered in the Society's garden, without producing anything of value. That received in 1846 under the name of *hexangular* variety, so long and vainly sought for, all proved to be the old myrtle leaved sort (Camellia japonica myrtifolia introduced to the Kew Gardens before 1811); as did also that named the 'Star' by Fortune". — Bot. Cab. tab. 354 (1819).

In his Journ. Bot. IV, 1866, p. 1. Seemann gives a coloured plate of *C. japonica,* var. *variegata,* single flowers and pretty variegated leaves which, he says, had been introduced from China by Fortune.

In his article, Notes on some Chinese Plants recently introduced into England, Gard. Chron. 1860, 169, Fortune states that he introduced two beautiful varieties of Camellia, which he found cultivated in the gardens of Ning po and Shang hai. He named them: *Princess Frederick William* and *Cup of Beauty.* They are striped kinds; their flowers are very double, most perfect in form, and when the plants are a few years old both striped and self-coloured blossoms are produced on the same specimen. See also Gard. Chron. 1859, 948, 972.

Camellia reticulata, Lindl. (single flowered) was introduced from China in 1820 (v. s. p. 272, 282). The Bot. Mag. tab. 4976 (1857) figures a variety of this plant with large double flowers, which was from Standish & Noble, Bagshot nursery. It had been sent home, years ago by Fortune from China. — See also Fl. d. Serres, XII (1857) 185. — Illustr. Hort. 1861, tab. 305.

Yellow Camellia. In 1870, Fortune wrote in Journ. Hort. Soc., new ser. II, 46: "I well remember how in my instructions drawn up by the late Dr. Lindley, I was especially desired to look out for some extraordinary productions of China, as the blue Paeony, the yellow Camellia, the kumquat (Citrus japonica) and other remarkable plants". As to the yellow Camellia, he was not successful during his first expedition. It was in vain that he inquired about this variety. (W. 82). But in April 1850, when inspecting the nursery gardens 10 or 12 miles E. of Shang hai, he met in one of them with a yellow Camellia in bloom and bought it. It was certainly a most curious plant, although not very handsome and single-flowered. The flowers belong to the Anemone or Warratah class; the outer petals are of a French white, and the inner ones are of a primrose yellow. It appears to be a very distinct species in foliage. (T. 339).

Seemann, Syn. Camellia and Thea, Trans. Linn. Soc. XXII (1859) 352, states that in that year a fine specimen of Fortune's Yellow Camellia flowered at Kew. It proved to be a variety of *Camellia sasanqua,* Thb., of the Warratah or Anemone class. It flowers in autumn. (But Fortune saw it in bloom in April, at Shang hai).

The plant is figured as *Camellia sasanqua*, var. *anemoniflora*, in Bot. Mag. tab. 5152 (1859). The outer petals are white, the stamina, which almost all are changed into petals, are yellow.

Tea, Thea sinensis, Lin. Two tea-plants, considered to be distinct varieties, are met with in China, both of which have been imported into Europe. One, the Canton variety, is called *Thea Bohea;* the other, the northern variety is called *Thea viridis*. The former produces the inferior green and black teas which are made about Canton, and from the latter are made all the fine green teas in the great Hwuy chow tea country (in southern An hui) and in the adjoining province, Che kiang (Ning po, Chusan). Until a few years back it was generally supposed that the fine black teas of the Bohea Hills were also made from the Canton variety, and hence its name. Such, however, is not the case. F. has proved that the Woo-e-shan (Bohea) plant is closely allied to the Thea viridis and originally identical with that variety, but slightly altered by climate. The Woo-e plant shows less inclination to throw out branches than the Hwuy chow one, and its leaves are sometimes rather darker and more finely serrated. — The tea-plants of China have been common enough for many years in the gardens of England. In the Royal Bot. Garden at Kew they have been growing in the open air for some years. They are also to be met with in many other gardens, and almost in every nursery. They are pretty evergreen shrubs, and produce a profusion of single white flowers in the winter and spring, about the time that camellias are in bloom. — It was at one time commonly supposed that the two well-marked sorts of Tea, Black and Green were the produce of distinct species or varieties. But F. has proved, that the black and green teas of the northern districts of China (those districts in which the greater part of the teas for the foreign markets are made) are both produced from the same variety, the Thea viridis, or what is commonly called the green tea plant. On the other hand, those black and green teas which are manufactured in considerable quantities in the vicinity of Canton are obtained from the (erroneously so called) Thea Bohea or black tea plant. The difference between green and black tea is caused solely by the diverse methods of preparation. For the manufacture of Black Tea the freshly gathered leaves, freed from extraneous moisture by a short exposure in the open air, are thrown in small quantities at a time into iron roasting pans and exposed to gentle fire-heat for about 5 minutes, which renders them soft and pliant, and causes them to give out a large quantity of moisture. After this the rolling process on the rattan worked table begins. The leaves are repeatedly squeezed and rolled in the hands, to give them their twist or curl. After having been placed for some time in the open air, they are finally exposed in iron pans to a slow and steady fire-heat until completely dried, and the leaves have assumed their dark colour. The chief difference in the manufacture of genuine Green Tea consists in the leaves being exposed so long to the air after rolling that fermentation does not take place, and in not beeing subjected to such a high temperature in the final drying. The leaves are not allowed to become black but the firing is stopped after they have assumed a dullishgreen colour, which becomes brighter afterwards. Green tea thus retains far more of the peculiar oil and sap in the leaves than the black. — The greater part, if not the whole of the Green Tea consumed in Europe and America is coloured artificially by the Chinese to suit foreign trade. At Canton as well as in the Hwuy chow green-tea country this colouring is practised. For this purpose the Chinese there use Prussian blue and gypsum. The Chinese never use these dyed teas themselves. (W. 186—89, 201. — T. 93; 273—78, 283).

The mode of scenting tea is only understood and practised at Canton with teas destined for the foreign markets. The flowers of various plants are used for this purpose:

1. *Rose*, scented, *(Tsing moi qui-hwa)*.
2. *Plum*, double, *(Moi hwa)*.
3. *Jasminum Sambac*, *(Mo le hwa)*.
4. *Jasminum paniculatum*, *(Sieu hing hwa)*.
5. *Aglaia odorata*, *(Lan hwa* or *Yu chu lan)*.
6. *Olea fragrans*, *(Kwei hwa)*.
7. *Orange*, *(Chang hwa)*.
8. *Gardenia florida*, *(Pak shin hwa)*.

It has been frequently stated that the Chloranthus inconspicuus *(Chu lan)* is largely used. F. found it to be a mistake. (W. 213. — T. 197—206).

F. met with the tea-shrub in cultivation in China, from Canton in the south up to the 31st degree of n. lat., and Mr. Reeves says it is found in the province of Shan tung near the city of Tang chow foo, (?) in lat. 36° 30 north. The principal tea districts of China, however, and those which supply the greater portion of the teas exported to Europe and America, lie between the 25th and 31st degrees, and the best districts are those between 27° and 31°. The districts where Black Tea is manufactured are situated in the provinces of Fu kien and Kiang si, and where the Green is made, in Che kiang and An hui. (W. 213. — T. 197—206, 272).

The tea-shrub cultivated in various parts of the Canton province is

the Tea Bohea. The teas here produced are of a very inferior quality and
are scented for the foreign market. The best known are grown in and near
a place called Tai shan. (W. 187, 214. — R. 202, 203).

The finest Black Tea is manufactured at Woo-e-shan (Bohea), in the
province of Fu kien. That which comes from the mountains near Fu chou
fu and the An koy teas (Eastern Fu kien) are considered greatly inferior.
W. 219, 382).

The best Green Tea is manufactured in the vicinity of the Sung lo Hill
situated in southern An hui, district of Hieu ning (T. 86, seqq.). After this
follow the Green Teas of Che kiang, especially in the districts near Ning po
(W. 155). The green tea shrub is extensively cultivated in the islands of
Chusan and Kin tang. (W. 57. — T. 117).

The tea plant is multiplied by seeds. The seeds are ripe in October.
When gathered they are put into a basket, and mixed up with sand and
earth in a damp state, and in this condition they are kept until the spring.
In the month of March they are taken out and placed in the ground. When
the young plants are a year old they are transplanted. The first crop of
leaves is taken from these plants in the third year. When under cultivation
they rarely attain a greater height than 3 or 4 feet. (T. 93).

Hibiscus syriacus, L. var. *chinensis*, Lindley, in Journ. Hort. Soc. VIII
(1853). Raised from seeds presented to the Society by J. Reeves in 1844
(v. s. p. 265). Large violet flowers with a crimson eye. Fortune saw it in
its wild state on the hill sides on Poo to san and other islands. (In the
Herbarium of the Bot. Garden St. Petersburg there is a specimen of this
plant labelled: Fortune China borealis, n. 58, 1845).

Hibiscus syriacus, var. Chinensis, the Chinese Althaea frutex is figured
in Paxt. Lindl. Flow. Gard. III, 1853, 165, tab. 106 (colour.). Lindley
states: "The common Althaea frutex is said upon no very good authority to
be a native of Palaestina and even of Carniolia, but it does not appear to
have been known to the Greeks. Forskähl expressly states that in Egypt it
is a garden plant. It seems to be very common in Eastern Asia, but always
cultivated".

In I. F. S., I, 88, it is noticed as wild and cultivated in China. See
also Lemaire, Jard. fleur. IV, t. 370 (colour).

Cotton. The vast level plain in which Shang hai is situated, is a deep
rich loam, and is without doubt the finest in China, if not in the world. The
land is exceedingly fertile and admirably adopted for Chinese cotton cul-

tivation, and consequently cotton is the staple summer production of the district. Both the white kind, and that called the "yellow cotton", from which the yellow Nan king cloth is made, are here cultivated. From this country large quantities of Nan king cotton are generally sent in junks to the north and south of China, as well as to the neighbouring islands. (W. 115, 233. — R. 333).

The Chinese or Nan king cotton plant is the *Gossypium herbaceum*, of botanists *(G. religiosum*, Roxb.), the *mie* (mien) *wha* of the Chinese. It is a branching annual, growing from one to 3 or 4 feet in height, according to the richness of the soil, and flowering from August to October. The flowers are of a dingy yellow colour, and, like the Hibiscus and Malva, which belong to the same tribe, remain expanded only for a few hours, in which time they perform the part allotted to them by nature, and then shrivel up and soon decay. At this stage the seed pod begins to swell rapidly, and when ripe, the outer coating bursts and exposes the pure white cotton in which the seeds lie imbedded.

The *Yellow Cotton,* from which the beautiful Nan king cloth is manufactured, is called *tze mie* (mien) *wha* by the Chinese, and differs but slightly in its structure and general appearance from the kind just noticed. F. often compared them in the cotton fields where they were growing, and although the yellow variety had a more stunted habit than the other, it had no characters which constitute a distinct species. The seeds frequently yield the white variety, and *vice versa.*

F. then gives a detailed account of the mode of cultivation, gathering etc. of cotton in China. The manure applied to the cotton lands is obtained from the canals, ponds and ditches which intersect the country in every direction, and consists of mud which has been formed by the decay of long grass, reeds and succulent water plants. In the end of April and beginning of May the sowing of the cotton seeds commences. The cotton plant produces its flowers in succession from August to the end of October. The pods are bursting every day and are gathered with great regularity. (W. 277—278).

Sterculia from North China (probably *St. plantanifolia)* introduced by Fortune, offered for sale. See Gard. Chron. 1863, 1040. In the Herbarium of the Bot. Gard. St. Petersb. there is a specimen of St. platanifolia labelled Fortune, China borealis, 1845.

Corchorus. A plant well known by the name of *Jute* in India — a species of *Corchorus* — is grown to a large extent in the plain of Ning po. In China this fibre is used in the manufacture of sacks and bays for holding rice and other grains. (R. 259).

Amongst the plants cultivated in the plain of Tien tsin for the sake of their fibre the Jute *(Corchorus* sp.) is the most important. It grows to a great height. (Y. 338, 340) *).

Linum stelleroides, Planchon, in Hook. Journ. Bot. (1848) 178, described. Hab. in China, Herb. Hooker, verosimiliter e collectione Fortune.— I. F. S., i, 95, Fortune, China.

Averrhoa Carambola, the Chinese gooseberry, cultivated at Hongkong, succeeds well. (T. 7). — I. F. S., i, 100.

Skimmia. In the spring of 1850, F. observed in a garden near Shang hai a fine new shrub, which he mistook for a holly. It turned out to be a species of *Skimmia* and was sent to England. F. observes that Lindley described it as *S. japonica*, but as F. found that it is quite a different plant from that known in English gardens under the above name, he proposes to call it *S. Reevesiana* in compliment to John Reeves, who has introduced many Chinese plants and who had been of great service to F. while in China **). It produces a profusion of whitish flowers, deliciously scented, and afterwards becomes covered during the winter with bunches of red berries like our common holly. Its glossy evergreen leaves and neat habit add greatly to its beauty. The Chinese call it the *Wang shan kwei*, and it is said to have been discovered on the celebrated mountain Wang shan in the district of Hwuy chow. (T. 329).

Skimmia japonica, Thunberg, is figured in Sieb. & Zucc. Fl. Japon. I (1839) 125, tab. 68, and in Bot. Mag. t. 4719 (1853), in the latter from specimens which flowered in March 1853, in Standish and Noble's nursery, where it had been introduced by Fortune from China. — Masters in Gard. Chron. 1889, I, 520, describes this plant, figured in the Botan. Mag. t. 4719, and which had been sent from China by Fortune under the name

---

*) It seems to me that the Jute plant observed by Fortune near Tien tsien was not Corchorus but *Abutilon Avicennae*, Gaert. *(Sida tiliaefolia*, Fischer), from which most of the Chinese Jute is derived. It is extensively cultivated at Peking. Dr. Cl. Abel (v. supra p. 228) notices this plant extremely cultivated on the banks of the Pe ho, between Tien tsin and Peking. Corchorus is not seen in this part of China. In the Kew Herbarium there is a specimen of A. Avicennae labelled Fortune, Che kiang. — I. F. S., i, 86.

**) Reeves Jun. V. s. 263.

of Ilex, as a new species, *Skimmia Fortunei*, distinguished from S. japonica by its dwarf habit, lanceolate leaves of a very dark green colour, and its invariably hermaphrodite flowers. Berries obovate, dark crimson. Seeds pointed at both ends. — Neither S. Fortuni nor S. Reevesii are noticed in the Index Kewensis. But in Kew Bull. 1897, App. I, 40, *Skimmia japonica*, Mast. figures in the list of trees cultivated at Kew.

Murraya exotica, Lin. and *Cookia punctata*, Sonn, cultivated in Hong kong, near Canton and near Foo chow. (W. 136, 142, 145, 384, 385. — T. 7).

The *Mandaria Orange* is much grown at the Fa tee Gardens near Canton, where the plants are kept in a dwarf state, and flower and fruit most profusely, producing large, flat, dark, red-skinned fruit.

Groves of *Orange* trees common near Chu chu foo (Kü chou fu) in Che kiang. (T. 177).

The *Kum-quat* *), *Citrus japonica*. The Chinese have a great variety of plants belonging to the orange tribe; and of one, which they call the *kum quat* — they make a most excellent preserve. This is a small species of Citrus, about the size of an oval gooseberry, with a sweet rind and sharp acid pulp. This fruit is well known in a preserved state by those who have any intercourse with Canton, and a small quantity is generally sent home as presents every year. Preserved in sugar it is excellent. In the island of Chusan groves of the kumquat are common on all the hill-sides. The bush grows from 3 to 6 feet high; and when covered with its orange-coloured fruit, is a very pretty object. (W. 58, 142).

In January 1849 F., when visiting Shang hai, saw the kumquat extensively grown there in pots. The trees were literally covered with their small orange-coloured fruit. It would be highly prized in England for decorative purposes during the winter months. It is much more hardy than any other of its tribe; it produces its flowers and fruit in great abundance, and would doubtless prove a plant of easy cultivation. In China it is grafted. (T. 122).

Living specimens of the kum quat were brought to England, for the Hort. Society when F. returned to England in May 1846. See Journ. Hort.

---

*) This name means «gold-orange», in the Mandarin dialect *Kin kü*.

Soc. I (1846) 223. Its cultivation was successful. It was figured in Journ. Hort. Soc. III (1848) 239, *Citrus japonica,* Thbg., where Fortune gives a notice of the plant. It is much cultivated in the island of Chusan, where good oranges, such as those known in the south as Mandarins and Coolies, are entirely unknown. In China the kum quat is propagated by grafting on a prickly wild species of Citrus, which seems of a more hardy nature than the kumquat itself *(Citrus trifoliata,* as F. says in Gard. Chron., 1874, II, 111, a plant which he likewise introduced to England). — In another earlier article on the kumquat, Journ. Hort. Soc. new ser. II (1870) 46, F. calls the plant on which the Chinese are accustomed to graft the kumquat, *Limonia trifoliata.*

I may observe that *Citrus trifoliata,* Lin. and *Limonia trifoliata,* Lin. are quite distinct plants, but have frequently been confounded. C. trifoliata is the same as *Aegle sepiaria,* DC., whilst Limonia trifoliata L. or Triphasia trifoliata, DC., is *Triphasia aurantiola,* Lour. (I. F. S., I, 111, 109). It seems therefore doubtful which of these plants was meant by Fortune. Sir J. Hooker in Bot. Mag. tab. 6128 (1874) says that the kumquat is grafted on Limonia trifoliata, but according to Mr. Nicholson, Dict. of Gard., Citrus trifoliata is used for this purpose. I think the latter is right. See also Gard. Chron. 1859, 508: Limonia trifoliata introduced by Fortune.

The kumquat, when cultivated in the south of China, does not succeed. — Comp. also Botan. Mag. tab. 6128 (1874), *Citrus aurantium,* var. *japonica.*

Pumeloes (Citrus decumana) — cultivated in the Min valley near Fu chou fu; F. in the summer of 1845. I may observe that Amoy is famed for its pumeloes.

The most striking plant in autumn or winter in the Fa-te Gardens near Canton is the curious *Fingered Citron,* which the Chinese gather and place in their dwellings or on their altars. — This plant is cultivated in great perfection in the nurseries near Foo chow foo. Fortune sent it to the Hort. Society in the beginning of 1844. (W. 142, 383. — Journ. Hort. Soc. I (1846) 215).

At Foo chow, in 1845, F. saw, for the first time, the tree commonly called the *Chinese Olive,* from the resemblance its fruit bears to the olive of Europe. (W. 384).

Europeans in China apply the above name to *Canarium album,* Raeusch and *C. Pimela,* Konig. There is a herbarium specimen of the latter in the Kew Herbarium. — I. F. S., ɪ, 113.

The sweet scented *Aglaja odorata,* cultivated in the Fa te Gardens near Canton and in the vicinity of Foo chow foo. The flowers are used for scenting tea and also for the mixing with and perfuming tobacco. (W. 142, 213, 385. — T. 146. — R. 201).

Ilex. Fortune introduced several new hollies from China, but in his books this name occurs only twice:

In enumerating the plants which he met with on his way to the Sung lo tea district, near the border of Hwuy chow prefecture, in Nov. 1848, he says: "The most interesting plant of all was a new evergreen Holly, with leaves somewhat like the Portugal laurel, very handsome and ornamental". F. procured seeds of this and sent them home to England. (T. 79). — Farther on, T. 88, Sung lo shan, we read: "A species of Holly not unlike the English is common here". — Besides these he notices in his "Notes of a Traveller in China", Gard. Chron. 1851, 5, a new species of Holly which he found in a Chinese nursery, 5 miles north of Shang hai, in April 1850. He named it *Ilex Reevesiana.* It is a dwarf shrub with entire, elliptical, acute, slightly wavy, dull green leaves, covered with dots, something like Elaeagnus. Flowers in short spikes, terminal. Fruit large, deep red. Very ornamental. It flowers in winter, and ripens its fruit during the following autumn, like our common English holly. The Chinese call it *Wang san qui wha.* It is said to be brought from Wang san, a celebrated mountain in Hwuy chow prefecture (v. s. p. 441). — Maximowicz, de Ilice (1881) 52, thinks that Fortune's I. Reevesiana is identical with *Ilex Fortunei,* Lindley, described in Gard. Chron. 1857, 868. "A handsome evergreen holly raised from seeds collected by Fortune at Hwuy chow". Glendinning's nursery of Turnham Green, 1853. In its young state it is much like I. cornuta, but in the adult condition it acquires quite another appearance, resembling a very broad-leaved, entire-leaved European holly. Flowers unknown, the specimen before Lindley being only in fruit. — I. F. S., ɪ, 116.

Ilex cornuta, Lindley in Paxt. Flow. Gard., I (1850) p. 43; short diagnosis, fig. 27 (uncoloured). F. found it in the neighbourhood of Shang hai and introduced it in 1849 to Standish & Co's nursery. A fine, hardy, evergreen holly. Leaves almost always furnished with three strong spines at the end, but when the plant is young there are added one or two more on each side. In the old plant the latter wholly disappear, while the endspines will occasionally extend, turn up their edges and assume the appearance of strong horns. When Fortune was in China in the service of the Hort. Society, he found this plant in flower, in April (1844), near Shang hai. During his last visit to that country, he again met with it, and in fruit, large berries, at a place called Kin tang (Kin tang or Silver Island, near Chusan, visited by F. end of 1848). It is a very remarkable plant. — Comp. also: Fl. d. Serres, VII (1852) 216 (uncoloured figure); ibidem IX (1853) 99, tab. 895 (colour.). — Ilustr. Hort. 1854, tab. 10. — I. F. S., I, 115: Staunton earlier.

Ilex leptacantha, Lindley in Paxt. Fl. Gard. III (1852) 72. Short diagnosis, no figure. A handsome evergreen shrub. Flowers and fruit unknown. Introduced from North China by Fortune. — I. F. S., I, 117.

Ilex microcarpa, Lindley, in Paxt. Fl. Gard. I (1850) 43, fig. 28, (uncoloured). Short diagnosis. Ovate, entire, acuminate leaves. Fortune introduced it from North China to the nursery of Standish & Co. He discovered it at Tein tung (near Ning po). — Maximowicz, l. c. 36, reduces this to *Ilex rotunda,* Thbg. — I. F. S., I, 118.

Ilex Terago, is noticed in Gard. Chron. 1859, 508, 948, 972, among Fortune's plants offered for sale by Stevens, Glendinning and others. — Comp. Linnaea XXVI (1853) 749. "*Ilex Tarayo,* hort. Angl., culta 1852, in hto bot. Vratislavensi. Patria ignota". — I may observe that *Terago,* is the Japanese name of *Ilex latifolia,* Thbg. which hitherto has not been recorded from China, although F. seems to have introduced it from that country. — Bot. Mag. tab. 5597 (1866).

Euonymus radicans, Sieb. var. *variegata,* mentioned in Gard. Chron. 1863, 1016 offered for sale amongst Fortune's plants. — Most probably F. introduced it from Japan, which country he visited in 1860 and 1861.

Elaeodendron (?) *Fortunei,* Turczaninov, in Bull. Mosc. 1863, I, 603. "China media, Fortune 1845, A. 46". — Maxim. Mél. XI (1881) 205, observes, that this is most likely an *Evonymus,* but the fruit is unknown. — I. F. S., I, 124.

At Foo chow in 1845 F. saw the *Chinese Date,* which produces a fruit not unlike the date imported into England. (W. 384). — He saw the fruit of *Zizyphus vulgaris*, Lam., Jujube.

Tien tsin 1861. The so called dates are the fruit of a Rhamnus not of a palm. (Y. 313). — Excellent large Jujubes are brought to Tien tsin and Peking from Shan tung.

The Mountains west of Peking abound in a species of stunted *Rhamnus, R. zizyphus* (?) (Y. 384). — This is *Z. vulgaris*, var. *spinosa*, Bunge.

A kind of *Rhamnus* is largely cultivated some miles to the westward of Hang chow foo. F. visited these fields in 1853. The Chinese called this plant *loh-zah* or *soh-loh-shoo*, and showed samples of cloth which had been dyed of a beautiful green with a dye-stuff obtained from this shrub. F. was told that two kinds were necessary — namely, the variety cultivated in these fields, and one which grew wild on the hills — in order to produce the dye in question. The former the farmers called the *yellow* kind, and the latter the *white* kind. The dye itself was not extracted by them, they were merely the growers. F. secured a good supply of plants and seeds of both kinds, which were afterwards sent to the Agricult. Society of India and England. — Fortune's further inquiries on the subject of the manufacture of this "green Indigo" were conducted in connection with **Dr. Lockhart,** and the **Rev. J. Edkins,** of Shang hai. They found that a considerable portion of this dye was made near the city of Kia hing foo, situated a few miles west (?) from Shang hai, and Edkins procured a bundle of chips there which exhibited the state in which the article is sold in the market. F. publishes the information procured by Edkins*):

"The bark of two kinds of the tree known as the "green shrub" *(luk chae)*, one wild, which is called the *white*, and another cultivated which is called the *yellow*, are used to obtain the dye. The white bark tree grows abundantly in the neighbourhoods of Kea hing and Ning po; the yellow is produced at Tsah kou pang, where the dye is manufactured. This place is 2 or 3 miles west from Wang tseen, a market-town situated a little to the south of Kea hing. The two kinds are placed together in iron pans and thoroughly boiled. With the residuum cotton cloth prepared with lime is dyed several times. Then the colouring matter is washed from the cloth with water and again boiled. Finally it is sprinkled on thin paper. When

---

*) **Edkins, Rev. Joseph,** D. D. This distinguished sinologist came out to China in 1848, as a missionary of the London Mission. He is now living at Shang hai, still engaged in sinological researches. His account of the Chinese Green Dye was originally communicated by Dr. W. Lockhart of Shang hai to the late D. Hanbury, in the early part of 1856, who published it in the same year in the Pharmac. Journ. See Hanb. Science Papers, 125.

half dry the paper is pasted on light screens and strongly exposed to the sun. The product is called *luk kaou*. In dyeing cotton cloth with it 10 parts are mixed with 3 of subcarbonate of potash in boiling water. All cotton fabrics, also grass cloths, take the colour readily, but it is not used to dye silk fabrics for it is only a rough surface which takes it readily. The dye does not fade with washing, which gives it a superiority over other greens. It is sent from Kea hing as far as Shan tung. It is also made in the province of Hoo nan and at Ning po. It has long been used by painters in water-colours, but the application of it to dye cloth was first made only about twenty years ago. (R. 166—170).

The two species of Rhamnus, from the bark of which the beautiful green dye (vert de Chine of the French) is prepared in China were described and figured as *Rh. chlorophorus* and *Rh. utilis* by Decaisne, in Comptes rend, Acad. Sc. XLIV, 1857, from herbarium specimens sent in 1854, 1855, by Montigny and Hélot. Fortune introduced the living plants. — Hemsley, I. F. S., I, 129, 130, reduces Rh. chlorophorus to *Rh. tinctorius,* Waldst. et Kit., and Rh. utilis to *Rh. davuricus,* Pall.

Grapes are plentiful at Tien tsin and may be had in perfection all the year round. On the approach of winter the stems of the vines are taken down from the trellis-work, and buried in the earth at a depth sufficient to protect them from the frost. Here they remain in safety during the winter, and are disinterred in spring. The vines seemed to be nearly all of one and the same variety, and produced large berries of a greenish colour, getting darker as they ripened, and covered with bloom. (Y. 321, 324, 325).

Foo chow, July 1845. The *lee chee* trees *(Nephelium Litchi,* Camb.) were covered with their fine red fruit, and were very beautiful, the fruit contrasting so well with the deep clear green foliage. (W. 384).

Min valley near Foo chow, 1849. Groves of *leechee* and *longan (Nephelium Longana,* Camb.), peach and plum trees are seen over all the plain.

Above Yen chow foo, Che kiang, November 1848. A species of *Maple* called by the Chinese the *fung gze,* the leaves of which had changed into a dark blood-red colour, was most picturesque. (T. 61).

Silver Island, December 1848. A species of maple, clothed in its autumnal hues; leaves of a clear blood-red colour. (T. 117).

Sabia japonica, Maximowicz, Mél. VI (1867) 202, and Fl. As. Fragm. (1879) 7. Maximowicz discovered this plant, in 1862, near Nagasaki. But Fortune gathered it much earlier in China. — I. F. S., i, 143.

The Agricult. and Hort. Society of India requested Fortune, when he performed his third expedition to China, to procure seeds of the Chinese *Varnish* tree. F. succeeded in introducing it into India, apparently from Ning po. The tree which yields the Chinese varnish is a species of *Rhus,* which, although producing an article of great value, is extremely danger-ous. The varnish is largely used in the country for giving a fine polish to tables and chairs used in the houses of the wealthy. The lacquer used in the manufacture of the beautiful lacquer-ware so extensively exported from Canton to foreign countries, and so well known and justly admired, is produced by this tree. It has the valuable property of being less liable than French-polish to be injured by a heated vessel which may be placed upon it; but it is very poisonous, and requires to be handled with great care by the workmen who use it. Indeed, after furniture is dry, it is very unsafe for certain constitutions until it has been in use for some time, and the smell entirely gone. A friend of F., the American Consul at Foo chow, Mr. Jones, used some furniture and was very ill for a long time from its effects. (R. 146, 148). — I. F. S., i, 148, *Rhus vernicifera,* DC.

On the hill-sides near Nagasaki, F. observed the Japan Waxtree, *Rhus succedaneum,* cultivated extensively. It occupies the same position on these hills as the Chinese Tallow-tree, *Stillingia sebifera,* does in Che kiang. It grows to about the same size, and, curiously enough, it produces the same effect upon the autumnal landscape by its leaves changing from green into a deep bloodred colour as they ripen, before falling off. (Y. 19).— I. F. S., i, 147.

Rhus microlepis, Turczan. in Bull. Soc. Nat. Mosc. 1858, 468. For-tune 97, A, China. — In I. F. S., i, 146, referred to *Rhus hypoleuca,* Champ., Hong kong. — Fortune evidently gathered it earlier.

The Mango, *Mangifera indica,* cultivated in Hong kong and near Canton. (W. 12, 136).

In the island of Chusan, and all over the rice country of Che kiang and Keang soo, there are two plants cultivated in the wintermonths, almost exclusively for manure, the one is a species of *Coronilla*, the other is *Trefoil* or Clover. Large ridges are thrown up on the wet rice fields in the autumn, and the seeds of the plants are dropt in. In a few days germination commences and long before the winter is past the tops of the ridges are covered with luxuriant herbage. In April, when the ground for the rice is prepared, the ridges are levelled, and the manure plants are scattered over the ground. The fields are flooded, plough and harrow are employed. The manure plants half buried in mud and water, commence to decay and give out a most disagreeable putrid smell. The young paddy doubtles derives strong nourishment from the ammonia given out by the fresh manure. — The large fresh leaves of the trefoil are also picked and used as a vegetable by the natives. (W. 54, 90, 311).

From seeds sent by Fortune to the Hort. Soc. his trefoil was raised and proved to be a fine broad-leaved variety of *Medicago denticulata,* Willd. See Gard. Chron. 1844, 815. — No Coronilla has hitherto been recorded for the Flora of China.

In the southern provinces of China a great deal of *Indigo (Indigofera)* is grown and manufactured, in addition to a large quantity which is annually imported from Manilla and the Straits. In the north, however, the plant which we call indigo is never met with, owing to the coldness of the winters. Its place is supplied by the *Isatis indigotica,* (v. supra p. 433). (W. 307). — I. F. S., ɪ, 157, *I. tinctoria,* Lin. Amoy, Fortune, A. 56.

In April 1850. In a garden near Shang hai F. saw the beautiful *Indigofera decora,* introduced by him to the Hort. Society between 1843 and 1846. Cultivated in pots. (T. 317).

I. decora, Lindley in Journ. Hort. Soc. I (1846) 8, and Bot. Reg. 1846, tab. 22. A very pretty bush, received from Fortune, 1st May 1845, who found it cultivated in the nursery gardens near Shang hai. Leaves pinnate in from 2 to 5 pairs and an odd one. Leaflets ovate. Flowers grow from the axils of the leaves in horizontal racemes. Very handsome light pink flowers. — Paxt. Mag. Bot. XVI (1849) 290, colour plate. — Bot. Mag. t. 5063 (1858). — I. F. S., ɪ, 156. Amoy, Fortune, A. 60.

Wistaria (Glycine) chinensis. — In the summer of 1844, F. met for the first time, in the island of Chusan, the beautiful *Glycine sinensis,* wild on the hills, where it climbs among the hedges and on trees, and its flowering branches hang in graceful festoons by the sides of the narrow roads

which lead over the mountains, or it dips its leaves and flowers in the
canals and mountain streams. In this place and all over the provinces of
Che kiang and Kiang soo, the Glycine seems to be at home. It grows wild
on every hill side. (W. 55. — T. 337).

Shang hai, September 1848. In the garden of the English Consul,
Mr. Alcock, there is a noble plant of the Glycine sinensis, which flowers
most profusely, and becomes covered with its long legumes, or pea-like
fruit, which ripen to perfection. (T. 15).

The Glycine or Wistaria chinensis has long been known in Europe,
and there are large trees of it on many of our house and garden walls. It
was introduced into England from a garden near Canton belonging to a
Chinese merchant named Consequa (v. s. p. 259). (J. Reeves sent it to the
Hort. Soc. in-1818). But it is not indigenous to the south of China, and
is rarely seen in perfection there. It is perfectly hardy in England. F.
observed a beautiful specimen of this plant in an abandoned garden in the
island of Koo-lung-su, near Amoy. But is was not found in a wild state
even at Amoy, and had evidently been brought from more northern latitudes. In the island of Chusan it grows most luxuriantly. By far the most
beautiful effect is produced when the Glycine attaches itself to the stems
and branches of other trees. One can scarcely imagine anything more gorgeous or beautiful than a large plant of this kind in full bloom. Its main
and larger branches are entwined round every branch and branchlet of the
tree, and from them hundreds of small ones hang down until they nearly
touch the ground. The whole of the branches are covered with flower-buds,
which a day or two of warm weather brings rapidly forward into bloom.
These fine long racemes of lilac flowers are produced before the plant puts
forth its leaves. There are some large specimens of the Glycine on the
island of Chusan. One in particular, was most striking. Not content with
monopolising one tree, it had scrambled over a whole clump, and formed
a pretty arbour underneath. When in full flower it has a charming appearance. The Chinese are fond of growing the Glycine on trellis-work, and
forming long covered walks in the garden, or arbours and porticos in front
of their doors. (T. 335—38).

In a garden near Shang hai F. saw a Wistaria chinensis in a dwarfed
state, growing in a pot. The tree was evidently aged, from the size of its
stem. It was about 6 feet high, the branches came out from the stem in
a regular and symmetrical manner. Every one of these branches was loaded
with long racemes of pendulous lilac blossoms. (T. 335).

Soo chow foo, 1844. Fortune observed a *white Glycine*. (W. 263).

The same variety was seen in a garden near Shang hai and in the garden of a mandarin at Ning po. In foliage and general habit it is like the typical form, but it bears long racemes of pure white flowers. (T. 317, 338).

Glycine sinensis is also a great favourite with the Japanese. (Y. 190).

Fortune brought living specimens of the white-flowered Glycine with him when he returned to England. It was cultivated in the garden of the Hort. Society. See Journ. Hort. Soc. I (1846) 223. — IV (1849) 221. — Illustr. Hort. 1858, tab. 166. — I. F. S., ɪ, 161.

Earth-nuts, *Arachis hypogaea*, a staple summer production in the hilly country of Che kiang. Earth-nuts are also grown most extensively during the summer season, in the southern provinces, more particularly in Fo kien. (W. 308, 309, 381).

The very sandy soil near the river (between Yen chow foo in Che kiang and the border of An hui) yields good crops of the ground-nut. It is also plentiful in the light sandy soil near Chu-chu-foo (Kü chou fu, Che kiang). (T. 84, 177).

Lespedeza ciliata, Bentham in Hook. Kew Journ. Bot. IV (1852) 48, in nota; Maxim. Acta Hti Petrop. II (1873) 351. Fortune Chin chew (1843) n. 31, Chusan n. 42. — I. F. S., ɪ, 180.

Soy bean, Soja hispida, much cultivated between Long yeou and Chu chu foo (Kü chou fu) in Che kiang; also in the Peking plain. (T. 177. — Y. 350).

Apios Fortunei, Max. Mél. IX (1873) 67. Fortune, A. 44. 1845. — I. F. S., ɪ, 189.

Sophora japonica, Lin. Common at Tien tsin and in the Peking plain. It yields a yellow dye called by the Chinese *whi (huai) hwa.* (Y. 349, 354).

Shang hai, 1853. F. saw pretty specimens of *S. japonica pendula,* grafted high, as we see the weeping-ash in England. (R. 139).

Che kiang, autumn of 1853. The *Soap-bean* tree, a kind of *Caesalpinia,* the fleshy pods of which, in all parts of China, are largely used as soap and may be bought in every market-town, was introduced by F. to the Agricult. and Hort. Society of India. (R. 146, 148). — I may observe that at Peking the large pods of *Gleditschia sinensis,* Lam. are used as soap, but the tree noticed by F. was perhaps *Gymnocladus chinensis*, Baillon, the fat black seeds of which are used by women in washing their head and hair, and which grows in Che kiang. — I. F. S., I, 207, 209.

Amongst the spring flowers observed in the gardens near Shang hai, in 1854, F. notices the pink *Judas-tree*, covered with blossoms. (R. 241).— F. evidently saw *Cercis chinensis,* Bunge. — I. F. S., I, 213. — The *C. japonica*, Siebold mss. hort. figured· by Planchon in Fl. d. Serres VIII (1852) 269, and introduced from Japan by Siebold, some years earlier, is, according to Index Kew = *C. chinensis.*

The *Peaches* at Foo chow foo are curiously shaped, but worthless. (W. 384. — T. 146). — Peaches are cultivated at Hang chow foo in Che kiang. (T. 30).

Amongst the more important of the acquisitions F. made in the vicinity of Shang hai, in the summer of 1844, was a fine and large variety of *Peach*, which comes into the markets there about the middle of August, and remains in perfection for about ten days. It is grown in the peach orchards, a few miles to the south of the city; and it is quite a usual thing to see peaches of this variety 11 inches in circumference and 12 ounces in weight. This is, probably, what some writers call the Peking peach, about which such exaggerated stories have been told. (W. 407). — The Peking peaches are indeed of very large size, and delicious.

This Shang hai Peach F. forwarded to the Hort. Society in the autumn of 1844. Along with a plant of it in a pot, he sent some of the peach stones. They were sown and came up abundantly. They flowered and produced fruit in 1852. Flowers large, petals deeply coloured. Fruit very large, 10 inches and more in circumference, roundish, very handsome, pale yellow where shaded and delicate, crimson red next the sun. Flesh pale yellow, next the skin, but very deep red at the stone. Juicy and rich, Lindley. See Journ. Hort. Soc. I (1846) 221, and VII (1852) 265.

Peach-flowers. During his first expedition to China, 1843—45, F. sent home to the Hort. Society beautiful double flowered varieties, which flowered in their Garden and proved to be a great acquisition.

Journ. Hort. Soc. II (1847) 311. *Double white Peach,* received from Fortune, from the north of China. Flowers white, full semi-double.

Ibid. III (1848) 246, 313. *Double crimson Peach*, black and white figure, young fruit. Received from Fortune. This plant has presented the peculiarity of producing very generally more than one fruit to each flower, where out of three flowers, one has two young peaches and another three. A very handsome plant. The flowers are dark crimson.

April 1850, Shang hai gardens. Beautiful peach-trees with double flowers. F. says that two of these have already been described by Dr. Lindley in the Journ. Hort. Soc., and named the *double white* and the *double crimson* peaches. But, fine as these undoubtedly are, there is a third far more beautiful and striking than either of them. This produces large double white flowers, which are striped with red or crimson lines like a carnation. A tree of this variety in full bloom is one of the most beautiful objects that can be imagined. Sometimes the branches "sport" and produce self-coloured flowers — the colours being in this case, either white or crimson. This fine tree then was safe in England. These double peach-trees seem to be particularly well adopted for forcing, as they form their flowerbuds fully in autumn, and are ready to burst into bloom with the first warm days in spring. A little artificial heat, therefore, will bring them into full flower about the new year, or any time from that period up to March. As spring flowers they are highly prized by the Chinese. Itinerant gardeners carry them about the streets for sale in the northern Chinese towns. The flower-buds are then just beginning to expand. In a day or two the little tree, having been put in a pot, is one mass of bloom. In this state all the three varieties are very beautiful, but the carnation-striped one is the handsomest of them all. (T. 319).

Shang hai, April 1854. The double-blossomed peaches, of which there are several very distinct varieties now in England, are perhaps the gayest of all things which flower in early spring. Fancy trees fully as large as our almond, literally loaded with rich-coloured blossoms, nearly as large and double as roses, and you will have some idea of the effect produced by these fine trees in this part of the world. (R. 139, 242).

In the Gard. Chron. 1860, 170, F. notices again the double flowered peaches introduced by him from China, and especially two varieties the *Camellia flowered* and the *Carnation flowered,* which he seems to have sent in 1854. Very remarkable trees, perfectly hardy. The carnation flowered peach has striped blooms resembling the carnation and, like the camellia flowered variety produces striped and selfcoloured flowers on the same tree. They are propagated by grafting. Small plants produce blossoms freely as well as large full grown trees.

Gard. Chron. 1859, 948, 972: Glendinning and Stevens offer for sale Fortune's camellia flowered, carnation flowered and rose flowered peaches.

The various varieties of Chinese peach flowers are found represented in the following horticultural periodicals:

Paxt. Flow. Gard. I (1850) 65: Large semidouble, crimson, rose and red — small semidouble red and white, altogether 5 varieties.

Lemaire, Jard. fleur. IV (1854) tab. 328. *Amygdalus persica,* var. *sinensis.*

Fl. d. Serres, X (1854) tab. 969, p. 1. Planchon: *Amygd. pers.* var. *sinensis:* a, *flore simplici albo* and b, *flore simpl. rubro.* Fortune 1843—48.

Ibid. XIII (1858) p. 17: *Amygd. pers.* var. *sinensis:* var. *camelliaeflora* and *dianthiflora,* both varieties introduced by Fortune to Glendinning.

Belg. Hort. VIII (1858) p. 97: *Amygd. pers. sin.,* var. *camelliaeflora.*

Illustr. Hort. 1858, tab. 165, and 1859, tab. 205. The same, cult. by Glendinning.

Revue Hort. 1863, 391, Carrière represents 4 new varieties of Chinese flowered peach trees: *rosaeflora, versicolor, caryophylliflora, camelliaeflora.*

Amongst fruits produced at Foo chow the *Plums* are good, but inferior to English plums. (W. 384). Groves of plum trees are seen over all the plain in the Min valey, near Foo chow. (T. 146).

Plums cultivated near Hang chow foo. (T. 30) and near T'ai hu Lake. (R. 340).

The pretty little *Prunus sinensis alba,* double flowered, seen by F. at Shang hai in flower, in January 1849. (T. 121). He had brought living specimens of it along with him when he returned from China in May 1846. Lindley in Journ. Hort. Soc., II (1847) 158, notices this plant, cultivated in the Society's Gardens, as *Cerasus japonica,* with double white flowers. A very nice plant, growing freely in any good sandy loam. It differs in no respect from the well known double dwarf Chinese Cherry or Plum, except in having perfectly white and very double flowers.

I. F. S., I, 219, sub *Prunus (Cerasus) japonica,* Thbg.

Prunus triloba, Lindley, Gard. Chron. 1857, 268, sent home from China by Fortune, flowered in the nursery of Mr. Glendinning at Chiswick. Most beautiful as a garden plant on account of its delicate, semi-double light rose coloured flowers, nearly $1\frac{1}{2}$ inch in diameter. Like Pr. trichocarpa it has the woolly ovary of the peach. Frequent tendency of the leaves to be wedge-shaped or three lobed.

F. seems to have introduced this plant during his third expedition to China, 1853—1855. In the Gard. Chron. 1860, 170, he states that *P. triloba*, is a very fine bush or dwarf tree, said to come from the province of Shan tung. It produces a profusion of semi-double, rose-coloured flowers, early in spring. — It is easily increased by budding and grafting, and no doubt perfectly hardy in England. — In 1861 he saw it cultivated at Tien tsin. (Y. 30). — This plant is much cultivated at Peking, where it was first discovered by Bunge (v. s. 332).

Carrière in Rev. Hort. 1862, 91, describes this plant as the type of a new genus, *Amygdalopsis Lindleyi*, and figures the hairy fruit. — Under the same name it is figured in Fl. d. Serres, XV (1862) p. 63 by Carrière. The latter states that Fortune introduced it into England in 1856. It came to France in 1859. — Illustr. Hort. 1861, t. 308. — I. F. S., i, 222.

Spiraea prunifolia, from Japan, was first described and figured by Sieb. & Zucc. in their Flora Jap. I (1840) 131, t. 70. — In the autumn of 1844 F. sent a double-flowered variety of this plant from China. Lindley in Journ. Hort. Soc. II (1847) 307, says that *Sp. prunifolia flore pleno*, received from Fortune, from North China, in 1845, is a very valuable ac-quisition. A bush with deciduous leaves and long, slender branches. Leaves oval, finely serrated, about $1\frac{1}{2}$ inch long. Flowers pure white, in clusters or 3 or 4 from the centre of the buds, and are agreeably relieved by 5 or 6 small green leaves that appear at the same time. The flowers are rather more than $\frac{1}{4}$ inch in diam., very regularly double and very ornamental.

Fortune procured the Sp. prunifolia probably from Shang hai. When in the spring of 1854 he visited this place he saw in the gardens there the pretty daisy-like Sp. prunifolia then covered with blossoms. (R. 241). — I. F. S., i, 226.

In the Journ. Hort. Soc. II (1847) 157, and in his Bot. Reg. 1847, t. 38, Lindley describes and figures a dwarf shrub which Fortune in 1843 had sent from Chusan and which Lindley believed to be *Sp. pubescens*, Turczan. It flowered in the garden of the Society. It has little hemisphaeri-cal umbels of small pure white, flowers, having a faint scent. Leaves $1\frac{1}{2}$ inch long, much wrinkled, wedge-shaped, entire at the base, unequally serrated towards the point, and covered beneath with wool, which becomes cinnamon-coloured as it grows old.

Maxim. Act. Hti Petrop. VI (1879) 193, found Fortune's plant to be a new species which he called *Spiraea chinensis*. — I. F. S., i, 224, both

Lindley's Sp. pubescens and Maxim.'s Sp. chinensis are reduced to *Spiraea dasyantha,* Bunge.

In Journ. Hort. Soc. IX (1855) 109, Lindley publishes a letter dated Ning po, August 14, 1853, written by Fortune, in which the latter states that during his last visit to Foo chow (in 1849) he discovered in a garden a *Spiraea,* which he took along with him to Shang hai, where it flowered, in 1850, in Beale's garden, a beautiful double variety. In the month of April it was loaded with daisy-like blossoms of the purest white. F. sent dried specimens and drawings of the plant to Lindley. It proved to be *Sp. Reevesiana,* described by the latter in Bot. Reg. 1844, t. 10. — F. saw it also cultivated in the gardens of Shang hai and Ning po, and met with it in a wild state in the mountains separating Che kiang from Fo kien, and above Yen chow foo in Che kiang (T. 57, 304, 320. — R. 28, 80, 241, 242).
Maxim. l. c. 195, reduces Sp. Reevesii to *Sp. cantoniensis,* Lour. — I. F. S., ɪ, 224.

Lindley in Paxt. Fl. Gard. II (1852) 113, fig. 191 (black), figures and describes *Spiraea callosa,* Thunb., introduced from North China by Fortune for Standish and Noble. A handsome, hardy, deciduous shrub with brilliant, rose-coloured flowers in branched cymes. Leaves dark green, nearly lanceolate, rugose, sharply serrate, tipped with little brown callosities.
Fortune saw it in a wild state in the mountains between Fo kien and Che kiang. (T. 304).
Planchon, in Bot. Mag. t. 5164 (1859) describes this plant, brought by Fortune as a new species, *Sp. Fortunei.* — See also Carrière in Rev. Hort. 1854, 21. Red flowers.
Maxim. l. c. 203, reduces both, Sp. callosa and Sp. Fortunei to *Sp. japonica,* Linn. — I. F. S., ɪ, 225.

Amelanchier racemosa, a handsome new plant discovered by Fortune in North China. Lindley notices it first in Bot. Reg. 1847, 38, note. Fortune mentions it, 1849, in the mountains near Ning po, as a shrub new to botanists, and scarcely yet known in Europe. It produces masses of flowers of the purest snowy white. (T. 154. — R. 28).
Hooker, in Bot. Mag. t. 4795 (1854) described and figured this plant as *Spiraea grandiflora.* Sent by Fortune from North China to Standish & Noble's nursery, Bagshot, where it lives in the open air, blossoming in May

1854. Conspicuous large white flowers. — See also Belg. Hort. V (1855) 65, col. plate.

Lindley in Gard. Chron. 1858, 925, calls it *Exochorda* (new genus) *grandiflora*. A handsome shrub with large flowers like those of a Philadelphus. Fortune's first specimens were marked: "a dwarf shrub, flowering in March, North China, 1845", his second: "Hills of Che kiang, May 1855", in young fruit. — I. F. S., I, 128.

Spiraea palmata, Thbg. Figured in Bot. Mag. t. 5726 (1868). Introduced by Fortune from Japan for Noble, Bagshot.

A species of *Fragaria*, is common in a wild state, on the banks and hill-sides, both in Japan and in China; but it has nothing to do with the species we cultivate in Europe, and is perfectly tasteless. (Y. 211). — Probably *Fragaria indica*, Andr. — I. F. S., I, 240.

Island of Koo lung soo, August 1843. Some very pretty *Roses*, producing small, double flowers of great neatness and beauty, although destitute of perfume. Fortune sent them home to the garden of the Hort. Society at Chiswick.

Rosa anemonaeflora, Lindley, Journ. Hort. Soc. II (1847) 316. Short diagnosis. Sent by Fortune (in the autumn of 1844, l. c. I (1846) 221), who found it in the gardens of Shang hai. It approaches R. moschata. Small clustered flowers, pale blush, of little beauty.

I. F. S., I, 247. Specimens labelled Fortune 61, Amoy, in the Kew Herb. Erroneously named *Rosa Fortunei* by Crépin. — Earlier Roxburgh.

F., when visiting Soo chow, in June 1844, saw there in the Chinese gardens a fine new *Double Yellow Rose*. Subsequently he found the same climbing double yellow Rose in the garden of a mandarin at Ning po. It was said that it was from the more northern districts of the empire. (W. 263, 322).

Lindley, in Journ. Hort. Soc. VI (1851) 52, mentions Fortune's double yellow Rose then cultivated in the garden of the society. It was brought home by F. on his return from China in 1846 (l. c. I, 223). F. speaks highly of its beauty in China, but in England it has little claim to notice. Flowers as large as those of the common China Rose, semi-double, solitary, dull buff, tinged with purple.

Lindley notices the same plant again in Paxt. Fl. Gard. III (1852) 156, and names it *Rosa Fortuniana*. It was first sent by Fortune, in 1845, to the Hort. Soc. (Journ. Hort. Soc. I, 218). Some years later he introduced it for Standish & Noble with whom it flowered. F. furnished the following memorandum: "I discovered it in the garden of a mandarin at Ning po, where it completely covered an old wall, and was then in full bloom. Masses of beautiful yellowish and salmon coloured flowers hung down in the greatest profusion. The Chinese call it *whang jang ve (huang ts'iang wei)* or yellow rose. The flowers vary a good deal in colour. It is quite distinct from any other known variety, and admirably adopted for covering walls". — Comp. also T. 318.

Bot. Mag. tab. 4679 (1852). Fortune's Double Yellow Rose. — Fl. d. Serres, VIII (1852) tab. 769. — Rev. Hort. 1854, 41, tab. 3. — Regel, Monogr. Ros. in Act. Hti Petrop. V (1877) 328, considers Fortune's Chin. yellow rose to be a variety of *R. sinica*, Murr.

There is yet another Chinese rose which bears Fortune's name: *Rosa Fortuniana* (Banksiana), Lindley in Paxt. Fl. Gard. II (1851) 71, fig. 171 (black), a scrambling evergreen hard shrub with large double, white, solitary flowers, 3 inches in diam., and ternate or quinate leaves. Introduced from China by Fortune, to the Hort. Society. — Regel, l. c. considers this likewise to be a variety of R. sinica, but according to Hemsley, I. F. S., I, 249, it is probably a hybrid between R. Banksiae and R. indica.

In 1844 Fortune met in a garden at Ning po with a rose, which the Chinese called the *Five coloured Rose*. It belongs to the section commonly called China Roses in England, but grows in a very strange and beautiful manner. Sometimes it produces self-coloured blooms — being either red or French white, and frequently having flowers of both on one plant at the same time — while at other times the flowers are striped with the two colours. (W. 322). Fortune sent it to the Hort. Society in the autumn of 1844. See Journ. Hort. Society I (1846) 221. — This rose is figured in Fl. d. Serres, IV (1848) 381: Rose de Thé, dite à cinq couleurs, Fortune.

Fortune sent from China a garden variety of *Rosa rugosa*, Thbg., with semi-double sweet-scented flowers of a very rich purple, about two inches across, which he discovered near Shang hai. Not very ornamental. Journ. Hort. Soc. III (1848) 313.

Near the city of Cading (Kia ting), in the Shang hai plain, F. observed orchards of *Apple*-trees, which is rather a rare sight in this district. The variety of apple was a small one, about as big as our golden pippin, but excellent in flavour; indeed, the only kind worth eating in this part of China. (R. 334).

Apples, pears and *Siberian Crabs* are cultivated in pots in the gardens at Tien tsin, and apparently with great success, for the little trees were all loaded with fruit (September). The Tientsin apples are very beautiful to look upon — the skin is thin and transparent, and the colour a delicate pink red, but the taste is sweet and almost insipid. (Y. 321).

Pears are perhaps the most abundant amongst all the autumnal fruits in Peking. They are exposed for sale in every direction, in shops, in stalls, on the pavement, as well as in the basket of the hawker. There were two or three kinds, and one of them was high-flavoured and melting. This is the first instance of a pear of this kind having been found in China, and it is a most welcome addition to the tables of the foreign residents of Peking. (Y. 363). — The Peking pear mentioned by F. is well known to me. The Chinese call it *pai li* or white pear. It is a middle-sized, roundish, apple-shaped very juicy fruit, most excellent in flavour. The skin is thin and of a very pale yellow colour.

The *Loquat, Eriobotrya japonica*, cultivated in Hong kong. F. saw also beautiful groves of this fruit-tree near Hang chow foo. (T. 7, 30).

Photinia Fortuneana, Maxim. Mél. IX (1873) 179. Fortune discovered this tree in 1843 near Amoy. Wenzig in Linnaea, 1874, 200, named it *Cotoneaster Fortunei*, — Decaisne, Pomac. 1874, 183: *Osteomeles pyracantha*. — I. F. S., i, 262. The tree is only known from the herbarium specimens sent by F.

Hills near the city of Tse kee (north of Ning po). *Photinia glabra*, a noble evergreen, which in the winter becomes covered with bunches of red berries. About the same hills F. met with *Ph. serrulata*, an evergreen tree with large glossy leaves, which came out from the trunk of a large juniper, about 12 feet from the ground, and appeared as if it had been grafted upon it. But the Photinia was no doubt rooted in the ground and had 12 feet of its stem cased in the decayed trunk of the juniper. (R. 52, 63). — I. F. S., i, 262, 263.

Raphiolepis indica, first described by Lindley in 1820 (v. s. p. 253). This was raised in the garden of the Hort. Society from seeds received from Fortune, as Lindley reports in the Journ. Hort. Soc. II (1847) 306. It proved to be a much finer plant than the R. rubra or the R. phaeostemon previously in cultivation (v. s. p. 253). It has a fine broad, evergreen foliage and flowers as large as those of the plum tree. The petals, however, are sharp pointed and either white or beautifully painted with rich rose colour. It is easily known from R. rubra by its broader leaves and narrower, shorter bracts. — Fortune sent the seeds probably from Canton or Hong kong. — I. F. S., ɪ, 264.

Saxifraga Fortunei, Hook. Bot. Mag. t. 5377 (1863). It differs from S. sarmentosa, Lin. particularly in the pure white flowers and the whole coloured (unicoloria) leaves. Fortune introduced it from Japan to Standish's nursery. — Illustr. Hort. 1864, tab. 398. — I. F. S., ɪ, 267.

In Journ. Hort. Soc. I (1846) 221, F. states that in the autumn of 1844 he sent home from China to the Hort. Society living specimens of a *Hydrangea* from the woods of Tein tung (near Ning po). — He saw handsome Hydrangeas cultivated in the nurseries near Foo chow, in 1845. They invariably produce there flowers of the deepest blue, much deeper than ever seen in England. (W. 383). — He notices a very beautiful species of Hydrangea met with in 1849 in the Bohea mountains. (T. 213). — In 1854 he found, in the mountains west of Ning po, a Hydrangea in a leafless state, which he introduced to Europe.

Hydrangea chinensis, Maxim. Rev. Hydrang. (1866) 7, was first discovered by F. at Amoy. He collected *Hydr. Hortensia,* DC. in Che kiang.— I. F. S., ɪ, 273.

Adamia versicolor, Fortune, in Journ. Hort. Society, I (1846) 298. Fortune found this plant in the island of Hong kong. The Society received living specimens from him in July 1844, and cultivated it. A fine bush with much the appearance of Hydrangea japonica so far as the foliage is concerned, but the flowers are quite different. They from a pyramidal panicle, nearly a foot in diameter, and when expanded are of the most brillant violet-blue; when in bud they are at first white but gradually change to purple and violet. When in full expansion they measure nearly an inch in diam. Petals 7 or 6, forming a star. Blue styles.

The plant is found figured in Paxt. Mag. Botan. XVI (1849) 322 (colour.), — in Paxt. Fl. Gard. I (1850) 19, plate 5 (colour.), — in Belg. Hort. V (1855) 1. — According to Bentham, Fl. Hgk. 128, this is the *Dichroa febrifuga,* Loureiro. — I. F. S., ı, 275.

In 1861 F. discovered in a temple near Yukuhama a new species of *Deutzia* with double rose-coloured flowers. It was in full bloom (June) and very beautiful. F. introduced it into England. (Y: 211).

Deutzia crenata, Sieb. & Zucc. var. *flore pleno,* figured in Illustr. Hort. XI (1864) 389. The plant had been introduced by Fortune (from Japan) to Standish. — The same figured in Belg. Hort. XIV (1864) 322. — Carrière in Rev. Hort. 1866, 338, named it *D. Fortunei.* — Fl. d. Serres, XVII (1867) 149. — Maxim. Hydrang. 1867, 45, reduces D. crenata to *D. scabra,* Thbg. — I. F. S., ı, 276.

Ribes fasciculatum, Sieb. & Zucc. Fl. Jap. Fam. nat. (1843) 189. Japan. — Maxim. Mél. IX (1873) 238, describes a variety of this plant, β *chinensis,* discovered by Fortune in China, during his first expedition. — I. F. S., ı, 289.

Corylopsis spicata, Sieb. & Zucc. Fl. Jap. I, 47, t. 19, and *C. pauciflora,* l. c. 48, tab. 20. Both introduced by F. from Japan. Fortune, Gard. Chron. 1880, I, 234.

Guavas cultivated near Canton. (W. 136). — The Guava is extensively cultivated for the sake of its fruit in the black tea country, Bohea hills. (T. 250, 254).

Barthea chinensis, Hook., gathered by Barthe, 1856, in Hong kong (v. supra p. 400), was earlier discovered, probably in the same island, by Fortune. — I. F. S., ı, 300.

Tien tsin. Here, as elsewhere the *Pomegranate* seemed an especial favourite, and was largely grown.

The *Trapa bicornis,* much cultivated in South China. (W. 307). — Tai hoo Lake. The water is very shallow, and a great part of it is covered with the Trapa bicornis, called *ling* by the Chinese. It produces a fruit of a very peculiar shape, resembling the head and horns of a bullock, and is highly esteemed in all parts of the empire. F. saw three distinct varieties, one of which has fruit of a beautiful red colour. Women and boys were sailing about on all parts of the lake, in tubs of the same size and form as our common washing tubs, gathering the ling fruit. F. gives a picture of this mode of gathering the Trapa. (T. 27. — R. 338).

Plain of Shang hai. *Melons* of several different kinds are extensively cultivated: when they are ripe the markets are literally crowded to over-flowing with them, and they are eaten by the natives much in the same way as apples are with us. (R. 334).

Water-melons abundant at Nan tsin in the silk district of Hoo chow, near Tai hoo Lake. (R. 340).

Tamsuy in Formosa, 1854, April. The far-famed Rice-paper plant, named by Sir W. Hooker *Aralia papyrifera,* was growing there apparently wild; but the site may have been an old plantation, which was now over-grown with weeds and brush wood. The largest specimens were about 5 or 6 feet in height, and from 6 to 8 inches in circumference at the base, but nearly of an equal thickness all up the stem. The stems were crowned at the top with a number of noble-looking palmate leaves, on long footstalks, which gave to the plant a very ornamental appearance. The under side of each leaf, its footstalk, and the top part of the stem, which was clasped by these stalks, was densely covered with down of a rich brown colour. The plants were not in flower then, for they flower and seed during the winter. F. dug up numerous small plants, took them with him to Shang hai, and deposited them in Beale's garden. They were finally sent to England and India. The proportion of pith in the stems is very great, and forms the pure white substance from which the beautiful article erroneously called "rice-paper" is prepared. The Chinese call this plant *tung tsao.* It is largely cultivated in many parts of Formosa, and with rice and camphor forms one of the chief articles of export. The Canton and Fo kien provinces are the chief consumers. This most beautiful and delicate substance is very cheap. (R. 197, 232—234).

One of Fortune's objects in visiting Japan, in 1861, was to procure the male variety of the common *Aucuba japonica* of our gardens, introduced in 1783 from Japan, a variety with variegated leaves. This is perhaps the most hardy and useful exotic evergreen shrub we possess. It is one of the

most common plants in the parks, squares and houses of London; but no one
in England has ever seen it covered with a profusion of crimson berries,
as it is met with in Japan. It belongs to a class of plants which have the
male and female flowers produced on different individuals. Curiously enough,
all the plants in Europe were females, and hence the absence of fruiting
specimens. At Yukuhama F. found at last the male plant in its native and
unvariegated form, and sent it home to Standish's nursery at Bagshot where
the mother plant was fertilized. (Y. 60, 146).

In 1864 Standish at Kensington exhibited this glorious shrub loaden
with bunches of its large oblong berries of the brightest coral red. (Gard.
Chron. 1864, 243).

The Aucuba in fruit was figured in Illustr. Hort. XI (1864) 399, —
in Bot. Mag. tab. 5512 (1865), — in Fl. d. Serres, XVI (1865) 5.

Viburnum plicatum, Thbg; Sieb. & Zucc. Fl. Jap. I (1839) 81, t. 38.
Journ. Hort. Soc. II (1847) 243. Lindley says: "a native of China, where
it is cultivated in gardens. Introduced by Fortune, cultivated in the Society's Garden. Leaves broad, coarsely serrate, somewhat plaited, dark
green, narrowed to the base, cuspidate. Flowers white, in large round
heads, most profuse." — Figured by Lindley in Bot. Reg. 1847, t. 51, —
also Lemaire, Jard. Fl. I, 1851, t. 88, — Paxt. Fl. Gard. I (1851) 147,
tab. 29 (colour.). — Illustr. Hort. 1860, tab. 249.

Maxim. Mél. X (1880) 661, considers V. plicatum to be a variety of
V. tomentosum, Thbg. — I. F. S., I, 356.

Viburnum macrocephalum, Fortune in Journ. Hort. Society, II (1847)
244. Deciduous shrub. Leaves 3 inches long, ovate, very blunt, on short
stalks, slightly toothed. Flowers in large compound cymes, which in the
neuter state (that in the Soc.'s garden) are as much as 8 inches in diam.,
not, however, globose like those of the Gueldres Rose, but rather pyramidal. Each flower is $1^3/_8$ inch in diam. snow-white. Fortune sent this fine
shrub from Chusan and Shang hai, where it is cultivated in the gardens of
the rich. F. saw a tree of it in the island of Chusan, 20 feet high. The
Chinese call it Hydrangea. It was received in the Society's Garden in June
1844. — Figured in Bot. Reg. 1847, tab. 43. — Carrière in Rev. Hort.
1863, 269, fig. 31, figured this plant under the name of *V. Keteleerii.* —
I. F. S., I, 353.

Viburnum dilatatum, Thbg. Fl. Jap. 124. Lindley in Journ. Hort.

Soc. III (1848) 247. Living plants received from Fortune, on his return from China, in May 1846. He found them at Tein tung near Ning po in May 1844. A small shrub. Bright green, plaited leaves, coarsely toothed, with rough hairs. Flowers in small, spreading cymes, white, not showy. — I. F. S., ɪ, 351.

Viburnum Hanceanum, Maxim. Mél. X (1880) 662. Sampson, 1870. Canton. — Fortune earlier, Amoy, 1843. — I. F. S., ɪ, 353. — Kew Bull. 1897, App. I, 41. Cult. Kew.

In October 1843, amongst the hills near Chimoo Bay, F. found a new species of *Abelia*, *A. rupestris*. He dug up living plants, and sent them to England. (W. 47).

Abelia rupestris, Lindley in Journ. Hort. Soc. I (1846) 63. Received from Fortune, June 20, 1844, who found the plant amongst the rocks of the Cha moo Hills. Flowered in the Society's Garden. It is a small bush. Branches very slender, covered with a fine down of a deep reddish brown. Leaves opposite, smooth, ovate, distantly serrated on short stalks. Flowers pure white, something like those of the honey-suckle, and come in pairs from the axils of the leaves, sweet-scented. Sepals of the calyx slightly stained rose colour. — Bot. Reg. 1846, tab. 8. — Paxt. Fl. Gard. 130, fig. 201 (black). — According to Hemsley, I. F. S., ɪ, 358, A. rupestris is a variety of *A. chinensis*, R. Br.

Fu kien side of the Bohea Mts, 1849. A fine species of *Abelia* was met with, which will probably be a favourite in English gardens. Its flowers are as large as those of Weigela rosea, of a blueish tinge, and bloom in great profusion for a long time. F., when he first saw this plant, took it to be *Abelia chinensis*, R. Br., discovered by Dr. Abel on the same mountains, about a hundred miles to the north-west of the spot, where F.'s plant was found — but Dr. Lindley, to whom it was sent for examination, calls it *A. uniflora*. (T. 213, 254).

The *Abelia uniflora*, named by R. Br. in Wallich Pl. As. rar. I (1830) 15, upon a specimen sent by J. Reeves (v. s. 261), was first described and figured in Paxt. Flow. Gard. II (1852) 145, fig. 208 (black). A small evergreen. Whitish flowers, which grow singly in the axils of the upper leaves. Fortune introduced living plants from the province of Fo kien near Ngan ke hyen *), in the black tea country, for Standish and Noble.

---

*) This is an erroneous statement. Reeves gathered A. uniflora in the tea district of An ki hien, in 1824, but Fortune never visited this district.

Bot. Mag. tab. 4694 (1853). It is there observed that the above name is unhappily chosen, for the living plants at Kew bear frequently 3 flowers from a peduncle. — Lemaire, Jard. fleur. IV (1854) tab. 380 (colour.). — I. F. S., ɪ, 359.

Lonicera fragrantissima, Lindley & Paxt. Flow. Gard. III (1852) fig. 268 (black). Short diagnosis. Evergreen shrub. Flowers whitish, very sweet-scented. Fortune sent living specimens from China, while in the service of the Hort. Society. They have not flowered in the Chiswick garden, but in January last the plant flowered in the garden of the Marquis of Salisbury at Hartfield. — Carrière in Fl. d. Serres, 1858, 563. — Figured in Gard. Chron. 1878, I, 106, 107, fig. 19, 20. — I. F. S., ɪ, 361.

Lonicera Maackii, Rupr., Mél. II (1857) 548, gathered by Maack, 1855, on the Amur River, was first discovered by Fortune, 1843, near Amoy. — I. F. S., ɪ, 364.

Lonicera Standishii, Bot. Mag. tab. 5709 (1868). Hooker fil. states: "This charming, early flowering honeysuckle has been for some time in cultivation in Europe, but hitherto it has borne no name in English gardens. M. Baillon assured us that in France it is cultivated under the above name, which I supposed to have been given to it by the late conductor of the Bot. Mag. in honour of the active and intelligent nurseryman, to whom many of Fortune's rich Chinese collections were consigned. Fortune informs us that it is common in gardens at Shang hai, and was sent home by him many years ago and distributed by the Hort. Society with L. fragrantissima. Flowers white, very sweet-scented".

L. Standishii, was first described and figured by Carrière in Fl. d. Serres, 1858, 63. It had been received from England. See also Carrière in Rev. Hort. 1873, 148 (black figure). — I. F. S., ɪ, 367.

Lonicera aureo-reticulata was introduced by Fortune from Japan. See Gard. Chron. 1863, 1040; — 1866, 588. A beautiful climber.

According to Nicholson, Dict. Gard., it is a variety of *Lonicera flexuosa,* Japan.

Weigela rosea, Lindley in Journ. Hort. Soc. I (1846) 65, 189, 215. Short diagnosis. Coloured plate, tab. 6. Fortune says it is a shrub like a Philadelphus. Old stems whitish, smooth, young ones green, slightly winged;

wings alternating with the leaves and covered with hairs. Leaves opposite, nearly sessile, elliptical, 1½ inch wide, 3·inches long, serrated above, nearly smooth below, on the midrib and veins hairy. Flowers axillary and terminal, 3 or 4, springing from each axil, rose-coloured. Peduncles short with green, short, thread like bracts at the base. Calyx cleft into 5 unequal segments, 3 above, 2 below, two-lipped. Corolla monopetalous, tubular, mouth reflexed and cleft into 5 equal segments. Stamens 5, shorter than the corolla and inserted to its sides, smooth above, but hairy from the point of union to the base of the corolla. Style 1, stigma capitate and little longer than the stamens. Germen inferior, rather more than an inch long, nearly sessile.

This beautiful shrub was discovered by Fortune in the island of Chusan, near Ting hai, in a mandarin's garden. Native name *noak chok wha*. It was literally loaded with its fine rose-coloured flowers. F. never met it in a wild state. It is unknown in the southern provinces. It was probably originally introduced from Japan to China. F. sent living specimens of the plant together with a drawing, in the beginning of 1844, to the Hort. Society, who cultivated it in their garden. It forms a neat middle-sized bush, deciduous in winter. It flowers in April and May. It has very large flowers 1½ inch in diam., more than an inch long. When the plant comes into blossom the branches are loaded with flowers and hang down in graceful natural festoons.

Fortune saw the plant cultivated in the gardens of Chusan, Shang hai, Tien tsin. (W. 320. — T. 16, 317. — R. 80, 242. — Y. 318).

Coloured figures of W. rosea are found in Paxt. Mag. Bot. 1848, 227, — in Bot. Mag. tab. 4396 (1848), — in Fl. d. Serres III (1847) 211, — Belg. Hort. I (1851) 377.

Maximowicz in Mél. XII (1886) 482, reduces W. rosea to *Diervilla florida*, Sieb. & Zucc. Fl. Jap. I, 75 (1835) t. 32. — I. F. S., I, 368.

Gardenia florida, Lin., var. *Fortuniana*, Lindley in Journ. Hort. Soc. I (1846) 226, 221. This plant was sent by Fortune from the north of China to the Hort. Society, in the autumn of 1844. The common single and double varieties of this species are well known. That which is now noticed differs merely in the extraordinary size of the flowers which are nearly 4 inches in diam., and in having fine, broad leaves, sometimes as much as 6 inches long. The flowers are pure white, changing to light buff as they go off, and not unlike a very large double Camellia, which it equals in the beauty of the flowers and leaves and infinitely excels in its delicious odour. — Figured by Lindley in the Bot. Reg. 1846, tab. 43, — see also Fl. d. Serres II (1846) 177.

Summer of 1844, Chinese Garden, Soo chow foo. A Gardenia with large white blossoms, like a Camellia. Sent to England. (W. 263). — The same observed in Shang hai gardens, in 1848, 1850. Some of the specimens 4 feet high and 15 feet in circumference. When covered with its large fragrant Camellia-looking blossoms it is extremely handsome, and at all times forms a pretty evergreen bush. (T. 17, 317).

The fruit of *G. radicans,* Thbg. is used in Japan as a yellow dye, in the same way as in China. (Y. 139).

I. F. S., i, 382. G. radicans is now generally united with G. florida.

Hong kong Flora. Higher up in the mountains we find the beautiful *Ixora coccinea,* flowering in profusion in the clefts of the rocks, and its scarlet heads of bloom under the Hong kong sun are of the most dazzling brightness. (W. 10). — I. F. S., i, 385.

Island of Koo lung soo (near Amoy) 1843. The hedges and crevices of the rocks abound in a little creeper called *Paederia foetida,* very pretty, but having a most disagreeable odour. (W. 32).

According to Maximowicz, Mél. XI (1883) 798, the Japanese and Chinese species of this genus is not P. foetida, Linn., but *P. tomentosa,* Blume, Bijdr. 968, of which latter there is in the Herbarium Hti Petrop. a specimen, labelled: Amoy Fortune, 93. — I. F. S., i, 389.

Vernonia Fortunei, Schultz-Bip. in Flora, 1852, p. 48. Only the name. — This is the *Vernonia solanifolia,* Benth. in Hook. Lond. Journ. Bot. I (1842) 486, first discovered in Hong kong by Hinds. — Fortune n. 175, gathered it a few years later. — I. F. S., i, 402.

Eupatorium Fortunei, Turcz. in Bull. Soc. Nat. Mosc. 1851, 170. Fortune, China. — In I. F. S., i, 403, this is reduced to *Eupatorium japonicum,* Thunb.

Aster turbinatus, S. Moore in Journ. Bot. 1878, 132. China, Fortune. — I. F. S., i, 417.

The *Sunflower (Helianthus)* grows to a very large size near Tientsin and attains a height of fourteen feet. (Y. 339).

When F. paid his first visit to Shang hai, towards the end of 1843, few plants were then in bloom except the *Chrysanthemum*, the varieties of which were as numerous there as in the south of China; and as the Chinese gardeners understand their cultivation well, they were, at this particular season, objects of great interest. (W. 125).

Shang hai, January 1849. The Chrysanthemum is the Chinese gardener's favourite winter flower, although it is generally past its full beauty at the Chinese new year. In the cultivation of the Chrysanthemum the Chinese stand unrivalled. F. found the plants trained in the form of animals, such as horses and deer, and at other times they were made to resemble the pagodas, so common in the country. Whether they were trained into these fanciful forms, or merely grown as simple bushes, they were always in high health, full of fresh green leaves, and never failed to bloom most profusely in the autumn and winter. Cuttings are struck every year from the young shoots of the Chrysanthemum. When they are rooted they are potted off at once into the pots in which they are to grow and bloom. The soil used in potting is of a very rich description. It is obtained from the bottom of lakes and ponds. The plants are trained each with a single stem and this is forced to send out numerous laterals which are tied down in a neat and regular manner, to have the plant clothed with branches.

About Shang hai and Ning po the Chrysanthemum is better managed than it is near Canton; but the success which attends it may be attributed, partly at least, to the more favourable nature of the climate, the plant being indigenous to the central and more northern parts of the empire. The Chinese are fond of having very large blooms, and, in order to obtain these, they generally pick off all small flower-buds.

Although we are indebted to China for the parents of those varieties of Chrysanthemums, which now enliven our gardens during the dull months of winter, yet, strange to say, the progeny is more numerous in Europe than in China itself. Some of the beautiful kinds raised by Salter in France would be much admired even by the Chinese florist. It is a curious fact, however, that many of those kinds, such as *formosum* and *lucidum*, which were originally raised from seed in Europe, are also met with in the north of China. (T. 123—126).

Chrysanthemums are largely cultivated in most of the gardens at Tien tsin. Some of them are trained as "standards" and somewhat resemble in form our dwarf standard rose-trees. In order that they may assume this appearance, they are grafted on the stout stems of a species of *Artemisia*. They grow with great vigour on this stock, and appear rather curious objects to those who have seen them only on their own stems. (Y. 320, 321).

Temple of "Ah-sax-saw" in the vicinity of Yedo. This place is most famed for the variety and beauty of its Chrysanthemums. F. found them in full bloom, in December 1860, and procured some extraordinary varieties, most peculiar in form and in colouring, and quite distinct from any of the kinds then known in Europe. One had petals like long, thick hairs, of a red colour, but tipped with yellow, looking like the fringe of a shawl or curtain; another had broad white petals striped with red like a carnation or camellia; while others were remarkable for their great size and brillant colouring. Most of these arrived safely in England. (Y. 126).

Fortune in his Notes on the Plants introduced by him from China and Japan, Gard. Chron. 1880, I, 73, notices a Chrysanthemum which he calls *Chusan Daisy*. He found this pretty little plant in a cottage garden on the island of Chusan, during his first expedition to China, and sent it home, presuming that great things might be done with this modest little flower in hybridising independently. It reached England alive, but for a time did not attract much attention. About that time the late Mr. Salter, of Hammersmith, had settled in Versailles, where he was making the culture of the Chrysanthemum a speciality. In his book on the Chrysanthemum, published in 1865, he says:

"In 1846 a new era commenced in the history of the Chrysanthemum, for at that time Fortune brought from China two small flowering varieties, known as the *Chusan Daisy* and *Chinese minimum*. In England they were considered too small and insignificant for English taste. But the French opinion was far different, for immediately after the introduction, in 1847, the little Chusan Daisy became a favourite in France. From these two varieties have sprung all the Pompons now in cultivation".

At that time the Chrysanthemums in England were far superior to those of China. F. could find nothing worthy of being introduced, except those two small varieties. A few years afterwards, however, in 1860 and 61, on visiting Japan, F. found himself in the very home of the Chrysanthemum. Here the varieties were altogether different from those met with in England. In so far as the culture is concerned, the Japanese are greatly assisted by the climate. What a glorious autumn they have in Japan. The sun shines from morning to evening in a clear sky, with scarcely a cloud to obscure its rays. Salter says:

"In 1862 Fortune introduced several Japanese varieties of Chrysanthemum, some of which were spotted and striped, others were of fantastic forms, called *Dragon,* and one laciniated, was a beautifully fringed white flower, most valuable for bouquets, having the appearance of a Japanese pink rather than of a chrysanthemum".

There is a curious circumstance about these plants, viz. that several distinct varieties are obtained from one plant.

There is a Chrysanthemum, introduced by Fortune, figured in Fl. d. Serres XV (1863) tab. 1574, Chr. ind. flore *roseo albo punctato.* F. introduced also the *Bronze Dragon, Yellow Dragon, grandiflorum, japonicum lanciniatum, striatum.*

Mosquito Tobacco, in Che kiang, Chusan, made of *Artemisia indica* and other species of wormwood, reduced to powder and mixed with sawings of resinous wood, Juniper or Pine tree *(pih hiang fun* or *sung shu),* Tobacco leaves, a mineral called *nu wang* and a small portion of arsenic. The whole is beaten up with water, and in the form of a thick paste rolled on a slip of bamboo, and then dried. This substance the Chinese burn to drive away mosquitos. On its being lighted the fumes rise slowly upwards and give out a fragrant small. In a few minutes all the mosquitos disappear. They also use for the same purposes the stems and leaves of the Artemisia twisted or plaited into ropes, and probably dipped in some preparation to make them burn. (T. 179. — R. 110—114).

For the *Moxa* burning operation in Japan the young leaves of *Artemisia* are used after having been reduced to powder. Little cones of this powder are placed in holes made in the skin and set on fire on the top. It burns slowly down and leaves a blister on the skin. (Y. 204, 205).

In the autumn of 1855, in a Chinese garden near Ning po F. met with a beautiful new herbaceous plant, having rich blotched or variegated leaves, which has since been named by Dr. Lindley *Farfugium grande.* The possessor of this rare plant informed F. that he had received it from Peking the year before. F. procured some plants from a nurseryman at Ning po, and sent them to England. (R. 420). — F. saw the same plant in Nagasaki. (Y. 12).

Farfugium grande, Lindley, in Gard. Chron. 1857, 4, 215 and 1860, 170. A charming plant cultivated by Mr. Glendinning. Fortune had sent it from China. Very large leaves, round, angular, heart-shaped, of a peculiar emerald green, copiously blotched with patches of clear yellow. The flowers are insignifiant, yellow ray with a dirty purple centre. — Figured in Fl. d. Serres, XII, 1857, 11. — Illustr. Hort. 1857, tab. 133.

This plant was known for a long time from Japan and has many scientific names. First noticed by Kaempfer. It is the *Tussilago japonica*, Lin.,— the *Senecio Kaempferi*, DC. Prod. VI (1839) 363, — the *Ligularia Kaempferi* and *L. gigantea*, Sieb. & Zucc. Fl. Jap. I (1839) 77, tab. 35 and 79, t. 36, — Bot. Mag. t. 5302 (1861) — the *Farfugium Kaempferi*, Benth. Fl. Hgk., 191. — I. F. S., I, 454.

The *hong wha (hung hua,* red flower*)* was found by F. for the first time, in 1855, in the fields near Cading (Kia ting hien) in the Shang hai plain. It is a variety of Safflower, *Carthamus tinctorius*. This dye is held in high esteem by the Chinese, and is used in dyeing the red and scarlet silks and crapes, which are so common in the country, and so much and justly admired by foreigners. F. was informed that large quantities of Safflower were annually produced in Che kiang, near Ning po. At this season (10th June) the crop of flowers had been gathered, and all the plants removed from the land, except some few here and there on the different farms, which had been left for seed. The seed was not yet ripe. Afterwards F. returned to the place, and secured ripe seed, which he sent to the Agricult. and Hort. Society of India, in order to compare the Chinese with the Indian Safflower. They turned out to be alike or nearly so. (R. 333).

Lindley in Journ. Hort. Soc. I (1846) 228, describes and figures a lettuce-like plant from Shang hai, the seeds of which were sent by F. from thence in 1844. The plant was cultivated in the Society's garden. The Chinese call it *hoo sung* or *oo sung*. It grows 2 or 3 feet high, and is of a light green colour with a succulent pith. Small yellow flowers in panicles. Perhaps it is *Youngea dentata*, DC. F. says it is a vegetable much esteemed by the Chinese. It is fit for use when the stem has grown to its full size, which is early in the spring at Shang hai. Mode of dressing: pare off the outer skin, cut off the leaves and take the stalk; either simply boil it with salt and eat it with pepper, or stew it with a few spoonfuls of soup, or with a little soy, salt and pepper. It would probably form a good preserve similar to that made of the stems of lettuces when running up and before they become hollow. — I. F. S., I, 480, *Lactuca denticulata*, Maxim.

Near Chimoo Bay, October 1843. *Campanula grandiflora* grows wild there. Fortune gathered several fine roots of it, and sent them to the Hort. Society at Chiswick, in the beginning of 1844. (R. 47. — Journ. Hort. Soc. I (1846) 215). — This is *Platycodon grandiflorus*, A DC.

A variety of the same with semidouble white flowers is recorded by Lindley, in Journ. Hort. Soc. I (1846) 305; figured, F. found it in gardens near Shang hai and sent it to the Society in April 1845; it flowered in the Chiswick garden. 5 lobed corolla placed within another so exactly that the two constitute a large white star of 10 points. — F. notices the white variety of *Pl. grandiflorus,* in T. 318, which he saw in full bloom in a garden near Shang hai, in April 1850. — Fortune's Pl. grandiflorus figured in Paxt. Mag. Bot. XIII (1847) 7. — Lindley in Paxt. Flow. Gard. II (1852) 121, t. 61 (colour.) describes and figures the same as a new plant, *Platycodon chinensis,* the finest herbaceous plant sent by Fortune from China. Very large, deep rich violet blue flowers, shaped like a balloon before expansion, and like a basin cut half way down into 5 regular sharp triangular lobes when expanded. The first knowledge we had of this plant was from finding it among some dried specimens collected by the Rev. G. H. Vatchell near Macao in December 1829. Fortune brought it from Cha moo. There is a semi-double white variety, figured in the Journ. Hort. Soc. Both produced seed in the garden of the Society.

In I. F. S., II, 5, P. chinensis is reduced to *P. grandiflorus.*

Campanula nobilis, Lindley, Journ. Hort. Soc., I (1846) 232. Short diagnosis. Herbaceous plant from Chusan and Shang hai, sent home by Fortune, received by the Society in June 1844. The root-leaves of this fine plant are deeply heart-shaped with foot-stalks from 6 to 9 inches long, forming a large tuft. The flowering stem rises more than twice the height, bears fine nodding flowers, which seem to be the largest yet seen among the genus Campanula, nearly 3 inches long, $1^1/_2$ in diam. Corolla pale purple on the outside, paler within, abundantly sprinkled with bright purple dots, and closely covered with long, delicate hairs. — The same in Bot. Reg. 1846, tab. 65. Fortune gives a short notice of the plant. It is a great favourite with the Chinese in the northern parts of the Empire. They call it *tze chung wha* or purple bell-flower. It flowers in summer. In the garden of the Society it grows freely. — Fl. d. Serres, tab. 247, 563. — Rev. Hort. 1846, 283.

I. F. S., II, 9. Reduced to *C. punctata,* Lam. Enc., I (1783) which was cultivated in 1815, by Lambert, who raised it from Siberian seeds.

Fortune on the Hong kong Flora. The *Enkianthus reticulatus (E. quinqueflorus)* is a plant very highly prized by the Chinese. It flowers in February and March, about the time of their new year, and they then

bring the branches down from the hills in great quantities for the decora-
.tion of their houses. The flowers are unexpanded when they are gathered,
but by being placed in water they very soon bloom in the houses, and
remain for more than a fortnight as fresh and beautiful as if they had been
taken up with their roots in the most careful manner. On the mountains
from 1000 to 2000 feet above the level of the sea, the Enkianthus is found
growing abundantly, and in great luxuriance. It is never seen in the valleys
or low lands, unless when brought down by the natives. (W. 11, 146. —
T. 128—130).

The plant was cultivated before Fortune in English gardens. Lindley's
E. reticulatus, in Bot. Reg. t. 885 (1825) and *E. quinqueflorus*, Lour. are
the same. — Rev. Hort. 1849, 221 (colour. plate). — I. F. S., II, 18.

Azalea, Rhododendron. The Azalea is indigenous to China. Every
mountain and hill in the southern and central provinces of China are covered with these beautiful plants. They are like the heath of our own country,
and quite as abundant. About Hongkong and Canton they are usually found
in a wild state high up on the sides of the mountains, from 1000 to ·2000
feet above the level of the sea. In lat. 25° north, in the province of Fo
kien azaleas are met with in less elevated situations, i. e. from 500 to
1000 feet; and when we reach Chusan, in lat. 30°, we find them growing
plentifully on the lower sides of all the hills, rarely at a high elevation.
Few can form any idea of the gorgeous and striking beauty of these azalea-
clad mountains, where on every side, as far as our vision extends, the eye
rests on masses of flowers of dazzling brightness and surpassing beauty.

Although this genus is thus found spreading itself over a vast tract of
country, yet the northern parts of China just indicated are evidently those
in which it is most at home. Most of the Chinese Azaleas flower early, in
March and April, only the section to which A. variegata (a variety of A.
indica) belongs, flowers in May. By far the finest are cultivated in gardens;
indeed it was only in gardens that Fortune could find any worthy of introduction into England. In these gardens the Azalea is propagated readily
and extensively. Layering is the common method employed, but grafting
and striking from cuttings are also resorted to with success. (T. 330—
334. — W. 56).

Azalea indica, Lin. Varieties noticed or introduced by Fortune. In
the Fa te Gardens near Canton, F. observed the varieties *alba, phoenicea,
lateritia, variegata*, all well known in Europe before. (W. 142).

Azalea obtusa, Lindley, Journ. Hort. Soc. I (1846) 152. Short diagnosis. Fortune sent it from Shang hai to the Society. Received July 26, 1844. This charming shrub may be regarded as the gayest of all the red Chinese Azaleas in cultivation. It is a little bush with very blunt leaves, small and narrow. Fine, deep red flowers, also smaller than in other Chinese species. Figured in Bot. Reg. 1846, t. 37. An acquisition of great importance. It is sweet-scented like Sweet Briar.

Fortune discovered it in the Pou shan Azalea gardens near Shang hai, where he also saw varieties of it with semi-double flowers. It is the most brilliant of all the red Azaleas, and seems to set itself and all things around it in a glow. (T. 327, 334). The colour of the flowers is much more brilliant in China than in the plants cultivated in England. — Maximowicz, Rhododendron, As. or. (1870) 40 = *Rhododendron indicum,* var. *obtusum.* Tilesius brought herbarium specimens from Macao, 1806 (v. s. p. 314). — I. F. S., II, 26.

Azalea ramentacea, Lindl., Journ. Hort. Soc. IV (1849) 291. Figured (black). Fortune brought it from Hong kong to the Society, in May 1846. Distinct species. Flowers not large. — Maxim. l. c. 30, reduces this to *Rhododendrum indicum,* var. *macranthum,* Sweet. — I. F. S., II, 26.

In 1850, F. discovered in the Azalea gardens near Shang hai a beautiful quite new variety with small semi-double pink flowers, which it produced in great profusion. F. named it *Azalea amoena* and sent living specimens to England. (T. 329). — *A. amoena,* Lindley, in Paxt. Flow. Gard. III (1852) 81, t. 89 (colour.). Dwarf crimson Chinese Azalea. Short diagnosis. A dwarf, evergreen shrub, resembling A. ferruginea in habit. The branches, when young, are closely covered with long, thin, white ramentaceous scales, when old rust-coloured. Leaves small, obovate, hairy, blunt. Flowers small, pentandrous, rich crimson, almost campanulate; no calyx. (Fortune). F. introduced it for Standish and Noble, from Shang hai. Originally brought from Soo chow. Perfectly hardy. — Bot. Mag. tab. 4728 (1853). — Fl. d. Serres, IX, 1853, 75, tab. 885. — Rev. Hort. 1854, 241. — Maxim. l. c. 40, *Rhod. indicum,* var. *amoenum.* — I. F. S., II, 26.

Azalea crispiflora, W. J. Hooker, in Bot. Mag. t. 4726 (1853). One of the many fine and showy plants introduced by Fortune from China. It flowered in April with Standish & Noble, Bagshot nursery. Corolla large, deep rose-coloured. Lobes rotundate, remarkably waved and crisped at the margin. — Lem. Jard. Fl. IV (1854) tab. 430. — Belg. Hort. IV (1854) 129. — According to Maximowicz, l. c. 41, near the variety *amoenum.* — I. F. S., II, 26.

Azalea ovata, Lindley, in Journ. Hort. Society, I (1846) 149. Short diagnosis. Among the early despatches from Fortune there was received a drawing of this beautiful shrub, which according to the Chinese artist has most delicate, pink flowers, of the size and form of Rhod. dauricum, growing in clusters at the ends of the branches. Of this F. found two varieties in Chusan, the one with white flowers, the other with pink or lilac. The original plants did not survive the voyage, but a packet of seeds has furnished an abundance of young plants. This species looks entirely different in foliage from all the other Chinese Azaleas, for instead of the pale green colour and abundant hairs, which characterise them all, this has perfectly hairless leaves of a very dark green. Their form too is quite distinct, for instead of tapering gradually to the stalk, they are abruptly ovate or even heart-shaped.

Ibidem II (1847) 126, colour. plate. Fortune met with it for the first time in the autumn of 1843, and sent home seeds. Subsequently he also sent living plants, which arrived in safety. It was found growing on the sides of the "Green Tea" Hills in Che kiang, and also in some of the islands of the Chusan Archipelago, where it is called *kin sze wha* or silver (gold) silk flower. One bears white flowers, the other is pink and both are beautifully dotted with dark spots on the under petals. This species is not met with in the south of China. — Azalea ovata, a fine plant grows wild on the hills in Chusan. (W. 320). — Bot. Mag. tab. 5064 (1858). — Ind. Fl. Sin., ii, 28, *Rhododendron ovatum,* Planchon.

Azalea squamata, Lindley, Journ. Hort. Soc. I (1846) 152. Short diagnosis. Sent by Fortune from the mountains of Hong kong, in the beginning of 1844, flowered in the Society's garden. The plant has been long known from dried specimens and a drawing sent by Reeves. In its natural state it blooms without leaves, produces at the end of every shoot one or two large flowers of a clear rose colour, distinctly spotted with crimson on one side, and guarded at the base by a large sheath of bright brown scales (whence the name). — Bot. Reg. 1847, tab. 3. — Maxim. l. c. 25 = *Rhododendron Farrerae,* Tate, introduced in 1829. — Ind. Fl. Sin., ii, 23.

Azalea narcissiflora, Fortune, described by Planchon in Fl. d. Serres, IX, 1853, and in Rev. Hort. III, 1854, 66. Fortune introduced it from China to Standish & Noble. — Maxim. l. c. 36 = *Rhodod. ledifolium,* Don. var. *narcissiflorum.* — I. F. S., ii, 27.

Azalea vittata, Fortune. This was first described and figured by Plan-chon in Fl. d. Serres IX (1853) 7, t. 887 (colour.) as *Azalea indica vittata,* introduced by Fortune from China, in 1844. In Rev. Hort. III (1854) 66, he names it *Rhododendron vittatum.*

Planchon notices in Fl. d. Serres IX (1853) 82, *Azalea Bealii,* which Fortune had introduced from China to Standish & Noble.

In Illustr. Hort, 1854, pl. 8, Azalea indica Bealii. — In Belg. Hort. XVI (1866) 1 (colour. plate) = *Azalea vittata,* var. *Bealii.*

In 1850, when Fortune visited the Azalea gardens near Shang hai, he found there the *Azalea vittata,* a most beautiful kind having the habit of A. indica and half deciduous. It had its flowers striped with pale blue or lilac lines, and sometimes blotched of the same colour upon a white ground. Another species allied to this, which he named *A. Bealii,* had red stripes, and a third was mottled and striped in its flowers, the colours being still the same. (T. 330).

The yellow *Azalea sinensis,* Lodd., first received, it seems, from Canton, is known in European gardens since 1824 (v. s. p. 281). It was again introduced from China by Fortune, in 1845. Bot. Mag. tab. 5905 (1871). — Fortune saw it cultivated in the Fa te Gardens near Canton. Afterwards he found it in a wild state amongst the hills near Ning po, where in spring it seemed to paint the hill sides, so large were its flowers and so vivid the colours. Its colours are far more brilliant, and its trus-ses of flowers much larger than they are ever seen in any of our exhi-bitions in Europe. (W. 90, 143. — T. 154. — R. 28). — Ind. Fl. Sin., ii, 30.

Rhododendron Fortunei, Lindley, Gard. Chron. 1859, 868. It was discovered by Fortune amongst the mountains of Che kiang, about 3000 feet above the level of the sea, the only Rhododendron met in those parts of China, where the Azalea is one of the most common plants. The only other species of Rhododendron in China is R. Championae in Hong kong. F. brought ripe seeds of the Rh. Fortunei, which germinated at Chiswick. The plants have not yet flowered.

In R. 412, F. states that in 1855, end of October, when he visited the high mountains S. W. of Ning po, he came upon a remarkably fine looking Rhododendron covered with ripe seeds, which he sent to Glendin-ning, Chiswick, who raised a good stock of young plants. — When after-wards these plants flowered the *Rh. Fortunei,* was described and figured in

Bot. Mag. t. 5596 (1866). Corolla of a fine pale rose colour, very beautiful. Fragrant odour. — I. F. S., II, 23.

Statice Fortunei, Lindley, Bot. Reg. 1845, tab. 63, and Journ. Hort. Soc. I (1846) 70. The seeds of this yellow flowered Statice, a very interesting species, perennial, were sent home from China by Fortune, who gathered them on the Bay of Chin chew (in October 1843) in sandy soil. Fortune's wild plants are not more than a foot high, while those which have flowered in the garden of the Society, have been twice or thrice as large, or even more. But F.'s herbarium species are far prettier, the small yellow flowers being more compact. Short, close, one-sided racemes. — Fl. d. Serres, II (1846) 28, tab. 9. — According to Bentham, Fl. Hgk. 281 = *Statice sinensis,* Girard, gathered in 1793, by Staunton. — Ind. Fl. Sin., II, 35.

Plumbago Larpentae, Lindley, Gard. Chron. 1847, ·732, figured. Lady's Larpent's Leadwort. This charming little plant is a native of Shang hai, where Fortune found it growing on the ruined ramparts. He sent herbarium specimens (and, it seems, seed). The Hort. Soc., however, did not succeed in raising it. Its introduction is owing to Sir George Larpent, who first cultivated it in his garden at Rochampton. It had been sent to him by Mr. Smith, with the following memorandum, dated Canton, May 16, 1846: "Mr. Fortune tried to get a plant of it but failed; yours is therefore the only one in England. It is very rare even at Shang hai, and I found it on the city wall. It is one of the most ornamental plants I have seen in China". — Flowers of an intense violet with a little red in the throat. Leaves obovate with minute scales on each side, finely serrated and fringed. Stem slender, covered with scales and close pressed hairs. — Coloured drawing in Fl. d. Serres, IV (1848) tab. 307.

This plant was afterwards found to be identical with *Ceratostigma plumbaginoides,* Bunge, who discovered it near Peking in 1831 (v. supra p. 336). — I. F. S., II, 36.

Primula japonica, A. Gray, discovered in 1855, by Ch. Wright in Japan (v. s. 398), was introduced by Fortune from Japan, who procured seeds in 1870. — See Bot. Mag. tab. 5916 (1871).

Lysimachia candida, Lindley, Journ. Hort. Soc. I (1846) 301. This plant was raised from the soil contained in one of the boxes sent from China by Fortune and received by the Society in April 1846. Description. Flowers white in close racemes. — I. F. S., ɪɪ, 48. This plant was earlier known from India.

Lysimachia Fortunei, Maxim. Mél. VI (1867) 270. Fortune, China borealis, 1845. According to I. F. S., ɪɪ, 52, Fortune, Amoy.

Lysimachia Klattiana, Hance, Journ. Bot. 1878, 236. First discovered by Fortune in China, Amoy. — I. F. S., ɪɪ, 53.

Symplocos congesta, Benth. Fl. Hong kong, 211. Fortune, Chusan. — I. F. S., ɪɪ, 72.

Halesia Fortunei, Hemsley in I. F. S., ɪɪ, 75. — Fortune, Amoy.

Styrax Fortuni, Hce, in Journ. Bot. 1882, 36. — I. F. S., ɪɪ, 77 = *Styrax serrulatum,* Roxb., Fortune, Amoy.

Jasminum nudiflorum, Lindley, in Journ. Hort. Soc. I (1846) 153, and Bot. Reg. 1846, tab. 48. Received from Fortune from Nin kin (?) July 24, 1844. A shrub with angular, deep green, trailing branches. Leaves shining, deep green, and each consisting of 3 sessile, ovate leaflets, which fall off early in the autumn, soon after which they are succeeded by large yellow scentless flowers, which grow singly from the buds formed in the axils of the leaves that have previously dropped. Limb of the corolla about one inch in diam. and divided into 6 broad, oblong, blunt, flat segments. The plant was earlier discovered (1831) in Peking by Bunge and described under the erroneous name of J. angulare (v. s. 336). Fortune says that he discovered this species in gardens and nurseries in the north of China, particularly about Shang hai, Soo chow, Nan king (F. never visited Nan king). It is a very ornamental dwarf shrub and seems to be perfectly hardy. The flower-buds expand in early spring, often when the snow is on the ground — and look like little prime roses. It is easily multiplied by cuttings or layers, as it has a tendency to throw out roots at the joints on the stem. The Chinese often graft it on the more common kinds of Jasminum. — Bot. Mag. tab. 4649 (1852). — Fl. d. Serres, VIII (1852) 31, tab. 762. — Jasminum nudiflorum at Shang hai and Tien tsin. (R. 241. — Y. 317). — I. F. S., ɪɪ, 79.

Jasminum paniculatum, Roxb. and *J. Sambac,* Ait., used at Canton for scenting tea. (R. 201).

J. Sambac, mo le wha, is much cultivated in China especially in Fo kien. In the fields in the vicinity of Foo chow F. observed large quantities of it cultivated. It is used to decorate the hair of the ladies, and to garnish the tables of the wealthy. Fortune saw it also in the Amoy gardens and at Tien tsin. (W. 385, 32. — T. 147, 175. — Y. 317.) — I. F. S., ii, 80.

Forsythia viridissima, Lindl., Journ. Hort. Soc. I (1846) 226, 215; — II (1847) 157; — Bot. Reg. 1847, tab. 39. Fortune sent this fine plant from North China to the Society, in the beginning of 1844. It flowered in their garden in 1847. It is a bush with a very rich green colour, something like a Viburnum. A very distinct species, very ornamental. Leaves oblong or oblong lanceolate (not ovate as in the original Forsythia suspensa). Branches 4 cornered (not terete), perfectly erect. Yellow flowers as large as those of Chimonanthus fragrans. Fortune says it is a deciduous shrub with very dark green leaves, which are prettily serrated at the margin. It grows about 8 or 10 feet high and sheds its leaves in autumn. It then remains dormant, but is remarkable for the number of large prominent buds, which are scattered along the young stems produced in the previous summer. Early in spring the flowerbuds gradually unfold themselves, before the leaves, and present a profusion of bright yellow blossoms all over the shrub. F. first discovered it in a Mandarin's garden in the island of Chusan. Afterwards he found it wild amongst the mountains of the interior in Che kiang, where he thought it even more ornamental in its natural state than when cultivated in gardens. — Fl. d. Serres, III (1847) 261.

F. viridissima was seen in the Shang hai gardens, on the hills near Yen chow in Che kiang, among the Ning po Hills, in hedges, in the Ning po gardens. (T. 17, 57, 317. — R. 27, 80, 241). — I. F. S., ii, 82.

Forsythia Fortunei, Lindley, in Gard. Chron. 1864, 412. Fortune discovered this species, which is quite different from F. viridissima, 1861, near Peking and introduced it to the Hort. Society. — Gard. Chron. 1884, I, 312: It was in fine bloom at Chiswick. As an ornamental plant it is a great improvement on Fortune viridissima. See also Y. 320, Tien tsin, new Forsythia. — According to Nicholson, Dict. Gard. F. Fortunei is nothing but the old *Forsythia suspensa,* for a long time known from Japan. — I. F. S., ii, 82.

Syringa oblata, Lindl. in Gard. Chron. 1859, 868. Fortune discovered it first in a garden near Shang hai and was informed by the Chinese there, that is was introduced from the north, and that it is common in the gardens of Peking. Full grown specimens are about the size of our common lilac but more tree-like in their form. The leaves are very striking, being large, rather fleshy and oblately cordate. It blooms profusely. Purple flowers. There is also a white variety. Both are handsome ornamental trees, and both were introduced by Fortune. S. oblata, the North China Lilac, differs from the common Lilac in its leaves being as broad and even broader than they are long, and in the flowers, which are not more than half the size, forming a thin, loose panicle. Mr. Glendinning's nursery contains the only specimens of the purple variety, the white is cultivated by Messrs Henderson, St. John's wood. Fortune says that the Chinese nurserymen propagate these trees by grafting on the Ligustrum lucidum.

S. oblata, both varieties, are much cultivated at Peking, where Bunge saw it in 1831. He believed that it was S. chinensis, Willd., (an obscure plant). — I. F. S., ɪɪ, 83. — A good figure of S. oblata is found in Mr. Sargent's Garden and Forest, 1888, 221.

The *Wax-insect* tree is no doubt a species of Ash, *Fraxinus.* It grows abundantly on the banks of ponds and canals in Che kiang; and a small quantity of wax is also produced in this province. F. received from Dr. Mc. Cartee, of Ning po, some beautiful specimens of the fresh insect upon the branches of this tree. This insect has been named *Coccus pela* by Mr. Westwood (see Gard. Chron. 1853, 484, 532). When fully developed on the trees it has a most remarkable appearance; they seem as if covered with flakes of snow. The wax is an article of great value in Chinese commerce, and a small portion is exported. F. gives a figure of the tree covered with wax. (R. 147, 148).

In Kew Journ. Bot. IV (1852) 153, we read that Fortune brought from China a deciduous tree as the true plant which yields the wax in question, and that this tree then was living at the garden of the Hort. Society, but was not in a condition to enable the genus or family of the plant to be determined. — D. Hanbury, in his Notes on Chin. Materia Med. 1862 (see his Science Papers 271) states that Fortune's wax-insect tree is *Fraxinus chinensis,* Roxb. — I. F. S., ɪɪ, 85.

Fraxinus retusa, Champion, Kew J. B. IV (1852) 330. — Amoy, Fortune, 84. — I. F. S., ɪɪ, 86.

Fontanesia Fortunei, Carrière, Rev. Hort. 1859, 43, fig. 9. This plant was introduced to the Muséum d'hist. nat. Paris in 1854, from Shang hai, by Montigny. Carrière found in the Herbarium of the Museum a specimen gathered by Fortune in China.

Debeaux, in his Fl. Shang hai, 1875, 40, publishes a plant, *Fontanesia phillyreoides,* Labillardière, Pl. Syriae, I (1791) 91, tab. 1, var. *sinensis,* which he had gathered near Shang hai, 1861. Maximowicz, Fl. As. Or. Fragm. 1879, 32, records that in the Herb. Hti Petrop. there is a specimen of the same plant labelled: Fortune, n. 45, a. 1846, probably gathered at Shang hai. Hemsley, I. F. S., II, 87, has ascertained that Carrière's F. Fortunei is the same plant, and retains Debeaux's denomination.

Osmanthus aquifolius, a fine ornamental evergreen shrub or small tree, which F. in 1860, saw in some gardens near Yedo, and which he introduced to England. This genus is closely allied to the olive; it produces sweet-scented white flowers, and has dark-green prickly leaves like the holly. Curiously enough, the leaves on the upper branches and shoots of the Osmanthus are produced without spines, exactly as we see on old holly trees. All the species of Osmanthus have variegated varieties in Japan, many of which are very beautiful objects for garden decoration. (Y. 59, 145). — I. F. S., II, 87. This is the *O. ilicifolius* or *Olea ilicifolia* of English gardeners and also the *Osm. Fortunei,* Carrière in Rev. Hort. 1864, 69, said there to have been sent to England, a few years earlier.

Olea fragrans, Thunberg, the *kwei wha* of the Chinese, is cultivated throughout the Chinese empire for its fragrant flowers. F. observed it in the Fa te gardens near Canton, in Hong kong, at Amoy, Shang hai, Ning po, Tien tsin, largely cultivated, and in autumn, when the trees are in bloom, perfuming the air with the most delicious fragrance. It seems to succeed much better at Shang hai than in the south of China. The kwei wha is one of the most favourite flowers with the Chinese. It forms a good-sized bush, about as large as a lilac, and flowers in the autumn. There are three or four varieties, the main difference between them consisting in the colour of their blossoms. Those kinds which produce brownish-yellow flowers are the finest and are most highly esteemed by the natives. In the north-eastern provinces (Kiang su, Che kiang) the bushes are seen growing near all the villages, and are plentiful in gardens and nurseries. When they are in flower in the autumnal months, the air in their vicinity is literally loaded with the most delicious perfume. One tree is enough to scent a whole garden. In England we know nothing of the beauty of this charming plant.

The flowers of the kwei wha are a source of great profit to the Chinese cottagers, as well as to the nurserymen, who produce them in large quantities for the market. There is a great demand for them in all the large towns. Ladies are fond of wearing wreaths of them in their hair; they are also dried and placed in ornamental jars, in the same way as we do rose-leaves in Europe, and they are used largely for mixing with the finer kind of tea, in order to give it an agreeable perfume. The kwei wha is used in the northern districts as a scent for a rare and expensive kind of "Hyson Pekoe" — a tea which forms a most delicious and refreshing beverage when taken *à la chinoise* without sugar and milk. The Olea fragrans tea will only keep well for one year; at the end of two years it has either become scentless, or has a peculiar disagreable oily odour. (W. 32, 143, 213. — T. 9, 17, 79, 331, 333. — R. 80, 81, 201, 202. — Y. 317). — I. F. S., II, 88.

Chionanthus retusus, Lindley in Paxt. Fl. Gard. III (1852) 86, fig. 273 (black). Fortune obtained the plant in a garden near Foo chow, and introduced it for Standish and Noble. The Chinese call it *ting heang* (a name properly applied to Syringa). A very pretty bush. Flowers snow-white, singularly fragrant, and on this account it is much prized by the natives of Fo kien. Fortune suspects it has been brought from a more northern latitude, perhaps from Japan or Lew kew. The Chinese propagate it by grafting on Olea fragrans. The flowers are produced in slender, terminal, somewhat whorled panicles. The corolla has a distinct tube, rather longer than the subulate sepals, and is 4—5 lobed.

Maximowicz in Mél. IX (1874) 393, described *Ch. chinensis*, a new species based upon specimens gathered by Tatarinov 1847, near Peking and one, received from the Hort. Soc. London and labelled: China borealis, Fortune, A 37, a. 1845. According to Hemsley, I. F. S., II, 88, this plant is identical with Lindley's Ch. retusus, which latter name has many years priority, and Fortune's herbarium specimens of the plant are from Fu chou and Amoy.

The Kew Journ. Bot. IV (1852) 153, refers to a communication received from Fortune regarding *Ligustrum lucidum*, Ait. With reference to a statement, found in the Bot. Mag. sub tab. 2565 (1825) that from the berries of L. lucidum in China a vegetable wax is procured, F. reports that after careful inquiry on the matter, in districts where this shrub abounds (Che kiang, Kiang nan), he could not learn that any such substance is yielded by it. F. further told D. Hanbury (Science Pap. 67), that he also never observed the wax-insect upon L. lucidum, which the Chinese call *tung tsing*. Hanbury, however, is wrong in assuming that the wax-insect never feeds upon Ligustrum, for in Sze ch'uan it is one of the insect-wax-trees.

Decaisne in Monogr. Ligustrum (1878) 20, notices a herbarium specimen of L. lucidum received from the Hort. Society London and labelled: Shang hai, Fortune, 1856. The I. F. S., I, 92, sub L. lucidum, does not mention Fortune.

Ligustrum coriaceum, Carrière, Rev. Hort. 1874, 418, fig. 56 (black) and 1888, 440, fig. 101. Introduced from Japan by Fortune to Standish. — I. F. S., II, 90. Between 1861 and 63 Oldham gathered the same plant in Japan or in the Corean Archipelago.

Ligustrum sinense, Lour. — Lindley in Gard. Chron. 1858, 621, figured, and 1878, II, 364, figured. Fortune introduced it to Glendinning's nursery (1855), probably from Shang hai, 1855. It has small fragrant, white flowers, succeeded by small pear-shaped apiculate purplish berries, the size of small shot.

Rhynchospermum jasminoides, Lindley in Journ. Hort. Soc. I (1846) 74. Black and white figure. Ibidem p. 221. F. sent living plants, taken in Shang hai, to the Society in the autumn of 1844. A slender, climbing, evergreen shrub, rooting along its branches wherever it touches a damp surface, like ivy. When wounded the branches discharge a milky fluid. Leaves opposite, oval, smooth, sharp pointed. Flowers white, deliciously sweet-scented, and produced in irregular corymbs on the ends of peduncles, considerably longer than the leaves. Corolla about $^3/_4$ inch long, pure white, salver-shaped. — The same figured in Paxt. Flow. Gard. II (1851) 26, fig. 147 (black). — Fl. d. Serres, VI (1850) 263, tab. 615 (colour.). — Bot. Mag. tab. 4737 (1853). — Lemaire, Jard. Fleur. I (1851) tab. 61 (colour.), named it *Trachelospermum* (new genus) *jasminoides*, under which name it figures in I. F. S., II, 99. — This plant has been for a long time known from Japan, Thunberg's Nerium divaricatum.

Buddleia Lindleyana, F. in literis, Lindley in Bot. Reg. XXX, 1844, Miscellaneous n. 25 and XXXII, 1846, t. 4. This was one of the earliest plants found by F. upon his arrival at Chusan. He immediately (Nov. 13, 1843) sent home seeds with a particular request that this species, if new, might bear the above name. In little more than three months after they were posted in Chusan, plants were growing in the garden of the Hort. Soc. F. at the same time transmitted a Chinese drawing of the plant, which represented it as of considerable beauty. F. also sent dried specimens of it. It is a Scrophulariacea. F. describes it as a handsome small bush. Flowers in close terminal racemes, about 2 inches long, and are themselves $^3/_4$ of an inch in length. Their colour is a rich violet or lilac. Lindley says: Flowers in spikes of a deep, rich violet, a little verging upon gray, on account of the numerous short hairs with which they are closely covered. — The plant is also figured in Fl. d. Serres, II (1846, May) tab. 9 (colour.); — in Paxt. Mag. Bot. XIV (1848) 5 (colour.).

F. met with this graceful plant first in Chusan; its long spikes of purple flowers hung in profusion from the hedges on the hill sides. He introduced it from Chusan to Hong kong, where it is almost always in bloom, although the flower spikes are not so fine as they are in a colder climate. He found the B. Lindleyana in a wild state on the hills above Yen chow foo in Che kiang, and in the Bohea (Woo-e-shan) Hills. — I. F. S., II, 119.

Gentiana Fortunei, W. J. Hooker in Bot. Mag. tab. 4776 (1854). A tall robust growing plant. Large blue flowers, a native of North China, exact locality not stated. It was sent by Fortune, in 1849, to Standish & Noble, Bagshot nursery. The blossoms expanded in December 1853. — Fl. d. Serres, IX (1854) 231, t. 947. — Illustr. Hort. 1854, t. 36. — In Gard. Chron. 1867, 212, G. Fortunei is styled the "queen of the genus". — In the autumn of 1853, F. saw a fine species of Gentian in the Snowy Valley, S. W. of Ning po. According to the Chinese an excellent stomachic. (R. 184). — According to I. F. S., II, 134, Fortune's plant is to be reduced to *G. scabra*, Bunge (1832), Siberia.

Convolvulus reptans, Lin., much cultivated as a vegetable in the southern provinces of China. (W. 307).

Calystegia pubescens, flore *pleno*, Lindl., in Journ. Hort. Soc. I (1846) 70 (see also 221). Figured black and white. Raised from a small portion of a root, which F. sent in the autumn of 1844, from Shang hai, as a double Convolvulus, which was supposed to be dead when received at the garden.

This curious plant approaches very nearly to the common Convolvulus sepium, from which it differs in having firmer and smaller leaves. Double flowers, as large as those of a double Anemone, but the petals are arranged with the irregularity of the Rose. They are of a very delicate pale pink. No trace of stamens or pistil. It is the first plant of its order, that has been mentioned as producing double flowers. It flowered in 1845. — The same figured in Bot. Reg. 1846, t. 42. — Fl. d. Serres, II (1846) t. 172.

In I. F. S., II, 164, Lindley's C. pubescens is reduced to *Calystegia hederacea*, Wall., known for a long time from India.

Sweet Potatoes (Ipomoea Batatas, Lam.) are largely cultivated in China. F. observed them in the neigbourhood of Foo chow, amongst the mountains south-west of Ning po, in Chusan, near Tien tsin. They belong to the summer crops of the most fertile hilly country from the province of Fo kien northward to the great valley of the Yang tze kiang, where they form the staple production, if we except rice. In the southern provinces, when the winters are mild, the roots of the sweet potato frequently remain in the ground all the winter. In the north the cold is too severe, and consequently the natives are obliged to dig up and protect the roots. In April those roots which have been saved for "seed" are planted thickly in beds near the houses or in the corners of the fields. They begin to push out their young shoots immediately. About the 10th or 12th of May they are planted in the ground, which in the meantime has been prepared on the hill sides, and then grow as readily as couch grass. It is astonishing how well they succeed, considering the little care spent upon them. But it must be kept in mind that this is the commencement of the rainy season at the change of the monsoon. The sweet potato succeeds better farther north than in the southern provinces. In China the sweet potato never produces seeds. (W. 53, 308, 381. — R. 27, 188. — Y. 350).

Egg-aples or *Brinjals (Solanum Melongena)* grow to a gigantic size at Tien tsin, some of them measuring 18 inches in circumference. (Y. 339, 350).

Tobacco is extensively grown in the Bohea Mountains, as it is in all parts of the province of Fo kien. (W. 381, 384. — T. 251).

Campsis Fortunei, Seem. Journ. Bot. 1867, 373. China, Fortune. Hemsley, I. F. S., II, 180 = *Paulownia Fortunei.*

Pterostigma grandiflorum, Benth. Scroph. Ind. (1835) 21. This plant is noticed by Lindley in Journ. Hort. Soc. I (1846) 66 and figured in Bot. Reg. 1846, tab. 16. It was cultivated in the Hort. Soc.'s garden. Received July 30, 1843, from Fortune, who sent it from Hong kong as a blue-flowered herbaceous plant, growing on hill-sides and near streams. Leaves opposite, stalked, ovate, crenated, very much marked with sunken veins and deep green. Flowers as large as those of a Digitalis, grow singly in the axils of the leaves. This is Loureiro's *Digitalis sinensis,* Linnaeus' *Gerardia glutinosa,* common about Macao and Canton.

Maxim. Mél. IX (1874) 405, notices a specimen in the herbarium of the Bot. garden, St. Petersb., labelled: Pu toy Island, Fortune 91. This is a small island south of Hong kong where Wright gathered the same plant (v. supra p. 395, note). But perhaps F. means Pu tu I., Chusan group. — I. F. S., II, 185 = *Adenosma grandiflora,* Bentham, China, Fortune, 91.

Torenia concolor, Lindley in Bot. Reg. 1846, tab. 62. This plant is probably regarded in herbaria as T. asiatica, Lin., but living specimens forbid its union with that species. The leaves are roundish-ovate, or even cordate, and by no means ovate-lanceolate; the flowers have no side spots. It was received from China in the garden of the Hort. Soc. in July 1844, and Fortune, who sent it, wrote the following memorandum: "This plant was found growing in marshy ground, on the mountain of Hong kong, nearly 2000 feet above the level of the sea. It flowers in the autumn. After the flowering season is past, the dry weather sets in, the stems and leaves shrivel up, and remain in this state during the winter. During the hot and damp summer months, it grows again with great vigour and forms its flowers in profusion". — Bot. Mag. t. 6797 (1883). — I. F. S., II, 187.

Monochasma Savatieri, Franchet was first discovered by Fortune, near Amoy. — I. F. S., II, 203.

Cyrtandra Fortunei, Clarke in DC. Monogr. V (1883) 251. Fortune, China. — I. F. S., II, 224.

Lysionotus pauciflorus, Maxim. Mél. IX (1874) 366. Maximowicz, Japan, 1863. — I. F. S., II, 225: Fortune, China.

Chirita sinensis, Lindley, Bot. Reg. 1844, t. 59. This charming little greenhouse plant is one of the first results of any importance, from the voyage to China, by Fortune, on account of the Hort. Society. It was sent home (in the beginning of 1844), and its beautiful large lilac, foxglove-like flowers were open, when it arrived. — This plant was found wild and cultivated in Hong kong. (W. 10. — T. 8). — Fl. d. Serres, I (1845) 41 (colour.). — Bot. Mag. tab. 4284 (1847). — Paxt. Mag. Bot. XIV (1848) 243). — I. F. S., II, 232.

The *Oily Grain, Sesamum orientale,* is extensively grown on the plain of Tien tsin, from the Gulf of Pechele to the mountains beyond Peking, and is fully twice as large (5 feet) and productive as that grown in the south. (Y. 338, 350).

In 1853, F. discovered in the province of Che kiang an interesting tinctorial plant. In the mountains south and south-west of Ning po and also amongst the Fung hwa Mountains to the westward of Ning po, there are large quantities of a *blue dye* produced, which is in fact the indigo of this part of the country, like that yielded by the Isatis indigotica (supra p. 433) is the indigo of Shang hai. It is made from a species of *Ruellia* for which F. proposes the name *R. indigotica.* It is a curious circumstance that the same plant, apparently, has lately been discovered in the Assam country in India, where it is also cultivated for the blue dye it affords. This Ruellia succeeds admirably well as a summer crop in Che kiang. It is planted in the end of April or beginning of May, and is cleared from the ground in October. During this period it attains a height of a foot or $1\frac{1}{2}$, becomes very bushy, and is densely covered with large green leaves. It is cut down before any flowers are formed. The roots which are left in the ground after the gathering season are all destroyed by the first frost of winter. The Chinese cultivator does not depend upon these for the crop of the following year. Cuttings are found to be much more vigorous and productive than the old roots. When the stems are cut down for the manufacture of indigo, a sufficient quantity have their leaves stripped off and the stems, tied in large bundles, are carried to a dry shed, where they are packed closely and firmly together, and banked round with very dry loam. When the winter has passed, a few roots are found formed. In this state they are taken to the fields and planted, and soon grow luxuriantly. For preparing the dye-stuff large quantities of stems and leaves are thrown into one of the tanks on the edges of the fields, and allowed to remain for about five days, when they become partially decomposed, and are removed by means of large brooms made of bamboo twigs. The water is then well stirred with the brooms, and kept in a rapid circular motion, and about 40 pounds of lime thoroughly mixed with it, after which it is beaten with bamboo rakes and then allowed to settle for three or four days. By this time the colouring matter has separated itself from the water, which is now entirely drained off — the dye occupying three or four inches of the bottom in the form of a thick paste and of a beautiful blue colour. In this state it is packed in baskets, and exposed for sale in all the country towns in this part of China. Like the Shang hai indigo it is called *tien ching* by the Chinese. During the season of its preparation every mountain stream in the highland valleys in the Ning po district is coloured and polluted with the refuse liquid drawn off from the tanks, and the stench which fills the air is almost unendurable. (R. 158—163).

I. F. S., II, 239, Fortune's plant is the *Strobilanthes flaccidifolius*, Nees in DC. Prod. XI (1847) 194, first discovered by Griffith in Assam. Figured in Bot. Mag. tab. 6947 (1887).

Polycoelium chinense, Alph. DC. Prodr. XI (1847) 706. China, Fortune, n. 12. — I. F. S., II, 250 = *Myoporum bontioides*, Asa Gray in Proc. Am. Acad. VI (1862) 52. — Before Fortune known from Japan.

Premna microphylla, Turczan. Bull. Nat. Mosc. 1863, II, 217. China, Fortune, A 23. — I. F. S., II, 256. Earlier known from Japan.

Clerodendron cyrtophyllum, Turczaninov, l. c. 222. China borealis, Fortune. — I. F. S., II, 259. Cantor a few years earlier, Chusan (v. supra p. 360).

Clerodendron foetidum, Bunge (nec. Don., v. supra p. 338), figured in Bot. Mag. tab. 4880 (1855). A charming species from North China first discovered by Bunge (1831) at Peking, and more recently introduced by Fortune (about 1850) to Standish & Noble. Large terminal corymbs, compound, hemispherical. Flowers rose-coloured, very numerous, compact. — Decaisne, in Rev. Hort. 1851, 405. — *Clerod. Bungei*, Steud. Nom. Bot. Under the latter name figured in Fl. d. Serres, IX (1853) 17, t. 863. — Rev. Hort. 1866, 470, Carrière. — I. F. S., II, 259.

Clerodendron? Fortunei, Hemsley in Ind. Fl. Sin., II, 259. China, Fortune, 20.

Mastacanthus sinensis, Lindley in Bot. Reg. 1846, tab. 2. This is an autumn-flowering herbaceous plant, growing from $1\frac{1}{2}$ to 2 feet high, and forming neat little, bushy tufts. It furnishes an abundance of rich, violet blossoms. Fortune sent it from China to the Hort. Society in whose garden it flowered in October 1845. Fortune found it wild in the neighbourhood of Canton, in Chusan and at Koo lung soo. It was known much earlier. It is the *Barbula sinensis* of Loureiro. — I. F. S., ii, 265 = *Caryopteris Mastacanthus,* Schauer in DC. Prod. XI, 625.

? *Teucrium nepetaefolium,* Bentham, in DC. Prod. XII (1848) 580. China, Fortune, A 73. — Maxim. in Mél. IX (1876) 830, named the plant *Caryopteris nepetaefolia.* — I. F. S., ii, 264.

Salvia Fortunei, Bentham in DC. Prod. XII (1848) 354. China, Fortune, 82. — I. F. S., ii, 284, reduced to *S. japonica,* Thunb., Japan.

Lamium chinense, Benth., l. c. (1848) 512. China, Fortune, A 83. — I. F. S., ii, 303.

Teucrium Fortunei, Benth. l. c. (1848) 583. China, Fortune, 71. — Maxim. Mél. IX (1876) 827, reduces this to *T. quadrifarium,* Hamilton, India, 1825. — I. F. S., ii, 314.

In Jan. 1845, the Hort. Society received from Fortune the seeds of a Chinese vegetable cultivated in Chusan under the name of *han tsi* and used as spinage. He also sent a drawing. Lindley cultivated, described and figured it, in Journ. Hort. Soc. I (1846) 72. It proved to be *Amarantus oleraceus,* Lin.

Fortune also sent seeds of a variety of this, cultivated near Shang hai, under the same name.

Buckwheat cultivated near Ning po in the mountains, and in other parts of Che kiang. (T. 177. — R. 188). — DC. Prodr. XIV, 143. *Fagopyrum esculentum,* Moench, Fortune, China, 111. — I. F. S., ii, 339.

Aristolochia Sinarum, Lindley in Gard. Chron. 1859, 708. Short diagnosis. We received this from Glendinning, to whom it (the living plant) had been sent from China by Fortune. It is a hardy perennial with long, branching, scrambling stems, deep green leaves, a little glaucous beneath, and dull green flowers about 2 inches long, with a dark dull purple stain at the mouth of the tube. The whole plant when bruised has the heavy oppressive smell of the genus. — I. F. S., II, 363.

Saururopsis chinensis, Turcz. Bull. Nat. Mosc. XXI, (1848) I, 590. China borealis, Fortune, 102, A.

Turczaninov's new genus Saururopsis is now reduced to *Saururus*, and according to Maxim., Fl. As. Fr. (1879) 54, Fortune's plant is *S. Loureiri*, Decaisne *(Spathium chinense*, Lour.). — I. F. S., II, 363.

In W. 213, 385, — T. 146, F. states that *Chloranthus inconspicuus* is cultivated near Foo chow and in other places for its blossoms, which are used for mixing with tea. But in R. 201, he declares this latter statement to be a mistake, the Chloranthus never being used for this purpose.

Camphor trees, are abundant in the province of Che kiang and in the island of Chusan where they attain a very large size. Fortune observed fine specimens in the valley of Yen chow foo, in the Bohea mountains, in the Ning po hills, where the Camphor tree is the most striking of all the trees and quite the monarch of the woods. But no camphor is extracted or exported from this part of China. (W. 57. — T. 56, 61, 165, 170, 176, 195, 224, 249. — R. 29, 51). — Some camphor trees of enormous size were observed about the temples on the outskirts of Nagasaki, in Japan. (Y. 20).

Chinese Cinnamon, cultivated in gardens at Hong kong. (T. 7).

Daphne Fortunei, Lindley, Journ. Hort. Society, I (1846) 147, and II (1847) 34, with a coloured plate. A small, downy, branched bush with thin deciduous ovate-oblong leaves covered with soft fine hairs. Flowers pale bluish lilac, arranged in clusters of 4 upon branches scarcely beginning to put forth their leaves. They are more than an inch long, covered externally with soft, closely pressed hairs, and divided on the border into 4 lobes. Fortune found it wild amongst the Chusan and Ning po Hills. The plant sent by him to the Hort. Society he took from a nursery garden near Shang hai, in the winter of 1843. It was then leafless. When taken down to the south of China, it flowered before it was sent to England. It reached the Soc.'s garden in good condition, July 26, 1844, and flowered in England for the first time in January 1846. According to F. the Chinese call this plant *nu lan ee* and use the bark, which is acrid and poisonous, to produce blisters on the skin in cases of rheumatism. Like the English Mezereum it is the harbinger of spring. In April the flowerbuds expand, and then the whole of the hill sides are tinged with its beautiful lilac-coloured blossoms. — See also W. 320, — T. 57, — R. 241.

Flore d. Serres, III (1847) t. 208. — According to Maxim. Mél. XI (1881) 310, D. Fortunei is only a variety of the Japanese *D. Genkwa*, Sieb. & Zucc. Fl. Jap. I (1840) 137, t. 75. — I. F. S., II, 395.

Edgeworthia chrysantha, Lindley, Journ. Hort. Society, I (1846) 148, and Bot. Reg. 1847, tab. 48. This plant was received by the Hort. Soc. on the 9th of April 1845. It had been sent by F. together with a Chinese drawing from Chusan as a deciduous shrub, producing bunches of yellow sweet-scented flowers. In the Soc.'s garden it flowered for the first time in Febr. 1847. In order to induce the plant to flower, the Chinese bind the stems so as to form a loop; and this practice has been followed in the Chiswick garden. — This is a dwarf soft-wooded shrub, throwing up rod-like dull-green stems from its base, and bearing the leaves exclusively near the ends. The leaves are about 8 or 9 inches long, oblong lanceolate, stalked, very dull green and covered with fine hairs. Flowers bright golden yellow, something less than an inch long, covered with exceedingly thick hair on the outside and collected into balls, about 2 inches in diameter, at the ends of the shoots. They are sweet-scented and appear in Chusan in July. — F. found this plant also in the gardens of Shang hai, and in a wild state amongst the hills above Yen chow foo in Che kiang. (T. 57, 316, 317). — Fl. d. Serres, III, (1847) t. 289. — E. papyrifera, Sieb. & Zucc. Fl. Jap. Fam. nat. (1846) n. 694, from Japan, is the same. — I. F. S., II, 396, 401.

Wikstroemia nutans, Champ. (v. s. p. 378). China, Fortune, 174. — I. F. S., II, 400.

Elaeagnus tenuiflora, Bentham in Kew Journ. Bot. V (1853) 197. China, Fortune, 114. — Maxim. Mél. VII (1870) 561, reduces this to the Japanese *E. glabra*, Thbg. — I. F. S., II, 402.

Buxus Fortunei, Carrière, Rev. Hort. 1870, 519. — Gard. Chron. 1871, 1615. Introduced by Fortune from China. — I. F. S., II, 418 = *B. sempervirens*, Lin.

Glochidion Fortunei, Hance, Manip. (1863) 228. Fortune China, probably Amoy, 42, 129. — I. F. S., ɪɪ, 424.

The *tung eu* (yu), *Dryandra cordata,* Thunb. furnishes a valuable oil found in its seeds, which is much used by carpenters. The oil is mixed with the celebrated varnish of the country and hence this tree is often called the «Varnish tree». It is common in Silver Island (near Chusan) where much tung oil is produced and exported. F. saw the tree also in the Snowy valley near Ning po, and in the valley of Yen chow foo in Che kiang. (T. 57, 119, 170. — R. 184). — I. F. S., ɪɪ, 433, *Aleurites cordata,* Steud.

The Castor oil plant, *Ricinus communis,* Lin. is much cultivated in the southern provinces of China. Summer crop. (W. 307).

The *Tallow* tree, *Stillingia sebifera,* is abundant in the valleys of Chusan, and large quantities of tallow and oil are yearly extracted from its seeds: tallow mills are erected in several parts of the island. In Silver Island also vegetable tallow is an article of export. On the sides of the roads and scattered over the hills F. observed large quantities of the tallow tree. The seeds are carefully gathered by the natives, and are valuable for the oil and tallow which they contain. — In the Ning po district the tallow tree is very abundant and occupies a prominent place on the edges of the fields and canals, as well as on the hill sides. It is a most important tree to the Chinese. When F. in the autumn of 1848 proceeded to the green tea district of Hwuy chow, in ascending the Tsien tang River and its northwestern affluent, he found the tallow tree extensively cultivated in the valleys especially at Yen chow foo, and at this season of the year the trees were clothed in their autumnal hues, producing a striking effect upon the varied landscape. The leaves had changed from a light green to a dark blood-red colour. In the next year on his visit to the Bohea tea districts he again saw large quantities of the same tree in southern Che kiang and in the Bohea Mountains. The tallow tree is indigenous to Hong kong, but no use is made there of its fruit. (W. 11, 56. — T. 48, 58, 61, 112, 113, 117, 119, 165, 169, 176, 249, 254, 298).

In W. 65, Fortune gives a valuable account of the Chinese method of extracting the tallow from the seeds of the Stillingia sebifera, for which he was indebted to Dr. Rawes, of the Madras army, who was some time resident in Chusan: "The seeds are picked at the commencement of the cold weather, in November and December, when all the leaves have fallen from the trees. Having been separated from the stalks, they are then put into a wooden cylinder, open at the top, but with a perforated bottom. This is placed over an iron vessel containing water, by which means the seeds are well steamed, for the purpose of softening the tallow and causing it more readily to separate. When the seeds have steamed 10 minutes or more, they are thrown into a large stone mortar and are gently beaten with stone mallets for the purpose of detaching the tallow from the other parts of the seed. They are then thrown upon a sieve, heated over the fire, and sifted, by which process the tallow is separated. The other part of the seed is ground and pressed for oil. The tallow now resembles coarse linseed meal, but with more white spots in it, and derives its brown colour from the thin covering over the seed, between it and the tallow, which is separated by the pounding and sifting. In this state it is put between circles of twisted straw, 5 or 6 of which are laid upon each other, and thus forming a hollow cylinder for its reception. When this straw cylinder has been filled, it is placed in the press, which is a very rude and simple contrivance, but answers the purpose remarkably. The tallow is pressed out by means of wedges driven in very tightly and passes through a hole in the bottom of the press into a tub, which is sunk there to receive it. It is now freed from all impurities, and is a semifluid of a beautiful white colour, but soon gets solid, and in cold weather is very brittle. The tallow is then removed in a solid state from the tubs, and in this condition the cakes are exposed for sale in the market. As the candles made from this vegetable tallow have a tendency to get soft and to melt in hot weather, they are commonly dipped in wax of various colours, as red, green, yellow. The cake, or refuse, which remains after the tallow has been pressed out of it, is used for fuel, or to manure the land, and so is the refuse from the other part of the seeds from which oil is extracted".

I. F. S., ii, 445, *Sapium sebiferum*, Roxb.

In R. 259, Fortune states that a gigantic species of *Hemp*, *Cannabis*, growing from 10 to 15 feet in height, is one of the staple summer crops in the district of Ning po. It is chiefly used in making ropes and string of various sizes, such articles being in great demand for tracking the boats up rivers, and in the canals of the country.

This is evidently the *C. gigantea*, Delile, Ind. sem. hti Monspel. 1849, raised from seeds received from China, 1846.

During his travels in China, Fortune passed three times, and at various seasons, through the famous silk districts in the northernmost part of Che kiang, where the *Mulberry* tree is the principal object of cultivation, viz: in May 1845, when proceeding from Cha poo to Shang hai, in October 1848, when he travelled by boat from Shang hai via Kea hing fu to Hang chow foo; and finally, in June 1855, he visited the great silk country of Hoo chow foo, south and south-east of the Tai hoo Lake. In his books he gives interesting accounts of the cultivation of this useful tree in China, and the gathering of its leaves employed for feeding the silkworm. Over all this part of the country the mulberry tree is observed in great abundance on the banks of the canals and in patches. The country, when viewed from a distance, resembles a vast mulberry garden, and when the trees are in full leaf, it has a very rich appearance. The whole surface of the country is cut up, and embankments formed for the cultivation of the mulberry. It appears to grow better upon the surface and sides of these embankments than upon level land. It is on the banks of canals, ricefields, small lakes and ponds, where the mulberry is generally cultivated, and where it is most at home. The variety of mulberry cultivated in these districts appears to be quite distinct from that which is grown in the southern parts of China, and in the silk districts of India. Its leaves are much larger, more glossy, and have more firmness and substance than any other variety. It may be that this circumstance has something to do with the superior quality of the silk produced in these districts, which is considered as being the finest in China. In 1845, F. sent a living mulberry plant of this variety to England, in order to determine it there, and this safely arrived *). This peculiar variety is not reproduced by seed, and hence all the plantations are formed of grafted trees.

The trees, or rather bushes are planted in rows, the banks of the canals being a favourite situation, and they are not allowed to grow more than from 4 to 6 feet in height, for the convenience of gathering the leaves. Before the leaves have been taken off, the plants seem in a high state of health, producing vigorous shoots and fine large and thick shining leaves. The young shoots with their leaves are then cut off with strong scissors close by the stump and carried home to the farm-yard to be plucked and prepared for the worms. The bushes now look like a collection of dead stumps, and in the middle of summer have a curious wintry appearance. But the rain, which falls copiously, and the fertility of the soil revive a

---

*) It is unknown to me whether Fortune's plant has ever been examined by competent botanists, but it is probably identical with the variety *latifolia* of *Morus alba*, Lin., described by Bureau in DC. Prodr. XVII (1873) 244, from specimens gathered in Che kiang by Hedde.

succulent plant like the mulberry. The silkworms are fed in the cottages, and commonly kept in dark rooms, in round bamboo sieves placed upon shelves. (W. 361—363. — T. 26. — R. 339, 343—347, 367). In R. 345, 347, F. gives two figures of the Chinese mulberry tree to show its habit and form. — I. F. S., II, 455.

Japanese paper is made chiefly out of the bark of the Paper mulberry, *Broussonetia papyrifera*. It is particularly well suited for decorative purposes, such as the papering of rooms. It has a glossy, silky, and comfortable appearance. For some reasons it is made in very small sheets. (Y. 122).

Ficus nitida, the Chinese Banyan is common in the south of China. Fortune observed it in Hong kong, Canton, Amoy and Foo chow. It is always seen near villages and temples, where its dark green leaves and white spreading branches afford an agreeable shade from the fierce rays of the sun. It attains a large size and sends down long roots, forming a very ornamental tree. It grows rapidly with but little care. But this tree, so common around all the houses and temples in the south, is unknown in Chusan. (W. 11, 26, 56, 136, 371, 381. — T. 6. — R. 217). — Thunberg's *F. nitida*, Japan in the same as *F. retusa*, Lin., India.

There is a plant in Chusan called *Urtica nivea*, both wild and cultivated, which grows about 3 or 4 feet in height, and produces a strong fibre in the bark, which is prepared by the natives, and sold for the purpose of making ropes and cables. The same species is said to furnish a very fine fibre, which is used in the manufacture of grass cloth. (W. 53). — Urtica nivea, the plant which is supposed to produce that beautiful fabric made in the Canton province, and largely exported to Europe and America, is also abundantly grown in the western part of Che kiang and in the adjoining province of Kiang si. Fabrics of various degrees of fineness are made from this fibre, and sold in these provinces; but Fortune did not see any so fine as that made about Canton. It is also spun into thread for sewing purposes, and is found to be very strong and durable. There are two very distinct varieties of this plant common in Che kiang — one the cultivated, the other the wild. The cultivated variety has larger leaves; on the upper side they are of a lighter green, and on the under they are much more downy. The stems also are lighter in colour, and the whole plant has a silky feel about it, which the wild one lacks. The wild variety grows plentifully on sloping banks, on city walls, and other old and ruinous buildings. It is not prized by the natives, who say its fibre is not so fine, and more broken and confused in its structure than the other kind. The cultivated kind yields three crops a year. (R. 259). — *Bochmeria nivea*, Hook. & Arn.

Fortunea chinensis, Lindley in Journ. Hort. Soc. I (1846) 150, 221. Genus nov. Black and white figure. Fortune sent seeds of this plant and herbarium specimens (in the autumn of 1844) to the Hort. Soc. from the hills of Chusan and Ning po. An empty cone of this singular plant was received some years earlier from Dr. Cantor. It was at that time supposed to belong to some unknown conifer. This is a plant like Rhus in aspect but a most curious genus of the order Juglands. A walnut reduced to the size and texture of a seed of the Alder tree, and then many collected into a small cone composed of hard, brittle, sharp pointed scales. This shrub or tree is perfectly distinct from all the other genera of Juglands. — Paxt. Fl. Gard. II (1851) 98, (no figure). The shrub has pinnated leaves and cones of green flowers. The Chinese use the fruit to dye the black colour of their cloths. — Fl. d. Serres, IV (1848) t. 331 (black). — Revue Hort. 1863, 13. The plant flowered with Leroy.

DC. Prod. XVI, II (1864) 145, identifies this plant with *Platycarya strobilacea*, Sieb. & Zucc. first described and figured in Abh. Münch. Acad. III (1843) 742, tab. 5, from Japan; also figured in Fl. Japan II (1870) 87, tab. 149.

There is a fruit called *yang mae* much cultivated in the island of Chusan. It is of a scarlet colour, not unlike an arbutus or strawberry, but having a stone like a plum in the centre. The yang mae is also cultivated in various parts of Che kiang. F. saw groves of the tree near Hang chow foo and in the Ning po district. It is a species of Myrica, allied to the Himalayan *M. sapida,* noticed by Frazer, Royle and others. The latter is the *kaiphul* of the hill tribes of India, but far inferior to the Chinese kind. The yang mae fruit is brought to the market in Chusan, in June and is sold at a very cheap rate. The natives consider it a great luxury. In the summer of 1850, F. visited one of the yang mae plantations on the hills in the centre of the island. The trees are bushy, round-headed, and from 15 to 20 feet high, and were at this time loaded with dark red fruit, not unlike the fruit of our arbutus, although very differently formed and much larger. F. observed two kinds, one with red fruit, and the other with fruit of a yellowish colour. The natives were busily engaged in gathering the fruit and packing it in baskets for the markets. Large quantities are consumed in Ting hae, the capital of Chusan, and a great deal is taken across to the mainland. The streets of Ning po are crowded with it during the season. It is a very fine fruit. — In 1853, F. saw yang mae plantations amongst the hills 10 or 12 miles N. W. of Ning po, and procured some grafted plants for India, which succeeded well in the North-West Provinces. The Chinese graft the fine variety of yang mae upon the wild kind, which they call the *san* or hill variety. In R. p. 65, the yang mae tree is figured. (W. 58. — T. 30, 345. — R. 52, 65).

DC. Prodr. XVI, ɪ (1864) 151, *Myrica Nagi*, Thunb. This is the species noticed by Fortune. It is much cultivated for its fruit in Japan and in China. — Bot. Mag. tab. 5727 (1868). — Gard. Chron. 1869, 136. Fortune introduced it into England.

Quercus inversa, Lindley in Paxt. Flow. Gard. I (1850) 58, fig. 36 (black). An evergreen oak from the north of China introduced by Fortune to Standish & Noble, Bagshot nursery (1849). Leaves deep green, shining on the upper side, coriaceous, obovate, petiolate, cuspidate, only at the apex serrate. — Seemann, Botany Herald, 414, reduces this species to *Q. Thalassica*, Hance, discovered in Hong kong before 1848, as does also DC. Prodr. XVI, ɪɪ (1864) 84.

Quercus sclerophylla, Lindl., l. c. 59, fig. 37 (black). Sent by Fortune from China (together with the last mentioned) to Standish & Noble. A much finer oak than the last. Large leaves, coarsely serrated. — DC. Prod. l. c. 81. — Fortune in T. 250, 304, states, that he discovered this and the Q. inversa, both with evergreen large glossy leaves, not unlike the Portugal Laurel at a distance, highly ornamental, amongst the Bohea Mountains in 1849 (July or August).

Quercus bambusaefolia, Fortune in Gard. Chron. 1860, 170. This fine evergreen oak, sold under this name in England, was found wild by F. in the mountains of Che kiang. Full-sized trees are from 30 to 50 feet high, and are very ornamental. The beautiful and rare *Dicronocephalus Wallichii* is generally found on this species. This oak is supposed to be hardy in England, and, if so, it will be a valuable introduction. It is probably distinct from the Hong kong species, which has been published under this name. (Hance).

Mr. F. B. Forbes, in Journ. Bot. 1884, 85, has proved that Fortune's

Q. bambusaefolia, which he brought, in 1854, from the Che kiang Hills, and gave to Glendinning, is identical with the Japanese oak described by Franchet, Enum. Jap. II (1876) 498, as *Q. Vibráyeana*. Hance's Quercus bambusaefolia is quite different (v. s. 369).

In 1861, September, F., met amongst the hills west of Peking, a new oak tree, *Quercus sinensis,* of great interest and beauty. It grows to a goodly size — 60 to 80 feet and probably higher — has large glossy leaves, and its bark is rough, somewhat resembling the corktree of the south of Europe. Its acorns were just ripe, and were lying in heaps in all the temple-courts. They are used in the manufacture of some kind of dye. F. secured a large quantity of these acorns, which he sent to Standish, in whose nursery at Bagshot they were subsequently growing luxuriantly. Y. 382).
This is Bunge's *Q. chinensis* (v. supra p. 340).

In the summer of 1853, when visiting the hills about 12 miles north-westward from Ning po, F. for the first time in China met there amongst the woods with the *Chesnut.* This discovery was of great importance, as he was most anxious to introduce this to the Himalayan Mountains in India. Many attempts had been made to introduce it from Europe, but they had not succeeded. The seeds of such trees as oaks, chesnuts, tea,, etc, retain their vitality for a very short time after they are gathered if they are not sown and allowed to vegetate. It is therefore useless to attempt to send these seeds in dry paper parcels or in hermetically sealed bottles from Europe to the north of India. The chesnuts which F. had met with in the markets of China, although excellent for the dessert, were generally too old for vegetating. But now, when he had discovered the locality where they grew, there was no longer any difficulty in procuring them quite fresh. There are *two* species cultivated in these hills. One is somewhat like the Spanish, and, although probably a different variety, it produces fruit quite equal in quality, if not superior, to the Spanish chesnut. The other is a delicious *little* kind, bearing fruit about the size and form of our common hazelnut. F. found chesnuts also cultivated amongst the mountains S. W. of Ning po. In the autumn of the same year F. returned to the hills N. W. of Ning po and procured large quantities of the two kinds of chesnut. Upwards of 300 (young) chesnut trees were sent to Calcutta, and arrived at their destination in excellent condition. (R. 51, 144, 196, 266, 277). These Chinese chesnuts were also introduced into England. See Fortune in Gard. Chron. 1860, 170.

Dr. Cl. Abel, in 1816, observed small chesnuts at Ta t'ung hien, on the Yang tze (v. s. 230), and Father A. David saw the same at Kiu kiang, in 1868. Comp. Franchet Pl. Dav. I, 277, *Castanea vulgaris*, Lam., var. *japonica*, A. DC.

The *Weeping-willow*, apparently the same species as we possess in England, is common in the neighbourhood of Shang hai, on the sides of all the rivers and canals, as well as in the gardens of the Chinese. Near Whampoa (Canton) F. notices a short of weeping willow very much like our own, and frequently met with there. The Chinese call it "sighing" willow. Near Ning po weeping willows are sometimes used for planting round the graves. (W. 118, 136. — R. 52).

Salix babylonica, Lin. seems to be a common tree all over China. Near Peking it is generally met with in its typical upright form. — We cannot agree with Bunge's statement (Enum. Chin.) that at Peking the female plant is very seldom met with.

In the gardens of Tien tsin F. observed a *Poplar*, which grows to a large size and is a tree of considerable beauty. Whether it be indigenous or introduced from some other country is unknown. (Y. 324). — The poplar noticed by F. was probably *P. alba*, Lin., which in the neighbourhood of Peking sometimes attains an enormous size.

At Che foo and amongst the mountains west of Peking F. observed an *Arbor vitae*, apparently distinct from the *Thuja orientalis*, which grows about Shang hai and in the more southern provinces of the empire. In the Peking mountains it attained a gigantic size. (Y. 307, 382). — The Peking arbor-vitae is the typical form of *Thuja orientalis*, Lin. According to Masters, Journ. Linn. Soc. IX (1867) 488, there is (in the Kew herbarium?) a specimen of Thuja orientalis, var. *pendula*, labelled Fortune, China.

Thuyopsis Standishii, Gordon, Pin. Suppl. (1862) 100. Introd. from Japan by F. Coloured drawing in Rev. Hort. 1896, 160. — Franchet, Enum. Jap. I, 469, considers it a variety of the American *Thuja gigantea*, Nutt.

Cupressus funebris, Endlicher. Lindley in Gard. Chron. 1849, 243, writes, that cones of this tree, which was first mentioned by Staunton in 1793 (v. s. 181), were sent by Fortune to Standish, Bagshot nursery, who raised the plant from seed. Fortune procured these cones from a place 200 miles north *) of Shang hai. — Figured (black) in Paxt. Fl. Gard. I (1850) 47, t. 31. — Fl. d. Serres, VI (1850) 89, 90 (black and white figures).

In the Gard. Chron. 1850, 228, there is a notice by Fortune regarding the discovery of this interesting tree. He first met with it, in November 1848, in the district of Hwuy chow, in the green tea country, not far from the town of Tun che. He observed a noble-looking fir tree, about 60 feet high with a stem as straight as the Norfolk Island pine and pendulous branches like the weeping willow. The branches grew at first horizontally with the main stem, then described a graceful curve upwards and drooped again at the points. From the main branches others, long and slender, hung down towards the ground, and gave the whole tree a graceful weeping form. The leaves were formed like those of the well-known arbor vitae. The tree was covered with ripe seeds, of which F. secured a quantity and sent them to England. The seed arrived safely and young plants were raised. Afterwards, as F. journeyed westward (towards the Hwuy chow green tea district, Sung lo shan) this tree became more common, and was frequently seen in clumps on the sides of the hills, generally near villages or amongst the graves. Lindley to whom F. sent a dried specimen has named it *Cupressus funebris*. — In 1853, F. sent a messenger to Hwuy chow, who brought him a good supply of the seeds of the funeral cypress. In T. 63, this beautiful tree is found figured. F. saw it cultivated at Shang hai and in 1849, he met with it again near Nan che and Yen chow foo in Che kiang. In 1853, he found it growing in great abundance amongst the mountains S. W. of Ning po. (T. 16, 61—64, 74, 107, 170, 314. — R. 145, 177).

Juniperus sphaerica, Lindley in Paxt. Fl. Gard. I (1850) 58, fig. 35 (black). An evergreen from the north of China, introduced by Fortune to Standish & Noble. It is a tree from 30 to 40 feet high. Young branches 4 cornered. Leaves minute, scaly, with a circular pit at the back. Fruit quite round, about as large as the ball of a pocket pistol. It differs from J. chinensis apparently in not having acicular leaves, and, very decidedly in the size and form of its fruit, which is twice as large as in that species and not at all depressed at the end.

Fortune notices the J. sphaerica in the hills north-west of Ning po and near Shang hai. The Chinese are fond of planting this tree round their graves. (R. 63, 140). — DC. Prod. XVI, II, 488.

---

*) This statement is erroneus. We have to understand south-west of Shang hai.   •

Cryptomeria japonica, Don, Japan Cedar. Described and figured in Sieb. & Zucc. Fl. Jap. I (1836) 43, t. 124. Gordon in Journ. Hort. Soc. I (1846) 57, states, that nothing was known of the living plant in Europe until Fortune succeeded in obtaining seeds of it at Shang hai for the Hort. Society. They reached the Chiswick garden in 1844, and from them the first plants were raised.

In 1843, F. met for the first time in the plain of Shang hai, with the beautiful *Cryptomeria japonica*, a species of pine not unlike an Araucaria. When growing luxuriantly it is highly ornamental, rising from the ground as straight as a larch, and sending out numerous side branches almost horizontally from the main stem, which again droop towards the ground in a graceful and "weeping" manner. The wood of the tree has a kind of twisted grain, and possesses great strenght and durability. It is highly valued by the Chinese, and from its beauty and straightness is often used by the mandarins and priests for those long poles, which are generally seen in front of their houses and temples. It is also well known and highly prized by the natives of Japan. At Nagasaki on the hill sides it is a very common tree. In the next year F. observed some very beautiful specimens of this new fir in the mountains S. W. of Ning po, and obtained some plants and seeds of it, which were sent to the Hortic. Gardens at Chiswick. This tree, together with Cryptomeria and other trees, forms dense woods in these regions, and yields an excellent timber. A great number of the wooden vessels at Ning po are made of the wood of Cryptomeria japonica, which is remarkable for the number of beautiful rings and veins which show to great advantage when the wood is polished. The Chinese name of the tree is *lew san*. Amongst the Ning po mountains it is even seen at high elevations. Fortune also met with beautiful specimens of the Japan Cedar in the Bohea mountains, when he crossed them in 1849, proceeding from the north-west. He notices especially a fine solitary tree, at least 120 feet in height, which stood near the celebrated gates in the highest range, which divides the provinces of Fo kien and Kiang si. On his way back to Che kiang, when he crossed the same chain more to the north-east, he again saw some noble specimens of the Cryptomeria. (W. 117, 159. — T. 155, 212, 304. — R. 145, 184, 189, 256, 266, 277, 412).

Fortune observed at Whampoa, near Canton, a species of pine, called the *water pine* by the Chinese, from its always growing by the sides of the rivers and canals. (W. 136). — *Shui sung* or water pine is the Chinese name for *Glyptostrobus heterophyllus*, Endl. V. s. 181, Staunton, 1793.

Cephalotaxus Fortunei, W. J. Hooker, Bot. Mag. tab. 4499 (1850). This interesting plant was detected by Fortune, during his present second visit to the north of China (1848—50), 200 miles north of Shang see *), at the same place where he discovered the Cupressus funebris. The present plant is of the Yew tribe of Coniferae. Large size of the foliage, with pectinated arrangement on the branches. — Lindley in Paxt. Fl. Gard. I (1850) 58, fig. 34 (black). Introduced by Fortune from North China, to Standish & Co. Young plants. — Fl. d. Serres, VI (1850) 51, t. 555.

In R. 412, F. notices C. Fortunei amongst the high mountains S. W. of Ning po (1855). — In Gard. Chron. 1880, I, p. 11, he states that he introduced C. Fortunei, the male as well as the female plant to the Hort. Soc. between 1843 to 46.

Siebold & Zucc. in Flora Jap. II (1870) p. 66, 67, tab. 131, 132, described and figured two Japanese species, *Cephalotaxus drupacea* and *C. pedunculata,* which by some botanists are considered to be only varieties of the same plant. Masters, in Gard. Chron. 1884, I, 113, states that C. Fortunei is really the female plant of C. pedunculata. The Kew Index reduces C. Fortunei to C. drupacea.

Torreya grandis, Fortune, in Gard. Chron. 1857, 788, and 1860, 170. — R. 411—15. When on an expedition for seeds and plants of Abies Kaempferi in the high mountains S. W. of Ning po, in the autumn of 1855, Fortune observed for the first time in a garden at the mountain village of Poo- in chee two fine Yew trees, apparently quite new. They were too young to have seeds upon them, and too large to dig up and carry away. The owner of the garden informed F. that these trees had been raised from seeds received from a place about 10 or 15 miles distant amongst the mountains, where the trees grew to a great size and produced seeds annually in considerable abundance. It is called *fee shoo* by the natives, and its seeds are to be found in a dry state in all the doctor's shops in Chinese towns. They are considered valuable in cases of cough, asthma etc. Higher up in the mountains, F. came to the «Valley of the Nine Stones» and a little town of the same name, about 4000 feet high. Here numerous fine trees of the new yew were met with, growing on the sides of the hill above the town. Many of them were from 60 to 80 feet in height with fine round heads, and altogether had a striking and ornamental appearance. There

---

*) Erroneous statement. Comp. supra, note on p. 500.

were no seeds to be seen on them. F. was informed that they had been lately gathered and could be purchased in the town. Fortune succeeded in obtaining from and old farmer a goodly supply of the seeds, and sent them to Glendinning's nursery, Chiswick, where they vegetated freely. The plant was found to be a new Torreya, and was subsequently much admired in English gardens. The tree resembles the Torreya nucifera of Japan, figured by Kaempfer in Amoen. exot. 815. — *Torreya grandis* figured in Rev. Hort. 1879, 173. — Gard. Chron. 1884, II, 682.

Taxus cuspidata, Sieb. & Zucc. Fl. Jap. Fam. nat. (1846) n. 814, and Fl. Jap. II (1870) 61, tab. 128, Japan. — Fortune, in Gard. Chron. 1860, states that he introduced this plant into England. He had received it from the late Mr. Beale at Shang hai (v. supra 408, before 1856), to whom it had been brought from Japan. But in Y. 110, he says that he introduced it from China.

The *Salisburia adiantifolia* or Maiden-hair tree grows to a very large size in the Shang hai district and in northern Che kiang. Its leaves resemble those of a fern, Adiantum. Its fruit is sold in the markets in all Chinese towns by the name of *pa kwo*, and is not unlike dried almonds, only whiter, fuller, and rounder. The natives seem very fond of it, although it is rarely eaten by Europeans. This noble tree seems to be indigenous to this part of China. The Chinese like to dwarf it, and it is often seen in that state in their gardens. F. observed fine specimens of the maiden-hair tree near Hoo chow foo in northern Che kiang. It is common also about the temples in Japan, where it is called *ging ko*. (W. 118, 251. — R. 140, 348, 363. — Y. 59). — It is the *Gingko biloba* of Linnaeus.

Cunninghamia sinensis or *C. lanceolata*, the lance-leaved pine, is found in abundance in the island of Chusan and on Poo too shan Island. On the island of Hong kong it is rare, although frequently met with on the mainland. In the mountainous districts near Foo chow the forest trees consist chiefly of Pinus sinensis and C. sinensis. In the Bohea Mountains F. saw, in 1849, beautiful forests of the tree, especially in the more northern part of the range, where the sides of the mountains were clothed with dense woods of the lance-leaved pine, which is of great value as a timber-tree in this part of China. Many of the specimens were at least 80 feet in height, and perfectly straight. There was a richness too in the appearance of the foliage, which F. had never seen before; sometimes it was of a deep green colour, while at others it was of a bluish tint. There are doubtless many varieties of this tree amongst these hills. Above Yen chow foo F. saw plantations of C. lanceolata. In the snowy valley S. W. of Ning po and in the high mountains farther west F. met with beautiful dense woods of this and other coniferous trees, but the trees were generally young and not remarkable for size. (W. 11, 57, 173, 282. — T. 147, 215, 248, 249. — R. 189, 277).

Sciadopitys verticillata, Sieb. & Zucc. Fl. Jap. II (1870) tab. 101, 102, Umbrella Pine. Japan. Introduced by Fortune from Japan in 1861. Gard. Chron. 1863, p. 1016, 1872, p. 1526. — Figured Y. 47.

The Chinese Pine, *Pinus sinensis,* which seems to be found all over China, in every degree of latitude, is frequently mentioned in Fortune's books. It is common in Hong kong, as it is all along the coast of China, and also a common forest tree in the mountains of Fo kien, and amongst the Bohea Mountains. It is much exported as timber for building houses, from Foo chow, to which place it is floated down the Min River together with other timber. It abounds in Chusan, in the mountains S. W. of Ning po, where the tree attains to a great size. F. saw it also in other parts of Che kiang and also in Japan. (W. 2, 11, 57, 158, 173, 375, 383, 388. — T. 58, 74, 142, 147, 155, 213, 249, 254. — R. 51, 189, 362. — Y. 14).

In the mountains west of Peking Fortune observed, in the autumn of 1861, a species of Pine-tree, having a peculiar habit and most striking appearance. It had a thick trunk, which rose from the ground to the height of 3 or 4 feet only. At this point some 8 or 10 branches sprang out, not branching or bending in the usual way, but rising perpendicularly, as straight as a larch, to the height of 80 or 100 feet. The bark of the main-stem and the secondary stems was of a milky white colour, peeling like that of the Arbutus, and the leaves, which were chiefly on the top of the tree, were of a lighter green than those of the common Pine. The trees were covered with cones. It proved to be *Pinus Bungeana*, Zucc. (discovered 1831 by Bunge). F. had already met with this tree in a young state in the country near Shang hai and introduced it into England [*]), but, until then,

---

[*]) To Glendinnings. See Gard. Chron. 1857, 216.

he had not the slightest idea of its extraordinary appearance when full grown. One of the trunks, which F. measured at 3 feet from the ground, was 12 feet in circumference. (Y. 378, with a figure). — Figured also in Gard. Chron. 1882, II, 8.

In 1850, in Paxt. Fl. Gard. I, 42, fig. 26 (black) Lindley described a cone and leaves of a magnificent evergreen coniferous tree introduced, as he says, by Standish & Co. from Japan. He determined it as the *Abies jezoensis*, Sieb. Leaves brilliant green, cones 6 inches long. This account and the figure were reproduced in Gard. Chron. 1850, 311, and in Fl. d. Serres, 1851, 223, and 1853, 7, tab. 858. Cone of a blue colour. But as Murray, "Pines and Firs of Japan", 1863, 43, has proved, that the tree in question did not come from Japan, but from China, and was not A. jezoenis, Fortune told Murray that he was the discoverer and had sent from China the cones and the seeds to Standish. It was near Foo chow foo that F. found it. A single tree in the grounds of a famous temple, named Koo shan, in the mountains, there struck his attention. It was an aged fir, stretching out its branches in a tabulated form, like a cedar of Lebanon, and on these were growing the magnificent purple cones, which Lindley figured, standing erect (not pendent) and thickly grouped like rows of soldiers. It was the only tree of the kind he saw, and from it he obtained the seeds and specimens, which he sent to Standish & Noble. It may have been introduced to Foo chow from some other country. — This is the noble species of *Abies* noticed by F., in T. 142, (February? 1849) near the celebrated temple of Koo shan situated amongst the mountains a few miles to the eastward of Foo chow, about 2000 feet high. — Murray, l. c. named the tree *Abies Fortunei*. — Carrière, in Rev. Hort. 1866, 449, c. ic., proposed for it a new genus: *Keteleeria Fortunei*.

Abies Kaempferi, Lindley in Penny Cyclop. I, 34, and Gard. Chron. 1854, 255, 455 (a cone figured). Lindley notices under the above name a plant sent by Fortune, which he believed was the seosi or karametz nomi, a kind of Larch in Japan, mentioned by Kaempfer, Amoen. exot. 883, but which after Kaempfer no traveller had seen and of which Siebold takes no notice. Fortune in a letter, dated Hong kong, Febr. 10, 1854, announced the discovery of a Larch found in the central, northern and eastern provinces of China, a very beautiful tree. The cones, when young are very pretty, but very brittle. From specimens sent by F. and from seeds of this tree transmitted by him to Messrs Glendinning of Turnham Green, Lindley

was able to ascertain that this tree was identical with that noticed by
Kaempfer for Japan. Its branches are exactly those of the common larch,
leaves very slender, clustered, deciduous. Cones pendulous, about 3 inches
long, $2\frac{1}{2}$ in diam., with excessively deciduous scales, diverging like those
of an artichoke head. Each scale is woody, flat, cordate, and more than an
inch long. The seeds are exactly the size of the scales, two of them occupying the whole inner face with their wings. It is unlike any coniferous plant
previously discovered.

Fortune discovered the *Abies Kaempferi* in 1853, in the mountains
S. W. of Ning po. He had been acquainted with this interesting tree for
several years in China, but only in gardens, and as a pot plant in a dwarfed
state. The Chinese by their favourite system of dwarfing, contrive to make
it, when only $1\frac{1}{2}$ or 2 feet high, have all the characters of an aged cedar
of Lebanon. It is called by them the *kin le (lo) sung*, or Golden Pine (more
correctly translated: golden pine with deciduous leaves). The long, green,
silky leaves, when first unfolded in spring are singularly beautiful, and so
they are again in the autumn, when they change into a golden yellow
colour; hence the Chinese name. Although F. had often made enquiries
after it, and endeavoured to get the natives to bring him some cones, or to
take him to a place where such cones could be procured, he met with no
success until the autumn of 1853, when he unexpectedly came upon some
fine specimens of full grown trees near the monastery of Tsan tsin, at an
elevation of 1000 or 1500 feet in the mountains, S. W. of Ning po. Their
stems were fully 5 feet in circumference, 2 feet from the ground, and
carried this size, with a slight diminution, to a height of 50 feet, that
being the height of the lower branches. The total height he estimated
about 120 or 130 feet. The stems were perfectly straight throughout, the
branches symmetrical, slightly inclined to the horizontal form, and having
the appearance of something between the cedar and larch. Fortune procured a large supply of the curious cones, which he sent to England in
the winter of 1853. In the autumn of 1854, F. visited the same place,
but could not detect a single cone on the trees. When proceeding westward, beyond the temple of Quan ting, near the top of the mountain range,
at an elevation of 4000 feet, he met with a plantation of the tree. One
tree was standing all alone, and seemed the queen of the forest, from its
great size and beauty. It measured 8 feet in circumference, was fully 130
feet high, and its lower branches were nearly touching the ground. They
came out almost at right angles with the stem, but the upper part of
the tree was of a conical shape. There were, however, no cones even on

this or on any of the others, although the natives informed him they had
been loaded with them on the previous year. F. had, therefore, to content
himself with digging up a few self-sown young plants; these were afterwards sent to England and arrived in good condition. F. does not agree
with Lindley in calling this an Abies. It is a plant intermediate between
the cedar and the larch; that is, it has deciduous leaves like the larch and
deciduous scales like the cedar. It is a noble tree, produces excellent timber,
and would be very ornamental in park scenery in England. At the end of
October 1855, F. visited these mountains again, and came upon some fine
examples of the golden pine tree, and, to his delight, they were loaded
with ripe cones. F. looks upon the Abies Kaempferi as the most important
of all his Chinese introductions. He figures the tree in R. 274. It was
subsequently growing in Glendinning's nursery at Chiswick. (R. 266,
274, 275, 286, 287, 412—16. — Fortune in Gard. Chron. 1855, 242, —
1860, 170). — *Larix Kaempferi,* Carrière, Fl. d. Serres, XI (1856) 97. —
Illustr. Hort. XIV (1867) 506 (colour. fig.). — *Pseudo-Larix Kaempferi,*
Gordon, Pin. 1858, 292, et suppl. 91. — Fl. d. Serres, XVII, 1867, 111
(colour. fig.).

In Gard. Chron. 1884, I, 581, 584, figures, it is stated that Lindley
in naming the tree after Kaempfer, was mistaken; it is not Japanese but
a native of China, and was first discovered by Fortune.

Spathoglottis Fortunei, Lindley, Bot. Reg. 1845, tab. 19. One of the
first plants, which Fortune met with on the granitic mountains of Hong
kong, was this pretty little Bletia-like plant figured in this plate. From
some corms of it which he sent home, in the beginning of 1844, this
specimen sprang up in the Hort. Soc.'s garden. Like the Bletias it has thin
plaited leaves and fleshy corms. Yellow flowers. (W. 11. — T. 8). — Benth.
Fl. Hgk. 355. Only known from Hong kong.

Pholidota chinensis, Lindl., Journ. Hort. Soc. II (1847) 308. Received
from Fortune as Coelogyne species with yellow flowers, from the mountain
sides, Hong kong. This species has long been known from Chinese drawings,
in which it is represented with yellow flowers. It is a small neat species,
but not very ornamental. A creeping epiphyte, bearing ovate, wrinkled
quadrangular bulbs on a short, stout rhizome. Leaves coriaceous. Flowers
greenishwhite, in drooping racemes, not more than 2 or 3 inches long. —
Benth. Fl. Hgk. 354. Only known from Hong kong.

Arundina chinensis, Blume (v. s. 309). Fortune in Journ. Hort. Soc. I (1846) 215, states, that he sent this plant from Hong kong to the Society in the beginning of 1844. — The tops of the highest hills in Hong kong are covered in the summer and autumn months with the purple *A. chinensis,* and the yellow *Spathoglottis Fortunei.* (W. 11). — Bentham, Fl. Hong kong, 355.

Glossaspis antennifera, Reichb. Fil. in Linnaea, XXV (1852) 225. China, Fortune, 182. — According to Bentham, l. c. 361, this is the same as *G. tentaculata,* Lindley, brought in 1824, from China by J. D. Parks.

Platanthera Galeandra, Reichb. Fil. l. c. 226. China, Fortune. — *Habenaria Galeandra,* Benth. l. c. 363.

Platanthera Mandarinorum, Reich. Fil. l. c. 226. China, Fortune. — Not noticed by Bentham.

Spiranthes Stylites, Lindley Orchid. Ind. Journ. Linn. Soc. I (1857) 178. Fortune, Che kiang.

A considerable quantity of *Ginger* is produced on the mainland near Amoy. It is also grown to a great extent near Foo chow. (W. 29, 381, 384).

Fortune observed, when in 1845 he visited Manila, that a kind of hemp, the produce of a species of *Musa (M. textilis),* is made there into ropes and cables; it is highly prized and in much demand amongst the shipping in the East. The beautiful cloth, generally known by the name of *Pinia,* and made from the fibre of the *Pine-apple* plant, is manufactured and embroidered by the natives and is sold in the shops. (W. 336).

Near Kanagava in Japan. Fortune observed a species of *Iris,* "shobu" which was growing thickly on the flattened ridges of the roofs of the houses. (W. 56, 57, figured). — This is the *Iris tectorum,* Maxim. Mél. VII (1870) 563.

The common *Jonquil* is a great favourite amongst the Chinese; and in the streets of Canton one meets with thousands of bulbs growing in small pans amongst water and a few white stones. (W. 146).

At Shang hai a pretty bulbous plant, a species of *Lycoris*, covers the graves in autumn with masses of brilliant purple. (W. 333). — This was probably *L. radiata*, Herbert. There is in the herbarium of the Bot. Garden, St. Petersb. a specimen of this labelled Fortune, Chin chew 148.

Lycoris straminea, Lindl., Journ. Hort. Soc. III (1848) 71. Received from Fortune, in 1845, from China. Pale, straw-coloured flowers with a pink line along the middle of the segments, and a few sattered dots. A pretty bulbous plant, very nearly allied to *L*. aurea.

The sweet-scented *Italian Tuberose*, much cultivated near Foo chow. (T. 147).

In Gard. Chron. 1855, 318, F. notices the *Chinese Yam, Dioscorea Batatas*, which is very abundant in China, but is not a staple crop, like the sweet potato. This esculent seems to be attracting a good deal of notice in England as well as in France and is by some persons considered equal, if not superior to the potato. But perhaps the summers in England are not hot enough for the Yam. Like the sweet potato the cultivated Dioscoreas (Igname) never produce seeds, but form axillary bulbils.

Funkia Fortunei, Baker, Gard. Chron. 1876, II, 36. Japan. It resembles F. ovata and Sieboldiana.

Caloscordum exsertum, is noticed by Herbert in Bot. Reg. 1847, sub t. 5, as a new species discovered by Fortune in Chusan, 102. It has small heads of flowers and only 10 in number, long styles. — Baker in Journ. Bot. 1874, 294, calls it *Allium exsertum*. It was earlier discovered by Staunton.

Lilium sinicum, Lindley in Paxt. Fl. Gard. II (1851) 384, fig. 384 (black). A handsome Chinese green house bulbous plant with scarlet flowers, only 1 foot high. This plant was earlier introduced, and flowered in the Hort. Soc.'s garden, in 1824. Recently it was reintroduced from China by Fortune for Messrs Standish & Noble. — Fl. d. Serres, VII (1852) 218 (black figure). — Illustr. Hort. 1856, tab. 100.

Lilium auratum, Lindley, Gard. Chron. 1862, 644, was introduced by J. G. Veitch from Japan (v. infra), and at the same time (1860 or 61) Fortune found the same plant in Japan and gave it to Standish & Noble, Bagshot. See Gard. Chron. 1880, I, p. 11, and Bot. Mag. tab. 5338 (1862). Fortune remarks that he usually saw this plant 4 feet high in Japan, and often with 3 or 4 of these large flowers on the same stem. He suspects that it is the great-grandmother of L. speciosum, which may be a hybrid offspring.

Lilium Brownii, Miellez (v. s. p. 200) was first found in a wild state near Tam sui, by Fortune in 1854. (R. 232).

Lilium Fortunei, Lindley, Gard. Chron. 1862, 212. Sent by Fortune from Japan to Standish, who exhibited it. It grows 18 inches high, has narrow grassy leaves, solitary flowers, orange yellow, dotted with brown purple. It resembles L. tenuifolium, Fisch.

Near Ning po, amongst the hills to the S. W. large tracts of land are planted with the bulbs of a liliaceous plant — probably a *Fritillaria* — which are used in medicine. This is planted in November, and dug up again in April and May. In March these lily-fields are in full blossom, and give quite a character to the country. The flowers are of a dingy greyish white, and not very ornamental. (R. 262).

I may observe that Baker, Tulipeae, Journ. Linn. Soc. XIV (1875) 258, reports on a herb. specimen of *F. verticillata*, Willd. var. *Thunbergii*, labelled: Fortune, China, 27. Another collector sent the same plant from the hills of Che kiang to Dr. Hance.

Tricyrtis hirta, Hook. Bot. Mag. tab. 5355 (1863) or Uvularia hirta, Thunberg. Fortune sent it from Japan to Standish, in whose nursery it flowered in November 1862. It grows 4 to 5 feet high, and the copious blossoms which appear on the axils of all the upper leaves, and which are of a pearly white, dotted with clear purple, render it very singular looking and beautiful.

Juncus effusus, used in Japan in the manufacture of the mats which are so common in the country. (Y. 278).

There is a palm tree, a *Chamaerops,* found in the provinces of Che kiang and Kiang su, indigenous to this part of China, where it is much cultivated on account of the large quantity of strong and useful fibre formed on its stem at the base of the leaf-stalks. These large, brown, hair-like fibres are cut and pulled off by the natives at certain seasons of the year, and used for many purposes. Ropes and cables for their junks are made out of this substance, and seem to last, even under water, for a very long time. It is probably better and stronger for those purposes than the fibre of the cocoa-nut, which it resembles to a certain extent. Bed-bottoms are wrought out of this, and are largely used in the country by all classes of the natives. Agricultural labourers are fond of wearing hats and cloaks made out of the same substance, and called *so-e* or garment of leaves, which in wet weather keeps out a good deal of rain. In the south of China the so-e is made from the leaves of the bamboo and other broad-leaved grasses. — The Chinese call this valuable tree *sung (tsung).* Fortune names it the *Hemp palm,* also the *Chusan palm,* for he first observed it, in 1843, in the island of Chusan, growing on the hill sides, cultivated. Subsequently he found it also on Silver Island and on the hills S. E. and N. W. of Ning po. In 1848, he met with specimens in a high state of perfection near Yen chow foo in Che kiang, and Sung lo shan. The tree is perfectly hardy about Shang hai, and thrives there unprotected throughout the severest winters. It is most ornamental in the country where it grows. Its tropical looking appearance has a curious effect in a landscape not unlike our own, in so far as the vegetation is concerned. Fortune gives a figure of the Hemp palm, in T. 59, and of the rain cloak and hat made of the fibres, in R. 145. — Fortune sent some plants of the Chinese Chamaerops (in 1848 or 49) to Sir William Hooker at Kew and requested him to forward some to H. R. H. the Prince Consort for the Royal Gardens at Osborne House, Isle of Wight. In 1853, Fortune obtained seeds of the tree near Tse kee, N. W. of Ning po. (W. 53, 54. — T. 58, 88, 117, 318. — R. 5, 145, 189. — Fortune in Gard. Chron. 1850, 757; — 1860, 170; — 1880, I, 234). Sir William Hooker, in the Bot. Mag. 1850, sub tab. 4499, writes: A palm *Chamaerops excelsa* (?) sent to the Royal gardens by Fortune, has braved unarmed, and unprotected the severe winter 1849—50. — In Bot. Mag. tab. 5221 (1860). Hooker describes and figures Fortune's Chusan palm as a new species, *Ch. Fortunei.* It approaches closely the Ch. excelsa of Japan, but is a more robust species. Pendulous apices of the segments of the leaves. It is the most hardy of all the palms, and stands almost unprotected in winter, in the latitude of London. At Osborne it had attained (in 1860) a height of 10 feet in the open air, with no protection in winter. It has blossomed for the last three years. The plants at Kew were introduced by Fortune, in 1849 and had attained 8 feet in height, in 1860. The fruit was then unknown.

Scirpus tuberosus, much cultivated in South China. (W. 307).

The *Sugar cane* is extensively grown both in the provinces of Quan tung and Fo kien, and probably in other parts of the Empire. In its raw state it is an article in great demand amongst the Chinese. It is manufact- ured into sugar candy and brown sugar; many kinds of the latter being particularly fine, though not much used by the foreigners residing in the country, who generally prefer the candy reduced to powder, in which state it is very fine and white. Our loaf-sugar is not made in China. F. observed great sugar plantations near Whampoa (Canton) in the flat country near Foo chow, especially in the valley of the Min, near Amoy, on the mainland. (W. 29, 136, 307, 381, 384. — T. 146).

Among the *Cereals* cultivated in China *Rice* is the most important, for it is the chief article of food and, of course, the staple production of the country, more particularly in the south, where two crops of it can easily be raised in the hot months. The ground is prepared in spring for the first crop, as soon as the winter green crops are removed from the fields.

The land is always flooded before it is ploughed. The plough, which is commonly drawn by a buffalo or bullock, is a rude implement. The opera- tion of ploughing turns up a layer of mud and water, 6 or 8 inches deep, which lies on a solid floor of hard clay. After the plough comes the harrow. The labourer stands on the top of it and presses it down upon the muddy soil while it is drawn along. Previously to the preparation of the fields the rice seed is sown thickly in small patches of highly manured ground, and the young plants are transplanted in the fields overflowed with water to the depth of 3 inches. They are put in, in patches and in rows. In the south the first crop is fit to cut by the end of June. Before it is quite ripe, another crop of seedlings is raised on the beds or corners of the fields. The second crop is ready for cutting in November. In the latitude of Ning po the summers are too short to have the land cropped in the same way in which it is done in the south. The farmers here manage to have two crops of paddy in the summer by planting the second crop two or three weeks after the first, in alternate rows. The second crop is ready for the middle of November. About one hundred miles further north, in the Shang hai district, the summers are too short and the husbandman is therefore obliged to content himself with one crop of rice. During the growth of the rice, the fields are always kept flooded, when water can be obtained, until the crops are nearly ripe. The terraces near the base of the hills are supplied by the mountain streams, and the fields which are above the level of any adjoining river or canal are flooded by the celebrated water-wheel, which is in use all over the country. When ripe the crops are cut with a small instrument, not very unlike our own reaping-hook, and are generally thrashed out at once in the fields where they have grown. (W. 299—300. — R. 300, 343).

Wheat and *Barley*, with various other green crops (Cabbage oil plant) are cultivated in winter and reaped in spring or during the early summer months. In Fo kien wheat and barley ripen in April, and in the Shang hai district about the middle of May. About Chin chew and Amoy the wheat crops are so poor that the labourers pull them up by the hand. They are of course much better in the rich district of Shang hai, but the varieties both of wheat and barley are far inferior to ours; and as the Chinese sow them too thickly, they are generally much drawn, and the heads and corn small. These two cereals are also cultivated in the Ning po plain. (W. 270, 310.— R. 261, 188).

Near Che foo and in the plain between Peking and Tien tsin Fortune observed a kind of millet cultivated there, which is not met with in the more southern parts of China. It grows to the height of from 12 to 15 feet, the Chinese call it *kow-leang*. F. believed this to be the *Sorghum*, which then had been introduced to Europe, as a good substitute for the sugar cane. (Y. 306, 316, 346).

The *Indian Corn (Zea Mays)* is much cultivated as a summer crop in the hills near Ning po, in the valleys of the upper Tsien tang River, and in the Hwuy chow district. (T. 74, 177. — R. 27, 188, 189).

Various kinds of *Bamboo* were met with in all the Chinese provinces visited by Fortune. The bamboo is one of the most valuable trees in China, and is used for almost every conceivable purpose. It is employed in making soldier's hats and shields, umbrellas, soles of shoes, scaffolding poles, measures, baskets, ropes, paper, pencil-holders, brooms, sedan-chairs, pipes, flower-stakes and trellis-work in gardens; pillows are made of the shavings; a kind of rush cloak for wet weather is made from the leaves and is called a *so-e* or "garment of leaves". On the water it is used in making sails and covers for boats, for fishing-rods and fish-baskets, fishing stakes and buoys; catamarans are rude boats, or rather floats, formed of a few logs of bamboo lashed firmly together. In agriculture the bamboo is used in making aqueducts for conveying water to the land; it forms part of the celebrated water-wheel, as well as of the plough, the harrow, and other implements of husbandry. Excellent water-pipes are made of it for conveying springs from the hills, to supply houses and temples in the valleys with pure water. Its roots are often cut into the most grotesque figures, and its stems finely carved into ornaments for the curious, or into incense-burners for the temples. The Ning po furniture, the most beautiful in China, is often inlaid with figures of people, houses, temples, and pagodas in bamboo. The young shoots, in the state when they come out from the ground are highly prized as a vegetable for the table. They are for this purpose boiled; sweetmeats are also made of them. A substance found in the joints, called *tabasheer* is used in medicine. And last, though not least, the celebrated chop-sticks, the most important articles in domestic use — are made of it. It is in universal demand, in the houses and in the fields, on water and on land. — For making paper from the bamboo it is soaked for some time, then split up and saturated with lime and water until it becomes quite soft. It is then beaten up into a pulp in mortars. When the mass has been reduced to a fine pulpy substance, it is then taken to a furnace and well boiled until it has become perfectly fine, and of the proper consistency. It is then formed into sheets of paper. Bamboo-paper is made of various degrees of fineness according to the purposes for which it is intended. It is not only used for writing upon, and for packing with, but a large quantity of a coarse description is made for the sole purpose of mixing with the mortar used by bricklayers. (W. 166. — T. 155. — R. 335).

In the south of China, that is about Hong kong and Canton, several kinds of the bamboo are very common. There is a *yellow* variety with beautiful green stripes, painted on its stem as if done by the hand of a most delicate artist. But all these kinds resemble the Indian varieties, that is, they grow in dense bushes, their stems are not remarkable for their straightness, and the large joints, and branches, which are produced on all parts of the stem, give it a rough surface and consequently render it unsuitable for fine work. These tropical, jungley looking bamboos disappear as we go to the more northern latitudes; and in their places we have the *mow-chok,* the *long-sin chok,* the *hoo-chok,* and one or two others, all with clean stems and feathery branches, suited for the most delicate kinds of work, and all "good for food". These trees are well worth the attention of people who inhabit temperate climates, such as the south of France, Italy and other parts of the south of Europe. No doubt they would be well worth introduction. In Che kiang the bamboos invariably grow in a rich yellow loam on the slopes of the hills or in ravines. (W. 158, 166. — R. 189—191).

In T. 8, F. speaks again of the beautiful *yellow Bamboo* of Hong kong with a straight stem of a fine yellow colour, striped with green, of which he sent a plant to the Hort. Society in 1844. This is, it seems the dwarf variegated Bamboo, he notices in Y. 12, as met with in Japan, in 1860, and as having been previously introduced by him into England from China. — The beautiful *yellow Bamboo* of Chusan with a clean and straight stem, and graceful top and branches waving in the breeze, mentioned in W. 57, is evidently a different kind. — In R. 372, F. records, that in the mountains west of Hoo chow foo, in northern Che kiang, fine bamboos are produced, which are sent down to the low country, where they are made into paper.

In 1853, F. discovered amongst the mountains S. W. of Ning po the *mow-chuk*, a graceful bamboo, which he declares to be the most beautiful in the world, and which he had never seen before in any other part of the globe. It is largely cultivated in the central and eastern provinces of China, particularly on the sides mountains, where the soil is rich, and in the vicinity of temples and other monastic buildings. Its stems are straight, smooth, and clean, the joints are small, it grows to the height of from 60 to 80 feet. Twenty or thirty feet of the lower part of its stem are generally free from branches. These are produced on the upper portion of the tree, and then they are so light and feathery that they do not affect the cleanness of the main stem. In addition, therefore, to the highly picturesque effect it produces upon the landscape, it is of great value in the arts, owing to the smoothness and fineness of its structure. It is used in the making of sieves for the manipulation of tea, rolling tables for the same purpose, baskets of all kinds, ornamental inlaid works, and for many other purposes, for which the bamboo found in India is wholly unsuitable. Like all other species of the same tribe, it grows with great rapidity and perfects its growth in a few months. To use a common expression "one could almost see it growing". F. measured the daily growth in the Chinese woods, and found that a healthy plant generally grew about 2 feet or $2\frac{1}{2}$ in the twenty four hours, and the greatest rate of growth was during the night. The young shoots just as they peep out of the ground are highly esteemed as food, and are taken to the markets in large quantities. F. was in the habit

of using them as a vegetable every day during the season, and latterly was
as fond of them as the Chinese are themselves. Sometimes he had them
split up, boiled, and dished by themselves; at other times they were used
in soup, like cabbage; and on one occasion Mr. Forbes, the United States'
Consul in China, taught him to make an excellent omelette, in which they
formed one of the ingredients. Among the high mountains S. W. of Ning
po, from 2000 to 3000 feet high, Fortune met with fine forests of the mow
chuk bamboo on each side of the road. Long trains of coolies — men and
boys — passed, loaded with young bamboo shoots brought from a district
further inland. F. succeeded in introducing the mow chuk to India. Several
plants were also sent to the Agricult. and Hort. Soc. of India, and reached
Calcutta in good condition. (R. 189—192, 266, 267, 277).

Fortune's *mow-chuk* is probably the *Bambusa edulis*, Carrière, Rev.
Hort. 1866, 380, or *Phyllostachys mitis*, Rivière, Bambous (1879) 231,
a tall tree, a native of China. — Fortune's yellow, striped Bamboo of Hong
kong, Canton, also observed in Japan, is a variety of the common bamboo,
Bambusa vulgaris, var. *vittata*, Rivière, l. c. 202. In our gardens it is
known under the name of *Bambusa variegata*, figured in Regel's Gartenflora, 1865, 362, tab. 490, fig. 5. This is the *Bambusa striata*, Lindley,
Loddiges (v. s. 281), figured in Bot. Mag. t. 6079 (1874). — Fl. d. Serres,
XXI (1875) 99. A graceful, tufted, very glabrous, slender species, from 6
to 20 feet high. Culms as thick as the thumb, striped yellow and green.

Bambusa Fortunei, foliis niveo-vittatis, Van Houtte, Fl. d. Serres,
1863, tab. 1535. Cultivated by Standish, who received the plant through
Fortune from China. A dwarf plant with long cilia at the margin of the
leaf. According to Miquel Prol. Fl. Jap. 173, it is the *Bambusa variegata*,
Siebold, mss. A common plant in Japan. — Rivière, Bambous (1879) 314,
named this plant *Arundinaria Fortunei*. — See also Bot. Mag. tab. 7146
(1890), sub A. Simonii. — According to Fortune, Gard. Chron., 1880, I,
11, he introduced this plant from China.

Bambusa dumetorum, Hance, Walp. Ann. III (1852) 781, Hong kong.
Munro in Seem. Bot. Her. 424, named it *Schizostachyum dumetorum*. First
discovered by Fortune, n. 164. — Rivière, l. c. 203.

Soon after his first arrival in China, in 1843, F., when travelling
amongst the hills of Hong kong, met with a most curious dwarf *Lycopodium*,
which he dug up and carried with him. This is much prized by the Chinese.

It is really very pretty, and often naturally takes the very form of a dwarf tree in miniature. (W. 84). Some years later he found in Braine's garden, Hong kong, a dwarf species of Lycopodium, highly prized by the Chinese, who call it *man neen chung*. (T. 8). — These two are probably the same. *Man nin tsung*, in the Canton dialect means "ten thousand years' fir". Bentham in Fl. Hgk. 437, records five species of Lycopodium for the Flora of the island, viz: *cernuum, involvens, atroviride, caudatum, flabellatum*. F.'s plant refers probably to the first or second. I may notice, that Dr. Abel in 1816, observed the same dwarf Lycopodium, resembling a fir tree growing in the Canton province (v. supra p. 233).

In the Journ. Hort. Soc. I, 1846, 221, 223, Fortune states, that among the plants he sent from China to the Society, in the autumn of 1844, there were *L. caesium* and *L. Willdenowii*. And the man neen chung he brought with him in 1846, when he returned from China. In R. 80 (1853) he notices, in a garden at Ning po, the pretty little *L. caesium*, introduced by him to England some years earlier. In the hills above Yen chow foo in Che kiang, F. observed, in 1848, many kinds of mosses and Lycopods, growing out of the crevices of the moist rocks; amongst the latter, and very abundant, was a fine species named *Lycopod. Willdenowii*. T. 57. — Spring, Monogr. Lycop. 1842, 109, identifies the L. caesium of English gardeners, of which he saw a specimen brought by Fortune from China, with *Lycopodium (Selaginella) uncinatum*, Desv., from India. The *L. Willdenowii*, India, China, a large climbing species is often grown in English gardens under the name of *Caesia arborea*.

Polypodium Fortunei, Kunze, in Mettenius, Polypod. (1857). China, Fortune A. 78. Herbarium Kunze. — Syn. Filic. 367, *P. (Drynaria) Fortunei*, Kunze.

In the Gard. Chron. 1855, 708, there is a notice of *Drynaria Fortunei*, Thomas Moore, Curator of the Chelsea Bot. Garden, Genera and Species of cultivated Ferns, ined. *). J. Smith in Seem. Bot. Herald, 425, cautions against confounding this with Kunze's P. Fortunei, and reduces it to *Pleopeltis nuda*, Hook. Exot. Fl. (1823) tab. 63. But E. J. Lowe, Ferns (1858) 42. B. calls it *Polypodium Fortunei*. At last Mettenius, Filic. Ind. Jap. in Miq. Ann. Mus. Bot. Lugd. II, 1866, 224, finds it to be identical with the Japanese *P. lineare*, Thunb. — Syn. Fil. 354.

---

*) Published 1858: J. Smith and Th. Moore, Exotic cultivated Ferns, p. 36. *D. Fortunei*. Fortune second journey.

Cyrtomium Fortunei, J. Smith, Ferns, British and Foreign, 1866. Syn. Fil. 257 = variety of *Aspidium (Cyrtomium) falcatum*, Swartz.

From the summary of Fortune's botanical discoveries in China and Japan, as given above, it appears that during his three visits to China, the last extending also to Japan, he introduced from those countries into English gardens, for the first time, nearly 190 species or varieties of plants, of which more than 120 were entirely new to botanists or horticulturists, whilst the remainder, up to that time, had been known in Europe only from herbarium specimens. In the collection of dried plants brought by him from China, about 25 proved to be novelties (not including those derived from the new plants introduced in a living state).

IV. French Embassy to China, 1844—46.

On the termination of the Anglo-Chinese war, in 1842, the French Government resolved upon sending a large Embassy to China, to form a treaty of commerce between France and the Chinese Government and to make investigations into the commerce, arts and industries of China, its resources and articles of export. M. de **Lagrené,** was at the head of this large mission *).

Lagrené, Théodore, Marie, Melchior Joseph de, was born near Amiens, in 1800. He entered the diplomatic service in 1823, was successively Secretary of Legation at St. Petersburg, Madrid, 1834, Chargé d'affaires at Darmstadt, 1835 French Minister for Greece. On August 9th 1843, he was appointed Envoyé extraordinaire et ministre plénipotentiaire to China. After his return from this mission, in 1846, he was made Pair de France; 1849 deputé de la Somme. He died in 1862.

Lagrené's mission was composed of 17 members **):

Marquis de Ferrière — le — Vayer, premier secrétaire.

Comte d'Harcourt, second secrétaire.

Callery, interprète.

X. Reymond, historiographe.

Yvan, médicin.

---

*) For the biographical notices regarding this French envoy and other French travellers and explorers in China, I am indebted to my respected friend Professor G. Devéria, at Paris.

**) Cordier, Bibl. Sin. 1227.

de Montigny, chancellier.

5 attachés: Macdonald de Tarente, Marey — Monge, F. Delahante, de la Guiche, de Charlus.

4 délégués du ministère du commerce, désignés par les chambres de commerce de Reims, de Mulhouse, de St. Etienne, de Lyon et de Paris: A. Haussmann, cotons, — N. Rondot, laines, — I. Hedde, soies, — Renard, articles dits de Paris.

2 représentants du ministère des finances: J. Itier, — Ch. Lavollée.

Several of the members of this mission, during their sojourn in China, rendered important services to natural science and especially botany, in elucidating the botanical origin of some Chinese vegetable productions appreciated in Europe.

M. de Lagrené, with his suite, embarked in December 1843, on board the Frigate "Sirène". Haussmann and his three colleagues took passage, at Brest, on board the Steamer "l'Archimède", on February 20, 1844; Cape 1844 of G. H., Isle Bourbon, Ceylon, Pondichery, Madras, Singapore, Manila, Hong kong. On August 24, the "Archimède" cast anchor at Macao. The Sirène with the minister had reached this port ten days earlier. Arrangements were immediately made for opening the negotiations with Ki ying, the imperial Chinese Commissioner. A Treaty of Commerce was signed, October 23, at Whampoa, near Canton. After this the French Minister and his suite left China to visit the Phillipines, Manila etc., and Java, whilst Haussmann and his colleagues remained at Whampoa till the 14 February, 1845, when they embarked on board the corvette "Alcmène" and set sail 1845 for Manila. From this expedition they returned to China on July 17. Meanwhile the Minister had also come back; Ferrière, Guiche, Itier, Marey-Monge and Montigny had left for Europe.

On Sept. 12, Lagrené with his suite and the delegates embarked on board the Frigate "Cléopâtre" to make a voyage to the north and on 2d October cast anchor at Chusan, at that time under British rule. On October 11, they embarked in Chinese junks, and paid a visit to Ning po, returning to Chusan October 18. Sir Francis Davis, Governor of Hong kong put at the disposal of the French mission the British Steamer "Nemesis", on which they proceeded to Shang hai, on October 25. Having spent at this place about ten days, and having visited Su chou fu, October 30, they returned to Chusan. On November 12, the Mission embarked again in the Cléopâtre and sailed for Amoy, which place they reached November 16, and remained

there till November 26. On Nov. 19, Chang chou fu was visited. Returned
to Macao, 1st December.

1846   On the 6th of January 1846, Haussmann and his colleagues embarked
on the corvette «Alcmène» to return to France and, after a favourable
voyage, the corvette anchored at Rochefort, May 13. M. de Lagrené and his
suite performed the voyage home by the shortest way. The Archimède
having brought them to Suez, they proceeded overland to Alexandria.

Haussmann, A., published an account of this expedition: VOYAGE EN
CHINE, COCHINCHINE, INDE ET MALAISIE PENDANT LES ANNÉES 1844—46.
Paris, 1847—48. 3 vol. From this the above details were taken.
    Lavolée, M. C. published a similar work, in 1852.

Rondot, Natalis, the most conspicuous amongst the specialists attached
to Lagrené's mission, was born at St. Quintin (Aisne), March 23, 1821.
After completing his education in a college, R. applied himself to the study
of the manufacturing of woollen goods. When the French mission was despatched to China he accompanied it, in consequence of the recommendation
of the Chamber of Commerce at Rheims, as a specialist for woollen goods,
but, during his sojourn in China, he investigated also other branches of
Chinese industry, especially sericulture, and inquired into Chinese articles
of export. After his return from China he wrote many articles and books
on the subject. He is now living at Chateau de Chamblon, near Yverdun in
Switzerland, and is still engaged in scientific pursuits. In 1855, he was
made «Officier de la Légion d'honneur».
    The following are the most important of R.'s publications:
    NOTICE SUR QUELQUES-UNES DES PLANTES TEXTILES DE CHINE. Séances
Acad. Reims, VI (1847) 15—17, — 505—8. Read before the Academy
in 1846. Translated into English in Chin. Repos. XVIII (1849) 209—16,
by S. W. Williams. *Boehmeria nivea* (Grass cloth), *Cannabis sativa, Sida
tiliaefolia, Corchorus*.
    EXCURSION À L'ÎLE DE POU TOU (PROVINCE DE TCHE KIANG) 7 et 8
Oct. 1845. 2 lithographies, Reims, 1846.
    CATALOGUE DES GRAINES DE PLANTES POTAGÈRES CULTIVÉES AU
JARDIN DE FAH TI, SUR LA RIVE DROITE DU FLEUVE TCHOU KIANG, PRÈS
CANTON. Séances Acad. Rheims, VI (1847) 15, 17, 505—8.
    APERÇU GÉOLOGIQUE SUR L'ÎLE TCHOU-SAN (TCHEKIANG). L. c.
(1847) 412.
    UNE PROMENADE DANS CANTON. LA MANUFACTURE DE LACQUES CHEZ

HIP-QUA ET L'ATELIER DE TABLETTERIE DE TA YU TONG. Journ. asiat. XI
(1848) 34, 64. Reproduced in Bazaine, Chine moderne, II, 631. Detailed
account of Chinese varnish and the manufacture of laquered-ware.

ÉTUDE PRATIQUE DU COMMERCE D'EXPORTATION DE LA CHINE par
I. Hedde, E. Renard, A. Haussmann et N. Rondot, revue et complétée par
N. Rondot, 1848. In this useful book the articles of export produced in
China are described, and interesting details given regarding their origin.
Many Chinese economic plants are mentioned.

LES COLONIES AGRICOLES DES CHINOIS. Paris, 1851.

NOTICE DU VERT DE CHINE ET DE LA TEINTURE EN VERT CHEZ LES
CHINOIS, par N. Rondot, suivie d'une ETUDE DES PROPRIÉTÉS CHIMIQUES ET
TINCTORIALES DU LO-KAO, par M. J. Persoz, et de RECHERCHES SUR LA
MATIERE COLORANTE DES NERPRUNS INDIGÈNES, par M. A. F. Michel. Im-
primé par ordre de la Chambre de commerce de Lyon, 1858. It was known
that the Chinese prepared a beautiful green dye from the bark of two
species of *Rhamnus*. Owing to the exertions of Rondot these plants were
introduced from China into France, and subsequently much cultivated in
that country. In 1857, Decaise described and figured them as *Rhamnus
chlorophorus*, and *Rh. utilis* (v. s. p. 447). Rondot reproduced this article
and the figures in his "Vert de Chine". Rondot made extensive experiments
with the Vert de Chine or Chinese green Indigo, of which considerable
quantities were then imported into Lyons and used for dyeing silk. As Mr.
Rondot informed me, in 1884, this dye-stuff was subsequently abandoned,
owing to its high price. In the same book R. gives an interesting account
of Chinese tinctorial plants.

But his principal work, which has rendered his name famous, was
published forty years after his visit to China. It bears the title: LES SOIES,
in 2 vols, 1885—1887, printed at the expense of the Chamber of Commerce at Lyons. It is the most complete and most trustworthy account
hitherto published on silk, its history, culture etc., an admirable monument
of research. R. has a remarkable ability in collecting useful information
from all quarters, sifting it critically, and making use of it for his purposes.
As silk originated in China, and sericulture in that country is still a very
important industry, Rondot was desirous to obtain all possible information
regarding silk directly from China, and in these pursuits he was most
liberally aided by Sir Robert Hart's effective exertions. Rondot's book contains a great deal of new information interesting also for botanists, especially with respect to the varieties of the Mulberry and other trees, upon the
leaves of which silkworms are fed in China.

Hedde, Isidore, born at Puy, May 12, 1801. He was one of the four delegates who in 1843, accompanied Lagrené's mission to China, sent by the Ministries of Agriculture and Commerce, on the recommendation of the Chamber of Commerce at St. Etienne.

In 1848, the Chamber of Commerce at St. Etienne published his work: Description méthodique des produits divers recueillis dans un voyage en Chine, 1843—46. This book contains a list of articles brought from the countries visited by the expedition, especially from China, and exhibited by the above-mentioned Chamber of Commerce. It abounds in curious information. Of Chinese articles noticed there, I may mention the following:

P. 56—60. Quatre-vingt-une espèce de graines de fleurs, fournies par Ha-Tchun, jardinier à Fa ti, près Canton, et envoyées à la Société d'Agriculture de Lyon. — Botanical names given.

P. 59. Douze variétés de graines potagères fournies à Canton p. M. Rondot, et remises à différentes personnes en France. — Chinese names given. — P. 60, le chou *pe-tsaï (Brassica chinensis)*. — Ibid. Dix variétés de Riz cultivées dans le Kwang-tong; — trois variétés de *ma*, dont les filaments sont employés à la fabrication des cordes: *Cannabis indica* (?), *Urtica nivea* (?), *Corchorus triumpheta* (?).

P. 62. Indigotier de Ko-long-sou, petite îsle en face d'Amoy.

P. 107—110. L'arbre *tché*, renommé pour la nourriture des vers à soie jaunes. — Description et dessin tirés de l'Encyclop. d'Agriculture chinoise. M. Hedde avait rapporté en France un plant d'arbre tché qui lui avait été procuré par M. Danicour, missionnaire à Chousan; il a péri avec beaucoup d'autres plantes intéressantes dans le trajet de Rochefort à Paris. The *tché* tree is well known now to botanists. It is the *Cudrania triloba,* Hance, introduced in 1862, to the Jardin des Plantes, Paris. Hook. Icon. Pl. tab. 1792 (1888).

P. 106—13, 116—20. Echantillons de différentes variétés du mûrier, de Canton, Macao, Chousan, Chang haï, Amoy. See DC. Prod. XVII, 243—245: *Morus alba,* var. *indica,* Bureau *(M. indica,* Lin.), Canton Hedde; — *Morus alba,* var. *latifolia,* Bureau. Hedde gathered it near Su chou fu in Kiang su, near Ning po, in Amoy, Ko lung su Island. It is considered the best for feeding silkworms.

P. 165—167. *Polygonum tinctorium,* near Canton, *Isatis indigotica,* Fortune, *Indigofera tinctoria.*

P. 167, 170. *Wei hwa,* substance dont les Chinois se servent pour teindre en jaune. Boutons du *Sophora japonica.*

P. 169. *Hwang-pei-pi,* écorce d'une espèce de cyprès (from the Chinese name) de la province de Kwang si. Employée pour teindre en jaune. — This is *Evodia glauca.*

P. 171. *Keou-hwa,* cônes d'une juglandée, qui sont employés dansle nord de la Chine, Chousan, Tche kiang, Kiang sou, pour teindre en noir. — This is *Fortunea chinensis,* v. supra p. 496.

P. 173. *Hong hwa, Carthamus tinctorius.*

P. 164, 182. *Pei tseu,* Galle de Chine. — Galls of *Rhus semialata.*

L'AGRICULTURE EN CHINE, par I. Hedde was published in 1850, 142 pages, comprenant 23 planches.

While still in China, Hedde communicated for the Chinese Repository a paper: AN EXCURSION TO THE CITY OF SU CHOW, MADE IN THE AUTUMN (Oct. 30, 1845), which appeared in that periodical, vol. XIV (1845) 584.

In vol. XVI (1847) 75, of the same, are NOTICES OF AN EXCURSION TO THE CHIEF CITY OF THE DEPARTMENT OF CHANG CHOW IN THE PROVINCE OF FU KIEN, by I. Hedde. He visited this place in Nov. 19, 1845.

Mr. Rondot informed me that Hedde died in 1880 or 81.

Itier, Jules, accompanied Lagrené's mission to China as a delegate of the Ministry of the Finances. Of China he saw only Canton and Macao, for he returned to France in July 1845.

He published: JOURNAL D'UN VOYAGE EN CHINE, 1843—46, 3 vols. 1848—53, and some papers on Chinese economic plants:

DE LA NATURALISATION EN FRANCE ET EN ALGÉRIE DE PLUSIEURS PLANTES TEXTILES ORIGINAIRES DE LA CHINE ET DE L'APPLICATION DES PROCÉDÉS CHINOIS À LA PRÉPARATION DES FILASSES. Extr. du Bull. de la Soc. d'Agr. de l'Hérault, Avril, May, Juin 1850. — Delile, in Ind. Sem. hti Monspell. 1849, states that Itier transmitted to him seeds of a kind of Chinese Hemp, *tsing ma,* which he raised. He named the plant *Cannabis chinensis.* In the same year M. Hébert presented seeds of another Chinese Hemp, from which plants were raised 4 mètres high. Delile called this *C. gigantea.* The same was cultivated in the Jard. d. Plantes, Paris. Vilmorin even obtained plants 7 mètres in height. Rev. Hort. 1847, 198, — 1851, 109—11.

Le Sorgho sucré, *Holcus saccharatus,* kao lien de la province de Kwang tong, Chine, par J. Itier. 1857.

Yvan, Dr. Melchior, Physician to Lagrené's mission. I learn from M. Rondot, who was on the most friendly terms with Yvan, that the latter

formed a herbarium while in China. A native from Canton collected plants
for him in the neighbourhood of this city and even as far as several days'
journey from it. After his return from China, Yvan first lived at Paris, but
many years ago retired to Digne, dép. des Basses-Alpes, his native city,
and since that time Rondot lost sight of him.

Dr. Yvan in 1847, published a small pamphlet: Lettre sur la
Pharmacie en Chine. — In 1850 appeared his: La Chine et la Pres-
qu'île Malaise. Relation d'un Voyage accompli 1843—46; — in 1852:
Souvenir de l'Ambassade française en Chine. Macao et ses environs; — in 1857: Canton, un coin du céleste empire.

Mr. A. Franchet of the Mus. d'Hist. nat., Paris, kindly communicated
to me, that in the vast herbarium of Mr. Drake del Castillo, Paris, which
contains about 85,000 species, — there is a collection of Chinese plants
labelled: Yvan, Pakoi, Kwang tong province. Mr. Franchet, who looked
over the phanerogamous plants, counted 850 species; he did not count the
ferns. This collection formed part of the herbarium of Count Franqueville,
and the labels show the handwriting of the latter. After his death, in 1891,
all his plants were bought by Mr. Drake. The formation of Franqueville's
herbarium dates from 1848 or 1850, when he began to buy all collections
offered to him. — Some of Yvan's plants are also found in the herbarium
of the Mus. d'Hist. nat. On the labels it is stated that these plants had
been presented by the Société de Commerce (unknown to Mr. Franchet).
But Mr. Rondot thinks that it is the Soc. de Comm. of Rouen, which
in about 1852, sent to China a mission, headed by Mr. Arnaudtizon.
There can be no doubt that the collection in question is that formed by
Dr. Yvan near Canton, between 1843—46. Pa koi is perhaps a clerical
error for Pak wan, name of some hills near Canton, famed for their
rare flowers.

I may finally notice, in this place, **Le Clancher,** Surgeon to the French
Corvette "La Favorite", who in 1844, presented to the Mus. d'Hist. nat.
herbarium specimens of the Nettles cultivated in China for textile purposes.
He had gathered them on the Yang tze kiang, 120 kilomètres from its
mouth; down the river from Nan king. They were examined by Decaisne,
who found that one of them was the *Urtica (Boehmeria) nivea,* Lin. The
other was the *Ramie,* which is much cultivated in Java, *Urtica (Boehmeria)
utilis,* Bl. See Decaisne in Journ. d'Agricult. prat. etc. 1845, 467—69,
and Revue Hort. 1854, 162—71, "Recherches sur la plante textile Ramie,
Boehmeria".

M. Franchet informs me that in the herbarium of the Museum, Paris, there is a botanical collection of about 230 species from South China, presented, in 1844 by Le Clancher (Voyage de "La Seine").

V. Botanical Collectors in the neighbourhood of Canton and Macao and in other localities in the Kuang tung Province, between 1840—60.

Callery, Joseph, Maxime Marie *), known as a distinguished sinologist as well as for his exploration of the Flora near Canton and Macao, was born at Turin, June 25, 1810. He was educated at Lyons and Chambery and originally destined for an industrial career, but finally he joined the Missions Etrangères, who determined to send him as a missionary to China and Corea. He embarked in March 1835, visited Batavia and the Philippines, where he explored the provinces of North Luzon, inhabited by the Igorrotes, and made an ornithological collection which he presented to the Museum of Turin. In 1836 he finally reached Macao and immediately began to study the Chinese language under the auspices of Gonçalves, the celebrated Portuguese sinologist. He soon acquired a thorough knowledge of Chinese and learned also the Corean language. In 1843 he wrote a Memoir on Corea. His leisure time he devoted to the geological and botanical exploration of the neighbourhood of Macao and Canton and published, in 1836, a memoir: Etat géologique des côtes méridionales de la Chine. He also made meteorological observations. Of his sinological works the most important is a Chinese-Latin Dictionary: Systema Phoneticum Scripturae Sinicae, Macao, 1841.

In April 1842, C. returned to France, and on February 13, 1843 he was appointed Interpreter to the French Consulate at Canton, and attached in the same year to Lagrené's Embassy, with which he went out again to China. On this occasion he had an opportunity of seeing Chusan, Shang hai, Ning po, Amoy and the other places visited by the mission. He subsequently wrote a Journal de l'Ambassade de M. de Lagrené, one half of which was printed in China; the other half still exists in MS. After his return to France, May 13, 1846, he lived in Paris. Subsequently he was naturalized as a French subject and, in 1847 was appointed Secrétaire-interprète du

*) The particulars regarding Callery's life, here given, I owe again to the kindness of Prof. G. Devéria, who obtained them from C.'s relatives and from the archives of the Foreign Office, Paris.

Roi. Put "en disponibilité" in 1849, he reassumed his function and held the post of "secrétaire interprète" till his death, which happened on June 5, 1862, at St. Martin les Boulangis (Seine et Marne). — He was Member of the Académie Royale des Sc., Turin, et Chevalier de la Légion d'honneur.

Prof. Devéria discovered in the archives of the Foreign Office at Paris a MS. memoir by Callery, in which he states, that he forwarded to to the Mus. d'Hist. nat. Paris, a collection of over 5000 specimens of Chinese plants, constituting 2000 distinct species, of which about 15 were entirely new plants. To the same Museum he transmitted a complete collection of Chinese drugs with their Chinese names.

Mr. Franchet writes me that Callery's collection of Chinese plants in the Muséum d'Hist. nat., most of them from Macao, was received in 1845, one part received in 1850 amounting to 450 species only. But he presented, besides these, a very interesting collection of about 800 species gathered in the island of Luzon. From A. De Candolle's Phytographie, 1880, 85, 158, we learn that Delessert, in 1838, received from A. Brongniart a collection of Chinese plants, gathered near Macao by the French missionary Callery. From this it appears that the latter first forwarded his plants to Brongniart. The Kew Herbarium also possesses Callery's plants, his name being not unfrequently quoted in Hemsley's Ind. Fl. Sin. Most of his Chinese specimens are labelled: China or Macao, but some of them also Chusan.

Callery's plants recorded as new:

Delphinium Calleryi, Franch. in Bull. Soc. Linn. 1882, 329. Macao. — I. F. S., I, 19.

Callerya (new genus) *tomentosa*, Endlicher, Gen. Pl. Suppl. III (1843) 104. — According to B. & H. Gen. Pl. I, 498, this is *Millettia nitida*, Benth., discovered earlier by Millett. — I. F. S., I, 159.

Rubus Hanceanus, O. Kuntze, Method. (1879) 72, 77. (R. Fordii, Hance). Macao, Callery. — I. F. S., I, 231.

Rosa Luciae, Franch. et Rochebr. was first discovered in China by Callery. See Bot. Mag. t. 7421 (1895). — I. F. S., I, 251.

Pyrus Calleryana, Decaisne, Jard. Fruit. du Mus. I (1858). China. — I. F. S., I, 256.

Pourthiaea (nov. gen.) *Calleryana*, Dcne, Nouv. Arch. Mus. X (1874) 147. China, Callery. — Hemsley in I. F. S., I, 263, reduces. Pourthiaea to Photinia, and unites all the species of Pourthiaea described by Decaisne, proposing a new name: *Photinia variabilis*, Hemsl.

Stransvaesia Calleryana, Decaisne, l. c. 179. Circa Cantonem. — I. F. S., I, 264.

Rhododendron Calleryi, Planchon, Fl. d. Serres, 1853, 81. China. — Maximowicz, Rhododend. (1870) 37 = *Rhododendron indicum*, Sweet. — I. F. S., ii, 26.

Ligustrum Calleryanum, Decaisne, Nouv. Arch. Mus. n. s. II (1878) 36. Macao. — I. F. S., ii, 89.

Tylophora hispida, Dcne in DC. Prod. VIII (1844) 610. Circa Cantonem. — I. F. S., ii, 113. Potts earlier.

Aporosa leptostachya, Benth. Fl. Hongkong, 317, Champion. It was earlier discovered by Callery, Macao. DC. Prod. XV, ii, 472. — Ind. Fl. Sin., ii, 429.

Alchornea trewioides, Müll. Arg. in DC. Prod. l. c. 901. — Ind. Fl. Sin., ii, 438.

Macaranga sinensis, Müller, Arg. l. c. 1001. China, 1844. — Ind. Fl. Sin., ii, 443.

Gironniera chinensis, Benth. Fl. Hongkong, 325. First discovered by Callery, China, 1844. See DC. Prod. XVII, 206. — I. F. S., ii, 452 = Variety of *G. subaequalis*, Planch. (1848) from Java and Ceylon.

Broussonetia Luzonica, Bureau in DC. Prod. XVII, 224. In Manillae montibus Igorrotes, Callery.

Pterocarya stenoptera, Cas. DC. Prod. XVI, ii (1864) 140. China, Callery, 1844. — Hance, Journ. Bot. 1873, 376.

Krone, Rev. Rudolph, German, protestant missionary of the Rhenish Mission. Soc. Dr. E. Faber of the same Mission informed me that Krone, who between 1855—1861 collected plants in the Kuang tung province, left China, in 1861, for Europe, visited Russia, embarked again for China, in 1863, died at a comparatively early age, on the way, at Aden, where he was buried. He possessed a thorough knowledge of botany.

The area of his botanical explorations comprises the districts of Hua hien (Fa yün in Cantonese), north of Canton, Tung kuan, Hui chou (Wai chou), both east of Canton, and Sin an (San on), south-east of Canton, in which districts the Rhenish Mission have several stations. Krone wrote several articles on his journeys in the regions east and south-east of Canton:

A Notice of the Sanon District, in Transactions of the China Br. R. Asiat. Soc. Part. VI, 1859 (resp. 1858) art. 5.

Reise der Missionäre Krone und Graves auf dem nördlichen Arme des East River (Tung kiang) aufwärts bis zur Stadt Shek lung im Jahre 1859. Rhein. Missionsblatt 1860. Peterm. Geogr. Mitth. 1862, 161.

BESTEIGUNG DES HEILIGEN BERGES LO FAU SHAN DURCH DIE MIS-
SIONÄRE KRONE UND GRAVES IM NOV. 1859. Rhein. Missionsblatt, Mai
1860. Peterm. l. c. 1860, 277, and 1864, 283—292, on account of the
same mountain published after Krone's death. The Lo fau shan, from 4000
to 5000 feet high, lies 14 (German) miles east of Canton. It does not seem
that Krone gathered plants there.

Some of his Kuang tung plants Krone forwarded to Dr. H. F. Hance,
then residing in Canton, who published the novelties in the Annales d. Sc.
nat. and in the Journ. of Botany. But the greater part of his collection he
seems to have sent to Europe. — Professor F. A. W. Miquel at Leyden
in Journal de Botanique néerlandaise, I (1861) 84—129, enumerates or
describes 220 plants, gathered by the Rev. Krone in the Kuang tung
province. This collection, Miquel states, he had received through the kindness of his colleague Mr. Millies. It was accompanied with some notes, by
Krone, regarding the localities, where these plants had been found, viz:

Samkok, la plus haute montagne de la contrée, 3000 feet. (On the
maps marked between the districts of Tung kuan and Sin an).

Tai thang, village sur le plateau (sometimes written Fai thang).

Su heang et Fuk wing, villages au bord de la mer. (Sin an district).

Pu kak et Ho an (some times written Ho au) dans le district de
Wai tchao (Hui chow). (On the maps S. W. of Hui chou).

Phu lu wei, montagne.

Pik than, plaine aride près de la mer.

Fa yün, dans l'intérieur de la province de Canton. (Distr. of Hua hien).

L'îsle de Hong kong.

Mr. Franchet informs me that in the Herbarium of Franqueville (v. s.
p. 524) he discovered a collection of 1200 phanerogamous plants (he did
not count the ferns) made by Krone in the Kwang tong province; six localities appear on the labels, viz: Pik thaio, Sa non, Pu kak, Tai wong
shan, Fuk ming, Sai heong.

Krone's plants recorded as novelties:

Uvaria synsepala, Miq. l. c. 128. Ho an. — I. F. S., I, 26.

Gynandropsis sinica, Miq. l. c. 128. Common. — I. F. S., I, 50 =
G. pentaphylla, DC. Common in warm countries.

Dianthus longicalyx, Miq. l. c. 127. Fuk wing. — I. F. S., I, 63.

Dianthus oreadum, Hance, Advers. 1864, 207. Ho au. — I. F. S., I,
64 = *Dianthus superbus*, Lin.

Stalagmites erosipetala, Miq. l. c. 126. — I. F. S., I, 75. Obscure.

Ilex oxyphylla, Miq. l. c. 124. Su heang. — I. F. S., I, 117.

Vitex sinica, Miq. l. c. 125. — I. F. S., I, 133 = *Vitex heterophylla,* Thunberg.?

Sophora Kronei, Hance, Manip. 1863, 220. S. China. — I. F. S., I, 202 = *S. flavescens,* Ait.

Amygdalus dasylepis, Miq. l. c. 122. Cultivé comme arbre fruitier. — I. F. S., I, 220 = variety of the Peach.

Quisqualis grandiflora, Miq. l. c. 119. Fuk wing. — I. F. S., I, 294= *Q. indica,* Lin.

Panax fallax, Miq. l. c. 118. — I. F. S., I, 118, obscure.

Scleromitrion sinense, Miq. l. c. 108. Su heang. — I. F. S., I, 375, obscure, probably a *Hedyotis.*

Mussaenda hirsutula, Miq. l. c. 109. Su heang. — I. F. S., I, 379, obscure.

Pavetta Kroneana, Miq. l. c. 107. Fuk wing, and *P. sinica,* ibid. — I. F. S., I, 385. Both perhaps = *Ixora chinensis,* Lam.

Vernonia Kroneana, Miq. l. c. 98. Su heang. — I. F. S., I, 402.

Vernonia exilis, Miq. l. c. 98. Ho an. — I. F. S., I, 402.

Eupatorium subtetragonum, Miq. l. c. 99. Su heang. — I. F. S., I, 404 = perhaps *E. Lindleyanum,* DC.

Aster scaberulus, Miq. l. c. 100. Pik than. — I. F. S., I, 416.

Conyza leucodasys, Miq. l. c. 103. Kwang tung. — I. F. S., I, 419. Does not seem a Conyza from description.

Emilia sinica, Miq. l. c. 105. Fuk wing. — I. F. S., I, 449 = *E. sonchifolia,* DC.

Youngia gracilis, Miq. l. c. 106. Su heang. — I. F. S., I, 476, probably a variety of *Crepis japonica,* Benth. *(Prenanthes japonica,* Lin.).

Youngia humilis, Miq. l. c. 106. Kwang tung. — I. F. S., I, 476 = probably a *Lactuca.*

Brachyramphus sinicus, Miq. l. c. 105. Su heang. — I. F. S., I, 481, reduced to *Lactuca denticulata,* Maxim. *(Prenanthes hastata,* Thunb.).

Lysimachia heteroganea, Klatt in Linn., XXXVII (1871) 501. Krone, Fuk wing. (Klatt examined the Primulaceae of Franqueville's herbarium).— I. F. S., II, 52.

Lysimachia inconspicua, Miq. l. c. 110. Fa yün. — I. F. S., II, 53.

Lysimachia sinica, Miq. l. c. 110. Fuk wing. — I. F. S., II, 57.

Olea ovalis, Miq. l. c. 111. Pu kak. — Ind. Fl. Sin., II, 88, perhaps *Osmanthus.*

Evolvulus sinicus, Miq. l. c. 112. Common. — I. F. S., II, 166, perhaps *E. alsinoides,* Lin.

Vandellia subcrenulata, Miq. l. c. 113. Fuk wing. — I. F. S., II, 191. Obscure.

Striga parvula, Miq. l. c. 113. Pu kak, Su heang. — I. F. S., II, 201 = *St. lutea,* Lour.?

Dipteracanthus subdenticulatus, Miq. l. c. 117. Kwang tung. — I. F. S., II, 249. Obscure.

Rostellularia trichochila, Miq. l. c. 118. Kwang tung. — I. F. S., II, 249. Obscure.

Hypoestes sinica, Miq. 117. Common. — I. F. S., II, 249.

Clerodendron amplius, Hce, Advers. 1864, 233. Kuang tung prov. — I. F. S., II, 259 = *Cl. cyrtophyllum,* Turcz. (v. s. p. 488).

Clerodend. oxysepalum, Miq. l. c. 114. Common. — I. F. S., II, 261.

Caryopteris ovata, Miq. l. c. 114. Fuk wing. — I. F. S., II, 265.

Plectranthus sinensis, Miq. l. c. 115. Fai thong. — I. F. S., II, 274.

Scutellaria adenophylla, Miq. l. c. 117. Fuk wing. — Ind. Fl. Sin., II, 293.

Scutellaria leucodasys, Miq. l. c. 116. Pu kak. — I. F. S., II, 296.

Polygonum subcordatum, Miq. l. c. 95. Su heang. — I. F. S., II, 350.

Melanthesa glaucescens, Miq. l. c. 97. Fuk wing. — Ind. Fl. Sin., II, 427 = *Breynia fruticosa,* Hook. *(Andrachne fruticosa,* Lin.)?

Croton Kroneanum, Miq. l. c. 97. Ho an. — I. F. S., II, 435, probably same as *C. lachnocarpus,* Benth. (Reeves).

Pinus canaliculata, Miq. l. c. 86.

Arundina pulchra, Miq. l. c. 90. Ho an.

Eulophia sinensis, Miq. l. c. 91. Fuk wing.

Habenaria endothrix, Miq. l. c. 92. Ho an.

Zingiber confine, Miq. 94. Fuk wing, cult.

Alpinia oxyphylla, Miq. 93. Ho an.

Phrynium sinicum, Miq. 94. Phu lu wei.

Smilax pteropus, Miq. l. c. 89. Su heang. — In DC. Monogr. I, 46= *Smilax China,* Lin.

Thysanotus chinensis, Benth. Fl. Hongk. 372. Krone, S. China. — Hance, Advers. 1864, 245.

Commelina ludens, Miq. l. c. 88. Fuk wing. — Clarce, Commel. in DC. Monogr. III, 171 = variety of *C. communis.*

Eriocaulon sinicum, Miq. 87. Fuk wing.

Park, a botanical collector in the neighbourhood of Canton and Macao, whose name appears frequently in DC. Prodromus, beginning with vol. XIV,

1856. The labels of his specimens, which are preserved in De Candolle's
herbarium, sometimes mention the Lappa Hills (Patera I., west of Macao)
and Mo ha Hills (unknown to me). The late Alph. De Candolle, to whom
I applied for information, kindly sent me a notice stating that in April
1856 he received 175 species of dried plants gathered by Park near
Canton. As he had not preserved Park's letter, accompanying the collection,
he was not able to say to what nationality he belonged. The great botanist
further informed me that in the Prodromus, the name is sometimes erroneously written Parker, and that Parkinson n. 108, quoted by Planchon in
DC. Prod. XVII, 161, sub *Ulmus parvifolia,* is a mistake for Park.

Tate, George Ralph, born at Alnwick 1835. In 1850 he entered the
University of Edinburg as a student of medicine. He gained the golden
medal for botany, in 1853, and in 1855, he passed his examination for a
Surgeon's degree. Some years later he took his degree of M. D., and in
March 1858 he joined the army as Assistant Surgeon in the Roy. Artillery.
He was then stationed for two years in Hong kong, and while there, went
on an excursion of some months into the interior of the country, and made
a collection of plants in the province of Shan tung. This collection, comprising about 800 specimens, is now in the Kew Herbarium. On his return he
was stationed in the Isle of Wight for some years. F. L. S. 1869. He died
Sept. 23, 1874. (Proc. Linn. Soc. 1874, p. LXIV).

I have some doubt whether T. ever gathered plants in Shan tung. The
plants from Tate's collection mentioned in the Ind. Fl. Sin., and by Munro,
Bamb. are stated to have been collected in Hong kong, near Macao and in
the Kwang tung province; localities given on the labels: Oo kai sa, Sim
chung, Wong yu, all unknown to me.

Tate discovered two new plants:

Sonerila peperomiaefolia, Oliver, in Hook. Icon. Pl. tab. 1814 (1889).
Tate, Kwang tung. — I. F. S., i, 301, Sonerila sp.? Kwang tung: mountain above Oo kai sa, at 2300 feet. Tate.

Plectranthus Tatei, Hemsl. Ind. Fl. Sin., ii, 274. Tate, Kuang tung:
Wong yu.

VI. British, French and Dutch Botanical Explorations, between about 1850—60, near the then opened Chinese Trading Ports (besides Canton), in Formosa, in the Corean Archipelago, on the Coasts of Corea and Manchuria, etc.

Let us first mention the names of three British officials in China, connected with the discovery of the famous Rice-paper plant, *Aralia papyrifera,* (v. s. p. 462) and its introduction into England.

Sir W. Hooker, in Kew Journ. Bot. II (1850) 27, reports that J. H. **Layton,** H. B. M. Consul at Amoy, sent him excellent specimens of the pith of the plant (then unknown), from which the so called Chinese Rice-paper is formed. It was known to be produced in Formosa. — In vol. IV (1852) 50, of the same Journal, Hooker states that, although Layton had died, his widow succeeded in procuring a living, rooted specimen of the plant, which was sent to England in 1851. It perished during the voyage, but Hooker was able to determine from the leaves that it was an araliaceous plant.

J. C. Bowring, in his article on the Chinese Rice-paper plant in 1852 (v. supra p. 381), records that in this year Mr. G. G. **Sullivan,** H. B. M. Consul at Amoy, had procured for him, from Formosa, six living specimens of this remarkable plant, four of which he (Bowring) transmitted to England. Of these, two arrived in a healthy state. Sullivan died in the same year.

In the same article Bowring published an interesting paper by Mr. **Sinclair,** interpreter to the Fu chou Consulate, on the Rice-Paper plant, its cultivation in Formosa, and the cutting of its pith into paper. — Sinclair was afterwards British Consul at Fu chou.

Hance, Dr. H. F. (v. s. p. 365). In the autumn of 1857, Hance visited Amoy, and spent there the months of October and November, collecting plants in the vicinity of this place. Some of these plants were gathered on Mount Lam tai wu, 2000 feet high. This mountain, in the Mandarin dialect Nan tai wu, is situated on the mainland south of Amoy. Hance notices the following new plants from his Amoy collection:

Coronilla buxifolia, Hce, Manipulus pl. nov. (1863) 219.

Rosa amoyensis, Hce, Journ. Bot. 1868, 297. — I. F. S., I, 252 = *R. microcarpa,* Lindley, (1821).

Abelia Hanceana, Martius in Hce, Advers. 1864, 216. Mons Lam tai wu. — I. F. S., I, 358 = *A. chinensis,* R. Brown (1818).

Galium miltorrhizum, Hance, Journ. Bot. 1868, 114. — Ind. Fl. Sin., I, 394 = *G. gracile,* Bunge (1833).

Gnaphalium amoyense, Hce, Journ. Bot. 1868, 174, — 1878, 108.— I. F. S., I, 426 = *G. hypoleucum,* DC. (1834).

Clerodendron amplius, Hance, Advers. 1864, 233. — I. F. S., II, 259 = *Cl. cyrtophyllum,* Turcz. (1863). Fortune, Amoy, earlier.

Cyperus amoyensis, Hance, Advers. 1864, 249.

Grijs, Dr. C. F. M. de, Dutch Military Surgeon and an experienced sinologist, subsequently Director of a school for training young Chinese at Amoy, to serve as interpreters in Java. He was also for some time Dutch Consul at Amoy. In 1863 he published a Dutch translation of the Sɪ ʏŭᴀɴ ʟᴜ, a Chinese work on Medical Jurisprudence.

From 1858 to 1862 G. collected plants for Dr. Hance in the neighbourhood of Amoy and amongst the tea-hills of An koe (An ki hien) in eastern Fu kien*). Hance described the novelties in Symbolae ad Floram sinicam, 1861, Adversaria 1864, and Journ. Bot.

The following novelties are recorded from Grijs collection:

Clematis caesariata, Hance, Journ. Bot. 1870, 71. Fo kien, 1861. — I. F. S., I, 3.

Camellia Edithae, Hance, Symbolae (1861) 221. An koe tea-district, 1861. — I. F. S., I, 81.

Camellia theiformis, Hance, Symb. 221. Prov. Fo kien, 1861. — I. F. S., I, 81 = *C. euryoides,* Lindley (1826).

Zanthoxylum simulans, Hce, Advers. 1864, 208. Fo kien, 1862. — I. F. S., I, 106 = *Z. Bungei,* Planch. (1831).

Prunus pogonostyla, Maxim. Fl. As. or. Fragm. 1879, 11, and Mél. XI (1883) 682. Mons Nam tai wu prope Amoy, de Grijs, 1862. Herb. Hance. — I. F. S., I, 221.

Prunus campanulata, Maxim. Mél. XI, 698. Fo kien, de Grijs, Herb. Hance. — I. F. S., I, 218.

Rubus althaeoides, Hce, Symb. 222. Fo kien. — I. F. S., I, 229.

Rubus jambosoides, Hce, l. c. Fo kien. — I. F. S., I, 232.

Corylopsis multiflora, Hce, Symb. 224. In collibus theiferis An koe.— I. F. S., I, 290.

---

*) Hance erroneously writes (Symb. 221). «Colles theiferi Ankoe, versus limites *occidentales* provinciae Fo kien». This district lies NW. of Ts'üan chou fu, not far from the sea. More than thirty years earlier J. Reeves sent to England plants gathered in the An koe tea-district (v. s. p. 261). The same district was visited in 1834 by G. J. Gordon in company with the Rev. Gutzlaff. Journ. As. Soc. Bengal, Febr. 1835, — Chin. Repos. IV (1835) 72.

Eugenia Grijsii, Hance, Journ. Bot. 1871, 5, — 1879, 10. Fo kien, 1861. — Ind. Fl. Sin., ɪ, 298, sub *Eugenia sinensis,* Hemsley. Earlier Abel (1817).

Patrinia graveolens, Hce, Symb. 224. Fo kien occidentalis. — I. F. S., ɪ, 396 = *P. heterophylla,* Bge?

Echinops Grijsii, Hce, Advers. 1864, 221. Fo kien. — I. F. S., ɪ, 459 = *E. dahuricus, Fischer* (1808).

Vaccinium iteophyllum, Hance, Manip. 1863, 223. Fo kien, 1861. — I. F. S., ɪɪ, 15.

Myrsine buxifolia, Hce, Symb. 225. — I. F. S., ɪɪ, 225.

Machilus Grijsii, Hance, Manip. 226. Fo kien, 1861. — Ind. Fl. Sin., ɪɪ, 375.

Machilus oreophila, Hce, l. c. 227. Fo kien. — I. F. S., ɪ, 377.

Elaeagnus Grijsii, Hce, Symb. 227. Fo kien. — I. F. S., ɪɪ, 403.

Buxus stenophylla, Hce, Journ. Bot. 1868, 331. — I. F. S., ɪɪ, 418 = *B. sempervirens,* Lin.

Pellionia Grijsii, Hce, Journ. Bot. 1868, 49. Fo kien, 1861.

Iris Grijsii, Maxim. Mél. X (1880) 702. Grijs, Fo kien. Noticed by Hce, Journ. Bot. 1870, 314, as *I. oxypetala,* Bge.

Chrysopogon pictus, Hce, Advers. 252. Fo kien.

Alsophila Metteniana, Hce, Journ. Bot. 1868, 175. Fo kien, 1861. — Syn. Fil. 459 = *A. glabra,* Hook. var. *Metteniana.*

Gregory, William, British Consular service. He went to China as interpreter in 1854, was employed in the Superintendency; proceeded to Fu chou fu with Mr. (afterwards Sir Walter) Medhurst, in March 1855; was appointed Post-office Agent, March 1856; was Acting Assistant in 1856 and 1857; appointed first Assistant at Fu chou, March 1859. Was Acting Interpreter at Swatow from 1860 to 1865. Appointed British Vice-Consul at Tamsuy, May 1865. Was transferred to Canton, in August 1871. Acting Consul at Tai wan, 1872—76. Promoted to be Consul at Swatow, May 1873; transferred to I chang, July 1886. Retired from the service, in April, 1890. (Foreign Office List).

G. gathered plants for Dr. Hance, in 1857, near Fu chou fu and after 1865 also near Tamsuy. It seems that he also sent dried plants to Kew. There are three of his specimens recorded as new by Hance:

Ixeris Gregorii, Hance, Advers. 1864, 222. Fo kien, May 1857. — I. F. S., ɪ, 485 = *Prenanthes dentata,* Thbg., or *Lactuca Thunbergiana,* Maximowicz.

Polygonum typhoniifolium, Hance, l. c. 239. Fo kien, 1857. — I. F.
S., ii, 349 = *P. senticosum,* Franchet, earlier known from Japan.

Woodwardia angustiloba, Hance, Journ. Bot. 1868, 176. Prope Fu
chou, May 1857. — According to M. Kuhn, Journ. Bot. 1868, 268, this
is *Woodwardia radicans,* Smith, in Act. Taur. V, 412, var. *auriculata,*
Bl. — Not in Syn. Filic.

Medhurst, Walter Henry (Junior), subsequently Sir Walter, not to be
confounded with his father, the well known sinologist, who had the same
Christian names. He came young to China, was attached to Sir H. Pottinger's suite as interpreter, in 1842. British Consul at Fu chou, in 1854,
where he remained for some years. In 1864 appointed to Han kow, in
1871 to Shang hai. He retired in 1878, after having been knighted, died
1886. — During his residence at Foo chow he collected some plants for
Dr. Hance. See Advers. 257, Journ. Bot. 1869, 235. — In the Kew
Report for 1875, 11, Medhurst's name is mentioned in connection with
Oeceoclades falcata, which he seems to have introduced from China.

Smith, J., Surgeon of H. M. S. "Cambrias", published in the Gard.
Chron. 1844, 220, a good article on Chinese Agriculture, from his
own observations.

Tarrant, William. He gathered some plants for Dr. Hance near Ning
po. One of them was described as new:
Fritillaria collicola, Hance, Journ. Bot. 1870, 71. In collibus prov.
Che kiang, Guil. Tarrant. — According to Baker, Tulipeae, 1873, 258,
this is *Fritillaria verticillata,* Willd. var. *Thunbergii,* or Uvularia cirrhosa,
Thunberg.

W. Tarrant was the editor of the periodical "The Friend of China",
which was edited in Hong kong, and after 1860 in Canton. In this he
published, in 1862, an Account of his Journey from Ning po to Shang hai,
in 1857, via the borders of An hui Province, Hu chou fu, Grand Canal.
T. died 1872 at Shang hai.

Fabre-Tonnerre, C., Dr. med., a French physician, who gathered for
Dr. Hance a few plants near Shang hai, in the province of Kiang su, of
which Hance described two as novelties:
Quercus Fabri, Hce, Journ. Linn. Soc. X (1869) 202. Prov. Kiang
su, no date.

Carex Fabri, Hance, l. c. XIII (1873, resp. 1871) 90. Juxta Shang hai. No date.

Among the Chinese plants noticed in Seemann's Bot. Herald, 1857, there are several collected by Dr. Fabre-Tonnerre near Shang hai (evidently from Hance's herbarium). One of them new:

Ischaemum cinerascens, Munro, in Seem. Botan. Herald, 424. Fabre-Tonnerre.

From this it appears that Fabre-Tonnerre's collection near Shang hai was made before 1857.

Montigny, Louis Charles Nicolas Maximilian, born at Hamburg, of French parents, 2d Aug. 1805. In 1843, he accompagnied the French Embassy (Lagrené) to China, in the capacity of *chancellier.* He returned from Canton to France, in July 1845, without having seen the more northern ports of China. But in January 1847, he was appointed Consular Agent at Shang hai and proceeded to that port. Having been promoted in 1850 to the rank of 2d class Consul, he went home on leave in July 1853. In October 1855 First Class Consul, returned to Shang hai, in July 1858 named Consul General at that port. In Febr. 1859 transferred to Canton, resigned the public service in August 1862, Commandeur de la Légion d'honneur, died September 14, 1868. See his obituary in Rev. Hort. 1869, 52.

During his residence in Shang hai, M. rendered great services to natural science and horticulture by introducing into France interesting beasts and birds*) and useful Chinese economic plants. These plants were cultivated and distributed by the Société d'Acclimatation of Paris, and in the pages of their Bulletin, 1855—1864, Montigny's name is frequently met with:

Bull. Soc. d'Accl. II (1855) p. 16: Chinese seeds brought by Montigny in 1854; — p. 17, Mountain Rice grown without irrigation; — p. 19, Gigantic Maize; — p. 20, Seeds of two varieties of the Oleaginous Pea of China, which contain 20% of oil; — p. 382, various Chinese oils; viz: Tea nut oil (obtained from tea seeds), Cabbage oil (from the seeds of *Brassica chinensis*), Cotton-seed oil, Pea oil, expressed from the oleaginous pea, which is *Soja hispida,* Moench. — Compare on the later plant,

---

*) When in 1853, he went on leave, he carried along with him, for the Société d'Acclimatation, twelve living yaks, *Bos grunniens* (domesticated) and living specimens of the beautiful Manchurian Crane named *Antigone Montignesii,* by Bonaparte, Comptes rend. Acad. Sc. XXXVIII (1854) 661.

Lachaume in Rev. Hort. 1857, 568 and 1859, 222; and Carrière in Rev. Hort. 1880, 153.

Bull. Soc. d'Accl., II (1855) 271, *l'Igname* de Chine. Of this plant, commonly grown in China for its edible roots, Montigny, in 1848, sent specimens to the Mus. d'Hist. nat., Paris, where the plant was cultivated. Decaisne found that it was a new species, which he named *Dioscorea Batatas*, and described and figured it in Fl. d. Serres, IX (1854) 167, and Rev. Hort. 1854, 243—52. Montigny says that the Chinese call it *sain in*, which is a typographical error for *san yü*.

Bull. Soc. d'Accl. II, 167, 298, 343, III (1857) 308, 314: In 1854 Montigny brought acorns of two species of *Oaks*, on the leaves of which the wild silk worms feed in Manchuria. Previously he had sent the cocoons of these silkworms, but they were dead when they arrived. The oaks were successfully cultivated in France. — See also Hance, Silkworm-oaks of N. China, in Journ. Linn. Soc. X (1868) 482, and XIII (1870) 8, *Quercus mongolica*, Fischer and *Q. dentata*, Thbg.

Bull. Soc. d'Accl. III (1856) 163, *Holcus saccharatus*, introduced from China by Montigny, successfully cultivated in the south of France. — Vilmorin, in Rev. Hort. 1854, 69, states that Montigny sent the seeds of this plant from Shang hai in 1850 or 51 under the name "Canne à sucre du nord de la Chine" to the French Geogr. Society, who distributed them. It is the *Sorghum saccharatum*.

Bull. Soc. d'Accl. IV (1857) 488. Two species of *Bamboo* from China cultivated in Algiers; the sprouts of one of them edible. Both of them had been sent from China by Montigny. The edible kind is *Phyllostachys mitis*, Rivière, Bambous, 231. V. s. p. 516.

Bull. Soc. d'Accl. VII (1860) 453. A cruciferous plant raised from seed sent from China by Montigny under the name of *ca tse* (probably *kie tze* = *Sinapis*). The leaves furnish an excellent salad.

Bull. Soc. d'Accl. n. sér. V (1868) 752—54: Three Chinese vegetables raised from seed sent 1864 by Montigny:

1. *Chou navet* (turnip) which has a taste of mustard. Comp. s. 49.

2. Chinese salad, a kind of *Chrysanthemum*. — Probably *Ch. coronarium*, L., which at Peking is much cultivated as a salad.

3. *Shan tung Cabbage, pe ts'aï*. — In Bull. n. sér. I, 621, there is a letter by Montigny, Aug. 1864, on this cabbage, which is cylindrical, does not form a head. It is *Brassica chinensis*, Lin.

Bull. Soc. d'Accl. II (1855) 349, and VIII (1861) 58. On April 28, 1855, Montigny received the great prize of 3000 francs, founded by the

Duc d'Orléans for having introduced the oleaginous pea, the Chinese Igname, the Sugar-Sorgho, the Chinese oaks, the Yak etc. On the medal which he was awarded on this occasion, was represented a wreath of the leaves of the Man chu oaks, the Igname and the Sorgho.

Montigny first discovered the two species of *Rhamnus*, *R. utilis* and *R. chlorophorus* (v. s. 447), about 1851, and cultivated them in the garden of the French Consulate at Shang hai, and in 1853, sent a number of living plants of both species to the Muséum d'hist. nat., Paris, but they all perished during the voyage. In Nov. 1853, Mr. Rémi (of the French Consulate Shang hai?) forwarded to N. Rondot ripe seed from Montigny's plants, a part of which germinated at Lyons, and the plants flowered in 1857. (Rondot, Vert de Chine, 43—45).

Fontanesia Fortunei, Carrière, (v. supra p. 481) was introduced by Montigny into the Jardin des Plantes, Paris, in 1854.

A peculiar Peach, called *Pêche Montigny* by Carrière, described and figured in Rev. Hort. 1861, 11, was introduced into the Jardin des Plantes by Montigny. According to Decaisne, Jardin fruitier, M. sent the stones from Shang hai in 1852, which germinated. The trees flowered several times, and in 1860 first produced fruit, almond-shaped, flesh adhering to the stone, edible but not very valuable. See also Rev. Hort. 1886.

When Montigny's Yaks were embarked at Shang hai, a considerable quantity of a large gourd, much cultivated in China, was sent along with them as food. It proved to be *Cucurbita melanosperma*, Al. Braun (1824) and was subsequently raised from seed and cultivated in France under the name of "Citrouille des yaks". Naudin in Ann. Sc. nat., 4 sér. VI (1856) 17, and in A. Paillieux and D. Bois, Potager d'un curieux (1892) 108. — According to Cogniaux, Cucurb., in DC. Monogr. III (1881) 548, probably indigenous to Eastern Asia.

Mr. A. Franchet informed me that in the Mus. d'hist. nat., Paris, there are herbarium specimens gathered by Montigny in China and labelled 1851, 1854. Shang hai, Kiang su, Chusan. We learn from Rondot (Vert de Chine, 43), that this collection was formed by Montigny's daughter.

Guierry, Edmond, French Missionary, Congrégation des Lazaristes, born 1825, arrived in China, in 1853; stationed at Ning po. From this place he sent in 1856 to the Société d'Acclimatation, ten different kinds of seeds of vegetables and trees, accompanied with interesting notices, which were published in the Bull. Soc. d'Accl. III (1857) 183. Among the seeds there were those of the *t'ung shu* tree *(Elaeococca verrucosa)* yielding an oil used for painting and caulking ships, and of the *wu t'ung shu (Sterculia platanifolia)*.

In 1864 G. was elected Bishop, in 1869 Vicaire apostolique of Che kiang. He died at Ning po Aug. 14, 1883.

Hélot, Louis, French, Jesuit missionary, born 1816, came to China in 1849, died at Shang hai, Sept. 22, 1867.

Hélot has much contributed to elucidate the question regarding the origin of the Chinese green dye known in France under the name of "Vert de Chine" and prepared from the bark of two species of Rhamnus (v. supra p. 447, 538). See his letters on the subject addressed to the Président du Conseil central de l'oeuve de la Propagation de la Foi, à Lyon, dated Shang hai April 6 and 27 1857, published in Rondot's Vert de Chine, 143—47, and Hélot's article, "Le Vert de Chine", in Etudes de Théol. Philos. et Hist., I (1857) 442—58.

Mr. A. Franchet informs me, that in the Muséum d'hist. nat., at Paris, there is a small collection of plants (about 134 species) gathered by the Jesuit Missionaries Hélot and d'Argy near Shang hai, received in 1865. The Rev. Father Heude of Shang hai wrote me, many years ago, that this collection is of little value, consisting of some plants gathered in the neighbourhood of Si ka wei, the great establishment of the Jesuits near Shang hai, for the greater part garden flowers. These plants, after having been for a long time kept in the house of the Jesuits at Paris, were finally handed to the Museum.

Hélot published also an article on CHINESE SPIRITUOUS LIQUORS, in Etudes religieuses, n. sér. VII (1865) 528—533:

Le *Tseu ia* ou *Matière alcoolisante des Chinois.*

Tseu tsiam ou Vin de *nou mi* (Oriza glutinosa).

Eau de vie des *Céréales*, procédés chinois.

Wilford, Charles, collector for the Royal Gardens, Kew, who from 1857 to 59 made botanical collections in various places on the coasts of China, Formosa, Corea, Manchuria, Japan. Sir W. Hooker in his Journ. Bot. IX (1857) 273, writes on Wilford's botanical mission to the Chinese Seas as follows:

"It was considered by the Director of the Royal Gardens, Kew, that the fact of the British Govt.'s sending a present of a splendid steam-yacht, "The Emperor", to the Emperor of Japan, might afford a good opportunity for making botanical researches among the numerous islands of the Japanese territories, a memorial was laid before the Lords of the Admiralty, requesting that a botanical collector might be permitted to be present at the arrival of the yacht in Japan. The application was liberally responded to. There was no difficulty in finding a suitable collector. The lot fell upon Mr. Ch. Wilford, Herbarium Assistant in the Roy. Gardens. On the 2d of May 1857, the latter embarked for Hong kong. He carried with him letters of recommendation to Sir John Bowring at Hong kong and to Sir Michael Seymour, the Admiral in the Indian Seas".

W. arrived at Hong kong in Nov. 1857 and rémained there till the beginning of June 1858 (v. s. p. 400). During his stay there he had an opportunity of visiting the quite isolated Pratas Island, in the China Sea, about 200 miles S. E. of Hong kong, in the end of April and beginning of May 1858. As we learn from Dr. W. Williams' Chin. Commercial Guide, 1863, Append. 176, Pratas and its reef were surveyed in April 1858 by H. M. S. "Saracen", and it was, we may suppose, on board this ship, that W. made the excursion. A small collection of plants were gathered there and presented to Dr. Hance, who in Journ. Bot. IX (1871) 202, enumerated them (no novelty):

Senebiera integrifolia, DC. — I. F. S., ɪ, 48.

Portulacca australis, Endl.

Triumfetta procumbens, Forst.

Scaevola Koenigii, Vahl. — I. F. S., ɪɪ, 2.

Tournefortia argentea, R. Br. — I. F. S., ɪɪ, 146.

Ipomoea congesta, R. Br. See Journ. Bot. 1878, 14.

Ipomoea pes caprae, Roth. — I. F. S., ɪɪ, 157.

Euphorbia Atoto, Forst. See Journ. Bot. 1878, 232.

Hance observes that all these species are found, without a single exception, in Queensland and seem primarily of Australian origin.

Hemsley, in I. F. S., ɪ, 384, 386, notices two more from W.'s Pratas plants, in the Kew herb., viz: Guettarda speciosa, L. and Morinda citrifolia, both also belonging to the Australian Flora.

From Hong kong W. proceeded, in the beginning of June, to Amoy, where he appears to have staid only a few days. The I. F. S., ɪ, 358, notices, that Abelia chinensis was gathered there by him.

At that time H. B. M. s/s. "Inflexible" had received order to sail for Formosa. W. was permitted to accompany the expedition, which started from Amoy on June 7th 1858. (See Biogr. Bot. sub Wilford). R. Swinhoe, the well known naturalist, of the Br. Consulate, Amoy, attended it as interpreter. He published an account of this cruise in the Journ. N. China, Br.

Asiat. Soc. May 1859, but does not mention Wilford's name. They visited T'ai wan fu, Ta kow, Lang kiao Bay, sailed round the southern point of the island, then along the eastern coast to Suao Bay, where the "Inflexible" arrived on the 18th. Two days later she anchored in the harbour of Ki lung. Coal-mines there visited. They started on an expedition to the Sulphur mines some forty miles from Ki lung and advanced as far as the trading place Banca not far from Tam suy. Then returned to Ki lung, which the Inflexible left on th 26th. Sailing round the north point, and then along the western coast of the island, they reached the entrance to T'ai wan fu, and visited that city again. After this they touched at the Pescadores, and on July 1st returned safely to Amoy. W. probably collected plants in every place where they landed. His Formosan plants in the Kew herbarium are generally labelled: W. coast, S. W. coast, N. E. coast of Formosa, also Ke lung. Tripterygium Wilfordii is labelled: "on banks of the river Sanar, Formosa", which we do not find marked on the best maps of the island, nor does Swinhoe mention it.

In the Report on the Progress of the Roy. Kew Gardens for 1853—59, dated 1st Jan. 1859, p. 12, we read: "Mr. C. Wilford, now on a botanical mission to Hong kong, North China, Japan, has sent us his collections from Hong kong and Formosa. He has now joined H. M. S. "Actaeon" and is about to proceed to the coast of Eastern Tartary in that survey vessel under the command of Lieutenant J. **Ward.** A special grant to the Gardens from the Treasury of 600 £, for three years. The Admiralty gives him a free passage and rations".

Sir W. Hooker in Spec. Fil. III (1860) 189, commemorated Ward's name in one of Wilford's novelties, *Asplenium Wardii*, acknowledging that John Ward, R. N. Commander of the "Actaeon" gave to Wilford, during a surveying voyage in the North China and Japan Seas, every facility and assistance that could be desired, in herborizing on shore. The result has been a very interesting herbarium from countries never before visited by botanists and among them some very interesting ferns.

We have not been able to ascertain how and where W., after his return to Amoy, on July 1st 1858, spent the rest of the year 1858 and the early part of 1859, for there is evidence that the Actaeon did not start for the Corean Archipelago and Manchurian coast before the spring of 1859. Commander Ward, from 1857 to 61 did much excellent work in the survey of the northern part of the Chinese and of the Corean and Manchurian coasts — as can be proved from the respective Admiralty Charts. The "Actaeon" belonged to the British expedition engaged in surveying the

Yang tze kiang from Nan king upwards to Han kou, in the autumn of 1858, and Comm. Ward is stated on the Charts representing the S. E. coast of Corea to have surveyed these regions in 1859. Most probably Wilford spent the winter $18^{58}/_{59}$ in Shang hai, and in the spring proceeded with the Actaeon towards the north. The names of Port Hamilton, Port Chusan, Tsusima, appear frequently on the labels of W.'s collection.

Sir Edw. Belcher, in June 1845, discovered in the Corean Archipelago a small island with a good harbour and named it Port Hamilton in compliment to the Secretary of the Admiralty. — In 1797, Oct. 13, Capt. Broughton, a British navigator, who visited these regions, anchored in a harbour on the S. E. coast of Corea, which, as he states, the natives called Cho san. For a long time this name figured on our maps. Wilford writes it Chusan. But these appellations are erroneous and originated in a mis-understanding. Chosan is the Corean and Japanese name for Corea. The place and harbour in question, first discovered by Broughton, are called Fu san. — The island of Tsu shima was also first visited by Broughton. It lies in the Corean Strait between Japan and Corea, consists properly of two islands separated by a narrow channel, and belongs to Japan.

Proceeding farther to the north, on board the Actaeon, W. botanized occasionally on the Manchurian coast, between 40°—45° N. Lat. as appears from some labels, namely on Bay Vladimir (44°), Bay St. Olga (43° 45′). In the Russian Morskoi Sbornik (Naval Magazine) for 1860, I find the following notice: "Capt. Derper of the Transport 'Baikal' reports from St. Olga, July 13, (25, new style), that on June 24, (July 6) the English sailing vessel "Actaeon", Capt. Ward, after having surveyed Port. Hamilton, Chusan, Harb., Tsu sima, arrived at St. Olga. On June 30 (July 12) she made sail for Hakodate.

We have not succeeded in making out how long Wilford staid at Hakodate, and by what way he returned to Europe. Commander Ward, in 1860, surveyed the Gulfs of Liao tung and Chili, and the coast of Shang tung.

Wilford's botanical collections are at Kew. Duplicates were distributed to the principal herbariums of Europe. The Bot. Garden, St. Petersb. received in 1859 from Kew 370 species of W.'s plants, gathered in Corea and Manchuria, and subsequently his plants from Hakodate. Maximowicz examined and described a considerable number of them. Wilford's Ferns were described in Hooker's Spec. Filic. since Vol. III (1860) and in Syn. Filic. by Hooker and Baker.

Plants from Wilford's collection (with exception of those from Hong kong) recorded as new:

Clematis orientalis, Lin., var. *Wilfordi,* Maxim. Mél. IX (1876) 583. Manchurian coast 44°—45°.

Cocculus cuneatus, Benth. MSS. in Herb. Kew Formosa, West coast.— I. F. S., ı, 28.

Corydalis Wilfordi, Regel in Pl. Radd. I (1861) 148, note. Corea.— Maxim. Mél. X (1877) 49 = *C. Pallida,* Pers. — I. F. S., ı, 38.

Hesperis lutea, Maxim. Mél. IX (1872) 12. Mandshurian Coast 1859, Wilford.

Viola pinnata, Lin. var. *chaerophylloides,* Regel, Pl. Radd. I (1861) 222. Corea. — I. F. S., ı, 54.

Lychnis Wilfordi, Maxim. Mél. VIII (1871) 419. Littus Manchuriae.

Geranium Wilfordi, Max. Mél. X (1880) 614. In Manchuria orient. littorali: inter 44 et 45°. — Cult. 1897 at Kew.

Euonymus sachalinensis, Maxim. Mél. IX (1873) 185. In Manchuria austro-orientali, littore inter 44 et 45°. — I. F. S., ı, 121.

Tripterygium (gen. nov.) *Wilfordi,* Hook. fil. in Benth. Hook. Gen. Pl. I (1862) 368. Formosa. — Regel, Gartenflora, 1869, 105, tab. 612. — I. F. S., ı, 125, Wilford, Formosa, banks of the river Sanar.

Sedum formosanum, N. E. Brown, Gard. Chron. 1885, II, 134. N. E. Formosa. — I. F. S., ı, 285.

Hydrocotyle Wilfordi, Maxim. Mél. XII (1886) 463. Korea, portu Chusan. — I. F. S., ı, 326.

Sanicula tuberculata, Maxim. Mél. VI (1867) 204. Korea, in aestuario Chusan. — I. F. S., ı, 327.

Lysimachia acroadenia, Maxim. Mél. VI (1867) 272. Tsu sima. — Ind. Fl. Sin., ıı, 51 = *Lysimachia decurrens,* G. Forster, Fl. Ins. Austr. Prod. (1786).

Cynoctonum Wilfordi, Maximovicz, Mél. IX (1876) 799. Korea, portu Chusan. — I. F. S., ıı, 109, *Cynanchum Wilfordi,* Franchet.

Ehretia formosama, Hemsley, — Ind. Fl. Sin., ıı, 144. West coast of Formosa.

Callicarpa formosana, Rolfe, Journ. Bot. 1882, 358. S. W. Formosa, Tam sui. — I. F. S., ıı, 252.

Nepeta Manchuriensis, M. Moore, J. B. 1880, 5.

Helicia formosana, Hemsley, I. F. S., ıı, 394. Formosa, N. E. district.

Gelonium aequorum, Hance, Journ. Bot. 1866, 173. Takow. — Ind. Fl. Sin., ıı, 444.

Pouzolzia elegans, Weddell in DC. Prodr. XVI, ɪ (1869) 230. Wilford, Formosa.

Woodsia Manchuriensis, Hook. 2d cent. Ferns (1861) t. 98. Manch. Wilf. — Syn. Fil. 48.

Leptocionium barbatum, van der Bosch, Ned. Kruitk. Arch. V, 62; *Hymenophyllum barbatum*, Baker, Syn. Fil. 68. Tsu sima, Wilford, (not Wright as Miquel, Prol. 347, states. Wright did not visit Tsu sima).

Davallia Wilfordii, Baker, Syn. Fil. (1868) 98. Japan.

Davallia pilosella, Hooker, 2d cent. Ferns (1861) tab. 96. Tsu sima, Wilf. — Syn. Fil. 98: reduced to *D. hirsuta*, Swartz (1806).

Asplenium Wilfordii, Mettenius; Kuhn in Rel. Mett., Linnaea XXXVI (1869) 94. Wilf. Tsu sima. — Syn. Fil. 487.

Asplenium Wardii, Hook. Spec. Fil. III (1860) 189. Wilford, Corea. Comp. supra p. 541.

Aspidium Tsusimense, Hook. Spec. Fil. IV (1862) 16. Corea, Wilf.— Syn. Fil. 252.

Nephrodium chinense, Baker, Syn. Fil. 278. Corea, Wilford.

Polypodium linearifolium, Hook. 2d cent. Ferns tab. 58. Tsu sima, Wilford. — Syn. Fil. 356.

Hemionitis Wilfordii, Hook. Filic. exot. (1859) tab. 93. Wilford, Formosa, Ke lung. — Syn. Fil. 399 = pinnatifid variety of *H. Griffithii*, Hook. fil. and Thomson, Khasya.

Mr. Hemsley of Kew, kindly informs me that Wilford died in 1893, at Wimble don, Surrey.

VII. Botanical Investigations in the Interior of China by French Missionaries, between 1840—60.

Verrolles, Emmanuel Jean François, French Missionary of the Missions Etrangères, born 1805, went out to China in 1830.

In the Annales de la Propag. de la Foi, IX (1836) 454—59, there is a letter by Verrolles to M. Dubois, dated Tchong king fou (Sze ch'uan) September 9, 1835. It contains interesting details regarding the Chinese Varnish tree *(Rhus vernix)* and the fruits and vegetables in that part of China.

In 1840 V. was elected bishop, Vicaire apostelique in Manchuria. He died in 1878 at Niu ch'uang.

Bertrand, Pierre Julien, French Missionary of the Missions Etrangè-res, went out to China, prov. of Sze ch'uan, in 1833.

He wrote to one of his friends in France a letter, giving interesting details on the management of the wild silkworms in the province of Kui chou, and the two *Oaks* upon the leaves of which the worms feed. This letter is dated Thong jin fou (in the north-eastern part of Kui chou), Juillet 19, 1842, and was first published in the Annales forestières, II (1843) 644. Bazin reproduced it in his Chine moderne, 1853, 583. An English translation of this memoir by Dr. Hance, is found in Linn. Soc. Journ. XIII (1870) 10.

More than 14 years later Father Bertrand furnished further informa-tion on the subject in two letters, one of them dated Sze tchouen, Sept. 11, 1856, and which were published in the Bull. Soc. d'Acclim. V (1858) 195, 272. He states that the wild silkworms and the two oaks are found also in Sze ch'uan. B. sent a supply of acorns of the two species to France from which a great number of trees were raised. One of these oaks has leaves resembling those of the chestnut. L. c. 139, 156.

Voisin, Joseph Etienne Polycarpe, born 1797, French Missionary of the Miss. Etr., went to China in 1824, was stationed in Sze ch'uan, returned to Europe previously to 1836, and became Director of the Seminary, Miss. Etr. Paris. He died in 1877. He introd. from China several economic plants, viz:

Red Chinese Radish (Radis rose d'hiver) introduced about 1836. Revue Hort. 1856, 181. Col. plate. — Fl. d. Serres, XII, 1857, 127, figured. — Bull. Soc. d'Acclim. VI, 1859, 521.

Pe tsaï or *Chinese Cabbage,* introduced at about the same time; Vil-morin in Rev. Hort. 1838, 86. Bull. Soc. d'Acclim. VI (1859) 776. — Paillieux et Bois, Potager d'un curieux, 1892, 410.

Voisin obtained from China the seeds of a textile plant, *ko,* which through Stan. Julien were presented to the Mus. d'Hist. nat. Paris. The plants raised from these seeds were examined by Prof. Jussieu, who found that it was a leguminous plant near *Dolichos bulbosus.* Rev. Hort. V (1843) 446. — In the Herb. of Bot. Gard. St. Petersburg, there is a specimen of this plant, received from Paris, and labelled *Pachyrhizus trilobus,* DC.? — It is known that the plant *ko,* of the fibres of which cloth is made in China and Japan, is the *P. Thunbergianus.*

In 1841, Voisin published in the Journ. d'Agriculture pratique an article: Divers procédés utiles des Chinois pour fabriquer des tissus d'été des fibres d'Urtica nivea. Revue Hort. V (1841) 59.

In the Revue de l'Orient, V (1844) 297—302, is a paper by Voisin: AGRICULTURE ET HORTICULTURE EN CHINE, Riz, Vin de Riz, Potiron, Pois et Haricots, Amarante épinard, Radis, Raves, Batates.

Perny, Paul, French Missionary of the Missions Etrangères, Provicaire apostolique in Kui chou, 1850—1860, the first botanical and zoological explorer of that Chinese province. At my request Mgr. Perny, in 1892 and 93, then living at Garches les St. Cloud, near Paris, furnished me with interesting information regarding his missionary life and his collections of natural objects made in that part of the empire. Destined by his superiors for Kui chou, he first arrived at Macao in 1845, and, as it was then impossible for missionaries to penetrate directly from the east to the interior of China, he embarked in a Chinese junk for Tung king, from whence he succeeded, travelling through Kuang si, dressed as a Chinese beggar, in reaching Kui chow. He was ordered to take up his residence in the northern part of the province, contiguous to Sze chʻuan. In 1850 he became Provicaire apostolique of Kui chou. Since 1848 he sent several times to Lyons, the seat of the Propagation de la Foi, cocoons of a silk moth living upon oaks, which subsequently was named *Bombyx Pernyi*. Towards the end of 1857 he went home on leave and carried with him for the Muséum d'Hist. nat. and the Société d'Acclimatation, Paris, a collection of dried plants, seeds, a living *Varnish* tree, living specimens of the trees upon which the Chinese wax-insect lives, and many interesting zoological specimens, among them the Chinese Pangolin *(Manis)*, the *Sciurus Pernyi*, the *Picus Pernyi*, etc. — In the Rev. de l'Orient et de l'Algérie, IX (1859) 330, 337, he published a paper on the "Province of Kui chow". In the same year he returned to China, and was obliged to wait a year in Canton, before he was allowed to proceed to Kui chou. In 1862 he was transferred to Chʻung kʻing fu in Sze chʻuan, where he was stationed till 1868, when he went to Shang hai, and in 1869 he returned home. From 1869—72 he published his Dictionnaire Français-Latin-Chinois, in 2 vols. with an Appendix "Histoire naturelle".

Guérin Mèneville in Bull. Soc. d'Acclim. I (1854) 49, reports that Perny in 1850 sent from China 500 living cocoons of the silkmoth living upon oaks to the Mus. d'Hist. nat. Lyons where they hatched in the winter $18^{50}/_{51}$. He described it in 1855 as *Bombyx Pernyi*.

In Bull. Soc. d'Accl. V (1858) 317, Perny published: Monographie du Ver-à-soie de Chêne au Kouy tcheou.

Ibidem, 111, 176, the *Varnish* and *Wax insect* trees brought from China by Perny noticed. See also X (1863) 151.

Ibidem, VI (1859) 520: Vilmorin, Essais de Culture sur les plantes de la Chine rapportées par Mgr. Perny. About 30 plants, cereals, vegetables. Chinese names added.

Perny writes that his herbarium specimens from Kui chow were transmitted in 1858, in good condition to the Muséum d'Hist. nat. Paris, a rich collection of several thousands of specimens. He was therefore disappointed, when after his last return from China, he found that these plants, gathered in a country never explored before, had been much neglected and partly lost in the Museum, and that his original labels, giving the Chinese names and other information had been thrown away.

Mr. A. Franchet of the Paris Museum informs me that Perny's botanical collection in the Museum, presented in 1858, consists of 264 species gathered in Kui chou, only a few in Sze ch'uan. It contains some very interesting plants. — The following have hitherto been examined and recorded as novelties:

Clematis urophylla, Franchet, Bull. Soc. Linn. Paris, 1884, 433. — I. F. S., I, 7.

Aconitum racemulosum, Fr., Journ. de Bot. 1894, 276. Kui chou prov.

Epimedium acuminatum, Franchet, Bull. Soc. Bot. France, XXXIII (1886) 10.

Ilex Pernyi, Franch. Pl. David. I (1884) 69. — Hook. Ic. Pl. t. 1539 (1886). — I. F. S., I, 117.

Euonymus acanthocarpa, Franch. Pl. Delav. 1889, 129.

Aria alnifolia, Decne, Pomac. in Nouv. Arch. Mus. X (1874) 166. — I. F. S., I, 255 = variety of the common *Pyrus Aria*.

Stransvaesia undulata, Decaisne, l. c. 179. — I. F. S., I, 264.

Chrysosplenium macrophyllum, Oliver, Ic. Pl. tab. 1744 (1888). A. Henry, Hupeh. — But according to Franchet, Nouv. Arch. Mus. II (1890) 103, the plant was earlier discovered by Perny.

Lysimachia paradiformis, Franch. Bull. Soc. Linn. Par. 1884, 433. — Bot. Mag. tab. 7226 (1892). — I. F. S., II, 55.

Pedicularis refracta, Max. Mél. XI (1881) 289. Maximowicz, Japan. 1861. — Mél. XII (1888), Perny earlier. — I. F. S., II, 214.

Pedicularis Rex, Clarke in shed. in Maxim. Mél. XII (1888) 875. — I. F. S., II, 214.

Didissandra Mihieri *), Franchet, Bull. Soc. Linn. Par. 1885, 450. Perny. — I. F. S., II, 227.

---

*) Mihières, Simon, French Missionary. Missions Etrangères, 1821—71.

Chirita Fauriei *), Franchet, l. c. — I. F. S., ɪɪ, 231.

Baea rufescens, Franch. l. c. 449. Simon 1858, Kouei tcheou. (Simon is evidently an error for Perny, for the former visited Kui chou several years later). — I. F. S., ɪɪ, 234.

Teucrium Pernyi, Franch. Pl. David. I, 245. — I. F. S., ɪɪ, 314.

Chloranthus Pernyanus, Solms in DC. Prodr. XVI, ɪ (1869) 474. — I. F. S., ɪɪ, 369.

Paris chinensis, Franch. Pl. David. II, 135. Perny, Mihières, Kouei tcheou. — Franchet, Monogr. Paris, 1888, 287.   ·

VIII. Illustrated English Botanical Periodicals, 1840—1860. Sir William J. Hooker. — Sir Joseph Paxton.

The BOTANICAL MAGAZINE, that long-lived useful periodical, in which since its foundation, more than a century ago, coloured figures are given of all the new exotic plants cultivated in the Kew Gardens, continued to flourish, during this period, under the able direction of the great botanist whose name has frequently appeared in the previous pages:

Hooker, Sir William Jackson, was born at Norwich, July 6, 1785, and received his education at the High School there. He early commenced to take interest in natural history and especially botany. Having at an early age inherited an ample competency from his godfather William Jackson, he formed the design of devoting his life to travelling and natural history. With the view of collecting plants he made expeditions to Scotland and its islands, France, Switzerland and Iceland. In 1806 he became F. L. S., in 1812 F. R. S. In 1815 he married the daughter of Dawson Turner, of Yarmouth, a distinguished botanist, a specialist for the knowledge of Algae, and settled at Halesworth in Suffolk. Here was laid the foundation of his herbarium, and here he began to write his long series of botanical works.

An increasing family and a decreasing income induced him, in 1820, to accept the Regius Professorship of Botany in Glasgow, at which place the next twenty years of his life were passed. — In 1836 he received the honour of Hanoverian knighthood, from William IV. In 1841 a new era of his life began with his appointment to the Directorship of the Roy. Bot. Gardens at Kew, which under his able direction soon became the most

---

*) **Faurie,** Louis, 1824—71, Vicaire apostolique du Kouei tcheou, Miss. Etr.

important seat of botanical research in the United Kingdom. — L. L. D. Glasgow, D. C. L. Oxon. 1845. — He died at Kew Aug. 12, 1865.

Although Sir W. Hooker had never been in China, he has eminently advanced our knowledge of the Flora of that country by cultivating in the Kew Gardens rare Chinese plants and describing them as well as the herbarium specimens, which he received from his correspondents in China.

Of the numerous botanical works and periodicals edited by him, I may mention those in which Chinese plants are described, and which are frequently quoted in these pages.

Exotic Flora, containing figures and descriptions of new, rare or otherwise interesting exotic plants. Edinburgh. 3 vols. 1823—1827. Col. tab. 232.

From 1826 Sir W. Hooker was connected with the Botanical Magazine (v. s. p. 201, 280) of which in 1845 he took the direction. In this there were given coloured plates of new or interesting plants cultivated in the Kew Gardens.

Companion to the Botanical Magazine, 1835—36. Two vols.

W. J. Hooker and R. K. Greville, Icones Filicum. Figures and descriptions of new or little-known Ferns, 2 vols. 1829—31.

Botanical Miscellany; containing figures and descriptions of such plants as recommend themselves by their novelty, rarity or history. 3 vols. 1830—33. — Second series, 4 vols, 1834—42 = Journal of Botany.

Icones Plantarum, or figures (black and white), with brief descriptive characters and remarks on new or rare plants selected from the author's herbarium. Vol. I, appeared in 1837. After Sir Williams' death continued by his son J. D. Hooker.

Hooker and G. A. Walker-Arnott, The Botany of Capt. Beechey's Voyage. 1841 (v. s. p. 289).

Hooker and Fr. Bauer, Genera Filicum, or Illustrations of Ferns, from the original coloured drawings of Francis Bauer, 1842.

The London Journal of Botany, Continuation of the Botanical Miscellany, 7 vols, 1842—48.

Journal of Botany and Kew Garden Miscellany, 9 vols. 1849—1857. In vol. IX (1857) 333—344, 353—363, Hooker enumerates all the Ferns then known from China, and describes the new species. Comp. supra p. 402.

Species Filicum; being descriptions of the known Ferns, particularly of such as exist in the author's herbarium. With black and white figures. 5 vols. 1846—1864.

A Century of Ferns; being figures with brief descriptions of one hundred new or rare, or imperfectly known species of Ferns from various parts of the world. A selection from the author's Icones Plantarum, 1854.— A second Century of Ferns appeared in 1860.

Filices Exoticae; or figures and descriptions of exotic ferns, chiefly of such as are cultivated in the Royal Gardens of Kew. 1859. 100 col. tab.

In 1865 Sir W. Hooker commenced the printing of his "Synopsis Filicum", a summary of his great work the "Species Filicum". But only a few sheets had passed through the press, when the venerable botanist died. Mr. J. G. Baker then took in hand the completion of the work, which appeared in 1868.

Hooker's immense herbarium was after his death acquired by the Government for the Kew Gardens.

J. E. Smith, in 1808, commemorated Hooker's name in a new genus of Mosses, *Hookeria* (England).

Paxton, Sir Joseph, born at Milton Bryant, Bedfordshire, Aug. 3d 1803, the son of humble parents. He commenced life as a gardener, working at Chiswick in the Hort. Society's garden. The Duke of Devonshire appointed him head gardener for his seat at Chatsworth, and subsequently his duties were extended to the management of the great Derbyshire estates of the duke. In 1831 P. commenced in conjunction with J. Harrison, the Horticultural Register and General Magazine, and in 1834 the Magazine of Botany and Register of Flowering Plants, of which 16 annual volumes appeared, 1834—1849; coloured figures. This was then somewhat remodelled and continued under the title of Paxton's Flower Garden, of which latter only two volumes appeared: I, 1850—51, — II, 1851—52. In all these serials descriptions and coloured figures of new Chinese plants are found, especially in the Flower Garden, in which, as we have seen, Dr. Lindley described many of Fortune's plants. — In 1851 Paxton was knighted. In 1853 he commenced the building of the Crystal Palace, at Sydenham, which was completed in June 1854. In the same year he became Member of Parliament for Coventry. He died at Sydenham, June 8, 1865.

IX. British Gardeners cultivating new Chinese and Japanese Plants, 1840—1860.

Messrs **Lucombe, Pince** & Co. of Exeter exhibited before the Hort. Soc. London, in July 1841, a Chinese plant which Lindley named *Quisqualis sinensis*, and figured and described it in Bot. Reg. 1844, t. 15. — In I. F. S., I, 294, it is reduced to *Q. indica*, Lin.

Veitch & Sons of the Roy. Exotic Nursery, Chelsea. The notices given below of this great and renowned Firm are chiefly derived from direct information kindly furnished by the present chiefs of it, Messrs James Veitch and Sons. The original founder of the firm was John **Veitch** who, born in Scotland, moved to Exeter, where he died in 1839, at the age of 84. John Veitch was helped by his son James **Veitch**, 1792—1863. The original establishment was commenced, at Budlake, about 6 miles from Exeter, and then later the nursery was formed close to Exeter. The two brothers Lobb travelled for James Veitch; — William in S. America, 1840—48, and in California, 1849—57; — Thomas, about 1847, in India and Malaya. — James Veitch of Exeter (son of John) was assisted by his son, also called James, born 1815, who afterwards, in 1853 removed to London and there bought the once famous establishment of Jos. Knight at Chelsea, which still bore the name of Roy. Exotic Nursery. James Veitch jun. died in 1869.

Veitch, John Gould, eldest son of the latter, was born at Exeter, in April 1839. He was like his father a devoted horticulturist, and was known as an intrepid voyager. He discovered and introduced, in the course of his travels, many valuable exotic plants. In April 1860 he started on his voyage to Japan and China, reached S h a n g h a i and from the latter place arrived at Nagasaki July 20, 1860, and after a stay of more than a month there proceeded to Kanagava. In September ascended with Mr. Alcock, British Consul at Yedo the celebrated mountain F u s i y a m a, visited the neighbourhood of Yukuhama. On Sept. 24, sailed for H a k o d a t e, spent a week there, returned to YEDO, where on November 12, he met with R. Fortune. Veitch seems to have left Japan in Nov. 1860. His letters on this country and its Flora were published in the Gard. Chron. 1860, 1104, 1126, and 1861, 22, 49, 97, 120. From Japan he proceeded to the Philippine Islands, and returned to Europe in the spring of 1862. — In 1864 we find him again *en route*, this time bound for Australia and the South Sea Islands, whence he returned in Febr. 1866, bringing with him some of the most beautiful plants of modern introduction. On Aug. 13,

1870, he died at this residence at Combewood from hemorrhage from the lungs. See Gard. Chron. 1870, 1117.

New Japanese plants discovered or introduced by J. G. Veitch:

Ampelopsis Veitchii hort. Gard. Chron. 1869, 838, and 1888, II, 664, fig. 126, introduced from Japan. — I. F. S., I, 133, same as *Cissus Thunbergii* or *Amp. tricuspidata*, Sieb. & Zucc., *Vitis inconstans*, Miq.

Corylopsis spicata, Sieb. & Zucc. Fl. Jap. I (1836) 47, t. 19. — Bot. Mag. t. 5458 (1864). Veitch introd. from Yukuhama. — I. F. S., I, 290.

Primula cortusoides, Lin., var. *amoena*, Bot. Mag. tab. 5528 (1865). Introduced.

Cryptomeria elegans, Veitch in Gord. Pinet. ed. 2, p. 73. Introd. — According to Masters, Conif. Jap. (1880) Journ. Linn. Soc. XVIII, 497, = *Cryptomeria japonica*, Don. var. *elegans*.

Picea Veitchii, Lindley, Gard. Chron. 1861, 23. Introd. — Masters l. c. 515, calls it *Abies Veitchii*.

Abies Alcockiana, J. G. Veitch in Gard. Chron. 1861, 23. Fusiama, Sept. 1860. Introd.— Hort. Soc. March 1862, cum icone. Masters l. c. 508.

Abies microsperma, Lindl., Gard. Chron. 1861, 22. Hakodate, Introduced. — Masters l. c. 509 = *Picea ajanensis*, Fisch. var. *microsperma*.

Lilium auratum, Lindl., V. s. p. 510. Introduced from Japan 1862.

Selaginella Veitchii, Mc Nab in Journ. Bot. 1867, 147.

Adiantum Veitchii, Hce, Symb. 1861, 229. Yokohama. J. G. Veitch.— Syn. Fil. 125 = *A. monochlamys*, Eaton. (Wright, v. s. 400).

Polypodium Veitchii, Baker, Gard. Chron. 1880, II, 494.

Among the new plants introduced by J. G. Veitch from Australia and the South Sea Islands, I may mention: 23 distinct kinds of *Croton*, many species of *Dracaena*, *Aralia Veitchii*, Hort., *Pandanus Veitchii*, Hort., *Coleus Veitchii*.

At Cape York in North Australia V. obtained a new palm, which has since been dedicated to his honour under the name of *Veitchia*[*]) *Johannis*, Wendl. in Seem. Flor. Vitiensis (1870).

Messrs **Standish** and **Noble** of Bagshot Nursery. They first cultivated most of the new Chinese plants discovered by Fortune during his second

---

[*]) The genus *Veitchia*, Lindley, Gard. Chron. 1861 (*V. japonica*) was founded upon a monstruous cone of a Japanese Picea. See Gen. Pl. III, 440.

visit to China, 1848—50. Fortune gave them also the living plants which, in 1860, he obtained in Japan.

Chinese Plants.

Berberis Bealii.	Azalea narcissiflora.
» consanguinea.	» Bealii.
» trifurca.	Chionanthus retusa.
Camellia reticulata.	Gentiana Fortunei.
Skimmia japonica.	Clerodendron foetidum.
Ilex cornuta.	Quercus inversa.
» microcarpa.	» sclerophylla.
» leptocantha.	» sinensis (Peking 1861).
Spiraea callosa.	Cupressus funebris.
Exorda grandiflora.	Juniperus sphaerica.
Abelia uniflora.	Cephalotaxus Fortunei.
Lonicera Standishii.	Abies Fortunei.
Azalea crispiflora.	Bambusa Fortunei.
» amoena.	

Japanese Plants.

Clematis lanuginosa.	Deutzia crenata.
» Fortunei.	Aucuba japonica.
Ilex Terago.	Ligustrum coriaceum.
Spiraea palmata.	Lilium Fortunei.
Saxifraga Fortunei.	Tricyrtis hirta.

Messrs **Glendinning**, nurserymen. They had their nurseries at Turnham Green and at Chiswick, and cultivated likewise Fortune's Chinese plants, but not the same as Standish and Noble. From Fortune's collection of living plants made during his second visit to China, they received and cultivated only one plant, *Ilex Fortunei*, but most of the new plants he discovered in China, during his third expedition, 1853—55, he sent to Glendinnings, viz:

Prunus triloba.	Aristolochia Sinarum.
Farfugium grande.	Quercus bambusaefolia.
Rhododendron Fortunei.	Torreya grandis.
Syringa oblata.	Pinus Bungeana.
Ligustrum sinense.	Abies Fortunei.

X. Botanical Information derived from Chinese Sources. Translations made by Sinologists.

Julien, Stanislas, the great French sinologist, was born 1797, succeeded in 1832 A. Rémusat as Professor of the Chinese Language at the Collège de France; 1833 Membre de l'Institut; died 1873. He translated several interesting Chinese treatises on economic plants, their cultivation and use, generally with a view to incite his countrymen to introduce these plants into France.

He wrote a book with the title: Résumé des principaux traités chinois sur la Culture des Mûriers et l'Education des Vers à Soie, which was published in 1838, by order of the Minister of Public Works, Agriculture and Commerce; 224 pages. It is translated from the Shou shi t'ung k'ao, a great Chinese work on Agriculture, and is followed by an account of the wild silkworms, which is a reproduction of a memoir written by Father d'Incarville (v. s. p. 51).

In Comptes rend. de L'Acad. Sc. VII, 1838, 703—4, he published: Procédés usités en Chine pour l'extraction de la matière colorante du Polygonum tinctorium, lan des Chinois. Translated from various Chinese works on botany and agriculture.

L. c. X (1840) 371—2: Sur une étoffe *(a pou* ou *hia pou)* recue de Canton, fabriquée en Chine avec les filaments de l'Urtica nivea. Observations de Stan. Julien et A. Brongniart.

Ibidem, 618—25. Sur une substance grasse produite par des insectes, la Cire d'arbre de Chine. — This is the *White Wax* or Insect wax of China. Stan Julien translates from various Chinese works on botany, agriculture. Account of the trees upon which the wax insect lives. See also Revue de l'Orient et de l'Algérie, V (1857) 218.

Ibidem, 697—703. Description des Procédés chinois pour la fabrication du Papier, traduite de l'ouvrage chinois Thien kong kai wei. Paper made of the *Bamboo*, of the bark of the *Mulberry* tree, of that of *Broussonetia papyrifera*. — See also Revue de l'Orient et de l'Algérie, XX, 1856, 74—78, and Industries de l'Empire chinois par. Stan. Julien et P. Champion, 1869, 140—52.

Comptes rend. Acad. Sc. XIV (1842) 40—42: Riz qui se cultive à sec dans la Mongolie. Stan. Julien had received from Father Gabet*), Missionary of the Lazarists near Jehol, Mongolia, his former pupil, a chest

---

*) The same who, a few years later, in company with Huc, performed the remarkable journey to Lhassa.

with rice of that sort, which is cultivated without requiring irrigation. Julien translated from the above-mentioned Shou shi t'ung k'ao the account there given of this kind of rice, and the mode of its cultivation.

Ibidem, XVII, 1843, 421—22. Sur la plante ko d'après le Shou shi t'ung k'ao. See also Stan. Julien et P. Champion, Industries de l'Emp. Chin. 1869, 170. This article was translated from the French in the Chinese Repository, XIV (1845) 216. The plant ko is the *Pachyrhizus Thunbergianus*. (V. s. p. 545).

Comptes rend. etc, XXVIII (1849) 394—400. Renseignements sur la plante textile tchou ma, Urtica nivea, extraits des livres chinois. See also Stan. Julien et P. Champion, l. c. 162.

Stan. Julien translated from Chinese works on botany and agriculture, all the information found there regarding the "Igname", *Dioscorea*, for Decaisne. See the article of the last, l'Igname Batate, in Rev. Hort. 1854, 248.

In the Bull. Soc. d'Acclim. II (1855) 225, Stan. Julien published a notice, Le Pois oléagineux de la Chine, you teou, translated from the Shou shi t'ung k'ao (v. s.). — This is the *Soja hispida*.

Williams, Dr. Samuel Wells. This accomplished American sinologist, long resident in China, was born at Utica, in the State of New York, Sept. 22, 1812, and was educated at the Rensselaer Institute at Troy in the same state. Like many literary men he began his career as a printer, and in this quality he was engaged by an American missionary society to go to China, to superintend a press there. He arrived at Canton, after a passage of 127 days from America, in Oct. 1833, and soon began to study Chinese.

In 1832 the Chinese Repository had been commenced by Dr. E. C. **Bridgman,** an American missionary (1801—61) and during the twenty years of the existence of this useful serial, the work of getting together and publishing it was done chiefly by Bridgman and Williams. It contains many valuable articles by these great sinologists. In vols. XI to XIX (1842—50) they published successively a Topography of the 18 Provinces of China, in which peculiar attention is paid to orography and hydrography. Williams had a great love for natural history, especially botany, and gradually acquired a considerable knowledge of this science. In the Chinese Repository we find several interesting papers relating to Chinese plants. We may notice the following:

Chin. Rep. III (1834) 83, sq. Natural History of China.

Ibid. 121, sq. Agriculture of China.

Ibid. 231, sq. RICE: its varieties; mode of cultivation, etc.

Ibid. 261, sq. DESCRIPTION OF THE BAMBOO IN CHINA; varieties and cultivation; its uses; mode of manufacturing paper. The COCOA NUT PALM, AND THE USES TO WHICH IT IS APPLIED IN CHINA.

Ibid. 458—460. DIET OF THE CHINESE. CHINESE CEREALS, VEGETABLES, FRUIT, OIL YIELDING PLANT, etc.

L. c. V (1836) 485 sq. DESCRIPTION OF THE AGRICULTURAL IMPLEMENTS USED BY THE CHINESE.

L. c. VIII (1839) 132—164. DESCRIPTION OF THE TEA PLANT; its name, cultivation, mode of curing the leaves, etc.

L. c. XIX (1850) 396. THE VARNISH TREE IN SZE CH'UAN.

In 1837, C. W. King of the firm of Oliphant & Co. determined to send the ship "Morrison" to Japan, for the purpose of restoring seven shipwrecked Japanese to their native land, and try to enter into connection with that country. Gutzlaff and Williams accompanied the vessel, which sailed (from Macao) 2 d July. The Morrison stopped at NAPA in LEW CHEW three days, and then proceeded to the Bay of Yedo. But they did not succeed in communicating with the Japanese authorities. The expedition failed in its object and all the shipwrecked Japanese were brought back. See Williams, Lecture on Japan, in Journ. N. Ch. Br. Asiat. Soc. 1859, 205. In the Chin. Rep. VI (1838) 406, Williams published: "NOTICES OF SOME OF THE SPECIMENS OF NATURAL HISTORY COLLECTED DURING THE VOYAGE OF THE MORRISON IN LEW CHEW".

In 1841 Bridgman published at Macao a CHINESE CHRESTOMATHY in the Canton dialect with an English translation and notes. Williams prepared the chapters on NATURAL HISTORY: XIII, Mineralogy, XIV, Botany, XV, Zoology. With respect to Botany we find there 445 Chinese names of plants, identified with the corresponding popular English or scientific appelations.

In 1844 W. returned to the United States, where the degree of L. L. D. was conferred upon him. In 1848 he published the MIDDLE KINGDOM, in 2 vols, one of his main works. In Sept. of the same year he rearrived at Canton as Superintendent of the Press.

In 1853, when the President of the United States sent an expedition to Japan under the command of Commodore Perry, Dr. Williams was appointed Interpreter to this mission. (V. s. p. 386). He returned with the commodore to Hong kong, after successfully completing negotiations, in August 1853, and accompanied him again, in the next year, to Japan. On March 31, 1854, the Treaty of Kanagava was signed, and in July W.

was back to Canton. The botanical collections he made during these two visits to Japan have been recorded in a previous chapter. (s. 388).

In 1856, Dr. W. severed his connection with the American Board of Foreign Missionaries and accepted office under the Government. In 1858 he went to Tien tsin, and was present at the formation of the Treaty between China and the United States. In 1860 he returned home to America, and came back to China in 1862.

After having brought out four editions (1844—56) of J. R. Morrison's CHINESE COMMERCIAL GUIDE, he published in 1863, the 5th, entirely remodeled, edition under his own name. William's Chinese Commercial Guide is replete with useful information with respect to China, and still still a standard work. In section V, 108—151, Description of Articles of Export, a mass of interesting notices are found regarding Chinese economic plants, drugs etc. Native names given in Chinese characters.

Mac Gowan, Dr. Daniel Jerome, American. He was born in Fall River, Mass. in 1814, came to Ning po as a medical missionary in 1843, and opened there a hospital for the Chinese. During the Civil war in the United States he served as a surgeon with the northern armies. He returned to China in 1865 as the agent of a syndicate, that proposed to build a telegraphic line to China by way of Bering's Strait, and from that time made Shang hai his headquarters. Sir Robert Hart gave him an appointment in the Chinese Marit. Customs, in 1879, and he served in Shang hai and Wen chow. He died at Shang hai July 20, 1893. He was a man of vast and various information and wrote many valuable articles on China of which the following present a botanical interest:

Chin. Repos. XVIII (1849) 354—358, NOTICES REGARDING THE PLANTS YIELDING THE FIBRE FROM WHICH GRASS CLOTH IS MANUFACTURED. This article appeared first in the Journ. of Agric. and Hort. Society, of India. It gives a short account of the textile plants of China and their Chinese names.

Chin. Repos. XX (1851) 422—26, USES OF THE STILLINGIA SEBIFERA OR TALLOW TREE, WITH A NOTICE OF THE PE LA OR INSECT WAX OF CHINA. This article was originally published in the Journ. Agric. and Hort. Soc. of India, VII (1850) Calcutta. — Another paper ON THE TALLOW TREE AND ITS USES by Mac Gowan, appeared in the Pharmac. Journ. 1872, p. 1034. — Compare also Hooker's Kew Journ. Bot. IV (1852) 150, sq.

Transact. China Br. Roy. Asiat. Soc. VI (1859) 41—52, NOTE ON CHINESE OPIUM, ITS INTRODUCTION INTO CHINA, information regarding

Opium in Chinese works. See also Notes and Queries on China and Japan, 1868, 158.

In the N. China Herald 1882, p. 306, Mac Gowan published an interesting paper On a peculiar sort of Bamboo, called the Square Bamboo, specimens of which he forwarded from Wen chou to the U. S. Consul General at Shang hai, destined for the Park at S. Francisco and for the Public Gardens Shang hai. The square bamboo is early mentioned in Chinese books. It grows in Che kiang, Ho nan, Sze ch'uan, Yün nan, Hu nan. The stem is square as if cut with a knife, abounding in small spines. Used for pipe stems, walking canes etc.

See also on the square bamboo of Wen chou, Oestreich. Monatsschrift f. d. Orient 1886, 113 (reproduced from the Chinese Recorder). — The Rev. Hort. 1876, 32, gives a figure of the square stemmed bamboo taken from a Japanese work and states that this bamboo is already cultivated in France. See also Gard. Chron. 1876, I, 147. I do not find it mentioned in Rivière's Bambous, 1879.

Shaw, Charles, British merchant, at Shang hai, published in the Chin. Repos. XVIII (1849) several articles on Chinese economic plants, translated from Chinese works:

P. 13, sq. Remarks on the Tea plant.

P. 449—69. Directions for the Cultivation of Cotton, especially in the District of Shang hai, translated from the Nung ching tsiuen shu or Encyclopaedia of Agriculture. (See Botan. sin. I, 82).

P. 303, sq. Cultivation of the Mulberry and Rearing the Silkworms. Translated from the Tsan sang ho pien.

Medhurst, Walter Henry (Senior), Protestant Missionary, British, (not to be confounded with his son Sir Walter Henry, v. s. p. 535). He was first stationed at Batavia, preaching the gospel to the Chinese there, and in 1843 came to Shang hai, where he lived till 1855, when he returned to England. He died at London, 1857. Medhurst was a distinguished sinologist. From 1842 to 48 he published two excellent Dictionaries, Chinese and English, and English and Chinese. In 1849 appeared at Shang hai, in the Chinese Miscellany his Dissertation on the Silk-Manufacture and the Cultivation of the Mulberry; translated from the above mentioned Nung ching tsiuen shu.

In November 1854 he performed from Shang hai an interesting journey to the famous mountain T'ien mu shan in the northern part of Che kiang, and published an account of it in the Shang hai Almanac 1855.

XI. Russian Botanical Collectors at Peking, 1840—60.

During this period the Herbarium of the Botanical Gardens, at St. Petersburg, continued to be supplied with valuable botanical collections formed by the members of the Russian Ecclesiastical Mission at Peking in the neighbourhood of the Chinese capital.

Tatarinov, Alexander Alexeyewich, M. D., was born about 1817 in the city of Penza, entered, in 1835 the Medico-chirurgical Academy, St. Petersburg, where, in 1839, he took his medical degree. In the same year he was appointed physician to the 12th Russian Eccles. Mission, which then was despatched to Peking under the conduct of N. I. Liubimov, to relieve the 11th Mission (v. s. 319, 324). The 12th Mission left St. Petersburg early in January 1840, started from Kiakhta, July 21, traversed Mongolia, and arrived at Peking, October 4. He spent ten years in the Chinese capital, and returned to Russia with his Mission, which left Peking in the first half of May 1850, and reached Kiakhta July 9. During his sojourn at Peking, Tatarinov devoted himself most ardently to the investigation of the Flora of the Peking plain, the mountains west, south-west, north and north-east of the capital. He extended his botanical excursions even to Jehol, beyond the Great Wall. He also gathered plants when passing through Mongolia in 1840 and 1850. He first sent his collections to Fischer and when in 1851 he returned to St. Petersburg, he handed over the bulk of his Pelïng plants, about 570 species, including many duplicates, to his preceptor and friend Horaninov, Paul (1796—1866) for many years Professor of Botany at the Medico-chirurgical Academy, St. Petersburg. Tatarinov also brought with him from Peking a valuable collection of Chinese drugs, upon which is founded his Catalogus Medicamentorum Sinensium, 1856. Horaninov determined the drugs. Besides Tatarinov brought home a beautiful set of coloured drawings representing 452 wild plants of the Peking Flora, and executed from nature by a Chinese artist under Tatarinov's direction, who also added the Chinese names in Chinese characters. In 1857 the Academy, St. Petersburg, bought Tatarinov's Herbarium of Peking plants, the collection of drugs and the coloured drawings, altogether at the price of 800 roubles.

Tatarinov's plants have never been described in a regular way. After Ruprecht and Regel had occasionally published some of the novelties, this collection was first thoroughly examined by Maximowicz, when he wrote his Primitiae Florae Amurensis, published 1859, at the end of which is found an Index Florae Pekinensis. Subsequently he described most of the novelties in various volumes of the Mélanges biologiques, 1866—93. See also his Flora Mongolica, 1889.

Tatarinov also brought from Peking a valuable entomological collection. Of this the Lepidoptera were described, in 1853, by O. Bremer and W. Grey: Schmetterlings Fauna des Nördlichen China's.

While still in Peking, Tatarinov wrote several interesting articles on Chinese Medicine (in Russian). Two of these were published in the Collection of Scientific Papers by the Members of the Peking Mission, viz:

II (1853) n. 5. The Chinese Art of Healing.

III (1857) n. 5. Remarks regarding the Application of Anaesthetics in operations, and Hydropathy among the Chinese.

In 1851, when Colonel Kovalevski was despatched by the Russian Government to Kuldja to negotiate a commercial treaty with China, T. accompanied him as interpreter, for which employment he was well qualified, owing to his thorough knowledge of the Chinese spoken language. After the cities of Kuldja (Ili) and Chuguchak (Tarbagatai) had been opened to Russian trade, T. was appointed Consul at the latter place. He held this post till 1855, when a troop of Chinese rebels attacked and destroyed the Russian Consulate there, and Tatarinov saved his life only through his presence of mind.

In 1857, T. accompanied Admiral Count Putiatin entrusted with a diplomatic mission to China (see farther on). After the treaty between Russia and China had been signed at Tien tsin, June $^1/_{13}$, 1858, T. returned, via Siberia, to St. Petersburg. Soon afterwards the Russian Government resolved upon sending General N. P. Ignatiev as ambassodor to Peking and T. was ordered to accompany him. The embassy, proceeding by the ordinary way, through Siberia and Mongolia, reached Peking In June 1859. On Nov. $^2/_{14}$, 1860, a new treaty was signed between Russia and China at Peking. In the negotiations with the Chinese T. displayed great ability. After his return to St. Petersburg he was attached to the Asiatic Department of the Foreign Office, of which Ignatiev became director. T.'s merits were rewarded with a pension for life. In 1866, he resigned the service and retired to Penza his native city, where he quietly spent the remaining years of his life. He died October 6th 1886.

Localities.

On the labels affixed to Tatarinov's plants, the collector always notices the localities where the plants were gathered, writing the Chinese names in accordance with the Russian mode of spelling Chinese sounds. Maximowicz, in attempting to render the Russian letters in Tatarinov's names by German equivalents has made them quite unintelligible. It may, therefore be well, that we give here all these names of localities, written according to the mode of transliterating Chinese sounds adopted in the present work. Most of the localities are marked on my Map of the Environs of Peking.

Rh cha. Second sluice on the canal leading from Peking eastward to T'ung chou. Örr tschsha, Maxim.

K'ang t'ai, outside the southern wall of Peking. V. s. p. 326.

Hai tien. Large village near the imperial summer palace Yüan ming yüan. V. s. p. 158.

Imperial Park, for deer. Tatarinov seems to mean the Nan hai tze, a large enclosure south of Peking.

Portuguese Cemetery. Near P'ing tze men, one of the western gates of Peking.

French Cemetery. Cheng fu sze, N. W. of Peking, S. of Wan shou shan.

Hun ho River, W. of Peking. Maxim. writes Chun che.

Shi king shan Hill, on the Hun ho River, W. of Peking.

Chang sin tien. A village S. W. of the Lu kou k'iao Bridge over the Hun ho.

Kün chu, a village in the western mountains, not far from the Hun ho.

Tsie t'ai sze, a large Buddhist Monastery in the mountains S. W. of Peking.

Shan a rh sze, a small temple near Tsie t'ai sze.

Fang shan hien, a district city about 30 miles S. W. of Peking.

Tou shuai sze, a monastery in the mountains W. of Fang shan hien.

Defile Ch'ang kou yü in the district of Fang shan hien.

Si ling, Western Cemetery of the Present Dynasty, about 75 miles S. W. of Peking.

Tsze king kuan, N. W. of the former. Gate of the Great Wall.

Yi tsien fang, E. of Si ling.

Ti men k'ou, near Si ling.

Si shan, Western Mountains, W. of Peking (v. s. p. 326).

Men t'ou k'ou, village W. of the Hun ho, and the road to the Po hua shan.

Ming ts'iao an, a hill beyond Men t'ou k'ou.

Miao an, temple among the mountains, on the road to the Po hua shan.

Wang p'ing k'ou, among the mountains on the road to the Po hua shan.

Ts'ien kün t'ai, on the road to the Po hua shan.

Tuan to nie, village and mountain on the road to the Po hua shan. Duan to oue, Maxim.

Tai ho (han) ling. Mountain ridge and defile on the road to the Po hua shan.

Po hua shan. Mountain, about 8000 feet, 60 miles W. of Peking. Maximowicz and Regel write erroneously Bo chuan tshuan.

Mei wo, village in the Western Mountains.

Shi t'ie rh, village in the Western Mountains.

Chai k'ou, village W. of Peking.

Wo fu sze, monastery N. W. of Peking.

Wen ts'üan sze, monastery with hot springs, N. W. of the Summer Palace.

Shui t'a sze, Villa, near the former.

Hei lung t'an, temple E. of Wen ts'üan.

Ta kio sze, large monastery W. of Wen ts'üan.

Miao feng shan, mountain and monastery N. W. of Ta kio sze.

T'ang shan. Hot springs and Imperial Gardens 18 miles N. of Peking (v. supra p. 327).

Kao li ying, village E. of T'ang shan.

Lung ts'üan sze, monastery N. of T'ang shan. V. s. p. 327.

Ts'ing ho, village, N. W. of Peking. V. s. p. 326.

Ch'ang p'ing chou, city N. W. of Peking. V. s. p. 327.

Nan k'ou, village, N. W. of the former. V. s. p. 327.

Kuan kou or Nan k'ou, Pass. V. s. p. 327.

Ming Tombs, Cemetery of the Ming Emperors, N. E. of Nan k'ou.

Sha ling rh, mountain on the road from the Ming Tombs to Mount Yin shan.

Mount Yin shan (Silver Mount), silver mine, N. E. of the Ming Tombs and N. W. from Lung ts'üan sze. Tatarinov's Yin shan has been erroneously identified by Maximowicz with the Yin shan of Przewalski, an ancient Chinese name meaning "dark mountain", for the mountain chain which separates the provinces of Shansi and Chili from the Mongolian plateau. The name is not in use now.

Ti tsan a rh, a small temple at the foot of the Yin shan, where Tatarinov spent the summer of 1847.

Hu men rh, village not far from the latter.

Rh tao kuan, E. of the Yin shan, outside of the Inner Great Wall.

Ho t'ao, mountain rivulet near Rh tao kuan.

Ma ts'üan tze ling, mountain beyond the (inner) Great Wall.

Ma tsi ling, mountain beyond the (inner) Great Wall.

Chang k'an, temple in the mountains N. of Peking. Maxim. writes Tschan chanj.

Po ho chai, village N. of Peking.

T'ao ling, village in the mountains N. of Peking.

Niu lan shan, hill N. E. of Peking, on the Pai ho River.

Men hien (error for Mi yün hien). District city N. E. of Peking, on the road to Ku pei k'ou.

Shi hia rh (ying), village N. E. of Mi yün hien.

Tung ling, Eastern Cemetery of the Present Dynasty, about 80 miles N. E. of Peking.

Ts'ao lu k'ou.

Kia tze men.   All these localities are near the

Kia rh k'ou.   Tung ling.

Mount Feng shui shan.

Ku pei k'ou (gate in the Great Wall), 85 miles N. E. of Peking.

River Pei ts'ao ho (Chao ho).

Mt. Si yü ling.

Mt. Lo shan.

Mt. Yin men shan.   All these locali-

River Yü rh ho.   ties are near Ku pei

Ki men ling.   k'ou.

Ki wu men.

Wu nan t'ien mien.

T'ou tao ling.

Sin ch'eng tze, village. Ssin tchen tsy, Maxim.

Je ho or Jehol, Imperial Summer Palace, about 55 miles N. E. from Ku pei k'ou (v. s. p. 158). — Sche che, Maxim.

Mt. La ma shan.

Ho kia kou.

Defile Hei k'u kou.   N. of Jehol.

Mt. Pai yü ling.

Mt. Lo han shan, N. E. of Jehol.

Mt. Wang p'ing ling.
Hu lu kou.
Wang lu k'ou.   } E. of Jehol.
San tsien fang.
Pai fu yü ling, 5 or 6 li (2 miles) S. of Jehol.
Siu yü tze t'an, W. of Jehol.
Defile Pei k'ou, near the city of Luan p'ing hien.
Ts'ao kia k'ou.   }
Yü t'ing ch'un.   } near Jehol.

New plants discovered by Tatarinov:

Clematis pinnata, Maximowicz, Mél. biol. IX (1876) 591. Near the temple Shan a rh. — I. F. S., I, 7.

Clematis Tatarinowii, Maximowicz, l. c. 590. Near Peking. — Ind. Fl. Sin., I, 7.

Paeonia obovata, Maxim. in Prim. Fl. Amur (1859) 29. Lower Amur. Tatarinov earlier, Peking. Herb. Mus. Acad. Petrop. See Max. Mél. XII (1886) 416. — I. F. S., I, 32.

Corydalis gamosepala, Max. Amur, 38, in nota. Hb. hti Petrop. Tatarinov, mountains W. of Peking. In Fl. As. Or. Fragm. (1879) 4, Maxim. reduces this to *C. solida*, Swartz (1819). — I. F. S., I, 36 = *C. bulbosa*, DC. Fl. Franc. 1805.

Corydalis Raddeana, Regel Plantae Radd. 1861, n. 169. Tatarinov, China borealis. — Comp. I. F. S., I, 37, sub *C. ochotensis*, Turcz.

Polygala Tatarinowii, Regel Pl. Radd. 1861, sub n. 274, in nota, c. icone. Near Peking. — I. F. S., I, 62.

Gypsophila acutifolia, Fischer, var. *chinensis*, Regel, l. c. n. 278. China bor., Tatarinov. — I. F. S., I, 64.

Silene melandryiformis, Max. Amur, 54, Amur, Ussuri. Earlier Tatarinov between Lung ts'üan and Yin shan. Hb. hti Petrop. — I. F. S., I, 64 = *S. aprica*, Turcz. (1835).

Silene Tatarinowii, Regel, Pl. Radd. 1861, n. 289, note. Herbarium Fischer in Hb. Hti. Petrop. Defilé Ch'ang kou yü. — I. F. S., I, 64.

Lychnis cognata, Max. Amur, 55, in nota. Herb. Fischer in Hb. hort. Petrop., Tatarinov Po hua shan. — I. F. S., I, 66 = *L. fulgens*, Fischer, 1819, Dahuria.

Stellaria chinensis, Regel, l. c. sub n. 329. — Maxim. Mél. IX (1873) 49. — I. F. S., I, 67.

Celastrus Tatarinowii, Rupr. in Mél. II (1857) 531. Hb. hti Petrop.,

Lung ts'üan, 1847. — Maximowicz, Mél. XI (1881) 200 = *C. articulata*, Thunb. — I. F. S., ɪ, 122.

Rhamnus argutus, Maximovicz, Rhamn. (1866) 7, fig. 48—51. Near Peking. — I. F. S., ɪ, 128.

Astragalus capillipes, Fischer ex Bunge, Astrag. II (1869) 21. Between Peking and Lung ts'üan. — Hb. Fischer in Hb. hti Petrop., Tatarinov Mongolia chinensis 1840, Yin shan.

Astragalus Koburensis, Bunge, l. c. 196. Mongolia australis prope Kobur.

Oxytropis mandshurica, Bge in Plant. Semenov. and in Gen. Oxytropis (1874) 39. Wilford, Manchuria (1859). Tatarinov earlier, Po hua shan. Hb. hti Petrop.

Oxytropis lasiopoda, Bge in Gen. Oxytropis, 151. Mongolia media prope Bussun chelo et Gashunhuduk.

Glycyrrhiza pallidiflora, Max. Amur (1859) 79, Amur. — Tatarinov earlier near Hei lung t'an. Max. l. c. Ind. Peking, 470. Hb. Mus. Acad. — I. F. S., ɪ, 169.

Lathyrus Davidiana, Hance, Journ. Bot. 1871, 130 (David). — Max. Mél. IX (1873) 63: Tatarinov earlier, Po hua shan. — I. F. S., ɪ, 186.

Hoteia chinensis, Max. Amur, 120: Maack 1855, Amur. Tatarinov earlier, Yin shan. — Franchet, Enum. Jap. I, 144 = *Astilbe chinensis*, following Bentham & Hooker, G. Pl., who reduce Hoteia to Astilbe. — I. F. S., ɪ, 265. — The Garden, 1892, 221: Recently introd. fr. China.

Hydrangea vestita, Wall. var. *pubescens*, Maxim. Hydrang. (1867) 10. First discovered by Kirilov (not yet in flower). Tatarinov first gathered flowering specimens near Tuan t'o nie. — I. F. S., ɪ, 274.

Ribes multiflorum, Kitab. Fl. Oestr., var. *mandshuricum*, Max. Mél. IX (1873) 229: Radde Manchuria. Earlier Tatarinov, Tuan t'o nie. — I. F. S., ɪ, 279.

Umbilicus ramosissimus, Max. Amur. (Ind. Peking) 472, in nota. Yin shan, Tatar. Herb. Fischer in Hb. hti Petrop. — In Mél. XI (1882) 728, Maximowicz reduces it to a variety of *U. fimbriatus*, Turcz. — I. F. S., ɪ, 281, sub *Cotyledon*.

Sedum drymarioides, Hce, Journ. Bot. 1865, 379 (Sampson 1865).— Maxim. Mél. XI (1882) 772: Tatarinov earlier, Tou shuai sze. Herb. hti Petrop. — I. F. S., ɪ, 283.

Sedum Tatarinowii, Max. l. c. 742. Yin shan. — I. F. S., ɪ, 287.

Mitrosicyos racemosus, Max. Amur, 112, in nota. Canal Rh cha. Herb. Fischer in Hb. hti Petrop. — The new genus *Mitrosicyos* proposed by

Maximowicz is reduced in the Gen. Plant. to *Actinostemma*. — I. F. S., I, 320, *A. racemosum.*

Mitrosicyos paniculatus, Max. l. c. 113, in nota. Monastery Tou shuai sze. Herb. Fischer in Hb. hti Petrop. — I. F. S., I, 320, *Actinostemma paniculatum.*

Begonia sinensis, DC. Prod. XV, I, (1864) 313. Tou shuai sze. Herb. Fischer in Hb. hti Petrop. — Hb. Mus. Acad. Petrop. — I. F. S., I, 323.

Angelica grosseserrata, Max. Mél. IX (1873) 253. Mountains near the Imp. Cemetery Tung ling. — I. F. S., I, 334.

Panax sessiliflorum, Maxim., Ruprecht in Mél. biol. II (1856) 426, and Max. Amur, 131. Herb. Fischer in Hb. hti Petrop. — Hb. Mus. Acad. Petrop. Tatarinov near Jehol. — *Acanthopanax sessiliflorum,* Seemann in Journ. Bot. 1867, 239. — I. F. S., I, 340.

Lonicera phyllocarpa, Max. Amur, 138, in nota. Mt. Ming ts'iao an. Max. Mél. X (1877) 71. — I. F. S., I, 365.

Lonicera Tatarinowii, Maxim. Amur, 138, in nota. Hb. Fischer in Hb. hti Petrop. Tatarinov on the way to the Po hua shan. Kirilov earlier (v. supra p. 349).

Rubia chinensis, Regel, Fl. Ussur. (1861) 76, c. icone. Amur, Ussuri, Maack. — Max. Mél. IX (1873) 266. Earlier Tatarinov near Peking. — I. F. S., I, 392.

Heteropappus decipiens, Maxim. Amur, 148, 473 (Ind. Pek.) Amur, Peking. Hb. hti Petrop. — I. F. S., I, 412 = *Aster hispidus,* Thbg.

Biotia discolor, Max. Amur, 146, Amur, Peking. — I. F. S., I, 415 = *Aster scaber,* Thbg.

Erigeron latisquamatus, Max. Amur, 473 (Ind. Pek.) in nota. Herb. Fischer in Hb. hti Petrop. China borealis, Tatarinov. — I. F. S., I, 417 = *Brachyactis ciliata,* Ledeb. Fl. Ross. II, 495 (according to Maxim. in litt.).

Artemisia igniaria, Maxim. Amur, 161, in nota. Near Peking. — Maxim. Mél. VIII (1872) 536 = *Artemisia vulgaris,* Lin. (A. indica). — I. F. S., I, 446.

Saussurea paleata, Maximowicz, Amur, 168. Herb. Fischer in Hb. hti Petrop. China borealis. — I. F. S., I, 467.

Youngia dentata, DC. var. *chinensis,* Max. Amur, 182, 473. Herb. Fischer in Hb. hti Petrop. Near Peking. In Mél. IX (1874) 359, Maxim. names it *Lactuca denticulata.* — I. F. S., I, 480.

Lactuca amurensis, Regel in Ind. sem. hti Petrop., 1857, 42, Maxim. Amur, 178. Lower Amur. Earlier Tatarinov near Peking. In Mél. IX (1874) 354, Maxim. identifies it with *Lactuca squarrosa,* Miq. *(Prenanthes squarrosa,* Thunberg). — I. F. S., ɪ, 479 = *Lactuca brevirostris,* Champ., which name is older than that given to the plant by Miquel.

Prenanthes Tatarinowii, Max. Amur, 474 (Ind. Pek.), in nota. Temple Chang kʻan. — I. F. S., ɪ, 486.

Glossocomia ussuriensis, Maxim. et Ruprecht, Mél. II (1856) 472, Max. Amur, 184. Lower Amur. Earlier Tatarinov near Tou shuai sze. Herb. Fischer in Hb. hti Petrop. — — I. F. S., ɪɪ, 7 = *Codonopsis ussuriensis,* Hemsl.

Adenophora trachelioides, Maxim. Amur, 186, in nota. Tsie tʻai sze. Herb. Fischer in Hb. hti Petrop. — I. F. S., ɪɪ, 13 = *Aden. remotiflora,* Miq. Earlier known from Japan.

Primula Maximowiczii, Regel, Acta Hti Petrop. III (1875) 139. Herb. Fischer in Hb. hti Petrop. Po hua shan. — I. F. S., ɪɪ, 40.

Chionanthus chinensis, Max. Mél. IX (1874) 393 = *Linociera chinensis,* Fischer, ined. in Hb. hti Petrop., Maxim. Ind. Pek. 474. Tatarinov 1847. Kuan kou. — I. F. S., ɪɪ, 89 = *Chionan. retusus,* Lindley (1852). Earlier Fortune (1844).

Pterygocalyx (gen. nov.) *volubilis,* Maxim. Amur, 198. Ussuri. Hb. Fischer in Hb. hti Petrop. Wang lu kʻou near Jehol, Aug. 1847, Tatarinov. — As Benth. and Hook. reduce this genus to Crawfurdia, Hemsley in I. F. S., ɪɪ, 123, calls the plant *Crawfurdia Pterygocalyx.*

Bothriospermum secundum, Maxim. Amur, 202. China borealis. — I. F. S., ɪɪ, 151.

Cuscuta colorans, Maxim. Amur, 202. Near Peking. — Engelmann, Cuscuta, 1859, 517, considers this to be a variety of *C. japonica,* Chois.— I. F. S., ɪɪ, 168.

Pedicularis Tatarinowii, Maxim. Mél. X (1877) 92. Herb. Fischer in Hb. hti Petrop. Jehol. — I. F. S., ɪɪ, 217.

Melampyrum roseum, Max. Amur, 210. Amur, Tsʻao kia kʻou near Jehol. — I. F. S., ɪɪ, 220.

Plectranthus glaucocalyx, Maxim. Amur, 212. Lower Amur. Tatarinov beyond Ku pei kʻou, village of Sin chʻen tze. — I. F. S., ɪɪ, 271.

Plectranthus pekinensis, Maxim. Amur, 213, in nota. Herb. Fischer in Hb. hti Petrop. Near Peking. — Maximowicz in Mél. IX (1874) 424, reduces this to *Plectranthus amethystoides,* Bentham, Lab. (1835). — Ind. Fl. Sin., ɪɪ, 269.

Salvia umbratica, Hance, Journ. Bot. 1870, 75 (David, Jehol). — Max. Mél. XI (1881) 303. Tatarinov, Jehol. — I. F. S., ɪɪ, 288.

Leonurus macranthus, Max. Amur, 476 (Ind. Pek.) in nota. Jehol. —

L. japonicus, Miq., gathered earlier by Siebold in Japan, is the same. — I. F. S., II, 302.

Salsola monoptera, Bunge in Salsol. Mongoliae, 1879, 296. Tatarinov, Mongolia.

Polygonum dentato-alatum, Fr. Schmidt in Max. Amur, 232. Herb. Fischer in Hb. hti Petrop. Mt Ma tsi ling. — I. F. S., II, 348 = *Polygonum scandens*, Lin.

Hemiptelea Davidii, Planch. (David, Jehol) was earlier, in 1847, discovered by Tatarinov between the Great Wall and the Yin shan. Max. Mél. IX (1872) 22.

Pteroceltis (gen. nov.) *Tatarinowii*, Max. Mél. IX (1872) 27. Peking.

Girardinia cuspidata, Weddell in DC. Prod. XVI, I, 103 (David, near Peking). — Max. Mél. IX (1876) 626. Tatarinov, 1847, Yin shan Mt.

Corylus mandschurica, Maxim. Amur, 241. Amur, North China. — *C. rostrata*, Ait. var. *mandschurica*, Regel. Flor. Ussur. 129. — Herb. Mus. Acad. Petrop. Tatarinov, Mountain Ma ts'tian tze ling. Figured in the Atlas appended to the Exploration of the Amur, 1859.

Smilacina hirta, Maxim. Amur, 277. Lower Amur. Peking, Western Mountains. — Maxim. Mél. XI (1883) 857, *Sm. japonica*, A. Gray, 1857 (Dr. Williams, 1854, v. s. p. 389), var. *mandshurica*.

Uvalaria viridescens, Maxim. Amur, 273. Lower Amur, North China. Herb. hti Petrop. Tatarinov, Tsie t'ai sze. — *Prosartes viridescens*, Regel, Fl. Ussur. (1861) 148. — Maximovicz, Mél. XI (1883) 859 = *Disporum smilacinum*, Asa Gray (1857). Dr. Williams 1854, Japan (v. s. p. 389), var. *viridescens*.

Monochoria Korsakowii, Regel, Ussur. 155, c. icone. Maack, Kengka Lake. — Herb. hti Petrop. Tatarinov near Peking and near Tou shuai sze.

Pinellia pedatisecta, Schott, Prod. Syst. Aroid., 1860, 3. Tatarinov near Shui t'a sze. Sinice *pan hia*. — According to Engler, Aroid. 1879, 566 (DC. Monogr. II) = variety of *P. tuberifera*, Tenore.

Arisaema Tatarinowii, Schott, Bonplandia, 1859, 27. Tatarinov, Yin shan. Herb. hti Petrop. Herb. Mus. Acad. Sinice *t'ien nan sing*. — Engler l. c. 559.

Typhonium giganteum, Engler. V. infra sub Skatchkov.

Acorus Tatarinowii, Schott, Oestr. bot. Zeitschr. 1859, 101. Hb. hti Petrop. T'ang shan. — Engler l. c. 217 = *Acorus calamus*, Linn. var. *terrestris*.

Carex neurocarpa, Max. Amur, 306. Lower Amur. N. China, Ti tsan a rh (Yin shan).

Panicum mandshuricum, Maxim., var. *pekinense,* Maxim. Amur, 329. N, China, Ku pei k'ou, Aug. 1847.

Cheilanthes Kuhnii, J. Milde Filices Europae, Asiae etc. 1867, 35. Herb. Fischer in Hb. hti Petrop. Tatarinov, temple Tsie t'ai sze. — Syn. Filic. 476.

Liubimov, N. I, conducted in 1840, the 12th Russian Eccles. Mission to Peking, and returned with the 11th Mission to Russia. He was subsequently Director of the Asiatic Department. He died in 1875.

In the History of the Imp. Bot. Gardens, St. Petersburg by R. Trautvetter (Acta Hti Petrop. II, 250) we read:

Mr. Liubimov, Councillor of State, on his return from Peking in 1842, transmitted to the Bot. Gardens 60 different living plants planted in 33 buckets. Notwithstanding their having been carried overland, they arrived in perfect condition.

Gorski, V. V., proceeded with the 12th Eccles. Mission to Peking, to learn there the Chinese and Manchurian languages. He died at Peking in 1849. In the Herb. of the Bot. Garden, St. Petersburg are some plants collected by Gorski in Mongolia, but the greater part of his botanical collections made in Mongolia and China were presented by the Asiatic Department to the Bot. Museum of the Academy.

Goshkewich, Joseph, of Polish origin, Mohilev, accompanied, in 1840, the 12th Eccles. Mission to Peking, to study Chinese there and make meteorological observations. He left Peking in 1850. He wrote several articles on Chinese economic plants in the Collection of Papers by the Members of the Mission (in Russian) viz:

I (1853) n. 4. THE MANUFACTURE OF INK AND ROUGE BY THE CHINESE.

III (1857) 119. ON THE CULTIVATION OF THE SHAN YAO OR DIOSCOREA.

Ibidem 125. ON THE IMPERIAL OR FRAGRANT RICE IN PEKING.

A collection of Lepidoptera formed at Peking by G. was described by O. Bremer and W. Grey (v. s. p. 560).

In 1852, Goshkewich proceeded with Admiral Putiatin (v. infra) to Japan, and in 1858 was sent as Russian Consul to Hakodate. He died 1874 in Wilna.

Basilevski, Stephan Ivanovich, entered the Medico-chirurgical Academy at St. Petersburg, took his degree of M. D., and was appointed physician to the 13th Russian Eccles. Mission which arrived at Peking Sept. 27, 1849, and left the Chinese capital in June 1859. B. took a great interest in natural history and during the first years of his sojourn at Peking formed a collection of Fishes, which he described and figured in the Nouv. Mém. Soc. natur. Moscou, vol. X, 1855, with the title of Ichthyographia Chinae borealis. Some years ago I discovered in the Botan. Museum of the Medico-chirurgical Academy, St. Petersburg a vast collection of Peking plants, consisting of twenty one large bundles which B. had presented to the Museum in 1860, and which bears no traces of ever having been touched since B. delivered it. The late C. Maximowicz was not aware of its existence. From the labels it appears that these plants were collected chiefly among the Peking Mountains, on the Po hua shan etc.

After his return from China, B. practised medicine for some years at St. Petersburg. In 1867 he was appointed Physician to the Russian Legation in Persia. He died a short time after his arrival at Teheran.

Skatchkov, Constantin Adrianovich, born 1821, completed his education at the Richelieu Lyceum Odessa, and came to St. Petersburg in 1847, where his studies were directed to agriculture and astronomy. In 1848 he was commissioned to accompany the 13th Russian Eccles. Mission to Peking in order to establish there a meteorological observatory. In 1857 the bad state of his health compelled him to return to Russia. Having recovered, he was in 1858 appointed Russian Consul at Chuguchak and held this post till 1863. From 1866 to 67 he was Professor of the Chinese Language at the St. Petersburg University, and then was again despatched to China to fill the post of Consul at Tien tsin. In 1875 he went home on leave and returned to China as Consul General at Shang hai. In 1879 he left for good and in the next year was appointed Interpreter to the Asiatic Department. He died March 26, 1883.

During the seven years he resided in Peking, S. applied himself to the study of Chinese agriculture and wrote several interesting papers (all in Russian) on Chinese economic plants and their cultivation.

In the Memoirs of the (Russian) Journal of Rural Economy, 1851, he published an article On the Chinese Vegetable Tallow *(Stillingia sebifera)*, and in 1857 in the same Periodical a valuable paper On the Cultivation of the Sweet Potato *(Batatas edulis)* in China, the history of its introduction from Manila in the 17th century, translated from Chinese books, and accompanied with drawings of the plant.

In 1858, S. transmitted to the Department of Agriculture in the Ministery of the State Domains, St. Petersburg a collection of about 500 kinds of seeds of Chinese cereals, vegetables, fruits and other economic plants, some of which were cultivated in the Botan. Gardens. They were also sent to the Société d'Acclimatation, Paris. (Regel's Gartenflora, 1860, 9). Dr. Merklin, Prof. of Botany at the Medico-chirurg. Academy determined them.

In the Gartenflora XI (1862) 407, tab. 383, is a description and a figure of an excellent large rose-coloured *Summer Radish* introduced by S. from Peking.

In the Moscow Journal of Agriculture (in Russian) III (1862) n. 7, 1—34, we find an article by Skatchkov. ON THE TREES UPON WHICH THE CHINESE USE TO REAR WILD SILKWORMS, and the mode of cultivating these trees: *Quercus chinensis (Bungeana), Ailantus glandulosa, Zanthoxylum Bungeanum.*

In 1864, S. published in conjunction with the French sinologist M. G. Pauthier: NOTICE SUR LA PLANTE MOU-SOU OU LUZERNE CHINOISE, *Medicago sativa.* (Revue de l'Orient etc. 1864, 69—77).

In the (Russian) Journal of Rural Economy, 1866, there is a paper by S. ON AGRICULTURE IN CHINA.

Two interesting articles prepared by S. for the press, which we have read, remained unpublished, viz:

ON THE TINCTORIAL PLANT, *Polygonum tinctorium,* its cultivation and use at Peking.

HYDROPYRUM LATIFOLIUM CULTIVTED AT PEKING FOR ITS SPROUTS, EATEN AS A VEGETABLE. Comp. Hance in Journ. Bot. 1872, 146.

We have seen also, in the possession of S. a collection of Chinese coloured drawings, about 50, representing Peking cultivated plants, beautifully executed by a Chinese artist, from nature.

There are in the Herbarium of the Bot. Gard. St. Petersb. about 200 herbarium specimens sent by S. from Peking, generally with the Chinese names on the labels, for the greater part cultivated plants. Among these there is a novelty:

Typhonium giganteum, first described and figured from S.'s specimen by Engler in his Jahrbücher IV (1883) 66. — There is, however, in the Herb. Mus. Acad., a specimen of this plant gathered earlier by Tatarinov. Its Chinese name is, according to these two collectors *tu küe lien.*

XII. Russian Occupation of the Amur River and Naval Expedition to Eastern Asia, 1852—55. Botanical Explorations connected with this Enterprise.

Before we proceed to record the important botanical discoveries made by Russian explorers in the middle of this century on the Amur River and its affluents, in Manchuria and in Japan, it will be necessary to give a short account of the memorable political events which led to the opening of these interesting regions to scientific research. Our information is chiefly borrowed from official Russian reports.

Down to the middle of our century Russia's possessions on the eastern coast of Asia did not extend farther south than the 55th parallel of N. lat. The Russians had two harbours in the inhospitable regions of the Siberian coast: Okhotsk and Ayan, both quite inaccessible to traders for the greater part of the year. There was besides these the Russian port Petropavlovsk in Avacha Bay in the south-eastern part of Kamtchatka, 53° N. lat., which presented more favourable conditions, but was too far off to be profitable. It was consequently natural for the Russians to cast their eyes southward, where there were excellent ports and harbours, and a country rich in timber and other unexplored resources and but loosely held in the feeble grasp of the Chinese Government.

For the acquisition of the Amur River and a part of Manchuria, Russia is indebted to the profound and energetic policy of N. Muraviov, the able Governor General of Eastern Siberia, who succeeded in accomplishing, without bloodshed, the annexation of this important province.

Muraviov, Nicolai Nicolayewich, born in 1810, of noble extraction, early entered the military service, distinguished himself in the Caucasian wars, rose to the rank of Major-General, was subsequently Governor of the province of Tula. In 1847, the Emperor Nicholas appointed him Governor General of Eastern Siberia. While in Petersburg, before his departure, Muraviov made the acquaintance of Captain-lieutenant Nevelski, Commander of the Transport "Baikal" of the Russo-American Trading Company, who had just received orders to sail from Kronstadt to Petropavlovsk with provisions for the Russian settlement there.

Nevelski, Ghenadi Ivanovich, was born about 1815. He rose subsequently to the rank of Admiral, died at St. Petersburg in 1876.

At the motion of Muraviov, Nevelski was commissioned by the Russian Admiralty to proceed, on his homeward voyage, from Petropavlovsk directly to the Manchurian coast in order to search for the still unknown mouth of the Amur River. Nevelski reached Petropavlovsk in May 1849 and, having discharged his ship, directed the course of the "Baikal" towards the Amur. Approaching from the Sea of Okhotsk, he discovered, on June 28, 1849, the liman (lagoon) before the mouth of the Amur. Captain Lieutenant Kazakevich, Peter Vassilievich, (afterwards Real-Admiral and Governor of the Sea coast Territory of E. Siberia, Commander of the Siberian Squadron. He died in 1887 at Kronstadt) who then was with Nevelski, explored the coast in a boat more closely, and succeeded in discovering the mouth of the great River, where it discharges itself into the liman. After this a survey of the coast of the mainland and the western coast of Sakhalin, forming the Gulf of Tatary, was drawn up by another naval officer **Orlov,** Dmitri Ivanovich, in 1849—50, by which survey it was ascertained that Sakhalin was not connected with the mainland, as had been presumed by Lapérouse, in 1787, and Krusenstern, in 1805, but that there was a narrow water passage between them, north of De Castries, by which the mouth of the Amur could be reached by ships coming from the south.

After possession had been taken of the mouth of the Amur, a Russian settlement was formed, in June 1850, for the Russo-American Trad. Comp. on the Sea of Okhotsk, on an inlet named Schastie (good luck) by the Russians. This settlement was called Petrovskoye zimoviye (Peter's winterquarters). In August 1851 the military post Nikolayevsk (subsequently city) was established on the left bank of the Amur, just where the river enters the liman. In June 1853, Nevelski occupied the Bay De Castries (discovered in 1787 by Lapérouse, who named it in honour of Marquis De Castries, then French Minister of the Navy), and established there the military post Alexandrovsk. At the same time another military post, Konstantinovsk, was formed farther south on Emperor's Harbour (Baracouta Harbour of English maps. Local name Hadji). Nevelski also sailed up the Amur and established the post of Marinsk, near the Kidzi Lake. A military post of 150 men was located on Aniva Bay in the south of Sakhalin.

When the news of the discovery and occupation of the mouth of the Amur reached St. Petersburg, the Emperor Nicholas immediately resolved upon sending a squadron to Eastern Asia. At the head of this Naval Expedition was placed Admiral Putiatin, who at the same time was entrusted with a diplomatic mission to Japan.

Putiatin, Yephimi Vasilievich, born 1803, was trained for the navy in the Imperial Naval School. He accompanied Captain M. P. Lazarev in his

circumnavigation of the globe, 1822—1825, and fought in the battle of
Navarino, Oct. 20, 1827. He was promoted to the rank of Rear-Admiral
in 1839. In 1842 he was sent out with a body of troops to chastise the
Turcomans near the south-eastern corner of the Caspian Sea, whose preda-
tory excursions troubled our trade with Persia. He was successful in this
expedition. After this he was despatched with a diplomatic mission to the
Shah of Persia. When in 1843 the news was received that the Chinese had
opened several ports to British trade, Putiatin, in a memorial, urged the
necessity for Russia to send a squadron to China to occupy a good harbour
on the Manchurian coast, and at the same time try to conclude commercial
treaties with China and Japan. This project met with the Emperor's ap-
proval, but owing to the objections made by the Minister of Finance, Count
Cancrin, it was postponed and not carried out till nine years later.

The Flag ship, the Frigate "Pallas", commanded by Captain-Lieutenant Unkovski, left the roads of Kronstadt October 7, 1852, and arrived,
October 31, at Portsmouth, to which place the admiral had proceeded in
advance. As the Pallas needed repairs, the departure was delayed. Mean
while Putiatin had bought at Bristol the Screw-steamer "Fearless" destined
to accompany the flag ship. She received the Russian name "Vostok" (the
East), the command of her was given to Capt. Lieut. Rimski-Korsakov. The
other two ships of the squadron, the Corvet "Olivutsa" and the Transport
Ship "Menshikov", belonging to the Russ.-Amer. Trading Company, had received orders to sail directly to Bonin sima Island and wait there for the
admiral's arrival.

1853   On January 6, 1853, the "Pallas" with the Admiral on board, and the
"Vostok" left Plymouth. On March 24, the flag ship arrived at False Bay
(Cape of G. H.), a week later the "Vostok". On April 12, the "Pallas"
again got under sail. The Vostok was ordered to visit the Cocos (Kee ling)
Islands, south of Sumatra, and then to repair to Hongkong. On May 17th
the "Pallas" anchored off Anjer (Java), reached Singapore on the 24th.
Here the admiral despatched his first officer Butakov to St. Petersburg, to
transmit to the Admiralty his report on the unseaworthiness of the Pallas,
and the request to send out, to reinforce his squadron, the Frigate "Diana",
recently built for the Baltic Fleet. Having quitted Singapore on June 1st,
the "Pallas" arrived at Hong kong on the 13th. The "Vostok" had cast
anchor there two days earlier. On June 18, Putiatin, on board the "Vostok",
proceeded up the Pearl River to Canton. His negotiations with the Chinese
authorities regarding the admission of Russian trading vessels to trade at
Canton having failed, he made sail for the Bonin Islands. The "Vostok" had

taken her departure a few days earlier. The Pallas on her way to Bonin was seriously damaged by a typhoon, but reached Port Lloyd on Peel Island in the Bonin Archipelago (v. s. p. 289) July 26, where all the other ships of the Russian squadron were found assembled. The "Olivutsa" had succeeded in surveying the southern part of the Bonin Group.

Then the squadron put to sea for Japan and on August 9, arrived at the roads of Nagasaki. Negotiations with the Japanese authorities, who finally consented to forward Putiatin's credentials as Russian Ambassador to the Japanese Court at Yedo. Unfortunately the Shogun died and the negotiations were interrupted.

On August 18, Putiatin despatched the "Vostok" to survey the Tatarian Gulf, and push on to the mouth of the Amur. Her commander Rimski-Korsakov, having reached Cape Crillon (the S. W. point of Sakhalin), sailed along the western coast of the island, landed on September 1st near Cape Bernizet, touched at Notosame and then proceeded to Jonquière Bay (near Due) which was surveyed. Then he steered to the Amur and landed, September 13, at the Pronghe Promontory, at the mouth of the Amur. The second half of September and the whole of October were spent in surveying the west coast of Sakhalin, where upon the "Vostok" returned to Nagasaki to join the squadron. The surveying operations on the west coast of Sakhalin were continued in July 1854.

On November 11, 1853, Admiral Putiatin with his squadron left Nagasaki and shaped his course to Shang hai. They anchored at Saddle Island, north of the Chusan Archipelago, opposite the southern entrance of the Yang tze. The Admiral proceeded, on board the "Vostok" to Shang hai. On Dec. 17, after his return from Shang hai, the Russian squadron left Saddle Island and on December 22, reached Nagasaki. Japanese plenipotentiaries arrived from Yedo to confer with the admiral, but these negotiations ended in nothing.

On January 24, 1854, the Admiral sent the "Vostok" to Shang hai 1854 to take the mail and carry it to Great Liu kiu, and with the remaining three ships made sail for the latter island. The "Vostok", having returned with the mail, the Admiral with three ships departed for Manila, whilst the "Vostok" was despatched to search for the island of Borodino, east of Liu kiu, previously discovered by Captain Ponafidin. The Admiral reached Manila, February 16, the "Vostok" joined the squadron a week later.

On February 27, the squadron left Manila. The "Menshikov" was sent to Shang hai for the mail with injunction to join the flag ship at Hamilton Island, south of Corea. The "Vostok" was despatched to Napa on Great

Liu kiu to inquire about the results of the negotiations of Commodore Perry with Japan (v. s. 387). The Americans had built a coal-dépôt at Napa.

The flag ship on her way from Manila to Hamilton, owing to the damaging of a mast, endured a delay of ten days in one of the islands of the Babuyan Group. At Hamilton Harbour the "Pallas" was joined by the other ships, and the Russian squadron for the third time proceeded to Nagasaki. After a short stay off this port they left it again. The "Vostok" was sent for the mail to Shang hai, the "Menshikov" to De Castries, whilst the "Pallas" passed through the Strait of Corea ànd from April 20, to May 11, the eastern coast of Corea was surveyed, beginning with Chusan (Fu sang) Harbour in the S. E. (v. supra p. 542). Two new and excellent harbours were discovered and named, Port Lazarev and Possiet Bay. All the inlets and promontories were named, mostly after the officers of the "Pallas". When entering the Gulf of Tatary, May 17, the flag ship was joined by the "Vostok" which brought the news that England and France, in March, had declared war upon Russia, and the order from St. Petersburg for all Russian ships in the Chinese and Japanese waters to collect in De Castries Bay. The "Vostok" was then sent to Okhotsk whence she returned in June, the "Menshikov" made sail for Petropavlosk. The "Pallas" anchored in De Castries Bay.

Let us now return to the occupation and exploration of the Amur River and its affluents by the Russians.

General Muraviov, who in 1853 was in St. Petersburg, obtained from the Emperor the permission to fit out an exploring expedition, and sail down the Amur River from the Russian frontier to its mouth. A small steamer (60 horse power) had already been built for this purpose at a great iron foundery established on the Shilka. Muraviov left St. Petersburg for Siberia in the beginning of 1854. When he arrived at the Shilka he found the steamer ready. She was named "Argun". The Governor General started at the head of the expedition on May 14. The steamer was accompanied by 75 barges and rafts with about a thousand men. They sailed down the Shilka, entered the Amur and on June 14th reached Marinsk. The "Pallas" was then already anchored in De Castries Bay. Marinsk is only about 28 miles distant from De Castries, by land.

In 1855 several scientific expeditions were sent down the Amur. One of them was headed by R. Maack (see farther on).

Thus a grand line of water communication connecting the interior of Siberia with the Pacific, was for the first time thrown open to Russia and the civilized world. Immediately after this success, the Russian Government began to colonize the northern part (left bank) of the Amur valley. In 1856 colonies of Russian peasants from the imperial domains were established between Marinsk and Nikolayevsk, and in the next year, Cossacks from the Baikal were transferred to the Amur, to form agricultural colonies along its left bank down to the mouth of the Zeya River. Nothing was neglected to take a firm hold of the Amur.

Hitherto Russia had invaded and taken possession of Chinese territory without any attempt to obtain a legal right to it. The Chinese were, of course, much alarmed by the progress of their northern neighbours, but unable to offer active resistance. In 1857, however, the Russian Government made ouvertures to get a formal cession of the annexed territory from the government of Peking. In consequence Admiral Putiatin, after returning from his first expedition to Eastern Asia, was despatched as ambassador to reach the Chinese Court by way of Siberia, but he was denied access (see farther on). Then Muraviov began to negotiate with the Chinese delegates who had been sent from Peking to Aigun, a place situated on the right bank of the Amur. These conferences finally led to the conclusion of the Aigun Convention, May 16, 1858, which secured to Russia the possession to the left bank of the Amur. In the same year several cities were founded on the river, viz: Blagoveshchensk, a little up stream from Aigun at the mouth of the Zeya River, — Khabarovka, at the mouth of the right branch of the Ussuri — and Sofisk on the lower Amur.

Muraviov, in recognition of the eminent services rendered, was created a Count and the title "Amurski" (of the Amur) was conferred upon him. He died at Paris in 1881.

On November 2nd 1860, after the termination of the war of the Western Powers with China, General N. P. Ignatiev, signed a definitive Treaty with China, according to which the latter ceded to Russia, besides the left bank of the Amur, the whole country situated between the Ussuri and the Sea up to the Corean frontier. A large and very important tract was added to the Russian Empire, with a seaboard of over 700 miles in extent and many excellent harbours.

The Frigate "Diana" sent out at the request of Admiral Putiatin to reinforce his squadron in the waters of China and Japan, commander Captain Lieutenant Lessovski, departed from the roads of Kronstadt, October 4, 1853, reached Kopenhagen 13th. — Madeira, stay November 8 to 18, —

1854 Rio Janeiro, stay December 13 to January 7, 1854. — Around Cape Horn,
where the "Diana" sustained a terrible storm. — Valparaiso, stay February
22, to March 11, — Honolulu, stay May 1 to 15. Then the "Diana" set
sail to search after the Frigate "Aurora", which had been despatched from
Kronstadt the previous August to Petropavlovsk, with orders to put into
Honolulu, but had not been seen in this port. After vain efforts to meet her
(the "Aurora" had passed directly from Calao to Petropavlovsk) the "Diana"
returned, May 29, to Honolulu, where meanwhile news had been received
of the declaration of war by the Western Powers against Russia, and that
the British Frigate "Pick" had been sent out to give chase to the "Diana".
The latter, therefore, got under sail with the least possible delay. Steering
directly to the Gulf of Tatary, she arrived, July 11, in safety in the Bay
De Castries, where the "Pallas" was found anchored.

As the appearance of the enemy's fleet in the Gulf of Tatary seemed
almost certain, measures had to be taken to secure the Russian ships there.
The "Pallas" was an old ship and, after the many storms sustained during
her way out, quite infit for a voyage. An attempt was made to bring her
into the Amur River, but, as it failed, it was deceided to destroy her. She
was conducted into Emperor's Harbour (v. s. 573), burnt there and sunk.
A part of her crew and officers were brought by the "Vostok" to Ayan and
returned to Russia overland. The rest were transferred to the "Diana",
then occupied by the Admiral and his staff.

On October 3, Putiatin with the only ship at his disposal, sailed for
Japan, in order to bring to an end the negotiations with the Japanese
Government. On October 9, 1854, the "Diana" reached Hakodate, where
the Admiral learned that Commodore Perry, in May last had concluded a
treaty with Japan, in consequence of which the ports of Hakodate and
Simoda were opened to American trade. Through the medium of the governor of Hakodate, Putiatin requested the Japanese Govt at Yedo to send
plenipotentiaries to Osaka, and, leaving Hakodate on October 16, and sailing along the eastern coast of Nippon and entering by Linschoten Strait
into the Inner Sea, reached Osaka, October 27. Here the "Diana" remained at anchor till November 10, when the answer from Yedo was
received inviting the Admiral to proceed to Simoda. The "Diana" reached this harbour on November 22. The Japanese plenipotentiaries arrived
ten days later. On December 11, a tremendous earth quake took place,
which destroyed the city of Simoda. Great damage was done to the "Diana",
her keel being torn off. The Japanese assented to the frigate being conducted into the great Bay of Taotomi (Saruga) situated N. W. of Simoda,

where she was sheltered from storm and not exposed to discovery by the enemy. Accordingly a great fleet of Japanese junks pulled her into the bay. But unhappily, a few miles before reaching the spot, where she had to be anchored, a sudden squall made her capsize and she sunk instantly, January 7, 1855. No life was lost.

1855

Then the Admiral, immediately, made up his mind to build a small vessel in order to be able to give notice at Petropavlovsk of the fatal loss. The Japanese most obliginly offered to furnish all the materials required and lend a helping hand. A wharf was established on an inlet at the village of Heda, near the north-eastern corner of the Saruga Bay, where excellent timber was obtainable from the neighbouring mountains. The building of the ship advanced rapidly.

On January 21, the Admiral's negotiations with the Japanese pleni-potentiaries were successfully terminated and a Treaty between Russia and Japan was signed, opening the ports of Hakodate, Simoda and Nagasaki to Russian trade.

Meanwhile the steam ship "Powhatan" of the American Squadron with the Commissary Mr. Adams on board, had arrived at Simoda, and when the latter left, February 10, he kindly offered to take care of the Admiral's report to his government.

In the beginning of March an American trading vessel, "Caroline E. Foote" came to Simoda. Profiting by this opportunity, Putiatin chartered her for the transportation of a part of his officers and crew to Petropav-lovsk. Accordingly, on March 30, Captain Lieut. Lessovski with 8 officers and 150 sailors left for that port.

On April 13th the new built schooner was ready and received the name "Heda". On the 26th the admiral with five officers and 40 sailors got under sail. They reached Petropavlovsk May 10, and, having learnt there that all Russian vessels in these waters had received orders to repair to De Castries, they departed the next day. The "Heda" had to pass the line of English ships posted in the Sea of Okhotsk, to watch the access to the Amur, but, succeeding during a dense fog in giving the enemy the slip, the admiral on June 8th reached Nikolayevsk. After the end of the war the "Heda" was returned to the Japanese.

Admiral Putiatin returned to St. Petersburg by the Amur and Siberia. He arrived at St. Petersburg in November 1855. The Emperor created him Count.

When the Admiral left Simoda there remained in this place 8 officers and 275 sailors of the "Diana". Mussin Pushkin, First Officer, upon whom involved the command, chartered the Bremen brig "Greta" to transport them all to Ayan. They left July 2d. On July 20, the "Greta" met, about 52° N. Lat., H. B. M. S. "Barracuta", was captured and declared a lawful price. The Russian captives were carried first to Ayan, then to Hakodate, Nagasaki, and finally to Hong kong, where they arrived on October 13. Baron N. Schilling, one of the officers of the "Diana", now Admiral, Adjutant to the Grand-duke Alexei, published in 1892, in the "Russian Archives" an anonymous article with the title: "Recollections of an Old Sailor" (in Russian), in which he gives a most interesting account of the voyage of the "Diana", the earthquake at Simoda, the destruction of the frigate and his captivity.

Scarcely 14 months had elapsed since the return of Admiral Putiatin from Japan, when, once more he was entrusted with a diplomatic mission to Eastern Asia. Emperor Alexander II resolved upon sending an Embassy to the Court of Peking at the head of which the Admiral was placed. To him was attached a young diplomatist of the Foreign Office, Baron Friedrich von der **Osten-Sacken,** subsequently well known among geographers as Secretary of the Imp. Russian Geographical Society, which much important post he most ably filled for many years. He was for a long time Director of one of the Departments of the Foreign Office.

The other members of this Embassy were:

Balluzek, Captain of Artillery. After the Treaty of Peking, 1860, he was for two years Resident Minister for Russia at Peking. He died a long time ago.

Archimandrite **Avakum** and Dr. A. **Tatarinov,** both in the capacity of interpreters. Avakum was a distinguished sinologist, from 1830—40, he belonged to the Russian Eccles. Mission, Peking. Dr. Tatarinov had been, from 1840—50, Physician to that Mission (v. s. p. 559).

Peshchurov, Alexei Alexeyewich, midshipman, subsequently Admiral, died in September 1891.

1857   This Embassy left St. Petersburg, Febr. 21, 1857, and proceeded through Siberia to Kiakhta at the Chinese frontier. Here they spent two months in vain negotiations, the Court of Peking refusing to receive so eminent a Russian official. Then the Admiral sailed down the Amur to Nikolayevsk, where he found the War-steamer "America", Capt. Chikhatchev, in readiness for him. The Embassy embarked on July 2d. Having visited De Castries, the ports of Due (Sakhalin), St. Vladimir, St. Olga, Hamilton, the "America" directed her course to Shan tung Cape, and then, sailing along the coast of Shan tung, appeared at the bar of the Pei (Pai) ho River on July 24th. Putiatin was denied access and the Chinese authorities even refused to receive the Emperor Alexander's letter to the Emperor of China. Leaving a letter for the Chinese ministers, Putiatin sailed for Shang hai. On September 2d he returned to the mouth of the Pei ho, and, as the Chinese peremptorily refused his request for a reception at Peking, he proceeded to Nagasaki, which he reached on the 9th. After a stay of 4 days there, the Admiral went to Shang hai, and returned to Nagasaki on the 29th. On October 12, he signed a new Treaty with Japan. On November 2d the "America" anchored off Hong kong. Putiatin conferred with the British, French and American plenipotentiaries, visited Canton, Macao. On February 19, 1858, the Admiral left Hong kong, 1858 visited Amoy, Fu chou, Ning po, and in the middle of March arrived at Shang hai. On March 28, left Shang hai for the mouth of the Pei ho. The America reached the bar on the first of April before the fleet of the Allies had arrived. Putiatin was present as a spectator at the taking of the forts of Ta ku, and, when Lord Elgin went up the Pei ho to Tien tsin, the "America" with the Russian Embassy followed, having on board the American plenipotentiary. After Putiatin had signed the Treaty of Tien tsin, 1st of June 1858, he sent the "America" back to the Amur, and embarked with his suite on board the lately arrived Man of war "Ascold" for Japan. Nagasaki, July 7; then went to Simoda, where they saw the "Heda" (v. s. p. 579), which had been handed over to the Japanese by the Russian Government, reached Kanagava, near Yedo. About the middle of August the Admiral left Japan for Shang hai. At last Putiatin and Baron Sacken returned to Europe by the English mail, Hong kong September 13. Baron Sacken spent a month in Ceylon.

As to the rest of the life of Admiral Putiatin, this remarkable man was, after his second return from Eastern Asia, attached to the Russian Embassy, London. In 1861, he was for a short time Minister of National Education, and then retired from service. He died in October 1883, at Paris by an accident, being scalded while taking a bath.

Maximovicz, Carl.

Deferring to a later chapter a full account of the life of this far famed botanist, we propose for the present to consider only the first period of his labours, in which he appears as botanical collector in Eastern Asia. He was a young man of 25 years, not long before appointed Conservator of the

Herbarium of the Botan. Gardens, St. Petersburg, when, in 1853, he was commissioned by the Director of the Gardens to accompany, as a botanical collector, the Frigate "Diana" (v. s. 577). As we have seen the "Diana"
1854 reached De Castries Bay on July 11, 1854. Owing to the war between the Western Powers and Russia, which then had broken out, M. was constrained to leave the frigate. He then resolved to explore the entirely unknown Flora of the Amur Country recently occupied by Russia. In the summer of 1854 he botanized on the Manchurian coast, between De Castries and the mouth of the Amur, and in the vicinity of Nikolayevsk. In September he went up the river, in a boat, to Marinsk and the Kidzi Lake, and thence returned by land to De Castries, where he passed the winter.

1855   In the spring of 1855 he again ascended the Amur, advanced as far as the mouth of the Dondon River and returned to Marinsk. At the end of July he started in company with his friend, the naturalist L. von Schrenck (see farther on), to explore the Ussuri River, which they sailed up as far as the mouth of the Nor River, an affluent from the west. In the beginning of September they were back in Marinsk, where they remained till the 8th
1856 of July 1856, gathering plants and other natural objects in the neighbourhood of this place and on the Kidzi Lake.

On July 8, 1856, M. set out on his homeward journey, and, ascending the great river in a boat, for three months reached the post Ust Strelka at the junction of the Argun and Shilka. Then he proceeded by the ordinary post route through Siberia and reached St. Petersburg, on March 17, 1857.

M. handed over to the Botan. Gardens a rich collection of herbarium specimens, among which there were many novelties, also seeds and living plants. A part of his collections had already been received in 1856, and had been described by F. I. Ruprecht, Botanist of the Imp. Academy in a paper published in Mél. biol. II (1856) 407—474, with the title: Die ersten botanischen Nachrichten über das Amurland. Beobachtungen von C. Maximowicz. — In Mél. biol. II (1857) 475—521, there is an article by Dr. E. Regel, Director of Bot. Gardens: Vegetations - Skizzen des Amurlandes, gesammelt von C. Maximowicz. Bäume und Sträucher.

After his return to St. Petersburg, M. at once set himself to work up his materials, and in 1859 he published the botanical results of his explorations, with the title:

Primitiae Florae Amurensis, with a Map and 10 Plates. It includes also some plants gathered on the Amur, on the Manchurian coast, and in Sakhalin by Orlov, Schrenck, Maack, Ditmar, Weyrich, Kuznetsov (see farther on).

M. enumerates 985 species, including 57 species of Mosses, gathered by himself and Maak. He describes 9 new genera and 112 new species. But 5 of these new genera have subsequently been reduced by the authors of the Gen. Plant. to already known genera.

At the end of the book, 468—79, is an Index Florae Pekinensis, and 479—486, an Index Florae Mongolicae, both compiled from the materials found in the Herb. Hti Petrop. and in Herb. Mus. Acad.

New Plants discovered by Maximowicz, 1854 to 56 in the Valleys of the Amur, Ussuri etc.

I may observe that some of these plants were gathered there at about the same time by Maack.

Thalictrum filamentosum, Maxim. Prim. Fl. Amur. 13.

Thalictrum amurense, Maxim. l. c. 15.

Adonis appennina, Lin., var. *daurica*, Ledeb. in Maxim. l. c. 19, is, according to Regel, Plant. Radd., 1861, n. 47, a new species = *A. amurensis*, Reg. — Bot. Mag. tab. 7490 (1896). A. amurensis introduced to Kew from Japan.

Ranunculus pleurocarpus, Maxim. l. c. 21.

Eranthis stellata, Max. l. c. 22.

Aconitum arcuatum, Max. l. c. 27.

Paeonia obovata, Maxim. l. c. 29. — Earlier Tatarinov, Peking (v. s. p. 564). — I. F. S., i, 22.

Maximowiczia (nov. gen. Schizandr.) *amurensis*, Rupr., Mél. II (1856) 412, 439, c. icone. Ibid. II (1857) 515, *Maximowiczia chinensis*, Rupr. — Max. Amur, 31, tab. 1. — Earlier Kirilov (v. s. 348). — Baillon reduced this genus to *Schizandra*. See I. F. S., i, 25.

Caulophyllum robustum, Maxim. l. c. 33.

Plagiorhegma (nov. gen. Berberid.) *dubium*, Max. l. c. 34, tab. 2. — I. F. S., i, 33 = *Jeffersonia dubia*, Bentham & Hooker, G. Pl. I, 44. — Dr. Hance, Journ. Botan. 1880, 258, changed the name into *Jeffersonia manchuriensis*.

Hylomecon (nov. gen. Papaverac.) *vernalis*, Max. l. c. 36, tab. 3. — Benth. & Hook. G. Pl. I, 52, reduce this genus to *Stylophorum*. — Earlier known from Japan. — I. F. S., i, 34.

Corydalis fumariaefolia, Max. l. c. 39.

Corydalis speciosa, Max. l. c. 39.

584 C. MAXIMOWICZ,

Viola brachysepala, Max. l. c. 50. — In Max. Mél. IX (1876) 742, reduced to *V. mirabilis*, Lin.

Silene foliosa, Max. l. c. 53.

Silene macrostyla, Max. l. c. 54.

Silene melandryiformis, Max. l. c. 54. — I. F. S., ɪ, 65 = *S. aprica*, Turczaninov.

Krasheninikowia sylvatica, Max. l. c. 57.

Prunus Kolomikta, Rupr. Maxim. Mél. II, 1856, 420. — *Kolomikta mandshurica*, Regel, Max. Mél. II (1857) 485. — *Trochostigma Kolomikta*, Rupr. Maack, Mél. II (1857) 517. — *Actinidia Kolomikta*, Maxim. Amur. 63. — It seems that L. v. Schrenck discovered this plant on the lower Amur before Maximowicz (June 28, (July 10), 1855). But in June 1855 Small gathered it at Yezo (v. s. p. 397).

Tilia mandshurica, Rupr. Maxim. Mél. II (1856) 413. Max. Amur. 63, Mél. X (1880) 586. — I. F. S., ɪ, 94.

Geranium Maximowiczii, Regel, Fl. Ussur. 38, tab. 3. In Max. Amur. 70, noticed as *Geranium Wlassowianum*, Fischer (1824). — Maxim. Mél. X (1880) 627.

Phellodendron (gen. nov. Zanthoxyl.) *amurense*, Rupr. et Maack, Mél. II (1857) 526. — Max. Amur. 72, tab. 4. — I. F. S., ɪ, 108.

Euonymus pauciflorus, Maxim. Amur. 74. — Regel Fl. Ussur. 41 = *E. verrucosus*, Scop. var. *pauciflorus*. — Max. Mél. XI (1881) 195.

Cissus brevipedunculata, Max. Amur. 68.

Acer Dedyle, Maxim. in Rupr. Mél. II (1856) 441. — Maxim. Amur. 65 = *A. spicatum*, Lam., var. *ukurunduense*, Middend.

Acer Ginnala, Max. in Rupr., Mél. II (1856) 415. Figured in Atlas explor. Amur. 1859. — Max. Amur. 67 = *A. tataricum*, Lin., var. *Ginnala*. — I. F. S., ɪ, 142.

Acer Mono, Maxim. in Rupr. Mél. II (1856) 416. — Maxim. Amur. 68. — Maximowicz, Mél. X (1880) 600 = *Acer pictum*, Thunberg. — I. F. S., ɪ, 141.

Acer tegmentosum, Maxim. in Rupr. Mél. II (1856) 415. — Maxim. Amur. 66.

Lespedeza stipulacea, Max. Amur, 85. — Max. Monogr. Lesp. (1873) 57 = *L. striata*, Hook. & Arn. (1827, v. s. 293). — I. F. S., ɪ, 182.

Orobus alatus, Maxim. Amur. 83. — Maxim. Mél. IX (1873) 59 = variety of *Lathyrus (Orobus) vernus*, Lin.

Orobus ramuliflorus, Max. Amur. 83. — Max. Mél. IX (1873) 67 = variety of *Vicia (Orobus) venosa*, Willd.

Prunus glandulifolia, Rupr. et Maxim., Mél. II (1856) 421. — Max.
Amur. 85. — Max. Mél. XI (1883) 700.

Prunus Maximowiczii, Rupr. Mél. II (1856) 422; Maxim. Amur. 89;
Max. Mél. XI (1883) 700. — I. F. S., ɪ, 219. — Gard. and Forest 1893,
195, figured. Introduced by Sargent from Japan, it seems.

Spiraea amurensis, Max. Amur. 90. — Max. Monogr. Spirac. (1879)
117 = *Physocarpus amurensis.*

Potentilla amurensis, Maxim. Amur, 98. Figured in Regel Fl. Ussur.
tab. 4. — I. F. S., ɪ, 240.

Pyrus ussuriensis, Maxim. in Rupr. Mél. II (1856) 424; Max. Amur.
102. Maximowicz August 1855. Schlippenbach earlier, 1854. — Maxim.
Mél. IX (1873) 168 = *P. sinensis,* Lindley (1824). — I. F. S., ɪ, 257.

Chrysosplenium pilosum, Maxim. Amur. 122. — Maxim. Mél. IX
(1876) 770.

Philadelphus tenuifolius, Rupr. et Maximowicz, Mél. II (1856) 425;
Maximowicz, Amur. 104. — Maxim. Hydrang. (1866) 36 = *Philadelphus
coronarius,* Lin. var.

Umbilicus erubescens, Maxim. Amur. 114. — Maxim. Mél. XI (1883)
725 = variety of *Cotyledon spinosa,* Lin.

Ludwigia epilobioides, Max. Amur. 104.

Mitrosicyos (nov. gen. Cucurb.) *lobatus,* Max. Amur. 112, tab. 7. —
Benth. and Hook. G. Pl. I, 838, reduce this genus to *Actinostemma.* —
I. F. S., ɪ, 320, *A. lobatum.*

Schizopepon (nov. gen. Cucurb.) *bryoniaefolius,* Max. Amur. 111, t. 6.

Sanicula rubriflora, Fr. Schmidt in Maxim. Amur. 123. — Ind. Fl.
Sin., ɪ, 327.

Osmorhiza amurensis, Fr. Schmidt in Maxim. Amur, 129. — Maxim.
Mél. XII (1886) 468.

Gomphopetalum Maximowiczii, Fr. Schmidt l. c. 126. — Max. Mél.
IX (1873) 253 = *Angelica Maximowiczii.* (G. Pl. I, 916).

Aralia mandshurica, Rupr. et Maxim. Mél. II (1856) 427. — *Dimorphanthus mandshuricus,* Maxim. Amur. 133. — I. F. S., ɪ, 338 = *Aralia
spinosa,* Lin. (Canada) or *A. chinensis,* Lin.

Viburnum burejanum, Herder. Pl. Radd. 1864, 7, and Bull. Soc. Nat.
Mosc. 1878, p. 11. — Maxim. Amur. 135, sub *V. dauricum,* Pall. —
Maxim. Mél. X (1880) 653. — Regel Gartenfl. 1862, 407, tab. 384. —
I. F. S., ɪ, 350.

Xylosteum gibbiflorum, Rupr. et Maximowicz, Mél. II (1856) 430. —
Max. Amur. 135 = *Lonicera chrysantha,* Turcz.

Xylosteum Maximowiczii, Rupr. in Max. Mél. II (1856) 431. — Max. Amur. 137 = *Lonicera Maximowiczii*.

Lonicera Ruprechtiana, Regel Gartenfl. 1870, 68, t. 645. (*Lonicera chrysantha*, var. *subtomentosa*, Max. Amur, 136).

Heteropappus decipiens, Maxim. Amur. 148. — Earlier Tatarinov. (V, supra p. 566).

Symphyllocarpus (nov. gen. Compos.) *exilis*, Max. Amur. 151, t. 8.— Benth. et Hook. G. Pl. II, 296.

Adenocaulon adhaerens, Maxim. Amur. 152. — I. F. S., ı, 432 = *A. bicolor*, Hook. Bot. Misc. (1830, India).

Achillea ptarmicoides, Max. Amur. 154. — I. F. S., ı, 436.

Chamaemelum limosum, Maxim. Amur. 156. — I. F. S., ı, 439 = *Matricaria inodora*, Lin.

Cirsium litorale, Maxim. l. c. 173. — In Max. Mél. IX (1874) 322, reduced to *Cnicus (Cirsium) japonicus*, DC. (1837). — I. F. S., ı, 461.

Saussurea grandifolia, Maximowicz, l. c. 169. — Maxim. Mél. IX (1874) 342.

Saussurea ussuriensis, Maxim. Amur. 167. — Max. Mél. IX (1874) 340. — I. F. S., ı, 468.

Hieracium hololeion, Max. Amur. 182.

Youngia chrysantha, Max. l. c. 181. — Max. Mél. IX (1874) 359 = *Lactuca denticulata*, Maxim. (*Youngia dentata*, DC., *Prenanthes hastata*, Thbg.). — I. F. S., ı, 481.

Lactuca triangulata, Max. Amur. 177. — Max. Mél. IX, 356.

Pyrola renifolia, Max. Amur. 190. — Max. Mél. VIII (1872) 624.

Fraxinus mandshurica, Rupr. in Maack, Mél. II (1857) 551. — Max. Amur, 194.

Vincetoxicum volubile, Max. Amur, 195. — Maximowicz and Maack, Amur. — I. F. S., ıı, 109, *Cynanchum volubile*, Hemsl.

Eritrichium myosotideum, Max. l. c. 203. — Max. Mél. VIII (1872) 549. Earlier Turczaninov, Argun (*E. radicans*). — Figured in Regel Fl. Ussur. tab. 9.

Cuscuta systyla, Maxim. Amur. 200. — Engelmann Cuscut. (1860) 79 = *C. japonica*, Chois. var. *thyrsoidea*.

Omphalotrix (nov. gen. Scrophul.) *longipes*, Max. l. c. 209, t. 10. — Benth. et Hook. G. Pl. II, 977.

Plectranthus excisus, Max. l. c. 213, tab. 8. — I. F. S., ıı, 270.

Lycopus parviflorus, Max. l. c. 216.

Scutellaria dependens, Max. l. c. 219. — I. F. S., ıı, 294.

Phlomis Maximowiczi, Regel, Acta Hti Petrop. IX (1886) 594 (Ph. umbrosa in Max. Amur, 476).

Chenopodium bryoniaefolium, Bge, Salsol. Chinae in Act. Hti Petrop. XIII (1893) 16, new species, noticed previously by Bunge, in Max. Amur. 222, as *Ch. ficifolium*, Smith.

Corispermum confertum, Bunge, in Max. Amur, 225.

Corispermum elongatum, Bunge, l. c. 224.

Corispermum macrocarpum, Bge, l. c. 226.

Rumex amurensis, Fr. Schmidt in Max. Amur. 228.

Euphorbia lucorum, Rupr. in Maxim. Amur. 239. — Maxim. Mél. XI (1883) 834.

Phyllanthus ussuriensis, Rupr. et Maximowicz, Mél. II (1856) 473; Max. Amur. 241. — DC. Prod. XVI, II, 392 = *Ph. simplex*, Müller, Arg. (India), var. *ussuriensis*. — I. F. S., II, 423.

Juglans mandshurica, Maxim. in Rupr. Mél. II (1856) 417; Maxim. Amur. 76. — Maxim. Mél. VIII (1872) 630.

Juglans stenocarpa, Max. Amur. 78. — Max. Mél. l. c. 632, 655 = variety of *J. mandshurica*.

Betula costata, Trautvetter in Maxim. Amur. 253. — Regel in DC. Prod. XVI, II, 176 = *Betula ulmifolia*, Sieb. & Zucc. (1846) var. *costata*.

Pinus mandshurica, Rupr. in Maack, Mél. II (1857) 567. — Maxim. Amur. 263. Amur, Maxim. in June, Maack in July 1855. — Regel, Fl. Ussur. 137. Maack, Ussuri. — Maxim. Mél. XI (1881) 349 = *Pinus koraensis*, Sieb. & Zucc.

Abies nephrolepis, Max. Mél. VI (1866) 21. — In Max. Amur. 260, first noticed by Trautvetter as *Abies sibirica*, Led. var. *nephrolepis*.

Polygonatum stenophyllum, Maxim. Amur, 274. — Maxim. Mél. XI (1883) 854.

Polygonatum Maximowiczii, Schmidt in Flora Sachalin (1868) 185. Max. Amur, 275, spec. indeterm. — Max. Mél. XI (1883) 847 = *Polygonatum officinale*, All. var. *Maximowiczii*. — Weyrich first Sakhalin.

Allium sacculiferum, Max. Amur. 281. — Regel, Monograph. Allium in Act. Hti Petrop. III (1875) 171. — Earlier (Kirilov or Tatarinov) Mongolia, Peking.

Allium Maximowiczii, Regel l. c. 153. — Noticed in Maxim. Amur. 283, as *A. prostratum*, Trev.

Juncus brachyspathus, Max. Amur. 293.

Arisaema amurensis, Maxim. l. c. 264. — Engler, Araceae in DC. Monogr. II (1879) 264.

Cyperus amuricus, Max. l. c. 296.

Cyperus limosus, Max. l. c. 294.

Fimbristylis leiocarpa, Max. l. c. 301.

Isolepis verrucifera, Max. l. c. 300.

Carex uda, Max. l. c. 303.

Carex neurocarpa, Max. l. c. 306.

Panicum mandshuricum, Max. l. c. 328.

Imperata sacchariflora, Max. l. c. 331.

Alopecurus longearistatus, Max. l. c. 327.

Calamagrostis varia, Host, l. c. 323, is according to Maxim. Fragm. 1879, 69, a new plant *C. robusta,* Franchet, Enum. Jap. II, 169, 600.

Diarrhena mandshurica, Maxim. Mél. XII (1888) 933. This is the gramen indeterminatum in Max. Amur. 332.

Cystopteris spinulosa, Maxim. Amur. 340. — *Asplenium spinulosum,* Baker, Syn. Fil. 225.

MAXIMOWICZ'S SECOND JOURNEY TO EASTERN ASIA: AMUR, MANCHURIA, JAPAN, 1859—64.

Maximowicz had but just finished his Primitiae Fl. Amur, when he was once more despatched to Eastern Asia, to continue there his successful botanical explorations, collect for the Bot. Gardens herbarium specimens and seeds, and procure living plants. The summary of his second journey to the Amur, extending to Manchuria and Japan, as given in the sequel, has been drawn from the traveller's official reports and private letters, viz:

Official Report (in Russian) in the Journal of the Ministery of the Domains of the Empire, LXXXVI, 1864, 434—45.

About the same account was published in German in Erman's Archiv für Wissenschaftliche Kunde von Russland, XIX (1860) 516, — XX (1861) 209, — (1862) 221, 553.

Bull. Acad. II (1860) 545; IV (1862) 225: Maximowicz's Letters to L. von Schrenck.

1859   Maximowicz left St. Petersburg March 15, 1859, and, travelling through Siberia by the ordinary post road, reached Irkutsk on May 1st. He left the capital of Eastern Siberia May 25, embarked near Nerchinsk, descended the Shilka, entered the Amur at Ust Strelka, on June 13. After 12 days Blagoveshchensk was reached. Departing from this city, July 2, M. arrived at the mouth of the Sungari, July 13, and then ascended this river for about 250 versts. Before reaching the city of San sïng ch'eng, he was compelled by the Chinese authorities to turn back.

On July 31, he again entered the great river, which he continued to descend to the mouth of the Ussuri. From August 18 to September 15, he explored the valley of this important affluent of the Amur upwards for about 400 versts, and then returned to the main river. Having been waiting in vain at Khabarovka for a steamer to take him down to Nikolayevsk, on Sept. 13, he finally got under way in a boat and reached Nikolayevsk, October 2. His intention was to embark as soon as possible on a Russian steamer for Japan. A vain attempt was made to reach De Castries Bay by sea, but the Amur beginning to freeze, he was constrained to return to Nikolayevsk. In the beginning of 1860 he set out for the Manchurian 1860 coast, and, proceeding in sledges drawn by dogs on the ice of the Amur, he arrived on February 8th at Khabarovka. Thence continuing in sledges on the frozen Ussuri, he reached March 6, the village (Russian settlement) of Busseva, 10 versts distant from the mouth of the Sungachi River. Here he waited for the melting of the snow in the mountains he had to cross. He employed his time usefully in making zoological collections. On the 6th of May he set out for the coast. This last part of the journey he performed on horseback over the mountain chain of Sikhota Alin, which separates the Ussury Valley from the Manchurian Coast, first ascending the valley of the Lifudin or Fudsi River, an affluent of the Ussuri, and then, after having passed the crest, descending into the valley of the Dadso shui or Lefule River, which led him to the Sea. He reached the Bay of St. Olga on June 1st 1860. Here he spent a month in making botanical excursions. He explored the valley of the Wai fu din or Avakumka River up to its sources. He visited also Port St. Vladimir, N. E. of St. Olga.

During the next three months M. had an opportunity of seeing all the new ports on the Manchurian Coast. On June 30, he went on board the Russian Steamer "America" in which Count Muraviov proceeded to inspect Possiet Harbour, situated near the Corean frontier. Here he was at leisure to botanize till July 23. Then he embarked on board the Corvette "Greden", and the other ports were successively visited:

Victoria Bay, also called the Bay of Peter the Great. It is divided into two parts by a long projecting peninsula on which in 1860 the city of Vladivostok was built.

From July 24, till August 2, stay in Port Bruce, now Sliavianski Zaliv, in the western part of Victoria Bay.

From August 6, to Sept. 10, in Port May, now called the "Golden Horn"; the harbour of Vladivostok. — Maxim. explored the valley of the Sui fun River, which discharges itself into Victoria Bay, N. W. of Vladivostok.

Port Deans Dondas, stay one week. Now named Inlet Novik on an island a little south of Vladivostok.

On Sept. 12, the "Greden" returned to Possiet, and finally directed her course to Japan with Maximowicz on board. She anchored off Hakodate Sept. 18, 1860.

In 1859 and 60 M. gathered about 800 species of plants. Besides these he sent to the Botan. Gardens 148 different kinds of seeds and a chest full of bulbs.

New plants discovered by Maximowicz, 1859—60 on the Amur, Sungari, Ussuri and in Russian Manchuria:

Thalictrum tuberiferum, Maximowicz, Mél. IX (1876) 607. Bay St. Olga, 1860.

Dentostemon hispidus, Maximowicz, Mél. IX (1872) 11. St. Olga Bay, 1860.

Viola phalacrocarpa, Maxim. Mél. IX (1876) 726. Manchuria 1860, Hakodate.

Melandrium Olgae, Max. Mél. VII (1870) 332. St. Olga Bay, 1860.

Lychnis laciniata, Maxim. Mél. VI (1867) 200. Var. *mandshurica*, Vladivostok, 1860, — var. *japonica*, Yedo.

Oxalis obtriangulata, Maxim. l. c. 260. Victoria Bay, 1860. — Ind. Fl. Sin., i, 99.

Euonymus ussuriensis, Max. Mél. XI (1881) 190. Ussuri, 1860.

Acer barbinerve, Maximowicz, Mél. VI (1867) 369. Bays St. Olga, Victoria, 1860.

Acer mandshuricum, Max. l. c. 371. S. E. Coast of Manchuria.

Lathyrus subrotundus, Maxim. Mél. IX (1873) 59. Port Bruce, Manchuria meridionalis.

Prunus armeniaca, Lin., var. *mandshurica*, Max. Mél. XI (1883) 675. Sungari, 1859.

Potentilla centigrana, Maxim. Mél. IX (1873) 156, var. *mandshurica*. Bay St. Olga, fl. Wai fu din, 1860, var. *japonica*. Maxim. Hakodate, 1861. Small earlier (v. supra p. 398).

Rosa Maximowiczciana, Regel, Monogr. Ros. (1878) 378. Possiet.

Philadelphus coronarius, Lin., var. *mandshurica*, Maxim. Hydrang. 1866, 41. Bays St. Olga, Victoria, 1860.

Sedum viviparum, Maxim. Mél. XI (1883) 747. Vladivostok, Deans Dundas, 1860.

Angelica coreana, Max. Mél. XII (1886) 471. Possiet Harb.

Triosteum sinuatum, Max. Mél. VII (1870) 553. Amur, 1859, Mandshuria rossica. — Icon. Plant. tab. 1586 (1887). James Manchuria.

Lonicera praeflorens, Batalin, Act. Hti Petrop. XII (1892) 169. Maximowicz Manchuria, 1860.

Galium paradoxum, Max. Mél. IX (1873) 263. Bay St. Olga, 1860.

Asperula Platygalium, Max. Mél. IX (1873) 267. Rivers Li fu din and Wai fu din, upper course. — I. F. S., i, 395.

Ligularia calthaefolia, Max. Mél. VII (1870) 554. Bays St. Olga and St. Vladimir.

Senecio otophorus, Max. Mél. VIII (1871) 11. Vladivostok, 1860.

Saussurea Maximowiczii, Herder Plant. Radd. 1869, n. 174. — Max. Mél. IX (1874) 337. Port Deans Dundas, 1860. Kiu siu.

Nabalus ochrolaceus, Max. Mél. VII (1870) 557. Vladivostok, Deans Dundas, 1860. — I. F. S., i, 486, *Prenanthes ochroleuca,* Hemsley. (See Gen. Pl. II, 527).

Chimaphila astyla, Max. Mél. VI (1867) 207. Vladivostok, 1860.

Vincetoxicum inamoenum, Maximowicz, Mél. IX (1876) 787. Bay St. Olga, 1860.

Brachybotrys paradiformis, Maxim. Oliver, in Icon. Plant. tab. 1254 (1878). Comp. Max. Mél. VIII (1872) 543, nov. genus Lithosperm. (fruiting specimens) Victoria Bay, 1860. — I. F. S., ii, 152.

Scrophularia amgunensis, Fr. Schmidt, Fl. Amgun. (1862) 57. — Maxim. As. orient. Fragm. (1879) 34. Maximowicz first, Bay St. Olga, Ussuri, 1859, 1860.

Scrophularia mandshurica, Maxim. l. c. 35. Amur australis, Porte Bruce 1859, 1860.

Pedicularis mandshurica, Maxim. Mél. X (1877) 120. Bays St. Olga, St. Vladimir.

Plectranthus serra, Max. Mél. IX (1874) 428. Sungari, 1859.

Euphorbia mandshurica, Max. Mél. XI (1883) 842. Sungari, 1859.

Abies holophylla, Maxim. Mél. VI (1866) 22. Victoria Bay, 1860. — Max. Mél. XI (1881) 349 = *Abies firma,* Sieb. & Zucc.

Fritillaria ussuriensis, Maxim. Decas Pl. nov. 1882, 9. Manchuria, Wai fu din, 1860.

Eriophorum japonicum, Maxim. Mél. XII (1886) 558. Sinus Possiet 1860, Nippon media, Tchon. A few years earlier Hook. & Thomson, Khasia.

Carex Maximowiczii, Böckler, Linnaea XLI, 237. Maxim., Amur.

Carex bostrychostigma, Maximowicz, Mél. XII (1886) 568. Bay St. Olga, 1860.

Carex capituliformis, Meinshausen in Maxim. l. c. 563. Upper Ussuri, Manchuria, 1860.

Carex oligostachys, Meinsh. l. c. 566. Upper Ussuri and his affluent Li fu din. End of May 1860.

Aspidium craspedosorum, Max. Mél. VII (1870) 341. Two varieties: var. *mandshurica,* Russian Manchuria, Upper Da dso shui River, 1860, var. *japonica,* Nippon, Mt. Hakone. — A. craspedosorum is figured in Icon. Plant. tab. 1655 (1866) Baker. — Syn. Fil. 492.

MAXIMOWICZ'S BOTANICAL EXPLORATIONS IN JAPAN, 1860—1864.

Having taken up his headquarters at Hakodate (v. supra p. 590), Maximowicz passed more than fourteen months there, from September 18, 1860 to the end of November 1861, and employed his time in exploring the almost unknown Flora of this part of Japan. He was allowed by the Japanese authorities to travel within the limits of 30 versts from the city. 1861 In 1861 M. sent from this place to St. Petersburg about 800 herbarium specimens, 250 different sorts of seeds and many bulbs. Comp. Gartenflora X (1861) M.'s article: Ueber die Vegetation von Hakodate.

Leaving Hakodate by steamer, November 28, he reached Yokohama on December first, and December 26, departed for Nagasaki, where 1862 he arrived January 4, 1862, to pass the winter. On March 30, he sailed to Yokohama and remained there from April 4 to December 21, 1862. Many excursions were made in the neighbourhood of Yokohama and Yedo.

In September 1862 M. despatched a native collector in his service to the mountains of Facone (Hakone) and to the volcano Fudzi yama, whence the latter returned at the end of October with a rich harvest.

Maximowicz, in Rhamn. As. or. (1866) 17, note, tells us that in Japan he met with an intelligent young Japanese, **Tchonoski** by name, who used to accompany him on his botanical excursions. When M. returned to Europe, Tchonoski offered his services to collect plants for him, which promise he kept in a most satisfactory manner. From 1866 to 1887 he sent to the Botan. Gardens, St. Petersburg herbarium specimens and seeds gathered in localities in the interior of Japan then not accessible to foreigners, as f. i. in the Nikko Alps, north of Yedo. In the Herbarium of the Garden there were in 1867 about 800 species sent by Tchonoski. M. named a number of new Japanese plants after him.

Maximowicz himself also visited the mountains of Hakone and the Fudzi yama. On December 21, 1862, he once more returned to Nagasaki, and spent more than a year there.

In the spring of 1863, M. despatched his native collector to the interior of the island of Kiu siu, the northern part of which the latter travelled over during six weeks. The southern part of the island, where the prince of Satsuma ruled, was not then accessible, even for Japanese from other parts of Japan. In the course of the summer this Japanese collector repeatedly visited the volcano Wun zen (Onzen), near Nagasaki, the mountains Kundsho san (Central Kiusiu) and other interesting spots of the island. From Maximowicz's labels it appears that he also himself visited the places just mentioned as well as Mt. Naga, near Nagasaki, and Simabara in the same island, east of Nagasaki.

On December 27, 1863, Maximowicz again proceeded to Yokohama, which place he reached January 9, 1864. After a month's stay there he embarked, February 11, for the homeward voyage and returned to Europe by the Cape of G. H. He arrived at St. Petersburg July 10, 1864.

Maxim. brought along with him very valuable collections from Japan: 72 chests with herbarium specimens, 300 different sorts of seeds, about 400 living plants. These collections, as well as the former, were determined by Maximowicz himself, and the duplicates distributed among the principal herbariums of the world.

New Plants discovered by Maximowicz in Japan, 1860—63 and described by himself, including new plants gathered by Tchonoski.

Clematis lasiandra, Maxim. Mél. IX (1876) 586. Mt Naga yama near Nagasaki. Maxim.

Anemone Nikoensis, Max. Mél. IX (1872) 1. In Nippon mediae alpe altissima Nikko. Tchonoski, 1864.

Anemone stolonifera, Maxim. Mél. XI (1876) 605. Tchonoski in silvis alpinis prov. Nambu. (Eastern coast of Nippon, 39°). Ito Keiske earlier.

Coptis quinquefolia, Maxim. Mél. VI (1867) 258. Nippon media et meridionalis. Tchonoski?

Coptis orientalis, Max. l. c. 259. Nippon merid.

Cercidiphyllum ovale, Max. Mél. VIII (1871) 369. Nippon media. Tchon.

Magnolia compressa, Maxim. Mél. VIII (1872) 506. Prope Nagasaki. Maxim. Culta, Yedo.

Schizandra nigra, Max. Mél. VIII (1871) 370. Kundsho san (Kiusiu), Fudziyama. Maxim. — Nambu, Nippon bor. Tchonoski.

Berberis Tchonoskii, Regel, Act. Hti Petrop. II (1873) 421. Nippon. Tchonoski.

Achlys japonica, Maxim. Mél. VI (1867) 260. Principatus Nambu, Nippon bor. Tchonoski, 1864.

Draba japonica, Max. Mél. IX (1876) 608. Nambu. Tchonoski.

Viola vaginata, Max. l. c. 734. Hakodate. Maxim.

Viola yezoensis, Max. l. c. 736. Hakodate. Maxim.

Idesia (nov. gen. Flacourt.) *polycarpa,* Max. Mél. VI (1866) 20. Mt Hikosan (Kiu siu). Max. — Known also as *Polycarpaea Maximowiczii,* Linden.

Silene Maximowiczii, Rohrb., Linnaea XXXVI (1870) 680. — Max. cult. Jap. — Maxim. Mél. XII (1888) 720. Tchonoski found it first wild, Sept. 1887 in the prov. of Kai, Nippon, Mt Motoyama.

Lychnis stellarioides, Max. Mél. IX (1873) 31. Nambu. Tchonoski. — Ibidem 656, same as *Silene inflata,* Smith in Miq. Prol. 10.

Cerastium schizopetalum, Maxim. Mél. XII (1888) 722. Nippon, provincia Kai, alpe Kumaga take. September 1887, legit Tchonoski, misit rev. pater Anatolius.

Stellaria diandra, Maxim. Mél. IX (1873) 43. In montibus Hakone. Maximowicz.

Stellaria tomentosa, Max. l. c. 45. Wunzen (Kiusiu). Maxim.

Krascheninikowia heterantha, Max. l. c. 38. With the preceding.

Moehringia platysperma, Max. l. c. 35. Hakodate. Maxim.

Hypericum electrocarpum, Maxim. Mél. VI (1867) 261. Nagasaki. Maxim. — I. F. S., I, 74; same as *H. Sampsoni,* Hance, discovered 1865, described in Journ. Bot. 1865, 378.

Stuartia pseudocamellia, Max. l. c. 201. Yedo culta.

Stuartia serrata, Max. l. c. 201. Kiu siu mts Hisokan. Maxim.

Phellodendron japonicum, Maxim. Mél. VIII (1871) 1. Mons Fudzi yama. Maximowicz.

Ilex geniculata, Max. Monogr. Ilex. (1881) 50. Fudzi yama. Tchon.

Euonymus nipponica, Maxim. Mél. XI (1881) 187. Alpes Nikko. Tchonoski, 1864.

Microrhamnus franguloides, Maxim. Rhamn. (1866) 4, fig. 15—23. Kiu siu: in principatu Higo, prope Wai fu. Maxim. — Maxim. Mél. XII (1886) 431; Siebold earlier.

Rhamnus japonica, Maxim. l. c. 11, fig. 52—60. Hakodate, 1860. Maximowicz.

Rhamnus costata, Maxim. l. c. 17, fig. 1—3. Nippon, prov. Senano. Tchonoski, 1864.

Vitis leeoides, Max. IX (1873) 148. Nagasaki. Maxim. — I. F. S., I, 131 = *V. contoniensis,* Seem. (1859).

Acer capillipes, Maxim. Mél. VI (1867) 367. — Mél. X (1880) 597. Nippon prov. Senana. Tchonoski.

Acer circumlobatum, Maxim. l. c. 368, Mél. X, 605. Same locality, Tchonoski.

Acer argutum, Maxim. l. c. 368. Nippon, in provinciis Senano et Nambu. Tchonoski.

Acer nicoense, Max. l. c. 370, Mél. X, 609. Kiu siu, Nippon. Tchonoski. — Gard. and Forest, 1893, 153, figured. Mr. Sargent introduced it 1892, from Japan.

Acer Tchonoskii, Max. XII (1886) 432. Prov. Nambu. Tchon.

Sabia japonica, Maxim. Mél. VI (1867) 202. Maxim. Nagasaki. — I. F. S., I, 143.

Meliosma tenuis, Max. l. c. 262. Nippon, prov. Senano. Tchon.

Smithia japonica, Maxim. Mél. IX (1873) 57. Kiu siu: mons Naga. Maximowicz.

Sophora platycaria, Max. l. c. 71. Fudzi yama. Nambu. Tchon.

Prunus armeniaca, Lin. var. *Ansu,* Max. Mél. XI (1883) 676. Hakodate, culta. Maxim.

Aruncus astilboides, Max. Spirac. (1879) 67. Prov. Nambu, Nippon. Tchonoski.

Spiraea bullata, Max. l. c. 100. Culta Yedo. Maxim.

Filipendula multijuga, Maxim. l. c. 143. Kiu siu: Kun dsho san, Nippon: Hakone. Maxim.

Rubus pectinellus, Maxim. Mél. VIII (1871) 374. Kiu siu: mons Higosan; Nippon: Fudzi yama. Tchon.

Rubus peltatus, Max. l. c. 384. Nippon med. Thon.

Rubus sorbifolius, Maxim. l. c. 390. Kiu siu: prov. Higo, alpe Higosan. — Earlier Hooker and Thomson, Khasia.

Rubus phoenicolasius, Max. l. c. 393. Insula Yezo: ad lacum Konoma (north of Hakodate). Max. — Nippon med. prov. Senano. Tchon.

Sanguisorba obtusa, Maxim. Mél. IX (1873) 152. Nippon bor., prov. Nambu. Tchon.

Pirus Tchonoskii, Max. l. c. 165. Fudzi yama. — Figured in Garden and Forest, 1894, 54. Mr. Sargent introduced this species from Japan into the Arnold Arboretum, in 1892.

Saxifraga sendaica, Max. Mél. VIII (1872) 601. Nippon borealis prov. Sendai. Maxim. knows this plant only from the figure in Somoku VIII, 17.

Mitella japonica, Max. Mél. VI (1867) 203. Kiu siu: mons Aso-san. Nippon: prov. Senano. Tchonoski.

Chrysosplenium Echinus, Maxim. Mél. IX (1876) 768. Kiu siu: in vulcano Wunzen. Nippon: in alpe Nikko.

Chrysosplenium rhabdospermum, Max. l. c. 768. Nagasaki. Maxim.

Chrysosplenium sphaerospermum, Maxim. l. c. 770. Kiu siu: Vulcan. Wunzen.

Parnassia nummularia, Maxim. Mél. VI (1867) 203. Kiu siu: mons Naga. Maximowicz. — I. F. S., I, 271 = *P. foliosa,* Hooker & Thomson (1858), Khasia.

Deinanthe (nov. gen. Hydrang.) *bifida,* Max. Hydrang. (1866) 2, t. 1. Mons Naga non procul a Nagasaki. Maxim. Nippon, prov. Senano. Tchon. 1864. Culta Yedo.

Ribes japonicum, Maxim. Mél. IX (1873) 221. Yezo merid. prope fodinas plumbeas Idzi Novatari. Maxim. Nippon bor. prov. Nambu. Tchon.

Cotyledon japonica, Maxim. Mél. XI (1883) 724. Yokohama culta. Maxim. About the same time (1862), Oldham, Japan.

Sedum sordidum, Max. l. c. 754. Yedo cultum.

Sedum Maximowiczii, Regel, Gartenflora, 1866, 355. — I. F. S., I, 282 = *S. Aizoon,* Linn.

Disanthus (nov. gen. Hamamel.) *cercidifolia,* Max. Mél. VI (1866) 21. Nippon interior. — Gard. and Forest, 1893, 214, figured. Introduced into Arnold Arbor. by Mr. Sargent from Japan, 1892.

Liquidambar acerifolia, Maxim. l. c. 21. Yedo culta. — I. F. S., I, 291 = *L. formosana,* Hance. Oldham, Formosa, 1864.

Hydrocotyle ramiflora, Max. Mél. XII (1886) 463. Circa Hakodate. Maximowicz.

Edosmia neurophyllum, Maxim. Mél. IX (1872) 16. Kiu siu: Kundsho san, Nippon merid.

Sium nipponicum, Max. l. c. 17. Nippon media. Tchon.

Selinum longeradiatum, Max. Mél. XII (1886) 469. Mons Naga prope Nagasaki. Maxim.

Angelica hakonensis, Max. Mél. IX (1873) 257. Montes Hakone.

Angelica inaequalis, Max. l. c. 186. Montes Hakone.

Angelica Miqueliana, Max. l. c. 255. Kiu siu interior: jugum Kundsho san. Yokohama. Prov. Nambu. Tchon.

Angelica polymorpha, Maximowicz, l. c. 187. Montes Hakone. Jugum Kundsho san.

Angelica kiusiana, Max. Mél. IX (1872) 14. Prope Nagasaki.

Angelica multisecta, Max. Mél. XII (1886) 470. Fudzi yama. Maxim. Tchonoski.

Peucedannm multivittatum, Max. l. c. 471. Nippon med. Tchon.

Lonicera Tchonoskii, Max. Mél. X (1877) 61. Alpe Nikko. Tchon.

Lonicera venulosa, Max. l. c. 63. Prov. Nambu. Tchon.

Lonicera pilosa, Max. l. c. 73. Ibid. Tchon.

Lonicera linderifolia, Maximowicz, l. c. 77. Nippon media. Tchonoski, 1866.

Pseudopyxis heterophylla, Maxim. Mél. XI (1883) 801. Mons Hakone. Maximowicz.

Galium brachypodion, Max. Mél. IX (1873) 260. Prope Hakodate.

Patrinia palmata, Max. Mél. VI (1867) 267. In montibus Hakone.

Patrinia gibbosa, Maxim. l. c. 267. Yezo circa sinum Vulcanarum. Maxim. — Prov. Nambu. Tchon.

Valeriana flaccidissima, Max. l. c. 372. Yokohama.

Aster rugulosus, Max. Mél. VII (1870) 333. Yokohama.

Aster spathulifolius, Max. Mél. VIII (1871) 7. Yedo cultus.

Carpesium glossophyllum, Max. Mél. IX (1874) 282. Yokohama.

Carpesium triste, Max. l. c. 287. Hakodate, Maxim.; prov. Nambu, in montibus Hakone. Tchon.

Pyrethrum seticuspe, Max. Mél. VIII (1872) 515. Yedo cultum.

Leucanthemum nipponicum, Franchet in litt. (nomen). Savatier Nippon media. Maximowicz saw it at Yedo, cultivated. Max. l. c. 511.

Ligularia clivorum, Maxim. Mél. VII (1870) 555. Japonia, frequens. Max. Mél. VIII (1871) 14. — I. F. S., I, 451.

Senecio stenocephalus, Maxim. Mél. VIII (1871) 10. In jugo Hakone. Tchon. — I. F. S., I, 458.

Cnicus purpuratus, Max. Mél. IX (1874) 305. Fudzi yama.

Cnicus nipponicus, Maxim. l. c. 311. Nippon bor., prov. Nambu, et media. Tchon.

Cnicus dipsacolepis, Max. l. c. 313. Yokohama, Fudzi yama. Maxim. Kundsho san.

Cnicus spicatus, Max. l. c. 318. In monte Naga non procul a Nagasaki. Maxim.

Cnicus effusus, Max. l. c. 317. In montibus Hakone. Tchon. Siebold 1861. Yedo.

Macroclinidium (nov. gen. Compos.) *robustum*, Max. Mél. VII (1870) 556. Prope Yokohama, Kiu siu: Kundsho san, Maxim.

Lampsana apogonoides, Max. Mél. IX (1872) 20. Nagasaki.

Nabalus acerifolius, Maxim. Mél. VII (1870) 557. Kiu siu interior, Nippon media.

Campanumoea japonica, Max. Mél. VI (1867) 268. Nagasaki.

Leucothoe Tchonoskii, Maxim. Mél. VIII (1872) 613. Nippon media. Tchonoski.

Epigaea asiatica, Maximowicz, Mél. VI (1867) 204. Prov. Nambu. Tchon. 1865.

Andromeda nana, Max. Mél. VIII (1872) 617. Prov. Nambu. Tchon.

Tripetaleia bracteata, Max. Mél. VI (1867) 206. Ins. Yezo, Maxim. Principatus Nambu. Tchon. 1865.

Rhododendron Albrechti, Max. Mél. VII (1870) 335. Ins. Yezo. Max. 1861. — Ibidem Albrecht 1862.

Rhododendron macrostemon, Maxim. Rhododendron (1870) 41, tab. 3. Cultum Yedo.

Rhododendron macrosepalum, Max. l. c. 31, and Mél. VII (1880) 335. Nikko Alps. Tchon.

Rhododendron semibarbatum, Maxim. Mél. VII (1870) 338. Nippon media. Tchon.

Rhododendron Tchonoskii, Max. l. c. 339. Max. Rhodod. (1870) 42, tab. 3. — Nippon media. Tchon. 1865, 1866.

Menziesia purpurea, Max. Mél. VI (1867) 187, 204. Kundsho san in Kiu siu. Figured in Maxim. Rhodod. 11, tab. 1.

Menziesia pentandra, Max. Mél. VI, 205, and Rhododend. 9. Yezo. Maxim. Prov. Nambu.

Menziesia multiflora, Maxim. Rhododend. 11, tab. 1. Nippon bor. et media. Tchon.

Pyrola subaphylla, Maxim. Mél. VI (1867) 206. Ins. Yezo. Fudzi yama, prov. Senano.

Schizocodon uniflorus, Max. l. c. 274. Nippon: provinciae Senano et Nambu. Tchon. — Max. Mél. VIII (1871) 20 = *Shortia uniflora,* Max.

Schizocodon ilicifolius, Max. l. c. 273. Montes Nikko.

Statice arbuscula, Max. Decas Plant. nov. 1882, 8. Yedo, culta.

Primula macrocarpa, Max. l. c. 269. In alpibus Nippon borealis.

Ligustrum Tchonoskii, Decaisne, Monogr. Ligustr. (1878) 18. Maximowicz, Nippon media et Tchonoski, 1866.

Vincetoxicum ambiguum, Maximowicz, Mél. IX (1876) 794. Kiu siu, vulc. Wunzen.

Golowninia japonica, Maxim. Mél. IV (1861) 40. Hakodate, same as *Crawfurdia japonica,* Sieb. & Zucc. (1846). Maxim. proposed a new genus in honour of the Russian navigator W. M. Golownin, who about 1818 visited Yezo. But in Gen. Pl. II, 815, this genus it not admitted. — I. F. S., II, 122 = *Crawfurdia fasciculata*, Wallich (India).

Gentiana japonica, Maxim. Mél. IX (1874) 396. Kiu siu centralis: alpe Aso san.

Ellisiophyllum (nov. gen. Hydrophyll.) *reptans*, Max. Mél. VIII (1871) 18. Nippon merid. Maxim. received the plant from a Japanese botanist.

Omphalodes sericea, Maxim. Mél. VIII (1872) 558. Kiu siu centralis: Kundsho san. — I. F. S., II, 148.

Eritrichium brevipes, Maxim. l. c. 547. Jugum Hakone in Nippon, Kundsho san in Kiu siu.

Ancistrocarya (nov. gen. Borrag.) *japonica*, Max. l. c. 544. In monte Naga, haud procul a Nagasaki.

Scopolia japonica, Max. l. c. 629. In alpe Nikko. Tchon.

Mimulus sessilifolius, Max. Mél. IX (1874) 401. Yezo: in vallis metallofodinae Idzi Nowatari. Max. At about the same time Albrecht, ibidem.

Pedicularis yezoensis, Max. Mél. X (1877) 106. Hakodate.

Pedicularis apodochila, Max. Mél. XII (1888) 907. Nippon media et borealis. Tchon.

Lysionotus pauciflorus, Max. Mél. IX (1874) 366. Kiu siu in monte Naga. — I. F. S., II, 225.

Mosla grosseserrata, Maxim. l. c. 432. Yokohama, Hakodate. — I. F. S., II, 280.

Perillula (nov. gen. Labiat) *reptans*, Maxim. l. c. 440. Kiu siu in monte Naga.

Calamintha multicaulis, Max. l. c. 444. — Kiu siu centralis, prope Ko-isi-wara.

Lamium humile, Maxim. Mél. XI (1883) cum icone. In Kiu siu interiore. Maxim.

Teucrium veronicoides, Max. Mél. IX (1876) 826. Yezo.

Ajuga yezoensis,. Max. Mél. XI (1883) 811, c. icone. Hakodate.

Polygonum suffultum, Max. Mél. IX (1876) 616. In monte Hakone, Nippon bor. prov. Nambu. Tchon. — I. F. S., II, 350.

Polygonum Maximowiczii, Regel, Ind. Sem. Hti Petrop. 1865. Yokoh.

Asarum albivenium, Regel, Gartenfl. 1864, 195, tab. 440. — Max. Mél. VIII (1871) 402.

Lindera hypoglauca, Max. Mél. VI (1867) 274. Nippon, in Montibus Hakone. Maxim.

Lindera membranacea, Max. l. c. 275. Prov. Senano. Tchon., 1864.

Daphne yezoensis, Max. in Grtfl. 1866, 34, t. 496. Maxim. Hakodate.

Urtica laetevirens, Max. Mél. IX (1876) 620. Hakodate.

Achudenia japonica, Maxim. Mél. l. c. 627. Kiu siu: in monte Naga, Nippon med: in monte Hakone.

Juglans cordiformis, Max. Mél. VIII (1872) 635. Yokohama, Hakodate. Ibidem, 655 = *J. mandshurica,* variety.

Betula Maximowicziana*), Regel, Bull. Nat. Mosc. I, 1865, II, 418, tab. 8; DC. Prodr. XVI, II, 180. Insula Yezo. — This beautiful tree was met with in the same island, in 1892, by Mr. C. S. Sargent and introduced into the Arnold Arboretum, Mass. See Garden and Forest, 1893, 343.

Betula corylifolia, Regel, DC. l. c., 178. Nippon, Tchon.

Carpinus yedoensis, Max. Mél. XI (1881) 314. Yedo cult. Maxim.

Carpinus Tchonoskii, Max. l. c. 313. Hakone, Fudzi yama. Tchon., 1864.

Fagus japonica, Maxim. Mél. XII (1886) 542. Nippon: Hakone et prov. Nambu. Tchonoski.

Thuja japonica, Max. Mél. VI (1866) 26. Yedo, culta.

Chamaecyparis breviramea, Maxim. l. c. 25. Yedo, culta.

Chamaecyparis pendula, Max. l. c. 26. Yedo, culta.

Juniperus nipponica, Maxim. Mél. VI (1867) 374. Nippon bor.: in alpibus prov. Nambu.

Juniperus litoralis, Maxim. l. c. 375. Hakodate, Nambu etc. Wright earlier (v. s. p. 398). — Max. XI (1881) 349 = *Juniperus conferta,* Parl. in Enum. Sem. Hti Florent., 1862, nomen antiquius et praeferendum.

Podocarpus caesia, Max. Mél. VII (1870) 561. Nagasaki, culta.

Podocarpus appressa, Maxim. l. c. 561. Yedo, culta.

Abies bicolor, Max. Mél. VI (1866) 24. Nippon alpibus, Fudzi yama. Same as *Abies Alcoquiana* (v. s. p. 552).

Abies brachyphylla, Maxim. l. c. 23. Nippon montibus altissimis, Fudzi yama.

Abies diversifolia, Maxim. Mél. VI (1867) 373. Nippon, alpinis; Kiu siu: in monte Naga.

Calanthe reflexa, Max. Mél. VIII (1872) 644. In monte Naga prope Nagasaki, vulc. Wunzen. Yedo, culta.

Goodyera velutina, Max. in Gartenfl. 1867, 36, t. 533, fig. 1. Japan.

Goodyera macrantha, Max. l. c. fig. 2. Japan.

Yoania (nov. gen. Orchid.) *japonica,* Maxim. l. c. 647. Nippon media. Tchonoski, 1864.

---

*) Not to be confounded with *Betula Maximowiczii,* Rupr. Mél., II (1856) 485, which Maximowicz brought from the Amur. It is now considered a variety of *B. daurica,* Pall.

Platanthera interrupta, Maxim. Mél. XII (1886) 550. Nagasaki and other places in Kiu siu. Maxim.

Lycoris squamigera, Max. Amaryll. (1884) 79: Kiu siu: prov. Simabara. — Siebold earlier.

Lycoris sanguinea, Max. l. c. 80. Yokohama. Siebold earlier.

Smilax stans, Max. Mél. VIII (1871) 407. Nippon: prov. Nambu, Tchon.

Allium japonicum, Regel, Monogr. Allium, Act. Hti Petrop. III (1875) 133. Maxim. Japan.

Allium Tchonoskianum, Rgl. l. c. 160. Tchonoski, Nippon.

Lilium Maximowiczii, Regel, Gartenfl. 1868, 322, tab. 596 — and 1870, 290, tab. 664, var. *tigrinum.* Japan.

Narthecium asiaticum, Maxim. Mél. VI (1867) 214. Nippon borealis principatu Nambu.

Metanarthecium (nov. gen. Melanthac.) *luteo-viride,* Maxim. Mél. VI (1867) 213. Japonia, frequens.

Metanarthecium foliatum, Maxim. Decas Plant. nov. 1882, Montes Hakone. Tchonoski.

Chionographis (nov. gen. Melanthac.) *japonica,* Max. Mél. VI (1867) 210. Kiu siu: Kundsho san, Kumamoto. Probably Thunberg's *Melanthium luteum.* Comp. Bot. Mag. tab. 6510 (1880). Raised in Kew from seeds sent by Maries.

Heloniopsis breviscapa, Max. l. c. 211. Kiu siu, vulc. Wunzen.

Tofieldia sordida, Max. l. c. 212. Yedo, culta.

Tofieldia nuda, Maxim. Mél. VIII (1871) 416. Nippon: prov. Owari, Yukumayu ssai (Jap. botanist).

Tricyrtis flava, Max. Mél. VI (1867) 208. Yedo culta.

Tricyrtis latifolia, Max. l. c. 208. Nippon borealis: principatu Nambu. Tchonoski.

Trillium Tchonoskii, Max. Mél. XI (1883) 863. Nippon, alpe Nikko. Tchonoski.

Veratrum Maximowiczii, Baker, Linn. Soc. Journ. XVII (1879) 472. Japan, Tchonoski.

Veratrum stamineum, Maxim. Mél. VIII (1870) 339. Jugo Hakone. Tchonoski, 1866.

Najas serristipula, Max. Mél. VI (1867) 275. Yokohama. Maxim.

Eriocaulon decemflorum, Maxim. Diagn. Pl. nov. Asiat. VIII (1892) 7. Nippon media. Tchon. 1866.

Eriocaulon kiusianum, Max. l. c. 22. Kiu siu, in principatu Simabara, Sept. 20, 1863. Maxim.

Scirpus fuirenoides, Maximowicz, Mél. XII (1886) 555. Yokohama. Maximowicz.

Carex grallatoria, Max. l. c. 560. Kiu siu: Kundsho san. Maxim.

Carex rhizopoda, Max. l. c. 561. Yezo, Kiu siu: vulc. Wuuzen. Max.

Carex scita, Max. l. c. 564. In montibus Hakone. Tchou.

Carex plocamostyla, Max. l. c. 565. Cum praecedente. Tchon.

Bambusa aureo-striata, Regel, Gartenfl. 1865, 362, t. 490, fig. 3, 4. A pretty small Bamboo with large leaves; yellow stripes. Maxim. brought living specimens from Japan.

Bambusa argenteo-striata, Regel, l. c. 363, tab. 490, fig. 5. Maxim. living plants from Japan. — Same as *Bambusa Fortunei*, or *variegata*, Fl. d. Serres, t. 1535 (v. s. 516). Known also as *B. Maximowiczii*.

Lycopodium cryptomerium, Maxim. Mél. VII (1870) 340. Kiu siu: in monte Naga, in jugo Kundsho san. Maxim. Yedo cultum.

Nephrodium Maximowiczii, Baker, Syn. Fil. 2d edition (1873) 499. Aspidium sanctum. Maxim. MSS. Yokohama, Maxim.

Polypodium Maximowiczii, Baker, l. c. 504. Nagasaki.

List of Plants introduced by Maximowicz from the Amur, Manchuria, Japan, into the Botanical Gardens, St. Petersburg,

This list has been extracted from the General Catalogue of Plants which have flowered in the Botan. Garden and from Regel's Gartenflora, where most of the plants introduced by Maximowicz are figured. Regel, in Gartenfl. 1891, 148, gives a list (incomplete) of the plants for the introduction of which to our gardens we are indebted to Maximowicz.

Clematis fusca, Turcz. var. *violacea*, Max. Amur, 10. — Gartenflora, 1864, 355, tab. 455.

Clematis paniculata, Thbg. Raised from seeds sent by Maxim. from Nagasaki. Flowered Oct. 1866.

Clematis japonica, Thunberg. Raised from seeds sent 1862 from Yokohama.

Clematis apiifolia, DC. Raised from seeds sent from Japan.

Clematis stans, Sieb. & Zucc. From seeds sent from Japan. Gartenfl. 1870, 203, tab. 657.

Illicum religiosum, Sieb. — Maxim. brought living plants from Japan, which flowered at St. Petersburg.

Trochodendron aralioides, Sieb. & Zucc. Fl. Jap. I, 83, tab. 39, 40. Introduced by M. from Japan in 1864. Flowered 1876.

Magnolia stellata, Maxim. Japan.

Maximowiczia chinensis, Rupr. From seeds sent by Maximowicz and Maack from the Amur. Flowered and produced fruit. Gartenfl. 1862, 406, tab. 382. — Fl. d. Serres, XV (1863) 173. Col. figure.

Stephania hernandifolia, Walp. From Japan.

Berberis Maximowiczii, Regel, Gartenfl. 1872, 238. Japan.

Berberis Thunbergii, DC. var. β *Maximowiczii,* l. c. 178. From seeds brought by Maximowicz. — Index Kew, both = *B. vulgaris,* L.

Berberis vulgaris, L. var. *amurensis,* Regel, Gartenflora, 1874, 176. From seeds sent by Maximowicz.

Epimedium Musschianum, Morr. & Decne. Max. brought living plants from Japan, 1864. Flowered 1866.

Chelidonium uniflorum, Sieb. & Zucc. (*Hylomecon vernalis,* Max. Amur, v. s. p. 583), Gartenflora, 1862, 88, tab. 355. Maxim. and Maack from the Amur.

Corydalis speciosa, Max. Amur. Gartenflora, 1858, 250.

Arabis japonica, Regel & Herder in Gartenfl. 1863, 308, tab. 414. From seeds, Hakodate, Maxim.

Crataeva falcata, DC. Introd. by Max. from Japan, 1864.

Idesia polycarpa, Maxim. (v. s. p. 594). Introd. by Max. from Japan. Gartenfl. 1883, p. 22; 1890, 63, cum iconibus. The plant was exhibited in 1867 by Linden under the (garden) name *Polycarpea Maximowiczii.* — Rev. Hort. 1868, 330; 1872, 174; black and white figures.

Hypericum patulum, Thbg. — Gartenfl. 1866, 195, tab. 513. From seeds, Japan, Maxim.

Cleyera japonica, Thunberg.

Eurya japonica, Thbg.

Actinidia Kolomikta, Max. (v. s. p. 584). It first flowered in the Bot. Gard. St. Petersb., in 1869. Gartenfl. 1880, 184, figured. Most probably first introduced by Maxim. from the Amur. — It was also cultivated at Paris, Mus. d'Hist. nat. See Carrière in Rev. Hort. 1872, 395, black and white figure.

Actinidia polygama, Planch.

Zanthoxylum piperitum, DC. From seeds, Japan, Maxim. Flowered in 1871.

Zanthoxylum planispinum, Sieb. & Zucc. From seeds, Maxim. Japan. Flowered 1871.

Zanthoxylum shinifolium, Sieb. & Zucc. From seeds, Maxim. Japan. Flowered 1871.

Phellodendron amurense, Rupr. (v. s. 584). Introd. from the Amur.

Phellodendron japonicum, Maximowicz, 1863 seeds from Japan.

Ilex integra, Thbg. Maxim. 1864 living plant from Japan.

Ilex crenata, Thbg. Introd. 1864 with the preceding.

Euonymus japonicus, Thbg. var. *ovata.* Introduced from Japan.

Euonymus alatus, Thbg.

Celastrus articulatus, Thbg. From seeds, Max., Japan. Flowered 1871.

Orixa japonica, Thbg. Introduced from Japan, 1864. — Gartenflora, 1886, 541, tab. 1232. Flowered in Breslau.

Tripterygium Wilfordi, Benth. & Hook. Gen. Pl. I, 368. — Gartenfl. 1869, 105, t. 612. From seeds, Maxim., Kiu siu, 1863. — Max. Mél. XI (1881) 206. Flowered 1872.

Rhamnus costata, Maxim. (v. s. 594). From Japan.

Vitis vinifera, Lin., var. *amurensis,* Regel, *(Vitis amurensis,* Rupr.). Gartenfl. 1861, 312, tab. 339. Introduced Maxim., Amur.

Vitis Thunbergii, Sieb. & Zucc. — Gartenflora, 1864, 34, tab. 424. Maxim. sent it from Japan.

Vitis heterophylla, Thbg. var. *Maximowiczii,* Regel, Gartenfl. 1873, 197, tab. 765. Introduced by Maxim. from Japan.

Ampelopsis serjaniaefolia, Bunge. — Gartenflora, 1867, 3, tab. 531. Introduced by Maxim. from Japan.

Acer palmatum, Thbg. — Gartenfl. 1871, 210. Introduced by Siebold and Maxim. from Japan.

Acer japonicum, Thunberg.
Acer crataegifolium, Sieb. & Zucc.
Acer micranthum, S. & Z.
Acer rufinerve, S. & Z.
Acer Sieboldianum, Miq.
Acer trifidum, Thbg.

}   All these species introduced from Japan into the Bot. Gard. St. Petersburg.

Acer nikoense, Maxim. (v. s. 595) was introduced later, about 1890. See Gartenfl. 1892, 149.

Staphylea Bumalda, S. & Z. — Maxim. from Japan.

Rhus semialata, Murr. var. *Osbeckii.* Raised from seeds sent by Max. from Japan, 1862. Flowered 1876.

Glycyrrhiza pallidiflora, Maxim. (v. s. 565). Introd. from the Amur.

Lespedeza bicolor, Turcz. (v. supra p. 175). — Gartenfl. 1858, 309; 1860, 268, tab. 299. From seeds sent by Maxim. from Amur. Flowered 1858, 1859.

Euchresta japonica, Bentham (1867) was described from specimens discovered near Nagasaki by Oldham (see farther on) in 1862. — Maximowicz found it in Kiu siu, in 1863, and introduced living plants, which flowered in 1864. Gartenflora, 1865, 320, tab. 487.

Maackia amurensis, Rupr. (v. infra sub Maack). Introduced by Maxim. from the Amur. Gartenflora, 1875, 152, c. icone. — Belgique hort. 1870, tab. 18. — Rev. hort. 1880, 452. — *Cladratis amurensis*, Bot. Mag. (1881) tab. 6551.

Spiraea amurensis, Maxim. — Gartenfl. 1865, 324, tab. 489. Introd. from the Amur.

Spiraea digitata, Willd. var. *glabra*, Maxim. Amur. Introduced with the preceding.

Spiraea bullata, Max. Spirac. (1879) 100, Japan. — Gartenfl. 1886, 65, tab. 1215. Introduced, it seems, by Maxim. — See also Gartenflora, 1892, 565, c. icone.

Stephanandra flexuosa, S. & Z. From Japan.

Rhodotypus kerrioides, S. & Z. — Gartenflora, 1866, 130, tab. 505. Raised from seeds sent by Maxim. from Yedo, 1862. Flowered 1865.

Rubus crataegifolius, Bge. — Gartenfl. 1868, 259, t. 591 (flowers); ibidem, 1878, 1, tab. 924 (in fruit). Flowered May 1868. From Japan. seeds, Maxim. 1863.

Rubus phoenicolasius, Maxim. Mél. VIII (1871) 393, Japan. Maxim. brought seeds, from which the plant was raised in Paris. — Bot. Mag. (1880) tab. 6479.

Rosa rugosa, Thbg. var. *typica*, Gartenflora, 1881, 197, tab. 1049. From seeds sent from Amur and Japan.

Rosa rugosa, Thbg. var. *purpurea plena*, Gartenfl. 1875, 321, t. 846. From seeds sent from Japan.

Crataegus pinnatifida, Bge. — Gartenfl. 1862, 204, tab. 366. From seeds sent by Maxim. and Maack from the Amur.

Rhaphiolepis japonica, Sieb. & Zucc. Introduced from Japan 1862. Flowered 1876.

Amelanchier asiatica, (*Aronia asiatica*, S. & Z.). Maxim. introduced 1864, a living plant from Japan.

Astilbe Thunbergii, Miq. Japan. From seeds sent by Maxim. Flowered in 1869. See also Gartenfl. 1863, 34, tab. 389.

Rodgersia podophylla, Asa Gray, Japan. — Gartenflora, 1871, 355, tab. 708. Introduced Maxim. — Bot. Mag. tab. 1691 (1883). Raised by Messrs Veitch from seeds sent by Maries from Japan.

Hydrangea stellata, S. & Z., var. *prolifera*. — Gartenfl. 1866, 291, tab. 521. Living plants introduced from Japan.

Hydrangea japonica, S. & Z., var. δ *macrosepala.* — Gartenfl. 1866, 289, tab. 520. Introduced from Japan.

Hydrangea paniculata, Sieb. var. *floribunda.* — Gartenflora 1867, 2, tab. 530. Introduced from Japan.

Hydrangea Hortensia, var. *otaksa*, blue flowers. — Gartfl. 1866, 289.

Deutzia crenata, S. & Z. — Introduced from Japan.

Philadelphus tenuifolius, Rupr. — Introduced from Amur.

Platycrater arguta, S. & Z. — Gartenfl. 1866, 229, t. 516. Maxim. introduced living plants from Japan. Flowered 1865.

Cardiandra alternifolia, S. & Z. — Gartenflora 1865, 292, tab. 486. Living plants introduced from Japan 1863.

Sedum japonicum, Sieb. — Gartenflora 1866, 196, tab. 513. From seeds sent from Yokohama. Flowered 1865.

Sedum Maximowiczii, Regel, Gartenfl. 1866, 355, tab. 528. — From seeds sent from Japan.

Distylium racemosum, Sieb. & Zucc. Introduced from Japan 1864. Flowered 1867.

Corylopsis pauciflora, S. & Z. — Introduced from Japan.

Trichosanthes japonica, Regel, Gartenfl. 1872, 35, tab. 714. Living plants Maximowicz from Yokohama.

Mitrosycios lobatus, Max. Amur. — Introd. by Maxim. and Maack.

Hydrocotyle rotundifolia, Roxb.
Torilis japonica, DC. } From seeds sent from Japan.

Dimorphanthus mandshuricus, Rupr. (v. supra p. 585). Introduced from the Amur.

Kalopanax ricinifolium, Miq. Raised from seeds sent by Maxim. from Japan. This is the *Aralia Maximowiczii*, van Houtte, Fl. d. Serres, XX (1874) 39, tab. 2067, and the *Acanthopanax ricinifolium*, S. & Z.

Helvingia rusciflora, Willd.

Eleutherococcus senticosus, Maxim. — Gartenfl. 1863, 84, tab. 393. Raised from seeds sent by Maximowicz from Manchuria and Maack from the Ussuri.

Viburnum Sandankwa, Hassk. Intr. 1864 from Japan. Flowered 1871.

Viburnum burejanum, Herder (v. s. p. 585). — Gartenfl. 1862, 407, tab. 384. Maxim. Bureya Mts.

Lonicera chrysantha, Turcz. — Gartenfl. 1863, 213, t. 404. Probably introduced by Maxim. from the Amur.

Lonicera Maximowiczii, Rupr. — Gartenflora 1868, 322, tab. 597. Probably introduced by Maxim. from the Amur.

Lonicera Ruprechtiana, Regel, Gartenfl. 1870, 68, tab. 645 (v. supra p. 586). Introduced by Maxim. or Maack.

Diervilla hortensis, Sieb. & Zucc. From seeds sent 1862 from Japan. Flowered in 1865.

Gardenia Maruba, Sieb. — Gartenflora 1866, tab. 494. Introduced from Japan.

Damnacanthus major. S. & Z. var. *submitis*, Max. — Gartenfl. 1868, 35, tab. 570. Living plants 1864 from Japan.

Patrinia villosa, S. & Z. From seeds sent from Japan. Flowered 1866.

Eupatorium japonicum, Thbg. From seeds sent from Yokohama.

Heteropappus decipiens, Maxim. (v. s. p. 586). — Gartenfl. 1864, 35, tab. 425. Introduced from the Amur.

Heteropappus hispidus, Less. From seeds sent 1863 from Nagasaki.

Boltonia indica, Benth. From seeds sent from Japan.

Emilia sonchifolia, DC. From seeds sent 1863 from Nagasaki. Flowered 1867.

Senecio flammeus, Turcz. — Gartenfl. 1863, 113, tab. 394. Introd. from the Amur.

Spanioptilon lineare, Less. From seeds sent 1863 from Nagasaki. Flowered 1865.

Saussurea japonica, DC. From seeds sent 1863 from Japan.

Lampsana parviflora, A. Gray. From seeds sent 1862 from Yokohama. Flowered 1866.

Lactuca amurensis, Regel. Introduced from the Amur.

Platycodon grandiflorum, DC. From seeds sent from Japan.

Glossocomia ussuriensis, Max. et Rupr. Introduced from the Amur.

Adenophora marsupiflora, Fischer, var. *dentata*, Regel. From seeds sent from the Amur.

Andromeda campanulata, Miq. From seeds sent from Hakodate.

Rhododendron linearifolium, S. & Z. Introduced from Japan 1864.

Rhododendron macrosepalum, Maximowicz (v. supra p. 598). Living specimens introduced from Japan. Also from seeds. Gartenflora 1870, 258, tab. 662.

Rhododendron semibarbatum, Max. — Gartenfl. 1870, 292, tab. 666. From seeds sent 1865 by Tchonoski.

Rhododendron brachycarpum, G. Don, Maxim. Rhododend. (1870) 22. Living plants introduced 1864 from Japan.

Rhododendron rhombicum, Miq. — Gartenflora 1868, 225, tab. 586; Maxim. Rhodod. 26. Introduced from Japan 1864.

Rhododendron sublanceolatum, Miq. — Max. Rhodod. 35. Introduced from Japan, 1864.

Clethra barbinervis, S. & Z. — Gartenflora 1870, 163, tab. 654. — From seeds sent by Maximowicz and Tchonoski.

Lysimachia lubinoides, S. & Z. From seeds sent from Hakodate.

Lysimachia japonica, Thbg. From seeds sent from Japan.

Lysimachia clethroides, Duby. — Gartenflora 1869, 163, tab. 618. Introduced from Japan.

Ardisia hortorum, Maxim. in Gartenfl. 1865, 363, tab. 491. Living plant introduced from Japan.

Diospyros Kaki, Lin.

Styrax japonica, S. & Z. — Gartfl. 1868, 193, tab. 583. Introduced from Japan. Flowered 1867.

Pterostyrax hispidum, Sieb. & Zucc.

Fraxinus mandshurica, Rupr. — Gartenflora 1879, 13. Introduced from the Amur.

Fraxinus longicuspis, Sieb. & Zucc. — Gartenfl. l. c. From Japan.

Olea aquifolia, S. & Z. foliis variegatis. Introduced 1864 from Japan. Flowered 1887.

Ligustrum japonicum, Thbg. var. *rotundifolium*, Bl. Introduced 1864 from Japan under the name of *L. ovalifolium*. Produced fruit 1883.

Parechites Thunbergii, Asa Gray, foliis variegatis.

Vincetoxicum purpurascens, Morr. & Decainse. Introduced 1864 from Japan. Flowered 1869.

Buddleya Lindleyana, Fortune. From seeds sent from Japan.

Solanum dulcamara, Lin. var. *macrocarpa*, Max. From seeds sent from Yezo, 1861. Flowered 1869.

Mazus rugosus, Lour. Introduced from the Amur and Japan.

Clerodendron trichotomum, Thbg.

Plectranthus glaucocalyx, Max. Introduced from Manchuria.

Hedeoma micrantha, Regel in Gartenfl. 1864, 357. From seeds sent from Yokohama.

Calamintha chinensis, Benth.

Salvia japonica, Thbg. From seeds sent from Japan.

Phlomis umbrosa, Turcz. From seeds sent from Manchuria.

Teloxys aristata, Moq. Tand. From seeds sent from the Amur.

Phytolacca acinosa, Roxb. From seeds sent from Japan.

Polygonum sachalinense, Fr. Schmidt. — Gartenfl. 1864, 68, t. 429. Probably introduced by Maxim. — Bot. Mag. tab. 6540 (1881). Sir Jos.

Hooker thinks that the plant was introduced into England by Wilford or Oldham, 1858 or 1863.

Polygonum Maximowiczii, Regel (v. s. 599). — Gartenfl. 1865, 99, tab. 468. Introduced from Yokohama.

Asarum albivenium, Rgl. (v. s. p. 599). Introduced from Japan.

Asarum variegatum, Al. Br. Living specimens 1863 from Japan.

Saururus Loureiri, Dcne. — Gartenfl. 1873, 1000, tab. 756. Introd. from Japan, probably Maxim.

Piper futocadsura, Sieb. & Zucc.

Cinnamomum pedunculatum, N. ab Es. ⎫
Cinnamomum sericeum, Sieb. ⎬ Introduced from Japan.
Cinnamomum brevifolium, Miq. ⎭

Lindera hypoglauca, Max. (v. s. p. 599). Living specimens introduced from Japan, 1864. Flowered 1867.

Helicia lancifolia, Sieb. & Zucc.

Daphne yezoensis, Max. (v. s. p. 600). Introduced from Hakodate.

Daphne Genkwa, Sieb. & Zucc. — Gartenflora 1866, 65, tab. 499. Living plants introduced from Japan.

Elaeagnus macrophylla, Thunberg. Introduced from Japan 1863. Flowered, 1867.

Elaeagnus pungens, Thbg, foliis variegatis. Introduced living plants from Japan, 1864.

Elaeagnus longipes, Asa Gray.

Phyllanthus ussuriensis, Rupr. et Maxim. (v. s. p. 587). From seeds sent from the Ussuri.

Geblera suffruticosa, Fischer et Meyer. From seeds from the Amur and Japan.

Acalypha gemina, Müll.

Juglans mandshurica, Max. From seeds sent from the Amur.

Quercus gilva, Bl. ⎫ Introd. living specimens from Japan 1864.
     » *serrata,* Thbg. ⎭

     » *lacera,* Bl. ⎫ Introd. living specimens from Yedo 1864.
     » *glauca,* Thbg. ⎭

     » *cuspidata,* Thbg. ⎫
     » *crispula,* Bl. ⎪
     » *thalassica,* Hance. ⎪
     » *glabra,* Thbg. ⎬ Introduced from Japan.
     » *acuta,* Thbg. ⎪
     » *lamellosa,* Smith. ⎭

Quercus mongolica, Fischer. Introduced from Manchuria.

Thuyopsis laetevirens, Lindl. Living specimens introd. from Japan.

Thuyopsis dolabrata, S. & Z.

Chamaecyparis pisifera, Sieb. & Zucc. From seeds sent from Japan, 1863.

Chamaecyparis obtusa, S. & Z. — Gartenflora 1876, 19. From seeds sent 1863 from Japan.

Chamaecyparis breviramea, Maxim. (v. s. p. 600). Living specimens introduced 1864 from Japan.

Juniperus rigida, S. & Z. var. *filiformis,* Maxim. Living specimens introduced 1864 from Japan. Flowered 1867.

Juniperus chinensis, Lin. var. *procumbens,* Sieb. Living plants introd. 1864 from Japan.

Juniperus conferta, Parl. (v. s. 600).

Podocarpus macrophylla, Don. var. *angustifolia,* Miq. Introd. 1864 from Japan.

Podocarpus chinensis, Wall. Living plants introduced from Japan.

Podocarpus appressa, Max. ⎱ V. s. 600. Living specimens introd.
Podocarpus caesia, Max. ⎰ from Japan 1863, 1864.

Pinus Massoniana, Lamb.

Pinus parviflora, S. & Z.

Abies diversifolia, Max. ⎱ V. s. p. 600. Living plants introduced
 »   *brachyphylla,* Max. ⎰ from Japan.
 »   *firma,* Sieb. & Zucc. ⎱
 »   *jesoensis,* S. & Z.   ⎪
 »   *Tsuga,* S. & Z.   ⎬ Introduced from Japan.
 »   *Alcockiana,* Parlat.   ⎭

Calanthe Sieboldi, Dcne, Rev. Hort. 1855, 381, tab. 20. — Gartenfl. 1869, 354, tab. 635. Living plants introduced from Japan.

Goodyera velutina, Maxim. (v. s. 600). Introduced from Japan, 1864. Flowered 1866.

Goodyera macrantha, Maximowicz (v. supra p. 600). Introduced with the preceding.

Iris tectorum, Maxim. Mél. VII (1870) 563. — Gartenfl. 1872, 65, tab. 716. Introduced from Japan.

Pardanthus dichotomus, Led. — Gartenfl. 1858, 309. From seeds received from the Amur.

Smilax biflora, Sieb. — Max. Mél. VIII (1871) 410. Introduced from Yedo, 1864. Flowered 1865.

Rohdea japonica, Roth, var. *aureo-variegata,* and var. *macrophylla,* foliis arg. marginatis. From Japan.

Hemerocallis fulva, Lin. var. *longituba,* Maxim. in Gartenfl. 1885, 98, tab. 1187. Introduced from Japan, 1864.

Lilium pulchellum, Fischer, Ind. Sem. VI (1835) 15, Turczaninov, Dauria. — Gartenfl. 1860, 81, tab. 284, fig. 2. Corolla red, spotted. From seeds sent from the Amur.

Lilium tenuifolium, Fischer Hort. Gorenki, 1812. Siberia, Amur. — Gartenfl. l. c. 80, t. 284, 1. Corolla red, not spotted. — From seeds sent from the Amur.

Lilium avenaceum, Fischer MS. Schmidt Fl. sachalin., 186. — Maxim. in Gartenfl. 1865, 290, tab. 485. Maxim. brought bulbs from Yeso, 1864. Flowered 1865.

Lilium Maximowiczii, Rgl. (v. s. 601). From bulbs sent from Japan.

Lilium auratum, Lindl. — Grtfl. 1873, 281. Bulbs introd. from Japan.

Maxim. had formed a great collection of bulbs of Japanese liliaceous plants. Unfortunately they were eaten by some pigs.

Tricyrtis hirta, Hooker, var. *japonica,* Miq. Introduced from Japan.

Tricyrtis macropoda, Miq. var. *crispa.* With the preceding.

Carex leucochlora, Bge. From seeds sent from Hakodate, 1861.

Imperata sacchariflora, Max. — Gartenfl. 1892, 92, tab. 357. From seeds sent from the Amur.

Spodiopogon sibericum, Trin. — From seeds sent from the Amur.

Hordeum hexastichon, Lin. var. *mandshuricum.* — Gartenfl. 1858, 45 and 1860, 156. — Maxim. sent seeds from the Amur. An excellent Barley, which has been successfully cultivated in several parts of Russia.

Bambusa aureo-striata, Regel. ⎫
Bambusa argenteo-striata, Rgl. ⎭ V. s. p. 602.

Bambusa viridi-striata, Sieb. MS. ex Illustr. Hort. XIX (1872) 319, tab. 108. Introduced from Japan by Siebold and Maximowicz.

Asplenium prolongatum, Hook.

Polypodium Lingua, Sw.

Polypodium decursivo-pinnatum, Hook.

Pteris japonica, Mett.

R. MAACK'S BOTANICAL EXPLORATIONS OF THE AMUR AND USSURI, 1855 and 1859.

General Muraviov's memorable opening of the navigation on the Amur, in 1854, was followed by several exploring expeditions on the great river and its affluents. To one of these expeditions, in 1855, R. Maack, a young naturalist, was attached.

Maack, Richard, born 1825, at Arensburg, in the Island of Oesel, studied natural sciences at the St. Petersburg University, took his degree of Candidate, in 1849, and in 1852 was appointed Professor of Natural Sciences at the Gymnasium of Irkutsk. Subsequently he became Director of that Gymnasium, and from 1868 to 1879, he was Superintendent of all schools in Eastern Siberia. He died at St. Petersburg, November 13, 1886.

Maack described his first expedition down the Amur and back in a book entitled: JOURNEY ON THE AMUR, IN 1855 (in Russian), published in 1859, accompanied with an Atlas containing maps, views and drawings of plants.

The expedition left Irkutsk in April 1855, and proceeded by the ordinary way to Nerchinsk. Here, at the discharging of the Nercha into the Shilka, they found a great raft prepared for them, on which they embarked on the 5th of May. Albazin, May 26, stay till 31st. — On August 8, the expedition arrived at the post Marinsk, near the Kidzi Lake and remained there till August 14. Then back up the Amur River, reached Aigun October 11, spent a month there. On November 12, started on horse back, for the Amur was frozen, following the river valley. Ust Strelka, December 30, Irkutsk January 16, 1856.

As on this river journey frequent stops were made, sometimes for several days, Maack had a favourable opportunity of making botanical and zoological collections. The plants gathered by him in the Amur valley, in 1855, were determined and described by Maximowicz in his Primitiae Florae Amurensis.

Maack had but just finished his account of the Amur journey, when he was commissioned by the Siberian Branch of the Imp. Geographical Society, to explore the valley of the Ussuri. This river journey was performed in 1859, and in 1861, Maack published a book with the title: JOURNEY IN THE VALLEY OF THE USSURI. The first volume (in Russian) gives an account of the journey, geological and zoological observations; the second (in German) is devoted to the Ussuri Flora (see farther on).

The traveller left St. Petersburg, February 16, 1859, and proceeded to Irkutsk, where he was joined by Mr. Brylkin, who took upon himself the ethnographical part of the exploration. They departed April 26, from Chita, sailed down the Ingoda and the Shilka rivers, reached, May 6, the Amur at Ust Strelka. Three days later they were at Albazin. On June 5, arrived at the mouth of the Ussuri. Khabarovka visited.

On June 16, they began to sail up the Ussuri. — June 28, mouth of the Nor River, — July 15, mouth of the Ima River, — Village (Russian settlement) Busseva, — July 25, mouth of the Sungachi Riv. Then they sailed up this river to the Keng ka (Hing ka) Lake, which was explored in a boat. August 28, they were back to the Ussuri, and, sailing down the river, arrived at Khabarovka. On Sept. 28, the expedition took passage in a steamer which went up the Amur. — Oct. 6, Blagoveshchensk. From this place they continued on horseback, arrived at Irkutsk in the beginning of January 1860, and reached St. Petersburg March 16.

As the river journey was frequently interrupted, and they sometimes went on shore for several days, M. was at leisure to botanize or collect zoological specimens.

New Plants discovered in 1855 and 1859 by Maack in the Amur and Ussuri valleys.

F. J. Ruprecht, DIE ERSTEN BOTANISCHEN NACHRICHTEN ÜBER DAS AMURLAND. Zweite Abtheilung: Bäume und Sträucher beobachtet by R. Maack*). Mél. biol. II (1857) 513—68.

E. Regel, TENTAMEN FLORAE USSURIENSIS, nach den von R. Maack (1859) gesammelten Pflanzen 1861. Accompanied with 12 plates. — Enumerated 643 species.

Clematis mandshurica, Rupr. in Maack, Mél. II 1857, 514. — Max. Amur, 10. Maack and Maxim. Amur, Ussuri. — Regel, Fl. Ussur. 1: Cultivated in Botan. Gard. St. Petersburg from seeds sent by Maxim. and Maack. Max. Mél. IX (1876) 594 = *Cl. recta*, Lin. var. *mandshurica*. — Earlier Staunton.

Delphinium Maackianum, Rgl. Fl. Ussur. 9. Sungachi. Grtfl. 1861, tab. 344. Cult. from seeds sent by Maack.

Berberis amurensis, Rupr. in Maack l. c. 517. — Maxim. Amur, 33. Maack, Maxim. Amur.

Nymphaea Wenzelii**), Maxim. Amur, 460. Maack, Lower Amur. Figured in Atlas appended to Maack, Amur.

Euonymus Maackii, Rupr. in Maack l. c. 532. — Max. Amur, 75.

Euonymus macropterus, Ruprecht in Maack, Mél. II (1857) 533; Maxim. Amur, 75. Discovered by Maack and Maximowicz at about the same time, in 1855, on the lower Amur. Also found in Japan, from which country the tree seems to have been introduced to the Arnold Arboretum

---

*) Many of Maack's Amur plants were gathered at about the same time by Maximowicz.
**) In honour of Mr. Wenzel, then Governor of Irkutsk.

by Mr. C. Sargent in 1892. Cultivated at Kew in 1896. See Kew Bull. 1897, App. I, 36.

Celastrus flagellaris, Rupr. in Maack, l. c. 531. — Max. Amur, 76. Amur near mouth of Sungari. Figured in Atlas, Maack, Amur.

Vitis amurensis, Rupr. in Maack, l. c. 524. — Maxim. Amur, 69. — I. F. S., I, 136 = a variety of *V. vinifera*, Lin.

Glycine ussuriensis, Rgl. Ussur. 50, tab. 7. — Max. Mél. IX (1873) 70 = *Glycine Soja*, Sieb. & Zucc. (1843). Siebold, Tatarinov earlier. Japan, Peking. — I. F. S., I, 189.

Maackia (nov. gen. Legumin.) *amurensis*, Rupr. in Maxim. Mél. II (1858) 440, 418, c. icone; Rupr. in Maack, Mél. II (1857) 534. — Max. Amur, 87, tab. 5. — Benth. et Hook. G. Pl. I, 154, reduce Maackia to *Cladastris*, Rafin. Introduced by Maximowicz. (V. s. p. 605).

Prunus (Padus) *Maackii*, Rupr. in Maack, Mél. II (1857) 536. — Maxim. Amur, 88.

Crataegus chlorosarca, Maxim. in Fl. Asiat. Fragm. (1879) 20. Cultivated in Buek's garden, St. Petersburg from. seeds received many years ago from the Ussuri under the name of *Crataegus mandshurica*. (Perhaps Maack?).

Sedum Selskianum, Rgl. Fl. Ussur., 66, t. 6, Maack, Kengka Lake.

Penthorum humile, Rgl. l. c. 65, tab. 6. Sungachi.

Myriophyllum ussuriense, Maxim. Mél. IX (1873) 183. In Regel Fl. Ussur. 60, t. 4, noticed as a variety of *M. verticillatum*, Lin. Kengka Lake.

Lonicera Ruprechtiana, Regel, Gartenfl., 1870, 68, tab. 645. First published by Rupr. in Maack, Mél. II (1857) 548, as *Xylosteum gibbiflorum*, var. *subtomentosum*, Maack, mouth of Sungari. — I. F. S., I, 366.

Lonicera Maackii, Max. Mél. X (1877) 66. First noticed by Rupr. in Maack, Mél. II (1857) 548, as *Xylosteum Maackii*. — Maxim. Amur, 136. Figured in Atlas, Maack, Amur. Earlier discovered by Fortune (v. supra p. 465). — Gartenfl. 1884, 225, tab. 1162. Flowered, in 1883, in Bot. Garden, St. Petersburg.

Aster Maackii, Regel, Fl. Ussur., 81, tab. 4. — Sungachi, Rev.

Galatella Meyendorfii, Regel, l. c. 81, tab. 5. — Kengka Lake. — I. F. S., I, 412 = *Aster hispidus*, Thbg.

Artemisia sylvatica, Max. Amur, 161. — Mouth of the Ussuri, Maack.

Cirsium Maackii, Maxim. Amur, 172. Maack, Amur. Max. Mél. IX (1874) 322 = *Cnicus japonicus*, DC. Prod. VI, 640. — I. F. S., I, 461.

Syringa (Ligustrina) amurensis, Ruprecht, in Maack, Mél. II (1857) 551. Maxim. Amur, 193. Maxim. and Maack both in 1855, on the Amur.

Earlier Kirilov, Tatarinov, Peking. Figured in Atlas, Maack, Amur. — I. F. S., II, 82.

Eritrichium Maackii, Max. Amur, 202. Maack, Upper, Amur. Max. Mél. VIII (1872) 546 = *Eritrichium pectinatum*, DC. (1846, Dahuria). — I. F. S., II, 150.

Calamintha ussuriensis, Regel, Fl. Ussur. 126, tab. 10. Sungachi, Kengka Lake.

Polygonum Maackianum, Rgl, l. c. 127, t. 10. Sungachi, Kenga Lake.

Chloranthus mandshuricus, Rupr. in Maack, Amur. Figured. Maxim. As. or. Fl. Fragm. 1879, 57 = *C. japonicus*, Sieb. — I. F. S., II, 368.

Geblera sungarensis, Rupr. in Maack, Mél. II (1857) 530. Maack, mouth of the Sungari. — Max. Amur, 240, var. of *G. suffruticosa*, Fisch. and Mey., or *Securinega ramiflora*, Müll. Arg. — I. F. S., II, 426.

Betula Maackii, Rupr. in Maack, Mél. II (1857) 564. Maack, Amur.— Regel in DC. Prod. XVI, II, 174 = *B. daurica*, Pallas.

Betula ovalifolia, Rupr. in Maack, Mél. II (1857) 560. Maack, Lower Amur. — Regel l. c. 174.

Betula reticulata, Rupr. in Maack, l. c. 561. Amur. Regel, Fl. Ussur., 135 = *B. humilis*, Schrank (1789) var. *reticulata*.

Habenaria linearifolia, Max. Amur, 269. Amur, Maack, 1855, Max. 1856. Figured in Regel Fl. Ussur. tab. 10.

Platanthera hologlottis, Max. Amur, 268. Maack, mouth of Sungari.— Earlier known from Dauria (Herb. Fischer).

Platanthera ussuriensis, Max. Mél. XII (1886) 551. Noticed in Regel Fl. Ussur. 142, t. 10, as *Pl. tipuloides*, Lindl. var. *ussuriensis*.

Iris Maackii, Maxim. Mél. X (1880) 740. — In Regel. Flor. Ussur. 148, as *I. Pseud-Acorus*, Lin.

Asparagus oligoclonos, Max. Amur, 286. Maack, Upper, Amur, 1855.

Veratrum Maackii, Regel, Fl. Ussur. 154, tab. 11. Maack, Lower, Ussuri. — Gartenfl., 1885, 5, tab. 1070. Raised in the Bot. Garden St. Petersburg from seeds sent from the Amur by Gueldenstaedt.

Monochoria Korsakowii*), Regel and Maack in Fl. Ussur., 155, t. 12. Kengka Lake. — Tatarinov earlier, Peking (v. s. 568).

Potamogeton serrulatus, Regel, Fl. Ussur. 139. Sungachi.

Potamogeton cristatus, Regel, l. c. 139, t. 10. Middle Ussuri.

Eriocaulon ussuriense, Körnicke in Fl. Ussur. 157. Keng ka Lake.

Carex Maackii, Maxim. Amur, 308. Middle Amur.

---

*) Named in honour of M. S. Korsakov, Governor General of Eastern Siberia, after Muraviov.

Carex capricornis, Meinshausen in Max. Mél. XII, 1886, 569. Noticed in Regel, Fl. Ussur. 165, t. 12, as *C. pseudocyperus*, Lin., var. *barystachys*. *Pleopeltis ussuriensis*, Rgl. Fl. Ussur. 175. Not found in Syn. Fil. *Mnium ussuriense*, Rgl. Fl. Ussur. 182.

Plants introduced by Maack from the Amur and Ussuri.
Clematis aethusaefolia, Turcz. (v. s. 343), var. *latisecta*, Max. Amur, 12. — Gartenfl. 1861, 342, tab. 342. Raised from seeds sent by Maack from the Amur. — Carrière in Rev. Hort. 1869, 10, c. icone. — Bot. Mag. (1881) tab. 6542.
Clematis mandshurica, Rupr. (v. supra p. 613).
Delphinium Maackianum, Regel (v. supra p. 613).
Pyrus ussuriensis, Maxim. (v. s. 585). Gartenfl. 1861, 374, tab. 345. Raised from seeds sent by Maack from the Amur.
Deutzia parviflora, Bge, var. *amurensis*, Regel, Gartenfl. 1862, 278, tab. 370. From seeds sent by Maack from Ussuri.
Sedum Selskianum, Rgl. (v. s. 614). — Gartenfl. 1862, 169, t. 361. From seeds sent by Maack from Ussuri.
Panax sessiliflorum, Rupr. (v. supra p. 566). — Gartenfl. 1862, 238, tab. 369. From seed sent by Maack from Amur.
Lonicera Maackii, Maxim. (v. s. 614). — Grtfl. 1884, 225, t. 1162. Introd. by Maack. Flower. 1883. The handsomest of the East. Asiat. species.
Syringa amurensis, Rupr. (v. s. 614). — *Ligustrina amurensis*, Grtfl. 1863, 115, tab. 396. Raised from seeds sent by Maack from the Ussuri.— Rev. Hort. 1861, 353, c. icone, also 1877, 453.
Lilium pumilum, Redouté (1802). — Grtfl. 1865, 65, t. 463. Redouté says: "a native of Dahuria", but no other botanist had seen the plant, when Maack sent seeds from the Amur, from which it was raised, and flowered.
Monochoria Korsakowii, Regel (v. s. 615). — Gartenfl. 1862, 312, tab. 374. Maack sent seeds from Keng ka Lake.

Schrenck, Dr. Leopold von *), of German extraction, was born near Kharkov, in Russia. He studied natural sciences at Dorpat, 1844—47; Candidatus Zoologiae, 1848; completed his studies at Berlin, 1852, Dr. Philos. Koenigsberg. In 1853 he was elected Member of the Imp. Academy, St. Petersburg, and commissioned by the Academy to explore the Amur Country. Schrenck embarked on board the Frigate "Aurora" (v. s. p. 578) which got under sail from Kronstadt on August 12, 1853. After this vessel

---

*) Cousin of A. Schrenck, who travelled in Dsungaria, 1840—43.

had reached Petropavlovsk, Schrenck in another Russian ship proceeded to the mouth of the Amur, where he arrived in July 1854. He remained in the Amur Country about two years. Although his investigations were chiefly directed to zoology, geography and ethnography, he also found time to make botanical collections. As Maximowicz states in his Prim. Fl. Amur. p. 5, his friend L. von Schrenck handed over to him plants gathered: July 26, 1854, at Hadji Bay (v. s. 573) — in August 1854, near Nikolayevsk, — in the spring of 1855, during a journey up the Amur to Pulsa, 70 versts from Kidzi Lake, — in June of the same year near De Castries Bay, — in the summer of 1856, during his journey home up the Amur River from Nikolayevsk to Ust Strelka. Schrenck's Amur plants were noticed by Maximowicz in the Prim. Fl. Amur. Most of them had been gathered at about the same time by Maximowicz himself and Maack.

Schrenck first discovered near Hadji Bay the *Ribes horridum*, Rupr., in Maxim. Amur, 117. In 1872 it was found also in Sakhalin. Maxim. Mél. IX, 1873, 226.   •

The *Philadelphus Schrenckii*, Rupr. in Maack, Mél. II, 1857, 542, discovered 1856 by Schrenck on the Lower Amur; Max. Amur, 109, — was reduced by Max. Hydrang. 1867, 36, to the common *Ph. coronarius*, L.

After his returning from the Amur expedition, in January 1857, Schrenck spent the remaining portion of his life at St. Petersburg, in working up his materials. He published the results of his explorations in the region of the Amur with the title: Reisen und Forschungen im Amur-lande, 1854—56, of which the first volume appeared 1858—60, the second 1859—67, the third 1881—95, the fourth 1876—77.

Schrenck died at St. Petersburg January 8, 1894.

Orlov, Dmitri Ivanovich, a Russian naval officer entrusted, 1849—50, with the hydrographical survey of the mouth of the Amur and the adjacent coasts of the mainland and of Sakhalin (v. s. 573). In Maxim. Amur it is stated, that there is in the Herbarium of the Bot. Gard. St. Petersburg a small collection of plants gathered by Orlov near the mouth of the Amur and on the Tugur River (which discharges itself into the Tugur Bay, N. W. of the Amur). O. was thus the first to bring plants from the Amur.

Radde, Gustav. This distinguished naturalist, of German extraction, was commissioned by the Geogr. Society, St. Peterb. in 1855, to explore Eastern Siberia. He travelled, from 1855 to 59 in the southern part of Eastern Siberia and in 1857 went down the Amur, and during the summer of this and in the next year explored its valley as far down as the mouth of the Ussuri. But his headquarters were in the Bureya Mountains, which range, stretching from N. E. to S. W., reaches the Amur at about 131° E. L. Greenw. There is still a post on the river which bears Radde's name. Radde made zoological collections for the Zool. Museum of the Academy, and collected plants and seeds for the Bot. Gardens.

The PLANTAE RADDEANAE (about 1000 species, East. Siberia and Amur) were partly worked up by Dr. E. Regel and Dr. Fr. von Herder of the Bot. Garden, St. Petersb. Regel took upon himself to describe the Polypetalous Dicotyledons, of which he published in 1861 and 1862 in the Bull. Soc. nat. Mosc., Ranunculaceae to inclus. Caryophyllaceae, and then discontinued.

Dr. Herder worked up the Monopetalous Dicotyledons and brought his task to an end. His Plantae Raddeanae appeared from 1864 to 1892, first in the Bull. Soc. nat. Mosc. finally in Act. Hti Petrop., not only Monopetalae but also Apetalae and Gymnospermeae.

The following new plants were gathered by Radde, 1858, in the Bureya Mountains.

Anemone Raddeana, Regel l. c. n. 29, tab. 1.

Enemion Raddeanum, Regel l. c. n. 94, tab. 2. — Maxim. Mél. XI (1883) 639 = *Isopyrum Raddeanum*.

Aconitum Raddeanum, Regel l. c. n. 113, tab. 3.

Corydalis Raddeana, Regel l. c. n. 169.

Viola Raddeana, Regel l. c. n. 256, tab. 7.

Lactuca Raddeana, Max. Mél. IX (1874) 355.

Mr. Radde, since 1867, Director of the Museum of Natural History etc. at Tiflis, has contributed much towards the knowledge of the Fauna and the Flora of the Caucasus.

Weyrich, Dr. Heinrich, of German extraction, born in Livonia, 1828, studied medicine at Dorpat, 1846—51, M. D. He entered the Russian Navy as a surgeon, accompanied Admiral Putiatin's naval expedition to Eastern Asia, in 1852, was first attached to the "Pallas", but after the steamer "Fearless" (Vostok) had been bought in England (v. s. 574), he was transferred to that ship and remained on it down to the end of the expedition. W. profited by every stay of the vessel to botanize. He transmitted his plants to the Bot. Gard. St. Petersburg.

In the middle of August 1853, Admiral Putiatin despatched the "Vostok", from Nagasaki, to survey the Gulf of Tatary. The surveying cruise lasted till the end of October.

Soon after leaving Nagasaki, W. had an opportunity of visiting two of the islands of the Gotto Group, one of them larger and wooded, the other smaller and sparsely covered with shrubs. In the former he gathered.

Rhododendron Weyrichii, Maxim. Rhod. (1870) 26, t. 2, and note.

In September he botanized at Hadji Bay, and at Pronghe, near the mouth of the Amur, and visited several spots on the western coast of Sakhalin, and discovered there the following new plants:

Angelica sachalinensis, Max. Amur, 127. Weyrich, W. coast of Sakhalin, beginning of Sept. 1854 (evidently an error for 1853).

Eleutherococcus senticosus, Maxim. Amur, 132. Weyrich, Notosama, Sept. 1853. Earlier Kirilov, Peking, v. s. p. 349.

Leucanthum Weyrichii, Max. Amur, 156. Weyrich, Dui, W. coast of Sakhalin, beginning of Sept. 1853.

Polygonum sachalinense, Fr. Schmidt in Maxim. Amur, 233. Weyrich, Notosama (v. s. p. 575). Sept. 1853.

Polygonatum Maximowiczii, Fr. Schmidt, Flora Sachal. 185. Maxim. Amur. 275. Weyrich, Notosama, Sept. 1853.

When in the summer of 1854 the «Vostok» had resumed the surveying of the coast of Sakhalin (western and northern) W. discovered there the following novelties:

Cirsium (Cnicus) Weyrichii, Max. Amur. 174. Weyrich, south of Dui, end of July 1854.

Polygonum Weyrichii, Max. Amur. 234. Weyrich, with the preceding, July 1854. Cultivated in the Bot. Gard. St. Petersburg.

Maximowicz records altogether 27 plants, gathered by Weyrich in 1853 and 1854 on the coast of Sakhalin *).

In Saddle Island, where Putiatin's squadron anchored from November 11th to December 17, 1853, Weyrich gathered a few plants **) of which Maximowicz records two:

Viola japonica, Langsdorff. Maxim. Mél. IX (1876) 724.

---

*) For nearly half a century *Artemisia sachalinensis*, gathered by Tilesius on the eastern coast of Sakhalin in 1805 (v. s. p. 316) had been the only plant known from that island. — The Flora of Sakhalin was for the first time thoroughly explored, in 1861 and 1862 by Magister Fr. Schmidt and P. von Glehn, Candidatus Bot. These botanists visited also in 1861 the coast of Manchuria and explored, in 1862 the valleys of the Amgun and the Bureya, two important affluents of the Amur, on its left side. The account given by Schmidt of this expedition, published in 1868, is entitled Reisen im Amurlande und auf der Insel Sachalin. Schmidt, Member of the Academy, now lives in St. Petersburg. Glehn died in 1876.

In 1872, M. Mitzul, a Russian Civil officer gathered plants in Sakhalin for the Bot. Gard.

**) I may notice here a curious misapprehension with respect to a botanical collection from South and Middle China in the Herbarium of the Botan. Gardens of St. Petersburg, which the botanists of this Garden attributed to a Russian collector, Seniavin.

Narzissus Tazetta, Liu. var. *chinensis*, Roemer, Amaryll. — Maxim. Amaryll. (1884) 75.

Napa, the principal port of Great Liu kiu was visited by the «Vostok» in February 1854. Weyrich here first discovered the *Wikstroemia retusa*,

---

The late C. Maximowicz recorded in various volumes of Mém. Acad. and the Mélanges biologiques, since 1866, twelve Chinese plants from a collection, said to have been received by Dr. Fischer, till 1850, Director of that Garden, from Seniavin, viz:

Hypericum Seniavini, Maxim. Mél. XI, 169. E. China australiore, a Seniavin datum. Herb. Fischer.

Rubus parvifolius, Linn. In China media, Seniavin. Max. Mél. VIII, 392.

Photinia villosa, DC. In China australiore, Seniavin, in Herb. Fischer. Max. Mél. IX, 176.

Hydrangea hortensis, var. In China australi, a Seniavin lecta. Max. Hydrang. 14.

Hydrangea chinensis, Maxim. l. c. 7. China australis. Herb. Fischer.

Viburnum dilatatum, Thbg. β. *formosanum*, Max. Mél. X, (1880) 666. In China australiori, Seniavin. — I. F. S., I, 351, same as *V. erosum*, Thbg. var. *formosana*, Hce, Advers. 1864, 216.

Lonicera japonica, Thbg. In China australi. Max. Mél. X, 56.

Rhododendron indicum, Lin., var. *Simsii*, Maxim. Rhodod. 38, China australis.

Rhododendron Seniavini, Max. l. c. 33, fig. t. 3. In China australiori.

Rhododendron ellipticum, Max. Mél. XII, 743. China australior.

Didymocarpus lanuginosus, Wall. China australis. Max. IX, 368. — According to Clarke, DC. Monogr. V, I, (1883) 63, this is a new species, which he names *Oreocharis Maximowiczii*. — I. F. S., II, 226.

Premna microphylla, Turcz. China australis. Max. Mél. XII, 510.

Premna japonica, Miq. In China australiori a Seniavin lecta. Herb. Fischer. Maxim. Fl. As. or Fragm. 40.

In an old inventory of the various collections preserved in the Botan. Museum of the Bot. Gardens St. Petersb., Sept. 1856, I found noticed: «Nine bundles of Chinese plants gathered by Seniavin and offered to Fischer». Time not given, but evidently previous to 1850. Now these plants are incorporated with the great Herbarium of Eastern Asiatic plants (Japan, Manchuria, Amur, Mongolia, China, Tibet) which is kept apart from the General Herbarium, and it is quite impossible to find them out without looking through the whole herbarium.

Seniavin is certainly a Russian family name, but considering the time which this collection implies, the view that it was made by a Russian cannot be accepted. As we have related on a previous page (v. s. 314) the Russian Admiral Krusenstern, on his circumnavigation of the globe, in November 1805, visited Macao. But from that time down to 1853 (Admiral Putiatin's Embassy to Japan and China) no Russian ship had been seen in the waters of China, and also no Russian traveller is known to have visited Middle or South China. As I was led to suppose that the botanical collection attributed to Seniavin might have found its way to St. Petersburg through the Russian Ecclesiastical Mission at Peking, I asked Professor V. P. Vassiliev whether he was able to give any explanation of the matter. This learned sinologist, who from 1840 to 50 has lived in Peking with the mission, favoured me with a most convincing conjecture regarding the origin of this interesting collection. He remembers that about 1845 Archimandrite Polycarpos, then at the head of the mission despatched an intelligent Chinaman, who was in his service, to the celebrated tea plantations in the province of Fu kien to procure some of the best tea produced there, and also commissioned him to purchase, on his way back (he travelled evidently overland and by rivers), at Nan king, some ham, which place is famed for this article. The Fu kien tea thus received directly, perhaps from the Bohea Hills, western borders of Fu kien, the Archimandrite forwarded to St. Petersb. and offered it as a present to L. G. **Seniavin**, who from 1838 to 1851 was Director of the Asiatic Department in the Foreign Office. Professor Vassiliev thinks it very likely that Dr. Tatarinov, who then likewise lived in Peking, had given instructions to the China man to gather plants in the regions visited, and these plants may have been sent to Seniavin, together with the tea and were finally transmitted to Fischer.

Asa Gray in shed. in Max. Mél. XII (1886) 538. Ch. Wright gathered the same at the same place in November 1854.

Maxim. Mél. XI (1883) 831, records a plant, *Euphorbia Sparmanni*, Boiss., brought by Weyrich from the Island of Borodino (v. s. 575), which the "Vostok" visited in the beginning of Febr. 1854. — I. F. S., II, 417.

After the dispersing of Putiatin's squadron, Weyrich went to Irkutsk, where he was appointed Chief Physician to the Military Hospital. He died at St. Petersburg, in 1863.

Schlippenbach, Baron Alexander von, a native of Curland. He was a naval officer on the steamer "Vostok", but seems to have been afterwards transferred to the "Pallas", which in April and May 1854, surveyed the eastern coast of Corea (v. s. 576), on which a promontory was named in honour of Schlippenbach.

Maximowicz notices some of the plants brought by Sch. from Eastern Asia. They were almost all gathered on the Corean coast. There is one novelty: *Rhododendron Schlippenbachii*, Maxim. Rhododend. 1870, 29, c. icone. Schlippenb. Corea in litore orientali, 1853 (error for 1854). — See also *Rubus Oldhami*, sub Oldham.

Kuznetsov, Dmitry, a naval officer on the Frigate "Diana" (v. s. 577). He was present at her foundering, near Simoda, January 7, 1855.

Maximowicz records two new plants gathered by K. near Simoda, in 1854 and 1855, viz:

Cnicus suffultus, Maxim. Mél. IX (1874) 314.

Pertya ovata, Max. Mél. VIII (1871) 8.

After the destruction of the "Diana", K. was subsequently transferred to the Corvet "Olivutsa". In 1856 he collected plants for the Bot. Garden St. Petersburg at Yokohama. See Acta Hti Petrop. II, 250.

See also supra 585, *Pyrus ussuriensis*, Maxim.

Admiral Baron Schilling told me that Kuznetsov died many years ago.

Yolkin, Peter (Maxim. writes his name Jolkin, to be pronounced as in German). Naval officer on the "Diana". There are in the Herbarium of the Bot. Gard. St. Petersburg 619 species from Japan, gathered by Yolkin, chiefly in the neighbourhood of Simoda, in 1854 and 55. See also Acta Hti Petrop. II, 250.

Euphorbia Jolkini, Boissier, Cent. Euphorb. (1860) 32, tab. 71. — I. F. S., II, 215. — Yolkin died a long time ago.

Albrecht, Dr. Michael, born in 1821 in Esthonia, studied medicine at Dorpat, 1843—49, entered the Russian naval service as a surgeon, served first at Kronstadt. In 1858 he was appointed Physician to the new Russian Consulate at Hakodate, and in the autumn of the same year he departed with Consul Goshkewicz and the Staff of the Consulate. He spent five or six years in Hakodate, where he formed valuable botanical collections, in 1862, 1863, which are now preserved in the Museums of the Academy and the Bot. Garden, St. Petersburg. From these Maximowicz described the following novelties:

Cardamine yezoensis, Maxim. Mél. IX (1872) 5.

Cnicus yezoensis, Maxim. Mél. IX (1874) 328.

Rhododendron Albrechti, v. supra p. 598.

Juglans cordiformis, Max. Mél. VIII (1872) 636, c. icone. Albrecht sent fruits of it, which germinated in the Bot. Gard., St. Petersb. and the plants grew well. — Max. Mél. IX (1876) 655 = variety of *J. mandshurica.*

I remember, that in 1865, I met with Dr. Albrecht at St. Petersburg. He died some years later.

XIII. The Imperial Botanical Garden, St. Petersburg.

After Dr. F. E. L. von Fischer, the first Director of this garden resigned his post, in 1850 (v. s. p. 318), the Directorship of this important institution was conferred upon his assistant C. A. Meyer.

Meyer, Carl, Anton. He was born at Witebsk in 1795, studied Pharmacy in Dorpat, 1813—14, travelled with Professor Ledebour in the Altai Mts and the Dsungarian Steppe, 1826—27, explored the flora of the Caucasus from 1829—30, and in 1832, was appointed Assistant Director of the Bot. Garden, St. Petersburg. In 1844, he was elected Member of the Russian Academy. He died Febr. 13, 1855.

Of his numerous botanical writings we may notice the following:

Meyer assisted Professor Ledebour in the elaboration of the Flora Altaica, 4 vols, 1829—34.

Verzeichniss der Pflanzen, welche C. A. Meyer in den Jahren 1829 und 1830 in Kaukasus und am westlichen Ufer des Kaspischen Meeres gesammelt, 1831.

In 1835 appeared the first Index Seminum, quae Hortus botanicus Petropolitanus pro mutua commutatione offert. Accedunt animadversiones botanicae nonnullae auctoribus F. E. L. Fischer et C. A. Meyer.

These Indices, published annually by the Botan. Garden, were edited by Fischer and Meyer in conjunction, from 1835—44, and then by Meyer alone till 1853. They contain occasionally descriptions of new Chinese plants, raised from seeds received from Peking.

Beschreibung einer neuen Art der Gattung Catalpa aus China (C. Bungei), C. A. Meyer. Bull. scient. de l'Acad. des sc. de St. Pétersb., II (1837) 49.

Verzeichniss der im J. 1838 (von Politkov) am Sai sang Nor und am Irtysh gesammelten Pflanzen, 1841.

A. von Schrenck's plants collected 1841—43 in Dsungaria etc., were worked up and described by Fischer and Meyer.

C. A. Meyer, über den Ginshen und die Arten der Gattung Panax, 1842. V. supra p. 349.

Sertum Petropolitanum seu icones et descriptiones plantarum, quae in Horto Botanico petropolitano floruerunt. Decas I, auctoribus Fischer et Meyer, 1846; Decas II, auct. Meyer, 1852; Decas III et IV auct. Rgl, 1869.

C. A. Meyer, Bemerkungen über Diervilla, Weigelia, Calysphyrum und eine neue verwandte Gattung. Bull. phys. math de l'Acad. d. sc. de St. Pétersb. XIII (1855) 216.

After Meyer's death, E. **Regel,** Chief Gardener at the Botan. Garden, Zuric, was appointed his successor. Of this eminent Gardener and botanist we shall have to speak in another chapter.

Ruprecht, Franz Joseph, Austrian, born 1814, studied medicine in Prague, 1830—36, took his degree of M. D. in 1838. He was much attached to the study of botany and wrote some valuable papers on Gramineae. On the recommendation of C. von Trinius, Member of the Russian Academy, he was appointed, in 1839, Keeper of the Botan. Museum of the Academy. In 1851 he became Assistant Director of the Botanical Garden, and held this post till 1855, when he succeeded C. A. Meyer in the Directorship of the Botanical Museum of the Academy. From 1860—62 he travelled in the Caucasus and made interesting botanical collections. Ruprecht died in July 1870.

In the following of his numerous botanical papers descriptions of East. Asiatic and Central-Asiatic plants are found:

Illustrationes Algarum in itinere circa orbem jussu Imperatoris Nicolai I atque auspiciis navarchi Friderici Lütke a. 1826—29, celoce Seniavin executo in oceano Pacifico imprimis septemtrionali ad littora ros-

sica asiatico-americana collectarum. Auctoribus Prof. A. Postels et Doct.
F. Ruprecht, 1840 (Comp. supra p. 322). Beautiful drawings from nature
by Postels.

In Mém. Acad. 6 sér. sciences nat. IV, ɪ, 78, 79, Ruprecht described
the LICHENES gathered in 1838 by Politkov on the SAI SANG NOR and
the IRTISH.

DISTRIBUTIO CRYPTOGAMARUM VASCULARIUM IN IMPERIO ROSSICO,
Ruprecht, in Beiträge zur Pflanzenkunde des Russischen Reiches (published
by the Academy) III 1845, 1—56. Some new Cryptogams from North
China described.

Ruprecht, DIE ERSTEN BOTANISCHEN NACHRICHTEN ÜBER DAS AMUR-
LAND. Erste Abtheilung: Beobachtungen von C. MAXIMOWICZ. Mélanges
biol. of the Acad., II (1856) 407—74.

Ruprecht, DIE ERSTEN BOTAN. NACHRICHTEN ÜBER DAS AMURLAND.
Zweite Abtheilung: Bäume und Sträucher beobachtet von R. MAACK. L. c.
II (1857) 513—68.

Ruprecht, DAS BOTANISCHE MUSEUM DER AKADEMIE DER WISSEN-
SCHAFTEN. Bull. Acad. VII (1864) Suppl. II.

Friedr. v. d. Osten-Sacken und Ruprecht, SERTUM TIANSCHANICUM.
Botanische Ergebnisse einer Reise im mittleren Tian-schan. Mém. Acad.
7 série, XIV (1869) № 4.

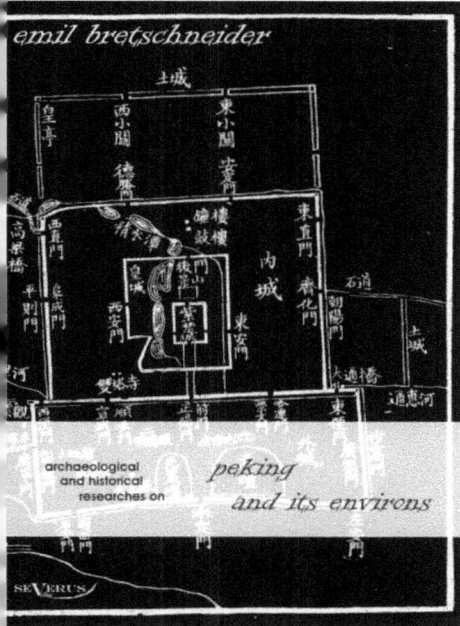

Emil Bretschneider
Archaeological and Historical
Researches on Peking and its Environs
SEVERUS 2011 / 80 Seiten / 24,50 Euro
ISBN 978-3-86347-164-4

„Almost all the celebrated capitals of the ancient kingdoms in Europe, western Asia and India, have been the subject of more or less extensive critical investigation by European antiquaries, and bulky works have been published relating to these matters. But with respect to China our scholars seem to be quite ignorant as to the remains of this ancient civilization; and even regarding Peking, one of the best-known places of the Middle Kingdom, and its classical soil, very little is known."

Emil Bretschneider (1833–1901) became famous among researchers for his valuable contributions to the field of sinology. His versatile approach – he was a physician and botanist as well as a sinologist – and his familiarity with Chinese literature distinguished him from his colleagues, many of whom were unable to read sources firsthand.

In this book, he presents his account of the history of Peking, where he was stationed as a physician from 1866–1883. Backed by historical sources as well as the reports of the first European to see the Chinese capital, Marco Polo, this work exemplifies Bretschneider's unique style of research.